area A width w base b
perimeter P surface area S circumference C area of base B
length l altitude (height) h radius r slant height s

 P9-DHQ-167

Rectangle

$A = lw \qquad P = 2l + 2w$

Triangle

$A = \dfrac{1}{2}bh$

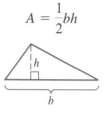

Square

$A = s^2 \qquad P = 4s$

Parallelogram

$A = bh$

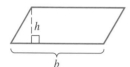

Trapezoid

$A = \dfrac{1}{2}h(b_1 + b_2)$

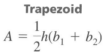

Circle

$A = \pi r^2 \qquad C = 2\pi r$

30°–60° Right Triangle

Right Triangle

$a^2 + b^2 = c^2$

Isosceles Right Triangle

Right Circular Cylinder

$V = \pi r^2 h \qquad S = 2\pi r^2 + 2\pi rh$

Sphere

$S = 4\pi r^2 \qquad V = \dfrac{4}{3}\pi r^3$

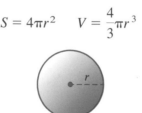

Right Circular Cone

$V = \dfrac{1}{3}\pi r^2 h \qquad S = \pi r^2 + \pi rs$

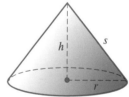

Pyramid

$V = \dfrac{1}{3}Bh$

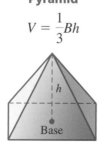

Prism

$V = Bh$

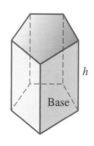

www.brookscole.com

www.brookscole.com is the World Wide Web site for Thomson Brooks/Cole and is your direct source to dozens of online resources.

At *www.brookscole.com* you can find out about supplements, demonstration software, and student resources. You can also send e-mail to many of our authors and preview new publications and exciting new technologies.

www.brookscole.com
Changing the way the world learns®

EIGHTH EDITION

Algebra for College Students

Jerome E. Kaufmann

Karen L. Schwitters
Seminole Community College

THOMSON

BROOKS/COLE

Australia • Brazil • Canada • Mexico • Singapore • Spain
United Kingdom • United States

THOMSON

———✦———™

BROOKS/COLE

Algebra for College Students, **Eighth Edition**
Jerome E. Kaufmann, Karen L. Schwitters

Editor: Gary Whalen
Assistant Editor: Rebecca Subity
Editorial Assistants: Katherine Cook and Dianna Muhammad
Technology Project Manager: Sarah Woicicki
Marketing Manager: Greta Kleinert
Marketing Assistant: Brian R. Smith
Marketing Communications Manager: Darlene Amidon-Brent
Project Manager, Editorial Production: Harold P. Humphrey
Art Director: Vernon T. Boes
Print Buyer: Barbara Britton

Permissions Editor: Stephanie Lee
Production Service: Susan Graham
Text Designer: John Edeen
Art Editor: Susan Graham
Photo Researcher: Sarah Evertson
Copy Editor: Susan Graham
Illustrator: Network Graphics and G&S Typesetters
Cover Designer: Lisa Henry
Cover Image: Doug Smock/Getty Images
Compositor: G & S Typesetters, Inc.
Text and Cover Printer: Transcontinental Printing/Interglobe

© 2007 Thomson Brooks/Cole, a part of The Thomson Corporation. Thomson, the Star logo, and Brooks/Cole are trademarks used herein under license.

ALL RIGHTS RESERVED. No part of this work covered by the copyright hereon may be reproduced or used in any form or by any means — graphic, electronic, or mechanical, including photocopying, recording, taping, web distribution, information storage and retrieval systems, or in any other manner — without the written permission of the publisher.

Printed in Canada.

1 2 3 4 5 6 7 10 09 08 07 06

Library of Congress Control Number: 2005936607

Student Edition ISBN 0-495-10510-4

Thomson Higher Education
10 Davis Drive
Belmont, CA 94002-3098
USA

For more information about our products, contact us at:
Thomson Learning Academic Resource Center
1-800-423-0563
For permission to use material from this text or product, submit a request online at **http://www.thomsonrights.com**.
Any additional questions about permissions can be submitted by e-mail to **thomsonrights@thomson.com**.

Contents

Chapter 4   Rational Expressions   165

Chapter 5   Exponents and Radicals   224

Chapter 6   Quadratic Equations and Inequalities   278

Chapter 7   Linear Equations and Inequalities in Two Variables   333

Chapter 8   Functions   391

Chapter 9   Polynomial and Rational Functions   463

Chapter 14   **Sequences and Mathematical Induction**   734

Chapter 15   **Counting Techniques, Probability, and the Binomial Theorem**   768

Preface

Algebra for College Students, Eighth Edition covers topics that are usually associated with intermediate algebra and college algebra. This text can be used in a one-semester course, but it contains ample material for a two-semester sequence.

In this book, we present the basic concepts of algebra in a simple, straight-forward way. Algebraic ideas are developed in a logical sequence and in an easy-to-read manner without excessive formalism. Concepts are developed through examples, reinforced through additional examples, and then applied in a variety of problem-solving situations. The examples show students how to use algebraic concepts to solve problems in a range of situations, and other situations have been provided in the problem sets for students to think about. In the examples, students are encouraged to organize their work and to decide when a meaningful shortcut can be used.

In preparing this edition, we made a special effort to incorporate ideas suggested by reviewers and by users of the earlier editions; at the same time, we have preserved the features of the book for which users have shown great enthusiasm.

■ New in This Edition

- Sections 7.1 and 7.2 have been reorganized so that only linear equations in two variables are graphed in Section 7.1. Then, in Section 7.2, the emphasis is on graphing nonlinear equations and using the graphs to motivate tests for x axis, y axis, and origin symmetry. These symmetry tests are used in Chapters 8, 9, 10, and 13, and will also be used in subsequent mathematics courses as students' graphing skills are enhanced.

- A focal point of every revision is the problem sets. Some users of the previous editions have suggested that the "very good" problem sets could be made even better by adding a few problems in different places. Based on these suggestions we have added approximately 90 new problems and distributed them among 15 different problem sets. For example, it was suggested that in Problem Set 6.6 we include a larger variety of quadratic inequalities. We inserted new problems 37–46 to satisfy this request. Likewise, in four problem sets in Chapter 13, "Conic Sections," we added problems to help students with the transition from the basic standard forms of equations of conics to the more general forms.

- In Section 10.2, some of the compound interest rates have been changed to be more in line with predictions for rates in the near future. However, in Section 10.2 and Problem Set 10.2 we have intentionally used a fairly wide range of interest rates. By varying the rates of interest, the number of compounding periods, and the amount of time, students can begin to see the effect that each variable has on the final result.

- The fact that logarithms are defined for only positive numbers does not imply that logarithmic equations cannot have negative solutions. We added an example at the end of Section 10.4 that shows a logarithmic equation that has a negative solution. We also added five new logarithmic equations in Problem Set 10.4 that have negative solutions or no solutions.

- As requested by a user of the previous edition, we have brought back a section on partial fractions that appeared in some earlier editions. It is now Section 11.6, and it is designated as an optional section. There are no problems pertaining to this section in the Chapter Review Problem Set or in the Chapter Test.

■ Other Special Features

- Throughout the book, students are encouraged to (a) learn a skill, (b) use the skill to help solve equations and inequalities, and then (c) use equations and inequalities to solve word problems. This focus has influenced some of the decisions we made in preparing and updating the text.

  1. Approximately 600 word problems are scattered throughout the text. These problems deal with a large variety of applications that show the connection between mathematics and its use in the real world.

  2. Many problem-solving suggestions are offered throughout the text, and there are special discussions in several sections. When appropriate, different methods for solving the same problem are shown. The problem-solving suggestions are demonstrated in more than 100 worked-out examples.

  3. Newly acquired skills are used as soon as possible to solve equations and inequalities, which are, in turn, used to solve word problems. Therefore, the concept of solving equations and inequalities is introduced early and reinforced throughout the text. The concepts of factoring, solving equations, and solving word problems are tied together in Chapter 3.

- As recommended by the American Mathematical Association of Two-Year Colleges, many basic geometric concepts are integrated into a problem-solving setting. This text contains 20 worked-out examples and 100 problems that connect algebra, geometry, and real world applications. Specific discussions of geometric concepts are contained in the following sections:

  Section 2.2: Complementary and supplementary angles; the sum of the measures of the angles of a triangle equals 180°
  Section 2.4: Area and volume formulas
  Section 3.4: More on area and volume formulas, perimeter, and circumference formulas
  Section 3.7: Pythagorean theorem
  Section 6.2: More on the Pythagorean theorem, including work with isosceles right triangles and 30°–60° right triangles.

- Specific graphing ideas (intercepts, symmetry, restrictions, asymptotes, and transformations) are introduced and used in Chapters 7, 8, 9, 10, and 13. In Section 8.5, the work with parabolas from Sections 8.3 and 8.4 is used to develop definitions for translations, reflections, stretchings, and shrinkings. These transformations are then applied to the graphs of

$$f(x) = x^3 \qquad f(x) = \frac{1}{x} \qquad f(x) = \sqrt{x} \qquad \text{and} \qquad f(x) = |x|$$

- Problems called **Thoughts into Words** are included in every problem set except the review exercises. These problems are designed to encourage students to express, in written form, their thoughts about various mathematical ideas. See, for examples, Problem Sets 2.1, 3.5, 4.7, 5.5, and 6.6.
- Many problem sets contain a special group of problems called **Further Investigations**, which lend themselves to small-group work. These problems encompass a variety of ideas: some are proofs, some show different approaches to topics covered in the text, some bring in supplementary topics and relationships, and some are more challenging problems. Although these problems add variety and flexibility to the problem sets, they can also be omitted without disrupting the continuity of the text. For examples, see Problem Sets 2.3, 2.7, 3.6, and 7.4.
- The graphing calculator is introduced in Section 7.1. From then on, many of the problem sets contain a group of problems called **Graphing Calculator Activities**. These activities, which are appropriate for either individual or small-group work, have been designed to reinforce concepts already presented and lay groundwork for concepts about to be discussed. In this text the use of a graphing calculator is considered optional.
- Photos and applications are used in the chapter openings to introduce some concepts presented in the chapter.
- Please note the exceptionally pleasing design features of the text, including the functional use of color. The open format makes for a continuous and easy flow of material instead of working through a maze of flags, caution symbols, reminder symbols, and so forth.
- *All* answers for Chapter Review Problem Sets, Chapter Tests, and Cumulative Review Problem Sets appear in the back of the text.

■ Additional Comments about Some of the Chapters

- Chapter 1 is written so that it can be covered quickly, and on an individual basis if necessary, by those who need only a brief review of some basic arithmetic and algebraic concepts. Appendix A is for students needing a more thorough review of operations with fractions.
- Chapter 2 presents an early introduction to the heart of an algebra course. We introduce problem solving and solving equations and inequalities early so that they can be used as unifying themes throughout the text.
- Chapter 6 is organized to give students the opportunity to learn, day by day, different techniques for solving quadratic equations. We treat completing the square

as a viable equation-solving process for certain types of quadratic equations. The emphasis on completing the square in this setting pays off in Chapters 8 and 13, when we graph parabolas, circles, ellipses, and hyperbolas. Section 6.5 offers some guidance about when to use a particular technique for solving quadratic equations. In addition, the often-overlooked relationships involving the sum and product of roots are discussed and used as an effective checking procedure.

- Chapter 8 is devoted entirely to functions, and the issue is not clouded by jumping back and forth between functions and relations that are not functions. Linear and quadratic functions are covered extensively and used in a variety of problem-solving situations.
- Chapters 14 and 15 have been written in a way that lends itself to individual or small-group work. Sequences, counting techniques, and some probability concepts are introduced and then used to solve problems.

■ Ancillaries

For the Instructor

Annotated Instructor's Edition. This special version of the complete student text contains a Resource Integration Guide with answers printed next to all respective exercises. Graphs, tables, and other answers appear in a special answer section at the back of the text. In every problem set, there are 20 problems that are available in electronic format through *iLrn*. These problems can be used by instructors to assign homework in an electronic format or to generate assessments for students. The *iLrn* problems are identified by a blue underline of the problem number.

Test Bank. The *Test Bank* includes eight tests per chapter as well as three final exams. The tests are made up of a combination of multiple-choice, free-response, true/false, and fill-in-the-blank questions.

Complete Solutions Manual. The *Complete Solutions Manual* provides worked-out solutions to all of the problems in the text.

iLrn™ Instructor Version. Providing instructors and students with unsurpassed control, variety, and all-in-one utility, *iLrn™* is a powerful and fully integrated teaching and learning system. *iLrn* ties together five fundamental learning activities: diagnostics, tutorials, homework, quizzing, and testing. Easy to use, *iLrn* offers instructors complete control when creating assessments in which they can draw from the wealth of exercises provided or create their own questions. *iLrn* features the greatest variety of problem types — allowing instructors to assess the way they teach. A real timesaver for instructors, *iLrn* offers automatic grading of homework, quizzes, and tests, with results flowing directly into the gradebook. The auto-enrollment feature also saves time with course setup as students self-enroll into the course gradebook. *iLrn* provides seamless integration with Blackboard™ and WebCT™.

Text-Specific Videotapes. These text-specific videotape sets, available at no charge to qualified adopters of the text, feature 10- to 20-minute problem-solving lessons that cover each section of every chapter.

For the Student

Student Solutions Manual. The *Student Solutions Manual* provides worked-out solutions to the odd-numbered problems, and all chapter review, chapter test, and cumulative review problems in the text.

Website (http://mathematics.brookscole.com). Instructors and students have access to a variety of teaching and learning resources. This website features everything from book-specific resources to newsgroups.

iLrn™ Tutorial Student Version. Featuring a variety of approaches that connect with all types of learners, *iLrn™ Tutorial* offers text-specific tutorials that require no setup by instructors. Students can begin exploring active examples from the text by using the access code packaged free with a new book. *iLrn Tutorial* supports students with explanations from the text, examples, step-by-step problem-solving help, unlimited practice, and chapter-by-chapter video lessons. With this self-paced system, students can even check their comprehension along the way by taking quizzes and receiving feedback. If they still are having trouble, students can easily access *vMentor™* for online help from a live math instructor. Students can ask any question and get personalized help through the interactive whiteboard and by using their computer microphones to speak with the instructor. While designed for self-study, instructors can also assign the individual tutorial exercises.

Interactive Video Skillbuilder CD-ROM. Think of it as portable instructor office hours. The *Interactive Video Skillbuilder CD-ROM* contains video instruction covering each chapter of the text. The problems worked during each video lesson are shown first so that students can try working them before watching the solution. To help students evaluate their progress, each section contains a 10-question web quiz (the results of which can be e-mailed to the instructor) and each chapter contains a chapter test with the answer to each problem on each test. A new learning tool on this CD-ROM is a graphing calculator tutorial for precalculus and college algebra, featuring examples, exercises, and video tutorials. Also new, English/Spanish closed caption translations can be selected to display along with the video instruction. This CD-ROM also features *MathCue* tutorial and testing software. Keyed to the text, *MathCue* offers these components:

- *MathCue Skill Builder*— Presents problems to solve, evaluates answers, and tutors students by displaying complete solutions with step-by-step explanations.
- *MathCue Quiz*— Allows students to generate large numbers of quiz problems keyed to problem types from each section of the book.
- *MathCue Chapter Test*— Also provides large numbers of problems keyed to problem types from each chapter.
- *MathCue Solution Finder*— This unique tool allows students to enter their own basic problems and receive step-by-step help as if they were working with a tutor.
- Score reports for any *MathCue* session can be printed and handed in for credit or extra credit.
- Print or e-mail score reports — Score reports for any *MathCue* session can be printed or sent to instructors via *MathCue's* secure e-mail score system.

vMentor™ Live, Online Tutoring. Packaged free with every text. Accessed seamlessly through *iLrn Tutorial, vMentor* provides tutorial help that can substantially improve student performance, increase test scores, and enhance technical aptitude. Students have access, via the web, to highly qualified tutors with thorough knowledge of our textbooks. When students get stuck on a particular problem or concept, they need only log on to *vMentor*, where they can talk (using their own computer microphones) to *vMentor* tutors who will skillfully guide them through the problem using the interactive whiteboard for illustration. Brooks/Cole also offers *Elluminate Live!*, an online virtual classroom environment that is customizable and easy to use. *Elluminate Live!* keeps students engaged with full two-way audio, instant messaging, and an interactive whiteboard — all in one, intuitive, graphical interface. For information about obtaining an *Elluminate Live!* site license, instructors may contact their Thomson representative. *For proprietary, college, and university adopters only. For additional information, instructors may consult their Thomson representative.*

Explorations in Beginning and Intermediate Algebra Using the TI-82/83/83-Plus/ 85/86 Graphing Calculator, Third Edition (0-534-40644-0)
Deborah J. Cochener and Bonnie M. Hodge, both of Austin Peay State University
This user-friendly workbook improves students' understanding and their retention of algebra concepts through a series of activities and guided explorations using the graphing calculator. An ideal supplement for any beginning or intermediate algebra course, *Explorations in Beginning and Intermediate Algebra, Third Edition* is an ideal tool for integrating technology without sacrificing course content. By clearly and succinctly teaching keystrokes, class time is devoted to investigations instead of how to use a graphing calculator.

The Math Student's Guide to the TI-83 Graphing Calculator (0-534-37802-1)
The Math Student's Guide to the TI-86 Graphing Calculator (0-534-37801-3)
The Math Student's Guide to the TI-83 Plus Graphing Calculator (0-534-42021-4)
The Math Student's Guide to the TI-89 Graphing Calculator (0-534-42022-2)
Trish Cabral of Butte College
These videos are designed for students who are new to the graphing calculator or for those who would like to brush up on their skills. Each instructional graphing calculator videotape covers basic calculations, the custom menu, graphing, advanced graphing, matrix operations, trigonometry, parametric equations, polar coordinates, calculus, Statistics I and one-variable data, and Statistics II with linear regression. These wonderful tools are each 105 minutes in length and cover all of the important functions of a graphing calculator.

Mastering Mathematics: How to Be a Great Math Student, Third Edition (0-534-34947-1)
Richard Manning Smith, Bryant College
Providing solid tips for every stage of study, *Mastering Mathematics* stresses the importance of a positive attitude and gives students the tools to succeed in their math course.

Activities for Beginning and Intermediate Algebra, Second Edition
Instructor Edition (0-534-99874-7); Student Edition (0-534-99873-9)
Debbie Garrison, Judy Jones, and Jolene Rhodes, all of Valencia Community College
Designed as a stand-alone supplement for any beginning or intermediate algebra text, *Activities in Beginning and Intermediate Algebra* is a collection of activities

written to incorporate the recommendations from the NCTM and from AMATYC's Crossroads. Activities can be used during class or in a laboratory setting to introduce, teach, or reinforce a topic.

Conquering Math Anxiety: A Self-Help Workbook, Second Edition (0-534-38634-2)
Cynthia Arem, Pima Community College
A comprehensive workbook that provides a variety of exercises and worksheets along with detailed explanations of methods to help "math-anxious" students deal with and overcome math fears. This edition now comes with a free relaxation CD-ROM and a detailed list of Internet resources.

Active Arithmetic and Algebra: Activities for Prealgebra and Beginning Algebra (0-534-36771-2)
Judy Jones, Valencia Community College
This activities manual includes a variety of approaches to learning mathematical concepts. Sixteen activities, including puzzles, games, data collection, graphing, and writing activities are included.

Math Facts: Survival Guide to Basic Mathematics, Second Edition (0-534-94734-4)
Algebra Facts: Survival Guide to Basic Algebra (0-534-19986-0)
Theodore John Szymanski, Tompkins-Cortland Community College
This booklet gives easy access to the most crucial concepts and formulas in algebra. Although it is bound, this booklet is structured to work like flash cards.

■ Acknowledgments

We would like to take this opportunity to thank the following people who served as reviewers for the new editions of this series of texts:

Yusuf Abdi
Rutgers University, Newark

Lynda Fish
St. Louis Community College at Forest Park

Cindy Fleck
Wright State University

James Hodge
Mountain State University

Barbara Laubenthal
University of North Alabama

Karolyn Morgan
University of Montevallo

Jayne Prude
University of North Alabama

Renee Quick
Wallace State Community College, Hanceville

We would like to express our sincere gratitude to the staff of Brooks/Cole, especially Gary Whalen, for his continuous cooperation and assistance throughout this project; and to Susan Graham and Hal Humphrey, who carry out the many details of production. Finally, very special thanks are due to Arlene Kaufmann, who spends numerous hours reading page proofs.

Jerome E. Kaufmann
Karen L. Schwitters

Basic Concepts and Properties

1.1 Sets, Real Numbers, and Numerical Expressions

1.2 Operations with Real Numbers

1.3 Properties of Real Numbers and the Use of Exponents

1.4 Algebraic Expressions

© Alden Pellett / The Image Works

Numbers from the set of integers are used to express temperatures that are below 0°F.

The temperature at 6 p.m. was −3°F. By 11 p.m. the temperature had dropped another 5°F. We can use the **numerical expression** −3 − 5 to determine the temperature at 11 p.m.

Justin has *p* pennies, *n* nickels, and *d* dimes in his pocket. The **algebraic expression** $p + 5n + 10d$ represents that amount of money in cents.

Algebra is often described as a **generalized arithmetic**. That description may not tell the whole story, but it does convey an important idea: A good understanding of arithmetic provides a sound basis for the study of algebra. In this chapter we use the concepts of **numerical expression** and **algebraic expression** to review some ideas from arithmetic and to begin the transition to algebra. Be sure that you thoroughly understand the basic concepts we review in this first chapter.

1.1   Sets, Real Numbers, and Numerical Expressions

In arithmetic, we use symbols such as 6, $\frac{2}{3}$, 0.27, and π to represent numbers. The symbols $+$, $-$, $\cdot$, and $\div$ commonly indicate the basic operations of addition, subtraction, multiplication, and division, respectively. Thus we can form specific **numerical expressions**. For example, we can write the indicated sum of six and eight as $6 + 8$.

In algebra, the concept of a variable provides the basis for generalizing arithmetic ideas. For example, by using x and y to represent any numbers, we can use the expression $x + y$ to represent the indicated sum of any two numbers. The x and y in such an expression are called **variables**, and the phrase $x + y$ is called an **algebraic expression**.

We can extend to algebra many of the notational agreements we make in arithmetic, with a few modifications. The following chart summarizes the notational agreements that pertain to the four basic operations.

Operation	Arithmetic	Algebra	Vocabulary
Addition	$4 + 6$	$x + y$	The **sum** of x and y
Subtraction	$14 - 10$	$a - b$	The **difference** of a and b
Multiplication	$7 \cdot 5$ or 7×5	$a \cdot b$, $a(b)$, $(a)b$, $(a)(b)$, or ab	The **product** of a and b
Division	$8 \div 4$, $\frac{8}{4}$, or $4\overline{)8}$	$x \div y$, $\frac{x}{y}$, or $y\overline{)x}$	The **quotient** of x and y

Note the different ways to indicate a product, including the use of parentheses. The ab form is the simplest and probably the most widely used form. Expressions such as abc, $6xy$, and $14xyz$ all indicate multiplication. We also call your attention to the various forms that indicate division. In algebra, we usually use the fractional form, $\frac{x}{y}$, although the other forms do serve a purpose at times.

■ Use of Sets

We can use some of the basic vocabulary and symbolism associated with the concept of sets in the study of algebra. A **set** is a collection of objects and the objects are called **elements** or **members** of the set. In arithmetic and algebra the elements of a set are usually numbers.

The use of set braces, { }, to enclose the elements (or a description of the elements) and the use of capital letters to name sets provide a convenient way to communicate about sets. For example, we can represent a set A, which consists of the vowels of the alphabet, in any of the following ways:

$A = \{$vowels of the alphabet$\}$ Word description

$A = \{a, e, i, o, u\}$ List or roster description

$A = \{x | x$ is a vowel$\}$ Set builder notation

We can modify the listing approach if the number of elements is quite large. For example, all of the letters of the alphabet can be listed as

$$\{a, b, c, \ldots, z\}$$

We simply begin by writing enough elements to establish a pattern; then the three dots indicate that the set continues in that pattern. The final entry indicates the last element of the pattern. If we write

$$\{1, 2, 3, \ldots\}$$

the set begins with the counting numbers 1, 2, and 3. The three dots indicate that it continues in a like manner forever; there is no last element. A set that consists of no elements is called the **null set** (written $\varnothing$).

   Set builder notation combines the use of braces and the concept of a variable. For example, $\{x | x$ is a vowel$\}$ is read "the set of all x such that x is a vowel." Note that the vertical line is read "such that." We can use set builder notation to describe the set $\{1, 2, 3, \ldots\}$ as $\{x | x > 0$ and x is a whole number$\}$.

   We use the symbol $\in$ to denote set membership. Thus if $A = \{a, e, i, o, u\}$, we can write $e \in A$, which we read as "e is an element of A." The slash symbol, $/$, is commonly used in mathematics as a negation symbol. For example, $m \notin A$ is read as "m is not an element of A."

   Two sets are said to be *equal* if they contain exactly the same elements. For example,

$$\{1, 2, 3\} = \{2, 1, 3\}$$

because both sets contain the same elements; the order in which the elements are written doesn't matter. The slash mark through the equality symbol denotes "is not equal to." Thus if $A = \{1, 2, 3\}$ and $B = \{1, 2, 3, 4\}$, we can write $A \neq B$, which we read as "set A is not equal to set B."

■ Real Numbers

We refer to most of the algebra that we will study in this text as the **algebra of real numbers**. This simply means that the variables represent real numbers. Therefore, it is necessary for us to be familiar with the various terms that are used to classify different types of real numbers.

$\{1, 2, 3, 4, \ldots\}$ Natural numbers, counting numbers, positive integers

$\{0, 1, 2, 3, \ldots\}$ Whole numbers, nonnegative integers

$\{\ldots -3, -2, -1\}$ Negative integers

$\{\ldots -3, -2, -1, 0\}$ Nonpositive integers

$\{\ldots -3, -2, -1, 0, 1, 2, 3, \ldots\}$ Integers

We define a **rational number** as any number that can be expressed in the form $\frac{a}{b}$, where a and b are integers and b is not zero. The following are examples of rational numbers.

$$-\frac{3}{4}, \quad \frac{2}{3}, \quad -4, \quad 0, \quad 0.3, \quad 6\frac{1}{2}$$

-4 because $-4 = \dfrac{-4}{1} = \dfrac{4}{-1}$ 0 because $0 = \dfrac{0}{1} = \dfrac{0}{2} = \dfrac{0}{3} = \ldots$

0.3 because $0.3 = \dfrac{3}{10}$ $6\dfrac{1}{2}$ because $6\dfrac{1}{2} = \dfrac{13}{2}$

We can also define a rational number in terms of a decimal representation. Before doing so, let's review the different possibilities for decimal representations. We can classify decimals as **terminating**, **repeating**, or **nonrepeating**. Some examples follow.

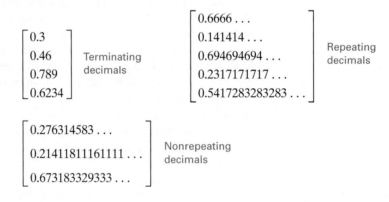

$$\begin{bmatrix} 0.3 \\ 0.46 \\ 0.789 \\ 0.6234 \end{bmatrix} \text{Terminating decimals} \qquad \begin{bmatrix} 0.6666\ldots \\ 0.141414\ldots \\ 0.694694694\ldots \\ 0.2317171717\ldots \\ 0.5417283283283\ldots \end{bmatrix} \text{Repeating decimals}$$

$$\begin{bmatrix} 0.276314583\ldots \\ 0.21411811161111\ldots \\ 0.673183329333\ldots \end{bmatrix} \text{Nonrepeating decimals}$$

A repeating decimal has a block of digits that repeats indefinitely. This repeating block of digits may be of any number of digits and may or may not begin immediately after the decimal point. A small horizontal bar (overbar) is commonly used to indicate the repeat block. Thus $0.6666\ldots$ is written as $0.\overline{6}$, and $0.2317171717\ldots$ is written as $0.23\overline{17}$.

In terms of decimals, we define a **rational number** as a number that has either a terminating or a repeating decimal representation. The following examples illustrate some rational numbers written in $\frac{a}{b}$ form and in decimal form.

$$\frac{3}{4} = 0.75 \qquad \frac{3}{11} = 0.\overline{27} \qquad \frac{1}{8} = 0.125 \qquad \frac{1}{7} = 0.\overline{142857} \qquad \frac{1}{3} = 0.\overline{3}$$

We define an **irrational number** as a number that *cannot* be expressed in $\frac{a}{b}$ form, where a and b are integers, and b is not zero. Furthermore, an irrational number has a nonrepeating and nonterminating decimal representation. Some examples of irrational numbers and a partial decimal representation for each follow.

$$\sqrt{2} = 1.414213562373095\ldots \qquad \sqrt{3} = 1.73205080756887\ldots$$

$$\pi = 3.14159265358979\ldots$$

The entire set of **real numbers** is composed of the rational numbers along with the irrationals. Every real number is either a rational number or an irrational number. The following tree diagram summarizes the various classifications of the real number system.

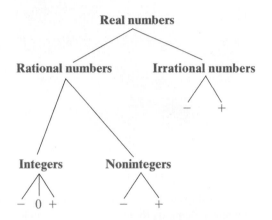

We can trace any real number down through the diagram as follows:

7 is real, rational, an integer, and positive.

$-\dfrac{2}{3}$ is real, rational, noninteger, and negative.

$\sqrt{7}$ is real, irrational, and positive.

0.38 is real, rational, noninteger, and positive.

Remark: We usually refer to the set of nonnegative integers, $\{0, 1, 2, 3, \ldots\}$, as the set of **whole numbers**, and we refer to the set of positive integers, $\{1, 2, 3, \ldots\}$, as the set of **natural numbers**. The set of whole numbers differs from the set of natural numbers by the inclusion of the number zero.

The concept of subset is convenient to use at this time. A set A is a **subset** of a set B if and only if every element of A is also an element of B. This is written as $A \subseteq B$ and read as "A is a subset of B." For example, if $A = \{1, 2, 3\}$ and $B = \{1, 2, 3, 5, 9\}$, then $A \subseteq B$ because every element of A is also an element of B. The slash mark again denotes negation, so if $A = \{1, 2, 5\}$ and $B = \{2, 4, 7\}$, we can say that A is not a subset of B by writing $A \not\subseteq B$. Figure 1.1 represents the subset

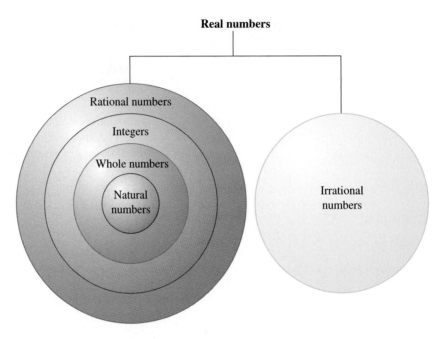

Figure 1.1

relationships for the set of real numbers. Refer to Figure 1.1 as you study the following statements that use subset vocabulary and subset symbolism.

1. The set of whole numbers is a subset of the set of integers.

$$\{0, 1, 2, 3, \ldots\} \subseteq \{\ldots, -2, -1, 0, 1, 2, \ldots\}$$

2. The set of integers is a subset of the set of rational numbers.

$$\{\ldots, -2, -1, 0, 1, 2, \ldots\} \subseteq \{x|x \text{ is a rational number}\}$$

3. The set of rational numbers is a subset of the set of real numbers.

$$\{x|x \text{ is a rational number}\} \subseteq \{y|y \text{ is a real number}\}$$

■ Equality

The relation **equality** plays an important role in mathematics — especially when we are manipulating real numbers and algebraic expressions that represent real numbers. An equality is a statement in which two symbols, or groups of symbols, are names for the same number. The symbol $=$ is used to express an equality. Thus we can write

$$6 + 1 = 7 \qquad 18 - 2 = 16 \qquad 36 \div 4 = 9$$

(The symbol $\neq$ means *is not equal to*.) The following four basic properties of equality are self-evident, but we do need to keep them in mind. (We will expand this list in Chapter 2 when we work with solutions of equations.)

■ Properties of Equality

Reflexive Property

For any real number a,

$a = a$

Examples: $14 = 14$ $x = x$ $a + b = a + b$

Symmetric Property

For any real numbers a and b,

if $a = b$, then $b = a$

Examples : If $13 + 1 = 14$, then $14 = 13 + 1$.
If $3 = x + 2$, then $x + 2 = 3$.

Transitive Property

For any real numbers a, b, and c,

if $a = b$ and $b = c$, then $a = c$

Examples: If $3 + 4 = 7$ and $7 = 5 + 2$, then $3 + 4 = 5 + 2$.
If $x + 1 = y$ and $y = 5$, then $x + 1 = 5$.

Substitution Property

For any real numbers a and b: If $a = b$, then a may be replaced by b, or b may be replaced by a, in any statement without changing the meaning of the statement.

Examples: If $x + y = 4$ and $x = 2$, then $2 + y = 4$.
If $a - b = 9$ and $b = 4$, then $a - 4 = 9$.

■ Numerical Expressions

Let's conclude this section by *simplifying some numerical expressions* that involve whole numbers. When simplifying numerical expressions, we perform the operations in the following order. Be sure that you agree with the result in each example.

1. Perform the operations inside the symbols of inclusion (parentheses, brackets, and braces) and above and below each fraction bar. Start with the innermost inclusion symbol.

2. Perform all multiplications and divisions in the order in which they appear from left to right.

3. Perform all additions and subtractions in the order in which they appear from left to right.

E X A M P L E   1

Simplify $20 + 60 \div 10 \cdot 2$

Solution

First do the division.

$$20 + 60 \div 10 \cdot 2 = 20 + 6 \cdot 2$$

Next do the multiplication.

$$20 + 6 \cdot 2 = 20 + 12$$

Then do the addition.

$$20 + 12 = 32$$

Thus $20 + 60 \div 10 \cdot 2$ simplifies to 32.   ∎

E X A M P L E   2

Simplify $7 \cdot 4 \div 2 \cdot 3 \cdot 2 \div 4$.

Solution

The multiplications and divisions are to be done from left to right in the order in which they appear.

$$7 \cdot 4 \div 2 \cdot 3 \cdot 2 \div 4 = 28 \div 2 \cdot 3 \cdot 2 \div 4$$
$$= 14 \cdot 3 \cdot 2 \div 4$$
$$= 42 \cdot 2 \div 4$$
$$= 84 \div 4$$
$$= 21$$

Thus $7 \cdot 4 \div 2 \cdot 3 \cdot 2 \div 4$ simplifies to 21.   ∎

E X A M P L E   3

Simplify $5 \cdot 3 + 4 \div 2 - 2 \cdot 6 - 28 \div 7$.

Solution

First we do the multiplications and divisions in the order in which they appear. Then we do the additions and subtractions in the order in which they appear. Our work may take on the following format.

$$5 \cdot 3 + 4 \div 2 - 2 \cdot 6 - 28 \div 7 = 15 + 2 - 12 - 4 = 1$$ ∎

EXAMPLE 4

Simplify $(4 + 6)(7 + 8)$.

Solution

We use the parentheses to indicate the *product* of the quantities $4 + 6$ and $7 + 8$. We perform the additions inside the parentheses first and then multiply.

$$(4 + 6)(7 + 8) = (10)(15) = 150$$ ∎

EXAMPLE 5

Simplify $(3 \cdot 2 + 4 \cdot 5)(6 \cdot 8 - 5 \cdot 7)$.

Solution

First we do the multiplications inside the parentheses.

$$(3 \cdot 2 + 4 \cdot 5)(6 \cdot 8 - 5 \cdot 7) = (6 + 20)(48 - 35)$$

Then we do the addition and subtraction inside the parentheses.

$$(6 + 20)(48 - 35) = (26)(13)$$

Then we find the final product.

$$(26)(13) = 338$$ ∎

EXAMPLE 6

Simplify $6 + 7[3(4 + 6)]$.

Solution

We use brackets for the same purposes as parentheses. In such a problem we need to simplify *from the inside out*; that is, we perform the operations in the innermost parentheses first. We thus obtain

$$6 + 7[3(4 + 6)] = 6 + 7[3(10)]$$
$$= 6 + 7[30]$$
$$= 6 + 210$$
$$= 216$$ ∎

EXAMPLE 7

Simplify $\dfrac{6 \cdot 8 \div 4 - 2}{5 \cdot 4 - 9 \cdot 2}$

Solution

First we perform the operations above and below the fraction bar. Then we find the final quotient.

$$\frac{6 \cdot 8 \div 4 - 2}{5 \cdot 4 - 9 \cdot 2} = \frac{48 \div 4 - 2}{20 - 18} = \frac{12 - 2}{2} = \frac{10}{2} = 5$$ ∎

Remark: With parentheses we could write the problem in Example 7 as $(6 \cdot 8 \div 4 - 2) \div (5 \cdot 4 - 9 \cdot 2)$.

Problem Set 1.1

For Problems 1–10, identify each statement as true or false.

1. Every irrational number is a real number.

2. Every rational number is a real number.

3. If a number is real, then it is irrational.

4. Every real number is a rational number.

5. All integers are rational numbers.

6. Some irrational numbers are also rational numbers.

7. Zero is a positive integer.

8. Zero is a rational number.

9. All whole numbers are integers.

10. Zero is a negative integer.

For Problems 11–18, from the list $0, 14, \frac{2}{3}, \pi, \sqrt{7}, -\frac{11}{14},$ $2.34, 3.2\overline{1}, \frac{55}{8}, -\sqrt{17}, -19,$ and -2.6, identify each of the following.

11. The whole numbers

12. The natural numbers

13. The rational numbers

14. The integers

15. The nonnegative integers

16. The irrational numbers

17. The real numbers

18. The nonpositive integers

For Problems 19–28, use the following set designations.

$N = \{x \mid x$ is a natural number$\}$

$Q = \{x \mid x$ is a rational number$\}$

$W = \{x \mid x$ is a whole number$\}$

$H = \{x \mid x$ is an irrational number$\}$

$I \ = \{x \mid x$ is an integer$\}$

$R = \{x \mid x$ is a real number$\}$

Place $\subseteq$ or $\not\subseteq$ in each blank to make a true statement.

19. R _____ N

20. N _____ R

21. I _____ Q

22. N _____ I

23. Q _____ H

24. H _____ Q

25. N _____ W

26. W _____ I

27. I _____ N

28. I _____ W

For Problems 29–32, classify the real number by tracing through the diagram in the text (see page 5).

29. -8

30. 0.9

31. $-\sqrt{2}$

32. $\frac{5}{6}$

For Problems 33–42, list the elements of each set. For example, the elements of $\{x \mid x$ is a natural number less than 4$\}$ can be listed as $\{1, 2, 3\}$.

33. $\{x \mid x$ is a natural number less than 3$\}$

34. $\{x \mid x$ is a natural number greater than 3$\}$

35. $\{n \mid n$ is a whole number less than 6$\}$

36. $\{y \mid y$ is an integer greater than $-4\}$

37. $\{y \mid y$ is an integer less than 3$\}$

38. $\{n \mid n$ is a positive integer greater than $-7\}$

39. $\{x \mid x$ is a whole number less than 0$\}$

40. $\{x \mid x$ is a negative integer greater than $-3\}$

41. $\{n \mid n$ is a nonnegative integer less than 5$\}$

42. $\{n \mid n$ is a nonpositive integer greater than 3$\}$

For Problems 43–50, replace each question mark to make the given statement an application of the indicated property of equality. For example, $16 = ?$ becomes $16 = 16$ because of the reflexive property of equality.

43. If $y = x$ and $x = -6$, then $y = ?$ (Transitive property of equality)

44. $5x + 7 = ?$ (Reflexive property of equality)

45. If $n = 2$ and $3n + 4 = 10$, then $3(?) + 4 = 10$ (Substitution property of equality)

46. If $y = x$ and $x = z + 2$, then $y = ?$ (Transitive property of equality)

47. If $4 = 3x + 1$, then $? = 4$ (Symmetric property of equality)

48. If $t = 4$ and $s + t = 9$, then $s + ? = 9$ (Substitution property of equality)

49. $5x = ?$ (Reflexive property of equality)

50. If $5 = n + 3$, then $n + 3 = ?$ (Symmetric property of equality)

For Problems 51–74, simplify each of the numerical expressions.

51. $16 + 9 - 4 - 2 + 8 - 1$

52. $18 + 17 - 9 - 2 + 14 - 11$

53. $9 \div 3 \cdot 4 \div 2 \cdot 14$

54. $21 \div 7 \cdot 5 \cdot 2 \div 6$

55. $7 + 8 \cdot 2$

56. $21 - 4 \cdot 3 + 2$

57. $9 \cdot 7 - 4 \cdot 5 - 3 \cdot 2 + 4 \cdot 7$

58. $6 \cdot 3 + 5 \cdot 4 - 2 \cdot 8 + 3 \cdot 2$

59. $(17 - 12)(13 - 9)(7 - 4)$

60. $(14 - 12)(13 - 8)(9 - 6)$

61. $13 + (7 - 2)(5 - 1)$

62. $48 - (14 - 11)(10 - 6)$

63. $(5 \cdot 9 - 3 \cdot 4)(6 \cdot 9 - 2 \cdot 7)$

64. $(3 \cdot 4 + 2 \cdot 1)(5 \cdot 2 + 6 \cdot 7)$

65. $7[3(6 - 2)] - 64$

66. $12 + 5[3(7 - 4)]$

67. $[3 + 2(4 \cdot 1 - 2)][18 - (2 \cdot 4 - 7 \cdot 1)]$

68. $3[4(6 + 7)] + 2[3(4 - 2)]$

69. $14 + 4\left(\dfrac{8 - 2}{12 - 9}\right) - 2\left(\dfrac{9 - 1}{19 - 15}\right)$

70. $12 + 2\left(\dfrac{12 - 2}{7 - 2}\right) - 3\left(\dfrac{12 - 9}{17 - 14}\right)$

71. $[7 + 2 \cdot 3 \cdot 5 - 5] \div 8$

72. $[27 - (4 \cdot 2 + 5 \cdot 2)][(5 \cdot 6 - 4) - 20]$

73. $\dfrac{3 \cdot 8 - 4 \cdot 3}{5 \cdot 7 - 34} + 19$

74. $\dfrac{4 \cdot 9 - 3 \cdot 5 - 3}{18 - 12}$

75. You must of course be able to do calculations like those in Problems 51–74 both with and without a calculator. Furthermore, different types of calculators handle the priority-of-operations issue in different ways. Be sure you can do Problems 51–74 with *your* calculator.

■ ■ ■ **THOUGHTS INTO WORDS**

76. Explain in your own words the difference between the reflexive property of equality and the symmetric property of equality.

77. Your friend keeps getting an answer of 30 when simplifying $7 + 8(2)$. What mistake is he making and how would you help him?

78. Do you think $3\sqrt{2}$ is a rational or an irrational number? Defend your answer.

79. Explain why every integer is a rational number but not every rational number is an integer.

80. Explain the difference between $1.\overline{3}$ and 1.3.

1.2 Operations with Real Numbers

Before we review the four basic operations with real numbers, let's briefly discuss some concepts and terminology we commonly use with this material. It is often helpful to have a geometric representation of the set of real numbers as indicated in Figure 1.2. Such a representation, called the **real number line**, indicates a one-to-one correspondence between the set of real numbers and the points on a line.

In other words, to each real number there corresponds one and only one point on the line, and to each point on the line there corresponds one and only one real number. The number associated with each point on the line is called the **coordinate** of the point.

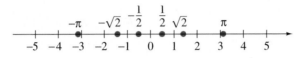

Figure 1.2

Many operations, relations, properties, and concepts pertaining to real numbers can be given a geometric interpretation on the real number line. For example, the addition problem $(-1) + (-2)$ can be depicted on the number line as in Figure 1.3.

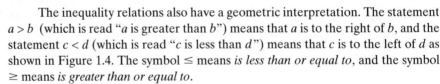

$$(-1) + (-2) = -3$$

Figure 1.3

Figure 1.4

The inequality relations also have a geometric interpretation. The statement $a > b$ (which is read "a is greater than b") means that a is to the right of b, and the statement $c < d$ (which is read "c is less than d") means that c is to the left of d as shown in Figure 1.4. The symbol $\leq$ means *is less than or equal to*, and the symbol $\geq$ means *is greater than or equal to*.

The property $-(-x) = x$ can be represented on the number line by following the sequence of steps shown in Figure 1.5.

1. Choose a point having a coordinate of x.

2. Locate its opposite, written as $-x$, on the other side of zero.

3. Locate the opposite of $-x$, written as $-(-x)$, on the other side of zero.

Therefore, we conclude that **the opposite of the opposite of any real number is the number itself**, and we symbolically express this by $-(-x) = x$.

Remark: The symbol -1 can be read "negative one," "the negative of one," "the opposite of one," or "the additive inverse of one." The opposite-of and additive-inverse-of terminology is especially meaningful when working with variables. For example, the symbol $-x$, which is read "the opposite of x" or "the additive inverse of x," emphasizes an important issue. Because x can be any real number, $-x$ (the opposite of x) can be zero, positive, or negative. If x is positive, then $-x$ is negative. If x is negative, then $-x$ is positive. If x is zero, then $-x$ is zero.

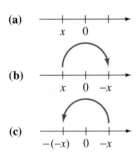

Figure 1.5

■ Absolute Value

We can use the concept of **absolute value** to describe precisely how to operate with positive and negative numbers. Geometrically, the absolute value of any number is the distance between the number and zero on the number line. For example, the absolute value of 2 is 2. The absolute value of -3 is 3. The absolute value of 0 is 0 (see Figure 1.6).

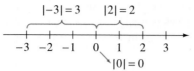

Figure 1.6

Symbolically, absolute value is denoted with vertical bars. Thus we write

$$|2| = 2 \qquad |-3| = 3 \qquad |0| = 0$$

More formally, we define the concept of absolute value as follows:

Definition 1.1

> For all real numbers a,
>
> **1.** If $a \geq 0$, then $|a| = a$.
> **2.** If $a < 0$, then $|a| = -a$.

According to Definition 1.1, we obtain

$|6| = 6$	By applying part 1 of Definition 1.1
$|0| = 0$	By applying part 1 of Definition 1.1
$|-7| = -(-7) = 7$	By applying part 2 of Definition 1.1

Note that the absolute value of a positive number is the number itself, but the absolute value of a negative number is its opposite. Thus the absolute value of any number except zero is positive, and the absolute value of zero is zero. Together, these facts indicate that the absolute value of any real number is equal to the absolute value of its opposite. We summarize these ideas in the following properties.

Properties of Absolute Value

> The variables a and b represent any real number.
>
> **1.** $|a| \geq 0$
> **2.** $|a| = |-a|$
> **3.** $|a - b| = |b - a|$ $\qquad$ $a - b$ and $b - a$ are opposites of each other.

■ Adding Real Numbers

We can use various physical models to describe the addition of real numbers. For example, profits and losses pertaining to investments: A loss of $25.75 (written as -25.75) on one investment, along with a profit of $22.20 (written as 22.20) on a second investment, produces an overall loss of $3.55. Thus $(-25.75) + 22.20 = -3.55$. Think in terms of profits and losses for each of the following examples.

$$50 + 75 = 125 \qquad\qquad 20 + (-30) = -10$$

$$-4.3 + (-6.2) = -10.5 \qquad -27 + 43 = 16$$

$$\frac{7}{8} + \left(-\frac{1}{4}\right) = \frac{5}{8} \qquad\qquad -3\frac{1}{2} + \left(-3\frac{1}{2}\right) = -7$$

Though all problems that involve addition of real numbers could be solved using the profit-loss interpretation, it is sometimes convenient to have a more precise description of the addition process. For this purpose we use the concept of absolute value.

Addition of Real Numbers

Two Positive Numbers   The sum of two positive real numbers is the sum of their absolute values.

Two Negative Numbers   The sum of two negative real numbers is the opposite of the sum of their absolute values.

One Positive and One Negative Number   The sum of a positive real number and a negative real number can be found by subtracting the smaller absolute value from the larger absolute value and giving the result the sign of the original number that has the larger absolute value. If the two numbers have the same absolute value, then their sum is 0.

Zero and Another Number   The sum of 0 and any real number is the real number itself.

Now consider the following examples in terms of the previous description of addition. These examples include operations with rational numbers in common fraction form. If you need a review on operations with fractions, see Appendix A.

$$(-6) + (-8) = -(|-6| + |-8|) = -(6 + 8) = -14$$

$$(-1.6) + (-7.7) = -(|-1.6| + |-7.7|) = -(1.6 + 7.7) = -9.3$$

$$6\frac{3}{4} + \left(-2\frac{1}{2}\right) = \left(\left|6\frac{3}{4}\right| - \left|-2\frac{1}{2}\right|\right) = \left(6\frac{3}{4} - 2\frac{1}{2}\right) = \left(6\frac{3}{4} - 2\frac{2}{4}\right) = 4\frac{1}{4}$$

$$14 + (-21) = -(|-21| - |14|) = -(21 - 14) = -7$$

$$-72.4 + 72.4 = 0 \qquad 0 + (-94) = -94$$

■ Subtracting Real Numbers

We can describe the subtraction of real numbers in terms of addition.

Subtraction of Real Numbers

If a and b are real numbers, then

$$a - b = a + (-b)$$

It may be helpful for you to read $a - b = a + (-b)$ as "a minus b is equal to a plus the opposite of b." In other words, every subtraction problem can be changed to an equivalent addition problem. Consider the following examples.

$$7 - 9 = 7 + (-9) = -2, \qquad -5 - (-13) = -5 + 13 = 8$$

$$6.1 - (-14.2) = 6.1 + 14.2 = 20.3, \qquad -16 - (-11) = -16 + 11 = -5$$

$$-\frac{7}{8} - \left(-\frac{1}{4}\right) = -\frac{7}{8} + \frac{1}{4} = -\frac{7}{8} + \frac{2}{8} = -\frac{5}{8}$$

It should be apparent that addition is a key operation. To simplify numerical expressions that involve addition and subtraction, we can first change all subtractions to additions and then perform the additions.

E X A M P L E   1

Simplify $7 - 9 - 14 + 12 - 6 + 4$.

Solution

$$7 - 9 - 14 + 12 - 6 + 4 = 7 + (-9) + (-14) + 12 + (-6) + 4$$

$$= -6 \qquad\qquad ■$$

E X A M P L E   2

Simplify $-2\frac{1}{8} + \frac{3}{4} - \left(-\frac{3}{8}\right) - \frac{1}{2}$

Solution

$$-2\frac{1}{8} + \frac{3}{4} - \left(-\frac{3}{8}\right) - \frac{1}{2} = -2\frac{1}{8} + \frac{3}{4} + \frac{3}{8} + \left(-\frac{1}{2}\right)$$

$$= -\frac{17}{8} + \frac{6}{8} + \frac{3}{8} + \left(-\frac{4}{8}\right) \qquad \begin{array}{l}\text{Change to equivalent}\\\text{fractions with a}\\\text{common denominator.}\end{array}$$

$$= -\frac{12}{8} = -\frac{3}{2} \qquad\qquad ■$$

It is often helpful to convert subtractions to additions *mentally*. In the next two examples, the work shown in the dashed boxes could be done in your head.

E X A M P L E   3

Simplify $4 - 9 - 18 + 13 - 10$.

Solution

$$4 - 9 - 18 + 13 - 10 = \boxed{4 + (-9) + (-18) + 13 + (-10)}$$
$$= -20 \qquad \blacksquare$$

E X A M P L E   4

Simplify $\left(\dfrac{2}{3} - \dfrac{1}{5}\right) - \left(\dfrac{1}{2} - \dfrac{7}{10}\right)$

Solution

$$\left(\frac{2}{3} - \frac{1}{5}\right) - \left(\frac{1}{2} - \frac{7}{10}\right) = \boxed{\left[\frac{2}{3} + \left(-\frac{1}{5}\right)\right] - \left[\frac{1}{2} + \left(-\frac{7}{10}\right)\right]}$$

$$= \left[\frac{10}{15} + \left(-\frac{3}{15}\right)\right] - \left[\frac{5}{10} + \left(-\frac{7}{10}\right)\right]$$

Within the brackets, change to equivalent fractions with a common denominator.

$$= \left(\frac{7}{15}\right) - \left(-\frac{2}{10}\right)$$

$$= \boxed{\left(\frac{7}{15}\right) + \left(+\frac{2}{10}\right)}$$

$$= \frac{14}{30} + \left(+\frac{6}{30}\right)$$

Change to equivalent fractions with a common denominator.

$$= \frac{20}{30} = \frac{2}{3} \qquad \blacksquare$$

■ Multiplying Real Numbers

We can interpret the multiplication of whole numbers as repeated addition. For example, $3 \cdot 2$ means three 2s; thus $3 \cdot 2 = 2 + 2 + 2 = 6$. This same repeated-addition interpretation of multiplication can be used to find the product of a positive number and a negative number, as shown by the following examples.

$$2(-3) = -3 + (-3) = -6, \qquad 3(-2) = -2 + (-2) + (-2) = -6$$

$$4(-1.2) = -1.2 + (-1.2) + (-1.2) + (-1.2) = -4.8$$

$$3\left(-\frac{1}{8}\right) = -\frac{1}{8} + \left(-\frac{1}{8}\right) + \left(-\frac{1}{8}\right) = -\frac{3}{8}$$

When we are multiplying whole numbers, the order in which we multiply two factors does not change the product. For example, $2(3) = 6$ and $3(2) = 6$. Using this idea, we can handle a negative number times a positive number as follows:

$$(-2)(3) = (3)(-2) = (-2) + (-2) + (-2) = -6$$

$$(-3)(4) = (4)(-3) = (-3) + (-3) + (-3) + (-3) = -12$$

$$\left(-\frac{3}{7}\right)(2) = (2)\left(-\frac{3}{7}\right) = -\frac{3}{7} + \left(-\frac{3}{7}\right) = -\frac{6}{7}$$

Finally, let's consider the product of two negative integers. The following pattern using integers helps with the reasoning.

$$4(-2) = -8 \qquad 3(-2) = -6 \qquad 2(-2) = -4$$

$$1(-2) = -2 \qquad 0(-2) = 0 \qquad (-1)(-2) = \;?$$

To continue this pattern, the product of -1 and -2 has to be 2. In general, this type of reasoning helps us realize that the product of any two negative real numbers is a positive real number. Using the concept of absolute value, we can describe the **multiplication of real numbers** as follows:

Multiplication of Real Numbers

1. The product of two positive or two negative real numbers is the product of their absolute values.
2. The product of a positive real number and a negative real number (either order) is the opposite of the product of their absolute values.
3. The product of zero and any real number is zero.

The following examples illustrate this description of multiplication. Again, the steps shown in the dashed boxes are usually performed mentally.

$$(-6)(-7) = |-6| \cdot |-7| = 6 \cdot 7 = 42$$

$$(8)(-9) = -(|8| \cdot |-9|) = -(8 \cdot 9) = -72$$

$$\left(-\frac{3}{4}\right)\left(\frac{1}{3}\right) = -\left(\left|-\frac{3}{4}\right| \cdot \left|\frac{1}{3}\right|\right) = -\left(\frac{3}{4} \cdot \frac{1}{3}\right) = -\frac{1}{4}$$

$$(-14.3)(0) = 0$$

The previous examples illustrated a step-by-step process for multiplying real numbers. In practice, however, the key is to remember that the product of two positive or two negative numbers is positive and that the product of a positive number and a negative number (either order) is negative.

■ Dividing Real Numbers

The relationship between multiplication and division provides the basis for dividing real numbers. For example, we know that $8 \div 2 = 4$ because $2 \cdot 4 = 8$. In other words, the quotient of two numbers can be found by looking at a related multiplication problem. In the following examples, we used this same type of reasoning to determine some quotients that involve integers.

$$\frac{6}{-2} = -3 \quad \text{because } (-2)(-3) = 6$$

$$\frac{-12}{3} = -4 \quad \text{because } (3)(-4) = -12$$

$$\frac{-18}{-2} = 9 \quad \text{because } (-2)(9) = -18$$

$$\frac{0}{-5} = 0 \quad \text{because } (-5)(0) = 0$$

$$\frac{-8}{0} \text{ is undefined} \qquad \text{Remember that division by zero is undefined!}$$

A precise description for **division of real numbers** follows.

Division of Real Numbers

1. The quotient of two positive or two negative real numbers is the quotient of their absolute values.
2. The quotient of a positive real number and a negative real number or of a negative real number and a positive real number is the opposite of the quotient of their absolute values.
3. The quotient of zero and any nonzero real number is zero.
4. The quotient of any nonzero real number and zero is undefined.

The following examples illustrate this description of division. Again, for practical purposes, the key is to remember whether the quotient is positive or negative.

$$\frac{-16}{-4} = \frac{|-16|}{|-4|} = \frac{16}{4} = 4 \qquad \frac{28}{-7} = -\left(\frac{|28|}{|-7|}\right) = -\left(\frac{28}{7}\right) = -4$$

$$\frac{-3.6}{4} = -\left(\frac{|-3.6|}{|4|}\right) = -\left(\frac{3.6}{4}\right) = -0.9 \qquad \frac{0}{\frac{7}{8}} = 0$$

Now let's simplify some numerical expressions that involve the four basic operations with real numbers. Remember that multiplications and divisions are done first, from left to right, before additions and subtractions are performed.

EXAMPLE 5

Simplify $-2\frac{1}{3} + 4\left(-\frac{2}{3}\right) - (-5)\left(-\frac{1}{3}\right)$

Solution

$$-2\frac{1}{3} + 4\left(-\frac{2}{3}\right) - (-5)\left(-\frac{1}{3}\right) = -2\frac{1}{3} + \left(-\frac{8}{3}\right) + \left(-\frac{5}{3}\right)$$

$$= -\frac{7}{3} + \left(-\frac{8}{3}\right) + \left(-\frac{5}{3}\right) \quad \text{Change to improper fraction.}$$

$$= -\frac{20}{3} \qquad \blacksquare$$

EXAMPLE 6

Simplify $-24 \div 4 + 8(-5) - (-5)(3)$.

Solution

$$-24 \div 4 + 8(-5) - (-5)(3) = -6 + (-40) - (-15)$$

$$= -6 + (-40) + 15$$

$$= -31 \qquad \blacksquare$$

EXAMPLE 7

Simplify $-7.3 - 2[-4.6(6 - 7)]$.

Solution

$$-7.3 - 2[-4.6(6 - 7)] = -7.3 - 2[-4.6(-1)] = -7.3 - 2[4.6]$$

$$= -7.3 - 9.2$$

$$= -7.3 + (-9.2)$$

$$= -16.5 \qquad \blacksquare$$

EXAMPLE 8

Simplify $[3(-7) - 2(9)][5(-7) + 3(9)]$.

Solution

$$[3(-7) - 2(9)][5(-7) + 3(9)] = [-21 - 18][-35 + 27]$$

$$= [-39][-8]$$

$$= 312 \qquad \blacksquare$$

Problem Set 1.2

For Problems 1–50, perform the following operations with real numbers.

1. $8 + (-15)$

2. $9 + (-18)$

3. $(-12) + (-7)$

4. $(-7) + (-14)$

5. $-8 - 14$

6. $-17 - 9$

7. $9 - 16$

8. $8 - 22$

9. $(-9)(-12)$

10. $(-6)(-13)$

11. $(5)(-14)$

12. $(-17)(4)$

13. $(-56) \div (-4)$

14. $(-81) \div (-3)$

15. $\dfrac{-112}{16}$

16. $\dfrac{-75}{5}$

17. $-2\dfrac{3}{8} + 5\dfrac{7}{8}$

18. $-1\dfrac{1}{5} + 3\dfrac{4}{5}$

19. $4\dfrac{1}{3} - \left(-1\dfrac{1}{6}\right)$

20. $1\dfrac{1}{12} - \left(-5\dfrac{3}{4}\right)$

21. $\left(-\dfrac{1}{3}\right)\left(\dfrac{2}{5}\right)$

22. $(-8)\left(\dfrac{1}{3}\right)$

23. $\dfrac{1}{2} \div \left(-\dfrac{1}{8}\right)$

24. $\dfrac{2}{3} \div \left(-\dfrac{1}{6}\right)$

25. $0 \div (-14)$

26. $(-19) \div 0$

27. $(-21) \div 0$

28. $0 \div (-11)$

29. $-21 - 39$

30. $-23 - 38$

31. $-17.3 + 12.5$

32. $-16.3 + 19.6$

33. $21.42 - 7.29$

34. $2.73 - 8.14$

35. $-21.4 - (-14.9)$

36. $-32.6 - (-9.8)$

37. $(5.4)(-7.2)$

38. $(-8.5)(-3.3)$

39. $\dfrac{-1.2}{-6}$

40. $\dfrac{-6.3}{0.7}$

41. $\left(-\dfrac{1}{3}\right) + \left(-\dfrac{3}{4}\right)$

42. $-\dfrac{5}{6} + \dfrac{3}{8}$

43. $-\dfrac{3}{2} - \left(-\dfrac{3}{4}\right)$

44. $\dfrac{5}{8} - \dfrac{11}{12}$

45. $-\dfrac{2}{3} - \dfrac{7}{9}$

46. $\dfrac{5}{6} - \left(-\dfrac{2}{9}\right)$

47. $\left(-\dfrac{3}{4}\right)\left(\dfrac{4}{5}\right)$

48. $\left(\dfrac{1}{2}\right)\left(-\dfrac{4}{5}\right)$

49. $\dfrac{3}{4} \div \left(-\dfrac{1}{2}\right)$

50. $\left(-\dfrac{5}{6}\right) \div \left(-\dfrac{7}{8}\right)$

For Problems 51–90, simplify each numerical expression.

51. $9 - 12 - 8 + 5 - 6$

52. $6 - 9 + 11 - 8 - 7 + 14$

53. $-21 + (-17) - 11 + 15 - (-10)$

54. $-16 - (-14) + 16 + 17 - 19$

55. $7\dfrac{1}{8} - \left(2\dfrac{1}{4} - 3\dfrac{7}{8}\right)$

56. $-4\dfrac{3}{5} - \left(1\dfrac{1}{5} - 2\dfrac{3}{10}\right)$

57. $16 - 18 + 19 - [14 - 22 - (31 - 41)]$

58. $-19 - [15 - 13 - (-12 + 8)]$

59. $[14 - (16 - 18)] - [32 - (8 - 9)]$

60. $[-17 - (14 - 18)] - [21 - (-6 - 5)]$

61. $4\dfrac{1}{12} - \dfrac{1}{2}\left(\dfrac{1}{3}\right)$

62. $-\dfrac{4}{5} - \dfrac{1}{2}\left(-\dfrac{3}{5}\right)$

63. $-5 + (-2)(7) - (-3)(8)$

64. $-9 - 4(-2) + (-7)(6)$

65. $\dfrac{2}{5}\left(-\dfrac{3}{4}\right) - \left(-\dfrac{1}{2}\right)\left(\dfrac{3}{5}\right)$

66. $-\dfrac{2}{3}\left(\dfrac{1}{4}\right) + \left(-\dfrac{1}{3}\right)\left(\dfrac{5}{4}\right)$

67. $(-6)(-9) + (-7)(4)$

68. $(-7)(-7) - (-6)(4)$

69. $3(5 - 9) - 3(-6)$

70. $7(8 - 9) + (-6)(4)$

71. $(6 - 11)(4 - 9)$

72. $(7 - 12)(-3 - 2)$

73. $-6(-3 - 9 - 1)$

74. $-8(-3 - 4 - 6)$

75. $56 \div (-8) - (-6) \div (-2)$

76. $-65 \div 5 - (-13)(-2) + (-36) \div 12$

77. $-3[5 - (-2)] - 2(-4 - 9)$

78. $-2(-7 + 13) + 6(-3 - 2)$

79. $\dfrac{-6 + 24}{-3} + \dfrac{-7}{-6 - 1}$

80. $\dfrac{-12 + 20}{-4} + \dfrac{-7 - 11}{-9}$

81. $14.1 - (17.2 - 13.6)$

82. $-9.3 - (10.4 + 12.8)$

83. $3(2.1) - 4(3.2) - 2(-1.6)$

84. $5(-1.6) - 3(2.7) + 5(6.6)$

85. $7(6.2 - 7.1) - 6(-1.4 - 2.9)$

86. $-3(2.2 - 4.5) - 2(1.9 + 4.5)$

87. $\dfrac{2}{3} - \left(\dfrac{3}{4} - \dfrac{5}{6}\right)$

88. $-\dfrac{1}{2} - \left(\dfrac{3}{8} + \dfrac{1}{4}\right)$

89. $3\left(\dfrac{1}{2}\right) + 4\left(\dfrac{2}{3}\right) - 2\left(\dfrac{5}{6}\right)$

90. $2\left(\dfrac{3}{8}\right) - 5\left(\dfrac{1}{2}\right) + 6\left(\dfrac{3}{4}\right)$

91. Use a calculator to check your answers for Problems 51–86.

92. A scuba diver was 32 feet below sea level when he noticed that his partner had his extra knife. He ascended 13 feet to meet his partner and then continued to dive down for another 50 feet. How far below sea level is the diver?

93. Jeff played 18 holes of golf on Saturday. On each of 6 holes he was 1 under par, on each of 4 holes he was 2 over par, on 1 hole he was 3 over par, on each of 2 holes he shot par, and on each of 5 holes he was 1 over par. How did he finish relative to par?

94. After dieting for 30 days, Ignacio has lost 18 pounds. What number describes his average weight change per day?

95. Michael bet $5 on each of the 9 races at the racetrack. His only winnings were $28.50 on one race. How much did he win (or lose) for the day?

96. Max bought a piece of trim molding that measured $11\dfrac{3}{8}$ feet in length. Because of defects in the wood, he had to trim $1\dfrac{5}{8}$ feet off one end, and he also had to remove $\dfrac{3}{4}$ of a foot off the other end. How long was the piece of molding after he trimmed the ends?

97. Natasha recorded the daily gains or losses for her company stock for a week. On Monday it gained 1.25 dollars; on Tuesday it gained 0.88 dollars; on Wednesday it lost 0.50 dollars; on Thursday it lost 1.13 dollars; on Friday it gained 0.38 dollars. What was the net gain (or loss) for the week?

98. On a summer day in Florida, the afternoon temperature was 96°F. After a thunderstorm, the temperature dropped 8°F. What would be the temperature if the sun came back out and the temperature rose 5°F?

99. In an attempt to lighten a dragster, the racing team exchanged two rear wheels for wheels that each weighed 15.6 pounds less. They also exchanged the crankshaft for one that weighed 4.8 pounds less. They changed the rear axle for one that weighed 23.7 pounds less but had to add an additional roll bar that weighed 10.6 pounds. If they wanted to lighten the dragster by 50 pounds, did they meet their goal?

100. A large corporation has five divisions. Two of the divisions had earnings of $2,300,000 each. The other three divisions had a loss of $1,450,000, a loss of $640,000, and a gain of $1,850,000, respectively. What was the net gain (or loss) of the corporation for the year?

■ ■ ■ THOUGHTS INTO WORDS

101. Explain why $\dfrac{0}{8} = 0$, but $\dfrac{8}{0}$ is undefined.

102. The following simplification problem is incorrect. The answer should be -11. Find and correct the error.

$$8 \div (-4)(2) - 3(4) \div 2 + (-1) = (-2)(2) - 12 \div 1$$
$$= -4 - 12$$
$$= -16$$

1.3   Properties of Real Numbers and the Use of Exponents

At the beginning of this section we will list and briefly discuss some of the basic properties of real numbers. Be sure that you understand these properties, for they not only facilitate manipulations with real numbers but also serve as the basis for many algebraic computations.

Closure Property for Addition

If a and b are real numbers, then $a + b$ is a unique real number.

Closure Property for Multiplication

If a and b are real numbers, then ab is a unique real number.

We say that the set of real numbers is *closed* with respect to addition and also with respect to multiplication. That is, the sum of two real numbers is a unique real number, and the product of two real numbers is a unique real number. We use the word *unique* to indicate *exactly one*.

Commutative Property of Addition

If a and b are real numbers, then

$$a + b = b + a$$

Commutative Property of Multiplication

If a and b are real numbers, then

$$ab = ba$$

We say that addition and multiplication are commutative operations. This means that the order in which we add or multiply two numbers does not affect the result. For example, $6 + (-8) = (-8) + 6$ and $(-4)(-3) = (-3)(-4)$. It is also important to realize that subtraction and division are *not* commutative operations; order *does* make a difference. For example, $3 - 4 = -1$ but $4 - 3 = 1$. Likewise, $2 \div 1 = 2$ but $1 \div 2 = \dfrac{1}{2}$.

Associative Property of Addition

If a, b, and c are real numbers, then

$$(a + b) + c = a + (b + c)$$

Associative Property of Multiplication

If a, b, and c are real numbers, then

$$(ab)c = a(bc)$$

Addition and multiplication are **binary operations**. That is, we add (or multiply) two numbers at a time. The associative properties apply if more than two numbers are to be added or multiplied; they are grouping properties. For example, $(-8 + 9) + 6 = -8 + (9 + 6)$; changing the grouping of the numbers does not affect the final sum. This is also true for multiplication, which is illustrated by $[(-4)(-3)](2) = (-4)[(-3)(2)]$. Subtraction and division are *not* associative operations. For example, $(8 - 6) - 10 = -8$, but $8 - (6 - 10) = 12$. An example showing that division is not associative is $(8 \div 4) \div 2 = 1$, but $8 \div (4 \div 2) = 4$.

Identity Property of Addition

If a is any real number, then

$$a + 0 = 0 + a = a$$

Zero is called the identity element for addition. This merely means that the sum of any real number and zero is identically the same real number. For example, $-87 + 0 = 0 + (-87) = -87$.

Identity Property of Multiplication

If a is any real number, then

$$a(1) = 1(a) = a$$

We call 1 the identity element for multiplication. The product of any real number and 1 is identically the same real number. For example, $(-119)(1) = (1)(-119) = -119$.

Additive Inverse Property

For every real number a, there exists a unique real number $-a$ such that

$$a + (-a) = -a + a = 0$$

The real number $-a$ is called the **additive inverse of a** or the **opposite of a**. For example, 16 and -16 are additive inverses, and their sum is 0. The additive inverse of 0 is 0.

Multiplication Property of Zero

If a is any real number, then

$$(a)(0) = (0)(a) = 0$$

The product of any real number and zero is zero. For example, $(-17)(0) = 0(-17) = 0$.

Multiplication Property of Negative One

If a is any real number, then

$$(a)(-1) = (-1)(a) = -a$$

The product of any real number and -1 is the opposite of the real number. For example, $(-1)(52) = (52)(-1) = -52$.

Multiplicative Inverse Property

For every nonzero real number a, there exists a unique real number $\dfrac{1}{a}$ such that

$$a\left(\frac{1}{a}\right) = \frac{1}{a}(a) = 1$$

The number $\dfrac{1}{a}$ is called the **multiplicative inverse of a** or the **reciprocal of a**. For example, the reciprocal of 2 is $\dfrac{1}{2}$ and $2\left(\dfrac{1}{2}\right) = \dfrac{1}{2}(2) = 1$. Likewise, the reciprocal of $\dfrac{1}{2}$ is $\dfrac{1}{\dfrac{1}{2}} = 2$. Therefore, 2 and $\dfrac{1}{2}$ are said to be reciprocals (or multiplicative inverses) of each other. Because division by zero is undefined, zero does not have a reciprocal.

Distributive Property

> If a, b, and c are real numbers, then
>
> $$a(b + c) = ab + ac$$

The distributive property ties together the operations of addition and multiplication. We say that **multiplication distributes over addition**. For example, $7(3 + 8) = 7(3) + 7(8)$. Because $b - c = b + (-c)$, it follows that **multiplication also distributes over subtraction**. This can be expressed symbolically as $a(b - c) = ab - ac$. For example, $6(8 - 10) = 6(8) - 6(10)$.

The following examples illustrate the use of the properties of real numbers to facilitate certain types of manipulations.

EXAMPLE 1

Simplify $[74 + (-36)] + 36$.

Solution

In such a problem, it is much more advantageous to group -36 and 36.

$$[74 + (-36)] + 36 = 74 + [(-36) + 36] \qquad \text{By using the associative property for addition}$$
$$= 74 + 0 = 74$$ ∎

EXAMPLE 2

Simplify $[(-19)(25)](-4)$.

Solution

It is much easier to group 25 and -4. Thus

$$[(-19)(25)](-4) = (-19)[(25)(-4)] \qquad \text{By using the associative property for multiplication}$$
$$= (-19)(-100)$$
$$= 1900$$ ∎

EXAMPLE 3

Simplify $17 + (-14) + (-18) + 13 + (-21) + 15 + (-33)$.

Solution

We could add in the order in which the numbers appear. However, because addition is commutative and associative, we could change the order and group in any convenient way. For example, we could add all of the positive integers and add all of the negative integers, and then find the sum of these two results. It might be convenient to use the vertical format as follows:

$$
\begin{array}{rrr}
 & -14 & \\
17 & -18 & \\
13 & -21 & -86 \\
\dfrac{15}{45} & \dfrac{-33}{-86} & \dfrac{45}{-41}
\end{array}
$$
■

E X A M P L E   4

Simplify $-25(-2 + 100)$.

Solution

For this problem, it might be easiest to apply the distributive property first and then simplify.

$$
\begin{aligned}
-25(-2 + 100) &= (-25)(-2) + (-25)(100) \\
&= 50 + (-2500) \\
&= -2450
\end{aligned}
$$
■

E X A M P L E   5

Simplify $(-87)(-26 + 25)$.

Solution

For this problem, it would be better not to apply the distributive property but instead to add the numbers inside the parentheses first and then find the indicated product.

$$
\begin{aligned}
(-87)(-26 + 25) &= (-87)(-1) \\
&= 87
\end{aligned}
$$
■

E X A M P L E   6

Simplify $3.7(104) + 3.7(-4)$.

Solution

Remember that the distributive property allows us to change from the form $a(b + c)$ to $ab + ac$ or from the form $ab + ac$ to $a(b + c)$. In this problem, we want to use the latter change. Thus

$$
\begin{aligned}
3.7(104) + 3.7(-4) &= 3.7[104 + (-4)] \\
&= 3.7(100) \\
&= 370
\end{aligned}
$$
■

Examples 4, 5, and 6 illustrate an important issue. Sometimes the form $a(b + c)$ is more convenient, but at other times the form $ab + ac$ is better. In these cases, as well as in the cases of other properties, you should *think first* and decide whether or not the properties can be used to make the manipulations easier.

■ Exponents

Exponents are used to indicate repeated multiplication. For example, we can write $4 \cdot 4 \cdot 4$ as 4^3, where the "raised 3" indicates that 4 is to be used as a factor 3 times. The following general definition is helpful.

Definition 1.2

If n is a positive integer and b is any real number, then

$$b^n = \underbrace{bbb \cdots b}_{n \text{ factors of } b}$$

We refer to the b as the **base** and to n as the **exponent**. The expression b^n can be read "b to the nth power." We commonly associate the terms *squared* and *cubed* with exponents of 2 and 3, respectively. For example, b^2 is read "b squared" and b^3 as "b cubed." An exponent of 1 is usually not written, so b^1 is written as b. The following examples illustrate Definition 1.2.

$$2^3 = 2 \cdot 2 \cdot 2 = 8 \qquad \left(\frac{1}{2}\right)^5 = \frac{1}{2} \cdot \frac{1}{2} \cdot \frac{1}{2} \cdot \frac{1}{2} \cdot \frac{1}{2} = \frac{1}{32}$$

$$3^4 = 3 \cdot 3 \cdot 3 \cdot 3 = 81 \qquad (0.7)^2 = (0.7)(0.7) = 0.49$$

$$-5^2 = -(5 \cdot 5) = -25 \qquad (-5)^2 = (-5)(-5) = 25$$

Please take special note of the last two examples. Note that $(-5)^2$ means that -5 is the base and is to be used as a factor twice. However, -5^2 means that 5 is the base and that after it is squared, we take the opposite of that result.

Simplifying numerical expressions that contain exponents creates no trouble if we keep in mind that exponents are used to indicate repeated multiplication. Let's consider some examples.

EXAMPLE 7

Simplify $3(-4)^2 + 5(-3)^2$.

Solution

$$3(-4)^2 + 5(-3)^2 = 3(16) + 5(9) \qquad \text{Find the powers.}$$

$$= 48 + 45$$

$$= 93$$

■

EXAMPLE 8   Simplify $(2 + 3)^2$

Solution

$$(2 + 3)^2 = (5)^2 \qquad \text{Add inside the parentheses before applying the exponent.}$$

$$= 25 \qquad \text{Square the 5.} \qquad\blacksquare$$

EXAMPLE 9   Simplify $[3(-1) - 2(1)]^3$.

Solution

$$[3(-1) - 2(1)]^3 = [-3 - 2]^3$$

$$= [-5]^3$$

$$= -125 \qquad\blacksquare$$

EXAMPLE 10   Simplify $4\left(\dfrac{1}{2}\right)^3 - 3\left(\dfrac{1}{2}\right)^2 + 6\left(\dfrac{1}{2}\right) + 2.$

Solution

$$4\left(\frac{1}{2}\right)^3 - 3\left(\frac{1}{2}\right)^2 + 6\left(\frac{1}{2}\right) + 2 = 4\left(\frac{1}{8}\right) - 3\left(\frac{1}{4}\right) + 6\left(\frac{1}{2}\right) + 2$$

$$= \frac{1}{2} - \frac{3}{4} + 3 + 2$$

$$= \frac{19}{4} \qquad\blacksquare$$

Problem Set 1.3

For Problems 1–14, state the property that justifies each of the statements. For example, $3 + (-4) = (-4) + 3$ because of the commutative property of addition.

1. $[6 + (-2)] + 4 = 6 + [(-2) + 4]$

2. $x(3) = 3(x)$

3. $42 + (-17) = -17 + 42$

4. $1(x) = x$

5. $-114 + 114 = 0$

6. $(-1)(48) = -48$

7. $-1(x + y) = -(x + y)$

8. $-3(2 + 4) = -3(2) + (-3)(4)$

9. $12yx = 12xy$

10. $[(-7)(4)](-25) = (-7)[4(-25)]$

11. $7(4) + 9(4) = (7 + 9)4$

12. $(x + 3) + (-3) = x + [3 + (-3)]$

13. $[(-14)(8)](25) = (-14)[8(25)]$

14. $\left(\dfrac{3}{4}\right)\left(\dfrac{4}{3}\right) = 1$

For Problems 15–26, simplify each numerical expression. Be sure to take advantage of the properties whenever they can be used to make the computations easier.

15. $36 + (-14) + (-12) + 21 + (-9) - 4$

16. $-37 + 42 + 18 + 37 + (-42) - 6$

17. $[83 + (-99)] + 18$

18. $[63 + (-87)] + (-64)$

19. $(25)(-13)(4)$

20. $(14)(25)(-13)(4)$

21. $17(97) + 17(3)$

22. $-86[49 + (-48)]$

23. $14 - 12 - 21 - 14 + 17 - 18 + 19 - 32$

24. $16 - 14 - 13 - 18 + 19 + 14 - 17 + 21$

25. $(-50)(15)(-2) - (-4)(17)(25)$

26. $(2)(17)(-5) - (4)(13)(-25)$

For Problems 27–54, simplify each of the numerical expressions.

27. $2^3 - 3^3$

28. $3^2 - 2^4$

29. $-5^2 - 4^2$

30. $-7^2 + 5^2$

31. $(-2)^3 - 3^2$

32. $(-3)^3 + 3^2$

33. $3(-1)^3 - 4(3)^2$

34. $4(-2)^3 - 3(-1)^4$

35. $7(2)^3 + 4(-2)^3$

36. $-4(-1)^2 - 3(2)^3$

37. $-3(-2)^3 + 4(-1)^5$

38. $5(-1)^3 - (-3)^3$

39. $(-3)^2 - 3(-2)(5) + 4^2$

40. $(-2)^2 - 3(-2)(6) - (-5)^2$

41. $2^3 + 3(-1)^3(-2)^2 - 5(-1)(2)^2$

42. $-2(3)^2 - 2(-2)^3 - 6(-1)^5$

43. $(3 + 4)^2$

44. $(4 - 9)^2$

45. $[3(-2)^2 - 2(-3)^2]^3$

46. $[-3(-1)^3 - 4(-2)^2]^2$

47. $2(-1)^3 - 3(-1)^2 + 4(-1) - 5$

48. $(-2)^3 + 2(-2)^2 - 3(-2) - 1$

49. $2^4 - 2(2)^3 - 3(2)^2 + 7(2) - 10$

50. $3(-3)^3 + 4(-3)^2 - 5(-3) + 7$

51. $3\left(\dfrac{1}{2}\right)^4 - 2\left(\dfrac{1}{2}\right)^3 + 5\left(\dfrac{1}{2}\right)^2 - 4\left(\dfrac{1}{2}\right) + 1$

52. $4(0.1)^2 - 6(0.1) + 0.7$

53. $-\left(\dfrac{2}{3}\right)^2 + 5\left(\dfrac{2}{3}\right) - 4$

54. $4\left(\dfrac{1}{3}\right)^3 + 3\left(\dfrac{1}{3}\right)^2 + 2\left(\dfrac{1}{3}\right) + 6$

C **55.** Use your calculator to check your answers for Problems 27–52.

C For Problems 56–64, use your calculator to evaluate each numerical expression.

56. 2^{10}

57. 3^7

58. $(-2)^8$

59. $(-2)^{11}$

60. -4^9

61. -5^6

62. $(3.14)^3$

63. $(1.41)^4$

64. $(1.73)^5$

The symbol, **C**, signals a problem that requires a calculator.

■ ■ ■ THOUGHTS INTO WORDS

65. State, in your own words, the multiplication property of negative one.

66. Explain how the associative and commutative properties can help simplify $[(25)(97)](-4)$.

67. Your friend keeps getting an answer of 64 when simplifying -2^6. What mistake is he making, and how would you help him?

68. Write a sentence explaining in your own words how to evaluate the expression $(-8)^2$. Also write a sentence explaining how to evaluate -8^2.

69. For what natural numbers n does $(-1)^n = -1$? For what natural numbers n does $(-1)^n = 1$? Explain your answers.

70. Is the set $\{0, 1\}$ closed with respect to addition? Is the set $\{0, 1\}$ closed with respect to multiplication? Explain your answers.

1.4 Algebraic Expressions

Algebraic expressions such as

$$2x, \quad 8xy, \quad 3xy^2, \quad -4a^2b^3c, \quad \text{and} \quad z$$

are called **terms**. A term is an indicated product that may have any number of factors. The variables involved in a term are called **literal factors**, and the numerical factor is called the **numerical coefficient**. Thus in $8xy$, the x and y are literal factors, and 8 is the numerical coefficient. The numerical coefficient of the term $-4a^2bc$ is -4. Because $1(z) = z$, the numerical coefficient of the term z is understood to be 1. Terms that have the same literal factors are called **similar terms** or **like terms**. Some examples of similar terms are

$$3x \quad \text{and} \quad 14x \qquad\qquad 5x^2 \quad \text{and} \quad 18x^2$$

$$7xy \quad \text{and} \quad -9xy \qquad\qquad 9x^2y \quad \text{and} \quad -14x^2y$$

$$2x^3y^2, \quad 3x^3y^2, \quad \text{and} \quad -7x^3y^2$$

By the symmetric property of equality, we can write the distributive property as

$$ab + ac = a(b + c)$$

Then the commutative property of multiplication can be applied to change the form to

$$ba + ca = (b + c)a$$

This latter form provides the basis for simplifying algebraic expressions by **combining similar terms**. Consider the following examples.

$$3x + 5x = (3 + 5)x \qquad\qquad -6xy + 4xy = (-6 + 4)xy$$
$$= 8x \qquad\qquad\qquad\qquad = -2xy$$
$$5x^2 + 7x^2 + 9x^2 = (5 + 7 + 9)x^2 \qquad 4x - x = 4x - 1x$$
$$= 21x^2 \qquad\qquad\qquad\qquad = (4 - 1)x = 3x$$

More complicated expressions might require that we first rearrange the terms by applying the commutative property for addition.

$$7x + 2y + 9x + 6y = 7x + 9x + 2y + 6y$$
$$= (7 + 9)x + (2 + 6)y \qquad \text{Distributive property}$$
$$= 16x + 8y$$

$$6a - 5 - 11a + 9 = 6a + (-5) + (-11a) + 9$$
$$= 6a + (-11a) + (-5) + 9 \qquad \text{Commutative property}$$
$$= (6 + (-11))a + 4 \qquad \text{Distributive property}$$
$$= -5a + 4$$

As soon as you thoroughly understand the various simplifying steps, you may want to do the steps mentally. Then you could go directly from the given expression to the simplified form, as follows:

$$14x + 13y - 9x + 2y = 5x + 15y$$

$$3x^2y - 2y + 5x^2y + 8y = 8x^2y + 6y$$

$$-4x^2 + 5y^2 - x^2 - 7y^2 = -5x^2 - 2y^2$$

Applying the distributive property to remove parentheses and then to combine similar terms sometimes simplifies an algebraic expression (as the next examples illustrate).

$$
\begin{aligned}
4(x + 2) + 3(x + 6) &= 4(x) + 4(2) + 3(x) + 3(6) \\
&= 4x + 8 + 3x + 18 \\
&= 4x + 3x + 8 + 18 \\
&= (4 + 3)x + 26 \\
&= 7x + 26
\end{aligned}
$$

$$
\begin{aligned}
-5(y + 3) - 2(y - 8) &= -5(y) - 5(3) - 2(y) - 2(-8) \\
&= -5y - 15 - 2y + 16 \\
&= -5y - 2y - 15 + 16 \\
&= -7y + 1
\end{aligned}
$$

$$
\begin{aligned}
5(x - y) - (x + y) &= 5(x - y) - 1(x + y) \qquad \text{Remember, } -a = -1(a). \\
&= 5(x) - 5(y) - 1(x) - 1(y) \\
&= 5x - 5y - 1x - 1y \\
&= 4x - 6y
\end{aligned}
$$

When we are multiplying two terms such as 3 and $2x$, the associative property for multiplication provides the basis for simplifying the product.

$$3(2x) = (3 \cdot 2)x = 6x$$

This idea is put to use in the following example.

$$
\begin{aligned}
3(2x + 5y) + 4(3x + 2y) &= 3(2x) + 3(5y) + 4(3x) + 4(2y) \\
&= 6x + 15y + 12x + 8y \\
&= 6x + 12x + 15y + 8y \\
&= 18x + 23y
\end{aligned}
$$

After you are sure of each step, a more simplified format may be used, as the following examples illustrate.

$$
\begin{aligned}
5(a + 4) - 7(a + 3) &= 5a + 20 - 7a - 21 \qquad \text{Be careful with this sign.} \\
&= -2a - 1
\end{aligned}
$$

$$3(x^2 + 2) + 4(x^2 - 6) = 3x^2 + 6 + 4x^2 - 24$$
$$= 7x^2 - 18$$

$$2(3x - 4y) - 5(2x - 6y) = 6x - 8y - 10x + 30y$$
$$= -4x + 22y$$

■ Evaluating Algebraic Expressions

An algebraic expression takes on a numerical value whenever each variable in the expression is replaced by a real number. For example, if x is replaced by 5 and y by 9, the algebraic expression $x + y$ becomes the numerical expression $5 + 9$, which simplifies to 14. We say that $x + y$ has a value of 14 when x equals 5 and y equals 9. If $x = -3$ and $y = 7$, then $x + y$ has a value of $-3 + 7 = 4$. The following examples illustrate the process of finding a value of an algebraic expression. We commonly refer to the process as **evaluating algebraic expressions**.

EXAMPLE 1 Find the value of $3x - 4y$ when $x = 2$ and $y = -3$.

Solution

$$3x - 4y = 3(2) - 4(-3), \quad \text{when } x = 2 \text{ and } y = -3$$
$$= 6 + 12$$
$$= 18 \qquad \blacksquare$$

EXAMPLE 2 Evaluate $x^2 - 2xy + y^2$ for $x = -2$ and $y = -5$.

Solution

$$x^2 - 2xy + y^2 = (-2)^2 - 2(-2)(-5) + (-5)^2, \quad \text{when } x = -2 \text{ and } y = -5$$
$$= 4 - 20 + 25$$
$$= 9 \qquad \blacksquare$$

EXAMPLE 3 Evaluate $(a + b)^2$ for $a = 6$ and $b = -2$.

Solution

$$(a + b)^2 = [6 + (-2)]^2, \quad \text{when } a = 6 \text{ and } b = -2$$
$$= (4)^2$$
$$= 16 \qquad \blacksquare$$

EXAMPLE 4

Evaluate $(3x + 2y)(2x - y)$ for $x = 4$ and $y = -1$.

Solution

$$(3x + 2y)(2x - y) = [3(4) + 2(-1)][2(4) - (-1)] \quad \text{when } x = 4$$
$$\text{and } y = -1$$

$$= (12 - 2)(8 + 1)$$

$$= (10)(9)$$

$$= 90 \qquad \blacksquare$$

EXAMPLE 5

Evaluate $7x - 2y + 4x - 3y$ for $x = -\dfrac{1}{2}$ and $y = \dfrac{2}{3}$.

Solution

Let's first simplify the given expression.

$$7x - 2y + 4x - 3y = 11x - 5y$$

Now we can substitute $-\dfrac{1}{2}$ for x and $\dfrac{2}{3}$ for y.

$$11x - 5y = 11\left(-\frac{1}{2}\right) - 5\left(\frac{2}{3}\right)$$

$$= -\frac{11}{2} - \frac{10}{3}$$

$$= -\frac{33}{6} - \frac{20}{6} \qquad \text{Change to equivalent fractions}$$
$$\text{with a common denominator.}$$

$$= -\frac{53}{6} \qquad \blacksquare$$

EXAMPLE 6

Evaluate $2(3x + 1) - 3(4x - 3)$ for $x = -6.2$.

Solution

Let's first simplify the given expression.

$$2(3x + 1) - 3(4x - 3) = 6x + 2 - 12x + 9$$

$$= -6x + 11$$

Now we can substitute -6.2 for x.

$$-6x + 11 = -6(-6.2) + 11$$

$$= 37.2 + 11$$

$$= 48.2 \qquad \blacksquare$$

Evaluate $2(a^2 + 1) - 3(a^2 + 5) + 4(a^2 - 1)$ for $a = 10$.

Solution

Let's first simplify the given expression.

$$2(a^2 + 1) - 3(a^2 + 5) + 4(a^2 - 1) = 2a^2 + 2 - 3a^2 - 15 + 4a^2 - 4$$

$$= 3a^2 - 17$$

Substituting $a = 10$, we obtain

$$3a^2 - 17 = 3(10)^2 - 17$$

$$= 3(100) - 17$$

$$= 300 - 17$$

$$= 283$$ ■

■ Translating from English to Algebra

To use the tools of algebra to solve problems, we must be able to translate from English to algebra. This translation process requires that we recognize key phrases in the English language that translate into algebraic expressions (which involve the operations of addition, subtraction, multiplication, and division). Some of these key phrases and their algebraic counterparts are listed in the following table. The variable n represents the number being referred to in each phrase. When translating, remember that the commutative property holds only for the operations of addition and multiplication. Therefore, order will be crucial to algebraic expressions that involve subtraction and division.

English phrase	Algebraic expression
Addition	
The sum of a number and 4	$n + 4$
7 more than a number	$n + 7$
A number plus 10	$n + 10$
A number increased by 6	$n + 6$
8 added to a number	$n + 8$
Subtraction	
14 minus a number	$14 - n$
12 less than a number	$n - 12$
A number decreased by 10	$n - 10$
The difference between a number and 2	$n - 2$
5 subtracted from a number	$n - 5$

English phrase	Algebraic expression
Multiplication	
14 times a number	$14n$
The product of 4 and a number	$4n$
$\frac{3}{4}$ of a number	$\frac{3}{4}n$
Twice a number	$2n$
Multiply a number by 12	$12n$
Division	
The quotient of 6 and a number	$\frac{6}{n}$
The quotient of a number and 6	$\frac{n}{6}$
A number divided by 9	$\frac{n}{9}$
The ratio of a number and 4	$\frac{n}{4}$
Mixture of operations	
4 more than three times a number	$3n + 4$
5 less than twice a number	$2n - 5$
3 times the sum of a number and 2	$3(n + 2)$
2 more than the quotient of a number and 12	$\frac{n}{12} + 2$
7 times the difference of 6 and a number	$7(6 - n)$

An English statement may not always contain a key word such as *sum, difference, product,* or *quotient.* Instead, the statement may describe a physical situation, and from this description we must deduce the operations involved. Some suggestions for handling such situations are given in the following examples.

EXAMPLE 8

Sonya can type 65 words per minute. How many words will she type in m minutes?

Solution

The total number of words typed equals the product of the rate per minute and the number of minutes. Therefore, Sonya should be able to type $65m$ words in m minutes. ∎

EXAMPLE 9

Russ has n nickels and d dimes. Express this amount of money in cents.

Solution

Each nickel is worth 5 cents and each dime is worth 10 cents. We represent the amount in cents by $5n + 10d$. ■

EXAMPLE 10

The cost of a 50-pound sack of fertilizer is d dollars. What is the cost per pound for the fertilizer?

Solution

We calculate the cost per pound by dividing the total cost by the number of pounds. We represent the cost per pound by $\dfrac{d}{50}$. ■

The English statement we want to translate into algebra may contain some geometric ideas. Tables 1.1 and 1.2 contain some of the basic relationships that pertain to linear measurement in the English and metric systems, respectively.

Table 1.1   English system

12 inches = 1 foot
3 feet = 1 yard
1760 yards = 1 mile
5280 feet = 1 mile

Table 1.2   Metric system

1 kilometer = 1000 meters
1 hectometer = 100 meters
1 dekameter = 10 meters
1 decimeter = 0.1 meter
1 centimeter = 0.01 meter
1 millimeter = 0.001 meter

EXAMPLE 11

The distance between two cities is k kilometers. Express this distance in meters.

Solution

Because 1 kilometer equals 1000 meters, the distance in meters is represented by $1000k$. ■

EXAMPLE 12

The length of a rope is y yards and f feet. Express this length in inches.

Solution

Because 1 foot equals 12 inches and 1 yard equals 36 inches, the length of the rope in inches can be represented by $36y + 12f$. ■

E X A M P L E   1 3   The length of a rectangle is l centimeters and the width is w centimeters. Express the perimeter of the rectangle in meters.

Solution

A sketch of the rectangle may be helpful (Figure 1.7).

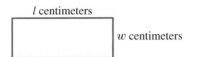

l centimeters

w centimeters

Figure 1.7

The perimeter of a rectangle is the sum of the lengths of the four sides. Thus the perimeter in centimeters is $l + w + l + w$, which simplifies to $2l + 2w$. Now, because 1 centimeter equals 0.01 meter, the perimeter, in meters, is $0.01(2l + 2w)$. This could also be written as $\dfrac{2l + 2w}{100} = \dfrac{2(l + w)}{100} = \dfrac{l + w}{50}$. ∎

Problem Set 1.4

Simplify the algebraic expressions in Problems 1–14 by combining similar terms.

1. $-7x + 11x$

2. $5x - 8x + x$

3. $5a^2 - 6a^2$

4. $12b^3 - 17b^3$

5. $4n - 9n - n$

6. $6n + 13n - 15n$

7. $4x - 9x + 2y$

8. $7x - 9y - 10x - 13y$

9. $-3a^2 + 7b^2 + 9a^2 - 2b^2$

10. $-xy + z - 8xy - 7z$

11. $15x - 4 + 6x - 9$

12. $5x - 2 - 7x + 4 - x - 1$

13. $5a^2b - ab^2 - 7a^2b$

14. $8xy^2 - 5x^2y + 2xy^2 + 7x^2y$

Simplify the algebraic expressions in Problems 15–34 by removing parentheses and combining similar terms.

15. $3(x + 2) + 5(x + 3)$

16. $5(x - 1) + 7(x + 4)$

17. $-2(a - 4) - 3(a + 2)$

18. $-7(a + 1) - 9(a + 4)$

19. $3(n^2 + 1) - 8(n^2 - 1)$

20. $4(n^2 + 3) + (n^2 - 7)$

21. $-6(x^2 - 5) - (x^2 - 2)$

22. $3(x + y) - 2(x - y)$

23. $5(2x + 1) + 4(3x - 2)$

24. $5(3x - 1) + 6(2x + 3)$

25. $3(2x - 5) - 4(5x - 2)$

26. $3(2x - 3) - 7(3x - 1)$

27. $-2(n^2 - 4) - 4(2n^2 + 1)$

28. $-4(n^2 + 3) - (2n^2 - 7)$

29. $3(2x - 4y) - 2(x + 9y)$

30. $-7(2x - 3y) + 9(3x + y)$

31. $3(2x - 1) - 4(x + 2) - 5(3x + 4)$

32. $-2(x - 1) - 5(2x + 1) + 4(2x - 7)$

33. $-(3x - 1) - 2(5x - 1) + 4(-2x - 3)$

34. $4(-x - 1) + 3(-2x - 5) - 2(x + 1)$

Evaluate the algebraic expressions in Problems 35–57 for the given values of the variables.

35. $3x + 7y$, $x = -1$ and $y = -2$

36. $5x - 9y$, $x = -2$ and $y = 5$

37. $4x^2 - y^2$, $x = 2$ and $y = -2$

38. $3a^2 + 2b^2$, $a = 2$ and $b = 5$

39. $2a^2 - ab + b^2$, $a = -1$ and $b = -2$

40. $-x^2 + 2xy + 3y^2$, $x = -3$ and $y = 3$

41. $2x^2 - 4xy - 3y^2$, $x = 1$ and $y = -1$

42. $4x^2 + xy - y^2$, $x = 3$ and $y = -2$

43. $3xy - x^2y^2 + 2y^2$, $x = 5$ and $y = -1$

44. $x^2y^3 - 2xy + x^2y^2$, $x = -1$ and $y = -3$

45. $7a - 2b - 9a + 3b$, $a = 4$ and $b = -6$

46. $-4x + 9y - 3x - y$, $x = -4$ and $y = 7$

47. $(x - y)^2$, $x = 5$ and $y = -3$

48. $2(a + b)^2$, $a = 6$ and $b = -1$

49. $-2a - 3a + 7b - b$, $a = -10$ and $b = 9$

50. $3(x - 2) - 4(x + 3)$, $x = -2$

51. $-2(x + 4) - (2x - 1)$, $x = -3$

52. $-4(2x - 1) + 7(3x + 4)$, $x = 4$

53. $2(x - 1) - (x + 2) - 3(2x - 1)$, $x = -1$

54. $-3(x + 1) + 4(-x - 2) - 3(-x + 4)$, $x = -\dfrac{1}{2}$

55. $3(x^2 - 1) - 4(x^2 + 1) - (2x^2 - 1)$, $x = \dfrac{2}{3}$

56. $2(n^2 + 1) - 3(n^2 - 3) + 3(5n^2 - 2)$, $n = \dfrac{1}{4}$

57. $5(x - 2y) - 3(2x + y) - 2(x - y)$, $x = \dfrac{1}{3}$ and $y = -\dfrac{3}{4}$

C For Problems 58–63, use your calculator and evaluate each of the algebraic expressions for the indicated values. Express the final answers to the nearest tenth.

58. πr^2, $\pi = 3.14$ and $r = 2.1$

59. πr^2, $\pi = 3.14$ and $r = 8.4$

60. $\pi r^2 h$, $\pi = 3.14$, $r = 1.6$, and $h = 11.2$

61. $\pi r^2 h$, $\pi = 3.14$, $r = 4.8$, and $h = 15.1$

62. $2\pi r^2 + 2\pi rh$, $\pi = 3.14$, $r = 3.9$, and $h = 17.6$

63. $2\pi r^2 + 2\pi rh$, $\pi = 3.14$, $r = 7.8$, and $h = 21.2$

For Problems 64–78, translate each English phrase into an algebraic expression and use n to represent the unknown number.

64. The sum of a number and 4

65. A number increased by 12

66. A number decreased by 7

67. Five less than a number

68. A number subtracted from 75

69. The product of a number and 50

70. One-third of a number

71. Four less than one-half of a number

72. Seven more than three times a number

73. The quotient of a number and 8

74. The quotient of 50 and a number

75. Nine less than twice a number

76. Six more than one-third of a number

77. Ten times the difference of a number and 6

78. Twelve times the sum of a number and 7

For Problems 79–99, answer the question with an algebraic expression.

79. Brian is n years old. How old will he be in 20 years?

80. Crystal is n years old. How old was she 5 years ago?

81. Pam is t years old, and her mother is 3 less than twice as old as Pam. What is the age of Pam's mother?

82. The sum of two numbers is 65, and one of the numbers is x. What is the other number?

83. The difference of two numbers is 47, and the smaller number is n. What is the other number?

84. The product of two numbers is 98, and one of the numbers is n. What is the other number?

85. The quotient of two numbers is 8, and the smaller number is y. What is the other number?

86. The perimeter of a square is c centimeters. How long is each side of the square?

87. The perimeter of a square is *m* meters. How long, in centimeters, is each side of the square?

88. Jesse has *n* nickels, *d* dimes, and *q* quarters in his bank. How much money, in cents, does he have in his bank?

89. Tina has *c* cents, which is all in quarters. How many quarters does she have?

90. If *n* represents a whole number, what represents the next larger whole number?

91. If *n* represents an odd integer, what represents the next larger odd integer?

92. If *n* represents an even integer, what represents the next larger even integer?

93. The cost of a 5-pound box of candy is *c* cents. What is the price per pound?

The symbol, $\boxed{\text{C}}$, signals a problem that requires a calculator.

94. Larry's annual salary is *d* dollars. What is his monthly salary?

95. Mila's monthly salary is *d* dollars. What is her annual salary?

96. The perimeter of a square is *i* inches. What is the perimeter expressed in feet?

97. The perimeter of a rectangle is *y* yards and *f* feet. What is the perimeter expressed in feet?

98. The length of a line segment is *d* decimeters. How long is the line segment expressed in meters?

99. The distance between two cities is *m* miles. How far is this, expressed in feet?

$\boxed{\text{C}}$ **100.** Use your calculator to check your answers for Problems 35–54.

■ ■ ■ **THOUGHTS INTO WORDS**

101. Explain the difference between simplifying a numerical expression and evaluating an algebraic expression.

102. How would you help someone who is having difficulty expressing *n* nickels and *d* dimes in terms of cents?

103. When asked to write an algebraic expression for "8 more than a number," you wrote $x + 8$ and another student wrote $8 + x$. Are both expressions correct? Explain your answer.

104. When asked to write an algebraic expression for "6 less than a number," you wrote $x - 6$ and another student wrote $6 - x$. Are both expressions correct? Explain your answer.

Chapter 1 Summary

(1.1) A **set** is a collection of objects; the objects are called **elements** or **members** of the set. Set A is a subset of set B if and only if every member of A is also a member of B. The sets of **natural numbers, whole numbers, integers, rational numbers**, and **irrational numbers** are all subsets of the set of **real numbers**.

We can evaluate **numerical expressions** by performing the operations in the following order.

1. Perform the operations inside the parentheses and above and below fraction bars.
2. Find all powers or convert them to indicated multiplication.
3. Perform all multiplications and divisions in the order in which they appear from left to right.
4. Perform all additions and subtractions in the order in which they appear from left to right.

(1.2) The **absolute value** of a real number a is defined as follows:

1. If $a \geq 0$, then $|a| = a$.
2. If $a < 0$, then $|a| = -a$.

■ Operations with Real Numbers

Addition

1. The sum of two positive real numbers is the sum of their absolute values.
2. The sum of two negative real numbers is the opposite of the sum of their absolute values.
3. The sum of one positive and one negative number is found as follows:
   a. If the positive number has the larger absolute value, then the sum is the difference of their absolute values when the smaller absolute value is subtracted from the larger absolute value.
   b. If the negative number has the larger absolute value, then the sum is the opposite of the difference of their absolute values when the smaller absolute value is subtracted from the larger absolute value.

Subtraction

Applying the principle that $a - b = a + (-b)$ changes every subtraction problem to an equivalent addition problem. Then the rules for addition can be followed.

Multiplication

1. The product of two positive numbers or two negative real numbers is the product of their absolute values.
2. The product of one positive and one negative real number is the opposite of the product of their absolute values.

Division

1. The quotient of two positive numbers or two negative real numbers is the quotient of their absolute values.
2. The quotient of one positive and one negative real number is the opposite of the quotient of their absolute values.

(1.3) The following basic properties of real numbers help with numerical manipulations and serve as a basis for algebraic computations.

■ Closure properties

$a + b$ is a real number

ab is a real number

■ Commutative properties

$a + b = b + a$

$ab = ba$

■ Associative properties

$(a + b) + c = a + (b + c)$

$(ab)c = a(bc)$

■ Identity properties

$a + 0 = 0 + a = a$

$a(1) = 1(a) = a$

■ **Additive inverse property**

$$a + (-a) = (-a) + a = 0$$

■ **Multiplication property of zero**

$$a(0) = 0(a) = 0$$

■ **Multiplication property of negative one**

$$-1(a) = a(-1) = -a$$

■ **Multiplicative inverse property**

$$a\left(\frac{1}{a}\right) = \left(\frac{1}{a}\right)a = 1$$

■ **Distributive properties**

$$a(b + c) = ab + ac$$

$$a(b - c) = ab - ac$$

(1.4) Algebraic expressions such as

$$2x, \quad 8xy, \quad 3xy^2, \quad -4a^2b^3c, \quad \text{and} \quad z$$

are called **terms**. A term is an indicated product and may have any number of factors. We call the variables in a term the **literal factors**, and we call the numerical factor the **numerical coefficient**. Terms that have the same literal factors are called **similar** or **like terms**.

The distributive property in the form $ba + ca = (b + c)a$ serves as the basis for **combining similar terms**. For example,

$$3x^2y + 7x^2y = (3 + 7)x^2y = 10x^2y$$

To translate English phrases into algebraic expressions, we must be familiar with the key phrases that signal whether we are to find a sum, difference, product, or quotient.

Chapter 1   Review Problem Set

1. From the list $0, \sqrt{2}, \dfrac{3}{4}, -\dfrac{5}{6}, \dfrac{25}{3}, -\sqrt{3}, -8, 0.34, 0.2\overline{3},$

67, and $\dfrac{9}{7}$, identify each of the following.

  a. The natural numbers

  b. The integers

  c. The nonnegative integers

  d. The rational numbers

  e. The irrational numbers

For Problems 2–10, state the property of equality or the property of real numbers that justifies each of the statements. For example, $6(-7) = -7(6)$ because of the commutative property for multiplication; and if $2 = x + 3$, then $x + 3 = 2$ is true because of the symmetric property of equality.

2. $7 + (3 + (-8)) = (7 + 3) + (-8)$

3. If $x = 2$ and $x + y = 9$, then $2 + y = 9$.

4. $-1(x + 2) = -(x + 2)$

5. $3(x + 4) = 3(x) + 3(4)$

6. $[(17)(4)](25) = (17)[(4)(25)]$

7. $x + 3 = 3 + x$

8. $3(98) + 3(2) = 3(98 + 2)$

9. $\left(\dfrac{3}{4}\right)\left(\dfrac{4}{3}\right) = 1$

10. If $4 = 3x - 1$, then $3x - 1 = 4$.

For Problems 11–22, simplify each of the numerical expressions.

11. $-8\dfrac{1}{4} + \left(-4\dfrac{5}{8}\right) - \left(-6\dfrac{3}{8}\right)$

12. $9\dfrac{1}{3} - 12\dfrac{1}{2} + \left(-4\dfrac{1}{6}\right) - \left(-1\dfrac{1}{6}\right)$

13. $-8(2) - 16 \div (-4) + (-2)(-2)$

41

14. $4(-3) - 12 \div (-4) + (-2)(-1) - 8$

15. $-3(2 - 4) - 4(7 - 9) + 6$

16. $[48 + (-73)] + 74$

17. $[5(-2) - 3(-1)][-2(-1) + 3(2)]$

18. $-4^2 - 2^3$

19. $(-2)^4 + (-1)^3 - 3^2$

20. $2(-1)^2 - 3(-1)(2) - 2^2$

21. $[4(-1) - 2(3)]^2$

22. $3 - [-2(3 - 4)] + 7$

For Problems 23–32, simplify each of the algebraic expressions by combining similar terms.

23. $3a^2 - 2b^2 - 7a^2 - 3b^2$

24. $4x - 6 - 2x - 8 + x + 12$

25. $\dfrac{1}{5}ab^2 - \dfrac{3}{10}ab^2 + \dfrac{2}{5}ab^2 + \dfrac{7}{10}ab^2$

26. $-\dfrac{2}{3}x^2y - \left(-\dfrac{3}{4}x^2y\right) - \dfrac{5}{12}x^2y - 2x^2y$

27. $3(2n^2 + 1) + 4(n^2 - 5)$

28. $-2(3a - 1) + 4(2a + 3) - 5(3a + 2)$

29. $-(n - 1) - (n + 2) + 3$

30. $3(2x - 3y) - 4(3x + 5y) - x$

31. $4(a - 6) - (3a - 1) - 2(4a - 7)$

32. $-5(x^2 - 4) - 2(3x^2 + 6) + (2x^2 - 1)$

For Problems 33–42, evaluate each of the algebraic expressions for the given values of the variables.

33. $-5x + 4y$ for $x = \dfrac{1}{2}$ and $y = -1$

34. $3x^2 - 2y^2$ for $x = \dfrac{1}{4}$ and $y = -\dfrac{1}{2}$

35. $-5(2x - 3y)$ for $x = 1$ and $y = -3$

36. $(3a - 2b)^2$ for $a = -2$ and $b = 3$

37. $a^2 + 3ab - 2b^2$ for $a = 2$ and $b = -2$

38. $3n^2 - 4 - 4n^2 + 9$ for $n = 7$

39. $3(2x - 1) + 2(3x + 4)$ for $x = 1.2$

40. $-4(3x - 1) - 5(2x - 1)$ for $x = -2.3$

41. $2(n^2 + 3) - 3(n^2 + 1) + 4(n^2 - 6)$ for $n = -\dfrac{2}{3}$

42. $5(3n - 1) - 7(-2n + 1) + 4(3n - 1)$ for $n = \dfrac{1}{2}$

For Problems 43–50, translate each English phrase into an algebraic expression and use n to represent the unknown number.

43. Four increased by twice a number

44. Fifty subtracted from three times a number

45. Six less than two-thirds of a number

46. Ten times the difference of a number and 14

47. Eight subtracted from five times a number

48. The quotient of a number and three less than the number

49. Three less than five times the sum of a number and 2

50. Three-fourths of the sum of a number and 12

For Problems 51–60, answer the question with an algebraic expression.

51. The sum of two numbers is 37 and one of the numbers is n. What is the other number?

52. Yuriko can type w words in an hour. What is her typing rate per minute?

53. Harry is y years old. His brother is 7 years less than twice as old as Harry. How old is Harry's brother?

54. If n represents a multiple of 3, what represents the next largest multiple of 3?

55. Celia has p pennies, n nickels, and q quarters. How much, in cents, does Celia have?

56. The perimeter of a square is i inches. How long, in feet, is each side of the square?

57. The length of a rectangle is y yards and the width is f feet. What is the perimeter of the rectangle expressed in inches?

58. The length of a piece of wire is d decimeters. What is the length expressed in centimeters?

59. Joan is f feet and i inches tall. How tall is she in inches?

60. The perimeter of a rectangle is 50 centimeters. If the rectangle is c centimeters long, how wide is it?

Chapter 1   Test

1. State the property of equality that justifies writing $x + 4 = 6$ for $6 = x + 4$.

2. State the property of real numbers that justifies writing $5(10 + 2)$ as $5(10) + 5(2)$.

For Problems 3–11, simplify each numerical expression.

3. $-4 - (-3) + (-5) - 7 + 10$

4. $7 - 8 - 3 + 4 - 9 - 4 + 2 - 12$

5. $5\left(-\dfrac{1}{3}\right) - 3\left(-\dfrac{1}{2}\right) + 7\left(-\dfrac{2}{3}\right) + 1$

6. $(-6) \cdot 3 \div (-2) - 8 \div (-4)$

7. $-\dfrac{1}{2}(3 - 7) - \dfrac{2}{5}(2 - 17)$

8. $[48 + (-93)] + (-49)$

9. $3(-2)^3 + 4(-2)^2 - 9(-2) - 14$

10. $[2(-6) + 5(-4)][-3(-4) - 7(6)]$

11. $[-2(-3) - 4(2)]^5$

12. Simplify $6x^2 - 3x - 7x^2 - 5x - 2$ by combining similar terms.

13. Simplify $3(3n - 1) - 4(2n + 3) + 5(-4n - 1)$ by removing parentheses and combining similar terms.

For Problems 14–20, evaluate each algebraic expression for the given values of the variables.

14. $-7x - 3y$ for $x = -6$ and $y = 5$

15. $3a^2 - 4b^2$ for $a = -\dfrac{3}{4}$ and $b = \dfrac{1}{2}$

16. $6x - 9y - 8x + 4y$ for $x = \dfrac{1}{2}$ and $y = -\dfrac{1}{3}$

17. $-5n^2 - 6n + 7n^2 + 5n - 1$ for $n = -6$

18. $-7(x - 2) + 6(x - 1) - 4(x + 3)$ for $x = 3.7$

19. $-2xy - x + 4y$ for $x = -3$ and $y = 9$

20. $4(n^2 + 1) - (2n^2 + 3) - 2(n^2 + 3)$ for $n = -4$

For Problems 21 and 22, translate the English phrase into an algebraic expression using *n* to represent the unknown number.

21. Thirty subtracted from six times a number

22. Four more than three times the sum of a number and 8

For Problems 23–25, answer each question with an algebraic expression.

23. The product of two numbers is 72 and one of the numbers is *n*. What is the other number?

24. Tao has *n* nickels, *d* dimes, and *q* quarters. How much money, in cents, does she have?

25. The length of a rectangle is *x* yards and the width is *y* feet. What is the perimeter of the rectangle expressed in feet?

2

Equations, Inequalities, and Problem Solving

Most shoppers take advantage of the discounts offered by retailers. When making decisions about purchases, it is beneficial to be able to compute the sale prices.

A retailer of sporting goods bought a putter for $18. He wants to price the putter to make a profit of 40% of the selling price. What price should he mark on the putter? The equation $s = 18 + 0.4s$ can be used to determine that the putter should be sold for $30.

Throughout this text, we develop algebraic skills, use these skills to help solve equations and inequalities, and then use equations and inequalities to solve applied problems. In this chapter, we review and expand concepts that are important to the development of problem-solving skills.

2.1 Solving First-Degree Equations

In Section 1.1, we stated that an equality (equation) is a statement where two symbols, or groups of symbols, are names for the same number. It should be further stated that an equation may be true or false. For example, the equation $3 + (-8) = -5$ is true, but the equation $-7 + 4 = 2$ is false.

Algebraic equations contain one or more variables. The following are examples of algebraic equations.

$$3x + 5 = 8 \qquad 4y - 6 = -7y + 9 \qquad x^2 - 5x - 8 = 0$$
$$3x + 5y = 4 \qquad x^3 + 6x^2 - 7x - 2 = 0$$

An algebraic equation such as $3x + 5 = 8$ is neither true nor false as it stands, and we often refer to it as an "open sentence." Each time that a number is substituted for x, the algebraic equation $3x + 5 = 8$ becomes a numerical statement that is true or false. For example, if $x = 0$, then $3x + 5 = 8$ becomes $3(0) + 5 = 8$, which is a false statement. If $x = 1$, then $3x + 5 = 8$ becomes $3(1) + 5 = 8$, which is a true statement. **Solving an equation** refers to the process of finding the number (or numbers) that make(s) an algebraic equation a true numerical statement. We call such numbers the **solutions** or **roots** of the equation, and we say that they **satisfy** the equation. We call the set of all solutions of an equation its **solution set**. Thus {1} is the solution set of $3x + 5 = 8$.

In this chapter, we will consider techniques for solving **first-degree equations in one variable**. This means that the equations contain only one variable and that this variable has an exponent of 1. The following are examples of first-degree equations in one variable.

$$3x + 5 = 8 \qquad \qquad \frac{2}{3}y + 7 = 9$$

$$7a - 6 = 3a + 4 \qquad \frac{x - 2}{4} = \frac{x - 3}{5}$$

Equivalent equations are equations that have the same solution set. For example,

1. $3x + 5 = 8$

2. $3x = 3$

3. $x = 1$

are all equivalent equations because {1} is the solution set of each.

The general procedure for solving an equation is to continue replacing the given equation with equivalent but simpler equations until we obtain an equation of the form *variable = constant* or *constant = variable*. Thus in the example above, $3x + 5 = 8$ was simplified to $3x = 3$, which was further simplified to $x = 1$, from which the solution set {1} is obvious.

To solve equations we need to use the various properties of equality. In addition to the reflexive, symmetric, transitive, and substitution properties we listed in Section 1.1, the following properties of equality play an important role.

Addition Property of Equality

> For all real numbers a, b, and c,
>
> $$a = b \quad \text{if and only if} \quad a + c = b + c$$

Multiplication Property of Equality

> For all real numbers a, b, and c, where $c \neq 0$,
>
> $$a = b \quad \text{if and only if} \quad ac = bc$$

The addition property of equality states that when the same number is added to both sides of an equation, an equivalent equation is produced. The multiplication property of equality states that we obtain an equivalent equation whenever we multiply both sides of an equation by the same *nonzero* real number. The following examples demonstrate the use of these properties to solve equations.

EXAMPLE 1   Solve $2x - 1 = 13$.

Solution

$$2x - 1 = 13$$

$$2x - 1 + 1 = 13 + 1 \qquad \text{Add 1 to both sides.}$$

$$2x = 14$$

$$\frac{1}{2}(2x) = \frac{1}{2}(14) \qquad \text{Multiply both sides by } \frac{1}{2}.$$

$$x = 7$$

The solution set is $\{7\}$. ■

To check an apparent solution, we can substitute it into the original equation and see if we obtain a true numerical statement.

✔ **Check**

$$2x - 1 = 13$$

$$2(7) - 1 \overset{?}{=} 13$$

$$14 - 1 \overset{?}{=} 13$$

$$13 = 13$$

Now we know that {7} is the solution set of $2x - 1 = 13$. We will not show our checks for every example in this text, but do remember that checking is a way to detect arithmetic errors.

EXAMPLE 2

Solve $-7 = -5a + 9$.

Solution

$$-7 = -5a + 9$$

$$-7 + (-9) = 5a + 9 + (-9) \qquad \text{Add } -9 \text{ to both sides.}$$

$$-16 = -5a$$

$$-\frac{1}{5}(-16) = -\frac{1}{5}(-5a) \qquad \text{Multiply both sides by } -\frac{1}{5}.$$

$$\frac{16}{5} = a$$

The solution set is $\left\{ \dfrac{16}{5} \right\}$. ∎

Note that in Example 2 the final equation is $\dfrac{16}{5} = a$ instead of $a = \dfrac{16}{5}$. Technically, the symmetric property of equality (if $a = b$, then $b = a$) would permit us to change from $\dfrac{16}{5} = a$ to $a = \dfrac{16}{5}$, but such a change is not necessary to determine that the solution is $\dfrac{16}{5}$. Note that we could use the symmetric property at the very beginning to change $-7 = -5a + 9$ to $-5a + 9 = -7$; some people prefer having the variable on the left side of the equation.

Let's clarify another point. We stated the properties of equality in terms of only two operations, addition and multiplication. We could also include the operations of subtraction and division in the statements of the properties. That is, we could think in terms of subtracting the same number from both sides of an equation and also in terms of dividing both sides of an equation by the same nonzero number. For example, in the solution of Example 2, we could subtract 9 from both sides rather than adding -9 to both sides. Likewise, we could divide both sides by -5 instead of multiplying both sides by $-\dfrac{1}{5}$.

EXAMPLE 3

Solve $7x - 3 = 5x + 9$.

Solution

$$7x - 3 = 5x + 9$$

$$7x - 3 + (-5x) = 5x + 9 + (-5x) \qquad \text{Add } -5x \text{ to both sides.}$$

$$2x - 3 = 9$$

$$2x - 3 + 3 = 9 + 3 \qquad \text{Add 3 to both sides.}$$

$$2x = 12$$

$$\frac{1}{2}(2x) = \frac{1}{2}(12) \qquad \text{Multiply both sides by } \frac{1}{2}.$$

$$x = 6$$

The solution set is $\{6\}$.   ■

E X A M P L E   4   Solve $4(y - 1) + 5(y + 2) = 3(y - 8)$.

Solution

$$4(y - 1) + 5(y + 2) = 3(y - 8)$$

$$4y - 4 + 5y + 10 = 3y - 24 \qquad \text{Remove parentheses by apply-ing the distributive property.}$$

$$9y + 6 = 3y - 24 \qquad \text{Simplify the left side by com-bining similar terms.}$$

$$9y + 6 + (-3y) = 3y - 24 + (-3y) \qquad \text{Add } -3y \text{ to both sides.}$$

$$6y + 6 = -24$$

$$6y + 6 + (-6) = -24 + (-6) \qquad \text{Add } -6 \text{ to both sides.}$$

$$6y = -30$$

$$\frac{1}{6}(6y) = \frac{1}{6}(-30) \qquad \text{Multiply both sides by } \frac{1}{6}.$$

$$y = -5$$

The solution set is $\{-5\}$.   ■

We can summarize the process of solving first-degree equations in one variable as follows:

Step 1   Simplify both sides of the equation as much as possible.

Step 2   Use the addition property of equality to isolate a term that contains the variable on one side of the equation and a constant on the other side.

Step 3   Use the multiplication property of equality to make the coefficient of the variable 1; that is, multiply both sides of the equation by the reciprocal of the numerical coefficient of the variable. The solution set should now be obvious.

Step 4   Check each solution by substituting it in the original equation and verifying that the resulting numerical statement is true.

■ Use of Equations to Solve Problems

To use the tools of algebra to solve problems, we must be able to translate back and forth between the English language and the language of algebra. More specifically, we need to translate English sentences into algebraic equations. Such translations allow us to use our knowledge of equation solving to solve word problems. Let's consider an example.

PROBLEM   1

If we subtract 27 from three times a certain number, the result is 18. Find the number.

Solution

Let n represent the number to be found. The sentence "If we subtract 27 from three times a certain number, the result is 18" translates into the equation $3n - 27 = 18$. Solving this equation, we obtain

$$3n - 27 = 18$$

$$3n = 45 \qquad \text{Add 27 to both sides.}$$

$$n = 15 \qquad \text{Multiply both sides by } \frac{1}{3}.$$

The number to be found is 15.   ■

We often refer to the statement "Let n represent the number to be found" as **declaring the variable**. We need to choose a letter to use as a variable and indicate what it represents for a specific problem. This may seem like an insignificant idea, but as the problems become more complex, the process of declaring the variable becomes even more important. Furthermore, it is true that you could probably solve a problem such as Problem 1 without setting up an algebraic equation. However, as problems increase in difficulty, the translation from English to algebra becomes a key issue. Therefore, even with these relatively easy problems, we suggest that you concentrate on the translation process.

The next example involves the use of integers. Remember that the set of integers consists of $\{\ldots -2, -1, 0, 1, 2, \ldots\}$. Furthermore, the integers can be classified as even, $\{\ldots -4, -2, 0, 2, 4, \ldots\}$, or odd, $\{\ldots -3, -1, 1, 3, \ldots\}$.

PROBLEM   2

The sum of three consecutive integers is 13 greater than twice the smallest of the three integers. Find the integers.

Solution

Because consecutive integers differ by 1, we will represent them as follows: Let n represent the smallest of the three consecutive integers; then $n + 1$ represents the second largest, and $n + 2$ represents the largest.

The sum of the three
consecutive integers   13 greater than twice the smallest

$$n + (n + 1) + (n + 2) = 2n + 13$$
$$3n + 3 = 2n + 13$$
$$n = 10$$

The three consecutive integers are 10, 11, and 12.   ■

To check our answers for Problem 2, we must determine whether or not they satisfy the conditions stated in the original problem. Because 10, 11, and 12 are consecutive integers whose sum is 33, and because twice the smallest plus 13 is also 33 $(2(10) + 13 = 33)$, we know that our answers are correct. (Remember, in checking a result for a word problem, it is *not* sufficient to check the result in the equation set up to solve the problem; the equation itself may be in error!)

In the two previous problems, the equation formed was almost a direct translation of a sentence in the statement of the problem. Now let's consider a situation where we need to think in terms of a guideline not explicitly stated in the problem.

PROBLEM   3

Khoa received a car repair bill for $106. This included $23 for parts, $22 per hour for each hour of labor, and $6 for taxes. Find the number of hours of labor.

Solution

See Figure 2.1. Let h represent the number of hours of labor. Then $22h$ represents the total charge for labor.

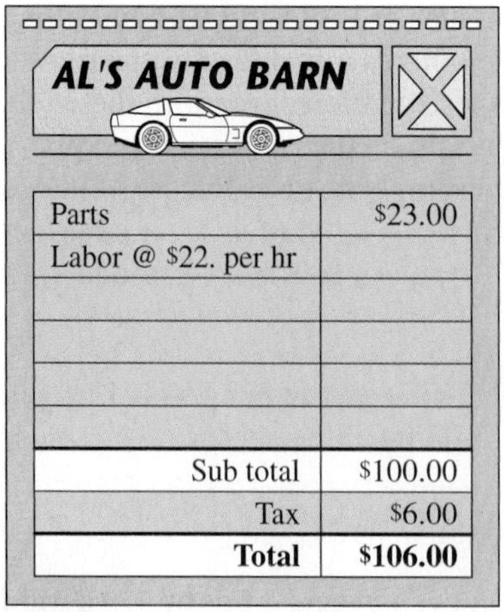

AL'S AUTO BARN	
Parts	$23.00
Labor @ $22. per hr	
Sub total	$100.00
Tax	$6.00
Total	**$106.00**

Figure 2.1

We can use a guideline of *charge for parts plus charge for labor plus tax equals the total bill* to set up the following equation.

Parts   Labor   Tax   Total bill

$$23 + 22h + 6 = 106$$

Solving this equation, we obtain

$$22h + 29 = 106$$
$$22h = 77$$
$$h = 3\frac{1}{2}$$

Khoa was charged for $3\frac{1}{2}$ hours of labor. ∎

Problem Set 2.1

For problems 1–50, solve each equation.

1. $3x + 4 = 16$

2. $4x + 2 = 22$

3. $5x + 1 = -14$

4. $7x + 4 = -31$

5. $-x - 6 = 8$

6. $8 - x = -2$

7. $4y - 3 = 21$

8. $6y - 7 = 41$

9. $3x - 4 = 15$

10. $5x + 1 = 12$

11. $-4 = 2x - 6$

12. $-14 = 3a - 2$

13. $-6y - 4 = 16$

14. $-8y - 2 = 18$

15. $4x - 1 = 2x + 7$

16. $9x - 3 = 6x + 18$

17. $5y + 2 = 2y - 11$

18. $9y + 3 = 4y - 10$

19. $3x + 4 = 5x - 2$

20. $2x - 1 = 6x + 15$

21. $-7a + 6 = -8a + 14$

22. $-6a - 4 = -7a + 11$

23. $5x + 3 - 2x = x - 15$

24. $4x - 2 - x = 5x + 10$

25. $6y + 18 + y = 2y + 3$

26. $5y + 14 + y = 3y - 7$

27. $4x - 3 + 2x = 8x - 3 - x$

28. $x - 4 - 4x = 6x + 9 - 8x$

29. $6n - 4 - 3n = 3n + 10 + 4n$

30. $2n - 1 - 3n = 5n - 7 - 3n$

31. $4(x - 3) = -20$

32. $3(x + 2) = -15$

33. $-3(x - 2) = 11$

34. $-5(x - 1) = 12$

35. $5(2x + 1) = 4(3x - 7)$

36. $3(2x - 1) = 2(4x + 7)$

37. $5x - 4(x - 6) = -11$

38. $3x - 5(2x + 1) = 13$

39. $-2(3x - 1) - 3 = -4$

40. $-6(x - 4) - 10 = -12$

41. $-2(3x + 5) = -3(4x + 3)$

42. $-(2x - 1) = -5(2x + 9)$

43. $3(x - 4) - 7(x + 2) = -2(x + 18)$

44. $4(x - 2) - 3(x - 1) = 2(x + 6)$

45. $-2(3n - 1) + 3(n + 5) = -4(n - 4)$

46. $-3(4n + 2) + 2(n - 6) = -2(n + 1)$

47. $3(2a - 1) - 2(5a + 1) = 4(3a + 4)$

48. $4(2a + 3) - 3(4a - 2) = 5(4a - 7)$

49. $-2(n - 4) - (3n - 1) = -2 + (2n - 1)$

50. $-(2n - 1) + 6(n + 3) = -4 - (7n - 11)$

For Problems 51–66, use an algebraic approach to solve each problem.

51. If 15 is subtracted from three times a certain number, the result is 27. Find the number.

52. If 1 is subtracted from seven times a certain number, the result is the same as if 31 is added to three times the number. Find the number.

53. Find three consecutive integers whose sum is 42.

54. Find four consecutive integers whose sum is −118.

55. Find three consecutive odd integers such that three times the second minus the third is 11 more than the first.

56. Find three consecutive even integers such that four times the first minus the third is six more than twice the second.

57. The difference of two numbers is 67. The larger number is three less than six times the smaller number. Find the numbers.

58. The sum of two numbers is 103. The larger number is one more than five times the smaller number. Find the numbers.

59. Angelo is paid double time for each hour he works over 40 hours in a week. Last week he worked 46 hours and earned $572. What is his normal hourly rate?

60. Suppose that a plumbing repair bill, not including tax, was $130. This included $25 for parts and an amount for 5 hours of labor. Find the hourly rate that was charged for labor.

61. Suppose that Maria has 150 coins consisting of pennies, nickels, and dimes. The number of nickels she has is 10 less than twice the number of pennies; the number of dimes she has is 20 less than three times the number of pennies. How many coins of each kind does she have?

62. Hector has a collection of nickels, dimes, and quarters totaling 122 coins. The number of dimes he has is 3 more than four times the number of nickels, and the number of quarters he has is 19 less than the number of dimes. How many coins of each kind does he have?

63. The selling price of a ring is $750. This represents $150 less than three times the cost of the ring. Find the cost of the ring.

64. In a class of 62 students, the number of females is one less than twice the number of males. How many females and how many males are there in the class?

65. An apartment complex contains 230 apartments each having one, two, or three bedrooms. The number of two-bedroom apartments is 10 more than three times the number of three-bedroom apartments. The number of one-bedroom apartments is twice the number of two-bedroom apartments. How many apartments of each kind are in the complex?

66. Barry sells bicycles on a salary-plus-commission basis. He receives a monthly salary of $300 and a commission of $15 for each bicycle that he sells. How many bicycles must he sell in a month to have a total monthly income of $750?

▪▪▪ THOUGHTS INTO WORDS

67. Explain the difference between a numerical statement and an algebraic equation.

68. Are the equations $7 = 9x − 4$ and $9x − 4 = 7$ equivalent equations? Defend your answer.

69. Suppose that your friend shows you the following solution to an equation.

$$17 = 4 − 2x$$
$$17 + 2x = 4 − 2x + 2x$$
$$17 + 2x = 4$$
$$17 + 2x − 17 = 4 − 17$$

$$2x = −13$$
$$x = \frac{−13}{2}$$

Is this a correct solution? What suggestions would you have in terms of the method used to solve the equation?

70. Explain in your own words what it means to declare a variable when solving a word problem.

71. Make up an equation whose solution set is the null set and explain why this is the solution set.

72. Make up an equation whose solution set is the set of all real numbers and explain why this is the solution set.

■ ■ ■ **FURTHER INVESTIGATIONS**

73. Solve each of the following equations.

(a) $5x + 7 = 5x - 4$

(b) $4(x - 1) = 4x - 4$

(c) $3(x - 4) = 2(x - 6)$

(d) $7x - 2 = -7x + 4$

(e) $2(x - 1) + 3(x + 2) = 5(x - 7)$

(f) $-4(x - 7) = -2(2x + 1)$

74. Verify that for any three consecutive integers, the sum of the smallest and largest is equal to twice the middle integer. [*Hint*: Use n, $n + 1$, and $n + 2$ to represent the three consecutive integers.]

75. Verify that no four consecutive integers can be found such that the product of the smallest and largest is equal to the product of the other two integers.

2.2   Equations Involving Fractional Forms

To solve equations that involve fractions, it is usually easiest to begin by **clearing the equation of all fractions**. This can be accomplished by multiplying both sides of the equation by the least common multiple of all the denominators in the equation. Remember that the least common multiple of a set of whole numbers is the smallest nonzero whole number that is divisible by each of the numbers. For example, the least common multiple of 2, 3, and 6 is 12. When working with fractions, we refer to the least common multiple of a set of denominators as the **least common denominator** (LCD). Let's consider some equations involving fractions.

E X A M P L E   1

Solve $\dfrac{1}{2}x + \dfrac{2}{3} = \dfrac{3}{4}$.

Solution

$$\frac{1}{2}x + \frac{2}{3} = \frac{3}{4}$$

$$12\left(\frac{1}{2}x + \frac{2}{3}\right) = 12\left(\frac{3}{4}\right)$$ Multiply both sides by 12, which is the LCD of 2, 3, and 4.

$$12\left(\frac{1}{2}x\right) + 12\left(\frac{2}{3}\right) = 12\left(\frac{3}{4}\right)$$ Apply the distributive property to the left side.

$$6x + 8 = 9$$

$$6x = 1$$

$$x = \frac{1}{6}$$

The solution set is $\left\{\dfrac{1}{6}\right\}$.

✔ **Check**

$$\frac{1}{2}x + \frac{2}{3} = \frac{3}{4}$$

$$\frac{1}{2}\left(\frac{1}{6}\right) + \frac{2}{3} \stackrel{?}{=} \frac{3}{4}$$

$$\frac{1}{12} + \frac{2}{3} \stackrel{?}{=} \frac{3}{4}$$

$$\frac{1}{12} + \frac{8}{12} \stackrel{?}{=} \frac{3}{4}$$

$$\frac{9}{12} \stackrel{?}{=} \frac{3}{4}$$

$$\frac{3}{4} = \frac{3}{4}$$

∎

EXAMPLE 2

Solve $\dfrac{x}{2} + \dfrac{x}{3} = 10$.

Solution

$$\frac{x}{2} + \frac{x}{3} = 10 \qquad \text{Recall that } \frac{x}{2} = \frac{1}{2}x.$$

$$6\left(\frac{x}{2} + \frac{x}{3}\right) = 6(10) \qquad \text{Multiply both sides by the LCD.}$$

$$6\left(\frac{x}{2}\right) + 6\left(\frac{x}{3}\right) = 6(10) \qquad \begin{array}{l}\text{Apply the distributive property to} \\ \text{the left side.}\end{array}$$

$$3x + 2x = 60$$

$$5x = 60$$

$$x = 12$$

The solution set is {12}. ∎

As you study the examples in this section, pay special attention to the steps shown in the solutions. There are no hard and fast rules as to which steps should be performed mentally; this is an individual decision. When you solve problems, show enough steps to allow the flow of the process to be understood and to minimize the chances of making careless computational errors.

EXAMPLE 3

Solve $\dfrac{x-2}{3} + \dfrac{x+1}{8} = \dfrac{5}{6}$.

Solution

$$\frac{x-2}{3} + \frac{x+1}{8} = \frac{5}{6}$$

$$24\left(\frac{x-2}{3}+\frac{x+1}{8}\right)=24\left(\frac{5}{6}\right)$$ Multiply both sides by the LCD.

$$24\left(\frac{x-2}{3}\right)+24\left(\frac{x+1}{8}\right)=24\left(\frac{5}{6}\right)$$ Apply the distributive property to the left side.

$$8(x-2)+3(x+1)=20$$

$$8x-16+3x+3=20$$

$$11x-13=20$$

$$11x=33$$

$$x=3$$

The solution set is {3}.   ■

EXAMPLE 4   Solve $\dfrac{3t-1}{5}-\dfrac{t-4}{3}=1$.

Solution

$$\frac{3t-1}{5}-\frac{t-4}{3}=1$$

$$15\left(\frac{3t-1}{5}-\frac{t-4}{3}\right)=15(1)$$ Multiply both sides by the LCD.

$$15\left(\frac{3t-1}{5}\right)-15\left(\frac{t-4}{3}\right)=15(1)$$ Apply the distributive property to the left side.

$$3(3t-1)-5(t-4)=15$$

$$9t-3-5t+20=15$$ Be careful with this sign!

$$4t+17=15$$

$$4t=-2$$

$$t=-\frac{2}{4}=-\frac{1}{2}$$ Reduce!

The solution set is $\left\{-\dfrac{1}{2}\right\}$.   ■

■ Problem Solving

As we expand our skills for solving equations, we also expand our capabilities for solving word problems. There is no one definite procedure that will ensure success at solving word problems, but the following suggestions can be helpful.

Suggestions for Solving Word Problems

1. Read the problem carefully and make certain that you understand the meanings of all of the words. Be especially alert for any technical terms used in the statement of the problem.
2. Read the problem a second time (perhaps even a third time) to get an overview of the situation being described. Determine the known facts as well as what is to be found.
3. Sketch any figure, diagram, or chart that might be helpful in analyzing the problem.
4. Choose a meaningful variable to represent an unknown quantity in the problem (perhaps t, if time is an unknown quantity) and represent any other unknowns in terms of that variable.
5. Look for a guideline that you can use to set up an equation. A guideline might be a formula, such as *distance equals rate times time*, or a statement of a relationship, such as "The sum of the two numbers is 28."
6. Form an equation that contains the variable and that translates the conditions of the guideline from English to algebra.
7. Solve the equation, and use the solution to determine all facts requested in the problem.
8. Check all answers back into the **original statement of the problem**.

Keep these suggestions in mind as we continue to solve problems. We will elaborate on some of these suggestions at different times throughout the text. Now let's consider some problems.

P R O B L E M   1

Find a number such that three-eighths of the number minus one-half of it is 14 less than three-fourths of the number.

Solution

Let n represent the number to be found.

$$\frac{3}{8}n - \frac{1}{2}n = \frac{3}{4}n - 14$$

$$8\left(\frac{3}{8}n - \frac{1}{2}n\right) = 8\left(\frac{3}{4}n - 14\right)$$

$$8\left(\frac{3}{8}n\right) - 8\left(\frac{1}{2}n\right) = 8\left(\frac{3}{4}n\right) - 8(14)$$

$$3n - 4n = 6n - 112$$

$$-n = 6n - 112$$

$$-7n = -112$$

$$n = 16$$

The number is 16. Check it!

PROBLEM 2

The width of a rectangular parking lot is 8 feet less than three-fifths of the length. The perimeter of the lot is 400 feet. Find the length and width of the lot.

Solution
Let l represent the length of the lot. Then $\frac{3}{5}l - 8$ represents the width (Figure 2.2).

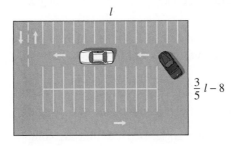

Figure 2.2

A guideline for this problem is the formula, *the perimeter of a rectangle equals twice the length plus twice the width* $(P = 2l + 2w)$. Use this formula to form the following equation.

$$P = 2l + 2w$$

$$400 = 2l + 2\left(\frac{3}{5}l - 8\right)$$

Solving this equation, we obtain

$$400 = 2l + \frac{6l}{5} - 16$$

$$5(400) = 5\left(2l + \frac{6l}{5} - 16\right)$$

$$2000 = 10l + 6l - 80$$

$$2000 = 16l - 80$$

$$2080 = 16l$$

$$130 = l.$$

The length of the lot is 130 feet, and the width is $\frac{3}{5}(130) - 8 = 70$ feet. ∎

In Problems 1 and 2, note the use of different letters as variables. It is helpful to choose a variable that has significance for the problem you are working on. For example, in Problem 2 the choice of l to represent the length seems natural and meaningful. (Certainly this is another matter of personal preference, but you might consider it.)

In Problem 2 a geometric relationship, ($P = 2l + 2w$), serves as a guideline for setting up the equation. The following geometric relationships pertaining to angle measure may also serve as guidelines.

1. Complementary angles are two angles the sum of whose measures is 90°.
2. Supplementary angles are two angles the sum of whose measures is 180°.
3. The sum of the measures of the three angles of a triangle is 180°.

PROBLEM 3

One of two complementary angles is 6° larger than one-half of the other angle. Find the measure of each of the angles.

Solution

Let a represent the measure of one of the angles. Then $\frac{1}{2}a + 6$ represents the measure of the other angle. Because they are complementary angles, the sum of their measures is 90°.

$$a + \left(\frac{1}{2}a + 6\right) = 90$$
$$2a + a + 12 = 180$$
$$3a + 12 = 180$$
$$3a = 168$$
$$a = 56$$

If $a = 56$, then $\frac{1}{2}a + 6$ becomes $\frac{1}{2}(56) + 6 = 34$. The angles have measures of 34° and 56°. ■

PROBLEM 4

Dominic's present age is 10 years more than Michele's present age. In 5 years Michele's age will be three-fifths of Dominic's age. What are their present ages?

Solution

Let x represent Michele's present age. Then Dominic's age will be represented by $x + 10$. In 5 years, everyone's age is increased by 5 years, so we need to add 5 to Michele's present age and 5 to Dominic's present age to represent their ages in 5 years. Therefore, in 5 years Michele's age will be represented by $x + 5$, and Dominic's age will be represented by $x + 15$. Thus we can set up the equation reflecting the fact that in 5 years, Michele's age will be three-fifths of Dominic's age.

$$x + 5 = \frac{3}{5}(x + 15)$$
$$5(x + 5) = 5\left[\frac{3}{5}(x + 15)\right]$$
$$5x + 25 = 3(x + 15)$$

$$5x + 25 = 3x + 45$$
$$2x + 25 = 45$$
$$2x = 20$$
$$x = 10$$

Because x represents Michele's present age, we know her age is 10. Dominic's present age is represented by $x + 10$, so his age is 20.   ■

Keep in mind that the problem-solving suggestions offered in this section simply outline a general algebraic approach to solving problems. You will add to this list throughout this course and in any subsequent mathematics courses that you take. Furthermore, you will be able to pick up additional problem-solving ideas from your instructor and from fellow classmates as you discuss problems in class. Always be on the alert for any ideas that might help you become a better problem solver.

Problem Set 2.2

For Problems 1–40, solve each equation.

1. $\dfrac{3}{4}x = 9$

2. $\dfrac{2}{3}x = -14$

3. $\dfrac{-2x}{3} = \dfrac{2}{5}$

4. $\dfrac{-5x}{4} = \dfrac{7}{2}$

5. $\dfrac{n}{2} - \dfrac{2}{3} = \dfrac{5}{6}$

6. $\dfrac{n}{4} - \dfrac{5}{6} = \dfrac{5}{12}$

7. $\dfrac{5n}{6} - \dfrac{n}{8} = \dfrac{-17}{12}$

8. $\dfrac{2n}{5} - \dfrac{n}{6} = \dfrac{-7}{10}$

9. $\dfrac{a}{4} - 1 = \dfrac{a}{3} + 2$

10. $\dfrac{3a}{7} - 1 = \dfrac{a}{3}$

11. $\dfrac{h}{4} + \dfrac{h}{5} = 1$

12. $\dfrac{h}{6} + \dfrac{3h}{8} = 1$

13. $\dfrac{h}{2} - \dfrac{h}{3} + \dfrac{h}{6} = 1$

14. $\dfrac{3h}{4} + \dfrac{2h}{5} = 1$

15. $\dfrac{x - 2}{3} + \dfrac{x + 3}{4} = \dfrac{11}{6}$

16. $\dfrac{x + 4}{5} + \dfrac{x - 1}{4} = \dfrac{37}{10}$

17. $\dfrac{x + 2}{2} - \dfrac{x - 1}{5} = \dfrac{3}{5}$

18. $\dfrac{2x + 1}{3} - \dfrac{x + 1}{7} = -\dfrac{1}{3}$

19. $\dfrac{n + 2}{4} - \dfrac{2n - 1}{3} = \dfrac{1}{6}$

20. $\dfrac{n - 1}{9} - \dfrac{n + 2}{6} = \dfrac{3}{4}$

21. $\dfrac{y}{3} + \dfrac{y - 5}{10} = \dfrac{4y + 3}{5}$

22. $\dfrac{y}{3} + \dfrac{y - 2}{8} = \dfrac{6y - 1}{12}$

23. $\dfrac{4x - 1}{10} - \dfrac{5x + 2}{4} = -3$

24. $\dfrac{2x - 1}{2} - \dfrac{3x + 1}{4} = \dfrac{3}{10}$

25. $\dfrac{2x - 1}{8} - 1 = \dfrac{x + 5}{7}$

26. $\dfrac{3x + 1}{9} + 2 = \dfrac{x - 1}{4}$

27. $\dfrac{2a - 3}{6} + \dfrac{3a - 2}{4} + \dfrac{5a + 6}{12} = 4$

28. $\dfrac{3a - 1}{4} + \dfrac{a - 2}{3} - \dfrac{a - 1}{5} = \dfrac{21}{20}$

29. $x + \dfrac{3x - 1}{9} - 4 = \dfrac{3x + 1}{3}$

30. $\dfrac{2x + 7}{8} + x - 2 = \dfrac{x - 1}{2}$

31. $\dfrac{x + 3}{2} + \dfrac{x + 4}{5} = \dfrac{3}{10}$

32. $\dfrac{x - 2}{5} - \dfrac{x - 3}{4} = -\dfrac{1}{20}$

33. $n + \dfrac{2n - 3}{9} - 2 = \dfrac{2n + 1}{3}$

34. $n - \dfrac{3n + 1}{6} - 1 = \dfrac{2n + 4}{12}$

35. $\dfrac{3}{4}(t - 2) - \dfrac{2}{5}(2t - 3) = \dfrac{1}{5}$

36. $\dfrac{2}{3}(2t + 1) - \dfrac{1}{2}(3t - 2) = 2$

37. $\dfrac{1}{2}(2x - 1) - \dfrac{1}{3}(5x + 2) = 3$

38. $\dfrac{2}{5}(4x - 1) + \dfrac{1}{4}(5x + 2) = -1$

39. $3x - 1 + \dfrac{2}{7}(7x - 2) = -\dfrac{11}{7}$

40. $2x + 5 + \dfrac{1}{2}(6x - 1) = -\dfrac{1}{2}$

For Problems 41–58, use an algebraic approach to solve each problem.

41. Find a number such that one-half of the number is 3 less than two-thirds of the number.

42. One-half of a number plus three-fourths of the number is 2 more than four-thirds of the number. Find the number.

43. Suppose that the width of a certain rectangle is 1 inch more than one-fourth of its length. The perimeter of the rectangle is 42 inches. Find the length and width of the rectangle.

44. Suppose that the width of a rectangle is 3 centimeters less than two-thirds of its length. The perimeter of the rectangle is 114 centimeters. Find the length and width of the rectangle.

45. Find three consecutive integers such that the sum of the first plus one-third of the second plus three-eighths of the third is 25.

46. Lou is paid $1\dfrac{1}{2}$ times his normal hourly rate for each hour he works over 40 hours in a week. Last week he worked 44 hours and earned $276. What is his normal hourly rate?

47. A board 20 feet long is cut into two pieces such that the length of one piece is two-thirds of the length of the other piece. Find the length of the shorter piece of board.

48. Jody has a collection of 116 coins consisting of dimes, quarters, and silver dollars. The number of quarters is 5 less than three-fourths of the number of dimes. The number of silver dollars is 7 more than five-eighths of the number of dimes. How many coins of each kind are in her collection?

49. The sum of the present ages of Angie and her mother is 64 years. In eight years Angie will be three-fifths as old as her mother at that time. Find the present ages of Angie and her mother.

50. Annilee's present age is two-thirds of Jessie's present age. In 12 years the sum of their ages will be 54 years. Find their present ages.

51. Sydney's present age is one-half of Marcus's present age. In 12 years, Sydney's age will be five-eighths of Marcus's age. Find their present ages.

52. The sum of the present ages of Ian and his brother is 45. In 5 years, Ian's age will be five-sixths of his brother's age. Find their present ages.

53. Aura took three biology exams and has an average score of 88. Her second exam score was 10 points better than her first, and her third exam score was 4 points better than her second exam. What were her three exam scores?

54. The average of the salaries of Tim, Maida, and Aaron is $24,000 per year. Maida earns $10,000 more than Tim, and Aaron's salary is $2000 more than twice Tim's salary. Find the salary of each person.

55. One of two supplementary angles is 4° more than one-third of the other angle. Find the measure of each of the angles.

56. If one-half of the complement of an angle plus three-fourths of the supplement of the angle equals 110°, find the measure of the angle.

57. If the complement of an angle is 5° less than one-sixth of its supplement, find the measure of the angle.

58. In $\triangle ABC$, angle B is 8° less than one-half of angle A and angle C is 28° larger than angle A. Find the measures of the three angles of the triangle.

■ ■ ■ **THOUGHTS INTO WORDS**

59. Explain why the solution set of the equation $x + 3 = x + 4$ is the null set.

60. Explain why the solution set of the equation $\dfrac{x}{3} + \dfrac{x}{2} = \dfrac{5x}{6}$ is the entire set of real numbers.

61. Why must potential answers to word problems be checked back into the original statement of the problem?

62. Suppose your friend solved the problem, *find two consecutive odd integers whose sum is 28* like this:

$$x + x + 1 = 28$$
$$2x = 27$$
$$x = \frac{27}{2} = 13\frac{1}{2}$$

She claims that $13\frac{1}{2}$ will check in the equation. Where has she gone wrong and how would you help her?

2.3 Equations Involving Decimals and Problem Solving

In solving equations that involve fractions, usually the procedure is to clear the equation of all fractions. For solving equations that involve decimals, there are two commonly used procedures. One procedure is to keep the numbers in decimal form and solve the equation by applying the properties. Another procedure is to multiply both sides of the equation by an appropriate power of 10 to clear the equation of all decimals. Which technique to use depends on your personal preference and on the complexity of the equation. The following examples demonstrate both techniques.

E X A M P L E 1

Solve $0.2x + 0.24 = 0.08x + 0.72$.

Solution

Let's clear the decimals by multiplying both sides of the equation by 100.

$$0.2x + 0.24 = 0.08x + 0.72$$
$$100(0.2x + 0.24) = 100(0.08x + 0.72)$$
$$100(0.2x) + 100(0.24) = 100(0.08x) + 100(0.72)$$
$$20x + 24 = 8x + 72$$
$$12x + 24 = 72$$
$$12x = 48$$
$$x = 4$$

✔ **Check**

$$0.2x + 0.24 = 0.08x + 0.72$$

$$0.2(4) + 0.24 \stackrel{?}{=} 0.08(4) + 0.72$$

$$0.8 + 0.24 \stackrel{?}{=} 0.32 + 0.72$$

$$1.04 = 1.04$$

The solution set is {4}.   ■

E X A M P L E   2   Solve $0.07x + 0.11x = 3.6$.

Solution

Let's keep this problem in decimal form.

$$0.07x + 0.11x = 3.6$$

$$0.18x = 3.6$$

$$x = \frac{3.6}{0.18}$$

$$x = 20$$

✔ **Check**

$$0.07x + 0.11x = 3.6$$

$$0.07(20) + 0.11(20) \stackrel{?}{=} 3.6$$

$$1.4 + 2.2 \stackrel{?}{=} 3.6$$

$$3.6 = 3.6$$

The solution set is {20}.   ■

E X A M P L E   3   Solve $s = 1.95 + 0.35s$.

Solution

Let's keep this problem in decimal form.

$$s = 1.95 + 0.35s$$

$$s + (-0.35s) = 1.95 + 0.35s + (-0.35s)$$

$$0.65s = 1.95 \qquad \text{Remember, } s = 1.00s.$$

$$s = \frac{1.95}{0.65}$$

$$s = 3$$

The solution set is {3}. Check it!   ■

EXAMPLE 4

Solve $0.12x + 0.11(7000 - x) = 790$.

Solution

Let's clear the decimals by multiplying both sides of the equation by 100.

$$0.12x + 0.11(7000 - x) = 790$$

$$100[0.12x + 0.11(7000 - x)] = 100(790) \qquad \text{Multiply both sides by 100.}$$

$$100(0.12x) + 100[0.11(7000 - x)] = 100(790)$$

$$12x + 11(7000 - x) = 79{,}000$$

$$12x + 77{,}000 - 11x = 79{,}000$$

$$x + 77{,}000 = 79{,}000$$

$$x = 2000$$

The solution set is $\{2000\}$. ∎

■ Back to Problem Solving

We can solve many consumer problems with an algebraic approach. For example, let's consider some discount sale problems involving the relationship, *original selling price minus discount equals discount sale price*.

Original selling price − Discount = Discount sale price

PROBLEM 1

Karyl bought a dress at a 35% discount sale for $32.50. What was the original price of the dress?

Solution

Let p represent the original price of the dress. Using the discount sale relationship as a guideline, we find that the problem translates into an equation as follows:

Original selling price	Minus	Discount	Equals	Discount sale price
p	−	$(35\%)(p)$	=	\$32.50

Switching this equation to decimal form and solving the equation, we obtain

$$p - (35\%)(p) = 32.50$$

$$(65\%)(p) = 32.50$$

$$0.65p = 32.50$$

$$p = 50$$

The original price of the dress was $50. ∎

PROBLEM   2

A pair of jogging shoes that was originally priced at $50 is on sale for 20% off. Find the discount sale price of the shoes.

Solution

Let s represent the discount sale price.

Original price	Minus	Discount	Equals	Sale price
↓		↓		↓
$50	−	(20%)($50)	=	s

Solving this equation we obtain

$$50 - (20\%)(50) = s$$
$$50 - (0.2)(50) = s$$
$$50 - 10 = s$$
$$40 = s$$

The shoes are on sale for $40.   ■

Remark: Keep in mind that if an item is on sale for 35% off, then the purchaser will pay $100\% - 35\% = 65\%$ of the original price. Thus in Problem 1 you could begin with the equation $0.65p = 32.50$. Likewise in Problem 2 you could start with the equation $s = 0.8(50)$.

Another basic relationship that pertains to consumer problems is *selling price equals cost plus profit*. We can state profit (also called markup, markon, and margin of profit) in different ways. Profit may be stated as a percent of the selling price, as a percent of the cost, or simply in terms of dollars and cents. We shall consider some problems for which the profit is calculated either as a percent of the cost or as a percent of the selling price.

$$\text{Selling price} = \text{Cost} + \text{Profit}$$

PROBLEM   3

A retailer has some shirts that cost $20 each. She wants to sell them at a profit of 60% of the cost. What selling price should be marked on the shirts?

Solution

Let s represent the selling price. Use the relationship *selling price equals cost plus profit* as a guideline.

Selling price	Equals	Cost	Plus	Profit
↓		↓		↓
s	=	$20	+	(60%)($20)

Solving this equation yields

$$s = 20 + (60\%)(20)$$
$$s = 20 + (0.6)(20)$$

$$s = 20 + 12$$

$$s = 32$$

The selling price should be $32.   ■

Remark: A profit of 60% of the cost means that the selling price is 100% of the cost plus 60% of the cost, or 160% of the cost. Thus in Problem 3 we could solve the equation $s = 1.6(20)$.

P R O B L E M   4

A retailer of sporting goods bought a putter for $18. He wants to price the putter such that he will make a profit of 40% of the selling price. What price should he mark on the putter?

Solution

Let s represent the selling price.

Selling price	Equals	Cost	Plus	Profit
↓		↓		↓
s	=	$18	+	$(40\%)(s)$

Solving this equation yields

$$s = 18 + (40\%)(s)$$

$$s = 18 + 0.4s$$

$$0.6s = 18$$

$$s = 30$$

The selling price should be $30.   ■

P R O B L E M   5

If a maple tree costs a landscaper $55.00, and he sells it for $80.00, what is his rate of profit based on the cost? Round the rate to the nearest tenth of a percent.

Solution

Let r represent the rate of profit, and use the following guideline.

Selling price	Equals	Cost	Plus	Profit
↓	↓	↓		↓
80.00	=	55.00	+	$r(55.00)$
25.00	=	$r(55.00)$		
$\dfrac{25.00}{55.00}$	=	r		
0.455	≈	r		

To change the answer to a percent, multiply 0.455 by 100. Thus his rate of profit is 45.5%.   ■

We can solve certain types of investment and money problems by using an algebraic approach. Consider the following examples.

Erick has 40 coins, consisting only of dimes and nickels, worth $3.35. How many dimes and how many nickels does he have?

Solution

Let x represent the number of dimes. Then the number of nickels can be represented by the total number of coins minus the number of dimes. Hence $40 - x$ represents the number of nickels. Because we know the amount of money Erick has, we need to multiply the number of each coin by its value. Use the following guideline.

Money from the dimes	Plus	Money from the nickels	Equals	Total money	
↓		↓		↓	
$0.10x$	$+$	$0.05(40 - x)$	$=$	3.35	
$10x$	$+$	$5(40 - x)$	$=$	335	Multiply both sides by 100.
$10x$	$+$	$200 - 5x$	$=$	335	
		$5x + 200$	$=$	335	
		$5x$	$=$	135	
		x	$=$	27	

The number of dimes is 27, and the number of nickels is $40 - x = 13$. So, Erick has 27 dimes and 13 nickels.   ∎

A man invests $8000, part of it at 11% and the remainder at 12%. His total yearly interest from the two investments is $930. How much did he invest at each rate?

Solution

Let x represent the amount he invested at 11%. Then $8000 - x$ represents the amount he invested at 12%. Use the following guideline.

Interest earned from 11% investment		Interest earned from 12% investment		Total amount of interest earned
↓		↓		↓
$(11\%)(x)$	$+$	$(12\%)(8000 - x)$	$=$	$\$930$

Solving this equation yields

$$(11\%)(x) + (12\%)(8000 - x) = 930$$

$$0.11x + 0.12(8000 - x) = 930$$

$$11x + 12(8000 - x) = 93,000 \qquad \text{Multiply both sides by 100.}$$

$$11x + 96,000 - 12x = 93,000$$

$$-x + 96,000 = 93,000$$

$$-x = -3000$$

$$x = 3000$$

Therefore, $3000 was invested at 11%, and $8000 − $3000 = $5000 was invested at 12%. ■

Don't forget to check word problems; determine whether the answers satisfy the conditions stated in the *original* problem. A check for Problem 7 follows.

✔ **Check**

We claim that $3000 is invested at 11% and $5000 at 12%, and this satisfies the condition that $8000 is invested. The $3000 at 11% produces $330 of interest, and the $5000 at 12% produces $600. Therefore, the interest from the investments is $930. The conditions of the problem are satisfied, and our answers are correct.

As you tackle word problems throughout this text, keep in mind that our primary objective is to expand your repertoire of problem-solving techniques. We have chosen problems that provide you with the opportunity to use a variety of approaches to solving problems. Don't fall into the trap of thinking "I will never be faced with this kind of problem." That is not the issue; the goal is to develop problem-solving techniques. In the examples that follow we are sharing some of our ideas for solving problems, but don't hesitate to use your own ingenuity. Furthermore, don't become discouraged — all of us have difficulty with some problems. Give each your best shot!

Problem Set 2.3

For Problems 1–28, solve each equation.

1. $0.14x = 2.8$

2. $1.6x = 8$

3. $0.09y = 4.5$

4. $0.07y = 0.42$

5. $n + 0.4n = 56$

6. $n - 0.5n = 12$

7. $s = 9 + 0.25s$

8. $s = 15 + 0.4s$

9. $s = 3.3 + 0.45s$

10. $s = 2.1 + 0.6s$

11. $0.11x + 0.12(900 - x) = 104$

12. $0.09x + 0.11(500 - x) = 51$

13. $0.08(x + 200) = 0.07x + 20$

14. $0.07x = 152 - 0.08(2000 - x)$

15. $0.12t - 2.1 = 0.07t - 0.2$

16. $0.13t - 3.4 = 0.08t - 0.4$

17. $0.92 + 0.9(x - 0.3) = 2x - 5.95$

18. $0.3(2n - 5) = 11 - 0.65n$

19. $0.1d + 0.11(d + 1500) = 795$

20. $0.8x + 0.9(850 - x) = 715$

21. $0.12x + 0.1(5000 - x) = 560$

22. $0.10t + 0.12(t + 1000) = 560$

23. $0.09(x + 200) = 0.08x + 22$

24. $0.09x = 1650 - 0.12(x + 5000)$

25. $0.3(2t + 0.1) = 8.43$

26. $0.5(3t + 0.7) = 20.6$

27. $0.1(x - 0.1) - 0.4(x + 2) = -5.31$

28. $0.2(x + 0.2) + 0.5(x - 0.4) = 5.44$

For Problems 29–50, use an algebraic approach to solve each problem.

29. Judy bought a coat at a 20% discount sale for $72. What was the original price of the coat?

30. Jim bought a pair of slacks at a 25% discount sale for $24. What was the original price of the slacks?

31. Find the discount sale price of a $64 item that is on sale for 15% off.

32. Find the discount sale price of a $72 item that is on sale for 35% off.

33. A retailer has some skirts that cost $30 each. She wants to sell them at a profit of 60% of the cost. What price should she charge for the skirts?

34. The owner of a pizza parlor wants to make a profit of 70% of the cost for each pizza sold. If it costs $2.50 to make a pizza, at what price should each pizza be sold?

35. If a ring costs a jeweler $200, at what price should it be sold to yield a profit of 50% on the selling price?

36. If a head of lettuce costs a retailer $0.32, at what price should it be sold to yield a profit of 60% on the selling price?

37. If a pair of shoes costs a retailer $24, and he sells them for $39.60, what is his rate of profit based on the cost?

38. A retailer has some skirts that cost her $45 each. If she sells them for $83.25 per skirt, find her rate of profit based on the cost.

39. If a computer costs an electronics dealer $300, and she sells them for $800, what is her rate of profit based on the selling price?

40. A textbook costs a bookstore $45, and the store sells it for $60. Find the rate of profit based on the selling price.

41. Mitsuko's salary for next year is $34,775. This represents a 7% increase over this year's salary. Find Mitsuko's present salary.

42. Don bought a used car for $15,794, with 6% tax included. What was the price of the car without the tax?

43. Eva invested a certain amount of money at 10% interest and $1500 more than that amount at 11%. Her total yearly interest was $795. How much did she invest at each rate?

44. A total of $4000 was invested, part of it at 8% interest and the remainder at 9%. If the total yearly interest amounted to $350, how much was invested at each rate?

45. A sum of $95,000 is split between two investments, one paying 6% and the other 9%. If the total yearly interest amounted to $7290, how much was invested at 9%?

46. If $1500 is invested at 6% interest, how much money must be invested at 9% so that the total return for both investments is $301.50?

47. Suppose that Javier has a handful of coins, consisting of pennies, nickels, and dimes, worth $2.63. The number of nickels is 1 less than twice the number of pennies, and the number of dimes is 3 more than the number of nickels. How many coins of each kind does he have?

48. Sarah has a collection of nickels, dimes, and quarters worth $15.75. She has 10 more dimes than nickels and twice as many quarters as dimes. How many coins of each kind does she have?

49. A collection of 70 coins consisting of dimes, quarters, and half-dollars has a value of $17.75. There are three times as many quarters as dimes. Find the number of each kind of coin.

50. Abby has 37 coins, consisting only of dimes and quarters, worth $7.45. How many dimes and how many quarters does she have?

■ ■ ■ **THOUGHTS INTO WORDS**

51. Go to Problem 39 and calculate the rate of profit based on cost. Compare the rate of profit based on cost to the rate of profit based on selling price. From a consumer's viewpoint, would you prefer that a retailer figure his profit on the basis of the cost of an item or on the basis of its selling price? Explain your answer.

52. Is a 10% discount followed by a 30% discount the same as a 30% discount followed by a 10% discount? Justify your answer.

53. What is wrong with the following solution and how should it be done?

$$1.2x + 2 = 3.8$$
$$10(1.2x) + 2 = 10(3.8)$$
$$12x + 2 = 38$$
$$12x = 36$$
$$x = 3$$

■ ■ ■ **FURTHER INVESTIGATIONS**

For Problems 54–63, solve each equation and express the solutions in decimal form. Be sure to check your solutions. Use your calculator whenever it seems helpful.

54. $1.2x + 3.4 = 5.2$

55. $0.12x - 0.24 = 0.66$

56. $0.12x + 0.14(550 - x) = 72.5$

57. $0.14t + 0.13(890 - t) = 67.95$

58. $0.7n + 1.4 = 3.92$

59. $0.14n - 0.26 = 0.958$

60. $0.3(d + 1.8) = 4.86$

61. $0.6(d - 4.8) = 7.38$

62. $0.8(2x - 1.4) = 19.52$

63. $0.5(3x + 0.7) = 20.6$

64. The following formula can be used to determine the selling price of an item when the profit is based on a percent of the selling price.

$$\text{Selling price} = \frac{\text{Cost}}{100\% - \text{Percent of profit}}$$

Show how this formula is developed.

65. A retailer buys an item for $90, resells it for $100, and claims that she is making only a 10% profit. Is this claim correct?

66. Is a 10% discount followed by a 20% discount equal to a 30% discount? Defend your answer.

2.4 Formulas

To find the distance traveled in 4 hours at a rate of 55 miles per hour, we multiply the rate times the time; thus the distance is 55(4) = 220 miles. We can state the rule *distance equals rate times time* as a formula: $d = rt$. Formulas are rules we state in symbolic form, usually as equations.

Formulas are typically used in two different ways. At times a formula is solved for a specific variable when we are given the numerical values for the other variables. This is much like evaluating an algebraic expression. At other times we need to change the form of an equation by solving for one variable in terms of the other variables. Throughout our work on formulas, we will use the properties of equality and the techniques we have previously learned for solving equations. Let's consider some examples.

EXAMPLE 1

If we invest P dollars at r percent for t years, the amount of simple interest i is given by the formula $i = Prt$. Find the amount of interest earned by \$500 at 7% for 2 years.

Solution

By substituting \$500 for P, 7% for r, and 2 for t, we obtain

$$i = Prt$$

$$i = (500)(7\%)(2)$$

$$i = (500)(0.07)(2)$$

$$i = 70$$

Thus we earn \$70 in interest.   ■

EXAMPLE 2

If we invest P dollars at a simple rate of r percent, then the amount A accumulated after t years is given by the formula $A = P + Prt$. If we invest \$500 at 8%, how many years will it take to accumulate \$600?

Solution

Substituting \$500 for P, 8% for r, and \$600 for A, we obtain

$$A = P + Prt$$

$$600 = 500 + 500(8\%)(t)$$

Solving this equation for t yields

$$600 = 500 + 500(0.08)(t)$$

$$600 = 500 + 40t$$

$$100 = 40t$$

$$2\frac{1}{2} = t$$

It will take $2\frac{1}{2}$ years to accumulate \$600.   ■

When we are using a formula, it is sometimes convenient first to change its form. For example, suppose we are to use the *perimeter* formula for a rectangle ($P = 2l + 2w$) to complete the following chart.

Perimeter (P)	32	24	36	18	56	80	
Length (l)	10	7	14	5	15	22	All in centimeters
Width (w)	?	?	?	?	?	?	

Because w is the unknown quantity, it would simplify the computational work if we first solved the formula for w in terms of the other variables as follows:

$$P = 2l + 2w$$

$$P - 2l = 2w \qquad \text{Add } -2l \text{ to both sides.}$$

$$\frac{P - 2l}{2} = w \qquad \text{Multiply both sides by } \frac{1}{2}.$$

$$w = \frac{P - 2l}{2} \qquad \text{Apply the symmetric property of equality.}$$

Now for each value for P and l, we can easily determine the corresponding value for w. Be sure you agree with the following values for w: 6, 5, 4, 4, 13, and 18. Likewise we can also solve the formula $P = 2l + 2w$ for l in terms of P and w. The result would be $l = \frac{P - 2w}{2}$.

Let's consider some other often-used formulas and see how we can use the properties of equality to alter their forms. Here we will be solving a formula for a specified variable in terms of the other variables. The key is to isolate the term that contains the variable being solved for. Then, by appropriately applying the multiplication property of equality, we will solve the formula for the specified variable. Throughout this section, we will identify formulas when we first use them. (Some geometric formulas are also given on the endsheets.)

EXAMPLE 3

Solve $A = \frac{1}{2}bh$ for h (area of a triangle).

Solution

$$A = \frac{1}{2}bh$$

$$2A = bh \qquad \text{Multiply both sides by 2.}$$

$$\frac{2A}{b} = h \qquad \text{Multiply both sides by } \frac{1}{b}.$$

$$h = \frac{2A}{b} \qquad \text{Apply the symmetric property of equality.} \qquad \blacksquare$$

EXAMPLE 4

Solve $A = P + Prt$ for t.

Solution

$$A = P + Prt$$

$$A - P = Prt \qquad \text{Add } -P \text{ to both sides.}$$

$$\frac{A - P}{Pr} = t \qquad \text{Multiply both sides by } \frac{1}{Pr}.$$

$$t = \frac{A - P}{Pr} \qquad \text{Apply the symmetric property of equality.} \qquad \blacksquare$$

EXAMPLE 5

Solve $A = P + Prt$ for P.

Solution

$$A = P + Prt$$

$$A = P(1 + rt) \qquad \text{Apply the distributive property to the right side.}$$

$$\frac{A}{1 + rt} = P \qquad \text{Multiply both sides by } \frac{1}{1 + rt}.$$

$$P = \frac{A}{1 + rt} \qquad \text{Apply the symmetric property of equality.} \qquad \blacksquare$$

EXAMPLE 6

Solve $A = \frac{1}{2}h(b_1 + b_2)$ for b_1 (area of a trapezoid).

Solution

$$A = \frac{1}{2}h(b_1 + b_2)$$

$$2A = h(b_1 + b_2) \qquad \text{Multiply both sides by 2.}$$

$$2A = hb_1 + hb_2 \qquad \text{Apply the distributive property to right side.}$$

$$2A - hb_2 = hb_1 \qquad \text{Add } -hb_2 \text{ to both sides.}$$

$$\frac{2A - hb_2}{h} = b_1 \qquad \text{Multiply both sides by } \frac{1}{h}.$$

$$b_1 = \frac{2A - hb_2}{h} \qquad \text{Apply the symmetric property of equality.} \qquad \blacksquare$$

In order to isolate the term containing the variable being solved for, we will apply the distributive property in different ways. In Example 5 you *must* use the distributive property to change from the form $P + Prt$ to $P(1 + rt)$. However, in Example 6 we used the distributive property to change $h(b_1 + b_2)$ to $hb_1 + hb_2$. In both problems the key is to isolate the term that contains the variable being solved for, so that an appropriate application of the multiplication property of equality will produce the desired result. Also note the use of subscripts to identify the two bases of a trapezoid. Subscripts enable us to use the same letter b to identify the bases, but b_1 represents one base and b_2 the other.

Sometimes we are faced with equations such as $ax + b = c$, where x is the variable and a, b, and c are referred to as *arbitrary constants*. Again we can use the properties of equality to solve the equation for x as follows:

$$ax + b = c$$

$$ax = c - b \qquad \text{Add } -b \text{ to both sides.}$$

$$x = \frac{c - b}{a} \qquad \text{Multiply both sides by } \frac{1}{a}.$$

In Chapter 7, we will be working with equations such as $2x - 5y = 7$, which are called equations of two variables in x and y. Often we need to change the form of such equations by solving for one variable in terms of the other variable. The properties of equality provide the basis for doing this.

EXAMPLE 7

Solve $2x - 5y = 7$ for y in terms of x.

Solution

$$2x - 5y = 7$$

$$-5y = 7 - 2x \qquad \text{Add } -2x \text{ to both sides.}$$

$$y = \frac{7 - 2x}{-5} \qquad \text{Multiply both sides by } \frac{1}{5}.$$

$$y = \frac{2x - 7}{5} \qquad \text{Multiply the numerator and denominator of the fraction on the right by } -1. \text{ (This final step is not absolutely necessary, but usually we prefer to have a positive number as a denominator.)} \blacksquare$$

Equations of two variables may also contain arbitrary constants. For example, the equation $\frac{x}{a} + \frac{y}{b} = 1$ contains the variables x and y and the arbitrary constants a and b.

EXAMPLE 8

Solve the equation $\frac{x}{a} + \frac{y}{b} = 1$ for x.

Solution

$$\frac{x}{a} + \frac{y}{b} = 1$$

$$ab\left(\frac{x}{a} + \frac{y}{b}\right) = ab(1) \qquad \text{Multiply both sides by } ab.$$

$$bx + ay = ab$$

$$bx = ab - ay \qquad \text{Add } -ay \text{ to both sides.}$$

$$x = \frac{ab - ay}{b} \qquad \text{Multiply both sides by } \frac{1}{b}. \qquad \blacksquare$$

Remark: Traditionally, equations that contain more than one variable, such as those in Examples 3–8, are called **literal equations**. As illustrated, it is sometimes necessary to solve a literal equation for one variable in terms of the other variable(s).

■ Formulas and Problem Solving

We often use formulas as guidelines for setting up an appropriate algebraic equation when solving a word problem. Let's consider an example to illustrate this point.

How long will it take $500 to double itself if we invest it at 8% simple interest?

Solution

For $500 to grow into $1000 (double itself), it must earn $500 in interest. Thus we let t represent the number of years it will take $500 to earn $500 in interest. Now we can use the formula $i = Prt$ as a guideline.

$$i = Prt$$

$$500 = 500(8\%)(t)$$

Solving this equation, we obtain

$$500 = 500(0.08)(t)$$

$$1 = 0.08t$$

$$100 = 8t$$

$$12\frac{1}{2} = t$$

It will take $12\frac{1}{2}$ years.   ■

Sometimes we use formulas in the analysis of a problem but not as the main guideline for setting up the equation. For example, uniform motion problems involve the formula $d = rt$, but the main guideline for setting up an equation for such problems is usually a statement about times, rates, or distances. Let's consider an example to demonstrate.

Mercedes starts jogging at 5 miles per hour. One-half hour later, Karen starts jogging on the same route at 7 miles per hour. How long will it take Karen to catch Mercedes?

Solution

First, let's sketch a diagram and record some information (Figure 2.3).

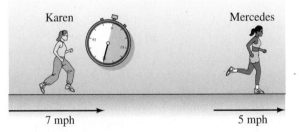

Karen   Mercedes

7 mph   5 mph

Figure 2.3

If we let t represent Karen's time, then $t + \frac{1}{2}$ represents Mercedes' time. We can use the statement *Karen's distance equals Mercedes' distance* as a guideline.

Solving this equation, we obtain

$$7t = 5t + \frac{5}{2}$$

$$2t = \frac{5}{2}$$

$$t = \frac{5}{4}$$

Karen should catch Mercedes in $1\frac{1}{4}$ hours.   ■

Remark: An important part of problem solving is the ability to sketch a meaningful figure that can be used to record the given information and help in the analysis of the problem. Our sketches were done by professional artists for aesthetic purposes. Your sketches can be very roughly drawn as long as they depict the situation in a way that helps you analyze the problem.

Note that in the solution of Problem 2 we used a figure and a simple arrow diagram to record and organize the information pertinent to the problem. Some people find it helpful to use a chart for that purpose. We shall use a chart in Problem 3. Keep in mind that we are not trying to dictate a particular approach; you decide what works best for you.

PROBLEM 3

Two trains leave a city at the same time, one traveling east and the other traveling west. At the end of $9\frac{1}{2}$ hours, they are 1292 miles apart. If the rate of the train traveling east is 8 miles per hour faster than the rate of the other train, find their rates.

Solution

If we let r represent the rate of the westbound train, then $r + 8$ represents the rate of the eastbound train. Now we can record the times and rates in a chart and then use the distance formula ($d = rt$) to represent the distances.

	Rate	Time	Distance ($d = rt$)
Westbound train	r	$9\frac{1}{2}$	$\frac{19}{2}r$
Eastbound train	$r + 8$	$9\frac{1}{2}$	$\frac{19}{2}(r + 8)$

Because the distance that the westbound train travels plus the distance that the eastbound train travels equals 1292 miles, we can set up and solve the following equation.

$$\begin{pmatrix}\text{Eastbound}\\\text{distance}\end{pmatrix} + \begin{pmatrix}\text{Westbound}\\\text{distance}\end{pmatrix} = \begin{pmatrix}\text{Miles}\\\text{apart}\end{pmatrix}$$

$$\frac{19r}{2} + \frac{19(r + 8)}{2} = 1292$$

$$19r + 19(r + 8) = 2584$$

$$19r + 19r + 152 = 2584$$

$$38r = 2432$$

$$r = 64$$

The westbound train travels at a rate of 64 miles per hour, and the eastbound train travels at a rate of $64 + 8 = 72$ miles per hour. ∎

Now let's consider a problem that is often referred to as a mixture problem. There is no basic formula that applies to all of these problems, but we suggest that you think in terms of a pure substance, which is often helpful in setting up a guideline. Also keep in mind that the phrase "a 40% solution of some substance" means that the solution contains 40% of that particular substance and 60% of something else mixed with it. For example, a 40% salt solution contains 40% salt, and the other 60% is something else, probably water. Now let's illustrate what we mean by suggesting that you think in terms of a pure substance.

P R O B L E M   4

Bryan's Pest Control stocks a 7% solution of insecticide for lawns and also a 15% solution. How many gallons of each should be mixed to produce 40 gallons that is 12% insecticide?

Solution

The key idea in solving such a problem is to recognize the following guideline.

$$\begin{pmatrix}\text{Amount of insecticide}\\\text{in the 7\% solution}\end{pmatrix} + \begin{pmatrix}\text{Amount of insecticide}\\\text{in the 15\% solution}\end{pmatrix} = \begin{pmatrix}\text{Amount of insecticide in}\\\text{40 gallons of 15\% solution}\end{pmatrix}$$

Let x represent the gallons of 7% solution. Then $40 - x$ represents the gallons of 15% solution. The guideline translates into the following equation.

$$(7\%)(x) + (15\%)(40 - x) = (12\%)(40)$$

Solving this equation yields

$$0.07x + 0.15(40 - x) = 0.12(40)$$

$$0.07x + 6 - 0.15x = 4.8$$

$$-0.08x + 6 = 4.8$$

$$-0.08x = -1.2$$

$$x = 15$$

Thus 15 gallons of 7% solution and $40 - x = 25$ gallons of 15% solution need to be mixed to obtain 40 gallons of 12% solution. ∎

P R O B L E M   5

How many liters of pure alcohol must we add to 20 liters of a 40% solution to obtain a 60% solution?

Solution

The key idea in solving such a problem is to recognize the following guideline.

$$\begin{pmatrix} \text{Amount of pure} \\ \text{alcohol in the} \\ \text{original solution} \end{pmatrix} + \begin{pmatrix} \text{Amount of} \\ \text{pure alcohol} \\ \text{to be added} \end{pmatrix} = \begin{pmatrix} \text{Amount of pure} \\ \text{alcohol in the} \\ \text{final solution} \end{pmatrix}$$

Let l represent the number of liters of pure alcohol to be added, and the guideline translates into the following equation.

$$(40\%)(20) + l = 60\%(20 + l)$$

Solving this equation yields

$$0.4(20) + l = 0.6(20 + l)$$

$$8 + l = 12 + 0.6l$$

$$0.4l = 4$$

$$l = 10$$

We need to add 10 liters of pure alcohol. (Remember to check this answer back into the original statement of the problem.) ∎

Problem Set 2.4

1. Solve $i = Prt$ for i, given that $P = \$300$, $r = 8\%$, and $t = 5$ years.

2. Solve $i = Prt$ for i, given that $P = \$500$, $r = 9\%$, and $t = 3\frac{1}{2}$ years.

3. Solve $i = Prt$ for t, given that $P = \$400$, $r = 11\%$, and $i = \$132$.

4. Solve $i = Prt$ for t, given that $P = \$250$, $r = 12\%$, and $i = \$120$.

5. Solve $i = Prt$ for r, given that $P = \$600$, $t = 2\frac{1}{2}$ years, and $i = \$90$. Express r as a percent.

6. Solve $i = Prt$ for r, given that $P = \$700$, $t = 2$ years, and $i = \$126$. Express r as a percent.

7. Solve $i = Prt$ for P, given that $r = 9\%$, $t = 3$ years, and $i = \$216$.

8. Solve $i = Prt$ for P, given that $r = 8\frac{1}{2}\%$, $t = 2$ years, and $i = \$204$.

9. Solve $A = P + Prt$ for A, given that $P = \$1000$, $r = 12\%$, and $t = 5$ years.

10. Solve $A = P + Prt$ for A, given that $P = \$850$, $r = 9\frac{1}{2}\%$, and $t = 10$ years.

11. Solve $A = P + Prt$ for r, given that $A = \$1372$, $P = \$700$, and $t = 12$ years. Express r as a percent.

12. Solve $A = P + Prt$ for r, given that $A = \$516$, $P = \$300$, and $t = 8$ years. Express r as a percent.

13. Solve $A = P + Prt$ for P, given that $A = \$326$, $r = 7\%$, and $t = 9$ years.

14. Solve $A = P + Prt$ for P, given that $A = \$720$, $r = 8\%$, and $t = 10$ years.

15. Use the formula $A = \frac{1}{2}h(b_1 + b_2)$ and complete the following chart.

A	98	104	49	162	$16\frac{1}{2}$	$38\frac{1}{2}$	square feet
h	14	8	7	9	3	11	feet
b_1	8	12	4	16	4	5	feet
b_2	?	?	?	?	?	?	feet

A = area, h = height, b_1 = one base, b_2 = other base

16. Use the formula $P = 2l + 2w$ and complete the following chart. (You may want to change the form of the formula.)

P	28	18	12	34	68	centimeters
w	6	3	2	7	14	centimeters
l	?	?	?	?	?	centimeters

P = perimeter, w = width, l = length

Solve each of the following for the indicated variable.

17. $V = Bh$ for h (Volume of a prism)

18. $A = lw$ for l (Area of a rectangle)

19. $V = \pi r^2 h$ for h (Volume of a circular cylinder)

20. $V = \frac{1}{3}Bh$ for B (Volume of a pyramid)

21. $C = 2\pi r$ for r (Circumference of a circle)

22. $A = 2\pi r^2 + 2\pi rh$ for h (Surface area of a circular cylinder)

23. $I = \dfrac{100M}{C}$ for C (Intelligence quotient)

24. $A = \frac{1}{2}h(b_1 + b_2)$ for h (Area of a trapezoid)

25. $F = \frac{9}{5}C + 32$ for C (Celsius to Fahrenheit)

26. $C = \frac{5}{9}(F - 32)$ for F (Fahrenheit to Celsius)

For Problems 27–36, solve each equation for x.

27. $y = mx + b$

28. $\dfrac{x}{a} + \dfrac{y}{b} = 1$

29. $y - y_1 = m(x - x_1)$

30. $a(x + b) = c$

31. $a(x + b) = b(x - c)$

32. $x(a - b) = m(x - c)$

33. $\dfrac{x - a}{b} = c$

34. $\dfrac{x}{a} - 1 = b$

35. $\dfrac{1}{3}x + a = \dfrac{1}{2}b$

36. $\dfrac{2}{3}x - \dfrac{1}{4}a = b$

For Problems 37–46, solve each equation for the indicated variable.

37. $2x - 5y = 7$ for x

38. $5x - 6y = 12$ for x

39. $-7x - y = 4$ for y

40. $3x - 2y = -1$ for y

41. $3(x - 2y) = 4$ for x

42. $7(2x + 5y) = 6$ for y

43. $\dfrac{y - a}{b} = \dfrac{x + b}{c}$ for x

44. $\dfrac{x - a}{b} = \dfrac{y - a}{c}$ for y

45. $(y + 1)(a - 3) = x - 2$ for y

46. $(y - 2)(a + 1) = x$ for y

Solve each of Problems 47–62 by setting up and solving an appropriate algebraic equation.

47. Suppose that the length of a certain rectangle is 2 meters less than four times its width. The perimeter of the rectangle is 56 meters. Find the length and width of the rectangle.

48. The perimeter of a triangle is 42 inches. The second side is 1 inch more than twice the first side, and the third side is 1 inch less than three times the first side. Find the lengths of the three sides of the triangle.

49. How long will it take $500 to double itself at 9% simple interest?

50. How long will it take $700 to triple itself at 10% simple interest?

51. How long will it take P dollars to double itself at 9% simple interest?

52. How long will it take P dollars to triple itself at 10% simple interest?

53. Two airplanes leave Chicago at the same time and fly in opposite directions. If one travels at 450 miles per hour and the other at 550 miles per hour, how long will it take for them to be 4000 miles apart?

54. Look at Figure 2.4. Tyrone leaves city A on a moped traveling toward city B at 18 miles per hour. At the same time, Tina leaves city B on a bicycle traveling toward city A at 14 miles per hour. The distance between the two cities is 112 miles. How long will it take before Tyrone and Tina meet?

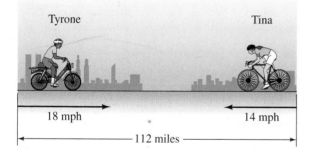

Figure 2.4

55. Juan starts walking at 4 miles per hour. An hour and a half later, Cathy starts jogging along the same route at 6 miles per hour. How long will it take Cathy to catch up with Juan?

56. A car leaves a town at 60 kilometers per hour. How long will it take a second car, traveling at 75 kilometers per hour, to catch the first car if it leaves 1 hour later?

57. Bret started on a 70-mile bicycle ride at 20 miles per hour. After a time he became a little tired and slowed down to 12 miles per hour for the rest of the trip. The entire trip of 70 miles took $4\frac{1}{2}$ hours. How far had Bret ridden when he reduced his speed to 12 miles per hour?

58. How many gallons of a 12%-salt solution must be mixed with 6 gallons of a 20%-salt solution to obtain a 15%-salt solution?

59. Suppose that you have a supply of a 30% solution of alcohol and a 70% solution of alcohol. How many quarts of each should be mixed to produce 20 quarts that is 40% alcohol?

60. How many cups of grapefruit juice must be added to 40 cups of punch that is 5% grapefruit juice to obtain a punch that is 10% grapefruit juice?

61. How many milliliters of pure acid must be added to 150 milliliters of a 30% solution of acid to obtain a 40% solution?

62. A 16-quart radiator contains a 50% solution of antifreeze. How much needs to be drained out and replaced with pure antifreeze to obtain a 60% antifreeze solution?

■ ■ ■ THOUGHTS INTO WORDS

63. Some people subtract 32 and then divide by 2 to estimate the change from a Fahrenheit reading to a Celsius reading. Why does this give an estimate and how good is the estimate?

64. One of your classmates analyzes Problem 56 as follows: "The first car has traveled 60 kilometers before the second car starts. Because the second car travels 15 kilometers per hour faster, it will take $\frac{60}{15} = 4$ hours for the second car to overtake the first car." How would you react to this analysis of the problem?

65. Summarize the new ideas relative to problem solving that you have acquired thus far in this course.

■ ■ ■ **FURTHER INVESTIGATIONS**

For Problems 66–73, use your calculator to help solve each formula for the indicated variable.

66. Solve $i = Prt$ for i, given that $P = \$875$, $r = 12\frac{1}{2}\%$, and $t = 4$ years.

67. Solve $i = Prt$ for i, given that $P = \$1125$, $r = 13\frac{1}{4}\%$, and $t = 4$ years.

68. Solve $i = Prt$ for t, given that $i = \$453.25$, $P = \$925$, and $r = 14\%$.

69. Solve $i = Prt$ for t, given that $i = \$243.75$, $P = \$1250$, and $r = 13\%$.

70. Solve $i = Prt$ for r, given that $i = \$356.50$, $P = \$1550$, and $t = 2$ years. Express r as a percent.

71. Solve $i = Prt$ for r, given that $i = \$159.50$, $P = \$2200$, and $t = 0.5$ of a year. Express r as a percent.

72. Solve $A = P + Prt$ for P, given that $A = \$1423.50$, $r = 9\frac{1}{2}\%$, and $t = 1$ year.

73. Solve $A = P + Prt$ for P, given that $A = \$2173.75$, $r = 8\frac{3}{4}\%$, and $t = 2$ years.

74. If you have access to computer software that includes spreadsheets, go to Problems 15 and 16. You should be able to enter the given information in rows. Then, when you enter a formula in a cell below the information and drag that formula across the columns, the software should produce all the answers.

2.5   Inequalities

We listed the basic inequality symbols in Section 1.2. With these symbols we can make various **statements of inequality**:

$a < b$ means a is less than b.

$a \leq b$ means a is less than or equal to b.

$a > b$ means a is greater than b.

$a \geq b$ means a is greater than or equal to b.

Here are some examples of **numerical statements of inequality**:

$$7 + 8 > 10 \qquad -4 + (-6) \geq -10$$

$$-4 > -6 \qquad 7 - 9 \leq -2$$

$$7 - 1 < 20 \qquad 3 + 4 > 12$$

$$8(-3) < 5(-3) \qquad 7 - 1 < 0$$

Note that only $3 + 4 > 12$ and $7 - 1 < 0$ are *false*; the other six are *true* numerical statements.

Algebraic inequalities contain one or more variables. The following are examples of algebraic inequalities.

$$x + 4 > 8 \qquad 3x + 2y \leq 4$$

$$3x - 1 < 15 \qquad x^2 + y^2 + z^2 \geq 7$$

$$y^2 + 2y - 4 \geq 0$$

An algebraic inequality such as $x + 4 > 8$ is neither true nor false as it stands, and we call it an **open sentence**. For each numerical value we substitute for x, the algebraic inequality $x + 4 > 8$ becomes a numerical statement of inequality that is true or false. For example, if $x = -3$, then $x + 4 > 8$ becomes $-3 + 4 > 8$, which is false. If $x = 5$, then $x + 4 > 8$ becomes $5 + 4 > 8$, which is true. **Solving an inequality** is the process of finding the numbers that make an algebraic inequality a true numerical statement. We call such numbers the *solutions* of the inequality; the solutions *satisfy* the inequality.

The general process for solving inequalities closely parallels the process for solving equations. We continue to replace the given inequality with equivalent, but simpler, inequalities. For example,

$$3x + 4 > 10 \tag{1}$$

$$3x > 6 \tag{2}$$

$$x > 2 \tag{3}$$

are all equivalent inequalities; that is, they all have the same solutions. By inspection we see that the solutions for (3) are all numbers greater than *2*. Thus (1) has the same solutions.

The exact procedure for simplifying inequalities so that we can determine the solutions is based primarily on two properties. The first of these is the addition property of inequality.

Addition Property of Inequality

For all real numbers a, b, and c,

$$a > b \quad \text{if and only if } a + c > b + c$$

The addition property of inequality states that we can add any number to both sides of an inequality to produce an equivalent inequality. We have stated the property in terms of $>$, but analogous properties exist for $<$, $\geq$, and $\leq$.

Before we state the multiplication property of inequality let's look at some numerical examples.

$2 < 5$	Multiply both sides by 4	$4(2) < 4(5)$	$8 < 20$
$-3 > -7$	Multiply both sides by 2	$2(-3) > 2(-7)$	$-6 > -14$
$-4 < 6$	Multiply both sides by 10	$10(-4) < 10(6)$	$-40 < 60$
$4 < 8$	Multiply both sides by -3	$-3(4) > -3(8)$	$-12 > -24$
$3 > -2$	Multiply both sides by -4	$-4(3) < -4(-2)$	$-12 < 8$
$-4 < -1$	Multiply both sides by -2	$-2(-4) > -2(-1)$	$8 > 2$

Notice in the first three examples that when we multiply both sides of an inequality by a *positive number*, we get an inequality of the *same sense*. That means that if the original inequality is *less than*, then the new inequality is *less than*; and if the original inequality is *greater than*, then the new inequality is *greater than*. The last three examples illustrate that when we multiply both sides of an inequality by a *negative number* we get an inequality of the *opposite sense*.

We can state the multiplication property of inequality as follows.

Multiplication Property of Inequality

(a) For all real numbers a, b, and c, with $c > 0$,

$$a > b \quad \text{if and only if } ac > bc$$

(b) For all real numbers a, b, and c, with $c < 0$,

$$a > b \quad \text{if and only if } ac < bc$$

Similar properties hold if we reverse each inequality or if we replace $>$ with $\geq$ and $<$ with $\leq$. For example, if $a \leq b$ and $c < 0$, then $ac \geq bc$.

Now let's use the addition and multiplication properties of inequality to help solve some inequalities.

E X A M P L E   1   Solve $3x - 4 > 8$.

Solution

$$3x - 4 > 8$$

$$3x - 4 + 4 > 8 + 4 \qquad \text{Add 4 to both sides.}$$

$$3x > 12$$

$$\frac{1}{3}(3x) > \frac{1}{3}(12) \qquad \text{Multiply both sides by } \frac{1}{3}.$$

$$x > 4$$

The solution set is $\{x \mid x > 4\}$. (Remember that we read the set $\{x \mid x > 4\}$ as "the set of all x such that x is greater than 4.") ■

In Example 1, once we obtained the simple inequality $x > 4$, the solution set $\{x \mid x > 4\}$ became obvious. We can also express solution sets for inequalities on a number line graph. Figure 2.5 shows the graph of the solution set for Example 1. The left-hand parenthesis at 4 indicates that 4 is *not* a solution, and the red part of the line to the right of 4 indicates that all numbers greater than 4 are solutions.

Figure 2.5

It is also convenient to express solution sets of inequalities using **interval notation**. For example, the notation $(4, \infty)$ also refers to the set of real numbers greater than 4. As in Figure 2.5, the left-hand parenthesis indicates that 4 is not to be included. The infinity symbol, ∞, along with the right-hand parenthesis, indicates that there is no right-hand endpoint. Following is a partial list of interval notations, along with the sets of graphs they represent (Figure 2.6). We will add to this list in the next section.

Set	Graph	Interval notation
$\{x \mid x > a\}$		(a, ∞)
$\{x \mid x \geq a\}$		$[a, \infty)$
$\{x \mid x < b\}$		$(-\infty, b)$
$\{x \mid x \leq b\}$	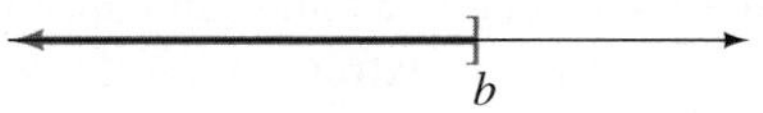	$(-\infty, b]$

Figure 2.6

Note the use of square brackets to indicate the inclusion of endpoints. From now on, we will express the solution sets of inequalities using interval notation.

EXAMPLE 2

Solve $-2x + 1 > 5$ and graph the solutions.

Solution

$$-2x + 1 > 5$$
$$-2x + 1 + (-1) > 5 + (-1) \qquad \text{Add } -1 \text{ to both sides.}$$
$$-2x > 4$$
$$-\frac{1}{2}(-2x) < -\frac{1}{2}(4) \qquad \text{Multiply both sides by } -\frac{1}{2}.$$
$$x < -2 \qquad \text{Note that the sense of the inequality has been reversed.}$$

The solution set is $(-\infty, -2)$, which can be illustrated on a number line as in Figure 2.7.

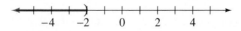

Figure 2.7 ∎

Checking solutions for an inequality presents a problem. Obviously, we cannot check all of the infinitely many solutions for a particular inequality. However, by checking at least one solution, especially when the multiplication property has been used, we might catch the common mistake of forgetting to change the sense of an inequality. In Example 2 we are claiming that all numbers less than -2 will satisfy the original inequality. Let's check one such number, say -4.

$$-2x + 1 > 5$$

$$-2(-4) + 1 \overset{?}{>} 5 \quad \text{when } x = -4$$

$$8 + 1 \overset{?}{>} 5$$

$$9 > 5$$

Thus -4 satisfies the original inequality. Had we forgotten to switch the sense of the inequality when both sides were multiplied by $-\dfrac{1}{2}$, our answer would have been $x > -2$, and we would have detected such an error by the check.

Many of the same techniques used to solve equations, such as removing parentheses and combining similar terms, may be used to solve inequalities. However, we must be extremely careful when using the multiplication property of inequality. Study each of the following examples very carefully. The format we used highlights the major steps of a solution.

EXAMPLE 3

Solve $-3x + 5x - 2 \geq 8x - 7 - 9x$.

Solution

$$-3x + 5x - 2 \geq 8x - 7 - 9x$$

$$2x - 2 \geq -x - 7 \qquad \text{Combine similar terms on both sides.}$$

$$3x - 2 \geq -7 \qquad \text{Add } x \text{ to both sides.}$$

$$3x \geq -5 \qquad \text{Add 2 to both sides.}$$

$$\frac{1}{3}(3x) \geq \frac{1}{3}(-5) \qquad \text{Multiply both sides by } \frac{1}{3}.$$

$$x \geq -\frac{5}{3}$$

The solution set is $\left[-\dfrac{5}{3}, \infty \right)$. ∎

EXAMPLE 4

Solve $-5(x - 1) \leq 10$ and graph the solutions.

Solution

$$-5(x - 1) \leq 10$$

$$-5x + 5 \leq 10 \qquad \text{Apply the distributive property on the left.}$$

$$-5x \leq 5 \qquad \text{Add } -5 \text{ to both sides.}$$

$$-\frac{1}{5}(-5x) \geq -\frac{1}{5}(5)$$

Multiply both sides by $-\frac{1}{5}$, which reverses the inequality.

$$x \geq -1$$

The solution set is $[-1, \infty)$, and it can be graphed as in Figure 2.8.

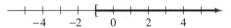

Figure 2.8

EXAMPLE 5

Solve $4(x - 3) > 9(x + 1)$.

Solution

$$4(x - 3) > 9(x + 1)$$

$$4x - 12 > 9x + 9$$ Apply the distributive property.

$$-5x - 12 > 9$$ Add $-9x$ to both sides.

$$-5x > 21$$ Add 12 to both sides.

$$-\frac{1}{5}(-5x) < -\frac{1}{5}(21)$$ Multiply both sides by $-\frac{1}{5}$, which reverses the inequality.

$$x < -\frac{21}{5}$$

The solution set is $\left(-\infty, -\frac{21}{5}\right)$.

The next example will solve the inequality without indicating the justification for each step. Be sure that you can supply the reasons for the steps.

EXAMPLE 6

Solve $3(2x + 1) - 2(2x + 5) < 5(3x - 2)$.

Solution

$$3(2x + 1) - 2(2x + 5) < 5(3x - 2)$$

$$6x + 3 - 4x - 10 < 15x - 10$$

$$2x - 7 < 15x - 10$$

$$-13x - 7 < -10$$

$$-13x < -3$$

$$-\frac{1}{13}(-13x) > -\frac{1}{13}(-3)$$

$$x > \frac{3}{13}$$

The solution set is $\left(\frac{3}{13}, \infty\right)$.

Problem Set 2.5

For Problems 1–8, express the given inequality in interval notation and sketch a graph of the interval.

1. $x > 1$

2. $x > -2$

3. $x \geq -1$

4. $x \geq 3$

5. $x < -2$

6. $x < 1$

7. $x \leq 2$

8. $x \leq 0$

For Problems 9–16, express each interval as an inequality using the variable x. For example, we can express the interval $[5, \infty)$ as $x \geq 5$.

9. $(-\infty, 4)$

10. $(-\infty, -2)$

11. $(-\infty, -7]$

12. $(-\infty, 9]$

13. $(8, \infty)$

14. $(-5, \infty)$

15. $[-7, \infty)$

16. $[10, \infty)$

For Problems 17–40, solve each of the inequalities and graph the solution set on a number line.

17. $x - 3 > -2$

18. $x + 2 < 1$

19. $-2x \geq 8$

20. $-3x \leq -9$

21. $5x \leq -10$

22. $4x \geq -4$

23. $2x + 1 < 5$

24. $2x + 2 > 4$

25. $3x - 2 > -5$

26. $5x - 3 < -3$

27. $-7x - 3 \leq 4$

28. $-3x - 1 \geq 8$

29. $2 + 6x > -10$

30. $1 + 6x > -17$

31. $5 - 3x < 11$

32. $4 - 2x < 12$

33. $15 < 1 - 7x$

34. $12 < 2 - 5x$

35. $-10 \leq 2 + 4x$

36. $-9 \leq 1 + 2x$

37. $3(x + 2) > 6$

38. $2(x - 1) < -4$

39. $5x + 2 \geq 4x + 6$

40. $6x - 4 \leq 5x - 4$

For Problems 41–70, solve each inequality and express the solution set using interval notation.

41. $2x - 1 > 6$

42. $3x - 2 < 12$

43. $-5x - 2 < -14$

44. $5 - 4x > -2$

45. $-3(2x + 1) \geq 12$

46. $-2(3x + 2) \leq 18$

47. $4(3x - 2) \geq -3$

48. $3(4x - 3) \leq -11$

49. $6x - 2 > 4x - 14$

50. $9x + 5 < 6x - 10$

51. $2x - 7 < 6x + 13$

52. $2x - 3 > 7x + 22$

53. $4(x - 3) \leq -2(x + 1)$

54. $3(x - 1) \geq -(x + 4)$

55. $5(x - 4) - 6(x + 2) < 4$

56. $3(x + 2) - 4(x - 1) < 6$

57. $-3(3x + 2) - 2(4x + 1) \geq 0$

58. $-4(2x - 1) - 3(x + 2) \geq 0$

59. $-(x - 3) + 2(x - 1) < 3(x + 4)$

60. $3(x - 1) - (x - 2) > -2(x + 4)$

61. $7(x + 1) - 8(x - 2) < 0$

62. $5(x - 6) - 6(x + 2) < 0$

63. $-5(x - 1) + 3 > 3x - 4 - 4x$

64. $3(x + 2) + 4 < -2x + 14 + x$

65. $3(x - 2) - 5(2x - 1) \geq 0$

66. $4(2x - 1) - 3(3x + 4) \geq 0$

67. $-5(3x + 4) < -2(7x - 1)$

68. $-3(2x + 1) > -2(x + 4)$

69. $-3(x + 2) > 2(x - 6)$

70. $-2(x - 4) < 5(x - 1)$

■ ■ ■ THOUGHTS INTO WORDS

71. Do the *less than* and *greater than* relations possess a symmetric property similar to the symmetric property of equality? Defend your answer.

72. Give a step-by-step description of how you would solve the inequality $-3 > 5 - 2x$.

73. How would you explain to someone why it is necessary to reverse the inequality symbol when multiplying both sides of an inequality by a negative number?

■ ■ ■ **FURTHER INVESTIGATIONS**

74. Solve each of the following inequalities.

(a) $5x - 2 > 5x + 3$

(b) $3x - 4 < 3x + 7$

(c) $4(x + 1) < 2(2x + 5)$

(d) $-2(x - 1) > 2(x + 7)$

(e) $3(x - 2) < -3(x + 1)$

(f) $2(x + 1) + 3(x + 2) < 5(x - 3)$

2.6   More on Inequalities and Problem Solving

When we discussed solving equations that involve fractions, we found that **clearing the equation of all fractions** is frequently an effective technique. To accomplish this, we multiply both sides of the equation by the least common denominator of all the denominators in the equation. This same basic approach also works very well with inequalities that involve fractions, as the next examples demonstrate.

EXAMPLE 1   Solve $\dfrac{2}{3}x - \dfrac{1}{2}x > \dfrac{3}{4}$.

Solution

$$\frac{2}{3}x - \frac{1}{2}x > \frac{3}{4}$$

$$12\left(\frac{2}{3}x - \frac{1}{2}x\right) > 12\left(\frac{3}{4}\right) \qquad \text{Multiply both sides by 12, which is the LCD of 3, 2, and 4.}$$

$$12\left(\frac{2}{3}x\right) - 12\left(\frac{1}{2}x\right) > 12\left(\frac{3}{4}\right) \qquad \text{Apply the distributive property.}$$

$$8x - 6x > 9$$

$$2x > 9$$

$$x > \frac{9}{2}$$

The solution set is $\left(\dfrac{9}{2}, \infty\right)$. ■

EXAMPLE 2   Solve $\dfrac{x + 2}{4} + \dfrac{x - 3}{8} < 1$.

Solution

$$\frac{x + 2}{4} + \frac{x - 3}{8} < 1$$

$$8\left(\frac{x + 2}{4} + \frac{x - 3}{8}\right) < 8(1) \qquad \text{Multiply both sides by 8, which is the LCD of 4 and 8.}$$

$$8\left(\frac{x+2}{4}\right) + 8\left(\frac{x-3}{8}\right) < 8(1)$$

$$2(x+2) + (x-3) < 8$$

$$2x + 4 + x - 3 < 8$$

$$3x + 1 < 8$$

$$3x < 7$$

$$x < \frac{7}{3}$$

The solution set is $\left(-\infty, \frac{7}{3}\right)$.   ▪

EXAMPLE 3   Solve $\frac{x}{2} - \frac{x-1}{5} \geq \frac{x+2}{10} - 4$.

Solution

$$\frac{x}{2} - \frac{x-1}{5} \geq \frac{x+2}{10} - 4$$

$$10\left(\frac{x}{2} - \frac{x-1}{5}\right) \geq 10\left(\frac{x+2}{10} - 4\right)$$

$$10\left(\frac{x}{2}\right) - 10\left(\frac{x-1}{5}\right) \geq 10\left(\frac{x+2}{10}\right) - 10(4)$$

$$5x - 2(x-1) \geq x + 2 - 40$$

$$5x - 2x + 2 \geq x - 38$$

$$3x + 2 \geq x - 38$$

$$2x + 2 \geq -38$$

$$2x \geq -40$$

$$x \geq -20$$

The solution set is $[-20, \infty)$.   ▪

The idea of **clearing all decimals** also works with inequalities in much the same way as it does with equations. We can multiply both sides of an inequality by an appropriate power of 10 and then proceed to solve in the usual way. The next two examples illustrate this procedure.

EXAMPLE 4   Solve $x \geq 1.6 + 0.2x$.

Solution

$$x \geq 1.6 + 0.2x$$

$$10(x) \geq 10(1.6 + 0.2x) \qquad \text{Multiply both sides by 10.}$$

$$10x \geq 16 + 2x$$
$$8x \geq 16$$
$$x \geq 2$$

The solution set is $[2, \infty)$. ∎

EXAMPLE 5

Solve $0.08x + 0.09(x + 100) \geq 43$.

Solution

$$0.08x + 0.09(x + 100) \geq 43$$
$$100(0.08x + 0.09(x + 100)) \geq 100(43) \qquad \text{Multiply both sides by 100.}$$
$$8x + 9(x + 100) \geq 4300$$
$$8x + 9x + 900 \geq 4300$$
$$17x + 900 \geq 4300$$
$$17x \geq 3400$$
$$x \geq 200$$

The solution set is $[200, \infty)$. ∎

■ Compound Statements

We use the words "and" and "or" in mathematics to form **compound statements**. The following are examples of compound numerical statements that use "and." We call such statements **conjunctions**. We agree to call a conjunction true only if all of its component parts are true. Statements 1 and 2 below are true, but statements 3, 4, and 5 are false.

1. $3 + 4 = 7$ and $-4 < -3$.   True

2. $-3 < -2$ and $-6 > -10$.   True

3. $6 > 5$ and $-4 < -8$.   False

4. $4 < 2$ and $0 < 10$.   False

5. $-3 + 2 = 1$ and $5 + 4 = 8$.   False

We call compound statements that use "or" **disjunctions**. The following are examples of disjunctions that involve numerical statements.

6. $0.14 > 0.13$ or $0.235 < 0.237$.   True

7. $\dfrac{3}{4} > \dfrac{1}{2}$ or $-4 + (-3) = 10$.   True

8. $-\dfrac{2}{3} > \dfrac{1}{3}$ or $(0.4)(0.3) = 0.12$.   True

9. $\dfrac{2}{5} < -\dfrac{2}{5}$ or $7 + (-9) = 16$.   False

A disjunction is true if at least one of its component parts is true. In other words, disjunctions are false only if all of the component parts are false. Thus statements 6, 7, and 8 are true, but statement 9 is false.

Now let's consider finding solutions for some compound statements that involve algebraic inequalities. Keep in mind that our previous agreements for labeling conjunctions and disjunctions true or false form the basis for our reasoning.

E X A M P L E   6

Graph the solution set for the conjunction $x > -1$ and $x < 3$.

Solution

The key word is "and," so we need to satisfy both inequalities. Thus all numbers between -1 and 3 are solutions, and we can indicate this on a number line as in Figure 2.9.

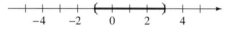

Figure 2.9

Using interval notation, we can represent the interval enclosed in parentheses in Figure 2.9 by $(-1, 3)$. Using set builder notation we can express the same interval as $\{x | -1 < x < 3\}$, where the statement $-1 < x < 3$ is read "Negative one is less than x, and x is less than three." In other words, x is between -1 and 3. ∎

Example 6 represents another concept that pertains to sets. The set of all elements common to two sets is called the **intersection** of the two sets. Thus in Example 6, we found the intersection of the two sets $\{x | x > -1\}$ and $\{x | x < 3\}$ to be the set $\{x | -1 < x < 3\}$. In general, we define the intersection of two sets as follows:

Definition 2.1

> The **intersection** of two sets A and B (written $A \cap B$) is the set of all elements that are in both A and in B. Using set builder notation, we can write
>
> $$A \cap B = \{x | x \in A \text{ and } x \in B\}$$

E X A M P L E   7

Solve the conjunction $3x + 1 > -5$ *and* $2x + 5 > 7$, and graph its solution set on a number line.

Solution

First, let's simplify both inequalities.

$$3x + 1 > -5 \quad \text{and} \quad 2x + 5 > 7$$
$$3x > -6 \quad \text{and} \quad 2x > 2$$
$$x > -2 \quad \text{and} \quad x > 1$$

Because this is a conjunction, we must satisfy both inequalities. Thus all numbers greater than 1 are solutions, and the solution set is $(1, \infty)$. We show the graph of the solution set in Figure 2.10.

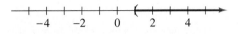

Figure 2.10   ∎

We can solve a conjunction such as $3x + 1 > -3$ and $3x + 1 < 7$, in which the same algebraic expression (in this case $3x + 1$) is contained in both inequalities, by using the **compact form** $-3 < 3x + 1 < 7$ as follows:

$$-3 < 3x + 1 < 7$$

$$-4 < 3x < 6 \qquad \text{Add } -1 \text{ to the left side, middle, and right side.}$$

$$-\frac{4}{3} < x < 2 \qquad \text{Multiply through by } \frac{1}{3}.$$

The solution set is $\left(-\dfrac{4}{3}, 2\right)$.

The word *and* ties the concept of a conjunction to the set concept of intersection. In a like manner, the word *or* links the idea of a disjunction to the set concept of **union**. We define the union of two sets as follows:

Definition 2.2

The **union** of two sets A and B (written $A \cup B$) is the set of all elements that are in A or in B, or in both. Using set builder notation, we can write

$$A \cup B = \{x \,|\, x \in A \text{ or } x \in B\}$$

EXAMPLE 8

Graph the solution set for the disjunction $x < -1$ *or* $x > 2$, and express it using interval notation.

Solution

The key word is "or," so all numbers that satisfy either inequality (or both) are solutions. Thus all numbers less than -1, along with all numbers greater than 2, are the solutions. The graph of the solution set is shown in Figure 2.11.

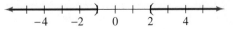

Figure 2.11

Using interval notation and the set concept of union, we can express the solution set as $(-\infty, -1) \cup (2, \infty)$.   ∎

Example 8 illustrates that in terms of set vocabulary, the solution set of a disjunction is the union of the solution sets of the component parts of the disjunction. Note that there is no compact form for writing $x < -1$ or $x > 2$ or for any disjunction.

E X A M P L E   9

Solve the disjunction $2x - 5 < -11$ or $5x + 1 \geq 6$, and graph its solution set on a number line.

Solution

First, let's simplify both inequalities.

$$
\begin{array}{ccc}
2x - 5 < -11 & \text{or} & 5x + 1 \geq 6 \\
2x < -6 & \text{or} & 5x \geq 5 \\
x < -3 & \text{or} & x \geq 1
\end{array}
$$

This is a disjunction, and all numbers less than -3, along with all numbers greater than or equal to 1, will satisfy it. Thus the solution set is $(-\infty, -3) \cup [1, \infty)$. Its graph is shown in Figure 2.12.

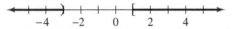

Figure 2.12

In summary, to solve a compound sentence involving an inequality, proceed as follows:

1. Solve separately each inequality in the compound sentence.

2. If it is a conjunction, the solution set is the intersection of the solution sets of each inequality.

3. If it is a disjunction, the solution set is the union of the solution sets of each inequality.

The following agreements on the use of interval notation (Figure 2.13) should be added to the list in Figure 2.6.

Set	Graph	Interval notation
$\{x \mid a < x < b\}$	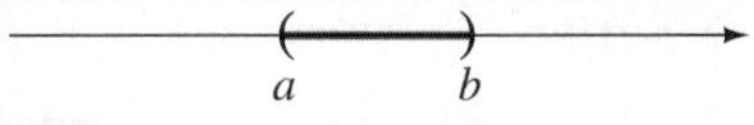	(a, b)
$\{x \mid a \leq x < b\}$	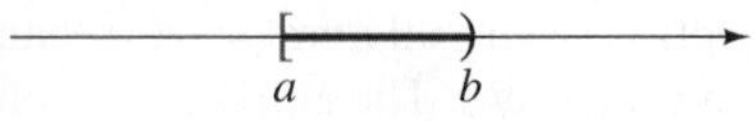	$[a, b)$
$\{x \mid a < x \leq b\}$		$(a, b]$
$\{x \mid a \leq x \leq b\}$	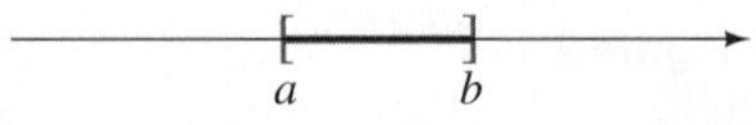	$[a, b]$

Figure 2.13

■ Problem Solving

We will conclude this section with some word problems that contain inequality statements.

Sari had scores of 94, 84, 86, and 88 on her first four exams of the semester. What score must she obtain on the fifth exam to have an average of 90 or better for the five exams?

Solution

Let s represent the score Sari needs on the fifth exam. Because the average is computed by adding all scores and dividing by the number of scores, we have the following inequality to solve.

$$\frac{94 + 84 + 86 + 88 + s}{5} \geq 90$$

Solving this inequality, we obtain

$$\frac{352 + s}{5} \geq 90$$

$$5\left(\frac{352 + s}{5}\right) \geq 5(90) \qquad \text{Multiply both sides by 5.}$$

$$352 + s \geq 450$$

$$s \geq 98$$

Sari must receive a score of 98 or better.   ■

An investor has $1000 to invest. Suppose she invests $500 at 8% interest. At what rate must she invest the other $500 so that the two investments together yield more than $100 of yearly interest?

Solution

Let r represent the unknown rate of interest. We can use the following guideline to set up an inequality.

Interest from 8% investment	+	Interest from r percent investment	>	$100
↓		↓		↓
(8%)($500)	+	r($500)	>	$100

Solving this inequality yields

$$40 + 500r > 100$$

$$500r > 60$$

$$r > \frac{60}{500}$$

$$r > 0.12 \qquad \text{Change to a decimal.}$$

She must invest the other $500 at a rate greater than 12%.   ■

PROBLEM 3

If the temperature for a 24-hour period ranged between 41°F and 59°F, inclusive (that is, $41 \le F \le 59$), what was the range in Celsius degrees?

Solution

Use the formula $F = \dfrac{9}{5}C + 32$, to solve the following compound inequality.

$$41 \le \frac{9}{5}C + 32 \le 59$$

Solving this yields

$$9 \le \frac{9}{5}C \le 27 \qquad \text{Add } -32.$$

$$\frac{5}{9}(9) \le \frac{5}{9}\left(\frac{9}{5}C\right) \le \frac{5}{9}(27) \qquad \text{Multiply by } \frac{5}{9}.$$

$$5 \le C \le 15$$

The range was between 5°C and 15°C, inclusive.   ■

Problem Set 2.6

For Problems 1–18, solve each of the inequalities and express the solution sets in interval notation.

1. $\dfrac{2}{5}x + \dfrac{1}{3}x > \dfrac{44}{15}$

2. $\dfrac{1}{4}x - \dfrac{4}{3}x < -13$

3. $x - \dfrac{5}{6} < \dfrac{x}{2} + 3$

4. $x + \dfrac{2}{7} > \dfrac{x}{2} - 5$

5. $\dfrac{x-2}{3} + \dfrac{x+1}{4} \ge \dfrac{5}{2}$

6. $\dfrac{x-1}{3} + \dfrac{x+2}{5} \le \dfrac{3}{5}$

7. $\dfrac{3-x}{6} + \dfrac{x+2}{7} \le 1$

8. $\dfrac{4-x}{5} + \dfrac{x+1}{6} \ge 2$

9. $\dfrac{x+3}{8} - \dfrac{x+5}{5} \ge \dfrac{3}{10}$

10. $\dfrac{x-4}{6} - \dfrac{x-2}{9} \le \dfrac{5}{18}$

11. $\dfrac{4x-3}{6} - \dfrac{2x-1}{12} < -2$

12. $\dfrac{3x+2}{9} - \dfrac{2x+1}{3} > -1$

13. $0.06x + 0.08(250 - x) \ge 19$

14. $0.08x + 0.09(2x) \ge 130$

15. $0.09x + 0.1(x + 200) > 77$

16. $0.07x + 0.08(x + 100) > 38$

17. $x \ge 3.4 + 0.15x$

18. $x \ge 2.1 + 0.3x$

For Problems 19–34, graph the solution set for each compound inequality, and express the solution sets in interval notation.

19. $x > -1$ and $x < 2$

20. $x > 1$ and $x < 4$

21. $x \le 2$ and $x > -1$

22. $x \le 4$ and $x \ge -2$

23. $x > 2$ or $x < -1$

24. $x > 1$ or $x < -4$

25. $x \le 1$ or $x > 3$

26. $x < -2$ or $x \ge 1$

27. $x > 0$ and $x > -1$

28. $x > -2$ and $x > 2$

29. $x < 0$ and $x > 4$

30. $x > 1$ or $x < 2$

31. $x > -2$ or $x < 3$

32. $x > 3$ and $x < -1$

33. $x > -1$ or $x > 2$

34. $x < -2$ or $x < 1$

For Problems 35–44, solve each compound inequality and graph the solution sets. Express the solution sets in interval notation.

35. $x - 2 > -1$ and $x - 2 < 1$

36. $x + 3 > -2$ and $x + 3 < 2$

37. $x + 2 < -3$ or $x + 2 > 3$

38. $x - 4 < -2$ or $x - 4 > 2$

39. $2x - 1 \geq 5$ and $x > 0$

40. $3x + 2 > 17$ and $x \geq 0$

41. $5x - 2 < 0$ and $3x - 1 > 0$

42. $x + 1 > 0$ and $3x - 4 < 0$

43. $3x + 2 < -1$ or $3x + 2 > 1$

44. $5x - 2 < -2$ or $5x - 2 > 2$

For Problems 45–56, solve each compound inequality using the compact form. Express the solution sets in interval notation.

45. $-3 < 2x + 1 < 5$

46. $-7 < 3x - 1 < 8$

47. $-17 \leq 3x - 2 \leq 10$

48. $-25 \leq 4x + 3 \leq 19$

49. $1 < 4x + 3 < 9$

50. $0 < 2x + 5 < 12$

51. $-6 < 4x - 5 < 6$

52. $-2 < 3x + 4 < 2$

53. $-4 \leq \dfrac{x - 1}{3} \leq 4$

54. $-1 \leq \dfrac{x + 2}{4} \leq 1$

55. $-3 < 2 - x < 3$

56. $-4 < 3 - x < 4$

For Problems 57–67, solve each problem by setting up and solving an appropriate inequality.

57. Suppose that Lance has $500 to invest. If he invests $300 at 9% interest, at what rate must he invest the remaining $200 so that the two investments yield more than $47 in yearly interest?

58. Mona invests $100 at 8% yearly interest. How much does she have to invest at 9% so that the total yearly interest from the two investments exceeds $26?

59. The average height of the two forwards and the center of a basketball team is 6 feet and 8 inches. What must

the average height of the two guards be so that the team average is at least 6 feet and 4 inches?

60. Thanh has scores of 52, 84, 65, and 74 on his first four math exams. What score must he make on the fifth exam to have an average of 70 or better for the five exams?

61. Marsha bowled 142 and 170 in her first two games. What must she bowl in the third game to have an average of at least 160 for the three games?

62. Candace had scores of 95, 82, 93, and 84 on her first four exams of the semester. What score must she obtain on the fifth exam to have an average of 90 or better for the five exams?

63. Suppose that Derwin shot rounds of 82, 84, 78, and 79 on the first four days of a golf tournament. What must he shoot on the fifth day of the tournament to average 80 or less for the five days?

64. The temperatures for a 24-hour period ranged between $-4°F$ and $23°F$, inclusive. What was the range in Celsius degrees? $\left(\text{Use } F = \dfrac{9}{5}C + 32. \right)$

65. Oven temperatures for baking various foods usually range between 325°F and 425°F, inclusive. Express this range in Celsius degrees. (Round answers to the nearest degree.)

66. A person's intelligence quotient (I) is found by dividing mental age (M), as indicated by standard tests, by chronological age (C) and then multiplying this ratio by 100. The formula $I = \dfrac{100M}{C}$ can be used. If the I range of a group of 11-year-olds is given by $80 \leq I \leq 140$, find the range of the mental age of this group.

67. Repeat Problem 66 for an I range of 70 to 125, inclusive, for a group of 9-year-olds.

■ ■ ■ THOUGHTS INTO WORDS

68. Explain the difference between a conjunction and a disjunction. Give an example of each (outside the field of mathematics).

69. How do you know by inspection that the solution set of the inequality $x + 3 > x + 2$ is the entire set of real numbers?

70. Find the solution set for each of the following compound statements, and in each case explain your reasoning.

(a) $x < 3$ and $5 > 2$

(b) $x < 3$ or $5 > 2$

(c) $x < 3$ and $6 < 4$

(d) $x < 3$ or $6 < 4$

2.7 Equations and Inequalities Involving Absolute Value

In Section 1.2, we defined the absolute value of a real number by

$$|a| = \begin{cases} a, & \text{if } a \geq 0 \\ -a, & \text{if } a < 0 \end{cases}$$

We also interpreted the absolute value of any real number to be the distance between the number and zero on a number line. For example, $|6| = 6$ translates to 6 units between 6 and 0. Likewise, $|-8| = 8$ translates to 8 units between -8 and 0.

The interpretation of absolute value as distance on a number line provides a straightforward approach to solving a variety of equations and inequalities involving absolute value. First, let's consider some equations.

EXAMPLE 1   Solve $|x| = 2$.

Solution

Think in terms of distance between the number and zero, and you will see that x must be 2 or -2. That is, the equation $|x| = 2$ is equivalent to

$$x = -2 \quad \text{or} \quad x = 2$$

The solution set is $\{-2, 2\}$. ∎

EXAMPLE 2   Solve $|x + 2| = 5$.

Solution

The number, $x + 2$, must be -5 or 5. Thus $|x + 2| = 5$ is equivalent to

$$x + 2 = -5 \quad \text{or} \quad x + 2 = 5$$

Solving each equation of the disjunction yields

$$x + 2 = -5 \quad \text{or} \quad x + 2 = 5$$
$$x = -7 \quad \text{or} \quad x = 3$$

The solution set is $\{-7, 3\}$.

✔ **Check**

$$|x + 2| = 5 \qquad |x + 2| = 5$$
$$|-7 + 2| \overset{?}{=} 5 \qquad |3 + 2| \overset{?}{=} 5$$
$$|-5| \overset{?}{=} 5 \qquad |5| \overset{?}{=} 5$$
$$5 = 5 \qquad 5 = 5$$

∎

The following general property should seem reasonable from the distance interpretation of absolute value.

Property 2.1

$|x| = k$ is equivalent to $x = -k$ or $x = k$, where k is a positive number.

Example 3 demonstrates our format for solving equations of the form $|x| = k$.

EXAMPLE 3

Solve $|5x + 3| = 7$.

Solution

$$|5x + 3| = 7$$

$$5x + 3 = -7 \quad \text{or} \quad 5x + 3 = 7$$

$$5x = -10 \quad \text{or} \quad 5x = 4$$

$$x = -2 \quad \text{or} \quad x = \frac{4}{5}$$

The solution set is $\left\{-2, \dfrac{4}{5}\right\}$. Check these solutions! ∎

The distance interpretation for absolute value also provides a good basis for solving some inequalities that involve absolute value. Consider the following examples.

EXAMPLE 4

Solve $|x| < 2$ and graph the solution set.

Solution

The number, x, must be less than two units away from zero. Thus $|x| < 2$ is equivalent to

$$x > -2 \quad \text{and} \quad x < 2$$

The solution set is $(-2, 2)$ and its graph is shown in Figure 2.14.

Figure 2.14 ∎

EXAMPLE   5

Solve $|x + 3| < 1$ and graph the solutions.

Solution

Let's continue to think in terms of distance on a number line. The number, $x + 3$, must be less than one unit away from zero. Thus $|x + 3| < 1$ is equivalent to

$$x + 3 > -1 \quad \text{and} \quad x + 3 < 1$$

Solving this conjunction yields

$$x + 3 \;>\; -1 \quad \text{and} \quad x + 3 \;<\; 1$$
$$x \;>\; -4 \quad \text{and} \quad x \;<\; -2$$

The solution set is $(-4, -2)$ and its graph is shown in Figure 2.15.

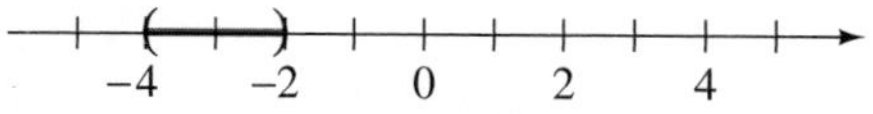

Figure 2.15   ∎

Take another look at Examples 4 and 5. The following general property should seem reasonable.

Property 2.2

$|x| < k$ is equivalent to $x > -k$ and $x < k$, where k is a positive number.

Remember that we can write a conjunction such as $x > -k$ and $x < k$ in the compact form $-k < x < k$. The compact form provides a very convenient format for solving inequalities such as $|3x - 1| < 8$, as Example 6 illustrates.

EXAMPLE   6

Solve $|3x - 1| < 8$ and graph the solutions.

Solution

$$|3x - 1| < 8$$
$$-8 < 3x - 1 < 8$$
$$-7 < 3x < 9 \qquad \text{Add 1 to left side, middle, and right side.}$$
$$\frac{1}{3}(-7) < \frac{1}{3}(3x) < \frac{1}{3}(9) \qquad \text{Multiply through by } \frac{1}{3}.$$
$$-\frac{7}{3} < x < 3$$

The solution set is $\left(-\dfrac{7}{3}, 3\right)$, and its graph is shown in Figure 2.16.

Figure 2.16   ■

  The distance interpretation also clarifies a property that pertains to *greater than* situations involving absolute value. Consider the following examples.

EXAMPLE 7

Solve $|x| > 1$ and graph the solutions.

Solution

The number, x, must be more than one unit away from zero. Thus $|x| > 1$ is equivalent to

$$x < -1 \qquad \text{or} \qquad x > 1$$

The solution set is $(-\infty, -1) \cup (1, \infty)$, and its graph is shown in Figure 2.17.

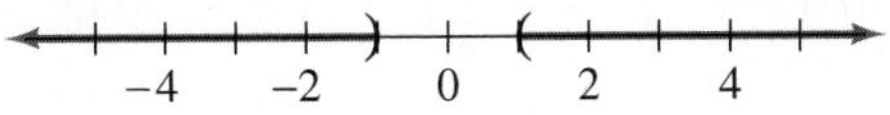

Figure 2.17   ■

EXAMPLE 8

Solve $|x - 1| > 3$ and graph the solutions.

Solution

The number, $x - 1$, must be more than three units away from zero. Thus $|x - 1| > 3$ is equivalent to

$$x - 1 < -3 \qquad \text{or} \qquad x - 1 > 3$$

Solving this disjunction yields

$$x - 1 < -3 \qquad \text{or} \qquad x - 1 > 3$$
$$x < -2 \qquad \text{or} \qquad x > 4$$

The solution set is $(-\infty, -2) \cup (4, \infty)$, and its graph is shown in Figure 2.18.

Figure 2.18   ■

Examples 7 and 8 illustrate the following general property.

Property 2.3

$|x| > k$ is equivalent to $x < -k$ or $x > k$, where k is a positive number.

Therefore, solving inequalities of the form $|x| > k$ can take the format shown in Example 9.

E X A M P L E   9

Solve $|3x - 1| > 2$ and graph the solutions.

Solution

$$|3x - 1| > 2$$

$$3x - 1 < -2 \qquad \text{or} \qquad 3x - 1 > 2$$

$$3x < -1 \qquad \text{or} \qquad 3x > 3$$

$$x < -\frac{1}{3} \qquad \text{or} \qquad x > 1$$

The solution set is $\left(-\infty, -\dfrac{1}{3}\right) \cup (1, \infty)$ and its graph is shown in Figure 2.19.

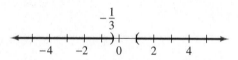

Figure 2.19

Properties 2.1, 2.2, and 2.3 provide the basis for solving a variety of equations and inequalities that involve absolute value. However, if at any time you become doubtful about what property applies, don't forget the distance interpretation. Furthermore, note that in each of the properties, k is a positive number. If k is a nonpositive number, we can determine the solution sets by inspection, as indicated by the following examples.

$|x + 3| = 0$ has a solution of $x = -3$, because the number $x + 3$ has to be 0. The solution set of $|x + 3| = 0$ is $\{-3\}$.

$|2x - 5| = -3$ has no solutions, because the absolute value (distance) cannot be negative. The solution set is $\varnothing$, the null set.

$|x - 7| < -4$ has no solutions, because we cannot obtain an absolute value less than -4. The solution set is $\varnothing$.

$|2x - 1| > -1$ is satisfied by all real numbers because the absolute value of $(2x - 1)$, regardless of what number is substituted for x, will always be greater than -1. The solution set is the set of all real numbers, which we can express in interval notation as $(-\infty, \infty)$.

Problem Set 2.7

For Problems 1–14, solve each inequality and graph the solutions.

1. $|x| < 5$

2. $|x| < 1$

3. $|x| \leq 2$

4. $|x| \leq 4$

5. $|x| > 2$

6. $|x| > 3$

7. $|x - 1| < 2$

8. $|x - 2| < 4$

9. $|x + 2| \leq 4$

10. $|x + 1| \leq 1$

11. $|x + 2| > 1$

12. $|x + 1| > 3$

13. $|x - 3| \geq 2$

14. $|x - 2| \geq 1$

For Problems 15–54, solve each equation and inequality.

15. $|x - 1| = 8$

16. $|x + 2| = 9$

17. $|x - 2| > 6$

18. $|x - 3| > 9$

19. $|x + 3| < 5$

20. $|x + 1| < 8$

21. $|2x - 4| = 6$

22. $|3x - 4| = 14$

23. $|2x - 1| \leq 9$

24. $|3x + 1| \leq 13$

25. $|4x + 2| \geq 12$

26. $|5x - 2| \geq 10$

27. $|3x + 4| = 11$

28. $|5x - 7| = 14$

29. $|4 - 2x| = 6$

30. $|3 - 4x| = 8$

31. $|2 - x| > 4$

32. $|4 - x| > 3$

33. $|1 - 2x| < 2$

34. $|2 - 3x| < 5$

35. $|5x + 9| \leq 16$

36. $|7x - 6| \geq 22$

37. $\left| x - \dfrac{3}{4} \right| = \dfrac{2}{3}$

38. $\left| x + \dfrac{1}{2} \right| = \dfrac{3}{5}$

39. $|-2x + 7| \leq 13$

40. $|-3x - 4| \leq 15$

41. $\left| \dfrac{x - 3}{4} \right| < 2$

42. $\left| \dfrac{x + 2}{3} \right| < 1$

43. $\left| \dfrac{2x + 1}{2} \right| > 1$

44. $\left| \dfrac{3x - 1}{4} \right| > 3$

45. $|2x - 3| + 2 = 5$

46. $|3x - 1| - 1 = 9$

47. $|x + 2| - 6 = -2$

48. $|x - 3| - 4 = -1$

49. $|4x - 3| + 2 = 2$

50. $|5x + 1| + 4 = 4$

51. $|x + 7| - 3 \geq 4$

52. $|x - 2| + 4 \geq 10$

53. $|2x - 1| + 1 \leq 6$

54. $|4x + 3| - 2 \leq 5$

For Problems 55–64, solve each equation and inequality *by inspection*.

55. $|2x + 1| = -4$

56. $|5x - 1| = -2$

57. $|3x - 1| > -2$

58. $|4x + 3| < -4$

59. $|5x - 2| = 0$

60. $|3x - 1| = 0$

61. $|4x - 6| < -1$

62. $|x + 9| > -6$

63. $|x + 4| < 0$

64. $|x + 6| > 0$

■ ■ ■ **THOUGHTS INTO WORDS**

65. Explain how you would solve the inequality $|2x + 5| > -3$.

66. Why is 2 the only solution for $|x - 2| \leq 0$?

67. Explain how you would solve the equation $|2x - 3| = 0$.

■ ■ ■ **FURTHER INVESTIGATIONS**

Consider the equation $|x| = |y|$. This equation will be a true statement if x is equal to y, or if x is equal to the opposite of y. Use the following format, $x = y$ or $x = -y$, to solve the equations in Problems 68–73.

For Problems 68–73, solve each equation.

68. $|3x + 1| = |2x + 3|$

69. $|-2x - 3| = |x + 1|$

70. $|2x - 1| = |x - 3|$

71. $|x - 2| = |x + 6|$

72. $|x + 1| = |x - 4|$

73. $|x + 1| = |x - 1|$

74. Use the definition of absolute value to help prove Property 2.1.

75. Use the definition of absolute value to help prove Property 2.2.

76. Use the definition of absolute value to help prove Property 2.3.

(2.1) Solving an algebraic equation refers to the process of finding the number (or numbers) that make(s) the algebraic equation a true numerical statement. We call such numbers the **solutions** or **roots** of the equation that **satisfy** the equation. We call the set of all solutions of an equation the **solution set**. The general procedure for solving an equation is to continue replacing the given equation with **equivalent but simpler** equations until we arrive at one that can be solved by inspection. Two properties of equality play an important role in the process of solving equations.

Addition Property of Equality $a = b$ if and only if $a + c = b + c$.

Multiplication Property of Equality   For $c \neq 0$, $a = b$ if and only if $ac = bc$.

(2.2) To solve an equation involving fractions, first **clear the equation of all fractions**. It is usually easiest to begin by multiplying both sides of the equation by the least common multiple of all of the denominators in the equation (by the least common denominator, or LCD).

Keep the following suggestions in mind as you solve word problems.

1. Read the problem carefully.
2. Sketch any figure, diagram, or chart that might be helpful.
3. Choose a meaningful variable.
4. Look for a guideline.
5. Form an equation or inequality.
6. Solve the equation or inequality.
7. Check your answers.

(2.3) To solve equations that contain decimals, you can clear the equation of all decimals by multiplying both sides by an appropriate power of 10, or you can keep the problem in decimal form and perform the calculations with decimals.

(2.4) We use equations to put rules in symbolic form; we call these rules **formulas**.

We can solve a formula such as $P = 2l + 2w$ for l $\left(l = \dfrac{P - 2w}{2} \right)$ or for w $\left(w = \dfrac{P - 2l}{2} \right)$ by applying the addition and multiplication properties of equality.

We often use formulas as **guidelines** for solving word problems.

(2.5) Solving an algebraic inequality refers to the process of finding the numbers that make the algebraic inequality a true numerical statement. We call such numbers the **solutions**, and we call the set of all solutions the **solution set**.

The general procedure for solving an inequality is to continue replacing the given inequality with **equivalent, but simpler**, inequalities until we arrive at one that we can solve by inspection. The following properties form the basis for solving algebraic inequalities.

1. $a > b$ if and only if $a + c > b + c$.   (Addition property)
2. **a.** For $c > 0$, $a > b$ if and only if $ac > bc$.

   b. For $c < 0$, $a > b$ if and only if $ac < bc$.   (Multiplication properties)

(2.6) To solve compound sentences that involve inequalities, we proceed as follows:

1. Solve separately each inequality in the compound sentence.
2. If it is a **conjunction**, the solution set is the **intersection** of the solution sets of each inequality.
3. If it is a **disjunction**, the solution set is the **union** of the solution sets of each inequality.

We define the intersection and union of two sets as follows.

Intersection $A \cap B = \{x | x \in A \quad and \quad x \in B\}$

Union $A \cup B = \{x | x \in A \quad or \quad x \in B\}$

The following are some examples of solution sets that we examined in Sections 2.5 and 2.6 (Figure 2.20).

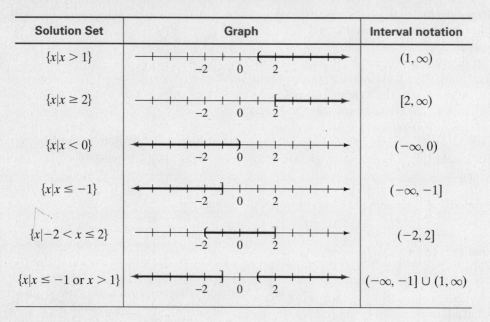

Solution Set	Graph	Interval notation
$\{x \mid x > 1\}$		$(1, \infty)$
$\{x \mid x \geq 2\}$		$[2, \infty)$
$\{x \mid x < 0\}$		$(-\infty, 0)$
$\{x \mid x \leq -1\}$		$(-\infty, -1]$
$\{x \mid -2 < x \leq 2\}$		$(-2, 2]$
$\{x \mid x \leq -1 \text{ or } x > 1\}$		$(-\infty, -1] \cup (1, \infty)$

Figure 2.20

(2.7) We can interpret the **absolute value** of a number on the number line as the distance between that number and zero. The following properties form the basis for solving equations and inequalities involving absolute value.

1. $|x| = k$ is equivalent to $x = -k$ or $x = k$
2. $|x| < k$ is equivalent to $x > -k$ and $x < k$ $\Big\}$ $k > 0$
3. $|x| > k$ is equivalent to $x < -k$ or $x > k$

Chapter 2   Review Problem Set

For Problems 1–15, solve each of the equations.

1. $5(x - 6) = 3(x + 2)$

2. $2(2x + 1) - (x - 4) = 4(x + 5)$

3. $-(2n - 1) + 3(n + 2) = 7$

4. $2(3n - 4) + 3(2n - 3) = -2(n + 5)$

5. $\dfrac{3t - 2}{4} = \dfrac{2t + 1}{3}$

6. $\dfrac{x + 6}{5} + \dfrac{x - 1}{4} = 2$

7. $1 - \dfrac{2x - 1}{6} = \dfrac{3x}{8}$

8. $\dfrac{2x + 1}{3} + \dfrac{3x - 1}{5} = \dfrac{1}{10}$

9. $\dfrac{3n - 1}{2} - \dfrac{2n + 3}{7} = 1$

10. $|3x - 1| = 11$

11. $0.06x + 0.08(x + 100) = 15$

12. $0.4(t - 6) = 0.3(2t + 5)$

13. $0.1(n + 300) = 0.09n + 32$

14. $0.2(x - 0.5) - 0.3(x + 1) = 0.4$

15. $|2n + 3| = 4$

For Problems 16–20, solve each equation for x.

16. $ax - b = b + 2$

17. $ax = bx + c$

18. $m(x + a) = p(x + b)$

19. $5x - 7y = 11$

20. $\dfrac{x - a}{b} = \dfrac{y + 1}{c}$

For Problems 21–24, solve each of the formulas for the indicated variable.

21. $A = \pi r^2 + \pi rs$ for s

22. $A = \dfrac{1}{2}h(b_1 + b_2)$ for b_2

23. $S_n = \dfrac{n(a_1 + a_2)}{2}$ for n

24. $\dfrac{1}{R} = \dfrac{1}{R_1} + \dfrac{1}{R_2}$ for R

For Problems 25–36, solve each of the inequalities.

25. $5x - 2 \geq 4x - 7$

26. $3 - 2x < -5$

27. $2(3x - 1) - 3(x - 3) > 0$

28. $3(x + 4) \leq 5(x - 1)$

29. $\dfrac{5}{6}n - \dfrac{1}{3}n < \dfrac{1}{6}$

30. $\dfrac{n - 4}{5} + \dfrac{n - 3}{6} > \dfrac{7}{15}$

31. $s \geq 4.5 + 0.25s$

32. $0.07x + 0.09(500 - x) \geq 43$

33. $|2x - 1| < 11$

34. $|3x + 1| > 10$

35. $-3(2t - 1) - (t + 2) > -6(t - 3)$

36. $\dfrac{2}{3}(x - 1) + \dfrac{1}{4}(2x + 1) < \dfrac{5}{6}(x - 2)$

For Problems 37–44, graph the solutions of each compound inequality.

37. $x > -1$ and $x < 1$

38. $x > 2$ or $x \leq -3$

39. $x > 2$ and $x > 3$

40. $x < 2$ or $x > -1$

41. $2x + 1 > 3$ or $2x + 1 < -3$

42. $2 \leq x + 4 \leq 5$

43. $-1 < 4x - 3 \leq 9$

44. $x + 1 > 3$ and $x - 3 < -5$

Solve each of Problems 45–56 by setting up and solving an appropriate equation or inequality.

45. The width of a rectangle is 2 meters more than one-third of the length. The perimeter of the rectangle is 44 meters. Find the length and width of the rectangle.

46. A total of $500 was invested, part of it at 7% interest and the remainder at 8%. If the total yearly interest from both investments amounted to $38, how much was invested at each rate?

47. Susan's average score for her first three psychology exams is 84. What must she get on the fourth exam so that her average for the four exams is 85 or better?

48. Find three consecutive integers such that the sum of one-half of the smallest and one-third of the largest is one less than the other integer.

49. Pat is paid time-and-a-half for each hour he works over 36 hours in a week. Last week he worked 42 hours for a total of $472.50. What is his normal hourly rate?

50. Marcela has a collection of nickels, dimes, and quarters worth $24.75. The number of dimes is 10 more than twice the number of nickels, and the number of quarters is 25 more than the number of dimes. How many coins of each kind does she have?

51. If the complement of an angle is one-tenth of the supplement of the angle, find the measure of the angle.

52. A retailer has some sweaters that cost her $38 each. She wants to sell them at a profit of 20% of her cost. What price should she charge for the sweaters?

53. How many pints of a 1% hydrogen peroxide solution should be mixed with a 4% hydrogen peroxide solution to obtain 10 pints of a 2% hydrogen peroxide solution?

54. Gladys leaves a town driving at a rate of 40 miles per hour. Two hours later, Reena leaves from the same place traveling the same route. She catches Gladys in 5 hours and 20 minutes. How fast was Reena traveling?

55. In $1\frac{1}{4}$ hours more time, Rita, riding her bicycle at 12 miles per hour, rode 2 miles farther than Sonya, who was riding her bicycle at 16 miles per hour. How long did each girl ride?

56. How many cups of orange juice must be added to 50 cups of a punch that is 10% orange juice to obtain a punch that is 20% orange juice?

For Problems 1–10, solve each equation.

1. $5x - 2 = 2x - 11$

2. $6(n - 2) - 4(n + 3) = -14$

3. $-3(x + 4) = 3(x - 5)$

4. $3(2x - 1) - 2(x + 5) = -(x - 3)$

5. $\dfrac{3t - 2}{4} = \dfrac{5t + 1}{5}$

6. $\dfrac{5x + 2}{3} - \dfrac{2x + 4}{6} = -\dfrac{4}{3}$

7. $|4x - 3| = 9$

8. $\dfrac{1 - 3x}{4} + \dfrac{2x + 3}{3} = 1$

9. $2 - \dfrac{3x - 1}{5} = -4$

10. $0.05x + 0.06(1500 - x) = 83.5$

11. Solve $\dfrac{2}{3}x - \dfrac{3}{4}y = 2$ for y

12. Solve $S = 2\pi r(r + h)$ for h

For Problems 13–20, solve each inequality and express the solution set using interval notation.

13. $7x - 4 > 5x - 8$

14. $-3x - 4 \le x + 12$

15. $2(x - 1) - 3(3x + 1) \ge -6(x - 5)$

16. $\dfrac{3}{5}x - \dfrac{1}{2}x < 1$

17. $\dfrac{x - 2}{6} - \dfrac{x + 3}{9} > -\dfrac{1}{2}$

18. $0.05x + 0.07(800 - x) \ge 52$

19. $|6x - 4| < 10$

20. $|4x + 5| \ge 6$

For Problems 21–25, solve each problem by setting up and solving an appropriate equation or inequality.

21. Dela bought a dress at a 20% discount sale for $57.60. Find the original price of the dress.

22. The length of a rectangle is 1 centimeter more than three times its width. If the perimeter of the rectangle is 50 centimeters, find the length of the rectangle.

23. How many cups of grapefruit juice must be added to 30 cups of a punch that is 8% grapefruit juice to obtain a punch that is 10% grapefruit juice?

24. Rex has scores of 85, 92, 87, 88, and 91 on the first five exams. What score must he make on the sixth exam to have an average of 90 or better for all six exams?

25. If the complement of an angle is $\dfrac{2}{11}$ of the supplement of the angle, find the measure of the angle.

3

Polynomials

A quadratic equation can be solved to determine the width of a uniform strip trimmed off both sides and ends of a sheet of paper to obtain a specified area for the sheet of paper.

© Tony Freeman /PhotoEdit

A strip of uniform width cut off of both sides and both ends of an 8-inch by 11-inch sheet of paper must reduce the size of the paper to an area of 40 square inches. Find the width of the strip. With the equation $(11 - 2x)(8 - 2x) = 40$, you can determine that the strip should be 1.5 inches wide.

The main object of this text is to help you develop algebraic skills, use these skills to solve equations and inequalities, and use equations and inequalities to solve word problems. The work in this chapter will focus on a class of algebraic expressions called **polynomials**.

Recall that algebraic expressions such as $5x$, $-6y^2$, $7xy$, $14a^2b$, and $-17ab^2c^3$ are called terms. A **term** is an indicated product and may contain any number of factors. The variables in a term are called **literal factors**, and the numerical factor is called the **numerical coefficient**. Thus in $7xy$, the x and y are literal factors, 7 is the numerical coefficient, and the term is in two variables (x and y).

Terms that contain variables with only whole numbers as exponents are called **monomials**. The previously listed terms, $5x$, $-6y^2$, $7xy$, $14a^2b$, and $-17ab^2c^3$, are all monomials. (We shall work later with some algebraic expressions, such as $7x^{-1}y^{-1}$ and $6a^{-2}b^{-3}$, that are not monomials.)

The **degree** of a monomial is the sum of the exponents of the literal factors.

$7xy$ is of degree 2.

$14a^2b$ is of degree 3.

$-17ab^2c^3$ is of degree 6.

$5x$ is of degree 1.

$-6y^2$ is of degree 2.

If the monomial contains only one variable, then the exponent of the variable is the degree of the monomial. The last two examples illustrate this point. We say that any nonzero constant term is of degree zero.

A **polynomial** is a monomial or a finite sum (or difference) of monomials. Thus

$$4x^2, \qquad 3x^2 - 2x - 4, \qquad 7x^4 - 6x^3 + 4x^2 + x - 1,$$

$$3x^2y - 2xy^2, \qquad \frac{1}{5}a^2 - \frac{2}{3}b^2, \qquad \text{and} \qquad 14$$

are examples of polynomials. In addition to calling a polynomial with one term a **monomial**, we also classify polynomials with two terms as **binomials**, and those with three terms as **trinomials**.

The **degree of a polynomial** is the degree of the term with the highest degree in the polynomial. The following examples illustrate some of this terminology.

The polynomial $4x^3y^4$ is a monomial in two variables of degree 7.

The polynomial $4x^2y - 2xy$ is a binomial in two variables of degree 3.

The polynomial $9x^2 - 7x + 1$ is a trinomial in one variable of degree 2.

■ Combining Similar Terms

Remember that *similar terms*, or *like terms*, are terms that have the same literal factors. In the preceding chapters, we have frequently simplified algebraic expressions by combining similar terms, as the next examples illustrate.

$$2x + 3y + 7x + 8y = \boxed{2x + 7x + 3y + 8y}$$
$$= \boxed{(2 + 7)x + (3 + 8)y}$$
$$= 9x + 11y$$

Steps in dashed boxes are usually done mentally.

$$4a - 7 - 9a + 10 = \boxed{4a + (-7) + (-9a) + 10}$$
$$= \boxed{4a + (-9a) + (-7) + 10}$$
$$= \boxed{(4 + (-9))a + (-7) + 10}$$
$$= -5a + 3$$

Both addition and subtraction of polynomials rely on basically the same ideas. The commutative, associative, and distributive properties provide the basis for rearranging, regrouping, and combining similar terms. Let's consider some examples.

EXAMPLE 1

Add $4x^2 + 5x + 1$ and $7x^2 - 9x + 4$.

Solution

We generally use the horizontal format for such work. Thus

$$(4x^2 + 5x + 1) + (7x^2 - 9x + 4) = (4x^2 + 7x^2) + (5x - 9x) + (1 + 4)$$
$$= 11x^2 - 4x + 5$$ ∎

EXAMPLE 2

Add $5x - 3$, $3x + 2$, and $8x + 6$.

Solution

$$(5x - 3) + (3x + 2) + (8x + 6) = (5x + 3x + 8x) + (-3 + 2 + 6)$$
$$= 16x + 5$$ ∎

EXAMPLE 3

Find the indicated sum: $(-4x^2y + xy^2) + (7x^2y - 9xy^2) + (5x^2y - 4xy^2)$.

Solution

$$(-4x^2y + xy^2) + (7x^2y - 9xy^2) + (5x^2y - 4xy^2)$$
$$= (-4x^2y + 7x^2y + 5x^2y) + (xy^2 - 9xy^2 - 4xy^2)$$
$$= 8x^2y - 12xy^2$$ ∎

The idea of subtraction as adding the opposite extends to polynomials in general. Hence the expression $a - b$ is equivalent to $a + (-b)$. We can form the opposite of a polynomial by taking the opposite of each term. For example, the opposite of $3x^2 - 7x + 1$ is $-3x^2 + 7x - 1$. We express this in symbols as

$$-(3x^2 - 7x + 1) = -3x^2 + 7x - 1$$

Now consider the following subtraction problems.

E X A M P L E   4

Subtract $3x^2 + 7x - 1$ from $7x^2 - 2x - 4$.

Solution

Use the horizontal format to obtain

$$(7x^2 - 2x - 4) - (3x^2 + 7x - 1) = (7x^2 - 2x - 4) + (-3x^2 - 7x + 1)$$
$$= (7x^2 - 3x^2) + (-2x - 7x) + (-4 + 1)$$
$$= 4x^2 - 9x - 3 \qquad ∎$$

E X A M P L E   5

Subtract $-3y^2 + y - 2$ from $4y^2 + 7$.

Solution

Because subtraction is not a commutative operation, be sure to perform the subtraction in the correct order.

$$(4y^2 + 7) - (-3y^2 + y - 2) = (4y^2 + 7) + (3y^2 - y + 2)$$
$$= (4y^2 + 3y^2) + (-y) + (7 + 2)$$
$$= 7y^2 - y + 9 \qquad ∎$$

The next example demonstrates the use of the vertical format for this work.

E X A M P L E   6

Subtract $4x^2 - 7xy + 5y^2$ from $3x^2 - 2xy + y^2$.

Solution

$$\begin{array}{l} 3x^2 - 2xy + \ \ y^2 \\ \underline{4x^2 - 7xy + 5y^2} \end{array}$$ Note which polynomial goes on the bottom and how the similar terms are aligned.

Now we can mentally form the opposite of the bottom polynomial and add.

$$\begin{array}{l} 3x^2 - 2xy + \ \ y^2 \\ \underline{4x^2 - 7xy + 5y^2} \\ -x^2 + 5xy - 4y^2 \end{array}$$ The opposite of $4x^2 - 7xy + 5y^2$ is $-4x^2 + 7xy - 5y^2$.

$∎$

We can also use the distributive property and the properties $a = 1(a)$ and $-a = -1(a)$ when adding and subtracting polynomials. The next examples illustrate this approach.

EXAMPLE 7

Perform the indicated operations: $(5x - 2) + (2x - 1) - (3x + 4)$.

Solution

$$
\begin{aligned}
(5x - 2) + (2x - 1) - (3x + 4) &= 1(5x - 2) + 1(2x - 1) - 1(3x + 4) \\
&= 1(5x) - 1(2) + 1(2x) - 1(1) - 1(3x) - 1(4) \\
&= 5x - 2 + 2x - 1 - 3x - 4 \\
&= 5x + 2x - 3x - 2 - 1 - 4 \\
&= 4x - 7 \quad \blacksquare
\end{aligned}
$$

We can do some of the steps mentally and simplify our format, as shown in the next two examples.

EXAMPLE 8

Perform the indicated operations: $(5a^2 - 2b) - (2a^2 + 4) + (-7b - 3)$.

Solution

$$
\begin{aligned}
(5a^2 - 2b) - (2a^2 + 4) + (-7b - 3) &= 5a^2 - 2b - 2a^2 - 4 - 7b - 3 \\
&= 3a^2 - 9b - 7 \quad \blacksquare
\end{aligned}
$$

EXAMPLE 9

Simplify $(4t^2 - 7t - 1) - (t^2 + 2t - 6)$.

Solution

$$
\begin{aligned}
(4t^2 - 7t - 1) - (t^2 + 2t - 6) &= 4t^2 - 7t - 1 - t^2 - 2t + 6 \\
&= 3t^2 - 9t + 5 \quad \blacksquare
\end{aligned}
$$

Remember that a polynomial in parentheses preceded by a negative sign can be written without the parentheses by replacing each term with its opposite. Thus in Example 9, $-(t^2 + 2t - 6) = -t^2 - 2t + 6$. Finally, let's consider a simplification problem that contains grouping symbols within grouping symbols.

EXAMPLE 10

Simplify $7x + [3x - (2x + 7)]$.

Solution

$$
\begin{aligned}
7x + [3x - (2x + 7)] &= 7x + [3x - 2x - 7] \qquad \text{Remove the innermost} \\
&= 7x + [x - 7] \qquad\qquad\; \text{parentheses first.} \\
&= 7x + x - 7 \\
&= 8x - 7 \quad \blacksquare
\end{aligned}
$$

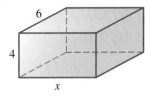

Figure 3.1

Sometimes we encounter polynomials in a geometric setting. For example, we can find a polynomial that represents the total surface area of the rectangular solid in Figure 3.1 as follows:

$$4x \quad + \quad 4x \quad + \quad 6x \quad + \quad 6x \quad + \quad 24 \quad + \quad 24$$

| Area of front | Area of back | Area of top | Area of bottom | Area of left side | Area of right side |

Simplifying $4x + 4x + 6x + 6x + 24 + 24$, we obtain the polynomial $20x + 48$, which represents the total surface area of the rectangular solid. Furthermore, by evaluating the polynomial $20x + 48$ for different positive values of x, we can determine the total surface area of any rectangular solid for which two dimensions are 4 and 6. The following chart contains some specific rectangular solids.

x	4 by 6 by x rectangular solid	Total surface area $(20x + 48)$
2	4 by 6 by 2	$20(2) + 48 = 88$
4	4 by 6 by 4	$20(4) + 48 = 128$
5	4 by 6 by 5	$20(5) + 48 = 148$
7	4 by 6 by 7	$20(7) + 48 = 188$
12	4 by 6 by 12	$20(12) + 48 = 288$

Problem Set 3.1

For Problems 1–10, determine the degree of the given polynomials.

1. $7xy + 6y$

2. $-5x^2y^2 - 6xy^2 + x$

3. $-x^2y + 2xy^2 - xy$

4. $5x^3y^2 - 6x^3y^3$

5. $5x^2 - 7x - 2$

6. $7x^3 - 2x + 4$

7. $8x^6 + 9$

8. $5y^6 + y^4 - 2y^2 - 8$

9. -12

10. $7x - 2y$

For Problems 11–20, add the given polynomials.

11. $3x - 7$ and $7x + 4$

12. $9x + 6$ and $5x - 3$

13. $-5t - 4$ and $-6t + 9$

14. $-7t + 14$ and $-3t - 6$

15. $3x^2 - 5x - 1$ and $-4x^2 + 7x - 1$

16. $6x^2 + 8x + 4$ and $-7x^2 - 7x - 10$

17. $12a^2b^2 - 9ab$ and $5a^2b^2 + 4ab$

18. $15a^2b^2 - ab$ and $-20a^2b^2 - 6ab$

19. $2x - 4$, $-7x + 2$, and $-4x + 9$

20. $-x^2 - x - 4$, $2x^2 - 7x + 9$, and $-3x^2 + 6x - 10$

For Problems 21–30, subtract the polynomials using the horizontal format.

21. $5x - 2$ from $3x + 4$

22. $7x + 5$ from $2x - 1$

23. $-4a - 5$ from $6a + 2$

24. $5a + 7$ from $-a - 4$

25. $3x^2 - x + 2$ from $7x^2 + 9x + 8$

26. $5x^2 + 4x - 7$ from $3x^2 + 2x - 9$

27. $2a^2 - 6a - 4$ from $-4a^2 + 6a + 10$

28. $-3a^2 - 6a + 3$ from $3a^2 + 6a - 11$

29. $2x^3 + x^2 - 7x - 2$ from $5x^3 + 2x^2 + 6x - 13$

30. $6x^3 + x^2 + 4$ from $9x^3 - x - 2$

For Problems 31–40, subtract the polynomials using the vertical format.

31. $5x - 2$ from $12x + 6$

32. $3x - 7$ from $2x + 1$

33. $-4x + 7$ from $-7x - 9$

34. $-6x - 2$ from $5x + 6$

35. $2x^2 + x + 6$ from $4x^2 - x - 2$

36. $4x^2 - 3x - 7$ from $-x^2 - 6x + 9$

37. $x^3 + x^2 - x - 1$ from $-2x^3 + 6x^2 - 3x + 8$

38. $2x^3 - x + 6$ from $x^3 + 4x^2 + 1$

39. $-5x^2 + 6x - 12$ from $2x - 1$

40. $2x^2 - 7x - 10$ from $-x^3 - 12$

For Problems 41–46, perform the operations as described.

41. Subtract $2x^2 - 7x - 1$ from the sum of $x^2 + 9x - 4$ and $-5x^2 - 7x + 10$.

42. Subtract $4x^2 + 6x + 9$ from the sum of $-3x^2 - 9x + 6$ and $-2x^2 + 6x - 4$.

43. Subtract $-x^2 - 7x - 1$ from the sum of $4x^2 + 3$ and $-7x^2 + 2x$.

44. Subtract $-4x^2 + 6x - 3$ from the sum of $-3x + 4$ and $9x^2 - 6$.

45. Subtract the sum of $5n^2 - 3n - 2$ and $-7n^2 + n + 2$ from $-12n^2 - n + 9$.

46. Subtract the sum of $-6n^2 + 2n - 4$ and $4n^2 - 2n + 4$ from $-n^2 - n + 1$.

For Problems 47–56, perform the indicated operations.

47. $(5x + 2) + (7x - 1) + (-4x - 3)$

48. $(-3x + 1) + (6x - 2) + (9x - 4)$

49. $(12x - 9) - (-3x + 4) - (7x + 1)$

50. $(6x + 4) - (4x - 2) - (-x - 1)$

51. $(2x^2 - 7x - 1) + (-4x^2 - x + 6) + (-7x^2 - 4x - 1)$

52. $(5x^2 + x + 4) + (-x^2 + 2x + 4) + (-14x^2 - x + 6)$

53. $(7x^2 - x - 4) - (9x^2 - 10x + 8) + (12x^2 + 4x - 6)$

54. $(-6x^2 + 2x + 5) - (4x^2 + 4x - 1) + (7x^2 + 4)$

55. $(n^2 - 7n - 9) - (-3n + 4) - (2n^2 - 9)$

56. $(6n^2 - 4) - (5n^2 + 9) - (6n + 4)$

For Problems 57–70, simplify by removing the inner parentheses first and working outward.

57. $3x - [5x - (x + 6)]$

58. $7x - [2x - (-x - 4)]$

59. $2x^2 - [-3x^2 - (x^2 - 4)]$

60. $4x^2 - [-x^2 - (5x^2 - 6)]$

61. $-2n^2 - [n^2 - (-4n^2 + n + 6)]$

62. $-7n^2 - [3n^2 - (-n^2 - n + 4)]$

63. $[4t^2 - (2t + 1) + 3] - [3t^2 + (2t - 1) - 5]$

64. $-(3n^2 - 2n + 4) - [2n^2 - (n^2 + n + 3)]$

65. $[2n^2 - (2n^2 - n + 5)] + [3n^2 + (n^2 - 2n - 7)]$

66. $3x^2 - [4x^2 - 2x - (x^2 - 2x + 6)]$

67. $[7xy - (2x - 3xy + y)] - [3x - (x - 10xy - y)]$

68. $[9xy - (4x + xy - y)] - [4y - (2x - xy + 6y)]$

69. $[4x^3 - (2x^2 - x - 1)] - [5x^3 - (x^2 + 2x - 1)]$

70. $[x^3 - (x^2 - x + 1)] - [-x^3 + (7x^2 - x + 10)]$

71. Find a polynomial that represents the perimeter of each of the following figures (Figures 3.2, 3.3, and 3.4).

(a)

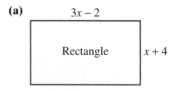

$3x - 2$

Rectangle $x + 4$

Figure 3.2

(b)

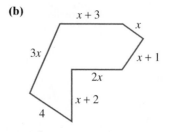

$x + 3$

x

$3x$

$x + 1$

$2x$

$x + 2$

4

Figure 3.3

(c)

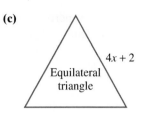

Figure 3.4

72. Find a polynomial that represents the total surface area of the rectangular solid in Figure 3.5.

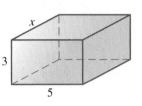

Figure 3.5

Now use that polynomial to determine the total surface area of each of the following rectangular solids.

(a) 3 by 5 by <u>4</u>   **(b)** 3 by 5 by <u>7</u>

(c) 3 by 5 by <u>11</u>   **(d)** 3 by 5 by <u>13</u>

73. Find a polynomial that represents the total surface area of the right circular cylinder in Figure 3.6. Now use that polynomial to determine the total surface area of each of the following right circular cylinders that have a base with a radius of 4. Use 3.14 for π, and express the answers to the nearest tenth.

(a) $h = 5$ **(b)** $h = 7$

(c) $h = 14$ **(d)** $h = 18$

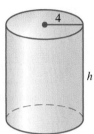

Figure 3.6

■ ■ ■ **THOUGHTS INTO WORDS**

74. Explain how to subtract the polynomial $-3x^2 + 2x - 4$ from $4x^2 + 6$.

75. Is the sum of two binomials always another binomial? Defend your answer.

76. Explain how to simplify the expression
$$7x - [3x - (2x - 4) + 2] - x$$

3.2   **Products and Quotients of Monomials**

Suppose that we want to find the product of two monomials such as $3x^2y$ and $4x^3y^2$. To proceed, use the properties of real numbers, and keep in mind that exponents indicate repeated multiplication.

$$(3x^2y)(4x^3y^2) = (3 \cdot x \cdot x \cdot y)(4 \cdot x \cdot x \cdot x \cdot y \cdot y)$$
$$= 3 \cdot 4 \cdot x \cdot x \cdot x \cdot x \cdot x \cdot y \cdot y \cdot y$$
$$= 12x^5y^3$$

You can use such an approach to find the product of any two monomials. However, there are some basic properties of exponents that make the process of multiplying monomials a much easier task. Let's consider each of these properties and illustrate its use when multiplying monomials. The following examples demonstrate the first property.

$$x^2 \cdot x^3 = (x \cdot x)(x \cdot x \cdot x) = x^5$$
$$a^4 \cdot a^2 = (a \cdot a \cdot a \cdot a)(a \cdot a) = a^6$$
$$b^3 \cdot b^4 = (b \cdot b \cdot b)(b \cdot b \cdot b \cdot b) = b^7$$

In general,

$$b^n \cdot b^m = \underbrace{(b \cdot b \cdot b \cdot \ldots b)}_{\substack{n \text{ factors} \\ \text{of } b}}\underbrace{(b \cdot b \cdot b \cdot \ldots b)}_{\substack{m \text{ factors} \\ \text{of } b}}$$

$$= \underbrace{b \cdot b \cdot b \cdot \ldots b}_{(n + m) \text{ factors of } b}$$

$$= b^{n+m}$$

We can state the first property as follows:

Property 3.1

If b is any real number, and n and m are positive integers, then

$$b^n \cdot b^m = b^{n+m}$$

Property 3.1 says that to find the product of two positive integral powers of the same base, we add the exponents and use this sum as the exponent of the common base.

$$x^7 \cdot x^8 = x^{7+8} = x^{15} \qquad\qquad y^6 \cdot y^4 = y^{6+4} = y^{10}$$
$$2^3 \cdot 2^8 = 2^{3+8} = 2^{11} \qquad\qquad (-3)^4 \cdot (-3)^5 = (-3)^{4+5} = (-3)^9$$
$$\left(\frac{2}{3}\right)^7 \cdot \left(\frac{2}{3}\right)^5 = \left(\frac{2}{3}\right)^{5+7} = \left(\frac{2}{3}\right)^{12}$$

The following examples illustrate the use of Property 3.1, along with the commutative and associative properties of multiplication, to form the basis for multiplying monomials. The steps enclosed in the dashed boxes could be performed mentally.

EXAMPLE 1

$$(3x^2y)(4x^3y^2) = \boxed{3 \cdot 4 \cdot x^2 \cdot x^3 \cdot y \cdot y^2}$$
$$= \boxed{12x^{2+3}y^{1+2}}$$
$$= 12x^5y^3$$

∎

EXAMPLE 2

$$(-5a^3b^4)(7a^2b^5) = \boxed{-5 \cdot 7 \cdot a^3 \cdot a^2 \cdot b^4 \cdot b^5}$$

$$= -35a^{3+2}b^{4+5}$$

$$= -35a^5b^9$$

∎

EXAMPLE 3

$$\left(\frac{3}{4}xy\right)\left(\frac{1}{2}x^5y^6\right) = \boxed{\frac{3}{4} \cdot \frac{1}{2} \cdot x \cdot x^5 \cdot y \cdot y^6}$$

$$= \frac{3}{8}x^{1+5}y^{1+6}$$

$$= \frac{3}{8}x^6y^7$$

∎

EXAMPLE 4

$$(-ab^2)(-5a^2b) = \boxed{(-1)(-5)(a)(a^2)(b^2)(b)}$$

$$= 5a^{1+2}b^{2+1}$$

$$= 5a^3b^3$$

∎

EXAMPLE 5

$$(2x^2y^2)(3x^2y)(4y^3) = \boxed{2 \cdot 3 \cdot 4 \cdot x^2 \cdot x^2 \cdot y^2 \cdot y \cdot y^3}$$

$$= 24x^{2+2}y^{2+1+3}$$

$$= 24x^4y^6$$

∎

The following examples demonstrate another useful property of exponents.

$$(x^2)^3 = x^2 \cdot x^2 \cdot x^2 = x^{2+2+2} = x^6$$

$$(a^3)^2 = a^3 \cdot a^3 = a^{3+3} = a^6$$

$$(b^4)^3 = b^4 \cdot b^4 \cdot b^4 = b^{4+4+4} = b^{12}$$

In general,

$$(b^n)^m = \underbrace{b^n \cdot b^n \cdot b^n \cdot \ldots b^n}_{m \text{ factors of } b^n}$$

$$= b^{\overbrace{n+n+n+\cdots+n}^{\text{adding } m \text{ of these}}}$$

$$= b^{mn}$$

We can state this property as follows:

Property 3.2

If b is any real number, and m and n are positive integers, then

$$(b^n)^m = b^{mn}$$

The following examples show how Property 3.2 is used to find "the power of a power."

$$(x^4)^5 = x^{5(4)} = x^{20} \qquad (y^6)^3 = y^{3(6)} = y^{18}$$

$$(2^3)^7 = 2^{7(3)} = 2^{21}$$

A third property of exponents pertains to raising a monomial to a power. Consider the following examples, which we use to introduce the property.

$$(3x)^2 = (3x)(3x) = 3 \cdot 3 \cdot x \cdot x = 3^2 \cdot x^2$$

$$(4y^2)^3 = (4y^2)(4y^2)(4y^2) = 4 \cdot 4 \cdot 4 \cdot y^2 \cdot y^2 \cdot y^2 = (4)^3(y^2)^3$$

$$(-2a^3b^4)^2 = (-2a^3b^4)(-2a^3b^4) = (-2)(-2)(a^3)(a^3)(b^4)(b^4)$$

$$= (-2)^2(a^3)^2(b^4)^2$$

In general,

$$(ab)^n = \underbrace{(ab)(ab)(ab) \cdot \ldots (ab)}_{n \text{ factors of } ab}$$

$$= \underbrace{(a \cdot a \cdot a \cdot a \cdot \ldots a)}_{\substack{n \text{ factors} \\ \text{of } a}} \underbrace{(b \cdot b \cdot b \cdot \ldots b)}_{\substack{n \text{ factors} \\ \text{of } b}}$$

$$= a^n b^n$$

We can formally state Property 3.3 as follows:

Property 3.3

If a and b are real numbers, and n is a positive integer, then

$$(ab)^n = a^n b^n$$

Property 3.3 and Property 3.2 form the basis for raising a monomial to a power, as in the next examples.

EXAMPLE 6

$(x^2y^3)^4 = (x^2)^4(y^3)^4$ Use $(ab)^n = a^nb^n$.

$= x^8y^{12}$ Use $(b^n)^m = b^{mn}$.   ∎

EXAMPLE 7

$(3a^5)^3 = (3)^3(a^5)^3$

$= 27a^{15}$ ∎

EXAMPLE 8

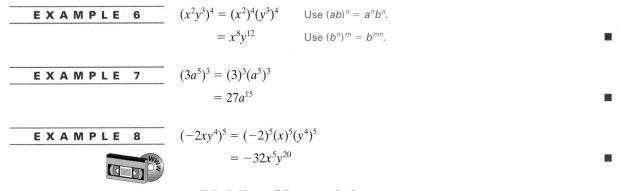

$(-2xy^4)^5 = (-2)^5(x)^5(y^4)^5$

$= -32x^5y^{20}$ ∎

■ Dividing Monomials

To develop an effective process for dividing by a monomial, we need yet another property of exponents. This property is a direct consequence of the definition of an exponent. Study the following examples.

$$\frac{x^4}{x^3} = \frac{x \cdot x \cdot x \cdot x}{x \cdot x \cdot x} = x \qquad\qquad \frac{x^3}{x^3} = \frac{x \cdot x \cdot x}{x \cdot x \cdot x} = 1$$

$$\frac{a^5}{a^2} = \frac{a \cdot a \cdot a \cdot a \cdot a}{a \cdot a} = a^3 \qquad\qquad \frac{y^5}{y^5} = \frac{y \cdot y \cdot y \cdot y \cdot y}{y \cdot y \cdot y \cdot y \cdot y} = 1$$

$$\frac{y^8}{y^4} = \frac{y \cdot y \cdot y \cdot y \cdot y \cdot y \cdot y \cdot y}{y \cdot y \cdot y \cdot y} = y^4$$

We can state the general property as follows:

Property 3.4

If b is any nonzero real number, and m and n are positive integers, then

1. $\dfrac{b^n}{b^m} = b^{n-m}$,   when $n > m$

2. $\dfrac{b^n}{b^m} = 1$,   when $n = m$

Applying Property 3.4 to the previous examples yields

$$\frac{x^4}{x^3} = x^{4-3} = x^1 = x \qquad \frac{x^3}{x^3} = 1$$

$$\frac{a^5}{a^2} = a^{5-2} = a^3 \qquad \frac{y^5}{y^5} = 1$$

$$\frac{y^8}{y^4} = y^{8-4} = y^4$$

(We will discuss the situation when $n < m$ in a later chapter.)

Property 3.4, along with our knowledge of dividing integers, provides the basis for dividing monomials. The following examples demonstrate the process.

$$\frac{24x^5}{3x^2} = 8x^{5-2} = 8x^3 \qquad \frac{-36a^{13}}{-12a^5} = 3a^{13-5} = 3a^8$$

$$\frac{-56x^9}{7x^4} = -8x^{9-4} = -8x^5 \qquad \frac{72b^5}{8b^5} = 9 \quad \left(\frac{b^5}{b^5} = 1\right)$$

$$\frac{48y^7}{-12y} = -4y^{7-1} = -4y^6 \qquad \frac{12x^4y^7}{2x^2y^4} = 6x^{4-2}y^{7-4} = 6x^2y^3$$

Problem Set 3.2

For Problems 1–36, find each product.

1. $(4x^3)(9x)$

2. $(6x^3)(7x^2)$

3. $(-2x^2)(6x^3)$

4. $(2xy)(-4x^2y)$

5. $(-a^2b)(-4ab^3)$

6. $(-8a^2b^2)(-3ab^3)$

7. $(x^2yz^2)(-3xyz^4)$

8. $(-2xy^2z^2)(-x^2y^3z)$

9. $(5xy)(-6y^3)$

10. $(-7xy)(4x^4)$

11. $(3a^2b)(9a^2b^4)$

12. $(-8a^2b^2)(-12ab^5)$

13. $(m^2n)(-mn^2)$

14. $(-x^3y^2)(xy^3)$

15. $\left(\frac{2}{5}xy^2\right)\left(\frac{3}{4}x^2y^4\right)$

16. $\left(\frac{1}{2}x^2y^6\right)\left(\frac{2}{3}xy\right)$

17. $\left(-\frac{3}{4}ab\right)\left(\frac{1}{5}a^2b^3\right)$

18. $\left(-\frac{2}{7}a^2\right)\left(\frac{3}{5}ab^3\right)$

19. $\left(-\frac{1}{2}xy\right)\left(\frac{1}{3}x^2y^3\right)$

20. $\left(\frac{3}{4}x^4y^5\right)(-x^2y)$

21. $(3x)(-2x^2)(-5x^3)$

22. $(-2x)(-6x^3)(x^2)$

23. $(-6x^2)(3x^3)(x^4)$

24. $(-7x^2)(3x)(4x^3)$

25. $(x^2y)(-3xy^2)(x^3y^3)$

26. $(xy^2)(-5xy)(x^2y^4)$

27. $(-3y^2)(-2y^2)(-4y^5)$

28. $(-y^3)(-6y)(-8y^4)$

29. $(4ab)(-2a^2b)(7a)$

30. $(3b)(-2ab^2)(7a)$

31. $(-ab)(-3ab)(-6ab)$

32. $(-3a^2b)(-ab^2)(-7a)$

33. $\left(\frac{2}{3}xy\right)(-3x^2y)(5x^4y^5)$

34. $\left(\frac{3}{4}x\right)(-4x^2y^2)(9y^3)$

35. $(12y)(-5x)\left(-\frac{5}{6}x^4y\right)$

36. $(-12x)(3y)\left(-\frac{3}{4}xy^6\right)$

For Problems 37–58, raise each monomial to the indicated power.

37. $(3xy^2)^3$

38. $(4x^2y^3)^3$

39. $(-2x^2y)^5$

40. $(-3xy^4)^3$

41. $(-x^4y^5)^4$

42. $(-x^5y^2)^4$

43. $(ab^2c^3)^6$

44. $(a^2b^3c^5)^5$

45. $(2a^2b^3)^6$

46. $(2a^3b^2)^6$

47. $(9xy^4)^2$

48. $(8x^2y^5)^2$

49. $(-3ab^3)^4$

50. $(-2a^2b^4)^4$

51. $-(2ab)^4$

52. $-(3ab)^4$

53. $-(xy^2z^3)^6$

54. $-(xy^2z^3)^8$

55. $(-5a^2b^2c)^3$

56. $(-4abc^4)^3$

57. $(-xy^4z^2)^7$

58. $(-x^2y^4z^5)^5$

For Problems 59–74, find each quotient.

59. $\dfrac{9x^4y^5}{3xy^2}$

60. $\dfrac{12x^2y^7}{6x^2y^3}$

61. $\dfrac{25x^5y^6}{-5x^2y^4}$

62. $\dfrac{56x^6y^4}{-7x^2y^3}$

63. $\dfrac{-54ab^2c^3}{-6abc}$

64. $\dfrac{-48a^3bc^5}{-6a^2c^4}$

65. $\dfrac{-18x^2y^2z^6}{xyz^2}$

66. $\dfrac{-32x^4y^5z^8}{x^2yz^3}$

67. $\dfrac{a^3b^4c^7}{-abc^5}$

68. $\dfrac{-a^4b^5c}{a^2b^4c}$

69. $\dfrac{-72x^2y^4}{-8x^2y^4}$

70. $\dfrac{-96x^4y^5}{12x^4y^4}$

71. $\dfrac{14ab^3}{-14ab}$

72. $\dfrac{-12abc^2}{12bc}$

73. $\dfrac{-36x^3y^5}{2y^5}$

74. $\dfrac{-48xyz^2}{2xz}$

For Problems 75–90, find each product. Assume that the variables in the exponents represent positive integers. For example,

$$(x^{2n})(x^{3n}) = x^{2n+3n} = x^{5n}$$

75. $(2x^n)(3x^{2n})$

76. $(3x^{2n})(x^{3n-1})$

77. $(a^{2n-1})(a^{3n+4})$

78. $(a^{5n-1})(a^{5n+1})$

79. $(x^{3n-2})(x^{n+2})$

80. $(x^{n-1})(x^{4n+3})$

81. $(a^{5n-2})(a^3)$

82. $(x^{3n-4})(x^4)$

83. $(2x^n)(-5x^n)$

84. $(4x^{2n-1})(-3x^{n+1})$

85. $(-3a^2)(-4a^{n+2})$

86. $(-5x^{n-1})(-6x^{2n+4})$

87. $(x^n)(2x^{2n})(3x^2)$

88. $(2x^n)(3x^{3n-1})(-4x^{2n+5})$

89. $(3x^{n-1})(x^{n+1})(4x^{2-n})$

90. $(-5x^{n+2})(x^{n-2})(4x^{3-2n})$

91. Find a polynomial that represents the total surface area of the rectangular solid in Figure 3.7. Also find a polynomial that represents the volume.

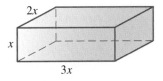
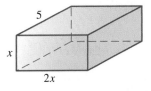

Figure 3.7

92. Find a polynomial that represents the total surface area of the rectangular solid in Figure 3.8. Also find a polynomial that represents the volume.

Figure 3.8

93. Find a polynomial that represents the area of the shaded region in Figure 3.9. The length of a radius of the larger circle is r units, and the length of a radius of the smaller circle is 6 units.

Figure 3.9

■ ■ ■ **THOUGHTS INTO WORDS**

94. How would you convince someone that $x^6 \div x^2$ is x^4 and not x^3?

95. Your friend simplifies $2^3 \cdot 2^2$ as follows:

$$2^3 \cdot 2^2 = 4^{3+2} = 4^5 = 1024$$

What has she done incorrectly and how would you help her?

3.3 Multiplying Polynomials

We usually state the distributive property as $a(b + c) = ab + ac$; however, we can extend it as follows:

$$a(b + c + d) = ab + ac + ad$$

$$a(b + c + d + e) = ab + ac + ad + ae \quad \text{etc.}$$

We apply the commutative and associative properties, the properties of exponents, and the distributive property together to find the product of a monomial and a polynomial. The following examples illustrate this idea.

E X A M P L E   1

$$3x^2(2x^2 + 5x + 3) = 3x^2(2x^2) + 3x^2(5x) + 3x^2(3)$$
$$= 6x^4 + 15x^3 + 9x^2 \qquad \blacksquare$$

E X A M P L E   2

$$-2xy(3x^3 - 4x^2y - 5xy^2 + y^3) = -2xy(3x^3) - (-2xy)(4x^2y)$$
$$-(-2xy)(5xy^2) + (-2xy)(y^3)$$
$$= -6x^4y + 8x^3y^2 + 10x^2y^3 - 2xy^4 \qquad \blacksquare$$

Now let's consider the product of two polynomials neither of which is a monomial. Consider the following examples.

E X A M P L E   3

$$(x + 2)(y + 5) = x(y + 5) + 2(y + 5)$$
$$= x(y) + x(5) + 2(y) + 2(5)$$
$$= xy + 5x + 2y + 10 \qquad \blacksquare$$

Note that each term of the first polynomial is multiplied by each term of the second polynomial.

E X A M P L E   4

$$(x - 3)(y + z + 3) = x(y + z + 3) - 3(y + z + 3)$$
$$= xy + xz + 3x - 3y - 3z - 9 \qquad \blacksquare$$

Multiplying polynomials often produces similar terms that can be combined to simplify the resulting polynomial.

E X A M P L E   5

$$(x + 5)(x + 7) = x(x + 7) + 5(x + 7)$$
$$= x^2 + 7x + 5x + 35$$
$$= x^2 + 12x + 35 \qquad \blacksquare$$

E X A M P L E   6

$$(x - 2)(x^2 - 3x + 4) = x(x^2 - 3x + 4) - 2(x^2 - 3x + 4)$$
$$= x^3 - 3x^2 + 4x - 2x^2 + 6x - 8$$
$$= x^3 - 5x^2 + 10x - 8 \qquad \blacksquare$$

In Example 6, we are claiming that

$$(x - 2)(x^2 - 3x + 4) = x^3 - 5x^2 + 10x - 8$$

for all real numbers. In addition to going back over our work, how can we verify such a claim? Obviously, we cannot try all real numbers, but trying at least one number gives us a partial check. Let's try the number 4.

$$(x - 2)(x^2 - 3x + 4) = (4 - 2)(4^2 - 3(4) + 4)$$
$$= 2(16 - 12 + 4)$$
$$= 2(8)$$
$$= 16$$

$$x^3 - 5x^2 + 10x - 8 = 4^3 - 5(4)^2 + 10(4) - 8$$
$$= 64 - 80 + 40 - 8$$
$$= 16$$

E X A M P L E   7

$$(3x - 2y)(x^2 + xy - y^2) = 3x(x^2 + xy - y^2) - 2y(x^2 + xy - y^2)$$
$$= 3x^3 + 3x^2y - 3xy^2 - 2x^2y - 2xy^2 + 2y^3$$
$$= 3x^3 + x^2y - 5xy^2 + 2y^3 \qquad \blacksquare$$

It helps to be able to find the product of two binomials without showing all of the intermediate steps. This is quite easy to do with the *three-step shortcut pattern* demonstrated by Figures 3.10 and 3.11 in the following examples.

E X A M P L E   8

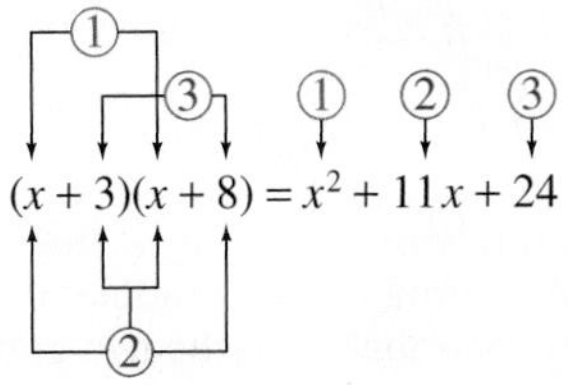

$$(x + 3)(x + 8) = x^2 + 11x + 24$$

Figure 3.10

Step ①. Multiply $x \cdot x$.

Step ②. Multiply $3 \cdot x$ and $8 \cdot x$ and combine.

Step ③. Multiply $3 \cdot 8$. $\qquad \blacksquare$

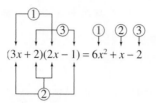

Figure 3.11 ∎

Now see if you can use the pattern to find the following products.

$$(x + 2)(x + 6) = \text{?}$$

$$(x - 3)(x + 5) = \text{?}$$

$$(2x + 5)(3x + 7) = \text{?}$$

$$(3x - 1)(4x - 3) = \text{?}$$

Your answers should be $x^2 + 8x + 12$, $x^2 + 2x - 15$, $6x^2 + 29x + 35$, and $12x^2 - 13x + 3$. Keep in mind that this shortcut pattern applies only to finding the product of two binomials.

We can use exponents to indicate repeated multiplication of polynomials. For example, $(x + 3)^2$ means $(x + 3)(x + 3)$, and $(x + 4)^3$ means $(x + 4)(x + 4) \cdot (x + 4)$. To square a binomial, we can simply write it as the product of two equal binomials and apply the shortcut pattern. Thus

$$(x + 3)^2 = (x + 3)(x + 3) = x^2 + 6x + 9$$

$$(x - 6)^2 = (x - 6)(x - 6) = x^2 - 12x + 36 \qquad \text{and}$$

$$(3x - 4)^2 = (3x - 4)(3x - 4) = 9x^2 - 24x + 16$$

When squaring binomials, be careful not to forget the middle term. That is to say, $(x + 3)^2 \neq x^2 + 3^2$; instead, $(x + 3)^2 = x^2 + 6x + 9$.

When multiplying binomials, there are some special patterns that you should recognize. We can use these patterns to find products, and later we will use some of them when factoring polynomials.

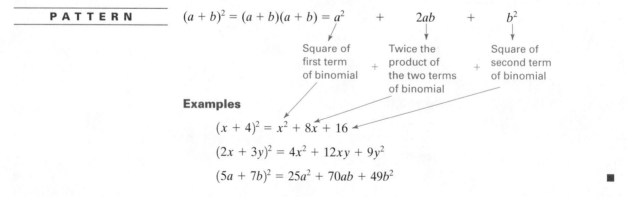

Examples

$$(x + 4)^2 = x^2 + 8x + 16$$

$$(2x + 3y)^2 = 4x^2 + 12xy + 9y^2$$

$$(5a + 7b)^2 = 25a^2 + 70ab + 49b^2$$

∎

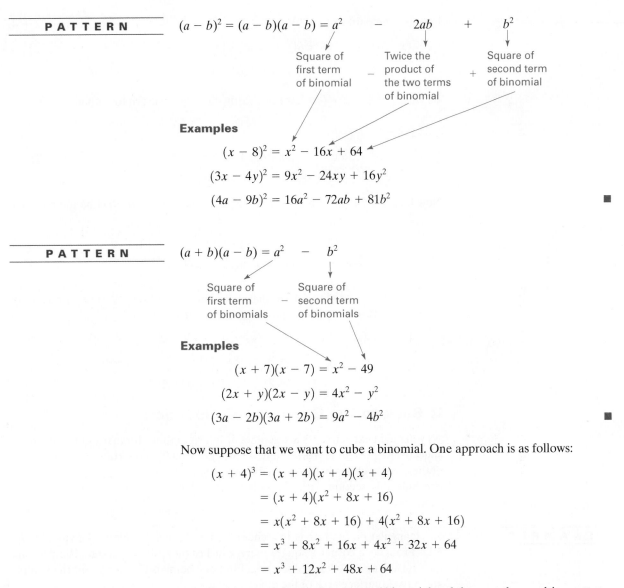

PATTERN

$$(a - b)^2 = (a - b)(a - b) = a^2 \quad - \quad 2ab \quad + \quad b^2$$

Square of
first term
of binomial

Twice the
product of
the two terms
of binomial

Square of
second term
of binomial

Examples

$$(x - 8)^2 = x^2 - 16x + 64$$

$$(3x - 4y)^2 = 9x^2 - 24xy + 16y^2$$

$$(4a - 9b)^2 = 16a^2 - 72ab + 81b^2$$ ∎

PATTERN

$$(a + b)(a - b) = a^2 \quad - \quad b^2$$

Square of
first term
of binomials

Square of
second term
of binomials

Examples

$$(x + 7)(x - 7) = x^2 - 49$$

$$(2x + y)(2x - y) = 4x^2 - y^2$$

$$(3a - 2b)(3a + 2b) = 9a^2 - 4b^2$$ ∎

Now suppose that we want to cube a binomial. One approach is as follows:

$$(x + 4)^3 = (x + 4)(x + 4)(x + 4)$$
$$= (x + 4)(x^2 + 8x + 16)$$
$$= x(x^2 + 8x + 16) + 4(x^2 + 8x + 16)$$
$$= x^3 + 8x^2 + 16x + 4x^2 + 32x + 64$$
$$= x^3 + 12x^2 + 48x + 64$$

Another approach is to cube a general binomial and then use the resulting pattern.

PATTERN

$$(a + b)^3 = (a + b)(a + b)(a + b)$$
$$= (a + b)(a^2 + 2ab + b^2)$$
$$= a(a^2 + 2ab + b^2) + b(a^2 + 2ab + b^2)$$
$$= a^3 + 2a^2b + ab^2 + a^2b + 2ab^2 + b^3$$
$$= a^3 + 3a^2b + 3ab^2 + b^3$$

Let's use the pattern $(a + b)^3 = a^3 + 3a^2b + 3ab^2 + b^3$ to cube the binomial $x + 4$.

$$(x + 4)^3 = x^3 + 3x^2(4) + 3x(4)^2 + 4^3$$
$$= x^3 + 12x^2 + 48x + 64 \qquad \blacksquare$$

Because $a - b = a + (-b)$, we can easily develop a pattern for cubing $a - b$.

PATTERN

$$(a - b)^3 = [a + (-b)]^3$$
$$= a^3 + 3a^2(-b) + 3a(-b)^2 + (-b)^3$$
$$= a^3 - 3a^2b + 3ab^2 - b^3$$

Now let's use the pattern $(a - b)^3 = a^3 - 3a^2b + 3ab^2 - b^3$ to cube the binomial $3x - 2y$.

$$(3x - 2y)^3 = (3x)^3 - 3(3x)^2(2y) + 3(3x)(2y)^2 - (2y)^3$$
$$= 27x^3 - 54x^2y + 36xy^2 - 8y^3 \qquad \blacksquare$$

Finally, we need to realize that if the patterns are forgotten or do not apply, then we can revert to applying the distributive property.

$$(2x - 1)(x^2 - 4x + 6) = 2x(x^2 - 4x + 6) - 1(x^2 - 4x + 6)$$
$$= 2x^3 - 8x^2 + 12x - x^2 + 4x - 6$$
$$= 2x^3 - 9x^2 + 16x - 6$$

■ Back to the Geometry Connection

As you might expect, there are geometric interpretations for many of the algebraic concepts we present in this section. We will give you the opportunity to make some of these connections between algebra and geometry in the next problem set. Let's conclude this section with a problem that allows us to use some algebra and geometry.

EXAMPLE 10

A rectangular piece of tin is 16 inches long and 12 inches wide as shown in Figure 3.12. From each corner a square piece x inches on a side is cut out. The flaps are then turned up to form an open box. Find polynomials that represent the volume and outside surface area of the box.

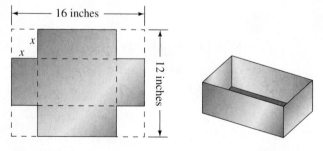

Figure 3.12

Solution

The length of the box will be $16 - 2x$, the width $12 - 2x$, and the height x. With the volume formula $V = lwh$, the polynomial $(16 - 2x)(12 - 2x)(x)$, which simplifies to $4x^3 - 56x^2 + 192x$, represents the volume.

    The outside surface area of the box is the area of the original piece of tin minus the four corners that were cut off. Therefore, the polynomial $16(12) - 4x^2$, or $192 - 4x^2$, represents the outside surface area of the box. ■

Remark: Recall that in Section 3.1 we found the total surface area of a rectangular solid by adding the areas of the sides, top, and bottom. Use this approach for the open box in Example 10 to check our answer of $192 - 4x^2$. Keep in mind that the box has no top.

Problem Set 3.3

For Problems 1–74, find each indicated product. Remember the shortcut for multiplying binomials and the other special patterns we discussed in this section.

1. $2xy(5xy^2 + 3x^2y^3)$

2. $3x^2y(6y^2 - 5x^2y^4)$

3. $-3a^2b(4ab^2 - 5a^3)$

4. $-7ab^2(2b^3 - 3a^2)$

5. $8a^3b^4(3ab - 2ab^2 + 4a^2b^2)$

6. $9a^3b(2a - 3b + 7ab)$

7. $-x^2y(6xy^2 + 3x^2y^3 - x^3y)$

8. $-ab^2(5a + 3b - 6a^2b^3)$

9. $(a + 2b)(x + y)$

10. $(t - s)(x + y)$

11. $(a - 3b)(c + 4d)$

12. $(a - 4b)(c - d)$

13. $(x + 6)(x + 10)$

14. $(x + 2)(x + 10)$

15. $(y - 5)(y + 11)$

16. $(y - 3)(y + 9)$

17. $(n + 2)(n - 7)$

18. $(n + 3)(n - 12)$

19. $(x + 6)(x - 6)$

20. $(t + 8)(t - 8)$

21. $(x - 6)^2$

22. $(x - 2)^2$

23. $(x - 6)(x - 8)$

24. $(x - 3)(x - 13)$

25. $(x + 1)(x - 2)(x - 3)$

26. $(x - 1)(x + 4)(x - 6)$

27. $(x - 3)(x + 3)(x - 1)$

28. $(x - 5)(x + 5)(x - 8)$

29. $(t + 9)^2$

30. $(t + 13)^2$

31. $(y - 7)^2$

32. $(y - 4)^2$

33. $(4x + 5)(x + 7)$

34. $(6x + 5)(x + 3)$

35. $(3y - 1)(3y + 1)$

36. $(5y - 2)(5y + 2)$

37. $(7x - 2)(2x + 1)$

38. $(6x - 1)(3x + 2)$

39. $(1 + t)(5 - 2t)$

40. $(3 - t)(2 + 4t)$

41. $(3t + 7)^2$

42. $(4t + 6)^2$

43. $(2 - 5x)(2 + 5x)$

44. $(6 - 3x)(6 + 3x)$

45. $(7x - 4)^2$

46. $(5x - 7)^2$

47. $(6x + 7)(3x - 10)$

48. $(4x - 7)(7x + 4)$

49. $(2x - 5y)(x + 3y)$

50. $(x - 4y)(3x + 7y)$

51. $(5x - 2a)(5x + 2a)$

52. $(9x - 2y)(9x + 2y)$

53. $(t + 3)(t^2 - 3t - 5)$

54. $(t - 2)(t^2 + 7t + 2)$

55. $(x - 4)(x^2 + 5x - 4)$

56. $(x + 6)(2x^2 - x - 7)$

57. $(2x - 3)(x^2 + 6x + 10)$

58. $(3x + 4)(2x^2 - 2x - 6)$

59. $(4x - 1)(3x^2 - x + 6)$

60. $(5x - 2)(6x^2 + 2x - 1)$

61. $(x^2 + 2x + 1)(x^2 + 3x + 4)$

62. $(x^2 - x + 6)(x^2 - 5x - 8)$

63. $(2x^2 + 3x - 4)(x^2 - 2x - 1)$

64. $(3x^2 - 2x + 1)(2x^2 + x - 2)$

65. $(x + 2)^3$

66. $(x + 1)^3$

67. $(x - 4)^3$

68. $(x - 5)^3$

69. $(2x + 3)^3$

70. $(3x + 1)^3$

71. $(4x - 1)^3$

72. $(3x - 2)^3$

73. $(5x + 2)^3$

74. $(4x - 5)^3$

For Problems 75–84, find the indicated products. Assume all variables that appear as exponents represent positive integers.

75. $(x^n - 4)(x^n + 4)$

76. $(x^{3a} - 1)(x^{3a} + 1)$

77. $(x^a + 6)(x^a - 2)$

78. $(x^a + 4)(x^a - 9)$

79. $(2x^n + 5)(3x^n - 7)$

80. $(3x^n + 5)(4x^n - 9)$

81. $(x^{2a} - 7)(x^{2a} - 3)$

82. $(x^{2a} + 6)(x^{2a} - 4)$

83. $(2x^n + 5)^2$

84. $(3x^n - 7)^2$

85. Explain how Figure 3.13 can be used to demonstrate geometrically that $(x + 2)(x + 6) = x^2 + 8x + 12$.

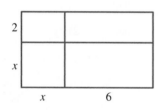

Figure 3.13

86. Find a polynomial that represents the sum of the areas of the two rectangles shown in Figure 3.14.

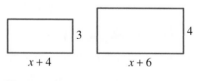

Figure 3.14

87. Find a polynomial that represents the area of the shaded region in Figure 3.15.

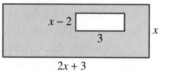

Figure 3.15

88. Explain how Figure 3.16 can be used to demonstrate geometrically that $(x + 7)(x - 3) = x^2 + 4x - 21$.

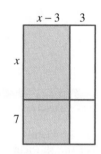

Figure 3.16

89. A square piece of cardboard is 16 inches on a side. A square piece x inches on a side is cut out from each corner. The flaps are then turned up to form an open box. Find polynomials that represent the volume and outside surface area of the box.

■ ■ ■ THOUGHTS INTO WORDS

90. How would you simplify $(2^3 + 2^2)^2$? Explain your reasoning.

91. Describe the process of multiplying two polynomials.

92. Determine the number of terms in the product of $(x + y)$ and $(a + b + c + d)$ without doing the multiplication. Explain how you arrived at your answer.

93. We have used the following two multiplication patterns.

$$(a + b)^2 = a^2 + 2ab + b^2$$

$$(a + b)^3 = a^3 + 3a^2b + 3ab^2 + b^3$$

By multiplying, we can extend these patterns as follows:

$$(a + b)^4 = a^4 + 4a^3b + 6a^2b^2 + 4ab^3 + b^4$$

$$(a + b)^5 = a^5 + 5a^4b + 10a^3b^2 + 10a^2b^3 + 5a^4 + b^5$$

On the basis of these results, see if you can determine a pattern that will enable you to complete each of the following without using the long-multiplication process.

(a) $(a + b)^6$ **(b)** $(a + b)^7$

(c) $(a + b)^8$ **(d)** $(a + b)^9$

94. Find each of the following indicated products. These patterns will be used again in Section 3.5.

(a) $(x - 1)(x^2 + x + 1)$ **(b)** $(x + 1)(x^2 - x + 1)$

(c) $(x + 3)(x^2 - 3x + 9)$ **(d)** $(x - 4)(x^2 + 4x + 16)$

(e) $(2x - 3)(4x^2 + 6x + 9)$

(f) $(3x + 5)(9x^2 - 15x + 25)$

95. Some of the product patterns can be used to do arithmetic computations mentally. For example, let's use the pattern $(a + b)^2 = a^2 + 2ab + b^2$ to compute 31^2 mentally. Your thought process should be "$31^2 = (30 + 1)^2 = 30^2 + 2(30)(1) + 1^2 = 961$." Compute each

of the following numbers mentally, and then check your answers.

(a) 21^2 **(b)** 41^2 **(c)** 71^2

(d) 32^2 **(e)** 52^2 **(f)** 82^2

96. Use the pattern $(a - b)^2 = a^2 - 2ab + b^2$ to compute each of the following numbers mentally, and then check your answers.

(a) 19^2 **(b)** 29^2 **(c)** 49^2

(d) 79^2 **(e)** 38^2 **(f)** 58^2

97. Every whole number with a units digit of 5 can be represented by the expression $10x + 5$, where x is a whole number. For example, $35 = 10(3) + 5$ and $145 = 10(14) + 5$. Now let's observe the following pattern when squaring such a number.

$$(10x + 5)^2 = 100x^2 + 100x + 25$$

$$= \overline{100x(x + 1) + 25}$$

The pattern inside the dashed box can be stated as "add 25 to the product of x, $x + 1$, and 100." Thus, to compute 35^2 mentally, we can think "$35^2 = 3(4)(100) + 25 = 1225$." Compute each of the following numbers mentally, and then check your answers.

(a) 15^2 **(b)** 25^2 **(c)** 45^2

(d) 55^2 **(e)** 65^2 **(f)** 75^2

(g) 85^2 **(h)** 95^2 **(i)** 105^2

3.4   Factoring: Use of the Distributive Property

Recall that 2 and 3 are said to be *factors* of 6 because the product of 2 and 3 is 6. Likewise, in an indicated product such as $7ab$, the 7, a, and b are called factors of the product. If a positive integer greater than 1 has no factors that are positive integers other than itself and 1, then it is called a **prime number**. Thus the prime numbers less than 20 are 2, 3, 5, 7, 11, 13, 17, and 19. A positive integer greater than 1 that is not a prime number is called a **composite number**. The composite numbers less than 20 are 4, 6, 8, 9, 10, 12, 14, 15, 16, and 18. Every composite number is the product of prime numbers. Consider the following examples.

$$4 = 2 \cdot 2 \qquad\qquad 63 = 3 \cdot 3 \cdot 7$$

$$12 = 2 \cdot 2 \cdot 3 \qquad\quad 121 = 11 \cdot 11$$

$$35 = 5 \cdot 7$$

The indicated product form that contains only prime factors is called the **prime factorization form** of a number. Thus the prime factorization form of 63 is $3 \cdot 3 \cdot 7$. We also say that the number has been **completely factored** when it is in the prime factorization form.

In general, factoring is the reverse of multiplication. Previously, we have used the distributive property to find the product of a monomial and a polynomial, as in the next examples.

$$3(x + 2) = 3(x) + 3(2) = 3x + 6$$

$$5(2x - 1) = 5(2x) - 5(1) = 10x - 5$$

$$x(x^2 + 6x - 4) = x(x^2) + x(6x) - x(4) = x^3 + 6x^2 - 4x$$

We shall also use the distributive property [in the form $ab + ac = a(b + c)$] to reverse the process — that is, to factor a given polynomial. Consider the following examples. (The steps in the dashed boxes can be done mentally.)

$$3x + 6 = \boxed{3(x) + 3(2)} = 3(x + 2),$$

$$10x - 5 = \boxed{5(2x) - 5(1)} = 5(2x - 1),$$

$$x^3 + 6x^2 - 4x = \boxed{x(x^2) + x(6x) - x(4)} = x(x^2 + 6x - 4)$$

Note that in each example a given polynomial has been factored into the product of a monomial and a polynomial. Obviously, polynomials could be factored in a variety of ways. Consider some factorizations of $3x^2 + 12x$.

$$3x^2 + 12x = 3x(x + 4) \qquad \text{or} \qquad 3x^2 + 12x = 3(x^2 + 4x) \qquad \text{or}$$

$$3x^2 + 12x = x(3x + 12) \qquad \text{or} \qquad 3x^2 + 12x = \frac{1}{2}(6x^2 + 24x)$$

We are, however, primarily interested in the first of the previous factorization forms, which we refer to as the **completely factored form**. A polynomial with integral coefficients is in completely factored form if

1. It is expressed as a product of polynomials with *integral coefficients*, and

2. No polynomial, other than a monomial, within the factored form can be further factored into polynomials with integral coefficients.

Do you see why only the first of the above factored forms of $3x^2 + 12x$ is said to be in completely factored form? In each of the other three forms, the polynomial inside the parentheses can be factored further. Moreover, in the last form, $\frac{1}{2}(6x^2 + 24x)$, the condition of using only integral coefficients is violated.

The factoring process that we discuss in this section, $ab + ac = a(b + c)$, is often referred to as **factoring out the highest common monomial factor**. The key idea in this process is to recognize the monomial factor that is common to all terms. For example, we observe that each term of the polynomial $2x^3 + 4x^2 + 6x$ has a factor of $2x$. Thus we write

$$2x^3 + 4x^2 + 6x = 2x(\qquad)$$

and insert within the parentheses the appropriate polynomial factor. We determine the terms of this polynomial factor by dividing each term of the original polynomial by the factor of $2x$. The final, completely factored form is

$$2x^3 + 4x^2 + 6x = 2x(x^2 + 2x + 3)$$

The following examples further demonstrate this process of factoring out the highest common monomial factor.

$$12x^3 + 16x^2 = 4x^2(3x + 4) \qquad 6x^2y^3 + 27xy^4 = 3xy^3(2x + 9y)$$

$$8ab - 18b = 2b(4a - 9) \qquad 8y^3 + 4y^2 = 4y^2(2y + 1)$$

$$30x^3 + 42x^4 - 24x^5 = 6x^3(5 + 7x - 4x^2)$$

Note that in each example, the common monomial factor itself is not in a completely factored form. For example, $4x^2(3x + 4)$ is not written as $2 \cdot 2 \cdot x \cdot x \cdot (3x + 4)$.

Sometimes there may be a common binomial factor rather than a common monomial factor. For example, each of the two terms of the expression $x(y + 2) + z(y + 2)$ has a binomial factor of $(y + 2)$. Thus we can factor $(y + 2)$ from each term, and our result is

$$x(y + 2) + z(y + 2) = (y + 2)(x + z)$$

Consider a few more examples that involve a common binomial factor.

$$a^2(b + 1) + 2(b + 1) = (b + 1)(a^2 + 2)$$

$$x(2y - 1) - y(2y - 1) = (2y - 1)(x - y)$$

$$x(x + 2) + 3(x + 2) = (x + 2)(x + 3)$$

It may be that the original polynomial exhibits no apparent common monomial or binomial factor, which is the case with $ab + 3a + bc + 3c$. However, by factoring a from the first two terms and c from the last two terms, we get

$$ab + 3a + bc + 3c = a(b + 3) + c(b + 3)$$

Now a common binomial factor of $(b + 3)$ is obvious, and we can proceed as before.

$$a(b + 3) + c(b + 3) = (b + 3)(a + c)$$

We refer to this factoring process as **factoring by grouping**. Let's consider a few more examples of this type.

$$ab^2 - 4b^2 + 3a - 12 = b^2(a - 4) + 3(a - 4)$$

Factor b^2 from the first two terms and 3 from the last two terms.

$$= (a - 4)(b^2 + 3)$$

Factor common binomial from both terms.

$$x^2 - x + 5x - 5 = x(x - 1) + 5(x - 1)$$

Factor x from the first two terms and 5 from the last two terms.

$$= (x - 1)(x + 5)$$

Factor common binomial from both terms.

$$x^2 + 2x - 3x - 6 = x(x + 2) - 3(x + 2)$$

Factor x from the first two terms and -3 from the last two terms.

$$= (x + 2)(x - 3)$$

Factor common binomial factor from both terms.

It may be necessary to rearrange some terms before applying the distributive property. Terms that contain common factors need to be grouped together, and this may be done in more than one way. The next example illustrates this idea.

$$4a^2 - bc^2 - a^2b + 4c^2 = 4a^2 - a^2b + 4c^2 - bc^2$$
$$= a^2(4 - b) + c^2(4 - b)$$
$$= (4 - b)(a^2 + c^2) \qquad \text{or}$$
$$4a^2 - bc^2 - a^2b + 4c^2 = 4a^2 + 4c^2 - bc^2 - a^2b$$
$$= 4(a^2 + c^2) - b(c^2 + a^2)$$
$$= 4(a^2 + c^2) - b(a^2 + c^2)$$
$$= (a^2 + c^2)(4 - b)$$

■ Equations and Problem Solving

One reason why factoring is an important algebraic skill is that it extends our techniques for solving equations. Each time we examine a factoring technique, we will then use it to help solve certain types of equations.

We need another property of equality before we consider some equations where the highest-common-factor technique is useful. Suppose that the product of two numbers is zero. Can we conclude that at least one of these numbers must itself be zero? Yes. Let's state a property that formalizes this idea. Property 3.5, along with the highest-common-factor pattern, provides us with another technique for solving equations.

Property 3.5

Let a and b be real numbers. Then

$$ab = 0 \quad \text{if and only if } a = 0 \text{ or } b = 0$$

E X A M P L E   1   Solve $x^2 + 6x = 0$.

Solution

$$x^2 + 6x = 0$$

$$x(x + 6) = 0 \qquad \text{Factor the left side.}$$

$$x = 0 \quad \text{or} \quad x + 6 = 0 \qquad ab = 0 \text{ if and only if } a = 0 \text{ or } b = 0$$

$$x = 0 \quad \text{or} \quad x = -6$$

Thus both 0 and -6 will satisfy the original equation, and the solution set is $\{-6,\ 0\}$. ∎

E X A M P L E   2   Solve $a^2 = 11a$.

Solution

$$a^2 = 11a$$

$$a^2 - 11a = 0 \qquad \text{Add } -11a \text{ to both sides.}$$

$$a(a - 11) = 0 \qquad \text{Factor the left side.}$$

$$a = 0 \quad \text{or} \quad a - 11 = 0 \qquad ab = 0 \text{ if and only if } a = 0 \text{ or } b = 0$$

$$a = 0 \quad \text{or} \quad a = 11$$

The solution set is $\{0,\ 11\}$. ∎

Remark: Note that in Example 2 we did *not* divide both sides of the equation by a. This would cause us to lose the solution of 0.

E X A M P L E   3   Solve $3n^2 - 5n = 0$.

Solution

$$3n^2 - 5n = 0$$

$$n(3n - 5) = 0$$

$$n = 0 \quad \text{or} \quad 3n - 5 = 0$$

$$n = 0 \quad \text{or} \quad 3n = 5$$

$$n = 0 \quad \text{or} \quad n = \frac{5}{3}$$

The solution set is $\left\{0, \dfrac{5}{3}\right\}$. ∎

EXAMPLE 4

Solve $3ax^2 + bx = 0$ for x.

Solution

$$3ax^2 + bx = 0$$

$$x(3ax + b) = 0$$

$$x = 0 \quad \text{or} \quad 3ax + b = 0$$

$$x = 0 \quad \text{or} \quad 3ax = -b$$

$$x = 0 \quad \text{or} \quad x = -\frac{b}{3a}$$

The solution set is $\left\{0, -\dfrac{b}{3a}\right\}$. ■

Many of the problems that we solve in the next few sections have a geometric setting. Some basic geometric figures, along with appropriate formulas, are listed in the inside front cover of this text. You may need to refer to them to refresh your memory.

PROBLEM 1

The area of a square is three times its perimeter. Find the length of a side of the square.

Solution

Let s represent the length of a side of the square (Figure 3.17). The area is represented by s^2 and the perimeter by $4s$. Thus

$$s^2 = 3(4s) \quad \text{The area is to be three times the perimeter.}$$

$$s^2 = 12s$$

$$s^2 - 12s = 0$$

$$s(s - 12) = 0$$

$$s = 0 \quad \text{or} \quad s = 12$$

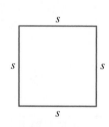

Figure 3.17

Because 0 is not a reasonable solution, it must be a 12-by-12 square. (Be sure to check this answer in the original statement of the problem!) ■

PROBLEM 2

Suppose that the volume of a right circular cylinder is numerically equal to the total surface area of the cylinder. If the height of the cylinder is equal to the length of a radius of the base, find the height.

Solution

Because $r = h$, the formula for volume $V = \pi r^2 h$ becomes $V = \pi r^3$, and the formula for the total surface area $S = 2\pi r^2 + 2\pi rh$ becomes $S = 2\pi r^2 + 2\pi r^2$, or $S = 4\pi r^2$. Therefore, we can set up and solve the following equation.

$$\pi r^3 = 4\pi r^2$$
$$\pi r^3 - 4\pi r^2 = 0$$
$$\pi r^2(r - 4) = 0$$
$$\pi r^2 = 0 \quad \text{or} \quad r - 4 = 0$$
$$r = 0 \quad \text{or} \quad r = 4$$

Zero is not a reasonable answer, therefore the height must be 4 units. ■

Problem Set 3.4

For Problems 1–10, classify each number as prime or composite.

1. 63

2. 81

3. 59

4. 83

5. 51

6. 69

7. 91

8. 119

9. 71

10. 101

For Problems 11–20, factor each of the composite numbers into the product of prime numbers. For example, $30 = 2 \cdot 3 \cdot 5$.

11. 28

12. 39

13. 44

14. 49

15. 56

16. 64

17. 72

18. 84

19. 87

20. 91

For Problems 21–46, factor completely.

21. $6x + 3y$

22. $12x + 8y$

23. $6x^2 + 14x$

24. $15x^2 + 6x$

25. $28y^2 - 4y$

26. $42y^2 - 6y$

27. $20xy - 15x$

28. $27xy - 36y$

29. $7x^3 + 10x^2$

30. $12x^3 - 10x^2$

31. $18a^2b + 27ab^2$

32. $24a^3b^2 + 36a^2b$

33. $12x^3y^4 - 39x^4y^3$

34. $15x^4y^2 - 45x^5y^4$

35. $8x^4 + 12x^3 - 24x^2$

36. $6x^5 - 18x^3 + 24x$

37. $5x + 7x^2 + 9x^4$

38. $9x^2 - 17x^4 + 21x^5$

39. $15x^2y^3 + 20xy^2 + 35x^3y^4$

40. $8x^5y^3 - 6x^4y^5 + 12x^2y^3$

41. $x(y + 2) + 3(y + 2)$

42. $x(y - 1) + 5(y - 1)$

43. $3x(2a + b) - 2y(2a + b)$

44. $5x(a - b) + y(a - b)$

45. $x(x + 2) + 5(x + 2)$

46. $x(x - 1) - 3(x - 1)$

For Problems 47–64, factor by grouping.

47. $ax + 4x + ay + 4y$

48. $ax - 2x + ay - 2y$

49. $ax - 2bx + ay - 2by$

50. $2ax - bx + 2ay - by$

51. $3ax - 3bx - ay + by$

52. $5ax - 5bx - 2ay + 2by$

53. $2ax + 2x + ay + y$

54. $3bx + 3x + by + y$

55. $ax^2 - x^2 + 2a - 2$

56. $ax^2 - 2x^2 + 3a - 6$

57. $2ac + 3bd + 2bc + 3ad$

58. $2bx + cy + cx + 2by$

59. $ax - by + bx - ay$

60. $2a^2 - 3bc - 2ab + 3ac$

61. $x^2 + 9x + 6x + 54$

62. $x^2 - 2x + 5x - 10$

63. $2x^2 + 8x + x + 4$

64. $3x^2 + 18x - 2x - 12$

For Problems 65–80, solve each of the equations.

65. $x^2 + 7x = 0$

66. $x^2 + 9x = 0$

67. $x^2 - x = 0$

68. $x^2 - 14x = 0$

69. $a^2 = 5a$

70. $b^2 = -7b$

71. $-2y = 4y^2$

72. $-6x = 2x^2$

73. $3x^2 + 7x = 0$

74. $-4x^2 + 9x = 0$

75. $4x^2 = 5x$

76. $3x = 11x^2$

77. $x - 4x^2 = 0$

78. $x - 6x^2 = 0$

79. $12a = -a^2$

80. $-5a = -a^2$

For Problems 81–86, solve each equation for the indicated variable.

81. $5bx^2 - 3ax = 0$ for x

82. $ax^2 + bx = 0$ for x

83. $2by^2 = -3ay$ for y

84. $3ay^2 = by$ for y

85. $y^2 - ay + 2by - 2ab = 0$ for y

86. $x^2 + ax + bx + ab = 0$ for x

For Problems 87–96, set up an equation and solve each of the following problems.

87. The square of a number equals seven times the number. Find the number.

88. Suppose that the area of a square is six times its perimeter. Find the length of a side of the square.

89. The area of a circular region is numerically equal to three times the circumference of the circle. Find the length of a radius of the circle.

90. Find the length of a radius of a circle such that the circumference of the circle is numerically equal to the area of the circle.

91. Suppose that the area of a circle is numerically equal to the perimeter of a square and that the length of a radius of the circle is equal to the length of a side of the square. Find the length of a side of the square. Express your answer in terms of π.

92. Find the length of a radius of a sphere such that the surface area of the sphere is numerically equal to the volume of the sphere.

93. Suppose that the area of a square lot is twice the area of an adjoining rectangular plot of ground. If the rectangular plot is 50 feet wide, and its length is the same as the length of a side of the square lot, find the dimensions of both the square and the rectangle.

94. The area of a square is one-fourth as large as the area of a triangle. One side of the triangle is 16 inches long, and the altitude to that side is the same length as a side of the square. Find the length of a side of the square.

95. Suppose that the volume of a sphere is numerically equal to twice the surface area of the sphere. Find the length of a radius of the sphere.

96. Suppose that a radius of a sphere is equal in length to a radius of a circle. If the volume of the sphere is numerically equal to four times the area of the circle, find the length of a radius for both the sphere and the circle.

■ ■ ■ THOUGHTS INTO WORDS

97. Is $2 \cdot 3 \cdot 5 \cdot 7 \cdot 11 + 7$ a prime or a composite number? Defend your answer.

98. Suppose that your friend factors $36x^2y + 48xy^2$ as follows:

$$36x^2y + 48xy^2 = (4xy)(9x + 12y)$$
$$= (4xy)(3)(3x + 4y)$$
$$= 12xy(3x + 4y)$$

Is this a correct approach? Would you have any suggestion to offer your friend?

99. Your classmate solves the equation $3ax + bx = 0$ for x as follows:

$$3ax + bx = 0$$
$$3ax = -bx$$
$$x = \frac{-bx}{3a}$$

How should he know that the solution is incorrect? How would you help him obtain the correct solution?

100. The total surface area of a right circular cylinder is given by the formula $A = 2\pi r^2 + 2\pi rh$, where r represents the radius of a base, and h represents the height of the cylinder. For computational purposes, it may be more convenient to change the form of the right side of the formula by factoring it.

$$A = 2\pi r^2 + 2\pi rh$$
$$= 2\pi r(r + h)$$

Use $A = 2\pi r(r + h)$ to find the total surface area of each of the following cylinders. Also, use $\dfrac{22}{7}$ as an approximation for π.

(a) $r = 7$ centimeters and $h = 12$ centimeters

(b) $r = 14$ meters and $h = 20$ meters

(c) $r = 3$ feet and $h = 4$ feet

(d) $r = 5$ yards and $h = 9$ yards

For Problems 101–106, factor each expression. Assume that all variables that appear as exponents represent positive integers.

101. $2x^{2a} - 3x^a$

102. $6x^{2a} + 8x^a$

103. $y^{3m} + 5y^{2m}$

104. $3y^{5m} - y^{4m} - y^{3m}$

105. $2x^{6a} - 3x^{5a} + 7x^{4a}$

106. $6x^{3a} - 10x^{2a}$

3.5	**Factoring: Difference of Two Squares and Sum or Difference of Two Cubes**

In Section 3.3, we examined some special multiplication patterns. One of these patterns was

$$(a + b)(a - b) = a^2 - b^2$$

This same pattern, viewed as a factoring pattern, is referred to as the difference of two squares.

Difference of Two Squares

$$a^2 - b^2 = (a + b)(a - b)$$

Applying the pattern is fairly simple, as these next examples demonstrate. Again, the steps in dashed boxes are usually performed mentally.

$$x^2 - 16 = (x)^2 - (4)^2 = (x + 4)(x - 4)$$

$$4x^2 - 25 = (2x)^2 - (5)^2 = (2x + 5)(2x - 5)$$

$$16x^2 - 9y^2 = (4x)^2 - (3y)^2 = (4x + 3y)(4x - 3y)$$

$$1 - a^2 = (1)^2 - (a)^2 = (1 + a)(1 - a)$$

Multiplication is commutative, so the order of writing the factors is not important. For example, $(x + 4)(x - 4)$ can also be written as $(x - 4)(x + 4)$.

You must be careful not to assume an analogous factoring pattern for the *sum* of two squares; *it does not exist*. For example, $x^2 + 4 \neq (x + 2)(x + 2)$ because $(x + 2)(x + 2) = x^2 + 4x + 4$. We say that a polynomial such as $x^2 + 4$ is a **prime polynomial** or that it is not factorable using integers.

Sometimes the difference-of-two-squares pattern can be applied more than once, as the next examples illustrate.

$$x^4 - y^4 = (x^2 + y^2)(x^2 - y^2) = (x^2 + y^2)(x + y)(x - y)$$
$$16x^4 - 81y^4 = (4x^2 + 9y^2)(4x^2 - 9y^2) = (4x^2 + 9y^2)(2x + 3y)(2x - 3y)$$

It may also be that the squares are other than simple monomial squares, as in the next three examples.

$$(x + 3)^2 - y^2 = ((x + 3) + y)((x + 3) - y) = (x + 3 + y)(x + 3 - y)$$
$$4x^2 - (2y + 1)^2 = (2x + (2y + 1))(2x - (2y + 1))$$
$$= (2x + 2y + 1)(2x - 2y - 1)$$
$$(x - 1)^2 - (x + 4)^2 = ((x - 1) + (x + 4))((x - 1) - (x + 4))$$
$$= (x - 1 + x + 4)(x - 1 - x - 4)$$
$$= (2x + 3)(-5)$$

It is possible to apply both the technique of factoring out a common monomial factor and the pattern of the difference of two squares to the same problem. In general, it is best to look first for a common monomial factor. Consider the following examples.

$$2x^2 - 50 = 2(x^2 - 25) \qquad\qquad 9x^2 - 36 = 9(x^2 - 4)$$
$$= 2(x + 5)(x - 5) \qquad\qquad = 9(x + 2)(x - 2)$$
$$48y^3 - 27y = 3y(16y^2 - 9)$$
$$= 3y(4y + 3)(4y - 3)$$

Word of Caution   The polynomial $9x^2 - 36$ can be factored as follows:

$$9x^2 - 36 = (3x + 6)(3x - 6)$$
$$= 3(x + 2)(3)(x - 2)$$
$$= 9(x + 2)(x - 2)$$

However, when one takes this approach, there seems to be a tendency to stop at the step $(3x + 6)(3x - 6)$. Therefore, remember the suggestion to *look first for a common monomial factor.*

The following examples should help you summarize all of the factoring techniques we have considered thus far.

$$7x^2 + 28 = 7(x^2 + 4)$$
$$4x^2y - 14xy^2 = 2xy(2x - 7y)$$

$$x^2 - 4 = (x + 2)(x - 2)$$

$$18 - 2x^2 = 2(9 - x^2) = 2(3 + x)(3 - x)$$

$y^2 + 9$ is not factorable using integers.

$5x + 13y$ is not factorable using integers.

$$x^4 - 16 = (x^2 + 4)(x^2 - 4) = (x^2 + 4)(x + 2)(x - 2)$$

■ Sum and Difference of Two Cubes

As we pointed out before, there exists no sum-of-squares pattern analogous to the difference-of-squares factoring pattern. That is, a polynomial such as $x^2 + 9$ is not factorable using integers. However, patterns do exist for both the sum and the difference of two cubes. These patterns are as follows:

Sum and Difference of Two Cubes

$$a^3 + b^3 = (a + b)(a^2 - ab + b^2)$$

$$a^3 - b^3 = (a - b)(a^2 + ab + b^2)$$

Note how we apply these patterns in the next four examples.

$$x^3 + 27 = (x)^3 + (3)^3 = (x + 3)(x^2 - 3x + 9)$$

$$8a^3 + 125b^3 = (2a)^3 + (5b)^3 = (2a + 5b)(4a^2 - 10ab + 25b^2)$$

$$x^3 - 1 = (x)^3 - (1)^3 = (x - 1)(x^2 + x + 1)$$

$$27y^3 - 64x^3 = (3y)^3 - (4x)^3 = (3y - 4x)(9y^2 + 12xy + 16x^2)$$

■ Equations and Problem Solving

Remember that each time we pick up a new factoring technique we also develop more power for solving equations. Let's consider how we can use the difference-of-two-squares factoring pattern to help solve certain types of equations.

EXAMPLE 1

Solve $x^2 = 16$.

Solution

$$x^2 = 16$$

$$x^2 - 16 = 0$$

$$(x + 4)(x - 4) = 0$$

$$x + 4 = 0 \quad \text{or} \quad x - 4 = 0$$

$$x = -4 \quad \text{or} \quad x = 4$$

The solution set is $\{-4, 4\}$. (Be sure to check these solutions in the original equation!) ■

EXAMPLE 2

Solve $9x^2 = 64$.

Solution

$$9x^2 = 64$$

$$9x^2 - 64 = 0$$

$$(3x + 8)(3x - 8) = 0$$

$$3x + 8 = 0 \quad \text{or} \quad 3x - 8 = 0$$

$$3x = -8 \quad \text{or} \quad 3x = 8$$

$$x = -\frac{8}{3} \quad \text{or} \quad x = \frac{8}{3}$$

The solution set is $\left\{-\frac{8}{3}, \frac{8}{3}\right\}$. ∎

EXAMPLE 3

Solve $7x^2 - 7 = 0$.

Solution

$$7x^2 - 7 = 0$$

$$7(x^2 - 1) = 0$$

$$x^2 - 1 = 0 \qquad \text{Multiply both sides by } \frac{1}{7}.$$

$$(x + 1)(x - 1) = 0$$

$$x + 1 = 0 \quad \text{or} \quad x - 1 = 0$$

$$x = -1 \quad \text{or} \quad x = 1$$

The solution set is $\{-1, 1\}$. ∎

In the previous examples we have been using the property $ab = 0$ if and only if $a = 0$ or $b = 0$. This property can be extended to any number of factors whose product is zero. Thus for three factors, the property could be stated $abc = 0$ if and only if $a = 0$ or $b = 0$ or $c = 0$. The next two examples illustrate this idea.

EXAMPLE 4

Solve $x^4 - 16 = 0$.

Solution

$$x^4 - 16 = 0$$

$$(x^2 + 4)(x^2 - 4) = 0$$

$$(x^2 + 4)(x + 2)(x - 2) = 0$$

$$x^2 + 4 = 0 \quad \text{or} \quad x + 2 = 0 \quad \text{or} \quad x - 2 = 0$$

$$x^2 = -4 \quad \text{or} \quad x = -2 \quad \text{or} \quad x = 2$$

The solution set is $\{-2, 2\}$. (Because no real numbers, when squared, will produce -4, the equation $x^2 = -4$ yields no additional real number solutions.) ■

EXAMPLE 5

Solve $x^3 - 49x = 0$.

Solution

$$x^3 - 49x = 0$$

$$x(x^2 - 49) = 0$$

$$x(x + 7)(x - 7) = 0$$

$$x = 0 \quad \text{or} \quad x + 7 = 0 \quad \text{or} \quad x - 7 = 0$$

$$x = 0 \quad \text{or} \quad x = -7 \quad \text{or} \quad x = 7$$

The solution set is $\{-7, 0, 7\}$. ■

The more we know about solving equations, the better we are at solving word problems.

PROBLEM 1

The combined area of two squares is 40 square centimeters. Each side of one square is three times as long as a side of the other square. Find the dimensions of each of the squares.

Solution

Let s represent the length of a side of the smaller square. Then $3s$ represents the length of a side of the larger square (Figure 3.18).

$$s^2 + (3s)^2 = 40$$

$$s^2 + 9s^2 = 40$$

$$10s^2 = 40$$

$$s^2 = 4$$

$$s^2 - 4 = 0$$

$$(s + 2)(s - 2) = 0$$

$$s + 2 = 0 \quad \text{or} \quad s - 2 = 0$$

$$s = -2 \quad \text{or} \quad s = 2$$

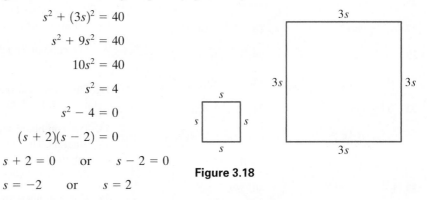

Figure 3.18

Because s represents the length of a side of a square, the solution -2 has to be disregarded. Thus the length of a side of the small square is 2 centimeters, and the large square has sides of length $3(2) = 6$ centimeters. ■

Problem Set 3.5

For Problems 1–20, use the difference-of-squares pattern to factor each of the following.

1. $x^2 - 1$

2. $x^2 - 9$

3. $16x^2 - 25$

4. $4x^2 - 49$

5. $9x^2 - 25y^2$

6. $x^2 - 64y^2$

7. $25x^2y^2 - 36$

8. $x^2y^2 - a^2b^2$

9. $4x^2 - y^4$

10. $x^6 - 9y^2$

11. $1 - 144n^2$

12. $25 - 49n^2$

13. $(x + 2)^2 - y^2$

14. $(3x + 5)^2 - y^2$

15. $4x^2 - (y + 1)^2$

16. $x^2 - (y - 5)^2$

17. $9a^2 - (2b + 3)^2$

18. $16s^2 - (3t + 1)^2$

19. $(x + 2)^2 - (x + 7)^2$

20. $(x - 1)^2 - (x - 8)^2$

For Problems 21–44, factor each of the following polynomials completely. Indicate any that are not factorable using integers. Don't forget to look first for a common monomial factor.

21. $9x^2 - 36$

22. $8x^2 - 72$

23. $5x^2 + 5$

24. $7x^2 + 28$

25. $8y^2 - 32$

26. $5y^2 - 80$

27. $a^3b - 9ab$

28. $x^3y^2 - xy^2$

29. $16x^2 + 25$

30. $x^4 - 16$

31. $n^4 - 81$

32. $4x^2 + 9$

33. $3x^3 + 27x$

34. $20x^3 + 45x$

35. $4x^3y - 64xy^3$

36. $12x^3 - 27xy^2$

37. $6x - 6x^3$

38. $1 - 16x^4$

39. $1 - x^4y^4$

40. $20x - 5x^3$

41. $4x^2 - 64y^2$

42. $9x^2 - 81y^2$

43. $3x^4 - 48$

44. $2x^5 - 162x$

For Problems 45–56, use the sum-of-two-cubes or the difference-of-two-cubes pattern to factor each of the following.

45. $a^3 - 64$

46. $a^3 - 27$

47. $x^3 + 1$

48. $x^3 + 8$

49. $27x^3 + 64y^3$

50. $8x^3 + 27y^3$

51. $1 - 27a^3$

52. $1 - 8x^3$

53. $x^3y^3 - 1$

54. $125x^3 + 27y^3$

55. $x^6 - y^6$

56. $x^6 + y^6$

For Problems 57–70, find all real number solutions for each equation.

57. $x^2 - 25 = 0$

58. $x^2 - 1 = 0$

59. $9x^2 - 49 = 0$

60. $4y^2 = 25$

61. $8x^2 - 32 = 0$

62. $3x^2 - 108 = 0$

63. $3x^3 = 3x$

64. $4x^3 = 64x$

65. $20 - 5x^2 = 0$

66. $54 - 6x^2 = 0$

67. $x^4 - 81 = 0$

68. $x^5 - x = 0$

69. $6x^3 + 24x = 0$

70. $4x^3 + 12x = 0$

For Problems 71–80, set up an equation and solve each of the following problems.

71. The cube of a number equals nine times the same number. Find the number.

72. The cube of a number equals the square of the same number. Find the number.

73. The combined area of two circles is 80π square centimeters. The length of a radius of one circle is twice the length of a radius of the other circle. Find the length of the radius of each circle.

74. The combined area of two squares is 26 square meters. The sides of the larger square are five times as long as the sides of the smaller square. Find the dimensions of each of the squares.

75. A rectangle is twice as long as it is wide, and its area is 50 square meters. Find the length and the width of the rectangle.

76. Suppose that the length of a rectangle is one and one-third times as long as its width. The area of the rectangle is 48 square centimeters. Find the length and width of the rectangle.

77. The total surface area of a right circular cylinder is 54π square inches. If the altitude of the cylinder is twice the length of a radius, find the altitude of the cylinder.

78. The total surface area of a right circular cone is 108π square feet. If the slant height of the cone is twice the length of a radius of the base, find the length of a radius.

79. The sum of the areas of a circle and a square is $(16\pi + 64)$ square yards. If a side of the square is twice the length of a radius of the circle, find the length of a side of the square.

80. The length of an altitude of a triangle is one-third the length of the side to which it is drawn. If the area of the triangle is 6 square centimeters, find the length of that altitude.

■ ■ ■ THOUGHTS INTO WORDS

81. Explain how you would solve the equation $4x^3 = 64x$.

82. What is wrong with the following factoring process?

$$25x^2 - 100 = (5x + 10)(5x - 10)$$

How would you correct the error?

83. Consider the following solution:

$$6x^2 - 24 = 0$$

$$6(x^2 - 4) = 0$$

$$6(x + 2)(x - 2) = 0$$

$$6 = 0 \quad \text{or} \quad x + 2 = 0 \quad \text{or} \quad x - 2 = 0$$

$$6 = 0 \quad \text{or} \quad x = -2 \quad \text{or} \quad x = 2$$

The solution set is $\{-2, \ 2\}$.

Is this a correct solution? Would you have any suggestion to offer the person who used this approach?

3.6   Factoring Trinomials

One of the most common types of factoring used in algebra is the expression of a trinomial as the product of two binomials. To develop a factoring technique, we first look at some multiplication ideas. Let's consider the product $(x + a)(x + b)$ and use the distributive property to show how each term of the resulting trinomial is formed.

$$(x + a)(x + b) = x(x + b) + a(x + b)$$

$$= x(x) + x(b) + a(x) + a(b)$$

$$= x^2 + (a + b)x + ab$$

Note that the coefficient of the middle term is the sum of a and b and that the last term is the product of a and b. These two relationships can be used to factor trinomials. Let's consider some examples.

EXAMPLE 1

Factor $x^2 + 8x + 12$.

Solution

We need to complete the following with two integers whose sum is 8 and whose product is 12.

$$x^2 + 8x + 12 = (x + \underline{})(x + \underline{})$$

The possible pairs of factors of 12 are $1(12)$, $2(6)$, and $3(4)$. Because $6 + 2 = 8$, we can complete the factoring as follows:

$$x^2 + 8x + 12 = (x + 6)(x + 2)$$

To check our answer, we find the product of $(x + 6)$ and $(x + 2)$. ■

EXAMPLE 2

Factor $x^2 - 10x + 24$.

Solution

We need two integers whose product is 24 and whose sum is -10. Let's use a small table to organize our thinking.

Factors	Product of the factors	Sum of the factors
$(-1)(-24)$	24	-25
$(-2)(-12)$	24	-14
$(-3)(-8)$	24	-11
$(-4)(-6)$	24	-10

The bottom line contains the numbers that we need. Thus

$$x^2 - 10x + 24 = (x - 4)(x - 6)$$ ■

EXAMPLE 3

Factor $x^2 + 7x - 30$.

Solution

We need two integers whose product is -30 and whose sum is 7.

Factors	Product of the factors	Sum of the factors
$(-1)(30)$	-30	29
$(1)(-30)$	-30	-29
$(2)(-15)$	-30	-13
$(-2)(15)$	-30	13
$(-3)(10)$	-30	7

No need to search any further

The numbers that we need are -3 and 10, and we can complete the factoring.

$$x^2 + 7x - 30 = (x + 10)(x - 3)$$ ■

E X A M P L E   4

Factor $x^2 + 7x + 16$.

Solution

We need two integers whose product is 16 and whose sum is 7.

Factors	Product of the factors	Sum of the factors
(1)(16)	16	17
(2)(8)	16	10
(4)(4)	16	8

We have exhausted all possible pairs of factors of 16 and no two factors have a sum of 7, so we conclude that $x^2 + 7x + 16$ *is not factorable using integers.* ■

The tables in Examples 2, 3, and 4 were used to illustrate one way of organizing your thoughts for such problems. Normally you would probably factor such problems mentally without taking the time to formulate a table. Note, however, that in Example 4 the table helped us to be absolutely sure that we tried all the possibilities. Whether or not you use the table, keep in mind that the key ideas are the product and sum relationships.

E X A M P L E   5

Factor $n^2 - n - 72$.

Solution

Note that the coefficient of the middle term is -1. Hence we are looking for two integers whose product is -72, and because their sum is -1, the absolute value of the negative number must be 1 larger than the positive number. The numbers are -9 and 8, and we can complete the factoring.

$$n^2 - n - 72 = (n - 9)(n + 8)$$ ■

E X A M P L E   6

Factor $t^2 + 2t - 168$.

Solution

We need two integers whose product is -168 and whose sum is 2. Because the absolute value of the constant term is rather large, it might help to look at it in prime factored form.

$$168 = 2 \cdot 2 \cdot 2 \cdot 3 \cdot 7$$

Now we can mentally form two numbers by using all of these factors in different combinations. Using two 2s and a 3 in one number and the other 2 and the 7 in the second number produces $2 \cdot 2 \cdot 3 = 12$ and $2 \cdot 7 = 14$. The coefficient of the middle term of the trinomial is 2, so we know that we must use 14 and -12. Thus we obtain

$$t^2 + 2t - 168 = (t + 14)(t - 12)$$ ■

■ Trinomials of the Form $ax^2 + bx + c$

We have been factoring trinomials of the form $x^2 + bx + c$—that is, trinomials where the coefficient of the squared term is 1. Now let's consider factoring trinomials where the coefficient of the squared term is not 1. First, let's illustrate an informal trial-and-error technique that works quite well for certain types of trinomials. This technique is based on our knowledge of multiplication of binomials.

EXAMPLE 7   Factor $2x^2 + 11x + 5$.

Solution

By looking at the first term, $2x^2$, and the positive signs of the other two terms, we know that the binomials are of the form

$$(x + \underline{\quad})(2x + \underline{\quad})$$

Because the factors of the last term, 5, are 1 and 5, we have only the following two possibilities to try.

$$(x + 1)(2x + 5) \qquad \text{or} \qquad (x + 5)(2x + 1)$$

By checking the middle term formed in each of these products, we find that the second possibility yields the correct middle term of $11x$. Therefore,

$$2x^2 + 11x + 5 = (x + 5)(2x + 1) \qquad ■$$

EXAMPLE 8   Factor $10x^2 - 17x + 3$.

Solution

First, observe that $10x^2$ can be written as $x \cdot 10x$ or $2x \cdot 5x$. Second, because the middle term of the trinomial is negative, and the last term is positive, we know that the binomials are of the form

$$(x - \underline{\quad})(10x - \underline{\quad}) \qquad \text{or} \qquad (2x - \underline{\quad})(5x - \underline{\quad})$$

The factors of the last term, 3, are 1 and 3, so the following possibilities exist.

$$(x - 1)(10x - 3) \qquad (2x - 1)(5x - 3)$$
$$(x - 3)(10x - 1) \qquad (2x - 3)(5x - 1)$$

By checking the middle term formed in each of these products, we find that the product $(2x - 3)(5x - 1)$ yields the desired middle term of $-17x$. Therefore,

$$10x^2 - 17x + 3 = (2x - 3)(5x - 1) \qquad ■$$

EXAMPLE 9   Factor $4x^2 + 6x + 9$.

Solution

The first term, $4x^2$, and the positive signs of the middle and last terms indicate that the binomials are of the form

$$(x + \underline{\quad})(4x + \underline{\quad}) \qquad \text{or} \qquad (2x + \underline{\quad})(2x + \underline{\quad}).$$

Because the factors of 9 are 1 and 9 or 3 and 3, we have the following five possibilities to try.

$$(x + 1)(4x + 9) \qquad (2x + 1)(2x + 9)$$

$$(x + 9)(4x + 1) \qquad (2x + 3)(2x + 3)$$

$$(x + 3)(4x + 3)$$

When we try all of these possibilities we find that none of them yields a middle term of $6x$. Therefore, $4x^2 + 6x + 9$ is not factorable using integers.   ∎

By now it is obvious that factoring trinomials of the form $ax^2 + bx + c$ can be tedious. The key idea is to organize your work so that you consider all possibilities. We suggested one possible format in the previous three examples. As you practice such problems, you may come across a format of your own. Whatever works best for you is the right approach.

There is another, more systematic technique that you may wish to use with some trinomials. It is an extension of the technique we used at the beginning of this section. To see the basis of this technique, let's look at the following product.

$$(px + r)(qx + s) = px(qx) + px(s) + r(qx) + r(s)$$

$$= (pq)x^2 + (ps + rq)x + rs$$

Note that the product of the coefficient of the x^2 term and the constant term is $pqrs$. Likewise, the product of the two coefficients of x, ps and rq, is also $pqrs$. Therefore, when we are factoring the trinomial $(pq)x^2 + (ps + rq)x + rs$, the two coefficients of x must have a sum of $(ps) + (rq)$ and a product of $pqrs$. Let's see how this works in some examples.

EXAMPLE 10

Factor $6x^2 - 11x - 10$

Solution

First, multiply the coefficient of the x^2 term, 6, and the constant term, -10.

$$(6)(-10) = -60$$

Now find two integers whose sum is -11 and whose product is -60. The integers 4 and -15 satisfy these conditions.

Rewrite the original problem, expressing the middle term as a sum of terms with these factors of -60 as their coefficients.

$$6x^2 - 11x - 10 = 6x^2 + 4x - 15x - 10$$

After rewriting the problem, we can factor by grouping — that is, factoring $2x$ from the first two terms and -5 from the last two terms.

$$6x^2 + 4x - 15x - 10 = 2x(3x + 2) - 5(3x + 2)$$

Now a common binomial factor of $(3x + 2)$ is obvious, and we can proceed as follows:

$$2x(3x + 2) - 5(3x + 2) = (3x + 2)(2x - 5)$$

Thus $6x^2 - 11x - 10 = (3x + 2)(2x - 5)$.   ∎

EXAMPLE 11

Factor $4x^2 - 29x + 30$

Solution

First, multiply the coefficient of the x^2 term, 4, and the constant term, 30.

$$(4)(30) = 120$$

Now find two integers whose sum is -29 and whose product is 120. The integers -24 and -5 satisfy these conditions.

Rewrite the original problem, expressing the middle term as a sum of terms with these factors of 120 as their coefficients.

$$4x^2 - 29x + 30 = 4x^2 - 24x - 5x + 30$$

After rewriting the problem, we can factor by grouping — that is, factoring $4x$ from the first two terms and -5 from the last two terms.

$$4x^2 - 29x - 5x + 30 = 4x(x - 6) - 5(x - 6)$$

Now a common binomial factor of $(x - 6)$ is obvious, and we can proceed as follows:

$$4x(x - 6) - 5(x - 6) = (x - 6)(4x - 5)$$

Thus $4x^2 - 29x + 30 = (x - 6)(4x - 5)$. ■

The technique presented in Examples 10 and 11 has concrete steps to follow. Examples 7 through 9 were factored by trial-and-error technique. Both of the techniques we used have their strengths and weaknesses. Which technique to use depends on the complexity of the problem and on your personal preference. The more that you work with both techniques, the more comfortable you will feel using them.

■ Summary of Factoring Techniques

Before we summarize our work with factoring techniques, let's look at two more special factoring patterns. In Section 3.3 we used the following two patterns to square binomials.

$$(a + b)^2 = a^2 + 2ab + b^2 \qquad \text{and} \qquad (a - b)^2 = a^2 - 2ab + b^2$$

These patterns can also be used for factoring purposes.

$$a^2 + 2ab + b^2 = (a + b)^2 \qquad \text{and} \qquad a^2 - 2ab + b^2 = (a - b)^2$$

The trinomials on the left sides are called **perfect-square trinomials**; they are the result of squaring a binomial. We can always factor perfect-square trinomials using the usual techniques for factoring trinomials. However, they are easily recognized by the nature of their terms. For example, $4x^2 + 12x + 9$ is a perfect-square trinomial because

1. The first term is a perfect square. $(2x)^2$

2. The last term is a perfect square. $(3)^2$

3. The middle term is twice the product of the quantities $2(2x)(3)$
being squared in the first and last terms.

Likewise, $9x^2 - 30x + 25$ is a perfect-square trinomial because

1. The first term is a perfect square. $\hspace{3cm}$ $(3x)^2$

2. The last term is a perfect square. $\hspace{3cm}$ $(5)^2$

3. The middle term is the negative of twice the product of $\hspace{0.5cm}$ $-2(3x)(5)$
the quantities being squared in the first and last terms.

Once we know that we have a perfect-square trinomial, the factors follow immediately from the two basic patterns. Thus

$$4x^2 + 12x + 9 = (2x + 3)^2 \hspace{1.5cm} 9x^2 - 30x + 25 = (3x - 5)^2$$

Here are some additional examples of perfect-square trinomials and their factored forms.

$$x^2 + 14x + 49 = (x)^2 + 2(x)(7) + (7) = (x + 7)^2$$

$$n^2 - 16n + 64 = (n)^2 - 2(n)(8) + (8)^2 = (n - 8)^2$$

$$36a^2 + 60ab + 25b^2 = (6a)^2 + 2(6a)(5b) + (5b)^2 = (6a + 5b)^2$$

$$16x^2 - 8xy + y^2 = (4x)^2 - 2(4x)(y) + (y)^2 = (4x - y)^2$$

Perhaps you will want to do this
step mentally after you feel
comfortable with the process.

As we have indicated, factoring is an important algebraic skill. We learned some basic factoring techniques one at a time, but you must be able to apply whichever is (or are) appropriate to the situation. Let's review the techniques and consider a variety of examples that demonstrate their use.

In this chapter, we have discussed

1. Factoring by using the distributive property to factor out a common monomial (or binomial) factor.

2. Factoring by applying the difference-of-two-squares pattern.

3. Factoring by applying the sum-of-two-cubes or the difference-of-two-cubes pattern.

4. Factoring of trinomials into the product of two binomials. (The perfect-square-trinomial pattern is a special case of this technique.)

As a general guideline, always look for a common monomial factor first and then proceed with the other techniques. Study the following examples carefully and be sure that you agree with the indicated factors.

$$2x^2 + 20x + 48 = 2(x^2 + 10x + 24) \hspace{1.5cm} 16a^2 - 64 = 16(a^2 - 4)$$

$$= 2(x + 4)(x + 6) \hspace{1.5cm} = 16(a + 2)(a - 2)$$

$$3x^3y^3 + 27xy = 3xy(x^2y^2 + 9) \qquad\qquad x^2 + 3x - 21 \text{ is not factorable using integers}$$

$$30n^2 - 31n + 5 = (5n - 1)(6n - 5) \qquad t^4 + 3t^2 + 2 = (t^2 + 2)(t^2 + 1)$$

$$2x^3 - 16 = 2(x^3 - 8) = 2(x - 2)(x^2 + 2x + 4)$$

Problem Set 3.6

For Problems 1–56, factor completely each of the polynomials and indicate any that are not factorable using integers.

1. $x^2 + 9x + 20$
2. $x^2 + 11x + 24$

3. $x^2 - 11x + 28$
4. $x^2 - 8x + 12$

5. $a^2 + 5a - 36$
6. $a^2 + 6a - 40$

7. $y^2 + 20y + 84$
8. $y^2 + 21y + 98$

9. $x^2 - 5x - 14$
10. $x^2 - 3x - 54$

11. $x^2 + 9x + 12$
12. $35 - 2x - x^2$

13. $6 + 5x - x^2$
14. $x^2 + 8x - 24$

15. $x^2 + 15xy + 36y^2$
16. $x^2 - 14xy + 40y^2$

17. $a^2 - ab - 56b^2$
18. $a^2 + 2ab - 63b^2$

19. $15x^2 + 23x + 6$
20. $9x^2 + 30x + 16$

21. $12x^2 - x - 6$
22. $20x^2 - 11x - 3$

23. $4a^2 + 3a - 27$
24. $12a^2 + 4a - 5$

25. $3n^2 - 7n - 20$
26. $4n^2 + 7n - 15$

27. $3x^2 + 10x + 4$
28. $4n^2 - 19n + 21$

29. $10n^2 - 29n - 21$
30. $4x^2 - x + 6$

31. $8x^2 + 26x - 45$
32. $6x^2 + 13x - 33$

33. $6 - 35x - 6x^2$
34. $4 - 4x - 15x^2$

35. $20y^2 + 31y - 9$
36. $8y^2 + 22y - 21$

37. $24n^2 - 2n - 5$
38. $3n^2 - 16n - 35$

39. $5n^2 + 33n + 18$
40. $7n^2 + 31n + 12$

41. $x^2 + 25x + 150$
42. $x^2 + 21x + 108$

43. $n^2 - 36n + 320$
44. $n^2 - 26n + 168$

45. $t^2 + 3t - 180$
46. $t^2 - 2t - 143$

47. $t^4 - 5t^2 + 6$
48. $t^4 + 10t^2 + 24$

49. $10x^4 + 3x^2 - 4$
50. $3x^4 + 7x^2 - 6$

51. $x^4 - 9x^2 + 8$
52. $x^4 - x^2 - 12$

53. $18n^4 + 25n^2 - 3$
54. $4n^4 + 3n^2 - 27$

55. $x^4 - 17x^2 + 16$
56. $x^4 - 13x^2 + 36$

Problems 57–94 should help you pull together all of the factoring techniques of this chapter. Factor completely each polynomial, and indicate any that are not factorable using integers.

57. $2t^2 - 8$
58. $14w^2 - 29w - 15$

59. $12x^2 + 7xy - 10y^2$
60. $8x^2 + 2xy - y^2$

61. $18n^3 + 39n^2 - 15n$
62. $n^2 + 18n + 77$

63. $n^2 - 17n + 60$
64. $(x + 5)^2 - y^2$

65. $36a^2 - 12a + 1$
66. $2n^2 - n - 5$

67. $6x^2 + 54$
68. $x^5 - x$

69. $3x^2 + x - 5$
70. $5x^2 + 42x - 27$

71. $x^2 - (y - 7)^2$
72. $2n^3 + 6n^2 + 10n$

73. $1 - 16x^4$
74. $9a^2 - 30a + 25$

75. $4n^2 + 25n + 36$
76. $x^3 - 9x$

77. $n^3 - 49n$
78. $4x^2 + 16$

79. $x^2 - 7x - 8$
80. $x^2 + 3x - 54$

81. $3x^4 - 81x$ **82.** $x^3 + 125$

83. $x^4 + 6x^2 + 9$ **84.** $18x^2 - 12x + 2$

85. $x^4 - 5x^2 - 36$ **86.** $6x^4 - 5x^2 - 21$

87. $6w^2 - 11w - 35$ **88.** $10x^3 + 15x^2 + 20x$

89. $25n^2 + 64$ **90.** $4x^2 - 37x + 40$

91. $2n^3 + 14n^2 - 20n$ **92.** $25t^2 - 100$

93. $2xy + 6x + y + 3$ **94.** $3xy + 15x - 2y - 10$

■ ■ ■ THOUGHTS INTO WORDS

95. How can you determine that $x^2 + 5x + 12$ is not factorable using integers?

96. Explain your thought process when factoring $30x^2 + 13x - 56$.

97. Consider the following approach to factoring $12x^2 + 54x + 60$.

$$12x^2 + 54x + 60 = (3x + 6)(4x + 10)$$
$$= 3(x + 2)(2)(2x + 5)$$
$$= 6(x + 2)(2x + 5)$$

Is this a correct factoring process? Do you have any suggestion for the person using this approach?

■ ■ ■ FURTHER INVESTIGATIONS

For Problems 98–103, factor each trinomial and assume that all variables that appear as exponents represent positive integers.

98. $x^{2a} + 2x^a - 24$ **99.** $x^{2a} + 10x^a + 21$

100. $6x^{2a} - 7x^a + 2$ **101.** $4x^{2a} + 20x^a + 25$

102. $12x^{2n} + 7x^n - 12$ **103.** $20x^{2n} + 21x^n - 5$

Consider the following approach to factoring $(x - 2)^2 + 3(x - 2) - 10$.

$(x - 2)^2 + 3(x - 2) - 10$

$= y^2 + 3y - 10$ Replace $x - 2$ with y.

$= (y + 5)(y - 2)$ Factor.

$= (x - 2 + 5)(x - 2 - 2)$ Replace y with $x - 2$.

$= (x + 3)(x - 4)$

Use this approach to factor Problems 104–109.

104. $(x - 3)^2 + 10(x - 3) + 24$

105. $(x + 1)^2 - 8(x + 1) + 15$

106. $(2x + 1)^2 + 3(2x + 1) - 28$

107. $(3x - 2)^2 - 5(3x - 2) - 36$

108. $6(x - 4)^2 + 7(x - 4) - 3$

109. $15(x + 2)^2 - 13(x + 2) + 2$

3.7   Equations and Problem Solving

The techniques for factoring trinomials that were presented in the previous section provide us with more power to solve equations. That is, the property "$ab = 0$ if and only if $a = 0$ or $b = 0$" continues to play an important role as we solve equations that contain factorable trinomials. Let's consider some examples.

EXAMPLE 1

Solve $x^2 - 11x - 12 = 0$.

Solution

$$x^2 - 11x - 12 = 0$$
$$(x - 12)(x + 1) = 0$$
$$x - 12 = 0 \quad \text{or} \quad x + 1 = 0$$
$$x = 12 \quad \text{or} \quad x = -1$$

The solution set is $\{-1, 12\}$. ■

EXAMPLE 2

Solve $20x^2 + 7x - 3 = 0$.

Solution

$$20x^2 + 7x - 3 = 0$$
$$(4x - 1)(5x + 3) = 0$$
$$4x - 1 = 0 \quad \text{or} \quad 5x + 3 = 0$$
$$4x = 1 \quad \text{or} \quad 5x = -3$$
$$x = \frac{1}{4} \quad \text{or} \quad x = -\frac{3}{5}$$

The solution set is $\left\{-\frac{3}{5}, \frac{1}{4}\right\}$. ■

EXAMPLE 3

Solve $-2n^2 - 10n + 12 = 0$.

Solution

$$-2n^2 - 10n + 12 = 0$$
$$-2(n^2 + 5n - 6) = 0$$
$$n^2 + 5n - 6 = 0 \qquad \text{Multiply both sides by } -\frac{1}{2}.$$
$$(n + 6)(n - 1) = 0$$
$$n + 6 = 0 \quad \text{or} \quad n - 1 = 0$$
$$n = -6 \quad \text{or} \quad n = 1$$

The solution set is $\{-6, 1\}$. ■

EXAMPLE 4

Solve $16x^2 - 56x + 49 = 0$.

Solution

$$16x^2 - 56x + 49 = 0$$
$$(4x - 7)^2 = 0$$

$$(4x - 7)(4x - 7) = 0$$

$$4x - 7 = 0 \quad \text{or} \quad 4x - 7 = 0$$

$$4x = 7 \quad \text{or} \quad 4x = 7$$

$$x = \frac{7}{4} \quad \text{or} \quad x = \frac{7}{4}$$

The only solution is $\frac{7}{4}$; thus the solution set is $\left\{\frac{7}{4}\right\}$.   ■

E X A M P L E   5

Solve $9a(a + 1) = 4$.

Solution

$$9a(a + 1) = 4$$

$$9a^2 + 9a = 4$$

$$9a^2 + 9a - 4 = 0$$

$$(3a + 4)(3a - 1) = 0$$

$$3a + 4 = 0 \quad \text{or} \quad 3a - 1 = 0$$

$$3a = -4 \quad \text{or} \quad 3a = 1$$

$$a = -\frac{4}{3} \quad \text{or} \quad a = \frac{1}{3}$$

The solution set is $\left\{-\frac{4}{3}, \frac{1}{3}\right\}$.   ■

E X A M P L E   6

Solve $(x - 1)(x + 9) = 11$.

Solution

$$(x - 1)(x + 9) = 11$$

$$x^2 + 8x - 9 = 11$$

$$x^2 + 8x - 20 = 0$$

$$(x + 10)(x - 2) = 0$$

$$x + 10 = 0 \quad \text{or} \quad x - 2 = 0$$

$$x = -10 \quad \text{or} \quad x = 2$$

The solution set is $\{-10, \ 2\}$.   ■

■ Problem Solving

As you might expect, the increase in our power to solve equations broadens our base for solving problems. Now we are ready to tackle some problems using equations of the types presented in this section.

A room contains 78 chairs. The number of chairs per row is one more than twice the number of rows. Find the number of rows and the number of chairs per row.

Solution

Let r represent the number of rows. Then $2r + 1$ represents the number of chairs per row.

$$r(2r + 1) = 78$$ The number of rows times the number of chairs per row yields the total number of chairs.

$$2r^2 + r = 78$$

$$2r^2 + r - 78 = 0$$

$$(2r + 13)(r - 6) = 0$$

$$2r + 13 = 0 \quad \text{or} \quad r - 6 = 0$$

$$2r = -13 \quad \text{or} \quad r = 6$$

$$r = -\frac{13}{2} \quad \text{or} \quad r = 6$$

The solution $-\dfrac{13}{2}$ must be disregarded, so there are 6 rows and $2r + 1$ or $2(6) + 1$ = 13 chairs per row.   ■

A strip of uniform width cut from both sides and both ends of an 8-inch by 11-inch sheet of paper reduces the size of the paper to an area of 40 square inches. Find the width of the strip.

Solution

Let x represent the width of the strip, as indicated in Figure 3.19.

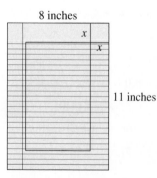

8 inches

11 inches

Figure 3.19

The length of the paper after the strips of width x are cut from both ends and both sides will be $11 - 2x$, and the width of the newly formed rectangle will be

$8 - 2x$. Because the area $(A = lw)$ is to be 40 square inches, we can set up and solve the following equation.

$$(11 - 2x)(8 - 2x) = 40$$

$$88 - 38x + 4x^2 = 40$$

$$4x^2 - 38x + 48 = 0$$

$$2x^2 - 19x + 24 = 0$$

$$(2x - 3)(x - 8) = 0$$

$$2x - 3 = 0 \quad \text{or} \quad x - 8 = 0$$

$$2x = 3 \quad \text{or} \quad x = 8$$

$$x = \frac{3}{2} \quad \text{or} \quad x = 8$$

The solution of 8 must be discarded because the width of the original sheet is only 8 inches. Therefore, the strip to be cut from all four sides must be $1\frac{1}{2}$ inches wide. (Check this answer!)   ■

The Pythagorean theorem, an important theorem pertaining to right triangles, can sometimes serve as a guideline for solving problems that deal with right triangles (see Figure 3.20). The Pythagorean theorem states that "in any right triangle, the square of the longest side (called the hypotenuse) is equal to the sum of the squares of the other two sides (called legs)." Let's use this relationship to help solve a problem.

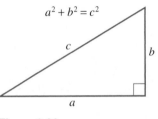

Figure 3.20

PROBLEM 3

One leg of a right triangle is 2 centimeters more than twice as long as the other leg. The hypotenuse is 1 centimeter longer than the longer of the two legs. Find the lengths of the three sides of the right triangle.

Solution

Let l represent the length of the shortest leg. Then $2l + 2$ represents the length of the other leg, and $2l + 3$ represents the length of the hypotenuse. Use the Pythagorean theorem as a guideline to set up and solve the following equation.

$$l^2 + (2l + 2)^2 = (2l + 3)^2$$

$$l^2 + 4l^2 + 8l + 4 = 4l^2 + 12l + 9$$

$$l^2 - 4l - 5 = 0$$

$$(l - 5)(l + 1) = 0$$

$$l - 5 = 0 \quad \text{or} \quad l + 1 = 0$$
$$l = 5 \quad \text{or} \quad l = -1$$

The negative solution must be discarded, so the length of one leg is 5 centimeters; the other leg is $2(5) + 2 = 12$ centimeters long, and the hypotenuse is $2(5) + 3 = 13$ centimeters long. ∎

Problem Set 3.7

For Problems 1–54, solve each equation. You will need to use the factoring techniques that we discussed throughout this chapter.

1. $x^2 + 4x + 3 = 0$

2. $x^2 + 7x + 10 = 0$

3. $x^2 + 18x + 72 = 0$

4. $n^2 + 20n + 91 = 0$

5. $n^2 - 13n + 36 = 0$

6. $n^2 - 10n + 16 = 0$

7. $x^2 + 4x - 12 = 0$

8. $x^2 + 7x - 30 = 0$

9. $w^2 - 4w = 5$

10. $s^2 - 4s = 21$

11. $n^2 + 25n + 156 = 0$

12. $n(n - 24) = -128$

13. $3t^2 + 14t - 5 = 0$

14. $4t^2 - 19t - 30 = 0$

15. $6x^2 + 25x + 14 = 0$

16. $25x^2 + 30x + 8 = 0$

17. $3t(t - 4) = 0$

18. $1 - x^2 = 0$

19. $-6n^2 + 13n - 2 = 0$

20. $(x + 1)^2 - 4 = 0$

21. $2n^3 = 72n$

22. $a(a - 1) = 2$

23. $(x - 5)(x + 3) = 9$

24. $3w^3 - 24w^2 + 36w = 0$

25. $16 - x^2 = 0$

26. $16t^2 - 72t + 81 = 0$

27. $n^2 + 7n - 44 = 0$

28. $2x^3 = 50x$

29. $3x^2 = 75$

30. $x^2 + x - 2 = 0$

31. $15x^2 + 34x + 15 = 0$

32. $20x^2 + 41x + 20 = 0$

33. $8n^2 - 47n - 6 = 0$

34. $7x^2 + 62x - 9 = 0$

35. $28n^2 - 47n + 15 = 0$

36. $24n^2 - 38n + 15 = 0$

37. $35n^2 - 18n - 8 = 0$

38. $8n^2 - 6n - 5 = 0$

39. $-3x^2 - 19x + 14 = 0$

40. $5x^2 = 43x - 24$

41. $n(n + 2) = 360$

42. $n(n + 1) = 182$

43. $9x^4 - 37x^2 + 4 = 0$

44. $4x^4 - 13x^2 + 9 = 0$

45. $3x^2 - 46x - 32 = 0$

46. $x^4 - 9x^2 = 0$

47. $2x^2 + x - 3 = 0$

48. $x^3 + 5x^2 - 36x = 0$

49. $12x^3 + 46x^2 + 40x = 0$

50. $5x(3x - 2) = 0$

51. $(3x - 1)^2 - 16 = 0$

52. $(x + 8)(x - 6) = -24$

53. $4a(a + 1) = 3$

54. $-18n^2 - 15n + 7 = 0$

For Problems 55–70, set up an equation and solve each problem.

55. Find two consecutive integers whose product is 72.

56. Find two consecutive even whole numbers whose product is 224.

57. Find two integers whose product is 105 such that one of the integers is one more than twice the other integer.

58. Find two integers whose product is 104 such that one of the integers is three less than twice the other integer.

59. The perimeter of a rectangle is 32 inches, and the area is 60 square inches. Find the length and width of the rectangle.

60. Suppose that the length of a certain rectangle is two centimeters more than three times its width. If the area of the rectangle is 56 square centimeters, find its length and width.

61. The sum of the squares of two consecutive integers is 85. Find the integers.

62. The sum of the areas of two circles is 65π square feet. The length of a radius of the larger circle is 1 foot less than twice the length of a radius of the smaller circle. Find the length of a radius of each circle.

63. The combined area of a square and a rectangle is 64 square centimeters. The width of the rectangle is 2 centimeters more than the length of a side of the square, and the length of the rectangle is 2 centimeters more than its width. Find the dimensions of the square and the rectangle.

64. The Ortegas have an apple orchard that contains 90 trees. The number of trees in each row is 3 more than twice the number of rows. Find the number of rows and the number of trees per row.

65. The lengths of the three sides of a right triangle are represented by consecutive whole numbers. Find the lengths of the three sides.

66. The area of the floor of the rectangular room shown in Figure 3.21 is 175 square feet. The length of the room is $1\frac{1}{2}$ feet longer than the width. Find the length of the room.

Area = 175 square feet

Figure 3.21

67. Suppose that the length of one leg of a right triangle is 3 inches more than the length of the other leg. If the length of the hypotenuse is 15 inches, find the lengths of the two legs.

68. The lengths of the three sides of a right triangle are represented by consecutive even whole numbers. Find the lengths of the three sides.

69. The area of a triangular sheet of paper is 28 square inches. One side of the triangle is 2 inches more than three times the length of the altitude to that side. Find the length of that side and the altitude to the side.

70. A strip of uniform width is shaded along both sides and both ends of a rectangular poster that measures 12 inches by 16 inches (see Figure 3.22). How wide is the shaded strip if one-half of the poster is shaded?

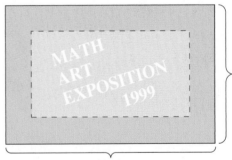

16 inches

Figure 3.22

■ ■ ■ **THOUGHTS INTO WORDS**

71. Discuss the role that factoring plays in solving equations.

72. Explain how you would solve the equation $(x + 6)(x - 4) = 0$ and also how you would solve $(x + 6)(x - 4) = -16$.

73. Explain how you would solve the equation $3(x - 1)(x + 2) = 0$ and also how you would solve the equation $x(x - 1)(x + 2) = 0$.

74. Consider the following two solutions for the equation $(x + 3)(x - 4) = (x + 3)(2x - 1)$.

Solution A

$$(x + 3)(x - 4) = (x + 3)(2x - 1)$$
$$(x + 3)(x - 4) - (x + 3)(2x - 1) = 0$$
$$(x + 3)[x - 4 - (2x - 1)] = 0$$
$$(x + 3)(x - 4 - 2x + 1) = 0$$
$$(x + 3)(-x - 3) = 0$$

$x + 3 = 0$ or $-x - 3 = 0$

$x = -3$ or $-x = 3$

$x = -3$ or $x = -3$

The solution set is $\{-3\}$.

Solution B

$$(x + 3)(x - 4) = (x + 3)(2x - 1)$$
$$x^2 - x - 12 = 2x^2 + 5x - 3$$
$$0 = x^2 + 6x + 9$$
$$0 = (x + 3)^2$$
$$x + 3 = 0$$
$$x = -3$$

The solution set is $\{-3\}$.

Are both approaches correct? Which approach would you use, and why?

Chapter 3   Summary

(3.1) A **term** is an indicated product and may contain any number of factors. The variables involved in a term are called **literal factors**, and the numerical factor is called the **numerical coefficient**. Terms that contain variables with only nonnegative integers as exponents are called **monomials**. The **degree** of a monomial is the sum of the exponents of the literal factors.

A **polynomial** is a monomial or a finite sum (or difference) of monomials. We classify polynomials as follows:

Polynomial with one term $\longrightarrow$ Monomial

Polynomial with two terms $\longrightarrow$ Binomial

Polynomial with three terms $\longrightarrow$ Trinomial

Similar terms, or **like terms**, have the same literal factors. The commutative, associative, and distributive properties provide the basis for rearranging, regrouping, and combining similar terms.

(3.2) The following properties provide the basis for multiplying and dividing monomials.

1. $b^n \cdot b^m = b^{n+m}$

2. $(b^n)^m = b^{mn}$

3. $(ab)^n = a^n b^n$

4. (a) $\dfrac{b^n}{b^m} = b^{n-m}, \quad if\ n > m$

   (b) $\dfrac{b^n}{b^m} = 1, \quad if\ n = m$

(3.3) The commutative and associative properties, the properties of exponents, and the distributive property work together to form a basis for multiplying polynomials. The following can be used as multiplication patterns.

$$(a + b)^2 = a^2 + 2ab + b^2$$
$$(a - b)^2 = a^2 - 2ab + b^2$$
$$(a + b)(a - b) = a^2 - b^2$$
$$(a + b)^3 = a^3 + 3a^2b + 3ab^2 + b^3$$
$$(a - b)^3 = a^3 - 3a^2b + 3ab^2 - b^3$$

(3.4) If a positive integer greater than 1 has no factors that are positive integers other than itself and 1, then it is called a **prime number**. A positive integer greater than 1 that is not a prime number is called a **composite number**.

The indicated product form that contains only prime factors is called the **prime factorization form** of a number.

An expression such as $ax + bx + ay + by$ can be factored as follows:

$$ax + bx + ay + by = x(a + b) + y(a + b)$$
$$= (a + b)(x + y)$$

This is called **factoring by grouping**.

The distributive property in the form $ab + ac = a(b + c)$ is the basis for **factoring out the highest common monomial factor**.

Expressing polynomials in factored form, and then applying the property $ab = 0$ if and only if $a = 0$ or $b = 0$, provides us with another technique for solving equations.

(3.5) The factoring pattern

$$a^2 - b^2 = (a + b)(a - b)$$

is called the **difference of two squares**.

The difference-of-two-squares factoring pattern, along with the property $ab = 0$ if and only if $a = 0$ or $b = 0$, provides us with another technique for solving equations. The factoring patterns

$$a^3 + b^3 = (a + b)(a^2 - ab + b^2) \quad \text{and}$$
$$a^3 - b^3 = (a - b)(a^2 + ab + b^2)$$

are called the **sum and difference of two cubes**.

(3.6) Expressing a trinomial (for which the coefficient of the squared term is 1) as a product of two binomials is based on the relationship

$$(x + a)(x + b) = x^2 + (a + b)x + ab$$

The coefficient of the middle term is the sum of a and b, and the last term is the product of a and b.

If the coefficient of the squared term of a trinomial does not equal 1, then the following relationship holds.

$$(px + r)(qx + s) = (pq)x^2 + (ps + rq)x + rs$$

The two coefficients of x, ps and rq, must have a sum of $(ps) + (rq)$ and a product of $pqrs$. Thus to factor something like $6x^2 + 7x - 3$, we need to find two integers whose product is $6(-3) = -18$ and whose sum is 7. The integers are 9 and -2, and we can factor as follows:

$$6x^2 + 7x - 3 = 6x^2 + 9x - 2x - 3$$

$$= 3x(2x + 3) - 1(2x + 3)$$

$$= (2x + 3)(3x - 1)$$

A **perfect-square trinomial** is the result of squaring a binomial. There are two basic perfect-square trinomial factoring patterns.

$$a^2 + 2ab + b^2 = (a + b)^2$$

$$a^2 - 2ab + b^2 = (a - b)^2$$

(3.7) The factoring techniques we discussed in this chapter, along with the property $ab = 0$ if and only if $a = 0$ or $b = 0$, provide the basis for expanding our repertoire of equation-solving processes.

The ability to solve more types of equations increases our capabilities for problem solving.

Chapter 3   Review Problem Set

For Problems 1–23, perform the indicated operations and simplify each of the following.

1. $(3x - 2) + (4x - 6) + (-2x + 5)$

2. $(8x^2 + 9x - 3) - (5x^2 - 3x - 1)$

3. $(6x^2 - 2x - 1) + (4x^2 + 2x + 5) - (-2x^2 + x - 1)$

4. $(-5x^2y^3)(4x^3y^4)$ **5.** $(-2a^2)(3ab^2)(a^2b^3)$

6. $5a^2(3a^2 - 2a - 1)$ **7.** $(4x - 3y)(6x + 5y)$

8. $(x + 4)(3x^2 - 5x - 1)$ **9.** $(4x^2y^3)^4$

10. $(3x - 2y)^2$ **11.** $(-2x^2y^3z)^3$

12. $\dfrac{-39x^3y^4}{3xy^3}$

13. $[3x - (2x - 3y + 1)] - [2y - (x - 1)]$

14. $(x^2 - 2x - 5)(x^2 + 3x - 7)$

15. $(7 - 3x)(3 + 5x)$ **16.** $-(3ab)(2a^2b^3)^2$

17. $\left(\dfrac{1}{2}ab\right)(8a^3b^2)(-2a^3)$ **18.** $(7x - 9)(x + 4)$

19. $(3x + 2)(2x^2 - 5x + 1)$ **20.** $(3x^{n+1})(2x^{3n-1})$

21. $(2x + 5y)^2$ **22.** $(x - 2)^3$

23. $(2x + 5)^3$

For Problems 24–45, factor each polynomial completely. Indicate any that are not factorable using integers.

24. $x^2 + 3x - 28$ **25.** $2t^2 - 18$

26. $4n^2 + 9$ **27.** $12n^2 - 7n + 1$

28. $x^6 - x^2$ **29.** $x^3 - 6x^2 - 72x$

30. $6a^3b + 4a^2b^2 - 2a^2bc$ **31.** $x^2 - (y - 1)^2$

32. $8x^2 + 12$ **33.** $12x^2 + x - 35$

34. $16n^2 - 40n + 25$ **35.** $4n^2 - 8n$

36. $3w^3 + 18w^2 - 24w$ **37.** $20x^2 + 3xy - 2y^2$

38. $16a^2 - 64a$ **39.** $3x^3 - 15x^2 - 18x$

40. $n^2 - 8n - 128$ **41.** $t^4 - 22t^2 - 75$

42. $35x^2 - 11x - 6$ **43.** $15 - 14x + 3x^2$

44. $64n^3 - 27$ **45.** $16x^3 + 250$

For Problems 46–65, solve each equation.

46. $4x^2 - 36 = 0$

47. $x^2 + 5x - 6 = 0$

48. $49n^2 - 28n + 4 = 0$

49. $(3x - 1)(5x + 2) = 0$

50. $(3x - 4)^2 - 25 = 0$

51. $6a^3 = 54a$

52. $x^5 = x$

53. $-n^2 + 2n + 63 = 0$

54. $7n(7n + 2) = 8$

55. $30w^2 - w - 20 = 0$

56. $5x^4 - 19x^2 - 4 = 0$

57. $9n^2 - 30n + 25 = 0$

58. $n(2n + 4) = 96$

59. $7x^2 + 33x - 10 = 0$

60. $(x + 1)(x + 2) = 42$

61. $x^2 + 12x - x - 12 = 0$

62. $2x^4 + 9x^2 + 4 = 0$

63. $30 - 19x - 5x^2 = 0$

64. $3t^3 - 27t^2 + 24t = 0$

65. $-4n^2 - 39n + 10 = 0$

For Problems 66–75, set up an equation and solve each problem.

66. Find three consecutive integers such that the product of the smallest and the largest is one less than 9 times the middle integer.

67. Find two integers whose sum is 2 and whose product is −48.

68. Find two consecutive odd whole numbers whose product is 195.

69. Two cars leave an intersection at the same time, one traveling north and the other traveling east. Some time later, they are 20 miles apart, and the car going east has traveled 4 miles farther than the other car. How far has each car traveled?

70. The perimeter of a rectangle is 32 meters, and its area is 48 square meters. Find the length and width of the rectangle.

71. A room contains 144 chairs. The number of chairs per row is two less than twice the number of rows. Find the number of rows and the number of chairs per row.

72. The area of a triangle is 39 square feet. The length of one side is 1 foot more than twice the altitude to that side. Find the length of that side and the altitude to the side.

73. A rectangular-shaped pool 20 feet by 30 feet has a sidewalk of uniform width around the pool (see Figure 3.23). The area of the sidewalk is 336 square feet. Find the width of the sidewalk.

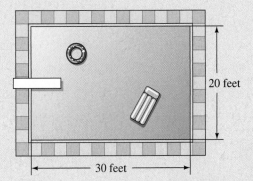

Figure 3.23

74. The sum of the areas of two squares is 89 square centimeters. The length of a side of the larger square is 3 centimeters more than the length of a side of the smaller square. Find the dimensions of each square.

75. The total surface area of a right circular cylinder is 32π square inches. If the altitude of the cylinder is three times the length of a radius, find the altitude of the cylinder.

For Problems 1–8, perform the indicated operations and simplify each expression.

1. $(-3x - 1) + (9x - 2) - (4x + 8)$

2. $(-6xy^2)(8x^3y^2)$

3. $(-3x^2y^4)^3$

4. $(5x - 7)(4x + 9)$

5. $(3n - 2)(2n - 3)$

6. $(x - 4y)^3$

7. $(x + 6)(2x^2 - x - 5)$

8. $\dfrac{-70x^4y^3}{5xy^2}$

For Problems 9–14, factor each expression completely.

9. $6x^2 + 19x - 20$

10. $12x^2 - 3$

11. $64 + t^3$

12. $30x + 4x^2 - 16x^3$

13. $x^2 - xy + 4x - 4y$

14. $24n^2 + 55n - 24$

For Problems 15–22, solve each equation.

15. $x^2 + 8x - 48 = 0$

16. $4n^2 = n$

17. $4x^2 - 12x + 9 = 0$

18. $(n - 2)(n + 7) = -18$

19. $3x^3 + 21x^2 - 54x = 0$

20. $12 + 13x - 35x^2 = 0$

21. $n(3n - 5) = 2$

22. $9x^2 - 36 = 0$

For Problems 23–25, set up an equation and solve each problem.

23. The perimeter of a rectangle is 30 inches, and its area is 54 square inches. Find the length of the longest side of the rectangle.

24. A room contains 105 chairs arranged in rows. The number of rows is one more than twice the number of chairs per row. Find the number of rows.

25. The combined area of a square and a rectangle is 57 square feet. The width of the rectangle is 3 feet more than the length of a side of the square, and the length of the rectangle is 5 feet more than the length of a side of the square. Find the length of the rectangle.

For Problems 1–10, evaluate each algebraic expression for the given values of the variables. Don't forget that in some cases it may be helpful to simplify the algebraic expression before evaluating it.

1. $x^2 - 2xy + y^2$ for $x = -2$ and $y = -4$

2. $-n^2 + 2n - 4$ for $n = -3$

3. $2x^2 - 5x + 6$ for $x = 3$

4. $3(2x - 1) - 2(x + 4) - 4(2x - 7)$ for $x = -1$

5. $-(2n - 1) + 5(2n - 3) - 6(3n + 4)$ for $n = 4$

6. $2(a - 4) - (a - 1) + (3a - 6)$ for $a = -5$

7. $(3x^2 - 4x - 7) - (4x^2 - 7x + 8)$ for $x = -4$

8. $-2(3x - 5y) - 4(x + 2y) + 3(-2x - 3y)$ for $x = 2$ and $y = -3$

9. $5(-x^2 - x + 3) - (2x^2 - x + 6) - 2(x^2 + 4x - 6)$ for $x = 2$

10. $3(x^2 - 4xy + 2y^2) - 2(x^2 - 6xy - y^2)$ for $x = -5$ and $y = -2$

For Problems 11–18, perform the indicated operations and express your answers in simplest form.

11. $4(3x - 2) - 2(4x - 1) - (2x + 5)$

12. $(-6ab^2)(2ab)(-3b^3)$

13. $(5x - 7)(6x + 1)$ **14.** $(-2x - 3)(x + 4)$

15. $(-4a^2b^3)^3$ **16.** $(x + 2)(5x - 6)(x - 2)$

17. $(x - 3)(x^2 - x - 4)$

18. $(x^2 + x + 4)(2x^2 - 3x - 7)$

For Problems 19–38, factor each of the algebraic expressions completely.

19. $7x^2 - 7$ **20.** $4a^2 - 4ab + b^2$

21. $3x^2 - 17x - 56$ **22.** $1 - x^3$

23. $xy - 5x + 2y - 10$ **24.** $3x^2 - 24x + 48$

25. $4n^4 - n^2 - 3$ **26.** $32x^4 + 108x$

27. $4x^2 + 36$ **28.** $6x^2 + 5x - 4$

29. $9x^2 - 30x + 25$ **30.** $2x^2 + 6xy + x + 3y$

31. $8a^3 + 27b^3$ **32.** $x^4 - 16$

33. $10m^4n^2 - 2m^3n^3 - 4m^2n^4$

34. $5x(2y + 7z) - 12(2y + 7z)$

35. $3x^2 - x - 10$ **36.** $25 - 4a^2$

37. $36x^2 + 60x + 25$ **38.** $64y^3 + 1$

For Problems 39–42, solve each equation for the indicated variable.

39. $5x - 2y = 6$ for x **40.** $3x + 4y = 12$ for y

41. $V = 2\pi rh + 2\pi r^2$ for h

42. $\dfrac{1}{R} = \dfrac{1}{R_1} + \dfrac{1}{R_2}$ for R_1

43. Solve $A = P + Prt$ for r, given that $A = \$4997$, $P = \$3800$, and $t = 3$ years.

44. Solve $C = \dfrac{5}{9}(F - 32)$ for C, given that $F = 5°$.

For Problems 45–62, solve each of the equations.

45. $(x - 2)(x + 5) = 8$

46. $(5n - 2)(3n + 7) = 0$

47. $-2(n - 1) + 3(2n + 1) = -11$

48. $x^2 + 7x - 18 = 0$

49. $8x^2 - 8 = 0$

50. $\dfrac{3}{4}(x - 2) - \dfrac{2}{5}(2x - 3) = \dfrac{1}{5}$

51. $0.1(x - 0.1) - 0.4(x + 2) = -5.31$

52. $\dfrac{2x - 1}{2} - \dfrac{5x + 2}{3} = 3$

53. $|3n - 2| = 7$

54. $|2x - 1| = |x + 4|$

55. $0.08(x + 200) = 0.07x + 20$

56. $2x^2 - 12x - 80 = 0$

57. $x^3 = 16x$

58. $x(x + 2) - 3(x + 2) = 0$

59. $-12n^2 + 5n + 2 = 0$

60. $3y(y + 1) = 90$

61. $2x^3 + 6x^2 - 20x = 0$

62. $(3n - 1)(2n + 3) = (n + 4)(6n - 5)$

For Problems 63–70, solve each of the inequalities.

63. $-5(3n + 4) < -2(7n - 1)$

64. $7(x + 1) - 8(x - 2) < 0$

65. $|2x - 1| > 7$

66. $|3x + 7| < 14$

67. $0.09x + 0.1(x + 200) > 77$

68. $\dfrac{2x - 1}{4} - \dfrac{x - 2}{6} \le \dfrac{3}{8}$

69. $-(x - 1) + 2(3x - 1) \ge 2(x + 4) - (x - 1)$

70. $\dfrac{1}{4}(x - 2) + \dfrac{3}{7}(2x - 1) < \dfrac{3}{14}$

For Problems 71–84, solve each problem by setting up and solving an appropriate equation or inequality.

71. Find three consecutive odd integers such that three times the first minus the second is one more than the third.

72. Inez has a collection of 48 coins consisting of nickels, dimes, and quarters. The number of dimes is one less than twice the number of nickels, and the number of quarters is ten greater than the number of dimes. How many coins of each denomination are there in the collection?

73. The sum of the present ages of Joey and his mother is 46 years. In 4 years, Joey will be 3 years less than one-half as old as his mother at that time. Find the present ages of Joey and his mother.

74. The difference of the measures of two supplementary angles is 56°. Find the measure of each angle.

75. Norm invested a certain amount of money at 8% interest and $200 more than that amount at 9%. His total yearly interest was $86. How much did he invest at each rate?

76. Sanchez has a collection of pennies, nickels, and dimes worth $9.35. He has five more nickels than pennies and twice as many dimes as pennies. How may coins of each kind does he have?

77. Sandy starts off with her bicycle at 8 miles per hour. Fifty minutes later, Billie starts riding along the same route at 12 miles per hour. How long will it take Billie to overtake Sandy?

78. How many milliliters of pure acid must be added to 150 milliliters of a 30% solution of acid to obtain a 40% solution?

79. A retailer has some carpet that cost him $18.00 a square yard. If he sells it for $30 a square yard, what is his rate of profit based on the selling price?

80. Brad had scores of 88, 92, 93, and 89 on his first four algebra tests. What score must he obtain on the fifth test to have an average better than 90 for the five tests?

81. Suppose that the area of a square is one-half the area of a triangle. One side of the triangle is 16 inches long, and the altitude to that side is the same length as a side of the square. Find the length of a side of the square.

82. A rectangle is twice as long as it is wide, and its area is 98 square meters. Find the length and width of the rectangle.

83. A room contains 96 chairs. The number of chairs per row is four more than the number of rows. Find the number of rows and the number of chairs per row.

84. One leg of a right triangle is 3 feet longer than the other leg. The hypotenuse is 3 feet longer than the longer leg. Find the lengths of the three sides of the right triangle.

Rational Expressions

Computers often work together to compile large processing jobs. Rational numbers are used to express the rate of the processing speed of a computer.

AP/Wide World Photos

It takes Pat 12 hours to complete a task. After he had been working on this task for 3 hours, he was joined by his brother, Liam, and together they finished the job in 5 hours. How long would it take Liam to do the job by himself? We can use the **fractional equation** $\dfrac{5}{12} + \dfrac{5}{h} = \dfrac{3}{4}$ to determine that Liam could do the entire job by himself in 15 hours.

Rational expressions are to algebra what rational numbers are to arithmetic. Most of the work we will do with rational expressions in this chapter parallels the work you have previously done with arithmetic fractions. The same basic properties we use to explain reducing, adding, subtracting, multiplying, and dividing arithmetic fractions will serve as a basis for our work with rational expressions. The techniques of factoring that we studied in Chapter 3 will also play an important role in our discussions. At the end of this chapter, we will work with some fractional equations that contain rational expressions.

4.1 Simplifying Rational Expressions

We reviewed the basic operations with rational numbers in an informal setting in Chapter 1. In this review, we relied primarily on your knowledge of arithmetic. At this time, we want to become a little more formal with our review so that we can use the work with rational numbers as a basis for operating with rational expressions. We will define a rational expression shortly.

You will recall that any number that can be written in the form $\frac{a}{b}$, where a and b are integers and $b \neq 0$, is called a rational number. The following are examples of rational numbers.

$$\frac{1}{2} \qquad \frac{3}{4} \qquad \frac{15}{7} \qquad \frac{-5}{6} \qquad \frac{7}{-8} \qquad \frac{-12}{-17}$$

Numbers such as $6, -4, 0, 4\frac{1}{2}, 0.7$, and 0.21 are also rational, because we can express them as the indicated quotient of two integers. For example,

$$6 = \frac{6}{1} = \frac{12}{2} = \frac{18}{3} \text{ and so on} \qquad 4\frac{1}{2} = \frac{9}{2}$$

$$-4 = \frac{4}{-1} = \frac{-4}{1} = \frac{8}{-2} \text{ and so on} \qquad 0.7 = \frac{7}{10}$$

$$0 = \frac{0}{1} = \frac{0}{2} = \frac{0}{3} \text{ and so on} \qquad 0.21 = \frac{21}{100}$$

Because a rational number is the quotient of two integers, our previous work with division of integers can help us understand the various forms of rational numbers. If the signs of the numerator and denominator are different, then the rational number is negative. If the signs of the numerator and denominator are the same, then the rational number is positive. The next examples and Property 4.1 show the equivalent forms of rational numbers. Generally, it is preferred to express the denominator of a rational number as a positive integer.

$$\frac{8}{-2} = \frac{-8}{2} = -\frac{8}{2} = -4 \qquad \frac{12}{3} = \frac{-12}{-3} = 4$$

Observe the following general properties.

Property 4.1

1. $\dfrac{-a}{b} = \dfrac{a}{-b} = -\dfrac{a}{b}$,   where $b \neq 0$

2. $\dfrac{-a}{-b} = \dfrac{a}{b}$,   where $b \neq 0$

Therefore, a rational number such as $\dfrac{-2}{5}$ can also be written as $\dfrac{2}{-5}$ or $-\dfrac{2}{5}$.

We use the following property, often referred to as the **fundamental principle of fractions**, to reduce fractions to lowest terms or express fractions in simplest or reduced form.

Property 4.2

If b and k are nonzero integers and a is any integer, then

$$\frac{a \cdot k}{b \cdot k} = \frac{a}{b}$$

Let's apply Properties 4.1 and 4.2 to the following examples.

EXAMPLE 1

Reduce $\dfrac{18}{24}$ to lowest terms.

Solution

$$\frac{18}{24} = \frac{3 \cdot 6}{4 \cdot 6} = \frac{3}{4}$$ ■

EXAMPLE 2

Change $\dfrac{40}{48}$ to simplest form.

Solution

$$\frac{\overset{5}{\cancel{40}}}{\underset{6}{\cancel{48}}} = \frac{5}{6}$$ A common factor of 8 was divided out of both numerator and denominator. ■

EXAMPLE 3

Express $\dfrac{-36}{63}$ in reduced form.

Solution

$$\frac{-36}{63} = -\frac{36}{63} = -\frac{4 \cdot 9}{7 \cdot 9} = -\frac{4}{7}$$ ■

EXAMPLE 4

Reduce $\dfrac{72}{-90}$ to simplest form.

Solution

$$\frac{72}{-90} = -\frac{72}{90} = -\frac{2 \cdot 2 \cdot 2 \cdot 3 \cdot 3}{2 \cdot 3 \cdot 3 \cdot 5} = -\frac{4}{5}$$ ■

Note the different terminology used in Examples 1–4. Regardless of the terminology, keep in mind that the number is not being changed, but the form of the numeral representing the number is being changed. In Example 1, $\frac{18}{24}$ and $\frac{3}{4}$ are equivalent fractions; they name the same number. Also note the use of prime factors in Example 4.

■ Rational Expressions

A **rational expression** is the indicated quotient of two polynomials. The following are examples of rational expressions.

$$\frac{3x^2}{5} \qquad \frac{x-2}{x+3} \qquad \frac{x^2+5x-1}{x^2-9} \qquad \frac{xy^2+x^2y}{xy} \qquad \frac{a^3-3a^2-5a-1}{a^4+a^3+6}$$

Because we must avoid division by zero, no values that create a denominator of zero can be assigned to variables. Thus the rational expression $\frac{x-2}{x+3}$ is meaningful for all values of x except $x = -3$. Rather than making restrictions for each individual expression, we will merely assume that all denominators represent nonzero real numbers.

Property 4.2 $\left(\frac{a \cdot k}{b \cdot k} = \frac{a}{b} \right)$ serves as the basis for simplifying rational expressions, as the next examples illustrate.

E X A M P L E   5

Simplify $\frac{15xy}{25y}$.

Solution

$$\frac{15xy}{25y} = \frac{3 \cdot \cancel{5} \cdot x \cdot \cancel{y}}{\cancel{5} \cdot 5 \cdot \cancel{y}} = \frac{3x}{5}$$

■

E X A M P L E   6

Simplify $\frac{-9}{18x^2y}$.

Solution

$$\frac{-9}{18x^2y} = -\frac{\overset{1}{\cancel{9}}}{\underset{2}{\cancel{18}}x^2y} = -\frac{1}{2x^2y} \qquad \text{A common factor of 9 was divided out of numerator and denominator.}$$

■

E X A M P L E   7

Simplify $\frac{-28a^2b^2}{-63a^2b^3}$.

Solution

$$\frac{-28a^2b^2}{-63a^2b^3} = \frac{4 \cdot \cancel{7} \cdot \cancel{a^2} \cdot \cancel{b^2}}{9 \cdot \cancel{7} \cdot \cancel{a^2} \cdot \underset{b}{\cancel{b^3}}} = \frac{4}{9b}$$

■

The factoring techniques from Chapter 3 can be used to factor numerators and/or denominators so that we can apply the property $\dfrac{a \cdot k}{b \cdot k} = \dfrac{a}{b}$. Examples 8–12 should clarify this process.

EXAMPLE 8

Simplify $\dfrac{x^2 + 4x}{x^2 - 16}$.

Solution

$$\frac{x^2 + 4x}{x^2 - 16} = \frac{x(x+4)}{(x-4)(x+4)} = \frac{x}{x-4}$$ ∎

EXAMPLE 9

Simplify $\dfrac{4a^2 + 12a + 9}{2a + 3}$.

Solution

$$\frac{4a^2 + 12a + 9}{2a + 3} = \frac{(2a+3)(2a+3)}{1(2a+3)} = \frac{2a+3}{1} = 2a + 3$$ ∎

EXAMPLE 10

Simplify $\dfrac{5n^2 + 6n - 8}{10n^2 - 3n - 4}$.

Solution

$$\frac{5n^2 + 6n - 8}{10n^2 - 3n - 4} = \frac{(5n-4)(n+2)}{(5n-4)(2n+1)} = \frac{n+2}{2n+1}$$ ∎

EXAMPLE 11

Simplify $\dfrac{6x^3y - 6xy}{x^3 + 5x^2 + 4x}$.

Solution

$$\frac{6x^3y - 6xy}{x^3 + 5x^2 + 4x} = \frac{6xy(x^2 - 1)}{x(x^2 + 5x + 4)} = \frac{6xy(x+1)(x-1)}{x(x+1)(x+4)} = \frac{6y(x-1)}{x+4}$$ ∎

Note that in Example 11 we left the numerator of the final fraction in factored form. This is often done if expressions other than monomials are involved. Either $\dfrac{6y(x-1)}{x+4}$ or $\dfrac{6xy - 6y}{x+4}$ is an acceptable answer.

Remember that the quotient of any nonzero real number and its opposite is -1. For example, $\dfrac{6}{-6} = -1$ and $\dfrac{-8}{8} = -1$. Likewise, the indicated quotient of any polynomial and its opposite is equal to -1; that is,

$$\frac{a}{-a} = -1 \quad \text{because } a \text{ and } -a \text{ are opposites}$$

$$\frac{a - b}{b - a} = -1 \quad \text{because } a - b \text{ and } b - a \text{ are opposites}$$

$$\frac{x^2 - 4}{4 - x^2} = -1 \quad \text{because } x^2 - 4 \text{ and } 4 - x^2 \text{ are opposites}$$

Example 12 shows how we use this idea when simplifying rational expressions.

EXAMPLE 12   Simplify $\dfrac{6a^2 - 7a + 2}{10a - 15a^2}$.

Solution

$$\frac{6a^2 - 7a + 2}{10a - 15a^2} = \frac{(2a - 1)(3a - 2)}{5a(2 - 3a)} \qquad \frac{3a - 2}{2 - 3a} = -1$$

$$= (-1)\left(\frac{2a - 1}{5a}\right)$$

$$= -\frac{2a - 1}{5a} \quad \text{or} \quad \frac{1 - 2a}{5a} \qquad \blacksquare$$

Problem Set 4.1

For Problems 1–8, express each rational number in reduced form.

1. $\dfrac{27}{36}$

2. $\dfrac{14}{21}$

3. $\dfrac{45}{54}$

4. $\dfrac{-14}{42}$

5. $\dfrac{24}{-60}$

6. $\dfrac{45}{-75}$

7. $\dfrac{-16}{-56}$

8. $\dfrac{-30}{-42}$

For Problems 9–50, simplify each rational expression.

9. $\dfrac{12xy}{42y}$

10. $\dfrac{21xy}{35x}$

11. $\dfrac{18a^2}{45ab}$

12. $\dfrac{48ab}{84b^2}$

13. $\dfrac{-14y^3}{56xy^2}$

14. $\dfrac{-14x^2y^3}{63xy^2}$

15. $\dfrac{54c^2d}{-78cd^2}$

16. $\dfrac{60x^3z}{-64xyz^2}$

17. $\dfrac{-40x^3y}{-24xy^4}$

18. $\dfrac{-30x^2y^2z^2}{-35xz^3}$

19. $\dfrac{x^2 - 4}{x^2 + 2x}$

20. $\dfrac{xy + y^2}{x^2 - y^2}$

21. $\dfrac{18x + 12}{12x - 6}$

22. $\dfrac{20x + 50}{15x - 30}$

23. $\dfrac{a^2 + 7a + 10}{a^2 - 7a - 18}$

24. $\dfrac{a^2 + 4a - 32}{3a^2 + 26a + 16}$

25. $\dfrac{2n^2 + n - 21}{10n^2 + 33n - 7}$

26. $\dfrac{4n^2 - 15n - 4}{7n^2 - 30n + 8}$

27. $\dfrac{5x^2 + 7}{10x}$

28. $\dfrac{12x^2 + 11x - 15}{20x^2 - 23x + 6}$

29. $\dfrac{6x^2 + x - 15}{8x^2 - 10x - 3}$

30. $\dfrac{4x^2 + 8x}{x^3 + 8}$

31. $\dfrac{3x^2 - 12x}{x^3 - 64}$

32. $\dfrac{x^2 - 14x + 49}{6x^2 - 37x - 35}$

33. $\dfrac{3x^2 + 17x - 6}{9x^2 - 6x + 1}$

34. $\dfrac{9y^2 - 1}{3y^2 + 11y - 4}$

35. $\dfrac{2x^3 + 3x^2 - 14x}{x^2y + 7xy - 18y}$

36. $\dfrac{3x^3 + 12x}{9x^2 + 18x}$

37. $\dfrac{5y^2 + 22y + 8}{25y^2 - 4}$

38. $\dfrac{16x^3y + 24x^2y^2 - 16xy^3}{24x^2y + 12xy^2 - 12y^3}$

39. $\dfrac{15x^3 - 15x^2}{5x^3 + 5x}$

40. $\dfrac{5n^2 + 18n - 8}{3n^2 + 13n + 4}$

41. $\dfrac{4x^2y + 8xy^2 - 12y^3}{18x^3y - 12x^2y^2 - 6xy^3}$

42. $\dfrac{3 + x - 2x^2}{2 + x - x^2}$

43. $\dfrac{3n^2 + 16n - 12}{7n^2 + 44n + 12}$

44. $\dfrac{x^4 - 2x^2 - 15}{2x^4 + 9x^2 + 9}$

45. $\dfrac{8 + 18x - 5x^2}{10 + 31x + 15x^2}$

46. $\dfrac{6x^4 - 11x^2 + 4}{2x^4 + 17x^2 - 9}$

47. $\dfrac{27x^4 - x}{6x^3 + 10x^2 - 4x}$

48. $\dfrac{64x^4 + 27x}{12x^3 - 27x^2 - 27x}$

49. $\dfrac{-40x^3 + 24x^2 + 16x}{20x^3 + 28x^2 + 8x}$

50. $\dfrac{-6x^3 - 21x^2 + 12x}{-18x^3 - 42x^2 + 120x}$

For Problems 51–58, simplify each rational expression. You will need to use factoring by grouping.

51. $\dfrac{xy + ay + bx + ab}{xy + ay + cx + ac}$

52. $\dfrac{xy + 2y + 3x + 6}{xy + 2y + 4x + 8}$

53. $\dfrac{ax - 3x + 2ay - 6y}{2ax - 6x + ay - 3y}$

54. $\dfrac{x^2 - 2x + ax - 2a}{x^2 - 2x + 3ax - 6a}$

55. $\dfrac{5x^2 + 5x + 3x + 3}{5x^2 + 3x - 30x - 18}$

56. $\dfrac{x^2 + 3x + 4x + 12}{2x^2 + 6x - x - 3}$

57. $\dfrac{2st - 30 - 12s + 5t}{3st - 6 - 18s + t}$

58. $\dfrac{nr - 6 - 3n + 2r}{nr + 10 + 2r + 5n}$

For Problems 59–68, simplify each rational expression. You may want to refer to Example 12 of this section.

59. $\dfrac{5x - 7}{7 - 5x}$

60. $\dfrac{4a - 9}{9 - 4a}$

61. $\dfrac{n^2 - 49}{7 - n}$

62. $\dfrac{9 - y}{y^2 - 81}$

63. $\dfrac{2y - 2xy}{x^2y - y}$

64. $\dfrac{3x - x^2}{x^2 - 9}$

65. $\dfrac{2x^3 - 8x}{4x - x^3}$

66. $\dfrac{x^2 - (y - 1)^2}{(y - 1)^2 - x^2}$

67. $\dfrac{n^2 - 5n - 24}{40 + 3n - n^2}$

68. $\dfrac{x^2 + 2x - 24}{20 - x - x^2}$

■ ■ ■ **THOUGHTS INTO WORDS**

69. Compare the concept of a rational number in arithmetic to the concept of a rational expression in algebra.

70. What role does factoring play in the simplifying of rational expressions?

71. Why is the rational expression $\dfrac{x + 3}{x^2 - 4}$ undefined for $x = 2$ and $x = -2$ but defined for $x = -3$?

72. How would you convince someone that $\dfrac{x - 4}{4 - x} = -1$ for all real numbers except 4?

4.2   Multiplying and Dividing Rational Expressions

We define multiplication of rational numbers in common fraction form as follows:

Definition 4.1

If a, b, c, and d are integers, and b and d are not equal to zero, then

$$\frac{a}{b} \cdot \frac{c}{d} = \frac{a \cdot c}{b \cdot d} = \frac{ac}{bd}$$

To multiply rational numbers in common fraction form, we merely **multiply numerators and multiply denominators**, as the following examples demonstrate. (The steps in the dashed boxes are usually done mentally.)

$$\frac{2}{3} \cdot \frac{4}{5} = \frac{2 \cdot 4}{3 \cdot 5} = \frac{8}{15}$$

$$\frac{-3}{4} \cdot \frac{5}{7} = \frac{-3 \cdot 5}{4 \cdot 7} = \frac{-15}{28} = -\frac{15}{28}$$

$$-\frac{5}{6} \cdot \frac{13}{3} = \frac{-5}{6} \cdot \frac{13}{3} = \frac{-5 \cdot 13}{6 \cdot 3} = \frac{-65}{18} = -\frac{65}{18}$$

We also agree, when multiplying rational numbers, to express the final product in reduced form. The following examples show some different formats used to multiply and simplify rational numbers.

$$\frac{3}{4} \cdot \frac{4}{7} = \frac{3 \cdot \cancel{4}}{\cancel{4} \cdot 7} = \frac{3}{7}$$

$$\frac{\overset{1}{\cancel{8}}}{\underset{1}{\cancel{9}}} \cdot \frac{\overset{3}{\cancel{27}}}{\underset{4}{\cancel{32}}} = \frac{3}{4}$$ A common factor of 9 was divided out of 9 and 27, and a common factor of 8 was divided out of 8 and 32.

$$\left(-\frac{28}{25}\right)\left(-\frac{65}{78}\right) = \frac{2 \cdot 2 \cdot 7 \cdot \cancel{5} \cdot \cancel{13}}{\cancel{5} \cdot 5 \cdot 2 \cdot 3 \cdot \cancel{13}} = \frac{14}{15}.$$ We should recognize that a negative times a negative is positive. Also, note the use of prime factors to help us recognize common factors.

Multiplication of rational expressions follows the same basic pattern as multiplication of rational numbers in common fraction form. That is to say, we multiply numerators and multiply denominators and express the final product in simplified or reduced form. Let's consider some examples.

$$\frac{3x}{4y} \cdot \frac{8y^2}{9x} = \frac{\overset{2}{\cancel{3}} \cdot \cancel{8} \cdot \cancel{x} \cdot \overset{y}{\cancel{y^2}}}{\cancel{4} \cdot \underset{3}{\cancel{9}} \cdot \cancel{x} \cdot \cancel{y}} = \frac{2y}{3}$$

Note that we use the commutative property of multiplication to rearrange the factors in a form that allows us to identify common factors of the numerator and denominator.

$$\frac{-4a}{6a^2b^2} \cdot \frac{9ab}{12a^2} = -\frac{\cancel{4} \cdot \overset{3}{\cancel{9}} \cdot \cancel{a^2} \cdot \cancel{b}}{\underset{2}{\cancel{6}} \cdot \underset{3}{\cancel{12}} \cdot \underset{a^2}{\cancel{a^4}} \cdot \underset{b}{\cancel{b^2}}} = -\frac{1}{2a^2b}$$

$$\frac{12x^2y}{-18xy} \cdot \frac{-24xy^2}{56y^3} = \frac{\overset{2}{\cancel{12}} \cdot \overset{3}{\cancel{24}} \cdot \overset{x^2}{\cancel{x^3}} \cdot \cancel{y^3}}{\underset{3}{\cancel{18}} \cdot \underset{7}{\cancel{56}} \cdot \cancel{x} \cdot \underset{y}{\cancel{y^4}}} = \frac{2x^2}{7y}$$

You should recognize that the first fraction is equivalent to $-\dfrac{12x^2y}{18xy}$ and the second to $-\dfrac{24xy^2}{56y^3}$; thus the product is positive.

If the rational expressions contain polynomials (other than monomials) that are factorable, then our work may take on the following format.

EXAMPLE 1

Multiply and simplify $\dfrac{y}{x^2 - 4} \cdot \dfrac{x + 2}{y^2}$.

Solution

$$\frac{y}{x^2 - 4} \cdot \frac{x + 2}{y^2} = \frac{\cancel{y}(\cancel{x + 2})}{\underset{y}{\cancel{y^2}}(\cancel{x + 2})(x - 2)} = \frac{1}{y(x - 2)}$$ ∎

In Example 1, note that we combined the steps of multiplying numerators and denominators and factoring the polynomials. Also note that we left the final answer in factored form. Either $\dfrac{1}{y(x - 2)}$ or $\dfrac{1}{xy - 2y}$ would be an acceptable answer.

EXAMPLE 2

Multiply and simplify $\dfrac{x^2 - x}{x + 5} \cdot \dfrac{x^2 + 5x + 4}{x^4 - x^2}$.

Solution

$$\frac{x^2 - x}{x + 5} \cdot \frac{x^2 + 5x + 4}{x^4 - x^2} = \frac{x(x - 1)}{x + 5} \cdot \frac{(x + 1)(x + 4)}{x^2(x - 1)(x + 1)}$$

$$= \frac{\cancel{x}(\cancel{x - 1})(\cancel{x + 1})(x + 4)}{(x + 5)(\cancel{x^2})(\cancel{x - 1})(\cancel{x + 1})} = \frac{x + 4}{x(x + 5)}$$ ∎

EXAMPLE 3 Multiply and simplify $\dfrac{6n^2 + 7n - 5}{n^2 + 2n - 24} \cdot \dfrac{4n^2 + 21n - 18}{12n^2 + 11n - 15}.$

Solution

$$\dfrac{6n^2 + 7n - 5}{n^2 + 2n - 24} \cdot \dfrac{4n^2 + 21n - 18}{12n^2 + 11n - 15}$$

$$= \dfrac{(3n + 5)(2n - 1)(4n - 3)(n + 6)}{(n + 6)(n - 4)(3n + 5)(4n - 3)} = \dfrac{2n - 1}{n - 4}$$ ∎

■ Dividing Rational Expressions

We define division of rational numbers in common fraction form as follows:

Definition 4.2

> If a, b, c, and d are integers, and b, c, and d are not equal to zero, then
>
> $$\dfrac{a}{b} \div \dfrac{c}{d} = \dfrac{a}{b} \cdot \dfrac{d}{c} = \dfrac{ad}{bc}$$

Definition 4.2 states that to divide two rational numbers in fraction form, we **invert the divisor and multiply**. We call the numbers $\dfrac{c}{d}$ and $\dfrac{d}{c}$ "reciprocals" or "multiplicative inverses" of each other, because their product is 1. Thus we can describe division by saying "to divide by a fraction, multiply by its reciprocal." The following examples demonstrate the use of Definition 4.2.

$$\dfrac{7}{8} \div \dfrac{5}{6} = \dfrac{7}{8} \cdot \dfrac{\overset{3}{\cancel{6}}}{5} = \dfrac{21}{20}, \qquad \dfrac{-5}{9} \div \dfrac{15}{18} = -\dfrac{\cancel{5}}{9} \cdot \dfrac{\overset{2}{\cancel{18}}}{\cancel{15}} = -\dfrac{2}{3}$$

$$\dfrac{14}{-19} \div \dfrac{21}{-38} = \left(-\dfrac{14}{19}\right) \div \left(-\dfrac{21}{38}\right) = \left(-\dfrac{\cancel{14}}{\cancel{19}}\right)\left(-\dfrac{\overset{2}{\cancel{38}}}{\cancel{21}}\right) = \dfrac{4}{3}$$

We define division of algebraic rational expressions in the same way that we define division of rational numbers. That is, the quotient of two rational expressions is the product we obtain when we multiply the first expression by the reciprocal of the second. Consider the following examples.

EXAMPLE 4 Divide and simplify $\dfrac{16x^2y}{24xy^3} \div \dfrac{9xy}{8x^2y^2}.$

Solution

$$\dfrac{16x^2y}{24xy^3} \div \dfrac{9xy}{8x^2y^2} = \dfrac{16x^2y}{24xy^3} \cdot \dfrac{8x^2y^2}{9xy} = \dfrac{16 \cdot 8 \cdot \overset{x^2}{\cancel{x^4}} \cdot \cancel{y^3}}{\underset{3}{\cancel{24}} \cdot 9 \cdot \cancel{x^2} \cdot \underset{y}{\cancel{y^4}}} = \dfrac{16x^2}{27y}$$ ∎

EXAMPLE 5

Divide and simplify $\dfrac{3a^2 + 12}{3a^2 - 15a} \div \dfrac{a^4 - 16}{a^2 - 3a - 10}$.

Solution

$$\frac{3a^2 + 12}{3a^2 - 15a} \div \frac{a^4 - 16}{a^2 - 3a - 10} = \frac{3a^2 + 12}{3a^2 - 15a} \cdot \frac{a^2 - 3a - 10}{a^4 - 16}$$

$$= \frac{3(a^2 + 4)}{3a(a - 5)} \cdot \frac{(a - 5)(a + 2)}{(a^2 + 4)(a + 2)(a - 2)}$$

$$= \frac{\overset{1}{\cancel{3}}(\cancel{a^2 + 4})(\cancel{a - 5})(\cancel{a + 2})}{\underset{1}{\cancel{3}}a(\cancel{a - 5})(\cancel{a^2 + 4})(\cancel{a + 2})(a - 2)}$$

$$= \frac{1}{a(a - 2)} \qquad \blacksquare$$

EXAMPLE 6

Divide and simplify $\dfrac{28t^3 - 51t^2 - 27t}{49t^2 + 42t + 9} \div (4t - 9)$.

Solution

$$\frac{28t^3 - 51t^2 - 27t}{49t^2 + 42t + 9} \div \frac{4t - 9}{1} = \frac{28t^3 - 51t^2 - 27t}{49t^2 + 42t + 9} \cdot \frac{1}{4t - 9}$$

$$= \frac{t(7t + 3)(4t - 9)}{(7t + 3)(7t + 3)} \cdot \frac{1}{(4t - 9)}$$

$$= \frac{t(\cancel{7t + 3})(\cancel{4t - 9})}{(\cancel{7t + 3})(7t + 3)(\cancel{4t - 9})}$$

$$= \frac{t}{7t + 3} \qquad \blacksquare$$

In a problem such as Example 6, it may be helpful to write the divisor with a denominator of 1. Thus we write $4t - 9$ as $\dfrac{4t - 9}{1}$; its reciprocal is obviously $\dfrac{1}{4t - 9}$.

Let's consider one final example that involves both multiplication and division.

EXAMPLE 7

Perform the indicated operations and simplify.

$$\frac{x^2 + 5x}{3x^2 - 4x - 20} \cdot \frac{x^2y + y}{2x^2 + 11x + 5} \div \frac{xy^2}{6x^2 - 17x - 10}$$

Solution

$$\frac{x^2 + 5x}{3x^2 - 4x - 20} \cdot \frac{x^2y + y}{2x^2 + 11x + 5} \div \frac{xy^2}{6x^2 - 17x - 10}$$

$$= \frac{x^2 + 5x}{3x^2 - 4x - 20} \cdot \frac{x^2y + y}{2x^2 + 11x + 5} \cdot \frac{6x^2 - 17x - 10}{xy^2}$$

$$= \frac{x(x + 5)}{(3x - 10)(x + 2)} \cdot \frac{y(x^2 + 1)}{(2x + 1)(x + 5)} \cdot \frac{(2x + 1)(3x - 10)}{xy^2}$$

$$= \frac{\cancel{x}(\cancel{x + 5})(\cancel{y})(x^2 + 1)(\cancel{2x + 1})(\cancel{3x - 10})}{(\cancel{3x - 10})(x + 2)(\cancel{2x + 1})(\cancel{x + 5})(\cancel{x})(\cancel{y^2})} = \frac{x^2 + 1}{y(x + 2)}$$ ∎

Problem Set 4.2

For Problems 1–12, perform the indicated operations involving rational numbers. Express final answers in reduced form.

1. $\dfrac{7}{12} \cdot \dfrac{6}{35}$

2. $\dfrac{5}{8} \cdot \dfrac{12}{20}$

3. $\dfrac{-4}{9} \cdot \dfrac{18}{30}$

4. $\dfrac{-6}{9} \cdot \dfrac{36}{48}$

5. $\dfrac{3}{-8} \cdot \dfrac{-6}{12}$

6. $\dfrac{-12}{16} \cdot \dfrac{18}{-32}$

7. $\left(-\dfrac{5}{7}\right) \div \dfrac{6}{7}$

8. $\left(-\dfrac{5}{9}\right) \div \dfrac{10}{3}$

9. $\dfrac{-9}{5} \div \dfrac{27}{10}$

10. $\dfrac{4}{7} \div \dfrac{16}{-21}$

11. $\dfrac{4}{9} \cdot \dfrac{6}{11} \div \dfrac{4}{15}$

12. $\dfrac{2}{3} \cdot \dfrac{6}{7} \div \dfrac{8}{3}$

For Problems 13–50, perform the indicated operations involving rational expressions. Express final answers in simplest form.

13. $\dfrac{6xy}{9y^4} \cdot \dfrac{30x^3y}{-48x}$

14. $\dfrac{-14xy^4}{18y^2} \cdot \dfrac{24x^2y^3}{35y^2}$

15. $\dfrac{5a^2b^2}{11ab} \cdot \dfrac{22a^3}{15ab^2}$

16. $\dfrac{10a^2}{5b^2} \cdot \dfrac{15b^3}{2a^4}$

17. $\dfrac{5xy}{8y^2} \cdot \dfrac{18x^2y}{15}$

18. $\dfrac{4x^2}{5y^2} \cdot \dfrac{15xy}{24x^2y^2}$

19. $\dfrac{5x^4}{12x^2y^3} \div \dfrac{9}{5xy}$

20. $\dfrac{7x^2y}{9xy^3} \div \dfrac{3x^4}{2x^2y^2}$

21. $\dfrac{9a^2c}{12bc^2} \div \dfrac{21ab}{14c^3}$

22. $\dfrac{3ab^3}{4c} \div \dfrac{21ac}{12bc^3}$

23. $\dfrac{9x^2y^3}{14x} \cdot \dfrac{21y}{15xy^2} \cdot \dfrac{10x}{12y^3}$

24. $\dfrac{5xy}{7a} \cdot \dfrac{14a^2}{15x} \cdot \dfrac{3a}{8y}$

25. $\dfrac{3x + 6}{5y} \cdot \dfrac{x^2 + 4}{x^2 + 10x + 16}$

26. $\dfrac{5xy}{x + 6} \cdot \dfrac{x^2 - 36}{x^2 - 6x}$

27. $\dfrac{5a^2 + 20a}{a^3 - 2a^2} \cdot \dfrac{a^2 - a - 12}{a^2 - 16}$

28. $\dfrac{2a^2 + 6}{a^2 - a} \cdot \dfrac{a^3 - a^2}{8a - 4}$

29. $\dfrac{3n^2 + 15n - 18}{3n^2 + 10n - 48} \cdot \dfrac{6n^2 - n - 40}{4n^2 + 6n - 10}$

30. $\dfrac{6n^2 + 11n - 10}{3n^2 + 19n - 14} \cdot \dfrac{2n^2 + 6n - 56}{2n^2 - 3n - 20}$

31. $\dfrac{9y^2}{x^2 + 12x + 36} \div \dfrac{12y}{x^2 + 6x}$

32. $\dfrac{7xy}{x^2 - 4x + 4} \div \dfrac{14y}{x^2 - 4}$

33. $\dfrac{x^2 - 4xy + 4y^2}{7xy^2} \div \dfrac{4x^2 - 3xy - 10y^2}{20x^2y + 25xy^2}$

34. $\dfrac{x^2 + 5xy - 6y^2}{xy^2 - y^3} \cdot \dfrac{2x^2 + 15xy + 18y^2}{xy + 4y^2}$

35. $\dfrac{5 - 14n - 3n^2}{1 - 2n - 3n^2} \cdot \dfrac{9 + 7n - 2n^2}{27 - 15n + 2n^2}$

36. $\dfrac{6 - n - 2n^2}{12 - 11n + 2n^2} \cdot \dfrac{24 - 26n + 5n^2}{2 + 3n + n^2}$

37. $\dfrac{3x^4 + 2x^2 - 1}{3x^4 + 14x^2 - 5} \cdot \dfrac{x^4 - 2x^2 - 35}{x^4 - 17x^2 + 70}$

38. $\dfrac{2x^4 + x^2 - 3}{2x^4 + 5x^2 + 2} \cdot \dfrac{3x^4 + 10x^2 + 8}{3x^4 + x^2 - 4}$

39. $\dfrac{3x^2 - 20x + 25}{2x^2 - 7x - 15} \div \dfrac{9x^2 - 3x - 20}{12x^2 + 28x + 15}$

40. $\dfrac{21t^2 + t - 2}{2t^2 - 17t - 9} \div \dfrac{12t^2 - 5t - 3}{8t^2 - 2t - 3}$

41. $\dfrac{10t^3 + 25t}{20t + 10} \cdot \dfrac{2t^2 - t - 1}{t^5 - t}$

42. $\dfrac{t^4 - 81}{t^2 - 6t + 9} \cdot \dfrac{6t^2 - 11t - 21}{5t^2 + 8t - 21}$

43. $\dfrac{4t^2 + t - 5}{t^3 - t^2} \cdot \dfrac{t^4 + 6t^3}{16t^2 + 40t + 25}$

44. $\dfrac{9n^2 - 12n + 4}{n^2 - 4n - 32} \cdot \dfrac{n^2 + 4n}{3n^3 - 2n^2}$

45. $\dfrac{nr + 3n + 2r + 6}{nr + 3n - 3r - 9} \cdot \dfrac{n^2 - 9}{n^3 - 4n}$

46. $\dfrac{xy + xc + ay + ac}{xy - 2xc + ay - 2ac} \cdot \dfrac{2x^3 - 8x}{12x^3 + 20x^2 - 8x}$

47. $\dfrac{x^2 - x}{4y} \cdot \dfrac{10xy^2}{2x - 2} \div \dfrac{3x^2 + 3x}{15x^2y^2}$

48. $\dfrac{4xy^2}{7x} \cdot \dfrac{14x^3y}{12y} \div \dfrac{7y}{9x^3}$

49. $\dfrac{a^2 - 4ab + 4b^2}{6a^2 - 4ab} \cdot \dfrac{3a^2 + 5ab - 2b^2}{6a^2 + ab - b^2} \div \dfrac{a^2 - 4b^2}{8a + 4b}$

50. $\dfrac{2x^2 + 3x}{2x^3 - 10x^2} \cdot \dfrac{x^2 - 8x + 15}{3x^3 - 27x} \div \dfrac{14x + 21}{x^2 - 6x - 27}$

■ ■ ■ **THOUGHTS INTO WORDS**

51. Explain in your own words how to divide two rational expressions.

52. Suppose that your friend missed class the day the material in this section was discussed. How could you draw on her background in arithmetic to explain to her how to multiply and divide rational expressions?

53. Give a step-by-step description of how to do the following multiplication problem.

$$\dfrac{x^2 + 5x + 6}{x^2 - 2x - 8} \cdot \dfrac{x^2 - 16}{16 - x^2}$$

4.3 Adding and Subtracting Rational Expressions

We can define addition and subtraction of rational numbers as follows:

Definition 4.3

If a, b, and c are integers, and b is not zero, then

$$\dfrac{a}{b} + \dfrac{c}{b} = \dfrac{a + c}{b} \qquad \text{Addition}$$

$$\dfrac{a}{b} - \dfrac{c}{b} = \dfrac{a - c}{b} \qquad \text{Subtraction}$$

We can add or subtract rational numbers with a common denominator by adding or subtracting the numerators and placing the result over the common denominator. The following examples illustrate Definition 4.3.

$$\frac{2}{9} + \frac{3}{9} = \frac{2+3}{9} = \frac{5}{9}$$

$$\frac{7}{8} - \frac{3}{8} = \frac{7-3}{8} = \frac{4}{8} = \frac{1}{2} \qquad \text{Don't forget to reduce!}$$

$$\frac{4}{6} + \frac{-5}{6} = \frac{4+(-5)}{6} = \frac{-1}{6} = -\frac{1}{6}$$

$$\frac{7}{10} + \frac{4}{-10} = \frac{7}{10} + \frac{-4}{10} = \frac{7+(-4)}{10} = \frac{3}{10}$$

We use this same *common denominator* approach when adding or subtracting rational expressions, as in these next examples.

$$\frac{3}{x} + \frac{9}{x} = \frac{3+9}{x} = \frac{12}{x}$$

$$\frac{8}{x-2} - \frac{3}{x-2} = \frac{8-3}{x-2} = \frac{5}{x-2}$$

$$\frac{9}{4y} + \frac{5}{4y} = \frac{9+5}{4y} = \frac{14}{4y} = \frac{7}{2y} \qquad \text{Don't forget to simplify the final answer!}$$

$$\frac{n^2}{n-1} - \frac{1}{n-1} = \frac{n^2-1}{n-1} = \frac{(n+1)(n-1)}{n-1} = n+1$$

$$\frac{6a^2}{2a+1} + \frac{13a+5}{2a+1} = \frac{6a^2+13a+5}{2a+1} = \frac{(2a+1)(3a+5)}{2a+1} = 3a+5$$

In each of the previous examples that involve rational expressions, we should technically restrict the variables to exclude division by zero. For example, $\frac{3}{x} + \frac{9}{x} = \frac{12}{x}$ is true for all real number values for x, except $x = 0$. Likewise, $\frac{8}{x-2} - \frac{3}{x-2} = \frac{5}{x-2}$ as long as x does not equal 2. Rather than taking the time and space to write down restrictions for each problem, we will merely assume that such restrictions exist.

If rational numbers that do not have a common denominator are to be added or subtracted, then we apply the fundamental principle of fractions $\left(\frac{a}{b} = \frac{ak}{bk} \right)$ **to obtain equivalent fractions with a common denominator.** Equivalent fractions are fractions such as $\dfrac{1}{2}$ and $\dfrac{2}{4}$ that name the same number. Consider the following example.

$$\frac{1}{2} + \frac{1}{3} = \frac{3}{6} + \frac{2}{6} = \frac{3+2}{6} = \frac{5}{6}$$

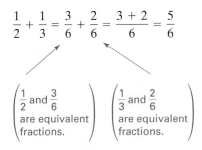

$\left(\begin{array}{l}\dfrac{1}{2}\text{ and }\dfrac{3}{6}\\ \text{are equivalent}\\ \text{fractions.}\end{array}\right)$ $\left(\begin{array}{l}\dfrac{1}{3}\text{ and }\dfrac{2}{6}\\ \text{are equivalent}\\ \text{fractions.}\end{array}\right)$

Note that we chose 6 as our common denominator, and 6 is the **least common multiple** of the original denominators 2 and 3. (The least common multiple of a set of whole numbers is the smallest nonzero whole number divisible by each of the numbers.) In general, we use the least common multiple of the denominators of the fractions to be added or subtracted as a **least common denominator** (LCD).

A least common denominator may be found by inspection or by using the prime-factored forms of the numbers. Let's consider some examples and use each of these techniques.

EXAMPLE   1

Subtract $\dfrac{5}{6} - \dfrac{3}{8}$.

Solution

By inspection, we can see that the LCD is 24. Thus both fractions can be changed to equivalent fractions, each with a denominator of 24.

$$\frac{5}{6} - \frac{3}{8} = \left(\frac{5}{6}\right)\left(\frac{4}{4}\right) - \left(\frac{3}{8}\right)\left(\frac{3}{3}\right) = \frac{20}{24} - \frac{9}{24} = \frac{11}{24}$$

$\quad\quad\quad\quad\quad\quad\uparrow\quad\quad\quad\quad\uparrow$
$\quad\quad\quad\quad$ Form of 1   Form of 1   ∎

In Example 1, note that the fundamental principle of fractions, $\dfrac{a}{b} = \dfrac{a \cdot k}{b \cdot k}$, can be written as $\dfrac{a}{b} = \left(\dfrac{a}{b}\right)\left(\dfrac{k}{k}\right)$. This latter form emphasizes the fact that 1 is the multiplication identity element.

EXAMPLE 2

Perform the indicated operations: $\dfrac{3}{5} + \dfrac{1}{6} - \dfrac{13}{15}$

Solution

Again by inspection, we can determine that the LCD is 30. Thus we can proceed as follows:

$$\frac{3}{5} + \frac{1}{6} - \frac{13}{15} = \left(\frac{3}{5}\right)\left(\frac{6}{6}\right) + \left(\frac{1}{6}\right)\left(\frac{5}{5}\right) - \left(\frac{13}{15}\right)\left(\frac{2}{2}\right)$$

$$= \frac{18}{30} + \frac{5}{30} - \frac{26}{30} = \frac{18 + 5 - 26}{30}$$

$$= \frac{-3}{30} = -\frac{1}{10} \qquad \text{Don't forget to reduce!} \qquad \blacksquare$$

EXAMPLE 3

Add $\dfrac{7}{18} + \dfrac{11}{24}$.

Solution

Let's use the prime-factored forms of the denominators to help find the LCD.

$$18 = 2 \cdot 3 \cdot 3 \qquad 24 = 2 \cdot 2 \cdot 2 \cdot 3$$

The LCD must contain three factors of 2 because 24 contains three 2s. The LCD must also contain two factors of 3 because 18 has two 3s. Thus the LCD = $2 \cdot 2 \cdot 2 \cdot 3 \cdot 3 = 72$. Now we can proceed as usual.

$$\frac{7}{18} + \frac{11}{24} = \left(\frac{7}{18}\right)\left(\frac{4}{4}\right) + \left(\frac{11}{24}\right)\left(\frac{3}{3}\right) = \frac{28}{72} + \frac{33}{72} = \frac{61}{72} \qquad \blacksquare$$

To add and subtract rational expressions with different denominators, follow the same basic routine that you follow when you add or subtract rational numbers with different denominators. Study the following examples carefully and note the similarity to our previous work with rational numbers.

EXAMPLE 4

Add $\dfrac{x + 2}{4} + \dfrac{3x + 1}{3}$.

Solution

By inspection, we see that the LCD is 12.

$$\frac{x + 2}{4} + \frac{3x + 1}{3} = \left(\frac{x + 2}{4}\right)\left(\frac{3}{3}\right) + \left(\frac{3x + 1}{3}\right)\left(\frac{4}{4}\right)$$

$$= \frac{3(x + 2)}{12} + \frac{4(3x + 1)}{12}$$

$$= \frac{3(x + 2) + 4(3x + 1)}{12}$$

$$= \frac{3x + 6 + 12x + 4}{12}$$

$$= \frac{15x + 10}{12} \qquad\blacksquare$$

Note the final result in Example 4. The numerator, $15x + 10$, could be factored as $5(3x + 2)$. However, because this produces no common factors with the denominator, the fraction cannot be simplified. Thus the final answer can be left as $\frac{15x + 10}{12}$. It would also be acceptable to express it as $\frac{5(3x + 2)}{12}$.

EXAMPLE 5

Subtract $\dfrac{a - 2}{2} - \dfrac{a - 6}{6}$.

Solution

By inspection, we see that the LCD is 6.

$$\frac{a - 2}{2} - \frac{a - 6}{6} = \left(\frac{a - 2}{2}\right)\left(\frac{3}{3}\right) - \frac{a - 6}{6}$$

$$= \frac{3(a - 2)}{6} - \frac{a - 6}{6}$$

$$= \frac{3(a - 2) - (a - 6)}{6} \qquad \text{Be careful with this sign as you move to the next step!}$$

$$= \frac{3a - 6 - a + 6}{6}$$

$$= \frac{2a}{6} = \frac{a}{3} \qquad \text{Don't forget to simplify.} \qquad\blacksquare$$

EXAMPLE 6

Perform the indicated operations: $\dfrac{x + 3}{10} + \dfrac{2x + 1}{15} - \dfrac{x - 2}{18}$.

Solution

If you cannot determine the LCD by inspection, then use the prime-factored forms of the denominators.

$$10 = 2 \cdot 5 \qquad 15 = 3 \cdot 5 \qquad 18 = 2 \cdot 3 \cdot 3$$

The LCD must contain one factor of 2, two factors of 3, and one factor of 5. Thus the LCD is $2 \cdot 3 \cdot 3 \cdot 5 = 90$.

$$\frac{x + 3}{10} + \frac{2x + 1}{15} - \frac{x - 2}{18} = \left(\frac{x + 3}{10}\right)\left(\frac{9}{9}\right) + \left(\frac{2x + 1}{15}\right)\left(\frac{6}{6}\right) - \left(\frac{x - 2}{18}\right)\left(\frac{5}{5}\right)$$

$$= \frac{9(x + 3)}{90} + \frac{6(2x + 1)}{90} - \frac{5(x - 2)}{90}$$

$$= \frac{9(x + 3) + 6(2x + 1) - 5(x - 2)}{90}$$

$$= \frac{9x + 27 + 12x + 6 - 5x + 10}{90}$$

$$= \frac{16x + 43}{90} \qquad \blacksquare$$

A denominator that contains variables does not create any serious difficulties; our approach remains basically the same.

E X A M P L E   7

Add $\dfrac{3}{2x} + \dfrac{5}{3y}$.

Solution

Using an LCD of $6xy$, we can proceed as follows:

$$\frac{3}{2x} + \frac{5}{3y} = \left(\frac{3}{2x}\right)\left(\frac{3y}{3y}\right) + \left(\frac{5}{3y}\right)\left(\frac{2x}{2x}\right)$$

$$= \frac{9y}{6xy} + \frac{10x}{6xy}$$

$$= \frac{9y + 10x}{6xy} \qquad \blacksquare$$

E X A M P L E   8

Subtract $\dfrac{7}{12ab} - \dfrac{11}{15a^2}$.

Solution

We can prime-factor the numerical coefficients of the denominators to help find the LCD.

$$\left.\begin{array}{l} 12ab = 2 \cdot 2 \cdot 3 \cdot a \cdot b \\ 15a^2 = 3 \cdot 5 \cdot a^2 \end{array}\right\} \longrightarrow \text{LCD} = 2 \cdot 2 \cdot 3 \cdot 5 \cdot a^2 \cdot b = 60a^2b$$

$$\frac{7}{12ab} - \frac{11}{15a^2} = \left(\frac{7}{12ab}\right)\left(\frac{5a}{5a}\right) - \left(\frac{11}{15a^2}\right)\left(\frac{4b}{4b}\right)$$

$$= \frac{35a}{60a^2b} - \frac{44b}{60a^2b}$$

$$= \frac{35a - 44b}{60a^2b}$$
■

EXAMPLE 9   Add $\dfrac{x}{x - 3} + \dfrac{4}{x}$.

Solution

By inspection, the LCD is $x(x - 3)$.

$$\frac{x}{x - 3} + \frac{4}{x} = \left(\frac{x}{x - 3}\right)\left(\frac{x}{x}\right) + \left(\frac{4}{x}\right)\left(\frac{x - 3}{x - 3}\right)$$

$$= \frac{x^2}{x(x - 3)} + \frac{4(x - 3)}{x(x - 3)}$$

$$= \frac{x^2 + 4(x - 3)}{x(x - 3)}$$

$$= \frac{x^2 + 4x - 12}{x(x - 3)} \qquad \text{or} \qquad \frac{(x + 6)(x - 2)}{x(x - 3)}$$
■

EXAMPLE 10   Subtract $\dfrac{2x}{x + 1} - 3$.

Solution

$$\frac{2x}{x + 1} - 3 = \frac{2x}{x + 1} - 3\left(\frac{x + 1}{x + 1}\right)$$

$$= \frac{2x}{x + 1} - \frac{3(x + 1)}{x + 1}$$

$$= \frac{2x - 3(x + 1)}{x + 1}$$

$$= \frac{2x - 3x - 3}{x + 1}$$

$$= \frac{-x - 3}{x + 1}$$
■

Problem Set 4.3

For Problems 1–12, perform the indicated operations involving rational numbers. Be sure to express your answers in reduced form.

1. $\dfrac{1}{4} + \dfrac{5}{6}$

2. $\dfrac{3}{5} + \dfrac{1}{6}$

3. $\dfrac{7}{8} - \dfrac{3}{5}$

4. $\dfrac{7}{9} - \dfrac{1}{6}$

5. $\dfrac{6}{5} + \dfrac{1}{-4}$

6. $\dfrac{7}{8} + \dfrac{5}{-12}$

7. $\dfrac{8}{15} + \dfrac{3}{25}$

8. $\dfrac{5}{9} - \dfrac{11}{12}$

9. $\dfrac{1}{5} + \dfrac{5}{6} - \dfrac{7}{15}$

10. $\dfrac{2}{3} - \dfrac{7}{8} + \dfrac{1}{4}$

11. $\dfrac{1}{3} - \dfrac{1}{4} - \dfrac{3}{14}$

12. $\dfrac{5}{6} - \dfrac{7}{9} - \dfrac{3}{10}$

For Problems 13–66, add or subtract the rational expressions as indicated. Be sure to express your answers in simplest form.

13. $\dfrac{2x}{x - 1} + \dfrac{4}{x - 1}$

14. $\dfrac{3x}{2x + 1} - \dfrac{5}{2x + 1}$

15. $\dfrac{4a}{a + 2} + \dfrac{8}{a + 2}$

16. $\dfrac{6a}{a - 3} - \dfrac{18}{a - 3}$

17. $\dfrac{3(y - 2)}{7y} + \dfrac{4(y - 1)}{7y}$

18. $\dfrac{2x - 1}{4x^2} + \dfrac{3(x - 2)}{4x^2}$

19. $\dfrac{x - 1}{2} + \dfrac{x + 3}{3}$

20. $\dfrac{x - 2}{4} + \dfrac{x + 6}{5}$

21. $\dfrac{2a - 1}{4} + \dfrac{3a + 2}{6}$

22. $\dfrac{a - 4}{6} + \dfrac{4a - 1}{8}$

23. $\dfrac{n + 2}{6} - \dfrac{n - 4}{9}$

24. $\dfrac{2n + 1}{9} - \dfrac{n + 3}{12}$

25. $\dfrac{3x - 1}{3} - \dfrac{5x + 2}{5}$

26. $\dfrac{4x - 3}{6} - \dfrac{8x - 2}{12}$

27. $\dfrac{x - 2}{5} - \dfrac{x + 3}{6} + \dfrac{x + 1}{15}$

28. $\dfrac{x + 1}{4} + \dfrac{x - 3}{6} - \dfrac{x - 2}{8}$

29. $\dfrac{3}{8x} + \dfrac{7}{10x}$

30. $\dfrac{5}{6x} - \dfrac{3}{10x}$

31. $\dfrac{5}{7x} - \dfrac{11}{4y}$

32. $\dfrac{5}{12x} - \dfrac{9}{8y}$

33. $\dfrac{4}{3x} + \dfrac{5}{4y} - 1$

34. $\dfrac{7}{3x} - \dfrac{8}{7y} - 2$

35. $\dfrac{7}{10x^2} + \dfrac{11}{15x}$

36. $\dfrac{7}{12a^2} - \dfrac{5}{16a}$

37. $\dfrac{10}{7n} - \dfrac{12}{4n^2}$

38. $\dfrac{6}{8n^2} - \dfrac{3}{5n}$

39. $\dfrac{3}{n^2} - \dfrac{2}{5n} + \dfrac{4}{3}$

40. $\dfrac{1}{n^2} + \dfrac{3}{4n} - \dfrac{5}{6}$

41. $\dfrac{3}{x} - \dfrac{5}{3x^2} - \dfrac{7}{6x}$

42. $\dfrac{7}{3x^2} - \dfrac{9}{4x} - \dfrac{5}{2x}$

43. $\dfrac{6}{5t^2} - \dfrac{4}{7t^3} + \dfrac{9}{5t^3}$

44. $\dfrac{5}{7t} + \dfrac{3}{4t^2} + \dfrac{1}{14t}$

45. $\dfrac{5b}{24a^2} - \dfrac{11a}{32b}$

46. $\dfrac{9}{14x^2y} - \dfrac{4x}{7y^2}$

47. $\dfrac{7}{9xy^3} - \dfrac{4}{3x} + \dfrac{5}{2y^2}$

48. $\dfrac{7}{16a^2b} + \dfrac{3a}{20b^2}$

49. $\dfrac{2x}{x - 1} + \dfrac{3}{x}$

50. $\dfrac{3x}{x - 4} - \dfrac{2}{x}$

51. $\dfrac{a - 2}{a} - \dfrac{3}{a + 4}$

52. $\dfrac{a + 1}{a} - \dfrac{2}{a + 1}$

53. $\dfrac{-3}{4n + 5} - \dfrac{8}{3n + 5}$

54. $\dfrac{-2}{n - 6} - \dfrac{6}{2n + 3}$

55. $\dfrac{-1}{x + 4} + \dfrac{4}{7x - 1}$

56. $\dfrac{-3}{4x + 3} + \dfrac{5}{2x - 5}$

57. $\dfrac{7}{3x - 5} - \dfrac{5}{2x + 7}$

58. $\dfrac{5}{x - 1} - \dfrac{3}{2x - 3}$

59. $\dfrac{5}{3x - 2} + \dfrac{6}{4x + 5}$

60. $\dfrac{3}{2x + 1} + \dfrac{2}{3x + 4}$

61. $\dfrac{3x}{2x + 5} + 1$

62. $2 + \dfrac{4x}{3x - 1}$

63. $\dfrac{4x}{x - 5} - 3$

64. $\dfrac{7x}{x + 4} - 2$

65. $-1 - \dfrac{3}{2x+1}$ **66.** $-2 - \dfrac{5}{4x-3}$

67. Recall that the indicated quotient of a polynomial and its opposite is -1. For example, $\dfrac{x-2}{2-x}$ simplifies to -1. Keep this idea in mind as you add or subtract the following rational expressions.

(a) $\dfrac{1}{x-1} - \dfrac{x}{x-1}$ (b) $\dfrac{3}{2x-3} - \dfrac{2x}{2x-3}$

(c) $\dfrac{4}{x-4} - \dfrac{x}{x-4} + 1$ (d) $-1 + \dfrac{2}{x-2} - \dfrac{x}{x-2}$

68. Consider the addition problem $\dfrac{8}{x-2} + \dfrac{5}{2-x}$. Note that the denominators are opposites of each other.

If the property $\dfrac{a}{-b} = -\dfrac{a}{b}$ is applied to the second fraction, we have $\dfrac{5}{2-x} = -\dfrac{5}{x-2}$. Thus we proceed as follows:

$$\dfrac{8}{x-2} + \dfrac{5}{2-x} = \dfrac{8}{x-2} - \dfrac{5}{x-2} = \dfrac{8-5}{x-2} = \dfrac{3}{x-2}$$

Use this approach to do the following problems.

(a) $\dfrac{7}{x-1} + \dfrac{2}{1-x}$ (b) $\dfrac{5}{2x-1} + \dfrac{8}{1-2x}$

(c) $\dfrac{4}{a-3} - \dfrac{1}{3-a}$ (d) $\dfrac{10}{a-9} - \dfrac{5}{9-a}$

(e) $\dfrac{x^2}{x-1} - \dfrac{2x-3}{1-x}$ (f) $\dfrac{x^2}{x-4} - \dfrac{3x-28}{4-x}$

■■■ **THOUGHTS INTO WORDS**

69. What is the difference between the concept of least common multiple and the concept of least common denominator?

70. A classmate tells you that she finds the least common multiple of two counting numbers by listing the multiples of each number and then choosing the smallest number that appears in both lists. Is this a correct procedure? What is the weakness of this procedure?

71. For which real numbers does $\dfrac{x}{x-3} + \dfrac{4}{x}$ equal $\dfrac{(x+6)(x-2)}{x(x-3)}$? Explain your answer.

72. Suppose that your friend does an addition problem as follows:

$$\dfrac{5}{8} + \dfrac{7}{12} = \dfrac{5(12)+8(7)}{8(12)} = \dfrac{60+56}{96} = \dfrac{116}{96} = \dfrac{29}{24}$$

Is this answer correct? If not, what advice would you offer your friend?

4.4 More on Rational Expressions and Complex Fractions

In this section, we expand our work with adding and subtracting rational expressions, and we discuss the process of simplifying complex fractions. Before we begin, however, this seems like an appropriate time to offer a bit of advice regarding your study of algebra. Success in algebra depends on having a good understanding of the concepts as well as on being able to perform the various computations. As for the computational work, you should adopt a carefully organized format that shows as many steps as you need in order to minimize the chances of making careless errors. Don't be eager to find shortcuts for certain computations before you have a thorough understanding of the steps involved in the process. This advice is especially appropriate at the beginning of this section.

Study Examples 1–4 very carefully. Note that the same basic procedure is followed in solving each problem:

Step 1   Factor the denominators.

Step 2   Find the LCD.

Step 3   Change each fraction to an equivalent fraction that has the LCD as its denominator.

Step 4   Combine the numerators and place over the LCD.

Step 5   Simplify by performing the addition or subtraction.

Step 6   Look for ways to reduce the resulting fraction.

E X A M P L E   1   Add $\dfrac{8}{x^2 - 4x} + \dfrac{2}{x}$.

Solution

$$\frac{8}{x^2 - 4x} + \frac{2}{x} = \frac{8}{x(x - 4)} + \frac{2}{x}$$ Factor the denominators.

The LCD is $x(x - 4)$.   Find the LCD.

$$= \frac{8}{x(x - 4)} + \left(\frac{2}{x}\right)\left(\frac{x - 4}{x - 4}\right)$$ Change each fraction to an equivalent fraction that has the LCD as its denominator.

$$= \frac{8 + 2(x - 4)}{x(x - 4)}$$ Combine numerators and place over the LCD.

$$= \frac{8 + 2x - 8}{x(x - 4)}$$ Simplify performing the addition or subtraction.

$$= \frac{2x}{x(x - 4)}$$

$$= \frac{2}{x - 4}$$ Reduce.   ∎

EXAMPLE 2

Subtract $\dfrac{a}{a^2 - 4} - \dfrac{3}{a + 2}$.

Solution

$$\frac{a}{a^2 - 4} - \frac{3}{a + 2} = \frac{a}{(a + 2)(a - 2)} - \frac{3}{a + 2}$$ Factor the denominators.

The LCD is $(a + 2)(a - 2)$.   Find the LCD.

$$= \frac{a}{(a + 2)(a - 2)} - \left(\frac{3}{a + 2}\right)\left(\frac{a - 2}{a - 2}\right)$$ Change each fraction to an equivalent fraction that has the LCD as its denominator.

$$= \frac{a - 3(a - 2)}{(a + 2)(a - 2)}$$ Combine numerators and place over the LCD.

$$= \frac{a - 3a + 6}{(a + 2)(a - 2)}$$ Simplify performing the addition or subtraction.

$$= \frac{-2a + 6}{(a + 2)(a - 2)} \qquad \text{or} \qquad \frac{-2(a - 3)}{(a + 2)(a - 2)}$$ ∎

EXAMPLE 3

Add $\dfrac{3n}{n^2 + 6n + 5} + \dfrac{4}{n^2 - 7n - 8}$.

Solution

$$\frac{3n}{n^2 + 6n + 5} + \frac{4}{n^2 - 7n - 8}$$

$$= \frac{3n}{(n + 5)(n + 1)} + \frac{4}{(n - 8)(n + 1)}$$ Factor the denominators.

The LCD is $(n + 5)(n + 1)(n - 8)$.   Find the LCD.

$$= \left(\frac{3n}{(n + 5)(n + 1)}\right)\left(\frac{n - 8}{n - 8}\right)$$

$$+ \left(\frac{4}{(n - 8)(n + 1)}\right)\left(\frac{n + 5}{n + 5}\right)$$ Change each fraction to an equivalent fraction that has the LCD as its denominator.

$$= \frac{3n(n - 8) + 4(n + 5)}{(n + 5)(n + 1)(n - 8)}$$ Combine numerators and place over the LCD.

$$= \frac{3n^2 - 24n + 4n + 20}{(n + 5)(n + 1)(n - 8)}$$ Simplify performing the addition or subtraction.

$$= \frac{3n^2 - 20n + 20}{(n + 5)(n + 1)(n - 8)}$$ ∎

Perform the indicated operations.

$$\frac{2x^2}{x^4 - 1} + \frac{x}{x^2 - 1} - \frac{1}{x - 1}$$

Solution

$$\frac{2x^2}{x^4 - 1} + \frac{x}{x^2 - 1} - \frac{1}{x - 1}$$

$$= \frac{2x^2}{(x^2 + 1)(x + 1)(x - 1)} + \frac{x}{(x + 1)(x - 1)} - \frac{1}{x - 1}$$ Factor the denominators.

The LCD is $(x^2 + 1)(x + 1)(x - 1)$. Find the LCD.

$$= \frac{2x^2}{(x^2 + 1)(x + 1)(x - 1)}$$ Change each fraction to an equivalent fraction that has the LCD as its denominator.

$$+ \left(\frac{x}{(x + 1)(x - 1)} \right) \left(\frac{x^2 + 1}{x^2 + 1} \right)$$

$$- \left(\frac{1}{x - 1} \right) \frac{(x^2 + 1)(x + 1)}{(x^2 + 1)(x + 1)}$$

$$= \frac{2x^2 + x(x^2 + 1) - (x^2 + 1)(x + 1)}{(x^2 + 1)(x + 1)(x - 1)}$$ Combine numerators and place over the LCD.

$$= \frac{2x^2 + x^3 + x - x^3 - x^2 - x - 1}{(x^2 + 1)(x + 1)(x - 1)}$$ Simplify performing the addition or subtraction.

$$= \frac{x^2 - 1}{(x^2 + 1)(x + 1)(x - 1)}$$

$$= \frac{(x + 1)(x - 1)}{(x^2 + 1)(x + 1)(x - 1)}$$

$$= \frac{1}{x^2 + 1}$$ Reduce. ■

■ Complex Fractions

Complex fractions are fractional forms that contain rational numbers or rational expressions in the numerators and/or denominators. The following are examples of complex fractions.

$$\frac{\dfrac{4}{x}}{\dfrac{2}{xy}} \qquad \frac{\dfrac{1}{2} + \dfrac{3}{4}}{\dfrac{5}{6} - \dfrac{3}{8}} \qquad \frac{\dfrac{3}{x} + \dfrac{2}{y}}{\dfrac{5}{x} - \dfrac{6}{y^2}} \qquad \frac{\dfrac{1}{x} + \dfrac{1}{y}}{2} \qquad \frac{-3}{\dfrac{2}{x} - \dfrac{3}{y}}$$

It is often necessary to **simplify** a complex fraction. We will take each of these five examples and examine some techniques for simplifying complex fractions.

E X A M P L E   5   Simplify $\dfrac{\dfrac{4}{x}}{\dfrac{2}{xy}}$.

Solution

This type of problem is a simple division problem.

$$\frac{\dfrac{4}{x}}{\dfrac{2}{xy}} = \frac{4}{x} \div \frac{2}{xy}$$

$$= \frac{\overset{2}{\cancel{4}}}{\cancel{x}} \cdot \frac{xy}{2} = 2y$$

■

E X A M P L E   6   Simplify $\dfrac{\dfrac{1}{2} + \dfrac{3}{4}}{\dfrac{5}{6} - \dfrac{3}{8}}$.

Let's look at two possible ways to simplify such a problem.

Solution A

Here we will simplify the numerator by performing the addition and simplify the denominator by performing the subtraction. Then the problem is a simple division problem like Example 5.

$$\frac{\dfrac{1}{2} + \dfrac{3}{4}}{\dfrac{5}{6} - \dfrac{3}{8}} = \frac{\dfrac{2}{4} + \dfrac{3}{4}}{\dfrac{20}{24} - \dfrac{9}{24}}$$

$$= \frac{\dfrac{5}{4}}{\dfrac{11}{24}} = \frac{5}{\cancel{4}} \cdot \frac{\overset{6}{\cancel{24}}}{11}$$

$$= \frac{30}{11}$$

Solution B

Here we find the LCD of all four denominators (2, 4, 6, and 8). The LCD is 24. Use this LCD to multiply the entire complex fraction by a form of 1, specifically $\dfrac{24}{24}$.

$$\frac{\dfrac{1}{2} + \dfrac{3}{4}}{\dfrac{5}{6} - \dfrac{3}{8}} = \left(\frac{24}{24}\right) \left(\frac{\dfrac{1}{2} + \dfrac{3}{4}}{\dfrac{5}{6} - \dfrac{3}{8}}\right)$$

$$= \frac{24\left(\dfrac{1}{2} + \dfrac{3}{4}\right)}{24\left(\dfrac{5}{6} - \dfrac{3}{8}\right)}$$

$$= \frac{24\left(\dfrac{1}{2}\right) + 24\left(\dfrac{3}{4}\right)}{24\left(\dfrac{5}{6}\right) - 24\left(\dfrac{3}{8}\right)}$$

$$= \frac{12 + 18}{20 - 9} = \frac{30}{11}$$

∎

E X A M P L E   7   Simplify $\dfrac{\dfrac{3}{x} + \dfrac{2}{y}}{\dfrac{5}{x} - \dfrac{6}{y^2}}$.

Solution A

Simplify the numerator and the denominator. Then the problem becomes a division problem.

$$\frac{\dfrac{3}{x} + \dfrac{2}{y}}{\dfrac{5}{x} - \dfrac{6}{y^2}} = \frac{\left(\dfrac{3}{x}\right)\left(\dfrac{y}{y}\right) + \left(\dfrac{2}{y}\right)\left(\dfrac{x}{x}\right)}{\left(\dfrac{5}{x}\right)\left(\dfrac{y^2}{y^2}\right) - \left(\dfrac{6}{y^2}\right)\left(\dfrac{x}{x}\right)}$$

$$= \frac{\dfrac{3y}{xy} + \dfrac{2x}{xy}}{\dfrac{5y^2}{xy^2} - \dfrac{6x}{xy^2}}$$

$$= \frac{\dfrac{3y + 2x}{xy}}{\dfrac{5y^2 - 6x}{xy^2}}$$

$$= \frac{3y + 2x}{xy} \div \frac{5y^2 - 6x}{xy^2}$$

$$= \frac{3y + 2x}{\cancel{xy}} \cdot \frac{\overset{y}{\cancel{xy^2}}}{5y^2 - 6x}$$

$$= \frac{y(3y + 2x)}{5y^2 - 6x}$$

Solution B

Here we find the LCD of all four denominators (x, y, x, and y^2). The LCD is xy^2. Use this LCD to multiply the entire complex fraction by a form of 1, specifically $\dfrac{xy^2}{xy^2}$.

$$\frac{\dfrac{3}{x} + \dfrac{2}{y}}{\dfrac{5}{x} - \dfrac{6}{y^2}} = \left(\frac{xy^2}{xy^2}\right)\left(\frac{\dfrac{3}{x} + \dfrac{2}{y}}{\dfrac{5}{x} - \dfrac{6}{y^2}}\right)$$

$$= \frac{xy^2\left(\dfrac{3}{x} + \dfrac{2}{y}\right)}{xy^2\left(\dfrac{5}{x} - \dfrac{6}{y^2}\right)}$$

$$= \frac{xy^2\left(\dfrac{3}{x}\right) + xy^2\left(\dfrac{2}{y}\right)}{xy^2\left(\dfrac{5}{x}\right) - xy^2\left(\dfrac{6}{y^2}\right)}$$

$$= \frac{3y^2 + 2xy}{5y^2 - 6x} \quad \text{or} \quad \frac{y(3y + 2x)}{5y^2 - 6x}$$ ∎

Certainly either approach (Solution A or Solution B) will work with problems such as Examples 6 and 7. Examine Solution B in both examples carefully. This approach works effectively with complex fractions where the LCD of all the denominators is easy to find. (Don't be misled by the length of Solution B for Example 6; we were especially careful to show every step.)

EXAMPLE 8   Simplify $\dfrac{\dfrac{1}{x} + \dfrac{1}{y}}{2}$.

Solution

The number 2 can be written as $\dfrac{2}{1}$; thus the LCD of all three denominators (x, y, and 1) is xy. Therefore, let's multiply the entire complex fraction by a form of 1, specifically $\dfrac{xy}{xy}$.

$$\left(\dfrac{\dfrac{1}{x} + \dfrac{1}{y}}{\dfrac{2}{1}}\right)\left(\dfrac{xy}{xy}\right) = \dfrac{xy\left(\dfrac{1}{x}\right) + xy\left(\dfrac{1}{y}\right)}{2xy}$$

$$= \dfrac{y + x}{2xy} \qquad \blacksquare$$

EXAMPLE 9   Simplify $\dfrac{-3}{\dfrac{2}{x} - \dfrac{3}{y}}$.

Solution

$$\left(\dfrac{\dfrac{-3}{1}}{\dfrac{2}{x} - \dfrac{3}{y}}\right)\left(\dfrac{xy}{xy}\right) = \dfrac{-3(xy)}{xy\left(\dfrac{2}{x}\right) - xy\left(\dfrac{3}{y}\right)}$$

$$= \dfrac{-3xy}{2y - 3x} \qquad \blacksquare$$

Let's conclude this section with an example that has a complex fraction as part of an algebraic expression.

EXAMPLE 10   Simplify $1 - \dfrac{n}{1 - \dfrac{1}{n}}$.

Solution

First simplify the complex fraction $\dfrac{n}{1 - \dfrac{1}{n}}$ by multiplying by $\dfrac{n}{n}$.

$$\left(\dfrac{n}{1 - \dfrac{1}{n}}\right)\left(\dfrac{n}{n}\right) = \dfrac{n^2}{n - 1}$$

Now we can perform the subtraction.

$$1 - \frac{n^2}{n-1} = \left(\frac{n-1}{n-1}\right)\left(\frac{1}{1}\right) - \frac{n^2}{n-1}$$

$$= \frac{n-1}{n-1} - \frac{n^2}{n-1}$$

$$= \frac{n-1-n^2}{n-1} \quad \text{or} \quad \frac{-n^2+n-1}{n-1} \quad \blacksquare$$

Problem Set 4.4

For Problems 1–40, perform the indicated operations, and express your answers in simplest form.

1. $\dfrac{2x}{x^2+4x} + \dfrac{5}{x}$

2. $\dfrac{3x}{x^2-6x} + \dfrac{4}{x}$

3. $\dfrac{4}{x^2+7x} - \dfrac{1}{x}$

4. $\dfrac{-10}{x^2-9x} - \dfrac{2}{x}$

5. $\dfrac{x}{x^2-1} + \dfrac{5}{x+1}$

6. $\dfrac{2x}{x^2-16} + \dfrac{7}{x-4}$

7. $\dfrac{6a+4}{a^2-1} - \dfrac{5}{a-1}$

8. $\dfrac{4a-4}{a^2-4} - \dfrac{3}{a+2}$

9. $\dfrac{2n}{n^2-25} - \dfrac{3}{4n+20}$

10. $\dfrac{3n}{n^2-36} - \dfrac{2}{5n+30}$

11. $\dfrac{5}{x} - \dfrac{5x-30}{x^2+6x} + \dfrac{x}{x+6}$

12. $\dfrac{3}{x+1} + \dfrac{x+5}{x^2-1} - \dfrac{3}{x-1}$

13. $\dfrac{3}{x^2+9x+14} + \dfrac{5}{2x^2+15x+7}$

14. $\dfrac{6}{x^2+11x+24} + \dfrac{4}{3x^2+13x+12}$

15. $\dfrac{1}{a^2-3a-10} - \dfrac{4}{a^2+4a-45}$

16. $\dfrac{6}{a^2-3a-54} - \dfrac{10}{a^2+5a-6}$

17. $\dfrac{3a}{8a^2-2a-3} + \dfrac{1}{4a^2+13a-12}$

18. $\dfrac{2a}{6a^2+13a-5} + \dfrac{a}{2a^2+a-10}$

19. $\dfrac{5}{x^2+3} - \dfrac{2}{x^2+4x-21}$

20. $\dfrac{7}{x^2+1} - \dfrac{3}{x^2+7x-60}$

21. $\dfrac{3x}{x^2-6x+9} - \dfrac{2}{x-3}$

22. $\dfrac{3}{x+4} + \dfrac{2x}{x^2+8x+16}$

23. $\dfrac{5}{x^2-1} + \dfrac{9}{x^2+2x+1}$

24. $\dfrac{6}{x^2-9} - \dfrac{9}{x^2-6x+9}$

25. $\dfrac{2}{y^2+6y-16} - \dfrac{4}{y+8} - \dfrac{3}{y-2}$

26. $\dfrac{7}{y-6} - \dfrac{10}{y+12} + \dfrac{4}{y^2+6y-72}$

27. $x - \dfrac{x^2}{x-2} + \dfrac{3}{x^2-4}$

28. $x + \dfrac{5}{x^2-25} - \dfrac{x^2}{x+5}$

29. $\dfrac{x+3}{x+10} + \dfrac{4x-3}{x^2+8x-20} + \dfrac{x-1}{x-2}$

30. $\dfrac{2x-1}{x+3} + \dfrac{x+4}{x-6} + \dfrac{3x-1}{x^2-3x-18}$

31. $\dfrac{n}{n-6} + \dfrac{n+3}{n+8} + \dfrac{12n+26}{n^2+2n-48}$

32. $\dfrac{n-1}{n+4} + \dfrac{n}{n+6} + \dfrac{2n+18}{n^2+10n+24}$

33. $\dfrac{4x-3}{2x^2+x-1} - \dfrac{2x+7}{3x^2+x-2} - \dfrac{3}{3x-2}$

34. $\dfrac{2x+5}{x^2+3x-18} - \dfrac{3x-1}{x^2+4x-12} + \dfrac{5}{x-2}$

35. $\dfrac{n}{n^2+1} + \dfrac{n^2+3n}{n^4-1} - \dfrac{1}{n-1}$

36. $\dfrac{2n^2}{n^4 - 16} - \dfrac{n}{n^2 - 4} + \dfrac{1}{n + 2}$

37. $\dfrac{15x^2 - 10}{5x^2 - 7x + 2} - \dfrac{3x + 4}{x - 1} - \dfrac{2}{5x - 2}$

38. $\dfrac{32x + 9}{12x^2 + x - 6} - \dfrac{3}{4x + 3} - \dfrac{x + 5}{3x - 2}$

39. $\dfrac{t + 3}{3t - 1} + \dfrac{8t^2 + 8t + 2}{3t^2 - 7t + 2} - \dfrac{2t + 3}{t - 2}$

40. $\dfrac{t - 3}{2t + 1} + \dfrac{2t^2 + 19t - 46}{2t^2 - 9t - 5} - \dfrac{t + 4}{t - 5}$

For Problems 41–64, simplify each complex fraction.

41. $\dfrac{\dfrac{1}{2} - \dfrac{1}{4}}{\dfrac{5}{8} + \dfrac{3}{4}}$

42. $\dfrac{\dfrac{3}{8} + \dfrac{3}{4}}{\dfrac{5}{8} - \dfrac{7}{12}}$

43. $\dfrac{\dfrac{3}{28} - \dfrac{5}{14}}{\dfrac{5}{7} + \dfrac{1}{4}}$

44. $\dfrac{\dfrac{5}{9} + \dfrac{7}{36}}{\dfrac{3}{18} - \dfrac{5}{12}}$

45. $\dfrac{\dfrac{5}{6y}}{\dfrac{10}{3xy}}$

46. $\dfrac{\dfrac{9}{8xy^2}}{\dfrac{5}{4x^2}}$

47. $\dfrac{\dfrac{3}{x} - \dfrac{2}{y}}{\dfrac{4}{y} - \dfrac{7}{xy}}$

48. $\dfrac{\dfrac{9}{x} + \dfrac{7}{x^2}}{\dfrac{5}{y} + \dfrac{3}{y^2}}$

49. $\dfrac{\dfrac{6}{a} - \dfrac{5}{b^2}}{\dfrac{12}{a^2} + \dfrac{2}{b}}$

50. $\dfrac{\dfrac{4}{ab} - \dfrac{3}{b^2}}{\dfrac{1}{a} + \dfrac{3}{b}}$

51. $\dfrac{\dfrac{2}{x} - 3}{\dfrac{3}{y} + 4}$

52. $\dfrac{1 + \dfrac{3}{x}}{1 - \dfrac{6}{x}}$

53. $\dfrac{3 + \dfrac{2}{n + 4}}{5 - \dfrac{1}{n + 4}}$

54. $\dfrac{4 + \dfrac{6}{n - 1}}{7 - \dfrac{4}{n - 1}}$

55. $\dfrac{5 - \dfrac{2}{n - 3}}{4 - \dfrac{1}{n - 3}}$

56. $\dfrac{\dfrac{3}{n - 5} - 2}{1 - \dfrac{4}{n - 5}}$

57. $\dfrac{\dfrac{-1}{y - 2} + \dfrac{5}{x}}{\dfrac{3}{x} - \dfrac{4}{xy - 2x}}$

58. $\dfrac{\dfrac{-2}{x} - \dfrac{4}{x + 2}}{\dfrac{3}{x^2 + 2x} + \dfrac{3}{x}}$

59. $\dfrac{\dfrac{2}{x - 3} - \dfrac{3}{x + 3}}{\dfrac{5}{x^2 - 9} - \dfrac{2}{x - 3}}$

60. $\dfrac{\dfrac{2}{x - y} + \dfrac{3}{x + y}}{\dfrac{5}{x + y} - \dfrac{1}{x^2 - y^2}}$

61. $\dfrac{\dfrac{3a}{2 - \dfrac{1}{a}} - 1}{}$

62. $\dfrac{\dfrac{a}{\dfrac{1}{a} + 4} + 1}{}$

63. $2 - \dfrac{x}{3 - \dfrac{2}{x}}$

64. $1 + \dfrac{x}{1 + \dfrac{1}{x}}$

■■■ THOUGHTS INTO WORDS

65. Which of the two techniques presented in the text would you use to simplify $\dfrac{\dfrac{1}{4} + \dfrac{1}{3}}{\dfrac{3}{4} - \dfrac{1}{6}}$? Which technique would you use to simplify $\dfrac{\dfrac{3}{8} - \dfrac{5}{7}}{\dfrac{7}{9} + \dfrac{6}{25}}$? Explain your choice for each problem.

66. Give a step-by-step description of how to do the following addition problem.

$$\dfrac{3x + 4}{8} + \dfrac{5x - 2}{12}$$

4.5 Dividing Polynomials

In Chapter 3, we saw how the property $\dfrac{b^n}{b^m} = b^{n-m}$, along with our knowledge of dividing integers, is used to divide monomials. For example,

$$\frac{12x^3}{3x} = 4x^2 \qquad \frac{-36x^4y^5}{4xy^2} = -9x^3y^3$$

In Section 4.3, we used $\dfrac{a}{b} + \dfrac{c}{b} = \dfrac{a+c}{b}$ and $\dfrac{a}{b} - \dfrac{c}{b} = \dfrac{a-c}{b}$ as the basis for adding and subtracting rational expressions. These same equalities, viewed as $\dfrac{a+b}{c} = \dfrac{a}{c} + \dfrac{b}{c}$ and $\dfrac{a-c}{b} = \dfrac{a}{b} - \dfrac{c}{b}$, along with our knowledge of dividing monomials, provide the basis for dividing polynomials by monomials. Consider the following examples.

$$\frac{18x^3 + 24x^2}{6x} = \frac{18x^3}{6x} + \frac{24x^2}{6x} = 3x^2 + 4x$$

$$\frac{35x^2y^3 - 55x^3y^4}{5xy^2} = \frac{35x^2y^3}{5xy^2} - \frac{55x^3y^4}{5xy^2} = 7xy - 11x^2y^2$$

To divide a polynomial by a monomial, we divide each term of the polynomial by the monomial. As with many skills, once you feel comfortable with the process, you may then want to perform some of the steps mentally. Your work could take on the following format.

$$\frac{40x^4y^5 + 72x^5y^7}{8x^2y} = 5x^2y^4 + 9x^3y^6 \qquad \frac{36a^3b^4 - 45a^4b^6}{-9a^2b^3} = -4ab + 5a^2b^3$$

In Section 4.1, we saw that a fraction like $\dfrac{3x^2 + 11x - 4}{x + 4}$ can be simplified as follows:

$$\frac{3x^2 + 11x - 4}{x + 4} = \frac{(3x - 1)(x + 4)}{x + 4} = 3x - 1$$

We can obtain the same result by using a dividing process similar to long division in arithmetic.

Step 1   Use the conventional long-division format, and arrange both the dividend and the divisor in descending powers of the variable.

$$x + 4 \overline{)3x^2 + 11x - 4}$$

Step 2   Find the first term of the quotient by dividing the first term of the dividend by the first term of the divisor.

$$\begin{array}{r} 3x \\ x + 4 \overline{)3x^2 + 11x - 4} \end{array}$$

Step 3   Multiply the entire divisor by the term of the quotient found in Step 2, and position the product to be subtracted from the dividend.

$$\begin{array}{r} 3x \\ x + 4 \overline{)3x^2 + 11x - 4} \\ \underline{3x^2 + 12x} \end{array}$$

Step 4   Subtract.

$$\dfrac{3x}{x+4{\overline{)3x^2+11x-4}}}$$
$$\dfrac{3x^2+12x}{}$$

Remember to add the opposite! ⟶ $-x-4$

$(3x^2+11x-4)-(3x^2+12x)=-x-4$ ⟶

Step 5   Repeat the process beginning with Step 2; use the polynomial that resulted from the subtraction in Step 4 as a new dividend.

$$\dfrac{3x\quad-1}{x+4{\overline{)3x^2+11x-4}}}$$
$$\dfrac{3x^2+12x}{}$$
$$\dfrac{-x-4}{}$$
$$-x-4$$

In the next example, let's *think* in terms of the previous step-by-step procedure but arrange our work in a more compact form.

Divide $5x^2+6x-8$ by $x+2$.

Solution   **Think Steps**

$$\dfrac{5x\;-4}{x+2{\overline{)5x^2+\;6x-8}}}$$
$$\dfrac{5x^2+10x}{}$$
$$-\;4x-8$$
$$\dfrac{-\;4x-8}{0}$$

1. $\dfrac{5x^2}{x}=5x$.

2. $5x(x+2)=5x^2+10x$.

3. $(5x^2+6x-8)-(5x^2+10x)=-4x-8$.

4. $\dfrac{-4x}{x}=-4$.

5. $-4(x+2)=-4x-8$.   ■

Recall that to check a division problem, we can multiply the divisor times the quotient and add the remainder. In other words,

Dividend = (Divisor)(Quotient) + (Remainder)

Sometimes the remainder is expressed as a fractional part of the divisor. The relationship then becomes

$$\dfrac{\text{Dividend}}{\text{Divisor}}=\text{Quotient}+\dfrac{\text{Remainder}}{\text{Divisor}}$$

Divide $2x^2-3x+1$ by $x-5$.

Solution

$$\dfrac{2x\;+7}{x-5{\overline{)2x^2-\;3x+\;1}}}$$
$$\dfrac{2x^2-10x}{}$$
$$7x+\;1$$
$$\dfrac{7x-35}{36}$$ ⟵ ——— Remainder

Thus

$$\dfrac{2x^2-3x+1}{x-5}=2x+7+\dfrac{36}{x-5}\qquad x\neq5$$

✔ **Check**

$$(x - 5)(2x + 7) + 36 \overset{?}{=} 2x^2 - 3x + 1$$

$$2x^2 - 3x - 35 + 36 \overset{?}{=} 2x^2 - 3x + 1$$

$$2x^2 - 3x + 1 = 2x^2 - 3x + 1 \qquad \blacksquare$$

Each of the next two examples illustrates another point regarding the division process. Study them carefully, and then you should be ready to work the exercises in the next problem set.

EXAMPLE 3

Divide $t^3 - 8$ by $t - 2$.

Solution

$$
\begin{array}{r}
t^2 + 2t + 4 \\
t - 2\overline{)t^3 + 0t^2 + 0t - 8} \\
\underline{t^3 - 2t^2} \\
2t^2 + 0t - 8 \\
\underline{2t^2 - 4t} \\
4t - 8 \\
\underline{4t - 8} \\
0
\end{array}
$$

⟵ Note the insertion of a "t-squared" term and a "t term" with zero coefficients.

Check this result!   ■

EXAMPLE 4

Divide $y^3 + 3y^2 - 2y - 1$ by $y^2 + 2y$.

Solution

$$
\begin{array}{r}
y + 1 \\
y^2 + 2y\overline{)y^3 + 3y^2 - 2y - 1} \\
\underline{y^3 + 2y^2} \\
y^2 - 2y - 1 \\
\underline{y^2 + 2y} \\
- 4y - 1
\end{array}
$$

⟵ Remainder of $-4y - 1$

(The division process is complete when the degree of the remainder is less than the degree of the divisor.) Thus

$$\frac{y^3 + 3y^2 - 2y - 1}{y^2 + 2y} = y + 1 + \frac{-4y - 1}{y^2 + 2y} \qquad \blacksquare$$

If the divisor is of the form $x - k$, where the coefficient of the x term is 1, then the format of the division process described in this section can be simplified by a procedure called **synthetic division**. This procedure is a shortcut for this type of polynomial division. If you are continuing on to study college algebra, then you will want to know synthetic division. If you are not continuing on to college algebra, then you probably will not need a shortcut and the long-division process will be sufficient.

First, let's consider an example and use the usual division process. Then, in step-by-step fashion, we can observe some shortcuts that will lead us into the synthetic-division procedure. Consider the division problem $(2x^4 + x^3 - 17x^2 + 13x + 2) \div (x - 2)$

$$
\require{enclose}
\begin{array}{r}
2x^3 + 5x^2 - 7x - 1 \\[-3pt]
x - 2 \overline{)2x^4 + x^3 - 17x^2 + 13x + 2} \\
\underline{2x^4 - 4x^3} \\
5x^3 - 17x^2 \\
\underline{5x^3 - 10x^2} \\
-7x^2 + 13x \\
\underline{-7x^2 + 14x} \\
-x + 2 \\
\underline{-x + 2}
\end{array}
$$

Note that because the dividend $(2x^4 + x^3 - 17x^2 + 13x + 2)$ is written in descending powers of x, the quotient $(2x^3 + 5x^2 - 7x - 1)$ is produced, also in descending powers of x. In other words, the numerical coefficients are the important numbers. Thus let's rewrite this problem in terms of its coefficients.

$$
\begin{array}{r}
2 + 5 - 7 - 1 \\[-3pt]
1 - 2 \overline{)2 + 1 - 17 + 13 + 2} \\
\underline{②- 4} \\
5 ⊖ 17 \\
\underline{⑤- 10} \\
-7 + ⑬ \\
\underline{⊖ + 14} \\
-1 + ② \\
\underline{⊖ + 2}
\end{array}
$$

Now observe that the numbers that are circled are simply repetitions of the numbers directly above them in the format. Therefore, by removing the circled numbers, we can write the process in a more compact form as

$$
\begin{array}{r}
2 \quad 5 - 7 - 1 \\[-3pt]
-2 \overline{)2 \quad 1 - 17 - 13 \quad 2} \\
\underline{-4 - 10 \quad 14 \quad 2} \\
5 - 7 - 1 \quad 0
\end{array}
\qquad
\begin{array}{l}
(1) \\
(2) \\
(3) \\
(4)
\end{array}
$$

where the repetitions are omitted and where 1, the coefficient of x in the divisor, is omitted.

Note that line (4) reveals all of the coefficients of the quotient, line (1), except for the first coefficient of 2. Thus we can begin line (4) with the first coefficient and then use the following form.

$$
\begin{array}{r}
-2 \overline{)2 \quad 1 - 17 \quad 13 \quad 2} \\
\underline{-4 - 10 \quad 14 \quad 2} \\
2 \quad 5 - 7 - 1 \quad 0
\end{array}
\qquad
\begin{array}{l}
(5) \\
(6) \\
(7)
\end{array}
$$

Line (7) contains the coefficients of the quotient, where the 0 indicates the remainder.

Finally, by changing the constant in the divisor to 2 (instead of -2), we can add the corresponding entries in lines (5) and (6) rather than subtract. Thus the final synthetic division form for this problem is

$$
\begin{array}{r}
2\overline{)2 \quad 1 \quad -17 \quad 13 \quad 2} \\
\underline{4 \quad 10 \quad -14 \quad -2} \\
2 \quad 5 \quad -7 \quad -1 \quad 0
\end{array}
$$

Now let's consider another problem that illustrates a step-by-step procedure for carrying out the synthetic-division process. Suppose that we want to divide $3x^3 - 2x^2 + 6x - 5$ by $x + 4$.

Step 1   Write the coefficients of the dividend as follows:

$$\overline{)3 \quad -2 \quad 6 \quad -5}$$

Step 2   In the divisor, $(x + 4)$, use -4 instead of 4 so that later we can add rather than subtract.

$$-4\overline{)3 \quad -2 \quad 6 \quad -5}$$

Step 3   Bring down the first coeffecient of the dividend (3).

$$
\begin{array}{r}
-4\overline{)3 \quad -2 \quad 6 \quad -5} \\
\hline
3
\end{array}
$$

Step 4   Multiply$(3)(-4)$, which yields -12; this result is to be added to the second coefficient of the dividend (-2).

$$
\begin{array}{r}
-4\overline{)3 \quad -2 \quad 6 \quad -5} \\
\underline{-12} \\
3 \quad -14
\end{array}
$$

Step 5   Multiply $(-14)(-4)$, which yields 56; this result is to be added to the third coefficient of the dividend (6).

$$
\begin{array}{r}
-4\overline{)3 \quad -2 \quad 6 \quad -5} \\
\underline{-12 \quad 56} \\
3 \quad -14 \quad 62
\end{array}
$$

Step 6   Multiply $(62)(-4)$, which yields -248; this result is added to the last term of the dividend (-5).

$$
\begin{array}{r}
-4\overline{)3 \quad -2 \quad 6 \quad -5} \\
\underline{-12 \quad 56 \quad -248} \\
3 \quad -14 \quad 62 \quad -253
\end{array}
$$

The last row indicates a quotient of $3x^2 - 14x + 62$ and a remainder of -253. Thus we have

$$\frac{3x^3 - 2x^2 + 6x - 5}{x + 4} = 3x^2 - 14x + 62 - \frac{253}{x + 4}$$

We will consider one more example, which shows only the final, compact form for synthetic division.

EXAMPLE 5   Find the quotient and remainder for $(4x^4 - 2x^3 + 6x - 1) \div (x - 1)$.

Solution

$$
\begin{array}{r|rrrr}
1) & 4 & -2 & 0 & 6 & -1 \\
  &   & 4 & 2 & 2 & 8 \\
\hline
  & 4 & 2 & 2 & 8 & 7 \\
\end{array}
$$

Note that a zero has been inserted as the coefficient of the missing x^2 term.

Therefore,

$$\frac{4x^4 - 2x^3 + 6x - 1}{x - 1} = 4x^3 + 2x^2 + 2x + 8 + \frac{7}{x - 1}$$ ∎

Problem Set 4.5

For Problems 1–10, perform the indicated divisions of polynomials by monomials.

1. $\dfrac{9x^4 + 18x^3}{3x}$

2. $\dfrac{12x^3 - 24x^2}{6x^2}$

3. $\dfrac{-24x^6 + 36x^8}{4x^2}$

4. $\dfrac{-35x^5 - 42x^3}{-7x^2}$

5. $\dfrac{15a^3 - 25a^2 - 40a}{5a}$

6. $\dfrac{-16a^4 + 32a^3 - 56a^2}{-8a}$

7. $\dfrac{13x^3 - 17x^2 + 28x}{-x}$

8. $\dfrac{14xy - 16x^2y^2 - 20x^3y^4}{-xy}$

9. $\dfrac{-18x^2y^2 + 24x^3y^2 - 48x^2y^3}{6xy}$

10. $\dfrac{-27a^3b^4 - 36a^2b^3 + 72a^2b^5}{9a^2b^2}$

For Problems 11–52, perform the indicated divisions.

11. $\dfrac{x^2 - 7x - 78}{x + 6}$

12. $\dfrac{x^2 + 11x - 60}{x - 4}$

13. $(x^2 + 12x - 160) \div (x - 8)$

14. $(x^2 - 18x - 175) \div (x + 7)$

15. $\dfrac{2x^2 - x - 4}{x - 1}$

16. $\dfrac{3x^2 - 2x - 7}{x + 2}$

17. $\dfrac{15x^2 + 22x - 5}{3x + 5}$

18. $\dfrac{12x^2 - 32x - 35}{2x - 7}$

19. $\dfrac{3x^3 + 7x^2 - 13x - 21}{x + 3}$

20. $\dfrac{4x^3 - 21x^2 + 3x + 10}{x - 5}$

21. $(2x^3 + 9x^2 - 17x + 6) \div (2x - 1)$

22. $(3x^3 - 5x^2 - 23x - 7) \div (3x + 1)$

23. $(4x^3 - x^2 - 2x + 6) \div (x - 2)$

24. $(6x^3 - 2x^2 + 4x - 3) \div (x + 1)$

25. $(x^4 - 10x^3 + 19x^2 + 33x - 18) \div (x - 6)$

26. $(x^4 + 2x^3 - 16x^2 + x + 6) \div (x - 3)$

27. $\dfrac{x^3 - 125}{x - 5}$

28. $\dfrac{x^3 + 64}{x + 4}$

29. $(x^3 + 64) \div (x + 1)$

30. $(x^3 - 8) \div (x - 4)$

31. $(2x^3 - x - 6) \div (x + 2)$

32. $(5x^3 + 2x - 3) \div (x - 2)$

33. $\dfrac{4a^2 - 8ab + 4b^2}{a - b}$

34. $\dfrac{3x^2 - 2xy - 8y^2}{x - 2y}$

35. $\dfrac{4x^3 - 5x^2 + 2x - 6}{x^2 - 3x}$

36. $\dfrac{3x^3 + 2x^2 - 5x - 1}{x^2 + 2x}$

37. $\dfrac{8y^3 - y^2 - y + 5}{y^2 + y}$

38. $\dfrac{5y^3 - 6y^2 - 7y - 2}{y^2 - y}$

39. $(2x^3 + x^2 - 3x + 1) \div (x^2 + x - 1)$

40. $(3x^3 - 4x^2 + 8x + 8) \div (x^2 - 2x + 4)$

41. $(4x^3 - 13x^2 + 8x - 15) \div (4x^2 - x + 5)$

42. $(5x^3 + 8x^2 - 5x - 2) \div (5x^2 - 2x - 1)$

43. $(5a^3 + 7a^2 - 2a - 9) \div (a^2 + 3a - 4)$

44. $(4a^3 - 2a^2 + 7a - 1) \div (a^2 - 2a + 3)$

45. $(2n^4 + 3n^3 - 2n^2 + 3n - 4) \div (n^2 + 1)$

46. $(3n^4 + n^3 - 7n^2 - 2n + 2) \div (n^2 - 2)$

47. $(x^5 - 1) \div (x - 1)$ **48.** $(x^5 + 1) \div (x + 1)$

49. $(x^4 - 1) \div (x + 1)$ **50.** $(x^4 - 1) \div (x - 1)$

51. $(3x^4 + x^3 - 2x^2 - x + 6) \div (x^2 - 1)$

52. $(4x^3 - 2x^2 + 7x - 5) \div (x^2 + 2)$

For problems 53–64, use synthetic division to determine the quotient and remainder.

53. $(x^2 - 8x + 12) \div (x - 2)$

54. $(x^2 + 9x + 18) \div (x + 3)$

55. $(x^2 + 2x - 10) \div (x - 4)$

56. $(x^2 - 10x + 15) \div (x - 8)$

57. $(x^3 - 2x^2 - x + 2) \div (x - 2)$

58. $(x^3 - 5x^2 + 2x + 8) \div (x + 1)$

59. $(x^3 - 7x - 6) \div (x + 2)$

60. $(x^3 + 6x^2 - 5x - 1) \div (x - 1)$

61. $(2x^3 - 5x^2 - 4x + 6) \div (x - 2)$

62. $(3x^4 - x^3 + 2x^2 - 7x - 1) \div (x + 1)$

63. $(x^4 + 4x^3 - 7x - 1) \div (x - 3)$

64. $(2x^4 + 3x^2 + 3) \div (x + 2)$

■ ■ ■ **THOUGHTS INTO WORDS**

65. Describe the process of long division of polynomials.

66. Give a step-by-step description of how you would do the following division problem.

$$(4 - 3x - 7x^3) \div (x + 6)$$

67. How do you know by inspection that $3x^2 + 5x + 1$ cannot be the correct answer for the division problem $(3x^3 - 7x^2 - 22x + 8) \div (x - 4)$?

4.6   Fractional Equations

The fractional equations used in this text are of two basic types. One type has only constants as denominators, and the other type contains variables in the denominators.

In Chapter 2, we considered fractional equations that involve only constants in the denominators. Let's briefly review our approach to solving such equations, because we will be using that same basic technique to solve any type of fractional equation.

Solve $\dfrac{x-2}{3} + \dfrac{x+1}{4} = \dfrac{1}{6}$.

Solution

$$\dfrac{x-2}{3} + \dfrac{x+1}{4} = \dfrac{1}{6}$$

$$12\left(\dfrac{x-2}{3} + \dfrac{x+1}{4}\right) = 12\left(\dfrac{1}{6}\right) \qquad \text{Multiply both sides by 12, which is the LCD of all of the denominators.}$$

$$4(x-2) + 3(x+1) = 2$$

$$4x - 8 + 3x + 3 = 2$$

$$7x - 5 = 2$$

$$7x = 7$$

$$x = 1$$

The solution set is {1}. Check it! ■

If an equation contains a variable (or variables) in one or more denominators, then we proceed in essentially the same way as in Example 1 **except that we must avoid any value of the variable that makes a denominator zero**. Consider the following examples.

Solve $\dfrac{5}{n} + \dfrac{1}{2} = \dfrac{9}{n}$.

Solution

First, we need to realize that n cannot equal zero. (Let's indicate this restriction so that it is not forgotten!) Then we can proceed.

$$\dfrac{5}{n} + \dfrac{1}{2} = \dfrac{9}{n}, \qquad n \neq 0$$

$$2n\left(\dfrac{5}{n} + \dfrac{1}{2}\right) = 2n\left(\dfrac{9}{n}\right) \qquad \text{Multiply both sides by the LCD, which is } 2n.$$

$$10 + n = 18$$

$$n = 8$$

The solution set is {8}. Check it! ■

Solve $\dfrac{35-x}{x} = 7 + \dfrac{3}{x}$.

Solution

$$\dfrac{35-x}{x} = 7 + \dfrac{3}{x}, \qquad x \neq 0$$

$$x\left(\frac{35 - x}{x}\right) = x\left(7 + \frac{3}{x}\right) \qquad \text{Multiply both sides by } x.$$

$$35 - x = 7x + 3$$

$$32 = 8x$$

$$4 = x$$

The solution set is {4}. ∎

E X A M P L E   4   Solve $\dfrac{3}{a - 2} = \dfrac{4}{a + 1}$.

Solution

$$\frac{3}{a - 2} = \frac{4}{a + 1}, \qquad a \neq 2 \text{ and } a \neq -1$$

$$(a - 2)(a + 1)\left(\frac{3}{a - 2}\right) = (a - 2)(a + 1)\left(\frac{4}{a + 1}\right) \qquad \begin{array}{l}\text{Multiply both sides}\\ \text{by } (a - 2)(a + 1).\end{array}$$

$$3(a + 1) = 4(a - 2)$$

$$3a + 3 = 4a - 8$$

$$11 = a$$

The solution set is {11}. ∎

Keep in mind that listing the restrictions at the beginning of a problem does not replace checking the potential solutions. In Example 4, the answer 11 needs to be checked in the original equation.

E X A M P L E   5   Solve $\dfrac{a}{a - 2} + \dfrac{2}{3} = \dfrac{2}{a - 2}$.

Solution

$$\frac{a}{a - 2} + \frac{2}{3} = \frac{2}{a - 2}, \qquad a \neq 2$$

$$3(a - 2)\left(\frac{a}{a - 2} + \frac{2}{3}\right) = 3(a - 2)\left(\frac{2}{a - 2}\right) \qquad \begin{array}{l}\text{Multiply both sides}\\ \text{by } 3(a - 2).\end{array}$$

$$3a + 2(a - 2) = 6$$

$$3a + 2a - 4 = 6$$

$$5a = 10$$

$$a = 2$$

Because our initial restriction was $a \neq 2$, we conclude that this equation has no solution. Thus the solution set is $\varnothing$. ∎

■ Ratio and Proportion

A **ratio** is the comparison of two numbers by division. We often use the fractional form to express ratios. For example, we can write the ratio of a to b as $\dfrac{a}{b}$. A statement of equality between two ratios is called a **proportion**. Thus if $\dfrac{a}{b}$ and $\dfrac{c}{d}$ are two equal ratios, we can form the proportion $\dfrac{a}{b} = \dfrac{c}{d}$ ($b \neq 0$ and $d \neq 0$). We deduce an important property of proportions as follows:

$$\frac{a}{b} = \frac{c}{d}, \qquad b \neq 0 \text{ and } d \neq 0$$

$$bd\left(\frac{a}{b}\right) = bd\left(\frac{c}{d}\right) \qquad \text{Multiply both sides by } bd.$$

$$ad = bc$$

Cross-Multiplication Property of Proportions

> If $\dfrac{a}{b} = \dfrac{c}{d}$ ($b \neq 0$ and $d \neq 0$), then $ad = bc$.

We can treat some fractional equations as proportions and solve them by using the cross-multiplication idea, as in the next examples.

E X A M P L E   6

Solve $\dfrac{5}{x + 6} = \dfrac{7}{x - 5}$.

Solution

$$\frac{5}{x + 6} = \frac{7}{x - 5}, \qquad x \neq -6 \text{ and } x \neq 5$$

$$5(x - 5) = 7(x + 6) \qquad \text{Apply the cross-multiplication property.}$$

$$5x - 25 = 7x + 42$$

$$-67 = 2x$$

$$-\frac{67}{2} = x$$

The solution set is $\left\{-\dfrac{67}{2}\right\}$.   ■

EXAMPLE   7   Solve $\dfrac{x}{7} = \dfrac{4}{x + 3}$.

Solution

$$\frac{x}{7} = \frac{4}{x + 3}, \qquad x \neq -3$$

$$x(x + 3) = 7(4) \qquad \text{Cross-multiplication property}$$

$$x^2 + 3x = 28$$

$$x^2 + 3x - 28 = 0$$

$$(x + 7)(x - 4) = 0$$

$$x + 7 = 0 \qquad \text{or} \qquad x - 4 = 0$$

$$x = -7 \qquad \text{or} \qquad x = 4$$

The solution set is $\{-7, 4\}$. Check these solutions in the original equation.   ■

■ Problem Solving

The ability to solve fractional equations broadens our base for solving word problems. We are now ready to tackle some word problems that translate into fractional equations.

PROBLEM   1   The sum of a number and its reciprocal is $\dfrac{10}{3}$. Find the number.

Solution

Let n represent the number. Then $\dfrac{1}{n}$ represents its reciprocal.

$$n + \frac{1}{n} = \frac{10}{3}, \qquad n \neq 0$$

$$3n\left(n + \frac{1}{n}\right) = 3n\left(\frac{10}{3}\right)$$

$$3n^2 + 3 = 10n$$

$$3n^2 - 10n + 3 = 0$$

$$(3n - 1)(n - 3) = 0$$

$$3n - 1 = 0 \qquad \text{or} \qquad n - 3 = 0$$

$$3n = 1 \qquad \text{or} \qquad n = 3$$

$$n = \frac{1}{3} \qquad \text{or} \qquad n = 3$$

If the number is $\dfrac{1}{3}$, then its reciprocal is $\dfrac{1}{\frac{1}{3}} = 3$. If the number is 3, then its reciprocal is $\dfrac{1}{3}$. ∎

Now let's consider a problem where we can use the relationship

$$\frac{\text{Dividend}}{\text{Divisor}} = \text{Quotient} + \frac{\text{Remainder}}{\text{Divisor}}$$

as a guideline.

PROBLEM 2

The sum of two numbers is 52. If the larger is divided by the smaller, the quotient is 9, and the remainder is 2. Find the numbers.

Solution

Let n represent the smaller number. Then $52 - n$ represents the larger number. Let's use the relationship we discussed previously as a guideline and proceed as follows:

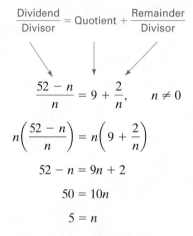

$$\frac{52 - n}{n} = 9 + \frac{2}{n}, \qquad n \neq 0$$

$$n\left(\frac{52 - n}{n}\right) = n\left(9 + \frac{2}{n}\right)$$

$$52 - n = 9n + 2$$

$$50 = 10n$$

$$5 = n$$

If $n = 5$, then $52 - n$ equals 47. The numbers are 5 and 47. ∎

We can conveniently set up some problems and solve them using the concepts of ratio and proportion. Let's conclude this section with two such examples.

PROBLEM 3

On a certain map $1\dfrac{1}{2}$ inches represents 25 miles. If two cities are $5\dfrac{1}{4}$ inches apart on the map, find the number of miles between the cities (see Figure 4.1).

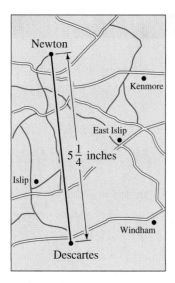

Figure 4.1

Solution

Let m represent the number of miles between the two cities. To set up the proportion, we will use a ratio of inches on the map to miles. Be sure to keep the ratio "inches on the map to miles" the same for both sides of the proportion.

$$\frac{1\frac{1}{2}}{25} = \frac{5\frac{1}{4}}{m}, \qquad m \neq 0$$

$$\frac{\frac{3}{2}}{25} = \frac{\frac{21}{4}}{m}$$

$$\frac{3}{2}m = 25\left(\frac{21}{4}\right) \qquad \text{Cross-multiplication property}$$

$$\frac{2}{3}\left(\frac{3}{2}m\right) = \frac{2}{\cancel{3}}(25)\left(\frac{\cancel{21}^{7}}{\cancel{4}_{2}}\right) \qquad \text{Multiply both sides by } \frac{2}{3}.$$

$$m = \frac{175}{2}$$

$$= 87\frac{1}{2}$$

The distance between the two cities is $87\frac{1}{2}$ miles.   ■

PROBLEM   4

A sum of \$750 is to be divided between two people in the ratio of 2 to 3. How much does each person receive?

Solution

Let d represent the amount of money that one person receives. Then $750 - d$ represents the amount for the other person.

$$\frac{d}{750 - d} = \frac{2}{3}, \qquad d \neq 750$$

$$3d = 2(750 - d)$$

$$3d = 1500 - 2d$$

$$5d = 1500$$

$$d = 300$$

If $d = 300$, then $750 - d$ equals 450. Therefore, one person receives \$300 and the other person receives \$450.   ■

Problem Set 4.6

For Problems 1–44, solve each equation.

1. $\dfrac{x+1}{4} + \dfrac{x-2}{6} = \dfrac{3}{4}$

2. $\dfrac{x+2}{5} + \dfrac{x-1}{6} = \dfrac{3}{5}$

3. $\dfrac{x+3}{2} - \dfrac{x-4}{7} = 1$

4. $\dfrac{x+4}{3} - \dfrac{x-5}{9} = 1$

5. $\dfrac{5}{n} + \dfrac{1}{3} = \dfrac{7}{n}$

6. $\dfrac{3}{n} + \dfrac{1}{6} = \dfrac{11}{3n}$

7. $\dfrac{7}{2x} + \dfrac{3}{5} = \dfrac{2}{3x}$

8. $\dfrac{9}{4x} + \dfrac{1}{3} = \dfrac{5}{2x}$

9. $\dfrac{3}{4x} + \dfrac{5}{6} = \dfrac{4}{3x}$

10. $\dfrac{5}{7x} - \dfrac{5}{6} = \dfrac{1}{6x}$

11. $\dfrac{47-n}{n} = 8 + \dfrac{2}{n}$

12. $\dfrac{45-n}{n} = 6 + \dfrac{3}{n}$

13. $\dfrac{n}{65-n} = 8 + \dfrac{2}{65-n}$

14. $\dfrac{n}{70-n} = 7 + \dfrac{6}{70-n}$

15. $n + \dfrac{1}{n} = \dfrac{17}{4}$

16. $n + \dfrac{1}{n} = \dfrac{37}{6}$

17. $n - \dfrac{2}{n} = \dfrac{23}{5}$

18. $n - \dfrac{3}{n} = \dfrac{26}{3}$

19. $\dfrac{5}{7x-3} = \dfrac{3}{4x-5}$

20. $\dfrac{3}{2x-1} = \dfrac{5}{3x+2}$

21. $\dfrac{-2}{x-5} = \dfrac{1}{x+9}$

22. $\dfrac{5}{2a-1} = \dfrac{-6}{3a+2}$

23. $\dfrac{x}{x+1} - 2 = \dfrac{3}{x-3}$

24. $\dfrac{x}{x-2} + 1 = \dfrac{8}{x-1}$

25. $\dfrac{a}{a+5} - 2 = \dfrac{3a}{a+5}$

26. $\dfrac{a}{a-3} - \dfrac{3}{2} = \dfrac{3}{a-3}$

27. $\dfrac{5}{x+6} = \dfrac{6}{x-3}$

28. $\dfrac{3}{x-1} = \dfrac{4}{x+2}$

29. $\dfrac{3x-7}{10} = \dfrac{2}{x}$

30. $\dfrac{x}{-4} = \dfrac{3}{12x-25}$

31. $\dfrac{x}{x-6} - 3 = \dfrac{6}{x-6}$

32. $\dfrac{x}{x+1} + 3 = \dfrac{4}{x+1}$

33. $\dfrac{3s}{s+2} + 1 = \dfrac{35}{2(3s+1)}$

34. $\dfrac{s}{2s-1} - 3 = \dfrac{-32}{3(s+5)}$

35. $2 - \dfrac{3x}{x-4} = \dfrac{14}{x+7}$

36. $-1 + \dfrac{2x}{x+3} = \dfrac{-4}{x+4}$

37. $\dfrac{n+6}{27} = \dfrac{1}{n}$

38. $\dfrac{n}{5} = \dfrac{10}{n-5}$

39. $\dfrac{3n}{n-1} - \dfrac{1}{3} = \dfrac{-40}{3n-18}$

40. $\dfrac{n}{n+1} + \dfrac{1}{2} = \dfrac{-2}{n+2}$

41. $\dfrac{-3}{4x+5} = \dfrac{2}{5x-7}$

42. $\dfrac{7}{x+4} = \dfrac{3}{x-8}$

43. $\dfrac{2x}{x-2} + \dfrac{15}{x^2-7x+10} = \dfrac{3}{x-5}$

44. $\dfrac{x}{x-4} - \dfrac{2}{x+3} = \dfrac{20}{x^2-x-12}$

For Problems 45–60, set up an algebraic equation and solve each problem.

45. A sum of $1750 is to be divided between two people in the ratio of 3 to 4. How much does each person receive?

46. A blueprint has a scale where 1 inch represents 5 feet. Find the dimensions of a rectangular room that measures $3\dfrac{1}{2}$ inches by $5\dfrac{3}{4}$ inches on the blueprint.

47. One angle of a triangle has a measure of 60° and the measures of the other two angles are in the ratio of 2 to 3. Find the measures of the other two angles.

48. The ratio of the complement of an angle to its supplement is 1 to 4. Find the measure of the angle.

49. The sum of a number and its reciprocal is $\dfrac{53}{14}$. Find the number.

50. The sum of two numbers is 80. If the larger is divided by the smaller, the quotient is 7, and the remainder is 8. Find the numbers.

51. If a home valued at $150,000 is assessed $2500 in real estate taxes, then how much, at the same rate, are the taxes on a home valued at $210,000?

52. The ratio of male students to female students at a certain university is 5 to 7. If there is a total of 16,200 students, find the number of male students and the number of female students.

53. Suppose that, together, Laura and Tammy sold $120.75 worth of candy for the annual school fair. If the ratio of Tammy's sales to Laura's sales was 4 to 3, how much did each sell?

54. The total value of a house and a lot is $168,000. If the ratio of the value of the house to the value of the lot is 7 to 1, find the value of the house.

55. The sum of two numbers is 90. If the larger is divided by the smaller, the quotient is 10, and the remainder is 2. Find the numbers.

56. What number must be added to the numerator and denominator of $\dfrac{2}{5}$ to produce a rational number that is equivalent to $\dfrac{7}{8}$?

57. A 20-foot board is to be cut into two pieces whose lengths are in the ratio of 7 to 3. Find the lengths of the two pieces.

58. An inheritance of $300,000 is to be divided between a son and the local heart fund in the ratio of 3 to 1. How much money will the son receive?

59. Suppose that in a certain precinct, 1150 people voted in the last presidential election. If the ratio of female voters to male voters was 3 to 2, how many females and how many males voted?

60. The perimeter of a rectangle is 114 centimeters. If the ratio of its width to its length is 7 to 12, find the dimensions of the rectangle.

■ ■ ■ THOUGHTS INTO WORDS

61. How could you do Problem 57 without using algebra?

62. Now do Problem 59 using the same approach that you used in Problem 61. What difficulties do you encounter?

63. How can you tell by inspection that the equation $\dfrac{x}{x+2} = \dfrac{-2}{x+2}$ has no solution?

64. How would you help someone solve the equation $\dfrac{3}{x} - \dfrac{4}{x} = \dfrac{-1}{x}$?

4.7 More Fractional Equations and Applications

Let's begin this section by considering a few more fractional equations. We will continue to solve them using the same basic techniques as in the previous section. That is, we will multiply both sides of the equation by the least common denominator of all of the denominators in the equation, with the necessary restrictions to avoid division by zero. Some of the denominators in these problems will require factoring before we can determine a least common denominator.

E X A M P L E   1

Solve $\dfrac{x}{2x-8} + \dfrac{16}{x^2-16} = \dfrac{1}{2}$.

Solution

$$\frac{x}{2x-8} + \frac{16}{x^2-16} = \frac{1}{2}$$

$$\frac{x}{2(x-4)} + \frac{16}{(x+4)(x-4)} = \frac{1}{2}, \qquad x \neq 4 \text{ and } x \neq -4$$

$$2(x-4)(x+4)\left(\frac{x}{2(x-4)} + \frac{16}{(x+4)(x-4)}\right) = 2(x+4)(x-4)\left(\frac{1}{2}\right) \qquad \begin{array}{l}\text{Multiply both}\\\text{sides by the LCD,}\\2(x-4)(x+4).\end{array}$$

$$x(x+4) + 2(16) = (x+4)(x-4)$$

$$x^2 + 4x + 32 = x^2 - 16$$

$$4x = -48$$

$$x = -12$$

The solution set is $\{-12\}$. Perhaps you should check it! ∎

In Example 1, note that the restrictions were not indicated until the denominators were expressed in factored form. It is usually easier to determine the necessary restrictions at this step.

E X A M P L E   2

Solve $\dfrac{3}{n-5} - \dfrac{2}{2n+1} = \dfrac{n+3}{2n^2-9n-5}$.

Solution

$$\frac{3}{n-5} - \frac{2}{2n+1} = \frac{n+3}{2n^2-9n-5}$$

$$\frac{3}{n-5} - \frac{2}{2n+1} = \frac{n+3}{(2n+1)(n-5)}, \qquad n \neq -\frac{1}{2} \text{ and } n \neq 5$$

$$(2n+1)(n-5)\left(\frac{3}{n-5} - \frac{2}{2n+1}\right) = (2n+1)(n-5)\left(\frac{n+3}{(2n+1)(n-5)}\right) \qquad \begin{array}{l}\text{Multiply both}\\\text{sides by the LCD,}\\(2n+1)(n-5).\end{array}$$

$$3(2n+1) - 2(n-5) = n+3$$

$$6n + 3 - 2n + 10 = n + 3$$

$$4n + 13 = n + 3$$

$$3n = -10$$

$$n = -\frac{10}{3}$$

The solution set is $\left\{-\dfrac{10}{3}\right\}$. ∎

E X A M P L E 3

Solve $2 + \dfrac{4}{x-2} = \dfrac{8}{x^2-2x}$.

Solution

$$2 + \frac{4}{x-2} = \frac{8}{x^2-2x}$$

$$2 + \frac{4}{x-2} = \frac{8}{x(x-2)}, \qquad x \neq 0 \text{ and } x \neq 2$$

$$x(x-2)\left(2 + \frac{4}{x-2}\right) = x(x-2)\left(\frac{8}{x(x-2)}\right) \qquad \text{Multiply both sides by the LCD, } x(x-2).$$

$$2x(x-2) + 4x = 8$$

$$2x^2 - 4x + 4x = 8$$

$$2x^2 = 8$$

$$x^2 = 4$$

$$x^2 - 4 = 0$$

$$(x+2)(x-2) = 0$$

$$x + 2 = 0 \quad \text{or} \quad x - 2 = 0$$

$$x = -2 \quad \text{or} \quad x = 2$$

Because our initial restriction indicated that $x \neq 2$, the only solution is -2. Thus the solution set is $\{-2\}$. ∎

In Section 2.4, we discussed using the properties of equality to change the form of various formulas. For example, we considered the simple interest formula $A = P + Prt$ and changed its form by solving for P as follows:

$$A = P + Prt$$

$$A = P(1 + rt)$$

$$\frac{A}{1+rt} = P \qquad \text{Multiply both sides by } \frac{1}{1+rt}.$$

If the formula is in the form of a fractional equation, then the techniques of these last two sections are applicable. Consider the following example.

E X A M P L E 4

If the original cost of some business property is C dollars and it is depreciated linearly over N years, then its value, V, at the end of T years is given by

$$V = C\left(1 - \frac{T}{N}\right)$$

Solve this formula for N in terms of V, C, and T.

Solution

$$V = C\left(1 - \frac{T}{N}\right)$$

$$V = C - \frac{CT}{N}$$

$$N(V) = N\left(C - \frac{CT}{N}\right) \qquad \text{Multiply both sides by } N.$$

$$NV = NC - CT$$

$$NV - NC = -CT$$

$$N(V - C) = -CT$$

$$N = \frac{-CT}{V - C}$$

$$N = -\frac{CT}{V - C} \qquad\qquad\qquad\qquad \blacksquare$$

■ Problem Solving

In Section 2.4 we solved some uniform motion problems. The formula $d = rt$ was used in the analysis of the problems, and we used guidelines that involve distance relationships. Now let's consider some uniform motion problems where guidelines that involve either times or rates are appropriate. These problems will generate fractional equations to solve.

PROBLEM 1

An airplane travels 2050 miles in the same time that a car travels 260 miles. If the rate of the plane is 358 miles per hour greater than the rate of the car, find the rate of each.

Solution

Let r represent the rate of the car. Then $r + 358$ represents the rate of the plane. The fact that the times are equal can be a guideline. Remember from the basic formula, $d = rt$, that $t = \dfrac{d}{r}$.

Time of plane	Equals	Time of car
↓		↓
$\dfrac{\text{Distance of plane}}{\text{Rate of plane}}$	$=$	$\dfrac{\text{Distance of car}}{\text{Rate of car}}$

$$\frac{2050}{r + 358} = \frac{260}{r}$$

$$2050r = 260(r + 358)$$

$$2050r = 260r + 93{,}080$$

$$1790r = 93{,}080$$

$$r = 52$$

If $r = 52$, then $r + 358$ equals 410. Thus the rate of the car is 52 miles per hour, and the rate of the plane is 410 miles per hour.   ■

P R O B L E M   2

It takes a freight train 2 hours longer to travel 300 miles than it takes an express train to travel 280 miles. The rate of the express train is 20 miles per hour greater than the rate of the freight train. Find the times and rates of both trains.

Solution

Let t represent the time of the express train. Then $t + 2$ represents the time of the freight train. Let's record the information of this problem in a table.

	Distance	Time	Rate $= \dfrac{\text{Distance}}{\text{Time}}$
Express train	280	t	$\dfrac{280}{t}$
Freight train	300	$t + 2$	$\dfrac{300}{t + 2}$

The fact that the rate of the express train is 20 miles per hour greater than the rate of the freight train can be a guideline.

Rate of express	Equals	Rate of freight train plus 20
$\dfrac{280}{t}$	$=$	$\dfrac{300}{t + 2} + 20$

$$t(t + 2)\left(\frac{280}{t}\right) = t(t + 2)\left(\frac{300}{t + 2} + 20\right)$$

$$280(t + 2) = 300t + 20t(t + 2)$$

$$280t + 560 = 300t + 20t^2 + 40t$$

$$280t + 560 = 340t + 20t^2$$

$$0 = 20t^2 + 60t - 560$$

$$0 = t^2 + 3t - 28$$

$$0 = (t + 7)(t - 4)$$

$$t + 7 = 0 \quad \text{or} \quad t - 4 = 0$$

$$t = -7 \quad \text{or} \quad t = 4$$

The negative solution must be discarded, so the time of the express train (t) is 4 hours, and the time of the freight train ($t + 2$) is 6 hours. The rate of the express train $\left(\dfrac{280}{t}\right)$ is $\dfrac{280}{4} = 70$ miles per hour, and the rate of the freight train $\left(\dfrac{300}{t + 2}\right)$ is $\dfrac{300}{6} = 50$ miles per hour. ∎

Remark: Note that to solve Problem 1 we went directly to a guideline without the use of a table, but for Problem 2 we used a table. Again, remember that this is a personal preference; we are merely acquainting you with a variety of techniques.

Uniform motion problems are a special case of a larger group of problems we refer to as **rate-time problems**. For example, if a certain machine can produce 150 items in 10 minutes, then we say that the machine is producing at a rate of $\dfrac{150}{10} = 15$ items per minute. Likewise, if a person can do a certain job in 3 hours, then, assuming a constant rate of work, we say that the person is working at a rate of $\dfrac{1}{3}$ of the job per hour. In general, if Q is the quantity of something done in t units of time, then the rate, r, is given by $r = \dfrac{Q}{t}$. We state the rate in terms of *so much quantity per unit of time*. (In uniform motion problems the "quantity" is distance.) Let's consider some examples of rate-time problems.

P R O B L E M 3

If Jim can mow a lawn in 50 minutes, and his son, Todd, can mow the same lawn in 40 minutes, how long will it take them to mow the lawn if they work together?

Solution

Jim's rate is $\dfrac{1}{50}$ of the lawn per minute, and Todd's rate is $\dfrac{1}{40}$ of the lawn per minute.

If we let m represent the number of minutes that they work together, then $\dfrac{1}{m}$ represents their rate when working together. Therefore, because the sum of the individual rates must equal the rate working together, we can set up and solve the following equation.

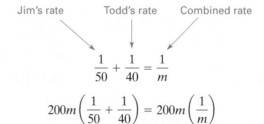

$$\underset{\text{Jim's rate}}{\dfrac{1}{50}} + \underset{\text{Todd's rate}}{\dfrac{1}{40}} = \underset{\text{Combined rate}}{\dfrac{1}{m}}$$

$$200m\left(\dfrac{1}{50} + \dfrac{1}{40}\right) = 200m\left(\dfrac{1}{m}\right)$$

$$4m + 5m = 200$$

$$9m = 200$$

$$m = \frac{200}{9} = 22\frac{2}{9}$$

It should take them $22\frac{2}{9}$ minutes.   ■

PROBLEM 4

Working together, Linda and Kathy can type a term paper in $3\frac{3}{5}$ hours. Linda can type the paper by herself in 6 hours. How long would it take Kathy to type the paper by herself?

Solution

Their rate working together is $\dfrac{1}{3\frac{3}{5}} = \dfrac{1}{\frac{18}{5}} = \dfrac{5}{18}$ of the job per hour, and Linda's rate

is $\dfrac{1}{6}$ of the job per hour. If we let h represent the number of hours that it would take

Kathy to do the job by herself, then her rate is $\dfrac{1}{h}$ of the job per hour. Thus we have

$$
\begin{array}{ccccc}
\text{Linda's rate} & & \text{Kathy's rate} & & \text{Combined rate}\\
\downarrow & & \downarrow & & \downarrow\\
\dfrac{1}{6} & + & \dfrac{1}{h} & = & \dfrac{5}{18}
\end{array}
$$

Solving this equation yields

$$18h\left(\frac{1}{6} + \frac{1}{h}\right) = 18h\left(\frac{5}{18}\right)$$

$$3h + 18 = 5h$$

$$18 = 2h$$

$$9 = h$$

It would take Kathy 9 hours to type the paper by herself.   ■

Our final example of this section illustrates another approach that some people find meaningful for rate-time problems. For this approach, think in terms of fractional parts of the job. For example, if a person can do a certain job in 5 hours, then at the end of 2 hours, he or she has done $\frac{2}{5}$ of the job. (Again, assume a constant rate of work.) At the end of 4 hours, he or she has finished $\frac{4}{5}$ of the job;

and, in general, at the end of h hours, he or she has done $\dfrac{h}{5}$ of the job. Then, just as in the motion problems where distance equals rate times the time, here the fractional part done equals the working rate times the time. Let's see how this works in a problem.

P R O B L E M   5

It takes Pat 12 hours to complete a task. After he had been working for 3 hours, he was joined by his brother Mike, and together they finished the task in 5 hours. How long would it take Mike to do the job by himself?

Solution

Let h represent the number of hours that it would take Mike to do the job by himself. The fractional part of the job that Pat does equals his working rate times his time. Because it takes Pat 12 hours to do the entire job, his working rate is $\dfrac{1}{12}$. He works for 8 hours (3 hours before Mike and then 5 hours with Mike). Therefore, Pat's part of the job is $\dfrac{1}{12}(8) = \dfrac{8}{12}$. The fractional part of the job that Mike does equals his working rate times his time. Because h represents Mike's time to do the entire job, his working rate is $\dfrac{1}{h}$; he works for 5 hours. Therefore, Mike's part of the job is $\dfrac{1}{h}(5) = \dfrac{5}{h}$. Adding the two fractional parts together results in 1 entire job being done. Let's also show this information in chart form and set up our guideline. Then we can set up and solve the equation.

	Time to do entire job	Working rate	Time working	Fractional part of the job done
Pat	12	$\dfrac{1}{12}$	8	$\dfrac{8}{12}$
Mike	h	$\dfrac{1}{h}$	5	$\dfrac{5}{h}$

Fractional part of the job that Pat does

Fractional part of the job that Mike does

$$\frac{8}{12} + \frac{5}{h} = 1$$

$$12h\left(\frac{8}{12} + \frac{5}{h}\right) = 12h(1)$$

$$12h\left(\frac{8}{12}\right) + 12h\left(\frac{5}{h}\right) = 12h$$

$$8h + 60 = 12h$$

$$60 = 4h$$

$$15 = h$$

It would take Mike 15 hours to do the entire job by himself. ∎

Problem Set 4.7

For Problems 1–30, solve each equation.

1. $\dfrac{x}{4x - 4} + \dfrac{5}{x^2 - 1} = \dfrac{1}{4}$ **2.** $\dfrac{x}{3x - 6} + \dfrac{4}{x^2 - 4} = \dfrac{1}{3}$

3. $3 + \dfrac{6}{t - 3} = \dfrac{6}{t^2 - 3t}$ **4.** $2 + \dfrac{4}{t - 1} = \dfrac{4}{t^2 - t}$

5. $\dfrac{3}{n - 5} + \dfrac{4}{n + 7} = \dfrac{2n + 11}{n^2 + 2n - 35}$

6. $\dfrac{2}{n + 3} + \dfrac{3}{n - 4} = \dfrac{2n - 1}{n^2 - n - 12}$

7. $\dfrac{5x}{2x + 6} - \dfrac{4}{x^2 - 9} = \dfrac{5}{2}$ **8.** $\dfrac{3x}{5x + 5} - \dfrac{2}{x^2 - 1} = \dfrac{3}{5}$

9. $1 + \dfrac{1}{n - 1} = \dfrac{1}{n^2 - n}$ **10.** $3 + \dfrac{9}{n - 3} = \dfrac{27}{n^2 - 3n}$

11. $\dfrac{2}{n - 2} - \dfrac{n}{n + 5} = \dfrac{10n + 15}{n^2 + 3n - 10}$

12. $\dfrac{n}{n + 3} + \dfrac{1}{n - 4} = \dfrac{11 - n}{n^2 - n - 12}$

13. $\dfrac{2}{2x - 3} - \dfrac{2}{10x^2 - 13x - 3} = \dfrac{x}{5x + 1}$

14. $\dfrac{1}{3x + 4} + \dfrac{6}{6x^2 + 5x - 4} = \dfrac{x}{2x - 1}$

15. $\dfrac{2x}{x + 3} - \dfrac{3}{x - 6} = \dfrac{29}{x^2 - 3x - 18}$

16. $\dfrac{x}{x - 4} - \dfrac{2}{x + 8} = \dfrac{63}{x^2 + 4x - 32}$

17. $\dfrac{a}{a - 5} + \dfrac{2}{a - 6} = \dfrac{2}{a^2 - 11a + 30}$

18. $\dfrac{a}{a + 2} + \dfrac{3}{a + 4} = \dfrac{14}{a^2 + 6a + 8}$

19. $\dfrac{-1}{2x - 5} + \dfrac{2x - 4}{4x^2 - 25} = \dfrac{5}{6x + 15}$

20. $\dfrac{-2}{3x + 2} + \dfrac{x - 1}{9x^2 - 4} = \dfrac{3}{12x - 8}$

21. $\dfrac{7y + 2}{12y^2 + 11y - 15} - \dfrac{1}{3y + 5} = \dfrac{2}{4y - 3}$

22. $\dfrac{5y - 4}{6y^2 + y - 12} - \dfrac{2}{2y + 3} = \dfrac{5}{3y - 4}$

23. $\dfrac{2n}{6n^2 + 7n - 3} - \dfrac{n - 3}{3n^2 + 11n - 4} = \dfrac{5}{2n^2 + 11n + 12}$

24. $\dfrac{x + 1}{2x^2 + 7x - 4} - \dfrac{x}{2x^2 - 7x + 3} = \dfrac{1}{x^2 + x - 12}$

25. $\dfrac{1}{2x^2 - x - 1} + \dfrac{3}{2x^2 + x} = \dfrac{2}{x^2 - 1}$

26. $\dfrac{2}{n^2 + 4n} + \dfrac{3}{n^2 - 3n - 28} = \dfrac{5}{n^2 - 6n - 7}$

27. $\dfrac{x + 1}{x^3 - 9x} - \dfrac{1}{2x^2 + x - 21} = \dfrac{1}{2x^2 + 13x + 21}$

28. $\dfrac{x}{2x^2 + 5x} - \dfrac{x}{2x^2 + 7x + 5} = \dfrac{2}{x^2 + x}$

29. $\dfrac{4t}{4t^2 - t - 3} + \dfrac{2 - 3t}{3t^2 - t - 2} = \dfrac{1}{12t^2 + 17t + 6}$

30. $\dfrac{2t}{2t^2 + 9t + 10} + \dfrac{1 - 3t}{3t^2 + 4t - 4} = \dfrac{4}{6t^2 + 11t - 10}$

For Problems 31– 44, solve each equation for the indicated variable.

31. $y = \dfrac{5}{6}x + \dfrac{2}{9}$ for x

32. $y = \dfrac{3}{4}x - \dfrac{2}{3}$ for x

33. $\dfrac{-2}{x - 4} = \dfrac{5}{y - 1}$ for y

34. $\dfrac{7}{y - 3} = \dfrac{3}{x + 1}$ for y

35. $I = \dfrac{100M}{C}$ for M

36. $V = C\left(1 - \dfrac{T}{N}\right)$ for T

37. $\dfrac{R}{S} = \dfrac{T}{S + T}$ for R

38. $\dfrac{1}{R} = \dfrac{1}{S} + \dfrac{1}{T}$ for R

39. $\dfrac{y - 1}{x - 3} = \dfrac{b - 1}{a - 3}$ for y

40. $y = -\dfrac{a}{b}x + \dfrac{c}{d}$ for x

41. $\dfrac{x}{a} + \dfrac{y}{b} = 1$ for y

42. $\dfrac{y - b}{x} = m$ for y

43. $\dfrac{y - 1}{x + 6} = \dfrac{-2}{3}$ for y

44. $\dfrac{y + 5}{x - 2} = \dfrac{3}{7}$ for y

Set up an equation and solve each of the following problems.

45. Kent drives his Mazda 270 miles in the same time that it takes Dave to drive his Nissan 250 miles. If Kent averages 4 miles per hour faster than Dave, find their rates.

46. Suppose that Wendy rides her bicycle 30 miles in the same time that it takes Kim to ride her bicycle 20 miles. If Wendy rides 5 miles per hour faster than Kim, find the rate of each.

47. An inlet pipe can fill a tank (see Figure 4.2) in 10 minutes. A drain can empty the tank in 12 minutes. If the tank is empty, and both the pipe and drain are open, how long will it take before the tank overflows?

Figure 4.2

48. Barry can do a certain job in 3 hours, whereas it takes Sanchez 5 hours to do the same job. How long would it take them to do the job working together?

49. Connie can type 600 words in 5 minutes less than it takes Katie to type 600 words. If Connie types at a rate of 20 words per minute faster than Katie types, find the typing rate of each woman.

50. Walt can mow a lawn in 1 hour, and his son, Malik, can mow the same lawn in 50 minutes. One day Malik started mowing the lawn by himself and worked for 30 minutes. Then Walt joined him and they finished the lawn. How long did it take them to finish mowing the lawn after Walt started to help?

51. Plane A can travel 1400 miles in 1 hour less time than it takes plane B to travel 2000 miles. The rate of plane B is 50 miles per hour greater than the rate of plane A. Find the times and rates of both planes.

52. To travel 60 miles, it takes Sue, riding a moped, 2 hours less time than it takes Doreen to travel 50 miles riding a bicycle. Sue travels 10 miles per hour faster than Doreen. Find the times and rates of both girls.

53. It takes Amy twice as long to deliver papers as it does Nancy. How long would it take each girl to deliver the papers by herself if they can deliver the papers together in 40 minutes?

54. If two inlet pipes are both open, they can fill a pool in 1 hour and 12 minutes. One of the pipes can fill the pool by itself in 2 hours. How long would it take the other pipe to fill the pool by itself?

55. Rod agreed to mow a vacant lot for $12. It took him an hour longer than he had anticipated, so he earned $1 per hour less than he had originally calculated. How long had he anticipated that it would take him to mow the lot?

56. Last week Al bought some golf balls for $20. The next day they were on sale for $0.50 per ball less, and he bought $22.50 worth of balls. If he purchased 5 more balls on the second day than on the first day, how many did he buy each day and at what price per ball?

57. Debbie rode her bicycle out into the country for a distance of 24 miles. On the way back, she took a much shorter route of 12 miles and made the return trip in one-half hour less time. If her rate out into the country was 4 miles per hour greater than her rate on the return trip, find both rates.

58. Felipe jogs for 10 miles and then walks another 10 miles. He jogs $2\frac{1}{2}$ miles per hour faster than he walks, and the entire distance of 20 miles takes 6 hours. Find the rate at which he walks and the rate at which he jogs.

■ ■ ■ THOUGHTS INTO WORDS

59. Why is it important to consider more than one way to do a problem?

60. Write a paragraph or two summarizing the new ideas about problem solving you have acquired thus far in this course.

(4.1) Any number that can be written in the form $\dfrac{a}{b}$, where a and b are integers and $b \neq 0$, is called a **rational number**.

A **rational expression** is defined as the indicated quotient of two polynomials. The following properties pertain to rational numbers and rational expressions.

1. $\dfrac{-a}{b} = \dfrac{a}{-b} = -\dfrac{a}{b}$

2. $\dfrac{-a}{-b} = \dfrac{a}{b}$

3. $\dfrac{a \cdot k}{b \cdot k} = \dfrac{a}{b}$ Fundamental principle of fractions

(4.2) Multiplication and division of rational expressions are based on the following definitions:

1. $\dfrac{a}{b} \cdot \dfrac{c}{d} = \dfrac{ac}{bd}$ Multiplication

2. $\dfrac{a}{b} \div \dfrac{c}{d} = \dfrac{a}{b} \cdot \dfrac{d}{c} = \dfrac{ad}{bc}$ Division

(4.3) Addition and subtraction of rational expressions are based on the following definitions:

1. $\dfrac{a}{b} + \dfrac{c}{b} = \dfrac{a + c}{b}$ Addition

2. $\dfrac{a}{b} - \dfrac{c}{b} = \dfrac{a - c}{b}$ Subtraction

(4.4) The following basic procedure is used to add or subtract rational expressions.

1. Factor the denominators.

2. Find the LCD.

3. Change each fraction to an equivalent fraction that has the LCD as its denominator.

4. Combine the numerators and place over the LCD.

5. Simplify by performing the addition or subtraction.

6. Look for ways to reduce the resulting fraction.

Fractional forms that contain rational numbers or rational expressions in the numerators and/or denominators are called **complex fractions**. The fundamental principle of fractions serves as a basis for simplifying complex fractions.

(4.5) To divide a polynomial by a monomial, we divide each term of the polynomial by the monomial. The procedure for dividing a polynomial by a polynomial, rather than a monomial, resembles the long-division process in arithmetic. (See the examples in Section 4.5.) Synthetic division is a shortcut to the long-division process when the divisor is of the form $x - k$.

(4.6) **To solve a fractional equation**, it is often easiest to begin by multiplying both sides of the equation by the LCD of all of the denominators in the equation. If an equation contains a variable in one or more denominators, then we must be careful to avoid any value of the variable that makes the denominator zero.

A **ratio** is the comparison of two numbers by division. A statement of equality between two ratios is a **proportion**.

We can treat some fractional equations as proportions, and we can solve them by applying the following property. This property is often called the **cross-multiplication** property:

$$\text{If } \frac{a}{b} = \frac{c}{d}, \quad \text{then } ad = bc.$$

(4.7) The techniques that we use to solve fractional equations can also be used to change the form of formulas containing rational expressions so that we can use those formulas to solve problems.

| **Chapter 4** | **Review Problem Set** |

For Problems 1–6, simplify each rational expression.

1. $\dfrac{26x^2y^3}{39x^4y^2}$

2. $\dfrac{a^2 - 9}{a^2 + 3a}$

3. $\dfrac{n^2 - 3n - 10}{n^2 + n - 2}$

4. $\dfrac{x^4 - 1}{x^3 - x}$

5. $\dfrac{8x^3 - 2x^2 - 3x}{12x^2 - 9x}$

6. $\dfrac{x^4 - 7x^2 - 30}{2x^4 + 7x^2 + 3}$

For Problems 7–10, simplify each complex fraction.

7. $\dfrac{\dfrac{5}{8} - \dfrac{1}{2}}{\dfrac{1}{6} + \dfrac{3}{4}}$

8. $\dfrac{\dfrac{3}{2x} + \dfrac{5}{3y}}{\dfrac{4}{x} - \dfrac{3}{4y}}$

9. $\dfrac{\dfrac{3}{x - 2} - \dfrac{4}{x^2 - 4}}{\dfrac{2}{x + 2} + \dfrac{1}{x - 2}}$

10. $1 - \dfrac{1}{2 - \dfrac{1}{x}}$

For Problems 11–22, perform the indicated operations, and express your answers in simplest form.

11. $\dfrac{6xy^2}{7y^3} \div \dfrac{15x^2y}{5x^2}$

12. $\dfrac{9ab}{3a + 6} \cdot \dfrac{a^2 - 4a - 12}{a^2 - 6a}$

13. $\dfrac{n^2 + 10n + 25}{n^2 - n} \cdot \dfrac{5n^3 - 3n^2}{5n^2 + 22n - 15}$

14. $\dfrac{x^2 - 2xy - 3y^2}{x^2 + 9y^2} \div \dfrac{2x^2 + xy - y^2}{2x^2 - xy}$

15. $\dfrac{2x + 1}{5} + \dfrac{3x - 2}{4}$

16. $\dfrac{3}{2n} + \dfrac{5}{3n} - \dfrac{1}{9}$

17. $\dfrac{3x}{x + 7} - \dfrac{2}{x}$

18. $\dfrac{10}{x^2 - 5x} + \dfrac{2}{x}$

19. $\dfrac{3}{n^2 - 5n - 36} + \dfrac{2}{n^2 + 3n - 4}$

20. $\dfrac{3}{2y + 3} + \dfrac{5y - 2}{2y^2 - 9y - 18} - \dfrac{1}{y - 6}$

21. $(18x^2 + 9x - 2) \div (3x + 2)$

22. $(3x^3 + 5x^2 - 6x - 2) \div (x + 4)$

For Problems 23–32, solve each equation.

23. $\dfrac{4x + 5}{3} + \dfrac{2x - 1}{5} = 2$

24. $\dfrac{3}{4x} + \dfrac{4}{5} = \dfrac{9}{10x}$

25. $\dfrac{a}{a - 2} - \dfrac{3}{2} = \dfrac{2}{a - 2}$

26. $\dfrac{4}{5y - 3} = \dfrac{2}{3y + 7}$

27. $n + \dfrac{1}{n} = \dfrac{53}{14}$

28. $\dfrac{1}{2x - 7} + \dfrac{x - 5}{4x^2 - 49} = \dfrac{4}{6x - 21}$

29. $\dfrac{x}{2x + 1} - 1 = \dfrac{-4}{7(x - 2)}$

30. $\dfrac{2x}{-5} = \dfrac{3}{4x - 13}$

31. $\dfrac{2n}{2n^2 + 11n - 21} - \dfrac{n}{n^2 + 5n - 14} = \dfrac{3}{n^2 + 5n - 14}$

32. $\dfrac{2}{t^2 - t - 6} + \dfrac{t + 1}{t^2 + t - 12} = \dfrac{t}{t^2 + 6t + 8}$

33. Solve $\dfrac{y - 6}{x + 1} = \dfrac{3}{4}$ for y.

34. Solve $\dfrac{x}{a} - \dfrac{y}{b} = 1$ for y.

For Problems 35–40, set up an equation, and solve the problem.

35. A sum of $1400 is to be divided between two people in the ratio of $\dfrac{3}{5}$. How much does each person receive?

36. Working together, Dan and Julio can mow a lawn in 12 minutes. Julio can mow the lawn by himself in 10 minutes less time than it takes Dan by himself. How long does it take each of them to mow the lawn alone?

37. Suppose that car A can travel 250 miles in 3 hours less time than it takes car B to travel 440 miles. The rate of car B is 5 miles per hour faster than that of car A. Find the rates of both cars.

38. Mark can overhaul an engine in 20 hours, and Phil can do the same job by himself in 30 hours. If they both work together for a time and then Mark finishes the job by himself in 5 hours, how long did they work together?

39. Kelly contracted to paint a house for $640. It took him 20 hours longer than he had anticipated, so he earned $1.60 per hour less than he had calculated. How long had he anticipated that it would take him to paint the house?

40. Nasser rode his bicycle 66 miles in $4\frac{1}{2}$ hours. For the first 40 miles he averaged a certain rate, and then for the last 26 miles he reduced his rate by 3 miles per hour. Find his rate for the last 26 miles.

For Problems 1–4, simplify each rational expression.

1. $\dfrac{39x^2y^3}{72x^3y}$

2. $\dfrac{3x^2 + 17x - 6}{x^3 - 36x}$

3. $\dfrac{6n^2 - 5n - 6}{3n^2 + 14n + 8}$

4. $\dfrac{2x - 2x^2}{x^2 - 1}$

For Problems 5–13, perform the indicated operations, and express your answers in simplest form.

5. $\dfrac{5x^2y}{8x} \cdot \dfrac{12y^2}{20xy}$

6. $\dfrac{5a + 5b}{20a + 10b} \cdot \dfrac{a^2 - ab}{2a^2 + 2ab}$

7. $\dfrac{3x^2 + 10x - 8}{5x^2 + 19x - 4} \div \dfrac{3x^2 - 23x + 14}{x^2 - 3x - 28}$

8. $\dfrac{3x - 1}{4} + \dfrac{2x + 5}{6}$

9. $\dfrac{5x - 6}{3} - \dfrac{x - 12}{6}$

10. $\dfrac{3}{5n} + \dfrac{2}{3} - \dfrac{7}{3n}$

11. $\dfrac{3x}{x - 6} + \dfrac{2}{x}$

12. $\dfrac{9}{x^2 - x} - \dfrac{2}{x}$

13. $\dfrac{3}{2n^2 + n - 10} + \dfrac{5}{n^2 + 5n - 14}$

14. Divide $3x^3 + 10x^2 - 9x - 4$ by $x + 4$.

15. Simplify the complex fraction $\dfrac{\dfrac{3}{2x} - \dfrac{1}{6}}{\dfrac{2}{3x} + \dfrac{3}{4}}$.

16. Solve $\dfrac{x + 2}{y - 4} = \dfrac{3}{4}$ for y.

For Problems 17–22, solve each equation.

17. $\dfrac{x - 1}{2} - \dfrac{x + 2}{5} = -\dfrac{3}{5}$

18. $\dfrac{5}{4x} + \dfrac{3}{2} = \dfrac{7}{5x}$

19. $\dfrac{-3}{4n - 1} = \dfrac{-2}{3n + 11}$

20. $n - \dfrac{5}{n} = 4$

21. $\dfrac{6}{x - 4} - \dfrac{4}{x + 3} = \dfrac{8}{x - 4}$

22. $\dfrac{1}{3x - 1} + \dfrac{x - 2}{9x^2 - 1} = \dfrac{7}{6x - 2}$

For Problems 23–25, set up an equation and solve the problem.

23. The denominator of a rational number is 9 less than three times the numerator. The number in simplest form is $\dfrac{3}{8}$. Find the number.

24. It takes Jodi three times as long to deliver papers as it does Jannie. Together they can deliver the papers in 15 minutes. How long would it take Jodi by herself?

25. René can ride her bike 60 miles in 1 hour less time than it takes Sue to ride 60 miles. René's rate is 3 miles per hour faster than Sue's rate. Find René's rate.

5

Exponents and Radicals

By knowing the time it takes for the pendulum to swing from one side to the other side and back, the formula, $T = 2\pi\sqrt{\dfrac{L}{32}}$, can be solved to find the length of the pendulum.

© Jonathan Nourok /PhotoEdit

How long will it take a pendulum that is 1.5 feet long to swing from one side to the other side and back? The formula $T = 2\pi\sqrt{\dfrac{L}{32}}$ can be used to determine that it will take approximately 1.4 seconds.

It is not uncommon in mathematics to find two separately developed concepts that are closely related to each other. In this chapter, we will first develop the concepts of exponent and root individually and then show how they merge to become even more functional as a unified idea.

5.1   Using Integers as Exponents

Thus far in the text we have used only positive integers as exponents. In Chapter 1 the expression b^n, where b is any real number and n is a positive integer, was defined by

$$b^n = b \cdot b \cdot b \cdot \ldots \cdot b \qquad n \text{ factors of } b$$

Then, in Chapter 3, some of the parts of the following property served as a basis for manipulation with polynomials.

Property 5.1

If m and n are positive integers, and a and b are real numbers (and $b \neq 0$ whenever it appears in a denominator), then

1. $b^n \cdot b^m = b^{n+m}$ **2.** $(b^n)^m = b^{mn}$

3. $(ab)^n = a^n b^n$

4. $\left(\dfrac{a}{b}\right)^n = \dfrac{a^n}{b^n}$

5. $\dfrac{b^n}{b^m} = b^{n-m}$ when $n > m$

$\dfrac{b^n}{b^m} = 1$ when $n = m$

$\dfrac{b^n}{b^m} = \dfrac{1}{b^{m-n}}$ when $n < m$

We are now ready to extend the concept of an exponent to include the use of zero and the negative integers as exponents.

First, let's consider the use of zero as an exponent. We want to use zero in such a way that the previously listed properties continue to hold. If $b^n \cdot b^m = b^{n+m}$ is to hold, then $x^4 \cdot x^0 = x^{4+0} = x^4$. In other words, x^0 *acts like* 1 because $x^4 \cdot x^0 = x^4$. This line of reasoning suggests the following definition.

Definition 5.1

If b is a nonzero real number, then

$$b^0 = 1$$

According to Definition 5.1, the following statements are all true.

$$5^0 = 1 \qquad\qquad (-413)^0 = 1$$

$$\left(\frac{3}{11}\right)^0 = 1 \qquad\qquad n^0 = 1, \quad n \neq 0$$

$$(x^3 y^4)^0 = 1, \qquad x \neq 0, y \neq 0$$

We can use a similar line of reasoning to motivate a definition for the use of negative integers as exponents. Consider the example $x^4 \cdot x^{-4}$. If $b^n \cdot b^m = b^{n+m}$ is to hold, then $x^4 \cdot x^{-4} = x^{4+(-4)} = x^0 = 1$. Thus x^{-4} must be the reciprocal of x^4, because their product is 1. That is,

$$x^{-4} = \frac{1}{x^4}$$

This suggests the following general definition.

Definition 5.2

If n is a positive integer, and b is a nonzero real number, then

$$b^{-n} = \frac{1}{b^n}$$

According to Definition 5.2, the following statements are all true.

$$x^{-5} = \frac{1}{x^5} \qquad\qquad 2^{-4} = \frac{1}{2^4} = \frac{1}{16}$$

$$10^{-2} = \frac{1}{10^2} = \frac{1}{100} \quad \text{or} \quad 0.01 \qquad \frac{2}{x^{-3}} = \frac{2}{\dfrac{1}{x^3}} = (2)\left(\frac{x^3}{1}\right) = 2x^3$$

$$\left(\frac{3}{4}\right)^{-2} = \frac{1}{\left(\dfrac{3}{4}\right)^2} = \frac{1}{\dfrac{9}{16}} = \frac{16}{9}$$

It can be verified (although it is beyond the scope of this text) that all of the parts of Property 5.1 hold for *all integers*. In fact, the following equality can replace the three separate statements for part (5).

$$\frac{b^n}{b^m} = b^{n-m} \quad \text{for all integers } n \text{ and } m$$

Let's restate Property 5.1 as it holds for all integers and include, at the right, a "name tag" for easy reference.

Property 5.2

If m and n are integers, and a and b are real numbers (and $b \neq 0$ whenever it appears in a denominator), then

1. $b^n \cdot b^m = b^{n+m}$ Product of two powers

2. $(b^n)^m = b^{mn}$ Power of a power

3. $(ab)^n = a^n b^n$ Power of a product

4. $\left(\dfrac{a}{b}\right)^n = \dfrac{a^n}{b^n}$ Power of a quotient

5. $\dfrac{b^n}{b^m} = b^{n-m}$ Quotient of two powers

Having the use of all integers as exponents enables us to work with a large variety of numerical and algebraic expressions. Let's consider some examples that illustrate the use of the various parts of Property 5.2.

E X A M P L E   1

Simplify each of the following numerical expressions.

(a) $10^{-3} \cdot 10^2$ **(b)** $(2^{-3})^{-2}$ **(c)** $(2^{-1} \cdot 3^2)^{-1}$

(d) $\left(\dfrac{2^{-3}}{3^{-2}}\right)^{-1}$ **(e)** $\dfrac{10^{-2}}{10^{-4}}$

Solution

(a) $10^{-3} \cdot 10^2 = 10^{-3+2}$ Product of two powers

$= 10^{-1}$

$= \dfrac{1}{10^1} = \dfrac{1}{10}$

(b) $(2^{-3})^{-2} = 2^{(-2)(-3)}$ Power of a power

$= 2^6 = 64$

(c) $(2^{-1} \cdot 3^2)^{-1} = (2^{-1})^{-1}(3^2)^{-1}$ Power of a product

$= 2^1 \cdot 3^{-2}$

$= \dfrac{2^1}{3^2} = \dfrac{2}{9}$

(d) $\left(\dfrac{2^{-3}}{3^{-2}}\right)^{-1} = \dfrac{(2^{-3})^{-1}}{(3^{-2})^{-1}}$ Power of a quotient

$$= \dfrac{2^3}{3^2} = \dfrac{8}{9}$$

(e) $\dfrac{10^{-2}}{10^{-4}} = 10^{-2-(-4)}$ Quotient of two powers

$$= 10^2 = 100$$ ■

EXAMPLE 2   Simplify each of the following; express final results without using zero or negative integers as exponents.

(a) $x^2 \cdot x^{-5}$ **(b)** $(x^{-2})^4$ **(c)** $(x^2y^{-3})^{-4}$

(d) $\left(\dfrac{a^3}{b^{-5}}\right)^{-2}$ **(e)** $\dfrac{x^{-4}}{x^{-2}}$

Solution

(a) $x^2 \cdot x^{-5} = x^{2+(-5)}$ Product of two powers

$$= x^{-3}$$

$$= \dfrac{1}{x^3}$$

(b) $(x^{-2})^4 = x^{4(-2)}$ Power of a power

$$= x^{-8}$$

$$= \dfrac{1}{x^8}$$

(c) $(x^2y^{-3})^{-4} = (x^2)^{-4}(y^{-3})^{-4}$ Power of a product

$$= x^{-4(2)}y^{-4(-3)}$$

$$= x^{-8}y^{12}$$

$$= \dfrac{y^{12}}{x^8}$$

(d) $\left(\dfrac{a^3}{b^{-5}}\right)^{-2} = \dfrac{(a^3)^{-2}}{(b^{-5})^{-2}}$ Power of a quotient

$$= \dfrac{a^{-6}}{b^{10}}$$

$$= \dfrac{1}{a^6b^{10}}$$

(e) $\dfrac{x^{-4}}{x^{-2}} = x^{-4-(-2)}$ Quotient of two powers

$= x^{-2}$

$= \dfrac{1}{x^2}$ ■

E X A M P L E   3

Find the indicated products and quotients; express your results using positive integral exponents only.

(a) $(3x^2 y^{-4})(4x^{-3} y)$ **(b)** $\dfrac{12a^3 b^2}{-3a^{-1} b^5}$ **(c)** $\left(\dfrac{15x^{-1} y^2}{5xy^{-4}} \right)^{-1}$

Solution

(a) $(3x^2 y^{-4})(4x^{-3} y) = 12x^{2+(-3)} y^{-4+1}$

$= 12x^{-1} y^{-3}$

$= \dfrac{12}{xy^3}$

(b) $\dfrac{12a^3 b^2}{-3a^{-1} b^5} = -4a^{3-(-1)} b^{2-5}$

$= -4a^4 b^{-3}$

$= -\dfrac{4a^4}{b^3}$

(c) $\left(\dfrac{15x^{-1} y^2}{5xy^{-4}} \right)^{-1} = (3x^{-1-1} y^{2-(-4)})^{-1}$ Note that we are first simplifying inside the parentheses.

$= (3x^{-2} y^6)^{-1}$

$= 3^{-1} x^2 y^{-6}$

$= \dfrac{x^2}{3y^6}$ ■

The final examples of this section show the simplification of numerical and algebraic expressions that involve sums and differences. In such cases, we use Definition 5.2 to change from negative to positive exponents so that we can proceed in the usual way.

E X A M P L E   4

Simplify $2^{-3} + 3^{-1}$.

Solution

$2^{-3} + 3^{-1} = \dfrac{1}{2^3} + \dfrac{1}{3^1}$

$$= \frac{1}{8} + \frac{1}{3}$$

$$= \frac{3}{24} + \frac{8}{24} \qquad \text{Use 24 as the LCD.}$$

$$= \frac{11}{24} \qquad\qquad\qquad ■$$

EXAMPLE 5

Simplify $(4^{-1} - 3^{-2})^{-1}$.

Solution

$$(4^{-1} - 3^{-2})^{-1} = \left(\frac{1}{4^1} - \frac{1}{3^2} \right)^{-1} \qquad \text{Apply } b^{-n} = \frac{1}{b^n} \text{ to } 4^{-1} \text{ and to } 3^{-2}.$$

$$= \left(\frac{1}{4} - \frac{1}{9} \right)^{-1}$$

$$= \left(\frac{9}{36} - \frac{4}{36} \right)^{-1} \qquad \text{Use 36 as the LCD.}$$

$$= \left(\frac{5}{36} \right)^{-1}$$

$$= \frac{1}{\left(\frac{5}{36} \right)^1} \qquad\qquad \text{Apply } b^{-n} = \frac{1}{b^n}.$$

$$= \frac{1}{\frac{5}{36}} = \frac{36}{5} \qquad\qquad ■$$

EXAMPLE 6

Express $a^{-1} + b^{-2}$ as a single fraction involving positive exponents only.

Solution

$$a^{-1} + b^{-2} = \frac{1}{a^1} + \frac{1}{b^2} \qquad\qquad \text{Use } ab^2 \text{ as the LCD.}$$

$$= \left(\frac{1}{a} \right) \left(\frac{b^2}{b^2} \right) + \left(\frac{1}{b^2} \right) \left(\frac{a}{a} \right) \qquad \begin{array}{l}\text{Change to equivalent fractions} \\ \text{with } ab^2 \text{ as the LCD.}\end{array}$$

$$= \frac{b^2}{ab^2} + \frac{a}{ab^2}$$

$$= \frac{b^2 + a}{ab^2} \qquad\qquad ■$$

Problem Set 5.1

For Problems 1–42, simplify each numerical expression.

1. 3^{-3}

2. 2^{-4}

3. -10^{-2}

4. 10^{-3}

5. $\dfrac{1}{3^{-4}}$

6. $\dfrac{1}{2^{-6}}$

7. $-\left(\dfrac{1}{3}\right)^{-3}$

8. $\left(\dfrac{1}{2}\right)^{-3}$

9. $\left(-\dfrac{1}{2}\right)^{-3}$

10. $\left(\dfrac{2}{7}\right)^{-2}$

11. $\left(-\dfrac{3}{4}\right)^{0}$

12. $\dfrac{1}{\left(\dfrac{4}{5}\right)^{-2}}$

13. $\dfrac{1}{\left(\dfrac{3}{7}\right)^{-2}}$

14. $-\left(\dfrac{5}{6}\right)^{0}$

15. $2^{7} \cdot 2^{-3}$

16. $3^{-4} \cdot 3^{6}$

17. $10^{-5} \cdot 10^{2}$

18. $10^{4} \cdot 10^{-6}$

19. $10^{-1} \cdot 10^{-2}$

20. $10^{-2} \cdot 10^{-2}$

21. $(3^{-1})^{-3}$

22. $(2^{-2})^{-4}$

23. $(5^{3})^{-1}$

24. $(3^{-1})^{3}$

25. $(2^{3} \cdot 3^{-2})^{-1}$

26. $(2^{-2} \cdot 3^{-1})^{-3}$

27. $(4^{2} \cdot 5^{-1})^{2}$

28. $(2^{-3} \cdot 4^{-1})^{-1}$

29. $\left(\dfrac{2^{-1}}{5^{-2}}\right)^{-1}$

30. $\left(\dfrac{2^{-4}}{3^{-2}}\right)^{-2}$

31. $\left(\dfrac{2^{-1}}{3^{-2}}\right)^{2}$

32. $\left(\dfrac{3^{2}}{5^{-1}}\right)^{-1}$

33. $\dfrac{3^{3}}{3^{-1}}$

34. $\dfrac{2^{-2}}{2^{3}}$

35. $\dfrac{10^{-2}}{10^{2}}$

36. $\dfrac{10^{-2}}{10^{-5}}$

37. $2^{-2} + 3^{-2}$

38. $2^{-4} + 5^{-1}$

39. $\left(\dfrac{1}{3}\right)^{-1} - \left(\dfrac{2}{5}\right)^{-1}$

40. $\left(\dfrac{3}{2}\right)^{-1} - \left(\dfrac{1}{4}\right)^{-1}$

41. $(2^{-3} + 3^{-2})^{-1}$

42. $(5^{-1} - 2^{-3})^{-1}$

For Problems 43–62, simplify each expression. Express final results without using zero or negative integers as exponents.

43. $x^{2} \cdot x^{-8}$

44. $x^{-3} \cdot x^{-4}$

45. $a^{3} \cdot a^{-5} \cdot a^{-1}$

46. $b^{-2} \cdot b^{3} \cdot b^{-6}$

47. $(a^{-4})^{2}$

48. $(b^{4})^{-3}$

49. $(x^{2}y^{-6})^{-1}$

50. $(x^{5}y^{-1})^{-3}$

51. $(ab^{3}c^{-2})^{-4}$

52. $(a^{3}b^{-3}c^{-2})^{-5}$

53. $(2x^{3}y^{-4})^{-3}$

54. $(4x^{5}y^{-2})^{-2}$

55. $\left(\dfrac{x^{-1}}{y^{-4}}\right)^{-3}$

56. $\left(\dfrac{y^{3}}{x^{-4}}\right)^{-2}$

57. $\left(\dfrac{3a^{-2}}{2b^{-1}}\right)^{-2}$

58. $\left(\dfrac{2xy^{2}}{5a^{-1}b^{-2}}\right)^{-1}$

59. $\dfrac{x^{-6}}{x^{-4}}$

60. $\dfrac{a^{-2}}{a^{2}}$

61. $\dfrac{a^{3}b^{-2}}{a^{-2}b^{-4}}$

62. $\dfrac{x^{-3}y^{-4}}{x^{2}y^{-1}}$

For Problems 63–74, find the indicated products and quotients. Express final results using positive integral exponents only.

63. $(2xy^{-1})(3x^{-2}y^{4})$

64. $(-4x^{-1}y^{2})(6x^{3}y^{-4})$

65. $(-7a^{2}b^{-5})(-a^{-2}b^{7})$

66. $(-9a^{-3}b^{-6})(-12a^{-1}b^{4})$

67. $\dfrac{28x^{-2}y^{-3}}{4x^{-3}y^{-1}}$

68. $\dfrac{63x^{2}y^{-4}}{7xy^{-4}}$

69. $\dfrac{-72a^{2}b^{-4}}{6a^{3}b^{-7}}$

70. $\dfrac{108a^{-5}b^{-4}}{9a^{-2}b}$

71. $\left(\dfrac{35x^{-1}y^{-2}}{7x^{4}y^{3}}\right)^{-1}$

72. $\left(\dfrac{-48ab^{2}}{-6a^{3}b^{5}}\right)^{-2}$

73. $\left(\dfrac{-36a^{-1}b^{-6}}{4a^{-1}b^{4}}\right)^{-2}$

74. $\left(\dfrac{8xy^{3}}{-4x^{4}y}\right)^{-3}$

For Problems 75–84, express each of the following as a single fraction involving positive exponents only.

75. $x^{-2} + x^{-3}$

76. $x^{-1} + x^{-5}$

77. $x^{-3} - y^{-1}$

78. $2x^{-1} - 3y^{-2}$

79. $3a^{-2} + 4b^{-1}$

80. $a^{-1} + a^{-1}b^{-3}$

81. $x^{-1}y^{-2} - xy^{-1}$

82. $x^2y^{-2} - x^{-1}y^{-3}$

83. $2x^{-1} - 3x^{-2}$

84. $5x^{-2}y + 6x^{-1}y^{-2}$

■ ■ ■ **THOUGHTS INTO WORDS**

85. Is the following simplification process correct?

$$(3^{-2})^{-1} = \left(\frac{1}{3^2}\right)^{-1} = \left(\frac{1}{9}\right)^{-1} = \frac{1}{\left(\frac{1}{9}\right)^1} = 9$$

Could you suggest a better way to do the problem?

86. Explain how to simplify $(2^{-1} \cdot 3^{-2})^{-1}$ and also how to simplify $(2^{-1} + 3^{-2})^{-1}$.

■ ■ ■ **FURTHER INVESTIGATIONS**

87. Use a calculator to check your answers for Problems 1–42.

88. Use a calculator to simplify each of the following numerical expressions. Express your answers to the nearest hundredth.

(a) $(2^{-3} + 3^{-3})^{-2}$

(b) $(4^{-3} - 2^{-1})^{-2}$

(c) $(5^{-3} - 3^{-5})^{-1}$

(d) $(6^{-2} + 7^{-4})^{-2}$

(e) $(7^{-3} - 2^{-4})^{-2}$

(f) $(3^{-4} + 2^{-3})^{-3}$

5.2 Roots and Radicals

To **square a number** means to raise it to the second power — that is, to use the number as a factor twice.

$$4^2 = 4 \cdot 4 = 16 \qquad \text{Read "four squared equals sixteen."}$$

$$10^2 = 10 \cdot 10 = 100$$

$$\left(\frac{1}{2}\right)^2 = \frac{1}{2} \cdot \frac{1}{2} = \frac{1}{4}$$

$$(-3)^2 = (-3)(-3) = 9$$

A **square root of a number** is one of its two equal factors. Thus 4 is a square root of 16 because $4 \cdot 4 = 16$. Likewise, -4 is also a square root of 16 because

$(-4)(-4) = 16$. In general, a is a square root of b if $a^2 = b$. The following generalizations are a direct consequence of the previous statement.

1. Every positive real number has two square roots; one is positive and the other is negative. They are opposites of each other.

2. Negative real numbers have no real number square roots because any real number except zero is positive when squared.

3. The square root of 0 is 0.

The symbol $\sqrt{}$, called a **radical sign**, is used to designate the nonnegative square root. The number under the radical sign is called the **radicand**. The entire expression, such as $\sqrt{16}$, is called a **radical**.

$\sqrt{16} = 4$ $\qquad\qquad$ $\sqrt{16}$ indicates the nonnegative or **principal square root** of 16.

$-\sqrt{16} = -4$ $\qquad\qquad$ $-\sqrt{16}$ indicates the negative square root of 16.

$\sqrt{0} = 0$ $\qquad\qquad$ Zero has only one square root. Technically, we could write $-\sqrt{0} = -0 = 0$.

$\sqrt{-4}$ is not a real number.

$-\sqrt{-4}$ is not a real number.

In general, the following definition is useful.

Definition 5.3

If $a \geq 0$ and $b \geq 0$, then $\sqrt{b} = a$ if and only if $a^2 = b$; a is called the **principal square root of b**.

To **cube a number** means to raise it to the third power — that is, to use the number as a factor three times.

$2^3 = 2 \cdot 2 \cdot 2 = 8$ $\qquad$ Read "two cubed equals eight."

$4^3 = 4 \cdot 4 \cdot 4 = 64$

$\left(\dfrac{2}{3}\right)^3 = \dfrac{2}{3} \cdot \dfrac{2}{3} \cdot \dfrac{2}{3} = \dfrac{8}{27}$

$(-2)^3 = (-2)(-2)(-2) = -8$

A **cube root of a number** is one of its three equal factors. Thus 2 is a cube root of 8 because $2 \cdot 2 \cdot 2 = 8$. (In fact, 2 is the only real number that is a cube root of 8.) Furthermore, -2 is a cube root of -8 because $(-2)(-2)(-2) = -8$. (In fact, -2 is the only real number that is a cube root of -8.)

In general, a is a cube root of b if $a^3 = b$. The following generalizations are a direct consequence of the previous statement.

1. Every positive real number has one positive real number cube root.

2. Every negative real number has one negative real number cube root.

3. The cube root of 0 is 0.

Remark: Technically, every nonzero real number has three cube roots, but only one of them is a real number. The other two roots are classified as complex numbers. We are restricting our work at this time to the set of real numbers.

The symbol $\sqrt[3]{}$ designates the cube root of a number. Thus we can write

$$\sqrt[3]{8} = 2 \qquad\qquad \sqrt[3]{\frac{1}{27}} = \frac{1}{3}$$

$$\sqrt[3]{-8} = -2 \qquad\qquad \sqrt[3]{-\frac{1}{27}} = -\frac{1}{3}$$

In general, the following definition is useful.

Definition 5.4

$$\sqrt[3]{b} = a \quad \text{if and only if } a^3 = b.$$

In Definition 5.4, if b is a positive number, then a, the cube root, is a positive number; whereas if b is a negative number, then a, the cube root, is a negative number. The number a is called the principal cube root of b or simply the cube root of b.

The concept of root can be extended to fourth roots, fifth roots, sixth roots, and, in general, nth roots.

Definition 5.5

The nth root of b is a,   if and only if $a^n = b$.

We can make the following generalizations.

If n is an even positive integer, then the following statements are true.

1. Every positive real number has exactly two real nth roots — one positive and one negative. For example, the real fourth roots of 16 are 2 and -2.

2. Negative real numbers do not have real nth roots. For example, there are no real fourth roots of -16.

If n is an odd positive integer greater than 1, then the following statements are true.

1. Every real number has exactly one real nth root.

2. The real nth root of a positive number is positive. For example, the fifth root of 32 is 2.

3. The real nth root of a negative number is negative. For example, the fifth root of -32 is -2.

The symbol $\sqrt[n]{}$ designates the principal nth root. To complete our terminology, the n in the radical $\sqrt[n]{b}$ is called the index of the radical. If $n = 2$, we commonly write $\sqrt{b}$ instead of $\sqrt[2]{b}$.

The following chart can help summarize this information with respect to $\sqrt[n]{b}$, where n is a positive integer greater than 1.

	If b is		
	Positive	**Zero**	**Negative**
n is even	$\sqrt[n]{b}$ is a positive real number	$\sqrt[n]{b} = 0$	$\sqrt[n]{b}$ is not a real number
n is odd	$\sqrt[n]{b}$ is a positive real number	$\sqrt[n]{b} = 0$	$\sqrt[n]{b}$ is a negative real number

Consider the following examples.

$$\sqrt[4]{81} = 3 \qquad\qquad \text{because } 3^4 = 81$$
$$\sqrt[5]{32} = 2 \qquad\qquad \text{because } 2^5 = 32$$
$$\sqrt[5]{-32} = -2 \qquad\qquad \text{because } (-2)^5 = -32$$
$$\sqrt[4]{-16} \text{ is not a real number} \qquad \text{because any real number, except zero, is positive when raised to the fourth power}$$

The following property is a direct consequence of Definition 5.5.

Property 5.3

1. $(\sqrt[n]{b})^n = b$ n is any positive integer greater than 1.

2. $\sqrt[n]{b^n} = b$ n is any positive integer greater than 1 if $b \geq 0$; n is an odd positive integer greater than 1 if $b < 0$.

Because the radical expressions in parts (1) and (2) of Property 5.3 are both equal to b, by the transitive property they are equal to each other. Hence $\sqrt[n]{b^n} = (\sqrt[n]{b})^n$.

The arithmetic is usually easier to simplify when we use the form $(\sqrt[n]{b})^n$. The following examples demonstrate the use of Property 5.3.

$$\sqrt{144^2} = (\sqrt{144})^2 = 12^2 = 144$$

$$\sqrt[3]{64^3} = (\sqrt[3]{64})^3 = 4^3 = 64$$

$$\sqrt[3]{(-8)^3} = (\sqrt[3]{-8})^3 = (-2)^3 = -8$$

$$\sqrt[4]{16^4} = (\sqrt[4]{16})^4 = 2^4 = 16$$

Let's use some examples to lead into the next very useful property of radicals.

$$\sqrt{4 \cdot 9} = \sqrt{36} = 6 \qquad \text{and} \qquad \sqrt{4} \cdot \sqrt{9} = 2 \cdot 3 = 6$$

$$\sqrt{16 \cdot 25} = \sqrt{400} = 20 \qquad \text{and} \qquad \sqrt{16} \cdot \sqrt{25} = 4 \cdot 5 = 20$$

$$\sqrt[3]{8 \cdot 27} = \sqrt[3]{216} = 6 \qquad \text{and} \qquad \sqrt[3]{8} \cdot \sqrt[3]{27} = 2 \cdot 3 = 6$$

$$\sqrt[3]{(-8)(27)} = \sqrt[3]{-216} = -6 \qquad \text{and} \qquad \sqrt[3]{-8} \cdot \sqrt[3]{27} = (-2)(3) = -6$$

In general, we can state the following property.

Property 5.4

$$\sqrt[n]{bc} = \sqrt[n]{b}\sqrt[n]{c} \qquad \sqrt[n]{b} \text{ and } \sqrt[n]{c} \text{ are real numbers}$$

Property 5.4 states that **the nth root of a product is equal to the product of the nth roots**.

■ Simplest Radical Form

The definition of nth root, along with Property 5.4, provides the basis for changing radicals to simplest radical form. The concept of **simplest radical form** takes on additional meaning as we encounter more complicated expressions, but for now it simply means that the radicand is not to contain any perfect powers of the index. Let's consider some examples to clarify this idea.

EXAMPLE 1   Express each of the following in simplest radical form.

(a) $\sqrt{8}$ (b) $\sqrt{45}$ (c) $\sqrt[3]{24}$ (d) $\sqrt[3]{54}$

Solution

(a) $\sqrt{8} = \sqrt{4 \cdot 2} = \sqrt{4}\sqrt{2} = 2\sqrt{2}$

$\uparrow$
4 is a
perfect
square.

(b) $\sqrt{45} = \sqrt{9 \cdot 5} = \sqrt{9}\sqrt{5} = 3\sqrt{5}$

9 is a
perfect
square.

(c) $\sqrt[3]{24} = \sqrt[3]{8 \cdot 3} = \sqrt[3]{8}\sqrt[3]{3} = 2\sqrt[3]{3}$

8 is a
perfect
cube.

(d) $\sqrt[3]{54} = \sqrt[3]{27 \cdot 2} = \sqrt[3]{27}\sqrt[3]{2} = 3\sqrt[3]{2}$

27 is a
perfect
cube.

■

The first step in each example is to express the radicand of the given radical as the product of two factors, one of which must be a perfect nth power other than 1. Also, observe the radicands of the final radicals. In each case, the radicand cannot have a factor that is a perfect nth power other than 1. We say that the final radicals $2\sqrt{2}, 3\sqrt{5}, 2\sqrt[3]{3}$, and $3\sqrt[3]{2}$ are in **simplest radical form**.

You may vary the steps somewhat in changing to simplest radical form, but the final result should be the same. Consider some different approaches to changing $\sqrt{72}$ to simplest form:

$$\sqrt{72} = \sqrt{9}\sqrt{8} = 3\sqrt{8} = 3\sqrt{4}\sqrt{2} = 3 \cdot 2\sqrt{2} = 6\sqrt{2} \quad \text{or}$$
$$\sqrt{72} = \sqrt{4}\sqrt{18} = 2\sqrt{18} = 2\sqrt{9}\sqrt{2} = 2 \cdot 3\sqrt{2} = 6\sqrt{2} \quad \text{or}$$
$$\sqrt{72} = \sqrt{36}\sqrt{2} = 6\sqrt{2}$$

Another variation of the technique for changing radicals to simplest form is to prime-factor the radicand and then to look for perfect nth powers in exponential form. The following example illustrates the use of this technique.

EXAMPLE 2   Express each of the following in simplest radical form.

(a) $\sqrt{50}$ **(b)** $3\sqrt{80}$ **(c)** $\sqrt[3]{108}$

Solution

(a) $\sqrt{50} = \sqrt{2 \cdot 5 \cdot 5} = \sqrt{5^2}\sqrt{2} = 5\sqrt{2}$

(b) $3\sqrt{80} = 3\sqrt{2 \cdot 2 \cdot 2 \cdot 2 \cdot 5} = 3\sqrt{2^4}\sqrt{5} = 3 \cdot 2^2\sqrt{5} = 12\sqrt{5}$

(c) $\sqrt[3]{108} = \sqrt[3]{2 \cdot 2 \cdot 3 \cdot 3 \cdot 3} = \sqrt[3]{3^3}\sqrt[3]{4} = 3\sqrt[3]{4}$

■

Another property of nth roots is demonstrated by the following examples.

$$\sqrt{\frac{36}{9}} = \sqrt{4} = 2 \quad \text{and} \quad \frac{\sqrt{36}}{\sqrt{9}} = \frac{6}{3} = 2$$

$$\sqrt[3]{\frac{64}{8}} = \sqrt[3]{8} = 2 \quad \text{and} \quad \frac{\sqrt[3]{64}}{\sqrt[3]{8}} = \frac{4}{2} = 2$$

$$\sqrt[3]{\frac{-8}{64}} = \sqrt[3]{-\frac{1}{8}} = -\frac{1}{2} \quad \text{and} \quad \frac{\sqrt[3]{-8}}{\sqrt[3]{64}} = \frac{-2}{4} = -\frac{1}{2}$$

In general, we can state the following property.

Property 5.5

$$\sqrt[n]{\frac{b}{c}} = \frac{\sqrt[n]{b}}{\sqrt[n]{c}} \qquad \sqrt[n]{b} \text{ and } \sqrt[n]{c} \text{ are real numbers, and } c \neq 0.$$

Property 5.5 states that **the nth root of a quotient is equal to the quotient of the nth roots**.

To evaluate radicals such as $\sqrt{\dfrac{4}{25}}$ and $\sqrt[3]{\dfrac{27}{8}}$, for which the numerator and denominator of the fractional radicand are perfect nth powers, you may use Property 5.5 or merely rely on the definition of nth root.

$$\sqrt{\frac{4}{25}} = \frac{\sqrt{4}}{\sqrt{25}} = \frac{2}{5} \quad \text{or} \quad \sqrt{\frac{4}{25}} = \frac{2}{5} \quad \text{because} \quad \frac{2}{5} \cdot \frac{2}{5} = \frac{4}{25}$$

$$\uparrow \qquad\qquad\qquad\qquad\qquad \uparrow$$

Property 5.5 $\qquad\qquad\qquad$ Definition of nth root

$$\downarrow \qquad\qquad\qquad\qquad\qquad \downarrow$$

$$\sqrt[3]{\frac{27}{8}} = \frac{\sqrt[3]{27}}{\sqrt[3]{8}} = \frac{3}{2} \quad \text{or} \quad \sqrt[3]{\frac{27}{8}} = \frac{3}{2} \quad \text{because} \quad \frac{3}{2} \cdot \frac{3}{2} \cdot \frac{3}{2} = \frac{27}{8}$$

Radicals such as $\sqrt{\dfrac{28}{9}}$ and $\sqrt[3]{\dfrac{24}{27}}$, in which only the denominators of the radicand are perfect nth powers, can be simplified as follows:

$$\sqrt{\frac{28}{9}} = \frac{\sqrt{28}}{\sqrt{9}} = \frac{\sqrt{28}}{3} = \frac{\sqrt{4}\sqrt{7}}{3} = \frac{2\sqrt{7}}{3}$$

$$\sqrt[3]{\frac{24}{27}} = \frac{\sqrt[3]{24}}{\sqrt[3]{27}} = \frac{\sqrt[3]{24}}{3} = \frac{\sqrt[3]{8}\sqrt[3]{3}}{3} = \frac{2\sqrt[3]{3}}{3}$$

Before we consider more examples, let's summarize some ideas that pertain to the simplifying of radicals. A radical is said to be in **simplest radical form** if the following conditions are satisfied.

---

1. No fraction appears with a radical sign. $\sqrt{\dfrac{3}{4}}$ violates this condition.

2. No radical appears in the denominator. $\dfrac{\sqrt{2}}{\sqrt{3}}$ violates this condition.

3. No radicand, when expressed in prime-factored form, contains a factor raised to a power equal to or greater than the index.

$\sqrt{2^3 \cdot 5}$ violates this condition.

Now let's consider an example in which neither the numerator nor the denominator of the radicand is a perfect nth power.

E X A M P L E   3

Simplify $\sqrt{\dfrac{2}{3}}$.

Solution

$$\sqrt{\dfrac{2}{3}} = \dfrac{\sqrt{2}}{\sqrt{3}} = \dfrac{\sqrt{2}}{\sqrt{3}} \cdot \dfrac{\sqrt{3}}{\sqrt{3}} = \dfrac{\sqrt{6}}{3}$$

$\uparrow$

Form of 1 ∎

We refer to the process we used to simplify the radical in Example 3 as **rationalizing the denominator**. Note that the denominator becomes a rational number. The process of rationalizing the denominator can often be accomplished in more than one way, as we will see in the next example.

E X A M P L E   4

Simplify $\dfrac{\sqrt{5}}{\sqrt{8}}$.

Solution A

$$\dfrac{\sqrt{5}}{\sqrt{8}} = \dfrac{\sqrt{5}}{\sqrt{8}} \cdot \dfrac{\sqrt{8}}{\sqrt{8}} = \dfrac{\sqrt{40}}{8} = \dfrac{\sqrt{4}\sqrt{10}}{8} = \dfrac{2\sqrt{10}}{8} = \dfrac{\sqrt{10}}{4}$$

Solution B

$$\dfrac{\sqrt{5}}{\sqrt{8}} = \dfrac{\sqrt{5}}{\sqrt{8}} \cdot \dfrac{\sqrt{2}}{\sqrt{2}} = \dfrac{\sqrt{10}}{\sqrt{16}} = \dfrac{\sqrt{10}}{4}$$

Solution C

$$\frac{\sqrt{5}}{\sqrt{8}} = \frac{\sqrt{5}}{\sqrt{4}\sqrt{2}} = \frac{\sqrt{5}}{2\sqrt{2}} = \frac{\sqrt{5}}{2\sqrt{2}} \cdot \frac{\sqrt{2}}{\sqrt{2}} = \frac{\sqrt{10}}{2\sqrt{4}} = \frac{\sqrt{10}}{2(2)} = \frac{\sqrt{10}}{4} \qquad \blacksquare$$

The three approaches to Example 4 again illustrate the need to think first and only then push the pencil. You may find one approach easier than another. To conclude this section, study the following examples and check the final radicals against the three conditions previously listed for **simplest radical form**.

EXAMPLE 5

Simplify each of the following.

(a) $\dfrac{3\sqrt{2}}{5\sqrt{3}}$ **(b)** $\dfrac{3\sqrt{7}}{2\sqrt{18}}$ **(c)** $\sqrt[3]{\dfrac{5}{9}}$ **(d)** $\dfrac{\sqrt[3]{5}}{\sqrt[3]{16}}$

Solution

(a) $\dfrac{3\sqrt{2}}{5\sqrt{3}} = \dfrac{3\sqrt{2}}{5\sqrt{3}} \cdot \dfrac{\sqrt{3}}{\sqrt{3}} = \dfrac{3\sqrt{6}}{5\sqrt{9}} = \dfrac{3\sqrt{6}}{15} = \dfrac{\sqrt{6}}{5}$

↑
Form of 1

(b) $\dfrac{3\sqrt{7}}{2\sqrt{18}} = \dfrac{3\sqrt{7}}{2\sqrt{18}} \cdot \dfrac{\sqrt{2}}{\sqrt{2}} = \dfrac{3\sqrt{14}}{2\sqrt{36}} = \dfrac{3\sqrt{14}}{12} = \dfrac{\sqrt{14}}{4}$

↑
Form of 1

(c) $\sqrt[3]{\dfrac{5}{9}} = \dfrac{\sqrt[3]{5}}{\sqrt[3]{9}} = \dfrac{\sqrt[3]{5}}{\sqrt[3]{9}} \cdot \dfrac{\sqrt[3]{3}}{\sqrt[3]{3}} = \dfrac{\sqrt[3]{15}}{\sqrt[3]{27}} = \dfrac{\sqrt[3]{15}}{3}$

↑
Form of 1

(d) $\dfrac{\sqrt[3]{5}}{\sqrt[3]{16}} = \dfrac{\sqrt[3]{5}}{\sqrt[3]{16}} \cdot \dfrac{\sqrt[3]{4}}{\sqrt[3]{4}} = \dfrac{\sqrt[3]{20}}{\sqrt[3]{64}} = \dfrac{\sqrt[3]{20}}{4}$

↑
Form of 1

■

■ Applications of Radicals

Many real-world applications involve radical expressions. For example, police often use the formula $S = \sqrt{30Df}$ to estimate the speed of a car on the basis of the length of the skid marks at the scene of an accident. In this formula, S represents the speed of the car in miles per hour, D represents the length of the skid marks in feet, and f represents a coefficient of friction. For a particular situation, the coefficient of friction is a constant that depends on the type and condition of the road surface.

E X A M P L E   6

Using 0.35 as a coefficient of friction, determine how fast a car was traveling if it skidded 325 feet.

Solution

Substitute 0.35 for f and 325 for D in the formula.

$$S = \sqrt{30Df} = \sqrt{30(325)(0.35)} = 58, \quad \text{to the nearest whole number}$$

The car was traveling at approximately 58 miles per hour.   ■

The **period** of a pendulum is the time it takes to swing from one side to the other side and back. The formula

$$T = 2\pi\sqrt{\frac{L}{32}}$$

where T represents the time in seconds and L the length in feet, can be used to determine the period of a pendulum (see Figure 5.1).

Figure 5.1

E X A M P L E   7

Find, to the nearest tenth of a second, the period of a pendulum of length 3.5 feet.

Solution

Let's use 3.14 as an approximation for π and substitute 3.5 for L in the formula.

$$T = 2\pi\sqrt{\frac{L}{32}} = 2(3.14)\sqrt{\frac{3.5}{32}} = 2.1, \quad \text{to the nearest tenth}$$

The period is approximately 2.1 seconds.   ■

Radical expressions are also used in some geometric applications. For example, the area of a triangle can be found by using a formula that involves a square root. If a, b, and c represent the lengths of the three sides of a triangle, the formula $K = \sqrt{s(s - a)(s - b)(s - c)}$, known as Heron's formula, can be used to determine the area (K) of the triangle. The letter s represents the semiperimeter of the triangle; that is, $s = \dfrac{a + b + c}{2}$.

EXAMPLE 8   Find the area of a triangular piece of sheet metal that has sides of lengths 17 inches, 19 inches, and 26 inches.

Solution

First, let's find the value of s, the semiperimeter of the triangle.

$$s = \frac{17 + 19 + 26}{2} = 31$$

Now we can use Heron's formula.

$$K = \sqrt{s(s - a)(s - b)(s - c)} = \sqrt{31(31 - 17)(31 - 19)(31 - 26)}$$

$$= \sqrt{31(14)(12)(5)}$$

$$= \sqrt{20{,}640}$$

$$= 161.4, \quad \text{to the nearest tenth}$$

Thus the area of the piece of sheet metal is approximately 161.4 square inches. ∎

Remark: Note that in Examples 6–8, we did not simplify the radicals. When one is using a calculator to approximate the square roots, there is no need to simplify first.

Problem Set 5.2

For Problems 1–20, evaluate each of the following. For example, $\sqrt{25} = 5$.

1. $\sqrt{64}$

2. $\sqrt{49}$

3. $-\sqrt{100}$

4. $-\sqrt{81}$

5. $\sqrt[3]{27}$

6. $\sqrt[3]{216}$

7. $\sqrt[3]{-64}$

8. $\sqrt[3]{-125}$

9. $\sqrt[4]{81}$

10. $-\sqrt[4]{16}$

11. $\sqrt{\dfrac{16}{25}}$

12. $\sqrt{\dfrac{25}{64}}$

13. $-\sqrt{\dfrac{36}{49}}$

14. $\sqrt{\dfrac{16}{64}}$

15. $\sqrt{\dfrac{9}{36}}$

16. $\sqrt{\dfrac{144}{36}}$

17. $\sqrt[3]{\dfrac{27}{64}}$

18. $\sqrt[3]{-\dfrac{8}{27}}$

19. $\sqrt[3]{8^3}$

20. $\sqrt[4]{16^4}$

For Problems 21–74, change each radical to simplest radical form.

21. $\sqrt{27}$

22. $\sqrt{48}$

23. $\sqrt{32}$

24. $\sqrt{98}$

25. $\sqrt{80}$

26. $\sqrt{125}$

27. $\sqrt{160}$

28. $\sqrt{112}$

29. $4\sqrt{18}$

30. $5\sqrt{32}$

31. $-6\sqrt{20}$

32. $-4\sqrt{54}$

33. $\dfrac{2}{5}\sqrt{75}$

34. $\dfrac{1}{3}\sqrt{90}$

35. $\dfrac{3}{2}\sqrt{24}$

36. $\dfrac{3}{4}\sqrt{45}$

37. $-\dfrac{5}{6}\sqrt{28}$

38. $-\dfrac{2}{3}\sqrt{96}$

39. $\sqrt{\dfrac{19}{4}}$

40. $\sqrt{\dfrac{22}{9}}$

41. $\sqrt{\dfrac{27}{16}}$

42. $\sqrt{\dfrac{8}{25}}$

43. $\sqrt{\dfrac{75}{81}}$

44. $\sqrt{\dfrac{24}{49}}$

45. $\sqrt{\dfrac{2}{7}}$

46. $\sqrt{\dfrac{3}{8}}$

47. $\sqrt{\dfrac{2}{3}}$

48. $\sqrt{\dfrac{7}{12}}$

49. $\dfrac{\sqrt{5}}{\sqrt{12}}$

50. $\dfrac{\sqrt{3}}{\sqrt{7}}$

51. $\dfrac{\sqrt{11}}{\sqrt{24}}$

52. $\dfrac{\sqrt{5}}{\sqrt{48}}$

53. $\dfrac{\sqrt{18}}{\sqrt{27}}$

54. $\dfrac{\sqrt{10}}{\sqrt{20}}$

55. $\dfrac{\sqrt{35}}{\sqrt{7}}$

56. $\dfrac{\sqrt{42}}{\sqrt{6}}$

57. $\dfrac{2\sqrt{3}}{\sqrt{7}}$

58. $\dfrac{3\sqrt{2}}{\sqrt{6}}$

59. $-\dfrac{4\sqrt{12}}{\sqrt{5}}$

60. $\dfrac{-6\sqrt{5}}{\sqrt{18}}$

61. $\dfrac{3\sqrt{2}}{4\sqrt{3}}$

62. $\dfrac{6\sqrt{5}}{5\sqrt{12}}$

63. $\dfrac{-8\sqrt{18}}{10\sqrt{50}}$

64. $\dfrac{4\sqrt{45}}{-6\sqrt{20}}$

65. $\sqrt[3]{16}$

66. $\sqrt[3]{40}$

67. $2\sqrt[3]{81}$

68. $-3\sqrt[3]{54}$

69. $\dfrac{2}{\sqrt[3]{9}}$

70. $\dfrac{3}{\sqrt[3]{3}}$

71. $\dfrac{\sqrt[3]{27}}{\sqrt[3]{4}}$

72. $\dfrac{\sqrt[3]{8}}{\sqrt[3]{16}}$

73. $\dfrac{\sqrt[3]{6}}{\sqrt[3]{4}}$

74. $\dfrac{\sqrt[3]{4}}{\sqrt[3]{2}}$

75. Use a coefficient of friction of 0.4 in the formula from Example 6 and find the speeds of cars that left skid marks of lengths 150 feet, 200 feet, and 350 feet. Express your answers to the nearest mile per hour.

76. Use the formula from Example 7, and find the periods of pendulums of lengths 2 feet, 3 feet, and 4.5 feet. Express your answers to the nearest tenth of a second.

77. Find, to the nearest square centimeter, the area of a triangle that measures 14 centimeters by 16 centimeters by 18 centimeters.

78. Find, to the nearest square yard, the area of a triangular plot of ground that measures 45 yards by 60 yards by 75 yards.

79. Find the area of an equilateral triangle, each of whose sides is 18 inches long. Express the area to the nearest square inch.

80. Find, to the nearest square inch, the area of the quadrilateral in Figure 5.2.

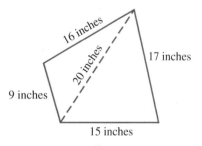

Figure 5.2

■■■ THOUGHTS INTO WORDS

81. Why is $\sqrt{-9}$ not a real number?

82. Why is it that we say 25 has two square roots (5 and −5), but we write $\sqrt{25} = 5$?

83. How is the multiplication property of 1 used when simplifying radicals?

84. How could you find a whole number approximation for $\sqrt{2750}$ if you did not have a calculator or table available?

■ ■ ■ **FURTHER INVESTIGATIONS**

85. Use your calculator to find a rational approximation, to the nearest thousandth, for (a) through (i).

(a) $\sqrt{2}$ **(b)** $\sqrt{75}$ **(c)** $\sqrt{156}$

(d) $\sqrt{691}$ **(e)** $\sqrt{3249}$ **(f)** $\sqrt{45{,}123}$

(g) $\sqrt{0.14}$ **(h)** $\sqrt{0.023}$ **(i)** $\sqrt{0.8649}$

86. Sometimes a fairly good estimate can be made of a radical expression by using whole number approximations. For example, $5\sqrt{35} + 7\sqrt{50}$ is approximately $5(6) + 7(7) = 79$. Using a calculator, we find that $5\sqrt{35} + 7\sqrt{50} = 79.1$, to the nearest tenth. In this case our whole number estimate is very good. For (a) through (f), first make a whole number estimate, and then use your calculator to see how well you estimated.

(a) $3\sqrt{10} - 4\sqrt{24} + 6\sqrt{65}$

(b) $9\sqrt{27} + 5\sqrt{37} - 3\sqrt{80}$

(c) $12\sqrt{5} + 13\sqrt{18} + 9\sqrt{47}$

(d) $3\sqrt{98} - 4\sqrt{83} - 7\sqrt{120}$

(e) $4\sqrt{170} + 2\sqrt{198} + 5\sqrt{227}$

(f) $-3\sqrt{256} - 6\sqrt{287} + 11\sqrt{321}$

5.3   Combining Radicals and Simplifying Radicals That Contain Variables

Recall our use of the distributive property as the basis for combining similar terms. For example,

$$3x + 2x = (3 + 2)x = 5x$$

$$8y - 5y = (8 - 5)y = 3y$$

$$\frac{2}{3}a^2 + \frac{3}{4}a^2 = \left(\frac{2}{3} + \frac{3}{4}\right)a^2 = \left(\frac{8}{12} + \frac{9}{12}\right)a^2 = \frac{17}{12}a^2$$

In a like manner, expressions that contain radicals can often be simplified by using the distributive property, as follows:

$$3\sqrt{2} + 5\sqrt{2} = (3 + 5)\sqrt{2} = 8\sqrt{2}$$

$$7\sqrt[3]{5} - 3\sqrt[3]{5} = (7 - 3)\sqrt[3]{5} = 4\sqrt[3]{5}$$

$$4\sqrt{7} + 5\sqrt{7} + 6\sqrt{11} - 2\sqrt{11} = (4 + 5)\sqrt{7} + (6 - 2)\sqrt{11} = 9\sqrt{7} + 4\sqrt{11}$$

Note that *in order to be added or subtracted, radicals must have the same index and the same radicand*. Thus we cannot simplify an expression such as $5\sqrt{2} + 7\sqrt{11}$.

Simplifying by combining radicals sometimes requires that you first express the given radicals in simplest form and then apply the distributive property. The following examples illustrate this idea.

EXAMPLE 1   Simplify $3\sqrt{8} + 2\sqrt{18} - 4\sqrt{2}$.

Solution

$$
\begin{aligned}
3\sqrt{8} + 2\sqrt{18} - 4\sqrt{2} &= 3\sqrt{4}\sqrt{2} + 2\sqrt{9}\sqrt{2} - 4\sqrt{2} \\
&= 3 \cdot 2 \cdot \sqrt{2} + 2 \cdot 3 \cdot \sqrt{2} - 4\sqrt{2} \\
&= 6\sqrt{2} + 6\sqrt{2} - 4\sqrt{2} \\
&= (6 + 6 - 4)\sqrt{2} = 8\sqrt{2}
\end{aligned}
$$

■

EXAMPLE 2   Simplify $\dfrac{1}{4}\sqrt{45} + \dfrac{1}{3}\sqrt{20}$.

Solution

$$
\begin{aligned}
\frac{1}{4}\sqrt{45} + \frac{1}{3}\sqrt{20} &= \frac{1}{4}\sqrt{9}\sqrt{5} + \frac{1}{3}\sqrt{4}\sqrt{5} \\[2mm]
&= \frac{1}{4} \cdot 3 \cdot \sqrt{5} + \frac{1}{3} \cdot 2 \cdot \sqrt{5} \\[2mm]
&= \frac{3}{4}\sqrt{5} + \frac{2}{3}\sqrt{5} = \left(\frac{3}{4} + \frac{2}{3}\right)\sqrt{5} \\[2mm]
&= \left(\frac{9}{12} + \frac{8}{12}\right)\sqrt{5} = \frac{17}{12}\sqrt{5}
\end{aligned}
$$

■

EXAMPLE 3   Simplify $5\sqrt[3]{2} - 2\sqrt[3]{16} - 6\sqrt[3]{54}$.

Solution

$$
\begin{aligned}
5\sqrt[3]{2} - 2\sqrt[3]{16} - 6\sqrt[3]{54} &= 5\sqrt[3]{2} - 2\sqrt[3]{8}\sqrt[3]{2} - 6\sqrt[3]{27}\sqrt[3]{2} \\
&= 5\sqrt[3]{2} - 2 \cdot 2 \cdot \sqrt[3]{2} - 6 \cdot 3 \cdot \sqrt[3]{2} \\
&= 5\sqrt[3]{2} - 4\sqrt[3]{2} - 18\sqrt[3]{2} \\
&= (5 - 4 - 18)\sqrt[3]{2} \\
&= -17\sqrt[3]{2}
\end{aligned}
$$

■

■ Radicals That Contain Variables

Before we discuss the process of simplifying radicals that contain variables, there is one technicality that we should call to your attention. Let's look at some examples to clarify the point. Consider the radical $\sqrt{x^2}$.

Let $x = 3$;   then $\sqrt{x^2} = \sqrt{3^2} = \sqrt{9} = 3$.

Let $x = -3$;   then $\sqrt{x^2} = \sqrt{(-3)^2} = \sqrt{9} = 3$.

Thus if $x \geq 0$, then $\sqrt{x^2} = x$, *but* if $x < 0$, then $\sqrt{x^2} = -x$. Using the concept of absolute value, we can state that for all real numbers, $\sqrt{x^2} = |x|$.

Now consider the radical $\sqrt{x^3}$. Because x^3 is negative when x is negative, we need to restrict x to the nonnegative reals when working with $\sqrt{x^3}$. Thus we can write, "if $x \geq 0$, then $\sqrt{x^3} = \sqrt{x^2}\sqrt{x} = x\sqrt{x}$," and no absolute-value sign is necessary. Finally, let's consider the radical $\sqrt[3]{x^3}$.

Let $x = 2$; then $\sqrt[3]{x^3} = \sqrt[3]{2^3} = \sqrt[3]{8} = 2.$

Let $x = -2$; then $\sqrt[3]{x^3} = \sqrt[3]{(-2)^3} = \sqrt[3]{-8} = -2.$

Thus it is correct to write, "$\sqrt[3]{x^3} = x$ for all real numbers," and again no absolute-value sign is necessary.

The previous discussion indicates that technically, every radical expression involving variables in the radicand needs to be analyzed individually in terms of any necessary restrictions imposed on the variables. To help you gain experience with this skill, examples and problems are discussed under Further Investigations in the problem set. For now, however, to avoid considering such restrictions on a problem-to-problem basis, we shall merely assume that all variables represent positive real numbers. Let's consider the process of simplifying radicals that contain variables in the radicand. Study the following examples, and note that the same basic approach we used in Section 5.2 is applied here.

E X A M P L E   4

Simplify each of the following.

    (a) $\sqrt{8x^3}$ **(b)** $\sqrt{45x^3y^7}$ **(c)** $\sqrt{180a^4b^3}$ **(d)** $\sqrt[3]{40x^4y^8}$

Solution

    (a) $\sqrt{8x^3} = \sqrt{4x^2}\sqrt{2x} = 2x\sqrt{2x}$

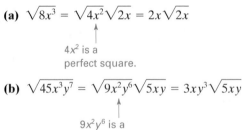

$4x^2$ is a
perfect square.

    (b) $\sqrt{45x^3y^7} = \sqrt{9x^2y^6}\sqrt{5xy} = 3xy^3\sqrt{5xy}$

$9x^2y^6$ is a
perfect square.

    (c) If the numerical coefficient of the radicand is quite large, you may want to look at it in the prime-factored form.

$$\sqrt{180a^4b^3} = \sqrt{2 \cdot 2 \cdot 3 \cdot 3 \cdot 5 \cdot a^4 \cdot b^3}$$
$$= \sqrt{36 \cdot 5 \cdot a^4 \cdot b^3}$$
$$= \sqrt{36a^4b^2}\sqrt{5b}$$
$$= 6a^2b\sqrt{5b}$$

(d) $\sqrt[3]{40x^4y^8} = \sqrt[3]{8x^3y^6}\sqrt[3]{5xy^2} = 2xy^2\sqrt[3]{5xy^2}$

$8x^3y^6$ is a
perfect cube.

■

Before we consider more examples, let's restate (in such a way as to include radicands containing variables) the conditions necessary for a radical to be in simplest radical form.

---

1. A radicand contains no polynomial factor raised to a power equal to or greater than the index of the radical.

$\sqrt{x^3}$ violates this condition.

2. No fraction appears within a radical sign.

$\sqrt{\dfrac{2x}{3y}}$ violates this condition.

3. No radical appears in the denominator.

$\dfrac{3}{\sqrt[3]{4x}}$ violates this condition.

EXAMPLE 5

Express each of the following in simplest radical form.

(a) $\sqrt{\dfrac{2x}{3y}}$ **(b)** $\dfrac{\sqrt{5}}{\sqrt{12a^3}}$ **(c)** $\dfrac{\sqrt{8x^2}}{\sqrt{27y^5}}$

(d) $\dfrac{3}{\sqrt[3]{4x}}$ **(e)** $\dfrac{\sqrt[3]{16x^2}}{\sqrt[3]{9y^5}}$

Solution

(a) $\sqrt{\dfrac{2x}{3y}} = \dfrac{\sqrt{2x}}{\sqrt{3y}} = \dfrac{\sqrt{2x}}{\sqrt{3y}} \cdot \dfrac{\sqrt{3y}}{\sqrt{3y}} = \dfrac{\sqrt{6xy}}{3y}$

Form of 1

(b) $\dfrac{\sqrt{5}}{\sqrt{12a^3}} = \dfrac{\sqrt{5}}{\sqrt{12a^3}} \cdot \dfrac{\sqrt{3a}}{\sqrt{3a}} = \dfrac{\sqrt{15a}}{\sqrt{36a^4}} = \dfrac{\sqrt{15a}}{6a^2}$

Form of 1

(c) $\dfrac{\sqrt{8x^2}}{\sqrt{27y^5}} = \dfrac{\sqrt{4x^2}\sqrt{2}}{\sqrt{9y^4}\sqrt{3y}} = \dfrac{2x\sqrt{2}}{3y^2\sqrt{3y}} = \dfrac{2x\sqrt{2}}{3y^2\sqrt{3y}} \cdot \dfrac{\sqrt{3y}}{\sqrt{3y}}$

$$= \dfrac{2x\sqrt{6y}}{(3y^2)(3y)} = \dfrac{2x\sqrt{6y}}{9y^3}$$

(d) $\dfrac{3}{\sqrt[3]{4x}} = \dfrac{3}{\sqrt[3]{4x}} \cdot \dfrac{\sqrt[3]{2x^2}}{\sqrt[3]{2x^2}} = \dfrac{3\sqrt[3]{2x^2}}{\sqrt[3]{8x^3}} = \dfrac{3\sqrt[3]{2x^2}}{2x}$

(e) $\dfrac{\sqrt[3]{16x^2}}{\sqrt[3]{9y^5}} = \dfrac{\sqrt[3]{16x^2}}{\sqrt[3]{9y^5}} \cdot \dfrac{\sqrt[3]{3y}}{\sqrt[3]{3y}} = \dfrac{\sqrt[3]{48x^2y}}{\sqrt[3]{27y^6}} = \dfrac{\sqrt[3]{8}\sqrt[3]{6x^2y}}{3y^2} = \dfrac{2\sqrt[3]{6x^2y}}{3y^2}$ ∎

Note that in part (c) we did some simplifying first before rationalizing the denominator, whereas in part (b) we proceeded immediately to rationalize the denominator. This is an individual choice, and you should probably do it both ways a few times to decide which you prefer.

Problem Set 5.3

For Problems 1–20, use the distributive property to help simplify each of the following. For example,

$$3\sqrt{8} - \sqrt{32} = 3\sqrt{4}\sqrt{2} - \sqrt{16}\sqrt{2}$$
$$= 3(2)\sqrt{2} - 4\sqrt{2}$$
$$= 6\sqrt{2} - 4\sqrt{2}$$
$$= (6 - 4)\sqrt{2} = 2\sqrt{2}$$

1. $5\sqrt{18} - 2\sqrt{2}$

2. $7\sqrt{12} + 4\sqrt{3}$

3. $7\sqrt{12} + 10\sqrt{48}$

4. $6\sqrt{8} - 5\sqrt{18}$

5. $-2\sqrt{50} - 5\sqrt{32}$

6. $-2\sqrt{20} - 7\sqrt{45}$

7. $3\sqrt{20} - \sqrt{5} - 2\sqrt{45}$

8. $6\sqrt{12} + \sqrt{3} - 2\sqrt{48}$

9. $-9\sqrt{24} + 3\sqrt{54} - 12\sqrt{6}$

10. $13\sqrt{28} - 2\sqrt{63} - 7\sqrt{7}$

11. $\dfrac{3}{4}\sqrt{7} - \dfrac{2}{3}\sqrt{28}$

12. $\dfrac{3}{5}\sqrt{5} - \dfrac{1}{4}\sqrt{80}$

13. $\dfrac{3}{5}\sqrt{40} + \dfrac{5}{6}\sqrt{90}$

14. $\dfrac{3}{8}\sqrt{96} - \dfrac{2}{3}\sqrt{54}$

15. $\dfrac{3\sqrt{18}}{5} - \dfrac{5\sqrt{72}}{6} + \dfrac{3\sqrt{98}}{4}$

16. $\dfrac{-2\sqrt{20}}{3} + \dfrac{3\sqrt{45}}{4} - \dfrac{5\sqrt{80}}{6}$

17. $5\sqrt[3]{3} + 2\sqrt[3]{24} - 6\sqrt[3]{81}$

18. $-3\sqrt[3]{2} - 2\sqrt[3]{16} + \sqrt[3]{54}$

19. $-\sqrt[3]{16} + 7\sqrt[3]{54} - 9\sqrt[3]{2}$

20. $4\sqrt[3]{24} - 6\sqrt[3]{3} + 13\sqrt[3]{81}$

For Problems 21–64, express each of the following in simplest radical form. All variables represent positive real numbers.

21. $\sqrt{32x}$

22. $\sqrt{50y}$

23. $\sqrt{75x^2}$

24. $\sqrt{108y^2}$

25. $\sqrt{20x^2y}$

26. $\sqrt{80xy^2}$

27. $\sqrt{64x^3y^7}$

28. $\sqrt{36x^5y^6}$

29. $\sqrt{54a^4b^3}$

30. $\sqrt{96a^7b^8}$

31. $\sqrt{63x^6y^8}$

32. $\sqrt{28x^4y^{12}}$

33. $2\sqrt{40a^3}$

34. $4\sqrt{90a^5}$

35. $\dfrac{2}{3}\sqrt{96xy^3}$

36. $\dfrac{4}{5}\sqrt{125x^4y}$

37. $\sqrt{\dfrac{2x}{5y}}$

38. $\sqrt{\dfrac{3x}{2y}}$

39. $\sqrt{\dfrac{5}{12x^4}}$

40. $\sqrt{\dfrac{7}{8x^2}}$

41. $\dfrac{5}{\sqrt{18y}}$

42. $\dfrac{3}{\sqrt{12x}}$

43. $\dfrac{\sqrt{7x}}{\sqrt{8y^5}}$

44. $\dfrac{\sqrt{5y}}{\sqrt{18x^3}}$

45. $\dfrac{\sqrt{18y^3}}{\sqrt{16x}}$

46. $\dfrac{\sqrt{2x^3}}{\sqrt{9y}}$

47. $\dfrac{\sqrt{24a^2b^3}}{\sqrt{7ab^6}}$

48. $\dfrac{\sqrt{12a^2b}}{\sqrt{5a^3b^3}}$

49. $\sqrt[3]{24y}$

50. $\sqrt[3]{16x^2}$

51. $\sqrt[3]{16x^4}$

52. $\sqrt[3]{54x^3}$

53. $\sqrt[3]{56x^6y^8}$

54. $\sqrt[3]{81x^5y^6}$

55. $\sqrt[3]{\dfrac{7}{9x^2}}$

56. $\sqrt[3]{\dfrac{5}{2x}}$

57. $\dfrac{\sqrt[3]{3y}}{\sqrt[3]{16x^4}}$

58. $\dfrac{\sqrt[3]{2y}}{\sqrt[3]{3x}}$

59. $\dfrac{\sqrt[3]{12xy}}{\sqrt[3]{3x^2y^5}}$

60. $\dfrac{5}{\sqrt[3]{9xy^2}}$

61. $\sqrt{8x + 12y}$ [*Hint:* $\sqrt{8x + 12y} = \sqrt{4(2x + 3y)}$]

62. $\sqrt{4x + 4y}$

63. $\sqrt{16x + 48y}$

64. $\sqrt{27x + 18y}$

For Problems 65–74, use the distributive property to help simplify each of the following. All variables represent positive real numbers.

65. $-3\sqrt{4x} + 5\sqrt{9x} + 6\sqrt{16x}$

66. $-2\sqrt{25x} - 4\sqrt{36x} + 7\sqrt{64x}$

67. $2\sqrt{18x} - 3\sqrt{8x} - 6\sqrt{50x}$

68. $4\sqrt{20x} + 5\sqrt{45x} - 10\sqrt{80x}$

69. $5\sqrt{27n} - \sqrt{12n} - 6\sqrt{3n}$

70. $4\sqrt{8n} + 3\sqrt{18n} - 2\sqrt{72n}$

71. $7\sqrt{4ab} - \sqrt{16ab} - 10\sqrt{25ab}$

72. $4\sqrt{ab} - 9\sqrt{36ab} + 6\sqrt{49ab}$

73. $-3\sqrt{2x^3} + 4\sqrt{8x^3} - 3\sqrt{32x^3}$

74. $2\sqrt{40x^5} - 3\sqrt{90x^5} + 5\sqrt{160x^5}$

■ ■ ■ THOUGHTS INTO WORDS

75. Is the expression $3\sqrt{2} + \sqrt{50}$ in simplest radical form? Defend your answer.

76. Your friend simplified $\dfrac{\sqrt{6}}{\sqrt{8}}$ as follows:

$$\dfrac{\sqrt{6}}{\sqrt{8}} \cdot \dfrac{\sqrt{8}}{\sqrt{8}} = \dfrac{\sqrt{48}}{8} = \dfrac{\sqrt{16}\sqrt{3}}{8} = \dfrac{4\sqrt{3}}{8} = \dfrac{\sqrt{3}}{2}$$

Is this a correct procedure? Can you show her a better way to do this problem?

77. Does $\sqrt{x + y}$ equal $\sqrt{x} + \sqrt{y}$? Defend your answer.

■ ■ ■ FURTHER INVESTIGATIONS

78. Use your calculator and evaluate each expression in Problems 1–16. Then evaluate the simplified expres-

sion that you obtained when doing these problems. Your two results for each problem should be the same.

Consider these problems, where the variables could represent any real number. However, we would still have the restriction that the radical would represent a real number. In other words, the radicand must be nonnegative.

$$\sqrt{98x^2} = \sqrt{49x^2}\,\sqrt{2} = 7|x|\sqrt{2}$$

An absolute-value sign is necessary to ensure that the principal root is nonnegative.

$$\sqrt{24x^4} = \sqrt{4x^4}\sqrt{6} = 2x^2\sqrt{6}$$

Because x^2 is nonnegative, there is no need for an absolute-value sign to ensure that the principal root is nonnegative.

$$\sqrt{25x^3} = \sqrt{25x^2}\sqrt{x} = 5x\sqrt{x}$$

Because the radicand is defined to be nonnegative, x must be nonnegative, and there is no need for an absolute-value sign to ensure that the principal root is nonnegative.

$$\sqrt{18b^5} = \sqrt{9b^4}\sqrt{2b} = 3b^2\sqrt{2b}$$

An absolute-value sign is not necessary to ensure that the principal root is nonnegative.

$$\sqrt{12y^6} = \sqrt{4y^6}\sqrt{3} = 2|y^3|\sqrt{3}$$

An absolute-value sign is necessary to ensure that the principal root is nonnegative.

79. Do the following problems, where the variable could be any real number as long as the radical represents a real number. Use absolute-value signs in the answers as necessary.

(a) $\sqrt{125x^2}$ (b) $\sqrt{16x^4}$

(c) $\sqrt{8b^3}$ (d) $\sqrt{3y^5}$

(e) $\sqrt{288x^6}$ (f) $\sqrt{28m^8}$

(g) $\sqrt{128c^{10}}$ (h) $\sqrt{18d^7}$

(i) $\sqrt{49x^2}$ (j) $\sqrt{80n^{20}}$

(k) $\sqrt{81h^3}$

5.4 Products and Quotients Involving Radicals

As we have seen, Property 5.4 ($\sqrt[n]{bc} = \sqrt[n]{b}\sqrt[n]{c}$) is used to express one radical as the product of two radicals and also to express the product of two radicals as one radical. In fact, we have used the property for both purposes within the framework of simplifying radicals. For example,

$$\frac{\sqrt{3}}{\sqrt{32}} = \frac{\sqrt{3}}{\sqrt{16}\sqrt{2}} = \frac{\sqrt{3}}{4\sqrt{2}} = \frac{\sqrt{3}}{4\sqrt{2}} \cdot \frac{\sqrt{2}}{\sqrt{2}} = \frac{\sqrt{6}}{8}$$

$$\uparrow \qquad \uparrow \qquad\qquad \uparrow \qquad \uparrow$$

$$\sqrt[n]{bc} = \sqrt[n]{b}\sqrt[n]{c} \qquad\qquad \sqrt[n]{b}\sqrt[n]{c} = \sqrt[n]{bc}$$

The following examples demonstrate the use of Property 5.4 to multiply radicals and to express the product in simplest form.

EXAMPLE 1 Multiply and simplify where possible.

(a) $(2\sqrt{3})(3\sqrt{5})$ (b) $(3\sqrt{8})(5\sqrt{2})$

(c) $(7\sqrt{6})(3\sqrt{8})$ (d) $(2\sqrt[3]{6})(5\sqrt[3]{4})$

Solution

(a) $(2\sqrt{3})(3\sqrt{5}) = 2 \cdot 3 \cdot \sqrt{3} \cdot \sqrt{5} = 6\sqrt{15}$

(b) $(3\sqrt{8})(5\sqrt{2}) = 3 \cdot 5 \cdot \sqrt{8} \cdot \sqrt{2} = 15\sqrt{16} = 15 \cdot 4 = 60.$

(c) $(7\sqrt{6})(3\sqrt{8}) = 7 \cdot 3 \cdot \sqrt{6} \cdot \sqrt{8} = 21\sqrt{48} = 21\sqrt{16}\sqrt{3}$

$$= 21 \cdot 4 \cdot \sqrt{3} = 84\sqrt{3}$$

(d) $(2\sqrt[3]{6})(5\sqrt[3]{4}) = 2 \cdot 5 \cdot \sqrt[3]{6} \cdot \sqrt[3]{4} = 10\sqrt[3]{24}$

$$= 10\sqrt[3]{8}\sqrt[3]{3}$$

$$= 10 \cdot 2 \cdot \sqrt[3]{3}$$

$$= 20\sqrt[3]{3} \qquad \blacksquare$$

Recall the use of the distributive property when finding the product of a monomial and a polynomial. For example, $3x^2(2x + 7) = 3x^2(2x) + 3x^2(7) = 6x^3 + 21x^2$. In a similar manner, the distributive property and Property 5.4 provide the basis for finding certain special products that involve radicals. The following examples illustrate this idea.

EXAMPLE 2 Multiply and simplify where possible.

(a) $\sqrt{3}(\sqrt{6} + \sqrt{12})$ **(b)** $2\sqrt{2}(4\sqrt{3} - 5\sqrt{6})$

(c) $\sqrt{6x}(\sqrt{8x} + \sqrt{12xy})$ **(d)** $\sqrt[3]{2}(5\sqrt[3]{4} - 3\sqrt[3]{16})$

Solution

(a) $\sqrt{3}(\sqrt{6} + \sqrt{12}) = \sqrt{3}\sqrt{6} + \sqrt{3}\sqrt{12}$

$$= \sqrt{18} + \sqrt{36}$$

$$= \sqrt{9}\sqrt{2} + 6$$

$$= 3\sqrt{2} + 6$$

(b) $2\sqrt{2}(4\sqrt{3} - 5\sqrt{6}) = (2\sqrt{2})(4\sqrt{3}) - (2\sqrt{2})(5\sqrt{6})$

$$= 8\sqrt{6} - 10\sqrt{12}$$

$$= 8\sqrt{6} - 10\sqrt{4}\sqrt{3}$$

$$= 8\sqrt{6} - 20\sqrt{3}$$

(c) $\sqrt{6x}(\sqrt{8x} + \sqrt{12xy}) = (\sqrt{6x})(\sqrt{8x}) + (\sqrt{6x})(\sqrt{12xy})$

$$= \sqrt{48x^2} + \sqrt{72x^2y}$$

$$= \sqrt{16x^2}\sqrt{3} + \sqrt{36x^2}\sqrt{2y}$$

$$= 4x\sqrt{3} + 6x\sqrt{2y}$$

(d) $\sqrt[3]{2}(5\sqrt[3]{4} - 3\sqrt[3]{16}) = (\sqrt[3]{2})(5\sqrt[3]{4}) - (\sqrt[3]{2})(3\sqrt[3]{16})$

$$= 5\sqrt[3]{8} - 3\sqrt[3]{32}$$

$$= 5 \cdot 2 - 3\sqrt[3]{8}\sqrt[3]{4}$$

$$= 10 - 6\sqrt[3]{4} \qquad \blacksquare$$

The distributive property also plays a central role in determining the product of two binomials. For example, $(x + 2)(x + 3) = x(x + 3) + 2(x + 3) = x^2 + 3x + 2x + 6 = x^2 + 5x + 6$. Finding the product of two binomial expressions that involve radicals can be handled in a similar fashion, as in the next examples.

EXAMPLE 3

Find the following products and simplify.

(a) $(\sqrt{3} + \sqrt{5})(\sqrt{2} + \sqrt{6})$ **(b)** $(2\sqrt{2} - \sqrt{7})(3\sqrt{2} + 5\sqrt{7})$

(c) $(\sqrt{8} + \sqrt{6})(\sqrt{8} - \sqrt{6})$ **(d)** $(\sqrt{x} + \sqrt{y})(\sqrt{x} - \sqrt{y})$

Solution

(a) $(\sqrt{3} + \sqrt{5})(\sqrt{2} + \sqrt{6}) = \sqrt{3}(\sqrt{2} + \sqrt{6}) + \sqrt{5}(\sqrt{2} + \sqrt{6})$

$$= \sqrt{3}\sqrt{2} + \sqrt{3}\sqrt{6} + \sqrt{5}\sqrt{2} + \sqrt{5}\sqrt{6}$$

$$= \sqrt{6} + \sqrt{18} + \sqrt{10} + \sqrt{30}$$

$$= \sqrt{6} + 3\sqrt{2} + \sqrt{10} + \sqrt{30}$$

(b) $(2\sqrt{2} - \sqrt{7})(3\sqrt{2} + 5\sqrt{7}) = 2\sqrt{2}(3\sqrt{2} + 5\sqrt{7})$

$$- \sqrt{7}(3\sqrt{2} + 5\sqrt{7})$$

$$= (2\sqrt{2})(3\sqrt{2}) + (2\sqrt{2})(5\sqrt{7})$$

$$- (\sqrt{7})(3\sqrt{2}) - (\sqrt{7})(5\sqrt{7})$$

$$= 12 + 10\sqrt{14} - 3\sqrt{14} - 35$$

$$= -23 + 7\sqrt{14}$$

(c) $(\sqrt{8} + \sqrt{6})(\sqrt{8} - \sqrt{6}) = \sqrt{8}(\sqrt{8} - \sqrt{6}) + \sqrt{6}(\sqrt{8} - \sqrt{6})$

$$= \sqrt{8}\sqrt{8} - \sqrt{8}\sqrt{6} + \sqrt{6}\sqrt{8} - \sqrt{6}\sqrt{6}$$

$$= 8 - \sqrt{48} + \sqrt{48} - 6$$

$$= 2$$

(d) $(\sqrt{x} + \sqrt{y})(\sqrt{x} - \sqrt{y}) = \sqrt{x}(\sqrt{x} - \sqrt{y}) + \sqrt{y}(\sqrt{x} - \sqrt{y})$

$$= \sqrt{x}\sqrt{x} - \sqrt{x}\sqrt{y} + \sqrt{y}\sqrt{x} - \sqrt{y}\sqrt{y}$$

$$= x - \sqrt{xy} + \sqrt{xy} - y$$

$$= x - y \qquad \blacksquare$$

Note parts (c) and (d) of Example 3; they fit the special-product pattern $(a + b)(a - b) = a^2 - b^2$. Furthermore, in each case the final product is in rational form. The factors $a + b$ and $a - b$ are called **conjugates**. This suggests a way of rationalizing the denominator in an expression that contains a binomial denominator with radicals. We will multiply by the conjugate of the binomial denominator. Consider the following example.

EXAMPLE 4

Simplify $\dfrac{4}{\sqrt{5} + \sqrt{2}}$ by rationalizing the denominator.

Solution

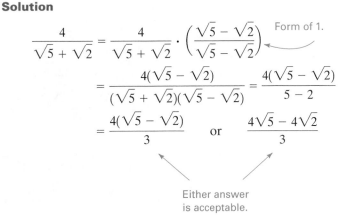

$$\frac{4}{\sqrt{5} + \sqrt{2}} = \frac{4}{\sqrt{5} + \sqrt{2}} \cdot \left(\frac{\sqrt{5} - \sqrt{2}}{\sqrt{5} - \sqrt{2}}\right) \quad \text{Form of 1.}$$

$$= \frac{4(\sqrt{5} - \sqrt{2})}{(\sqrt{5} + \sqrt{2})(\sqrt{5} - \sqrt{2})} = \frac{4(\sqrt{5} - \sqrt{2})}{5 - 2}$$

$$= \frac{4(\sqrt{5} - \sqrt{2})}{3} \quad \text{or} \quad \frac{4\sqrt{5} - 4\sqrt{2}}{3}$$

Either answer
is acceptable.

The next examples further illustrate the process of rationalizing and simplifying expressions that contain binomial denominators.

EXAMPLE 5

For each of the following, rationalize the denominator and simplify.

(a) $\dfrac{\sqrt{3}}{\sqrt{6} - 9}$ (b) $\dfrac{7}{3\sqrt{5} + 2\sqrt{3}}$

(c) $\dfrac{\sqrt{x} + 2}{\sqrt{x} - 3}$ (d) $\dfrac{2\sqrt{x} - 3\sqrt{y}}{\sqrt{x} + \sqrt{y}}$

Solution

(a) $\dfrac{\sqrt{3}}{\sqrt{6} - 9} = \dfrac{\sqrt{3}}{\sqrt{6} - 9} \cdot \dfrac{\sqrt{6} + 9}{\sqrt{6} + 9}$

$$= \frac{\sqrt{3}(\sqrt{6} + 9)}{(\sqrt{6} - 9)(\sqrt{6} + 9)}$$

$$= \frac{\sqrt{18} + 9\sqrt{3}}{6 - 81}$$

$$= \frac{3\sqrt{2} + 9\sqrt{3}}{-75}$$

$$= \frac{3(\sqrt{2} + 3\sqrt{3})}{(-3)(25)}$$

$$= -\frac{\sqrt{2} + 3\sqrt{3}}{25} \quad \text{or} \quad \frac{-\sqrt{2} - 3\sqrt{3}}{25}$$

(b) $\dfrac{7}{3\sqrt{5} + 2\sqrt{3}} = \dfrac{7}{3\sqrt{5} + 2\sqrt{3}} \cdot \dfrac{3\sqrt{5} - 2\sqrt{3}}{3\sqrt{5} - 2\sqrt{3}}$

$$= \dfrac{7(3\sqrt{5} - 2\sqrt{3})}{(3\sqrt{5} + 2\sqrt{3})(3\sqrt{5} - 2\sqrt{3})}$$

$$= \dfrac{7(3\sqrt{5} - 2\sqrt{3})}{45 - 12}$$

$$= \dfrac{7(3\sqrt{5} - 2\sqrt{3})}{33} \quad \text{or} \quad \dfrac{21\sqrt{5} - 14\sqrt{3}}{33}$$

(c) $\dfrac{\sqrt{x} + 2}{\sqrt{x} - 3} = \dfrac{\sqrt{x} + 2}{\sqrt{x} - 3} \cdot \dfrac{\sqrt{x} + 3}{\sqrt{x} + 3} = \dfrac{(\sqrt{x} + 2)(\sqrt{x} + 3)}{(\sqrt{x} - 3)(\sqrt{x} + 3)}$

$$= \dfrac{x + 3\sqrt{x} + 2\sqrt{x} + 6}{x - 9}$$

$$= \dfrac{x + 5\sqrt{x} + 6}{x - 9}$$

(d) $\dfrac{2\sqrt{x} - 3\sqrt{y}}{\sqrt{x} + \sqrt{y}} = \dfrac{2\sqrt{x} - 3\sqrt{y}}{\sqrt{x} + \sqrt{y}} \cdot \dfrac{\sqrt{x} - \sqrt{y}}{\sqrt{x} - \sqrt{y}}$

$$= \dfrac{(2\sqrt{x} - 3\sqrt{y})(\sqrt{x} - \sqrt{y})}{(\sqrt{x} + \sqrt{y})(\sqrt{x} - \sqrt{y})}$$

$$= \dfrac{2x - 2\sqrt{xy} - 3\sqrt{xy} + 3y}{x - y}$$

$$= \dfrac{2x - 5\sqrt{xy} + 3y}{x - y} \qquad ■$$

Problem Set 5.4

For Problems 1–14, multiply and simplify where possible.

1. $\sqrt{6}\sqrt{12}$

2. $\sqrt{8}\sqrt{6}$

3. $(3\sqrt{3})(2\sqrt{6})$

4. $(5\sqrt{2})(3\sqrt{12})$

5. $(4\sqrt{2})(-6\sqrt{5})$

6. $(-7\sqrt{3})(2\sqrt{5})$

7. $(-3\sqrt{3})(-4\sqrt{8})$

8. $(-5\sqrt{8})(-6\sqrt{7})$

9. $(5\sqrt{6})(4\sqrt{6})$

10. $(3\sqrt{7})(2\sqrt{7})$

11. $(2\sqrt[3]{4})(6\sqrt[3]{2})$

12. $(4\sqrt[3]{3})(5\sqrt[3]{9})$

13. $(4\sqrt[3]{6})(7\sqrt[3]{4})$

14. $(9\sqrt[3]{6})(2\sqrt[3]{9})$

For Problems 15–52, find the following products and express answers in simplest radical form. All variables represent nonnegative real numbers.

15. $\sqrt{2}(\sqrt{3} + \sqrt{5})$

16. $\sqrt{3}(\sqrt{7} + \sqrt{10})$

17. $3\sqrt{5}(2\sqrt{2} - \sqrt{7})$

18. $5\sqrt{6}(2\sqrt{5} - 3\sqrt{11})$

19. $2\sqrt{6}(3\sqrt{8} - 5\sqrt{12})$

20. $4\sqrt{2}(3\sqrt{12} + 7\sqrt{6})$

21. $-4\sqrt{5}(2\sqrt{5} + 4\sqrt{12})$

22. $-5\sqrt{3}(3\sqrt{12} - 9\sqrt{8})$

23. $3\sqrt{x}(5\sqrt{2} + \sqrt{y})$

24. $\sqrt{2x}(3\sqrt{y} - 7\sqrt{5})$

25. $\sqrt{xy}(5\sqrt{xy} - 6\sqrt{x})$ **26.** $4\sqrt{x}(2\sqrt{xy} + 2\sqrt{x})$

27. $\sqrt{5y}(\sqrt{8x} + \sqrt{12y^2})$ **28.** $\sqrt{2x}(\sqrt{12xy} - \sqrt{8y})$

29. $5\sqrt{3}(2\sqrt{8} - 3\sqrt{18})$ **30.** $2\sqrt{2}(3\sqrt{12} - \sqrt{27})$

31. $(\sqrt{3} + 4)(\sqrt{3} - 7)$ **32.** $(\sqrt{2} + 6)(\sqrt{2} - 2)$

33. $(\sqrt{5} - 6)(\sqrt{5} - 3)$ **34.** $(\sqrt{7} - 2)(\sqrt{7} - 8)$

35. $(3\sqrt{5} - 2\sqrt{3})(2\sqrt{7} + \sqrt{2})$

36. $(\sqrt{2} + \sqrt{3})(\sqrt{5} - \sqrt{7})$

37. $(2\sqrt{6} + 3\sqrt{5})(\sqrt{8} - 3\sqrt{12})$

38. $(5\sqrt{2} - 4\sqrt{6})(2\sqrt{8} + \sqrt{6})$

39. $(2\sqrt{6} + 5\sqrt{5})(3\sqrt{6} - \sqrt{5})$

40. $(7\sqrt{3} - \sqrt{7})(2\sqrt{3} + 4\sqrt{7})$

41. $(3\sqrt{2} - 5\sqrt{3})(6\sqrt{2} - 7\sqrt{3})$

42. $(\sqrt{8} - 3\sqrt{10})(2\sqrt{8} - 6\sqrt{10})$

43. $(\sqrt{6} + 4)(\sqrt{6} - 4)$

44. $(\sqrt{7} - 2)(\sqrt{7} + 2)$

45. $(\sqrt{2} + \sqrt{10})(\sqrt{2} - \sqrt{10})$

46. $(2\sqrt{3} + \sqrt{11})(2\sqrt{3} - \sqrt{11})$

47. $(\sqrt{2x} + \sqrt{3y})(\sqrt{2x} - \sqrt{3y})$

48. $(2\sqrt{x} - 5\sqrt{y})(2\sqrt{x} + 5\sqrt{y})$

49. $2\sqrt[3]{3}(5\sqrt[3]{4} + \sqrt[3]{6})$

50. $2\sqrt[3]{2}(3\sqrt[3]{6} - 4\sqrt[3]{5})$

51. $3\sqrt[3]{4}(2\sqrt[3]{2} - 6\sqrt[3]{4})$

52. $3\sqrt[3]{3}(4\sqrt[3]{9} + 5\sqrt[3]{7})$

For Problems 53–76, rationalize the denominator and simplify. All variables represent positive real numbers.

53. $\dfrac{2}{\sqrt{7} + 1}$ **54.** $\dfrac{6}{\sqrt{5} + 2}$

55. $\dfrac{3}{\sqrt{2} - 5}$ **56.** $\dfrac{-4}{\sqrt{6} - 3}$

57. $\dfrac{1}{\sqrt{2} + \sqrt{7}}$ **58.** $\dfrac{3}{\sqrt{3} + \sqrt{10}}$

59. $\dfrac{\sqrt{2}}{\sqrt{10} - \sqrt{3}}$ **60.** $\dfrac{\sqrt{3}}{\sqrt{7} - \sqrt{2}}$

61. $\dfrac{\sqrt{3}}{2\sqrt{5} + 4}$ **62.** $\dfrac{\sqrt{7}}{3\sqrt{2} - 5}$

63. $\dfrac{6}{3\sqrt{7} - 2\sqrt{6}}$ **64.** $\dfrac{5}{2\sqrt{5} + 3\sqrt{7}}$

65. $\dfrac{\sqrt{6}}{3\sqrt{2} + 2\sqrt{3}}$ **66.** $\dfrac{3\sqrt{6}}{5\sqrt{3} - 4\sqrt{2}}$

67. $\dfrac{2}{\sqrt{x} + 4}$ **68.** $\dfrac{3}{\sqrt{x} + 7}$

69. $\dfrac{\sqrt{x}}{\sqrt{x} - 5}$ **70.** $\dfrac{\sqrt{x}}{\sqrt{x} - 1}$

71. $\dfrac{\sqrt{x} - 2}{\sqrt{x} + 6}$ **72.** $\dfrac{\sqrt{x} + 1}{\sqrt{x} - 10}$

73. $\dfrac{\sqrt{x}}{\sqrt{x} + 2\sqrt{y}}$ **74.** $\dfrac{\sqrt{y}}{2\sqrt{x} - \sqrt{y}}$

75. $\dfrac{3\sqrt{y}}{2\sqrt{x} - 3\sqrt{y}}$ **76.** $\dfrac{2\sqrt{x}}{3\sqrt{x} + 5\sqrt{y}}$

■ ■ ■ THOUGHTS INTO WORDS

77. How would you help someone rationalize the denominator and simplify $\dfrac{4}{\sqrt{8} + \sqrt{12}}$?

78. Discuss how the distributive property has been used thus far in this chapter.

79. How would you simplify the expression $\dfrac{\sqrt{8} + \sqrt{12}}{\sqrt{2}}$?

80. Use your calculator to evaluate each expression in Problems 53–66. Then evaluate the results you obtained when you did the problems.

5.5 Equations Involving Radicals

We often refer to equations that contain radicals with variables in a radicand as **radical equations**. In this section we discuss techniques for solving such equations that contain one or more radicals. To solve radical equations, we need the following property of equality.

Property 5.6

Let a and b be real numbers and n be a positive integer.

   If $a = b$,   then $a^n = b^n$.

Property 5.6 states that we can raise both sides of an equation to a positive integral power. However, raising both sides of an equation to a positive integral power sometimes produces results that do not satisfy the original equation. Let's consider two examples to illustrate this point.

EXAMPLE 1   Solve $\sqrt{2x - 5} = 7$.

Solution

$$\sqrt{2x - 5} = 7$$
$$(\sqrt{2x - 5})^2 = 7^2 \qquad \text{Square both sides.}$$
$$2x - 5 = 49$$
$$2x = 54$$
$$x = 27$$

✔ **Check**

$$\sqrt{2x - 5} = 7$$
$$\sqrt{2(27) - 5} \overset{?}{=} 7$$
$$\sqrt{49} \overset{?}{=} 7$$
$$7 = 7$$

The solution set for $\sqrt{2x - 5} = 7$ is $\{27\}$. ■

EXAMPLE 2   Solve $\sqrt{3a + 4} = -4$.

Solution

$$\sqrt{3a + 4} = -4$$

$$(\sqrt{3a + 4})^2 = (-4)^2 \qquad \text{Square both sides.}$$

$$3a + 4 = 16$$

$$3a = 12$$

$$a = 4$$

✔ Check

$$\sqrt{3a + 4} = -4$$

$$\sqrt{3(4) + 4} \overset{?}{=} -4$$

$$\sqrt{16} \overset{?}{=} -4$$

$$4 \neq -4$$

Because 4 does not check, the original equation has no real number solution. Thus the solution set is $\varnothing$. ■

In general, raising both sides of an equation to a positive integral power produces an equation that has all of the solutions of the original equation, but it may also have some extra solutions that do not satisfy the original equation. Such extra solutions are called **extraneous solutions**. Therefore, when using Property 5.6, you *must* check each potential solution in the original equation.

Let's consider some examples to illustrate different situations that arise when we are solving radical equations.

EXAMPLE 3   Solve $\sqrt{2t - 4} = t - 2$.

Solution

$$\sqrt{2t - 4} = t - 2$$

$$(\sqrt{2t - 4})^2 = (t - 2)^2 \qquad \text{Square both sides.}$$

$$2t - 4 = t^2 - 4t + 4$$

$$0 = t^2 - 6t + 8$$

$$0 = (t - 2)(t - 4) \qquad \text{Factor the right side.}$$

$$t - 2 = 0 \quad \text{or} \quad t - 4 = 0 \qquad \text{Apply: } ab = 0 \text{ if and only if } a = 0 \text{ or } b = 0.$$

$$t = 2 \quad \text{or} \quad t = 4$$

✔ **Check**

$$\sqrt{2t-4} = t-2 \qquad\qquad \sqrt{2t-4} = t-2$$

$$\sqrt{2(2)-4} \overset{?}{=} 2-2, \quad \text{when } t=2 \quad \text{or} \quad \sqrt{2(4)-4} \overset{?}{=} 4-2, \quad \text{when } t=4$$

$$\sqrt{0} \overset{?}{=} 0 \qquad\qquad\qquad \sqrt{4} \overset{?}{=} 2$$

$$0 = 0 \qquad\qquad\qquad\qquad 2 = 2$$

The solution set is {2, 4}. ∎

EXAMPLE 4

Solve $\sqrt{y} + 6 = y$.

Solution

$$\sqrt{y} + 6 = y$$

$$\sqrt{y} = y - 6$$

$$(\sqrt{y})^2 = (y-6)^2 \qquad \text{Square both sides.}$$

$$y = y^2 - 12y + 36$$

$$0 = y^2 - 13y + 36$$

$$0 = (y-4)(y-9) \qquad \text{Factor the right side.}$$

$$y - 4 = 0 \quad \text{or} \quad y - 9 = 0 \qquad \text{Apply: } ab = 0 \text{ if and only if } a = 0 \text{ or } b = 0.$$

$$y = 4 \quad \text{or} \quad y = 9$$

✔ **Check**

$$\sqrt{y} + 6 = y \qquad\qquad \sqrt{y} + 6 = y$$

$$\sqrt{4} + 6 \overset{?}{=} 4, \quad \text{when } y=4 \quad \text{or} \quad \sqrt{9} + 6 \overset{?}{=} 9, \quad \text{when } y=9$$

$$2 + 6 \overset{?}{=} 4 \qquad\qquad\qquad 3 + 6 \overset{?}{=} 9$$

$$8 \neq 4 \qquad\qquad\qquad\qquad 9 = 9$$

The only solution is 9; the solution set is {9}. ∎

In Example 4, note that we changed the form of the original equation $\sqrt{y} + 6 = y$ to $\sqrt{y} = y - 6$ before we squared both sides. Squaring both sides of $\sqrt{y} + 6 = y$ produces $y + 12\sqrt{y} + 36 = y^2$, which is a much more complex equation that still contains a radical. Here again, it pays to think ahead before carrying out all the steps. Now let's consider an example involving a cube root.

EXAMPLE 5

Solve $\sqrt[3]{n^2 - 1} = 2$.

Solution

$$\sqrt[3]{n^2 - 1} = 2$$

$$(\sqrt[3]{n^2 - 1})^3 = 2^3 \qquad \text{Cube both sides.}$$

$$n^2 - 1 = 8$$
$$n^2 - 9 = 0$$
$$(n + 3)(n - 3) = 0$$
$$n + 3 = 0 \qquad \text{or} \qquad n - 3 = 0$$
$$n = -3 \qquad \text{or} \qquad n = 3$$

✔ **Check**

$$\sqrt[3]{n^2 - 1} = 2 \qquad\qquad \sqrt[3]{n^2 - 1} = 2$$
$$\sqrt[3]{(-3)^2 - 1} \overset{?}{=} 2, \quad \text{when } n = -3 \qquad \text{or} \qquad \sqrt[3]{3^2 - 1} \overset{?}{=} 2, \quad \text{when } n = 3$$
$$\sqrt[3]{8} \overset{?}{=} 2 \qquad\qquad\qquad\qquad \sqrt[3]{8} \overset{?}{=} 2$$
$$2 = 2 \qquad\qquad\qquad\qquad\qquad 2 = 2$$

The solution set is $\{-3, 3\}$. ■

It may be necessary to square both sides of an equation, simplify the resulting equation, and then square both sides again. The next example illustrates this type of problem.

EXAMPLE   6

Solve $\sqrt{x + 2} = 7 - \sqrt{x + 9}$.

Solution

$$\sqrt{x + 2} = 7 - \sqrt{x + 9}$$
$$(\sqrt{x + 2})^2 = (7 - \sqrt{x + 9})^2 \qquad\qquad \text{Square both sides.}$$
$$x + 2 = 49 - 14\sqrt{x + 9} + x + 9$$
$$x + 2 = x + 58 - 14\sqrt{x + 9}$$
$$-56 = -14\sqrt{x + 9}$$
$$4 = \sqrt{x + 9}$$
$$(4)^2 = (\sqrt{x + 9})^2 \qquad\qquad \text{Square both sides.}$$
$$16 = x + 9$$
$$7 = x$$

✔ **Check**

$$\sqrt{x + 2} = 7 - \sqrt{x + 9}$$
$$\sqrt{7 + 2} \overset{?}{=} 7 - \sqrt{7 + 9}$$
$$\sqrt{9} \overset{?}{=} 7 - \sqrt{16}$$
$$3 \overset{?}{=} 7 - 4$$
$$3 = 3$$

The solution set is $\{7\}$. ■

■ Another Look at Applications

In Section 5.1 we used the formula $S = \sqrt{30Df}$ to approximate how fast a car was traveling on the basis of the length of skid marks. (Remember that S represents the speed of the car in miles per hour, D represents the length of the skid marks in feet, and f represents a coefficient of friction.) This same formula can be used to estimate the length of skid marks that are produced by cars traveling at different rates on various types of road surfaces. To use the formula for this purpose, let's change the form of the equation by solving for D.

$$\sqrt{30Df} = S$$

$$30Df = S^2 \qquad \text{The result of squaring both sides of the original equation}$$

$$D = \frac{S^2}{30f} \qquad \begin{array}{l}\text{D, S, and f are positive numbers, so this final equation and}\\ \text{the original one are equivalent.}\end{array}$$

EXAMPLE 7   Suppose that for a particular road surface, the coefficient of friction is 0.35. How far will a car skid when the brakes are applied at 60 miles per hour?

Solution

We can substitute 0.35 for f and 60 for S in the formula $D = \dfrac{S^2}{30f}$.

$$D = \frac{60^2}{30(0.35)} = 343, \quad \text{to the nearest whole number}$$

The car will skid approximately 343 feet.   ■

Remark: Pause for a moment and think about the result in Example 7. The coefficient of friction 0.35 refers to a wet concrete road surface. Note that a car traveling at 60 miles per hour on such a surface will skid more than the length of a football field.

Problem Set 5.5

For Problems 1–56, solve each equation. Don't forget to check each of your potential solutions.

1. $\sqrt{5x} = 10$

2. $\sqrt{3x} = 9$

3. $\sqrt{2x} + 4 = 0$

4. $\sqrt{4x} + 5 = 0$

5. $2\sqrt{n} = 5$

6. $5\sqrt{n} = 3$

7. $3\sqrt{n} - 2 = 0$

8. $2\sqrt{n} - 7 = 0$

9. $\sqrt{3y + 1} = 4$

10. $\sqrt{2y - 3} = 5$

11. $\sqrt{4y - 3} - 6 = 0$

12. $\sqrt{3y + 5} - 2 = 0$

13. $\sqrt{3x - 1} + 1 = 4$

14. $\sqrt{4x - 1} - 3 = 2$

15. $\sqrt{2n + 3} - 2 = -1$

16. $\sqrt{5n + 1} - 6 = -4$

17. $\sqrt{2x - 5} = -1$

18. $\sqrt{4x - 3} = -4$

19. $\sqrt{5x + 2} = \sqrt{6x + 1}$

20. $\sqrt{4x + 2} = \sqrt{3x + 4}$

21. $\sqrt{3x + 1} = \sqrt{7x - 5}$

22. $\sqrt{6x + 5} = \sqrt{2x + 10}$

23. $\sqrt{3x - 2} - \sqrt{x + 4} = 0$

24. $\sqrt{7x - 6} - \sqrt{5x + 2} = 0$

25. $5\sqrt{t - 1} = 6$ **26.** $4\sqrt{t + 3} = 6$

27. $\sqrt{x^2 + 7} = 4$ **28.** $\sqrt{x^2 + 3} - 2 = 0$

29. $\sqrt{x^2 + 13x + 37} = 1$

30. $\sqrt{x^2 + 5x - 20} = 2$

31. $\sqrt{x^2 - x + 1} = x + 1$

32. $\sqrt{n^2 - 2n - 4} = n$

33. $\sqrt{x^2 + 3x + 7} = x + 2$

34. $\sqrt{x^2 + 2x + 1} = x + 3$

35. $\sqrt{-4x + 17} = x - 3$ **36.** $\sqrt{2x - 1} = x - 2$

37. $\sqrt{n + 4} = n + 4$ **38.** $\sqrt{n + 6} = n + 6$

39. $\sqrt{3y} = y - 6$ **40.** $2\sqrt{n} = n - 3$

41. $4\sqrt{x + 5} = x$ **42.** $\sqrt{-x - 6} = x$

43. $\sqrt[3]{x - 2} = 3$ **44.** $\sqrt[3]{x + 1} = 4$

45. $\sqrt[3]{2x + 3} = -3$ **46.** $\sqrt[3]{3x - 1} = -4$

47. $\sqrt[3]{2x + 5} = \sqrt[3]{4 - x}$

48. $\sqrt[3]{3x - 1} = \sqrt[3]{2 - 5x}$

49. $\sqrt{x + 19} - \sqrt{x + 28} = -1$

50. $\sqrt{x + 4} = \sqrt{x - 1} + 1$

51. $\sqrt{3x + 1} + \sqrt{2x + 4} = 3$

52. $\sqrt{2x - 1} - \sqrt{x + 3} = 1$

53. $\sqrt{n - 4} + \sqrt{n + 4} = 2\sqrt{n - 1}$

54. $\sqrt{n - 3} + \sqrt{n + 5} = 2\sqrt{n}$

55. $\sqrt{t + 3} - \sqrt{t - 2} = \sqrt{7 - t}$

56. $\sqrt{t + 7} - 2\sqrt{t - 8} = \sqrt{t - 5}$

57. Use the formula given in Example 7 with a coefficient of friction of 0.95. How far will a car skid at 40 miles per hour? at 55 miles per hour? at 65 miles per hour? Express the answers to the nearest foot.

58. Solve the formula $T = 2\pi\sqrt{\dfrac{L}{32}}$ for L. (Remember that in this formula, which was used in Section 5.2, T represents the period of a pendulum expressed in seconds, and L represents the length of the pendulum in feet.)

59. In Problem 58, you should have obtained the equation $L = \dfrac{8T^2}{\pi^2}$. What is the length of a pendulum that has a period of 2 seconds? of 2.5 seconds? of 3 seconds? Express your answers to the nearest tenth of a foot.

■ ■ ■ THOUGHTS INTO WORDS

60. Explain the concept of extraneous solutions.

61. Explain why possible solutions for radical equations *must* be checked.

62. Your friend makes an effort to solve the equation $3 + 2\sqrt{x} = x$ as follows:

$$(3 + 2\sqrt{x})^2 = x^2$$
$$9 + 12\sqrt{x} + 4x = x^2$$

At this step he stops and doesn't know how to proceed. What help would you give him?

5.6   Merging Exponents and Roots

Recall that the basic properties of positive integral exponents led to a definition for the use of negative integers as exponents. In this section, the properties of integral exponents are used to form definitions for the use of rational numbers as exponents. These definitions will tie together the concepts of exponent and root.

Let's consider the following comparisons.

From our study of radicals, we know that	If $(b^n)^m = b^{mn}$ is to hold when n equals a rational number of the form $\dfrac{1}{p}$, where p is a positive integer greater than 1, then
$(\sqrt{5})^2 = 5$	$\left(5^{\frac{1}{2}}\right)^2 = 5^{2\left(\frac{1}{2}\right)} = 5^1 = 5$
$(\sqrt[3]{8})^3 = 8$	$\left(8^{\frac{1}{3}}\right)^3 = 8^{3\left(\frac{1}{3}\right)} = 8^1 = 8$
$(\sqrt[4]{21})^4 = 21$	$\left(21^{\frac{1}{4}}\right)^4 = 21^{4\left(\frac{1}{4}\right)} = 21^1 = 21$

It would seem reasonable to make the following definition.

Definition 5.6

If b is a real number, n is a positive integer greater than 1, and $\sqrt[n]{b}$ exists, then
$$b^{\frac{1}{n}} = \sqrt[n]{b}$$

Definition 5.6 states that $b^{\frac{1}{n}}$ means the nth root of b. We shall assume that b and n are chosen so that $\sqrt[n]{b}$ exists. For example, $(-25)^{\frac{1}{2}}$ is not meaningful at this time because $\sqrt{-25}$ is not a real number. Consider the following examples, which demonstrate the use of Definition 5.6.

$$25^{\frac{1}{2}} = \sqrt{25} = 5 \qquad\qquad 16^{\frac{1}{4}} = \sqrt[4]{16} = 2$$

$$8^{\frac{1}{3}} = \sqrt[3]{8} = 2 \qquad\qquad \left(\frac{36}{49}\right)^{\frac{1}{2}} = \sqrt{\frac{36}{49}} = \frac{6}{7}$$

$$(-27)^{\frac{1}{3}} = \sqrt[3]{-27} = -3$$

The following definition provides the basis for the use of *all* rational numbers as exponents.

Definition 5.7

If $\dfrac{m}{n}$ is a rational number, where n is a positive integer greater than 1, and b is a real number such that $\sqrt[n]{b}$ exists, then

$$b^{\frac{m}{n}} = \sqrt[n]{b^m} = (\sqrt[n]{b})^m$$

In Definition 5.7, note that the denominator of the exponent is the index of the radical and that the numerator of the exponent is either the exponent of the radicand or the exponent of the root.

Whether we use the form $\sqrt[n]{b^m}$ or the form $(\sqrt[n]{b})^m$ for computational purposes depends somewhat on the magnitude of the problem. Let's use both forms on two problems to illustrate this point.

$$8^{\frac{2}{3}} = \sqrt[3]{8^2} \qquad \text{or} \qquad 8^{\frac{2}{3}} = (\sqrt[3]{8})^2$$
$$= \sqrt[3]{64} \qquad\qquad\qquad = 2^2$$
$$= 4 \qquad\qquad\qquad\quad = 4$$

$$27^{\frac{2}{3}} = \sqrt[3]{27^2} \qquad \text{or} \qquad 27^{\frac{2}{3}} = (\sqrt[3]{27})^2$$
$$= \sqrt[3]{729} \qquad\qquad\qquad = 3^2$$
$$= 9 \qquad\qquad\qquad\quad = 9$$

To compute $8^{\frac{2}{3}}$, either form seems to work about as well as the other one. However, to compute $27^{\frac{2}{3}}$, it should be obvious that $(\sqrt[3]{27})^2$ is much easier to handle than $\sqrt[3]{27^2}$.

E X A M P L E   1

Simplify each of the following numerical expressions.

(a) $25^{\frac{3}{2}}$ (b) $16^{\frac{3}{4}}$ (c) $(32)^{-\frac{2}{5}}$

(d) $(-64)^{\frac{2}{3}}$ (e) $-8^{\frac{1}{3}}$

Solution

(a) $25^{\frac{3}{2}} = (\sqrt{25})^3 = 5^3 = 125$

(b) $16^{\frac{3}{4}} = (\sqrt[4]{16})^3 = 2^3 = 8$

(c) $(32)^{-\frac{2}{5}} = \dfrac{1}{(32)^{\frac{2}{5}}} = \dfrac{1}{(\sqrt[5]{32})^2} = \dfrac{1}{2^2} = \dfrac{1}{4}$

(d) $(-64)^{\frac{2}{3}} = (\sqrt[3]{-64})^2 = (-4)^2 = 16$

(e) $-8^{\frac{1}{3}} = -\sqrt[3]{8} = -2$ ∎

The basic laws of exponents that we stated in Property 5.2 are true for all rational exponents. Therefore, from now on we will use Property 5.2 for rational as well as integral exponents.

Some problems can be handled better in exponential form and others in radical form. Thus we must be able to switch forms with a certain amount of ease. Let's consider some examples where we switch from one form to the other.

E X A M P L E   2

Write each of the following expressions in radical form.

(a) $x^{\frac{3}{4}}$ (b) $3y^{\frac{2}{5}}$ (c) $x^{\frac{1}{4}}y^{\frac{3}{4}}$ (d) $(x + y)^{\frac{2}{3}}$

Solution

(a) $x^{\frac{3}{4}} = \sqrt[4]{x^3}$ (b) $3y^{\frac{2}{5}} = 3\sqrt[5]{y^2}$

(c) $x^{\frac{1}{4}}y^{\frac{3}{4}} = (xy^3)^{\frac{1}{4}} = \sqrt[4]{xy^3}$ (d) $(x + y)^{\frac{2}{3}} = \sqrt[3]{(x + y)^2}$ ■

E X A M P L E   3

Write each of the following using positive rational exponents.

(a) $\sqrt{xy}$ (b) $\sqrt[4]{a^3b}$ (c) $4\sqrt[3]{x^2}$ (d) $\sqrt[5]{(x + y)^4}$

Solution

(a) $\sqrt{xy} = (xy)^{\frac{1}{2}} = x^{\frac{1}{2}}y^{\frac{1}{2}}$ (b) $\sqrt[4]{a^3b} = (a^3b)^{\frac{1}{4}} = a^{\frac{3}{4}}b^{\frac{1}{4}}$

(c) $4\sqrt[3]{x^2} = 4x^{\frac{2}{3}}$ (d) $\sqrt[5]{(x + y)^4} = (x + y)^{\frac{4}{5}}$ ■

The properties of exponents provide the basis for simplifying algebraic expressions that contain rational exponents, as these next examples illustrate.

E X A M P L E   4

Simplify each of the following. Express final results using positive exponents only.

(a) $\left(3x^{\frac{1}{2}}\right)\left(4x^{\frac{2}{3}}\right)$ (b) $\left(5a^{\frac{1}{3}}b^{\frac{1}{2}}\right)^2$ (c) $\dfrac{12y^{\frac{1}{3}}}{6y^{\frac{1}{2}}}$ (d) $\left(\dfrac{3x^{\frac{2}{5}}}{2y^{\frac{2}{3}}}\right)^4$

Solution

(a) $\left(3x^{\frac{1}{2}}\right)\left(4x^{\frac{2}{3}}\right) = 3 \cdot 4 \cdot x^{\frac{1}{2}} \cdot x^{\frac{2}{3}}$

$\qquad\qquad = 12x^{\frac{1}{2}+\frac{2}{3}}$ $\qquad$ $b^n \cdot b^m = b^{n+m}$

$\qquad\qquad = 12x^{\frac{3}{6}+\frac{4}{6}}$ $\qquad$ Use 6 as LCD.

$\qquad\qquad = 12x^{\frac{7}{6}}$

(b) $\left(5a^{\frac{1}{3}}b^{\frac{1}{2}}\right)^2 = 5^2 \cdot \left(a^{\frac{1}{3}}\right)^2 \cdot \left(b^{\frac{1}{2}}\right)^2$ $\qquad$ $(ab)^n = a^nb^n$

$\qquad\qquad = 25a^{\frac{2}{3}}b$ $\qquad$ $(b^n)^m = b^{mn}$

(c) $\dfrac{12y^{\frac{1}{3}}}{6y^{\frac{1}{2}}} = 2y^{\frac{1}{3}-\frac{1}{2}}$ $\qquad$ $\dfrac{b^n}{b^m} = b^{n-m}$

$\qquad\quad = 2y^{\frac{2}{6}-\frac{3}{6}}$

$\qquad\quad = 2y^{-\frac{1}{6}}$

$\qquad\quad = \dfrac{2}{y^{\frac{1}{6}}}$

(d) $\left(\dfrac{3x^{\frac{2}{5}}}{2y^{\frac{2}{3}}}\right)^4 = \dfrac{\left(3x^{\frac{2}{5}}\right)^4}{\left(2y^{\frac{2}{3}}\right)^4}$　　　$\left(\dfrac{a}{b}\right)^n = \dfrac{a^n}{b^n}$

$\qquad\qquad = \dfrac{3^4 \cdot \left(x^{\frac{2}{5}}\right)^4}{2^4 \cdot \left(y^{\frac{2}{3}}\right)^4}$　　$(ab)^n = a^n b^n$

$\qquad\qquad = \dfrac{81x^{\frac{8}{5}}}{16y^{\frac{8}{3}}}$　　$(b^n)^m = b^{mn}$　　∎

The link between exponents and roots also provides a basis for multiplying and dividing some radicals even if they have different indexes. The general procedure is as follows:

1. Change from radical form to exponential form.

2. Apply the properties of exponents.

3. Then change back to radical form.

The three parts of Example 5 illustrate this process.

EXAMPLE 5　　Perform the indicated operations and express the answers in simplest radical form.

$\qquad$ **(a)** $\sqrt{2}\,\sqrt[3]{2}$ 　　**(b)** $\dfrac{\sqrt{5}}{\sqrt[3]{5}}$ 　　**(c)** $\dfrac{\sqrt{4}}{\sqrt[3]{2}}$

Solution

(a) $\sqrt{2}\,\sqrt[3]{2} = 2^{\frac{1}{2}} \cdot 2^{\frac{1}{3}}$

$\qquad\qquad = 2^{\frac{1}{2}+\frac{1}{3}}$

$\qquad\qquad = 2^{\frac{3}{6}+\frac{2}{6}}$　　Use 6 as LCD.

$\qquad\qquad = 2^{\frac{5}{6}}$

$\qquad\qquad = \sqrt[6]{2^5} = \sqrt[6]{32}$

(b) $\dfrac{\sqrt{5}}{\sqrt[3]{5}} = \dfrac{5^{\frac{1}{2}}}{5^{\frac{1}{3}}}$

$\qquad\qquad = 5^{\frac{1}{2}-\frac{1}{3}}$

$\qquad\qquad = 5^{\frac{3}{6}-\frac{2}{6}}$　　Use 6 as LCD.

$\qquad\qquad = 5^{\frac{1}{6}} = \sqrt[6]{5}$

(c) $\dfrac{\sqrt{4}}{\sqrt[3]{2}} = \dfrac{4^{\frac{1}{2}}}{2^{\frac{1}{3}}}$

$\qquad\qquad = \dfrac{(2^2)^{\frac{1}{2}}}{2^{\frac{1}{3}}}$

$\qquad\qquad = \dfrac{2^1}{2^{\frac{1}{3}}}$

$\qquad\qquad = 2^{1-\frac{1}{3}}$

$\qquad\qquad = 2^{\frac{2}{3}} = \sqrt[3]{2^2} = \sqrt[3]{4}$　　∎

Problem Set 5.6

For Problems 1–30, evaluate each numerical expression.

1. $81^{\frac{1}{2}}$

2. $64^{\frac{1}{2}}$

3. $27^{\frac{1}{3}}$

4. $(-32)^{\frac{1}{5}}$

5. $(-8)^{\frac{1}{3}}$

6. $\left(-\dfrac{27}{8}\right)^{\frac{1}{3}}$

7. $-25^{\frac{1}{2}}$

8. $-64^{\frac{1}{3}}$

9. $36^{-\frac{1}{2}}$

10. $81^{-\frac{1}{2}}$

11. $\left(\dfrac{1}{27}\right)^{-\frac{1}{3}}$

12. $\left(-\dfrac{8}{27}\right)^{-\frac{1}{3}}$

13. $4^{\frac{3}{2}}$

14. $64^{\frac{2}{3}}$

15. $27^{\frac{4}{3}}$

16. $4^{\frac{7}{2}}$

17. $(-1)^{\frac{7}{3}}$

18. $(-8)^{\frac{4}{3}}$

19. $-4^{\frac{5}{2}}$

20. $-16^{\frac{3}{2}}$

21. $\left(\dfrac{27}{8}\right)^{\frac{4}{3}}$

22. $\left(\dfrac{8}{125}\right)^{\frac{2}{3}}$

23. $\left(\dfrac{1}{8}\right)^{-\frac{2}{3}}$

24. $\left(-\dfrac{1}{27}\right)^{-\frac{2}{3}}$

25. $64^{-\frac{7}{6}}$

26. $32^{-\frac{4}{5}}$

27. $-25^{\frac{3}{2}}$

28. $-16^{\frac{3}{4}}$

29. $125^{\frac{4}{3}}$

30. $81^{\frac{5}{4}}$

For Problems 31–44, write each of the following in radical form. For example,

$$3x^{\frac{2}{3}} = 3\sqrt[3]{x^2}$$

31. $x^{\frac{4}{3}}$

32. $x^{\frac{2}{5}}$

33. $3x^{\frac{1}{2}}$

34. $5x^{\frac{1}{4}}$

35. $(2y)^{\frac{1}{3}}$

36. $(3xy)^{\frac{1}{2}}$

37. $(2x - 3y)^{\frac{1}{2}}$

38. $(5x + y)^{\frac{1}{3}}$

39. $(2a - 3b)^{\frac{2}{3}}$

40. $(5a + 7b)^{\frac{3}{5}}$

41. $x^{\frac{2}{3}}y^{\frac{1}{3}}$

42. $x^{\frac{3}{7}}y^{\frac{5}{7}}$

43. $-3x^{\frac{1}{5}}y^{\frac{2}{5}}$

44. $-4x^{\frac{3}{4}}y^{\frac{1}{4}}$

For Problems 45–58, write each of the following using positive rational exponents. For example,

$$\sqrt{ab} = (ab)^{\frac{1}{2}} = a^{\frac{1}{2}}b^{\frac{1}{2}}$$

45. $\sqrt{5y}$

46. $\sqrt{2xy}$

47. $3\sqrt{y}$

48. $5\sqrt{ab}$

49. $\sqrt[3]{xy^2}$

50. $\sqrt[5]{x^2y^4}$

51. $\sqrt[4]{a^2b^3}$

52. $\sqrt[6]{ab^5}$

53. $\sqrt[5]{(2x - y)^3}$

54. $\sqrt[7]{(3x - y)^4}$

55. $5x\sqrt{y}$

56. $4y\sqrt[3]{x}$

57. $-\sqrt[3]{x + y}$

58. $-\sqrt[5]{(x - y)^2}$

For Problems 59–80, simplify each of the following. Express final results using positive exponents only. For example,

$$\left(2x^{\frac{1}{2}}\right)\left(3x^{\frac{1}{3}}\right) = 6x^{\frac{5}{6}}$$

59. $\left(2x^{\frac{2}{5}}\right)\left(6x^{\frac{1}{4}}\right)$

60. $\left(3x^{\frac{1}{4}}\right)\left(5x^{\frac{1}{3}}\right)$

61. $\left(y^{\frac{2}{3}}\right)\left(y^{-\frac{1}{4}}\right)$

62. $\left(y^{\frac{3}{4}}\right)\left(y^{-\frac{1}{2}}\right)$

63. $\left(x^{\frac{2}{5}}\right)\left(4x^{-\frac{1}{2}}\right)$

64. $\left(2x^{\frac{1}{3}}\right)\left(x^{-\frac{1}{2}}\right)$

65. $\left(4x^{\frac{1}{2}}y\right)^2$

66. $\left(3x^{\frac{1}{4}}y^{\frac{1}{5}}\right)^3$

67. $(8x^6y^3)^{\frac{1}{3}}$

68. $(9x^2y^4)^{\frac{1}{2}}$

69. $\dfrac{24x^{\frac{3}{5}}}{6x^{\frac{1}{3}}}$

70. $\dfrac{18x^{\frac{1}{2}}}{9x^{\frac{1}{3}}}$

71. $\dfrac{48b^{\frac{1}{3}}}{12b^{\frac{3}{4}}}$

72. $\dfrac{56a^{\frac{1}{6}}}{8a^{\frac{1}{4}}}$

73. $\left(\dfrac{6x^{\frac{2}{5}}}{7y^{\frac{2}{3}}}\right)^2$

74. $\left(\dfrac{2x^{\frac{1}{3}}}{3y^{\frac{1}{4}}}\right)^4$

75. $\left(\dfrac{x^2}{y^3}\right)^{-\frac{1}{2}}$

76. $\left(\dfrac{a^3}{b^{-2}}\right)^{-\frac{1}{3}}$

77. $\left(\dfrac{18x^{\frac{1}{3}}}{9x^{\frac{1}{4}}}\right)^2$ **78.** $\left(\dfrac{72x^{\frac{3}{4}}}{6x^{\frac{1}{2}}}\right)^2$

79. $\left(\dfrac{60a^{\frac{1}{5}}}{15a^{\frac{3}{4}}}\right)^2$ **80.** $\left(\dfrac{64a^{\frac{1}{3}}}{16a^{\frac{5}{9}}}\right)^3$

For Problems 81–90, perform the indicated operations and express answers in simplest radical form. (See Example 5.)

81. $\sqrt[3]{3}\sqrt{3}$ **82.** $\sqrt{2}\sqrt[4]{2}$

83. $\sqrt[4]{6}\sqrt{6}$ **84.** $\sqrt[3]{5}\sqrt{5}$

85. $\dfrac{\sqrt[3]{3}}{\sqrt{3}}$ **86.** $\dfrac{\sqrt{2}}{\sqrt[3]{2}}$

87. $\dfrac{\sqrt[3]{8}}{\sqrt[4]{4}}$ **88.** $\dfrac{\sqrt{9}}{\sqrt[3]{3}}$

89. $\dfrac{\sqrt[4]{27}}{\sqrt{3}}$ **90.** $\dfrac{\sqrt[3]{16}}{\sqrt[6]{4}}$

▪ ▪ ▪ THOUGHTS INTO WORDS

91. Your friend keeps getting an error message when evaluating $-4^{\frac{5}{2}}$ on his calculator. What error is he probably making?

92. Explain how you would evaluate $27^{\frac{2}{3}}$ without a calculator.

▪ ▪ ▪ FURTHER INVESTIGATIONS

93. Use your calculator to evaluate each of the following.

(a) $\sqrt[3]{1728}$ (b) $\sqrt[3]{5832}$

(c) $\sqrt[4]{2401}$ (d) $\sqrt[4]{65,536}$

(e) $\sqrt[5]{161,051}$ (f) $\sqrt[5]{6,436,343}$

94. Definition 5.7 states that

$$b^{\frac{m}{n}} = \sqrt[n]{b^m} = (\sqrt[n]{b})^m$$

Use your calculator to verify each of the following.

(a) $\sqrt[3]{27^2} = (\sqrt[3]{27})^2$ (b) $\sqrt[3]{8^5} = (\sqrt[3]{8})^5$

(c) $\sqrt[4]{16^3} = (\sqrt[4]{16})^3$ (d) $\sqrt[3]{16^2} = (\sqrt[3]{16})^2$

(e) $\sqrt[5]{9^4} = (\sqrt[5]{9})^4$ (f) $\sqrt[3]{12^4} = (\sqrt[3]{12})^4$

95. Use your calculator to evaluate each of the following.

(a) $16^{\frac{5}{2}}$ (b) $25^{\frac{7}{2}}$

(c) $16^{\frac{9}{4}}$ (d) $27^{\frac{5}{3}}$

(e) $343^{\frac{2}{3}}$ (f) $512^{\frac{4}{3}}$

96. Use your calculator to estimate each of the following to the nearest one-thousandth.

(a) $7^{\frac{4}{3}}$ (b) $10^{\frac{4}{5}}$

(c) $12^{\frac{3}{5}}$ (d) $19^{\frac{2}{5}}$

(e) $7^{\frac{3}{4}}$ (f) $10^{\frac{5}{4}}$

97. (a) Because $\dfrac{4}{5} = 0.8$, we can evaluate $10^{\frac{4}{5}}$ by evaluating $10^{0.8}$, which involves a shorter sequence of "calculator steps." Evaluate parts (b), (c), (d), (e), and (f) of Problem 96 and take advantage of decimal exponents.

(b) What problem is created when we try to evaluate $7^{\frac{4}{3}}$ by changing the exponent to decimal form?

| 5.7 | **Scientific Notation** |

Many applications of mathematics involve the use of very large or very small numbers.

1. The speed of light is approximately 29,979,200,000 centimeters per second.

2. A light year — the distance that light travels in 1 year — is approximately 5,865,696,000,000 miles.

3. A millimicron equals 0.000000001 of a meter.

Working with numbers of this type in standard decimal form is quite cumbersome. It is much more convenient to represent very small and very large numbers in **scientific notation**. The expression $(N)(10)^k$, where N is a number greater than or equal to 1 and less than 10, written in decimal form, and k is any integer, is commonly called scientific notation or the scientific form of a number. Consider the following examples, which show a comparison between ordinary decimal notation and scientific notation.

Ordinary notation	Scientific notation
2.14	$(2.14)(10)^0$
31.78	$(3.178)(10)^1$
412.9	$(4.129)(10)^2$
8,000,000	$(8)(10)^6$
0.14	$(1.4)(10)^{-1}$
0.0379	$(3.79)(10)^{-2}$
0.00000049	$(4.9)(10)^{-7}$

To switch from ordinary notation to scientific notation, you can use the following procedure.

> Write the given number as the product of a number greater than or equal to 1 and less than 10, and a power of 10. The exponent of 10 is determined by counting the number of places that the decimal point was moved when going from the original number to the number greater than or equal to 1 and less than 10. This exponent is (a) negative if the original number is less than 1, (b) positive if the original number is greater than 10, and (c) 0 if the original number itself is between 1 and 10.

Thus we can write

$$0.00467 = (4.67)(10)^{-3}$$

$$87,000 = (8.7)(10)^4$$

$$3.1416 = (3.1416)(10)^0$$

We can express the applications given earlier in scientific notation as follows:

Speed of light $29,979,200,000 = (2.99792)(10)^{10}$ centimeters per second.

Light year $5,865,696,000,000 = (5.865696)(10)^{12}$ miles.

Metric units   A millimicron is $0.000000001 = (1)(10)^{-9}$ meter.

To switch from scientific notation to ordinary decimal notation, you can use the following procedure.

> Move the decimal point the number of places indicated by the exponent of 10. The decimal point is moved to the right if the exponent is positive and to the left if the exponent is negative.

Thus we can write

$$(4.78)(10)^4 = 47,800$$

$$(8.4)(10)^{-3} = 0.0084$$

Scientific notation can frequently be used to simplify numerical calculations. We merely change the numbers to scientific notation and use the appropriate properties of exponents. Consider the following examples.

EXAMPLE 1

Perform the indicated operations.

(a) $(0.00024)(20,000)$

(b) $\dfrac{7,800,000}{0.0039}$

(c) $\dfrac{(0.00069)(0.0034)}{(0.0000017)(0.023)}$

(d) $\sqrt{0.000004}$

Solution

(a) $(0.00024)(20,000) = (2.4)(10)^{-4}(2)(10)^4$

$$= (2.4)(2)(10)^{-4}(10)^4$$

$$= (4.8)(10)^0$$

$$= (4.8)(1)$$

$$= 4.8$$

(b) $\dfrac{7,800,000}{0.0039} = \dfrac{(7.8)(10)^6}{(3.9)(10)^{-3}}$

$= (2)(10)^9$

$= 2,000,000,000$

(c) $\dfrac{(0.00069)(0.0034)}{(0.0000017)(0.023)} = \dfrac{(6.9)(10)^{-4}(3.4)(10)^{-3}}{(1.7)(10)^{-6}(2.3)(10)^{-2}}$

$= \dfrac{\overset{3}{(\cancel{6.9})}\overset{2}{(\cancel{3.4})}(10)^{-7}}{(\cancel{1.7})(\cancel{2.3})(10)^{-8}}$

$= (6)(10)^1$

$= 60$

(d) $\sqrt{0.00004} = \sqrt{(4)(10)^{-6}}$

$= ((4)(10)^{-6})^{\frac{1}{2}}$

$= 4^{\frac{1}{2}}((10)^{-6})^{\frac{1}{2}}$

$= (2)(10)^{-3}$

$= 0.002$ ∎

E X A M P L E   2

The speed of light is approximately $(1.86)(10^5)$ miles per second. When the earth is $(9.3)(10^7)$ miles away from the sun, how long does it take light from the sun to reach the earth?

Solution

We will use the formula $t = \dfrac{d}{r}$.

$t = \dfrac{(9.3)(10^7)}{(1.86)(10^5)}$

$t = \dfrac{(9.3)}{(1.86)}(10^2)$ Subtract exponents.

$t = (5)(10^2) = 500$ seconds

At this distance it takes light about 500 seconds to travel from the sun to the earth. To find the answer in minutes, divide 500 seconds by 60 seconds/minute. That gives a result of approximately 8.33 minutes. ∎

Many calculators are equipped to display numbers in scientific notation. The display panel shows the number between 1 and 10 and the appropriate exponent of 10. For example, evaluating $(3,800,000)^2$ yields

$\boxed{\text{1.444E13}}$

Thus $(3,800,000)^2 = (1.444)(10)^{13} = 14,440,000,000,000.$

Similarly, the answer for $(0.000168)^2$ is displayed as

$$2.8224\text{E-}8$$

Thus $(0.000168)^2 = (2.8224)(10)^{-8} = 0.000000028224$.

Calculators vary as to the number of digits displayed in the number between 1 and 10 when scientific notation is used. For example, we used two different calculators to estimate $(6729)^6$ and obtained the following results.

$$9.2833\text{E}22$$

$$9.283316768\text{E}22$$

Obviously, you need to know the capabilities of your calculator when working with problems in scientific notation. Many calculators also allow the entry of a number in scientific notation. Such calculators are equipped with an enter-the-exponent key (often labeled as $\boxed{\text{EE}}$ or $\boxed{\text{EEX}}$). Thus a number such as $(3.14)(10)^8$ might be entered as follows:

Enter	Press	Display		Enter	Press	Display
3.14	$\boxed{\text{EE}}$	3.14E	or	3.14	$\boxed{\text{EE}}$	3.14 ^{00}
8		3.14E8		8		3.14 ^{08}

A $\boxed{\text{MODE}}$ key is often used on calculators to let you choose normal decimal notation, scientific notation, or engineering notation. (The abbreviations Norm, Sci, and Eng are commonly used.) If the calculator is in scientific mode, then a number can be entered and changed to scientific form by pressing the $\boxed{\text{ENTER}}$ key. For example, when we enter 589 and press the $\boxed{\text{ENTER}}$ key, the display will show 5.89E2. Likewise, when the calculator is in scientific mode, the answers to computational problems are given in scientific form. For example, the answer for $(76)(533)$ is given as 4.0508E4.

It should be evident from this brief discussion that even when you are using a calculator, you need to have a thorough understanding of scientific notation.

Problem Set 5.7

For Problems 1–18, write each of the following in scientific notation. For example

$$27800 = (2.78)(10)^4$$

1. 89

2. 117

3. 4290

4. 812,000

5. 6,120,000

6. 72,400,000

7. 40,000,000

8. 500,000,000

9. 376.4

10. 9126.21

11. 0.347

12. 0.2165

13. 0.0214

14. 0.0037

15. 0.00005

16. 0.00000082

17. 0.00000000194

18. 0.0000000003

For Problems 19–32, write each of the following in ordinary decimal notation. For example,

$$(3.18)(10)^2 = 318$$

19. $(2.3)(10)^1$

20. $(1.62)(10)^2$

21. $(4.19)(10)^3$

22. $(7.631)(10)^4$

23. $(5)(10)^8$

24. $(7)(10)^9$

25. $(3.14)(10)^{10}$

26. $(2.04)(10)^{12}$

27. $(4.3)(10)^{-1}$

28. $(5.2)(10)^{-2}$

29. $(9.14)(10)^{-4}$

30. $(8.76)(10)^{-5}$

31. $(5.123)(10)^{-8}$

32. $(6)(10)^{-9}$

For Problems 33–50, use scientific notation and the properties of exponents to help you perform the following operations.

33. $(0.0037)(0.00002)$

34. $(0.00003)(0.00025)$

35. $(0.00007)(11,000)$

36. $(0.000004)(120,000)$

37. $\dfrac{360,000,000}{0.0012}$

38. $\dfrac{66,000,000,000}{0.022}$

39. $\dfrac{0.000064}{16,000}$

40. $\dfrac{0.00072}{0.0000024}$

41. $\dfrac{(60,000)(0.006)}{(0.0009)(400)}$

42. $\dfrac{(0.00063)(960,000)}{(3,200)(0.0000021)}$

43. $\dfrac{(0.0045)(60,000)}{(1800)(0.00015)}$

44. $\dfrac{(0.00016)(300)(0.028)}{0.064}$

45. $\sqrt{9,000,000}$

46. $\sqrt{0.00000009}$

47. $\sqrt[3]{8000}$

48. $\sqrt[3]{0.001}$

49. $(90,000)^{\frac{3}{2}}$

50. $(8000)^{\frac{2}{3}}$

51. Avogadro's number, 602,000,000,000,000,000,000,000, is the number of atoms in 1 mole of a substance. Express this number in scientific notation.

52. The Social Security program paid out approximately $33,200,000,000 in benefits in May 2000. Express this number in scientific notation.

53. Carlos's first computer had a processing speed of $(1.6)(10^6)$ hertz. He recently purchased a laptop computer with a processing speed of $(1.33)(10^9)$ hertz. Approximately how many times faster is the processing speed of his laptop than that of his first computer? Express the result in decimal form.

54. Alaska has an area of approximately $(6.15)(10^5)$ square miles. In 1999 the state had a population of approximately 619,000 people. Compute the population density to the nearest hundredth. Population density is the number of people per square mile. Express the result in decimal form rounded to the nearest hundredth.

55. In the year 2000 the public debt of the United States was approximately $5,700,000,000,000. For July 2000, the census reported that 275,000,000 people lived in the United States. Convert these figures to scientific notation, and compute the average debt per person. Express the result in scientific notation.

56. The space shuttle can travel at approximately 410,000 miles per day. If the shuttle could travel to Mars, and Mars was 140,000,000 miles away, how many days would it take the shuttle to travel to Mars? Express the result in decimal form.

57. Atomic masses are measured in atomic mass units (amu). The amu, $(1.66)(10^{-27})$ kilograms, is defined as $\dfrac{1}{12}$ the mass of a common carbon atom. Find the mass of a carbon atom in kilograms. Express the result in scientific notation.

58. The field of view of a microscope is $(4)(10^{-4})$ meters. If a single cell organism occupies $\dfrac{1}{5}$ of the field of view, find the length of the organism in meters. Express the result in scientific notation.

59. The mass of an electron is $(9.11)(10^{-31})$ kilogram, and the mass of a proton is $(1.67)(10^{-27})$ kilogram. Approximately how many times more is the weight of a proton than the weight of an electron? Express the result in decimal form.

60. A square pixel on a computer screen has a side of length $(1.17)(10^{-2})$ inches. Find the approximate area of the pixel in inches. Express the result in decimal form.

■ ■ ■ **THOUGHTS INTO WORDS**

61. Explain the importance of scientific notation.

62. Why do we need scientific notation even when using calculators and computers?

■ ■ ■ **FURTHER INVESTIGATIONS**

63. Sometimes it is more convenient to express a number as a product of a power of 10 and a number that is not between 1 and 10. For example, suppose that we want to calculate $\sqrt{640{,}000}$. We can proceed as follows:

$$\sqrt{640{,}000} = \sqrt{(64)(10)^4}$$
$$= ((64)(10)^4)^{\frac{1}{2}}$$
$$= (64)^{\frac{1}{2}}(10^4)^{\frac{1}{2}}$$
$$= (8)(10)^2$$
$$= 8(100) = 800$$

Compute each of the following without a calculator, and then use a calculator to check your answers.

(a) $\sqrt{49{,}000{,}000}$ **(b)** $\sqrt{0.0025}$

(c) $\sqrt{14{,}400}$ **(d)** $\sqrt{0.000121}$

(e) $\sqrt[3]{27{,}000}$ **(f)** $\sqrt[3]{0.000064}$

64. Use your calculator to evaluate each of the following. Express final answers in ordinary notation.

(a) $(27{,}000)^2$ **(b)** $(450{,}000)^2$

(c) $(14{,}800)^2$ **(d)** $(1700)^3$

(e) $(900)^4$ **(f)** $(60)^5$

(g) $(0.0213)^2$ **(h)** $(0.000213)^2$

(i) $(0.000198)^2$ **(j)** $(0.000009)^3$

65. Use your calculator to estimate each of the following. Express final answers in scientific notation with the number between 1 and 10 rounded to the nearest one-thousandth.

(a) $(4576)^4$ **(b)** $(719)^{10}$

(c) $(28)^{12}$ **(d)** $(8619)^6$

(e) $(314)^5$ **(f)** $(145{,}723)^2$

66. Use your calculator to estimate each of the following. Express final answers in ordinary notation rounded to the nearest one-thousandth.

(a) $(1.09)^5$ **(b)** $(1.08)^{10}$

(c) $(1.14)^7$ **(d)** $(1.12)^{20}$

(e) $(0.785)^4$ **(f)** $(0.492)^5$

(5.1) The following properties form the basis for manipulating with exponents.

1. $b^n \cdot b^m = b^{n+m}$ Product of two powers

2. $(b^n)^m = b^{mn}$ Power of a power

3. $(ab)^n = a^n b^n$ Power of a product

4. $\left(\dfrac{a}{b}\right)^n = \dfrac{a^n}{b^n}$ Power of a quotient

5. $\dfrac{b^n}{b^m} = b^{n-m}$ Quotient of two powers

(5.2) and **(5.3)** The **principal nth root of b** is designated by $\sqrt[n]{b}$, where n is the **index** and b is the **radicand**.

A radical expression is in **simplest radical form** if

1. A radicand contains no polynomial factor raised to a power equal to or greater than the index of the radical,

2. No fraction appears within a radical sign, and

3. No radical appears in the denominator.

The following properties are used to express radicals in simplest form.

$$\sqrt[n]{bc} = \sqrt[n]{b}\sqrt[n]{c} \qquad \sqrt[n]{\frac{b}{c}} = \frac{\sqrt[n]{b}}{\sqrt[n]{c}}$$

Simplifying by combining radicals sometimes requires that we first express the given radicals in simplest form and then apply the distributive property.

(5.4) The distributive property and the property $\sqrt[n]{b}\sqrt[n]{c} = \sqrt[n]{bc}$ are used to find products of expressions that involve radicals.

The special-product pattern $(a + b)(a - b) = a^2 - b^2$ suggests a procedure for **rationalizing the denominator** of an expression that contains a binomial denominator with radicals.

(5.5) Equations that contain radicals with variables in a radicand are called **radical equations**. The property "if $a = b$, then $a^n = b^n$" forms the basis for solving radical equations. Raising both sides of an equation to a positive integral power may produce **extraneous solutions** — that is, solutions that do not satisfy the original equation. Therefore, you must check each potential solution.

(5.6) If b is a real number, n is a positive integer greater than 1, and $\sqrt[n]{b}$ exists, then

$$b^{\frac{1}{n}} = \sqrt[n]{b}$$

Thus $b^{\frac{1}{n}}$ means **the nth root of b**.

If $\dfrac{m}{n}$ is a rational number, n is a positive integer greater than 1, and b is a real number such that $\sqrt[n]{b}$ exists, then

$$b^{\frac{m}{n}} = \sqrt[n]{b^m} = (\sqrt[n]{b})^m$$

Both $\sqrt[n]{b^m}$ and $(\sqrt[n]{b})^m$ can be used for computational purposes.

We need to be able to switch back and forth between **exponential form** and **radical form**. The link between exponents and roots provides a basis for multiplying and dividing some radicals even if they have different indexes.

(5.7) The **scientific form** of a number is expressed as

$$(N)(10)^k$$

where N is a number greater than or equal to 1 and less than 10, written in decimal form, and k is an integer. Scientific notation is often convenient to use with very small and very large numbers. For example, 0.000046 can be expressed as $(4.6)(10^{-5})$, and 92,000,000 can be written as $(9.2)(10)^7$.

Scientific notation can often be used to simplify numerical calculations. For example,

$$(0.000016)(30,000) = (1.6)(10)^{-5}(3)(10)^4$$

$$= (4.8)(10)^{-1} = 0.48$$

| Chapter 5 | **Review Problem Set** |

For Problems 1–12, evaluate each of the following numerical expressions.

1. 4^{-3}

2. $\left(\dfrac{2}{3}\right)^{-2}$

3. $(3^2 \cdot 3^{-3})^{-1}$

4. $\sqrt[3]{-8}$

5. $\sqrt[4]{\dfrac{16}{81}}$

6. $4^{\frac{5}{2}}$

7. $(-1)^{\frac{2}{3}}$

8. $\left(\dfrac{8}{27}\right)^{\frac{2}{3}}$

9. $-16^{\frac{3}{2}}$

10. $\dfrac{2^3}{2^{-2}}$

11. $(4^{-2} \cdot 4^2)^{-1}$

12. $\left(\dfrac{3^{-1}}{3^2}\right)^{-1}$

For Problems 13–24, express each of the following radicals in simplest radical form. Assume the variables represent positive real numbers.

13. $\sqrt{54}$

14. $\sqrt{48x^3y}$

15. $\dfrac{4\sqrt{3}}{\sqrt{6}}$

16. $\sqrt{\dfrac{5}{12x^3}}$

17. $\sqrt[3]{56}$

18. $\dfrac{\sqrt[3]{2}}{\sqrt[3]{9}}$

19. $\sqrt{\dfrac{9}{5}}$

20. $\sqrt{\dfrac{3x^3}{7}}$

21. $\sqrt[3]{108x^4y^8}$

22. $\dfrac{3}{4}\sqrt{150}$

23. $\dfrac{2}{3}\sqrt{45xy^3}$

24. $\dfrac{\sqrt{8x^2}}{\sqrt{2x}}$

For Problems 25–32, multiply and simplify. Assume the variables represent nonnegative real numbers.

25. $(3\sqrt{8})(4\sqrt{5})$

26. $(5\sqrt[3]{2})(6\sqrt[3]{4})$

27. $3\sqrt{2}(4\sqrt{6} - 2\sqrt{7})$

28. $(\sqrt{x} + 3)(\sqrt{x} - 5)$

29. $(2\sqrt{5} - \sqrt{3})(2\sqrt{5} + \sqrt{3})$

30. $(3\sqrt{2} + \sqrt{6})(5\sqrt{2} - 3\sqrt{6})$

31. $(2\sqrt{a} + \sqrt{b})(3\sqrt{a} - 4\sqrt{b})$

32. $(4\sqrt{8} - \sqrt{2})(\sqrt{8} + 3\sqrt{2})$

For Problems 33–36, rationalize the denominator and simplify.

33. $\dfrac{4}{\sqrt{7} - 1}$

34. $\dfrac{\sqrt{3}}{\sqrt{8} + \sqrt{5}}$

35. $\dfrac{3}{2\sqrt{3} + 3\sqrt{5}}$

36. $\dfrac{3\sqrt{2}}{2\sqrt{6} - \sqrt{10}}$

For Problems 37–42, simplify each of the following, and express the final results using positive exponents.

37. $(x^{-3}y^4)^{-2}$

38. $\left(\dfrac{2a^{-1}}{3b^4}\right)^{-3}$

39. $(4x^{\frac{1}{2}})(5x^{\frac{1}{5}})$

40. $\dfrac{42a^{\frac{3}{4}}}{6a^{\frac{1}{3}}}$

41. $\left(\dfrac{x^3}{y^4}\right)^{-\frac{1}{3}}$

42. $\left(\dfrac{6x^{-2}}{2x^4}\right)^{-2}$

For Problems 43–46, use the distributive property to help simplify each of the following.

43. $3\sqrt{45} - 2\sqrt{20} - \sqrt{80}$

44. $4\sqrt[3]{24} + 3\sqrt[3]{3} - 2\sqrt[3]{81}$

45. $3\sqrt{24} - \dfrac{2\sqrt{54}}{5} + \dfrac{\sqrt{96}}{4}$

46. $-2\sqrt{12x} + 3\sqrt{27x} - 5\sqrt{48x}$

For Problems 47 and 48, express each as a single fraction involving positive exponents only.

47. $x^{-2} + y^{-1}$

48. $a^{-2} - 2a^{-1}b^{-1}$

For Problems 49–56, solve each equation.

49. $\sqrt{7x - 3} = 4$

50. $\sqrt{2y + 1} = \sqrt{5y - 11}$

51. $\sqrt{2x} = x - 4$

52. $\sqrt{n^2 - 4n - 4} = n$

53. $\sqrt[3]{2x - 1} = 3$

54. $\sqrt{t^2 + 9t - 1} = 3$

55. $\sqrt{x^2 + 3x - 6} = x$

56. $\sqrt{x + 1} - \sqrt{2x} = -1$

59. $(0.000015)(400,000)$

60. $\dfrac{0.000045}{0.0003}$

61. $\dfrac{(0.00042)(0.0004)}{0.006}$

62. $\sqrt{0.000004}$

63. $\sqrt[3]{0.000000008}$

64. $(4,000,000)^{\frac{3}{2}}$

For Problems 57–64, use scientific notation and the properties of exponents to help perform the following calculations.

57. $(0.00002)(0.0003)$

58. $(120,000)(300,000)$

For Problems 1–4, simplify each of the numerical expressions.

1. $(4)^{-\frac{5}{2}}$

2. $-16^{\frac{5}{4}}$

3. $\left(\frac{2}{3}\right)^{-4}$

4. $\left(\frac{2^{-1}}{2^{-2}}\right)^{-2}$

For Problems 5–9, express each radical expression in simplest radical form. Assume the variables represent positive real numbers.

5. $\sqrt{63}$

6. $\sqrt[3]{108}$

7. $\sqrt{52x^4y^3}$

8. $\frac{5\sqrt{18}}{3\sqrt{12}}$

9. $\sqrt{\frac{7}{24x^3}}$

10. Multiply and simplify: $(4\sqrt{6})(3\sqrt{12})$

11. Multiply and simplify: $(3\sqrt{2} + \sqrt{3})(\sqrt{2} - 2\sqrt{3})$

12. Simplify by combining similar radicals:
$2\sqrt{50} - 4\sqrt{18} - 9\sqrt{32}$

13. Rationalize the denominator and simplify:
$$\frac{3\sqrt{2}}{4\sqrt{3} - \sqrt{8}}$$

14. Simplify and express the answer using positive exponents: $\left(\frac{2x^{-1}}{3y}\right)^{-2}$

15. Simplify and express the answer using positive exponents: $\dfrac{-84a^{\frac{1}{2}}}{7a^{\frac{4}{5}}}$

16. Express $x^{-1} + y^{-3}$ as a single fraction involving positive exponents.

17. Multiply and express the answer using positive exponents: $\left(3x^{-\frac{1}{2}}\right)\left(4x^{\frac{3}{4}}\right)$

18. Multiply and simplify:
$(3\sqrt{5} - 2\sqrt{3})(3\sqrt{5} + 2\sqrt{3})$

For Problems 19 and 20, use scientific notation and the properties of exponents to help with the calculations.

19. $\dfrac{(0.00004)(300)}{0.00002}$

20. $\sqrt{0.000009}$

For Problems 21–25, solve each equation.

21. $\sqrt{3x + 1} = 3$

22. $\sqrt[3]{3x + 2} = 2$

23. $\sqrt{x} = x - 2$

24. $\sqrt{5x - 2} = \sqrt{3x + 8}$

25. $\sqrt{x^2 - 10x + 28} = 2$

6

Quadratic Equations and Inequalities

The Pythagorean theorem is applied throughout the construction industry when right angles are involved.

© Jeff Greenberg/PhotoEdit

A page for a magazine contains 70 square inches of type. The height of the page is twice the width. If the margin around the type is 2 inches uniformly, what are the dimensions of a page? We can use the quadratic equation $(x - 4)(2x - 4) = 70$ to determine that the page measures 9 inches by 18 inches.

Solving equations is one of the central themes of this text. Let's pause for a moment and reflect on the different types of equations that we have solved in the last five chapters.

As the chart on the next page shows, we have solved second-degree equations in one variable, but only those for which the polynomial is factorable. In this chapter we will expand our work to include more general types of second-degree equations, as well as inequalities in one variable.

Type of Equation	Examples
First-degree equations in one variable	$3x + 2x = x - 4$; $5(x + 4) = 12$; $\dfrac{x + 2}{3} + \dfrac{x - 1}{4} = 2$
Second-degree equations in one variable *that are factorable*	$x^2 + 5x = 0$; $x^2 + 5x + 6 = 0$; $x^2 - 9 = 0$; $x^2 - 10x + 25 = 0$
Fractional equations	$\dfrac{2}{x} + \dfrac{3}{x} = 4$; $\dfrac{5}{a - 1} = \dfrac{6}{a - 2}$; $\dfrac{2}{x^2 - 9} + \dfrac{3}{x + 3} = \dfrac{4}{x - 3}$
Radical equations	$\sqrt{x} = 2$; $\sqrt{3x - 2} = 5$; $\sqrt{5y + 1} = \sqrt{3y + 4}$

6.1   Complex Numbers

Because the square of any real number is nonnegative, a simple equation such as $x^2 = -4$ has no solutions in the set of real numbers. To handle this situation, we can expand the set of real numbers into a larger set called the **complex numbers**. In this section we will instruct you on how to manipulate complex numbers.

To provide a solution for the equation $x^2 + 1 = 0$, we use the number i, such that

$$i^2 = -1$$

The number i is not a real number and is often called the **imaginary unit**, but the number i^2 is the real number -1. The imaginary unit i is used to define a complex number as follows:

Definition 6.1

> A **complex number** is any number that can be expressed in the form
>
> $\quad a + bi$
>
> where a and b are real numbers.

The form $a + bi$ is called the **standard form** of a complex number. The real number a is called the **real part** of the complex number, and b is called the **imaginary**

part. (Note that b is a real number even though it is called the imaginary part.) The following list exemplifies this terminology.

1. The number $7 + 5i$ is a complex number that has a real part of 7 and an imaginary part of 5.

2. The number $\dfrac{2}{3} + i\sqrt{2}$ is a complex number that has a real part of $\dfrac{2}{3}$ and an imaginary part of $\sqrt{2}$. (It is easy to mistake $\sqrt{2}i$ for $\sqrt{2i}$. Thus we commonly write $i\sqrt{2}$ instead of $\sqrt{2}i$ to avoid any difficulties with the radical sign.)

3. The number $-4 - 3i$ can be written in the standard form $-4 + (-3i)$ and therefore is a complex number that has a real part of -4 and an imaginary part of -3. [The form $-4 - 3i$ is often used, but we know that it means $-4 + (-3i)$.]

4. The number $-9i$ can be written as $0 + (-9i)$; thus it is a complex number that has a real part of 0 and an imaginary part of -9. (Complex numbers, such as $-9i$, for which $a = 0$ and $b \neq 0$ are called **pure imaginary numbers**.)

5. The real number 4 can be written as $4 + 0i$ and is thus a complex number that has a real part of 4 and an imaginary part of 0.

Look at item 5 in this list. We see that the set of real numbers is a subset of the set of complex numbers. The following diagram indicates the organizational format of the complex numbers.

Complex numbers $a + bi$,   where a and b are real numbers

Real numbers
$a + bi$,   where $b = 0$

Imaginary numbers
$a + bi$,   where $b \neq 0$

Pure imaginary numbers
$a + bi$,   where $a = 0$ and $b \neq 0$

Two complex numbers $a + bi$ and $c + di$ are said to be **equal** if and only if $a = c$ and $b = d$.

■ Adding and Subtracting Complex Numbers

To **add complex numbers**, we simply add their real parts and add their imaginary parts. Thus

$$(a + bi) + (c + di) = (a + c) + (b + d)i$$

The following examples show addition of two complex numbers.

1. $(4 + 3i) + (5 + 9i) = (4 + 5) + (3 + 9)i = 9 + 12i$

2. $(-6 + 4i) + (8 - 7i) = (-6 + 8) + (4 - 7)i$
$$= 2 - 3i$$

3. $\left(\dfrac{1}{2} + \dfrac{3}{4}i\right) + \left(\dfrac{2}{3} + \dfrac{1}{5}i\right) = \left(\dfrac{1}{2} + \dfrac{2}{3}\right) + \left(\dfrac{3}{4} + \dfrac{1}{5}\right)i$
$$= \left(\dfrac{3}{6} + \dfrac{4}{6}\right) + \left(\dfrac{15}{20} + \dfrac{4}{20}\right)i$$
$$= \dfrac{7}{6} + \dfrac{19}{20}i$$

The set of complex numbers is closed with respect to addition; that is, the sum of two complex numbers is a complex number. Furthermore, the commutative and associative properties of addition hold for all complex numbers. The addition identity element is $0 + 0i$ (or simply the real number 0). The additive inverse of $a + bi$ is $-a - bi$, because

$$(a + bi) + (-a - bi) = 0$$

To **subtract complex numbers**, $c + di$ from $a + bi$, add the additive inverse of $c + di$. Thus

$$(a + bi) - (c + di) = (a + bi) + (-c - di)$$
$$= (a - c) + (b - d)i$$

In other words, we subtract the real parts and subtract the imaginary parts, as in the next examples.

1. $(9 + 8i) - (5 + 3i) = (9 - 5) + (8 - 3)i$
$$= 4 + 5i$$

2. $(3 - 2i) - (4 - 10i) = (3 - 4) + (-2 - (-10))i$
$$= -1 + 8i$$

■ Products and Quotients of Complex Numbers

Because $i^2 = -1$, i is a square root of -1, so we let $i = \sqrt{-1}$. It should also be evident that $-i$ is a square root of -1, because

$$(-i)^2 = (-i)(-i) = i^2 = -1$$

Thus, in the set of complex numbers, -1 has two square roots, i and $-i$. We express these symbolically as

$$\sqrt{-1} = i \qquad \text{and} \qquad -\sqrt{-1} = -i$$

Let us extend our definition so that in the set of complex numbers every negative real number has two square roots. We simply define $\sqrt{-b}$, where b is a positive real number, to be the number whose square is $-b$. Thus

$$(\sqrt{-b})^2 = -b, \quad \text{for } b > 0$$

Furthermore, because $(i\sqrt{b})(i\sqrt{b}) = i^2(b) = -1(b) = -b$, we see that

$$\sqrt{-b} = i\sqrt{b}$$

In other words, a square root of any negative real number can be represented as the product of a real number and the imaginary unit i. Consider the following examples.

$$\sqrt{-4} = i\sqrt{4} = 2i$$

$$\sqrt{-17} = i\sqrt{17}$$

$$\sqrt{-24} = i\sqrt{24} = i\sqrt{4}\sqrt{6} = 2i\sqrt{6} \qquad \text{Note that we simplified the radical } \sqrt{24} \text{ to } 2\sqrt{6}.$$

We should also observe that $-\sqrt{-b}$, where $b > 0$, is a square root of $-b$ because

$$(-\sqrt{-b})^2 = (-i\sqrt{b})^2 = i^2(b) = -1(b) = -b$$

Thus in the set of complex numbers, $-b$ (where $b > 0$) has two square roots, $i\sqrt{b}$ and $-i\sqrt{b}$. We express these symbolically as

$$\sqrt{-b} = i\sqrt{b} \qquad \text{and} \qquad -\sqrt{-b} = -i\sqrt{b}$$

We must be very careful with the use of the symbol $\sqrt{-b}$, where $b > 0$. Some real number properties that involve the square root symbol do not hold if the square root symbol does not represent a real number. For example, $\sqrt{a}\sqrt{b} = \sqrt{ab}$ does not hold if a and b are both negative numbers.

Correct $\quad \sqrt{-4}\sqrt{-9} = (2i)(3i) = 6i^2 = 6(-1) = -6$

Incorrect $\quad \sqrt{-4}\sqrt{-9} = \sqrt{(-4)(-9)} = \sqrt{36} = 6$

To avoid difficulty with this idea, you should rewrite all expressions of the form $\sqrt{-b}$, where $b > 0$, in the form $i\sqrt{b}$ before doing any computations. The following examples further demonstrate this point.

1. $\sqrt{-5}\sqrt{-7} = (i\sqrt{5})(i\sqrt{7}) = i^2\sqrt{35} = (-1)\sqrt{35} = -\sqrt{35}$

2. $\sqrt{-2}\sqrt{-8} = (i\sqrt{2})(i\sqrt{8}) = i^2\sqrt{16} = (-1)(4) = -4$

3. $\sqrt{-6}\sqrt{-8} = (i\sqrt{6})(i\sqrt{8}) = i^2\sqrt{48} = (-1)\sqrt{16}\sqrt{3} = -4\sqrt{3}$

4. $\dfrac{\sqrt{-75}}{\sqrt{-3}} = \dfrac{i\sqrt{75}}{i\sqrt{3}} = \dfrac{\sqrt{75}}{\sqrt{3}} = \sqrt{\dfrac{75}{3}} = \sqrt{25} = 5$

5. $\dfrac{\sqrt{-48}}{\sqrt{12}} = \dfrac{i\sqrt{48}}{\sqrt{12}} = i\sqrt{\dfrac{48}{12}} = i\sqrt{4} = 2i$

Complex numbers have a binomial form, so we find the product of two complex numbers in the same way that we find the product of two binomials. Then, by replacing i^2 with -1, we are able to simplify and express the final result in standard form. Consider the following examples.

6. $(2 + 3i)(4 + 5i) = 2(4 + 5i) + 3i(4 + 5i)$

$$= 8 + 10i + 12i + 15i^2$$
$$= 8 + 22i + 15i^2$$
$$= 8 + 22i + 15(-1) = -7 + 22i$$

7. $(-3 + 6i)(2 - 4i) = -3(2 - 4i) + 6i(2 - 4i)$

$$= -6 + 12i + 12i - 24i^2$$
$$= -6 + 24i - 24(-1)$$
$$= -6 + 24i + 24 = 18 + 24i$$

8. $(1 - 7i)^2 = (1 - 7i)(1 - 7i)$

$$= 1(1 - 7i) - 7i(1 - 7i)$$
$$= 1 - 7i - 7i + 49i^2$$
$$= 1 - 14i + 49(-1)$$
$$= 1 - 14i - 49$$
$$= -48 - 14i$$

9. $(2 + 3i)(2 - 3i) = 2(2 - 3i) + 3i(2 - 3i)$

$$= 4 - 6i + 6i - 9i^2$$
$$= 4 - 9(-1)$$
$$= 4 + 9$$
$$= 13$$

Example 9 illustrates an important situation: The complex numbers $2 + 3i$ and $2 - 3i$ are conjugates of each other. In general, two complex numbers $a + bi$ and $a - bi$ are called **conjugates** of each other. *The product of a complex number and its conjugate is always a real number*, which can be shown as follows:

$$(a + bi)(a - bi) = a(a - bi) + bi(a - bi)$$
$$= a^2 - abi + abi - b^2i^2$$
$$= a^2 - b^2(-1)$$
$$= a^2 + b^2$$

We use conjugates to simplify expressions such as $\dfrac{3i}{5 + 2i}$ that indicate the quotient of two complex numbers. To eliminate i in the denominator and change the indicated quotient to the standard form of a complex number, we can multiply both the numerator and the denominator by the conjugate of the denominator as follows:

$$\frac{3i}{5 + 2i} = \frac{3i(5 - 2i)}{(5 + 2i)(5 - 2i)}$$

$$= \frac{15i - 6i^2}{25 - 4i^2}$$

$$= \frac{15i - 6(-1)}{25 - 4(-1)}$$

$$= \frac{15i + 6}{29}$$

$$= \frac{6}{29} + \frac{15}{29}i$$

The following examples further clarify the process of dividing complex numbers.

10. $\dfrac{2 - 3i}{4 - 7i} = \dfrac{(2 - 3i)(4 + 7i)}{(4 - 7i)(4 + 7i)}$ $\qquad$ 4 + 7*i* is the conjugate of 4 − 7*i*.

$$= \frac{8 + 14i - 12i - 21i^2}{16 - 49i^2}$$

$$= \frac{8 + 2i - 21(-1)}{16 - 49(-1)}$$

$$= \frac{8 + 2i + 21}{16 + 49}$$

$$= \frac{29 + 2i}{65}$$

$$= \frac{29}{65} + \frac{2}{65}i$$

11. $\dfrac{4 - 5i}{2i} = \dfrac{(4 - 5i)(-2i)}{(2i)(-2i)}$ $\qquad$ −2*i* is the conjugate of 2*i*.

$$= \frac{-8i + 10i^2}{-4i^2}$$

$$= \frac{-8i + 10(-1)}{-4(-1)}$$

$$= \frac{-8i - 10}{4}$$

$$= -\frac{5}{2} - 2i$$

In Example 11, where the denominator is a pure imaginary number, we can change to standard form by choosing a multiplier other than the conjugate. Consider the following alternative approach for Example 11.

$$\frac{4 - 5i}{2i} = \frac{(4 - 5i)(i)}{(2i)(i)}$$

$$= \frac{4i - 5i^2}{2i^2}$$

$$= \frac{4i - 5(-1)}{2(-1)}$$

$$= \frac{4i + 5}{-2}$$

$$= -\frac{5}{2} - 2i$$

Problem Set 6.1

For Problems 1–8, label each statement true or false.

1. Every complex number is a real number.

2. Every real number is a complex number.

3. The real part of the complex number $6i$ is 0.

4. Every complex number is a pure imaginary number.

5. The sum of two complex numbers is always a complex number.

6. The imaginary part of the complex number 7 is 0.

7. The sum of two complex numbers is sometimes a real number.

8. The sum of two pure imaginary numbers is always a pure imaginary number.

For Problems 9–26, add or subtract as indicated.

9. $(6 + 3i) + (4 + 5i)$

10. $(5 + 2i) + (7 + 10i)$

11. $(-8 + 4i) + (2 + 6i)$

12. $(5 - 8i) + (-7 + 2i)$

13. $(3 + 2i) - (5 + 7i)$

14. $(1 + 3i) - (4 + 9i)$

15. $(-7 + 3i) - (5 - 2i)$

16. $(-8 + 4i) - (9 - 4i)$

17. $(-3 - 10i) + (2 - 13i)$

18. $(-4 - 12i) + (-3 + 16i)$

19. $(4 - 8i) - (8 - 3i)$

20. $(12 - 9i) - (14 - 6i)$

21. $(-1 - i) - (-2 - 4i)$

22. $(-2 - 3i) - (-4 - 14i)$

23. $\left(\frac{3}{2} + \frac{1}{3}i\right) + \left(\frac{1}{6} - \frac{3}{4}i\right)$

24. $\left(\frac{2}{3} - \frac{1}{5}i\right) + \left(\frac{3}{5} - \frac{3}{4}i\right)$

25. $\left(-\frac{5}{9} + \frac{3}{5}i\right) - \left(\frac{4}{3} - \frac{1}{6}i\right)$

26. $\left(\frac{3}{8} - \frac{5}{2}i\right) - \left(\frac{5}{6} + \frac{1}{7}i\right)$

For Problems 27–42, write each of the following in terms of i and simplify. For example,

$$\sqrt{-20} = i\sqrt{20} = i\sqrt{4}\sqrt{5} = 2i\sqrt{5}$$

27. $\sqrt{-81}$

28. $\sqrt{-49}$

29. $\sqrt{-14}$

30. $\sqrt{-33}$

31. $\sqrt{-\frac{16}{25}}$

32. $\sqrt{-\frac{64}{36}}$

33. $\sqrt{-18}$

34. $\sqrt{-84}$

35. $\sqrt{-75}$

36. $\sqrt{-63}$

37. $3\sqrt{-28}$

38. $5\sqrt{-72}$

39. $-2\sqrt{-80}$

40. $-6\sqrt{-27}$

41. $12\sqrt{-90}$

42. $9\sqrt{-40}$

For Problems 43–60, write each of the following in terms of i, perform the indicated operations, and simplify. For example,

$$\sqrt{-3}\sqrt{-8} = (i\sqrt{3})(i\sqrt{8})$$

$$= i^2\sqrt{24}$$

$$= (-1)\sqrt{4}\sqrt{6}$$

$$= -2\sqrt{6}$$

43. $\sqrt{-4}\sqrt{-16}$ **44.** $\sqrt{-81}\sqrt{-25}$

45. $\sqrt{-3}\sqrt{-5}$ **46.** $\sqrt{-7}\sqrt{-10}$

47. $\sqrt{-9}\sqrt{-6}$ **48.** $\sqrt{-8}\sqrt{-16}$

49. $\sqrt{-15}\sqrt{-5}$ **50.** $\sqrt{-2}\sqrt{-20}$

51. $\sqrt{-2}\sqrt{-27}$ **52.** $\sqrt{-3}\sqrt{-15}$

53. $\sqrt{6}\sqrt{-8}$ **54.** $\sqrt{-75}\sqrt{3}$

55. $\dfrac{\sqrt{-25}}{\sqrt{-4}}$ **56.** $\dfrac{\sqrt{-81}}{\sqrt{-9}}$

57. $\dfrac{\sqrt{-56}}{\sqrt{-7}}$ **58.** $\dfrac{\sqrt{-72}}{\sqrt{-6}}$

59. $\dfrac{\sqrt{-24}}{\sqrt{6}}$ **60.** $\dfrac{\sqrt{-96}}{\sqrt{2}}$

For Problems 61–84, find each of the products and express the answers in the standard form of a complex number.

61. $(5i)(4i)$ **62.** $(-6i)(9i)$

63. $(7i)(-6i)$ **64.** $(-5i)(-12i)$

65. $(3i)(2 - 5i)$ **66.** $(7i)(-9 + 3i)$

67. $(-6i)(-2 - 7i)$ **68.** $(-9i)(-4 - 5i)$

69. $(3 + 2i)(5 + 4i)$ **70.** $(4 + 3i)(6 + i)$

71. $(6 - 2i)(7 - i)$ **72.** $(8 - 4i)(7 - 2i)$

73. $(-3 - 2i)(5 + 6i)$ **74.** $(-5 - 3i)(2 - 4i)$

75. $(9 + 6i)(-1 - i)$ **76.** $(10 + 2i)(-2 - i)$

77. $(4 + 5i)^2$ **78.** $(5 - 3i)^2$

79. $(-2 - 4i)^2$ **80.** $(-3 - 6i)^2$

81. $(6 + 7i)(6 - 7i)$ **82.** $(5 - 7i)(5 + 7i)$

83. $(-1 + 2i)(-1 - 2i)$ **84.** $(-2 - 4i)(-2 + 4i)$

For Problems 85–100, find each of the following quotients and express the answers in the standard form of a complex number.

85. $\dfrac{3i}{2 + 4i}$ **86.** $\dfrac{4i}{5 + 2i}$

87. $\dfrac{-2i}{3 - 5i}$ **88.** $\dfrac{-5i}{2 - 4i}$

89. $\dfrac{-2 + 6i}{3i}$ **90.** $\dfrac{-4 - 7i}{6i}$

91. $\dfrac{2}{7i}$ **92.** $\dfrac{3}{10i}$

93. $\dfrac{2 + 6i}{1 + 7i}$ **94.** $\dfrac{5 + i}{2 + 9i}$

95. $\dfrac{3 + 6i}{4 - 5i}$ **96.** $\dfrac{7 - 3i}{4 - 3i}$

97. $\dfrac{-2 + 7i}{-1 + i}$ **98.** $\dfrac{-3 + 8i}{-2 + i}$

99. $\dfrac{-1 - 3i}{-2 - 10i}$ **100.** $\dfrac{-3 - 4i}{-4 - 11i}$

101. Some of the solution sets for quadratic equations in the next sections will contain complex numbers such as $(-4 + \sqrt{-12})/2$ and $(-4 - \sqrt{-12})/2$. We can simplify the first number as follows.

$$\frac{-4 + \sqrt{-12}}{2} = \frac{-4 + i\sqrt{12}}{2} =$$

$$\frac{-4 + 2i\sqrt{3}}{2} = \frac{2(-2 + i\sqrt{3})}{2} = -2 + i\sqrt{3}$$

Simplify each of the following complex numbers.

(a) $\dfrac{-4 - \sqrt{-12}}{2}$ **(b)** $\dfrac{6 + \sqrt{-24}}{4}$

(c) $\dfrac{-1 - \sqrt{-18}}{2}$ **(d)** $\dfrac{-6 + \sqrt{-27}}{3}$

(e) $\dfrac{10 + \sqrt{-45}}{4}$ **(f)** $\dfrac{4 - \sqrt{-48}}{2}$

■ ■ ■ **THOUGHTS INTO WORDS**

102. Why is the set of real numbers a subset of the set of complex numbers?

103. Can the sum of two nonreal complex numbers be a real number? Defend your answer.

104. Can the product of two nonreal complex numbers be a real number? Defend your answer.

6.2 Quadratic Equations

A second-degree equation in one variable contains the variable with an exponent of 2, but no higher power. Such equations are also called **quadratic equations**. The following are examples of quadratic equations.

$$x^2 = 36 \qquad y^2 + 4y = 0 \qquad x^2 + 5x - 2 = 0$$

$$3n^2 + 2n - 1 = 0 \qquad 5x^2 + x + 2 = 3x^2 - 2x - 1$$

A quadratic equation in the variable x can also be defined as any equation that can be written in the form

$$ax^2 + bx + c = 0$$

where a, b, and c are real numbers and $a \neq 0$. The form $ax^2 + bx + c = 0$ is called the **standard form** of a quadratic equation.

In previous chapters you solved quadratic equations (the term *quadratic* was not used at that time) by factoring and applying the property, $ab = 0$ if and only if $a = 0$ or $b = 0$. Let's review a few such examples.

EXAMPLE 1   Solve $3n^2 + 14n - 5 = 0$.

Solution

$$3n^2 + 14n - 5 = 0$$

$$(3n - 1)(n + 5) = 0 \qquad \text{Factor the left side.}$$

$$3n - 1 = 0 \quad \text{or} \quad n + 5 = 0 \qquad \text{Apply: } ab = 0 \text{ if and only if } a = 0 \text{ or } b = 0.$$

$$3n = 1 \quad \text{or} \quad n = -5$$

$$n = \frac{1}{3} \quad \text{or} \quad n = -5$$

The solution set is $\left\{-5, \dfrac{1}{3}\right\}$.

■

EXAMPLE 2

Solve $x^2 + 3kx - 10k^2 = 0$ for x.

Solution

$$x^2 + 3kx - 10k^2 = 0$$

$$(x + 5k)(x - 2k) = 0 \qquad \text{Factor the left side.}$$

$$x + 5k = 0 \quad \text{or} \quad x - 2k = 0 \qquad \text{Apply: } ab = 0 \text{ if and}$$
$$\qquad\qquad\qquad\qquad\qquad\qquad \text{only if } a = 0 \text{ or } b = 0.$$

$$x = -5k \quad \text{or} \quad x = 2k$$

The solution set is $\{-5k, 2k\}$. ∎

EXAMPLE 3

Solve $2\sqrt{x} = x - 8$.

Solution

$$2\sqrt{x} = x - 8$$

$$(2\sqrt{x})^2 = (x - 8)^2 \qquad \text{Square both sides.}$$

$$4x = x^2 - 16x + 64$$

$$0 = x^2 - 20x + 64$$

$$0 = (x - 16)(x - 4) \qquad \text{Factor the right side.}$$

$$x - 16 = 0 \quad \text{or} \quad x - 4 = 0 \qquad \text{Apply: } ab = 0 \text{ if and}$$
$$\qquad\qquad\qquad\qquad\qquad\qquad \text{only if } a = 0 \text{ or } b = 0.$$

$$x = 16 \quad \text{or} \quad x = 4$$

✔ **Check**

$$2\sqrt{x} = x - 8 \qquad\qquad 2\sqrt{x} = x - 8$$

$$2\sqrt{16} \stackrel{?}{=} 16 - 8 \quad \text{or} \quad 2\sqrt{4} \stackrel{?}{=} 4 - 8$$

$$2(4) \stackrel{?}{=} 8 \qquad\qquad\qquad 2(2) \stackrel{?}{=} -4$$

$$8 = 8 \qquad\qquad\qquad\qquad 4 \neq -4$$

The solution set is $\{16\}$. ∎

We should make two comments about Example 3. First, remember that applying the property, if $a = b$, then $a^n = b^n$, might produce extraneous solutions. Therefore, we *must* check all potential solutions. Second, the equation $2\sqrt{x} = x - 8$ is said to be of **quadratic form** because it can be written as $2x^{\frac{1}{2}} = \left(x^{\frac{1}{2}}\right)^2 - 8$. More will be said about the phrase *quadratic form* later.

Let's consider quadratic equations of the form $x^2 = a$, where x is the variable and a is any real number. We can solve $x^2 = a$ as follows:

$$x^2 = a$$
$$x^2 - a = 0$$
$$x^2 - (\sqrt{a})^2 = 0 \qquad\qquad a = (\sqrt{a})^2$$
$$(x - \sqrt{a})(x + \sqrt{a}) = 0 \qquad\qquad \text{Factor the left side.}$$
$$x - \sqrt{a} = 0 \qquad \text{or} \qquad x + \sqrt{a} = 0 \qquad \text{Apply: } ab = 0 \text{ if and}$$
$$x = \sqrt{a} \qquad \text{or} \qquad x = -\sqrt{a}. \qquad \text{only if } a = 0 \text{ or } b = 0.$$

The solutions are $\sqrt{a}$ and $-\sqrt{a}$. We can state this result as a general property and use it to solve certain types of quadratic equations.

Property 6.1

> For any real number a,
>
> $$x^2 = a \quad \text{if and only if } x = \sqrt{a} \text{ or } x = -\sqrt{a}$$
>
> (The statement $x = \sqrt{a}$ or $x = -\sqrt{a}$ can be written as $x = \pm\sqrt{a}$.)

Property 6.1, along with our knowledge of square roots, makes it very easy to solve quadratic equations of the form $x^2 = a$.

E X A M P L E 4

Solve $x^2 = 45$.

Solution

$$x^2 = 45$$
$$x = \pm\sqrt{45}$$
$$x = \pm3\sqrt{5} \qquad \sqrt{45} = \sqrt{9}\sqrt{5} = 3\sqrt{5}$$

The solution set is $\{\pm3\sqrt{5}\}$.   ■

E X A M P L E 5

Solve $x^2 = -9$.

Solution

$$x^2 = -9$$
$$x = \pm\sqrt{-9}$$
$$x = \pm3i$$

Thus the solution set is $\{\pm3i\}$.   ■

EXAMPLE 6

Solve $7n^2 = 12$.

Solution

$$7n^2 = 12$$

$$n^2 = \frac{12}{7}$$

$$n = \pm\sqrt{\frac{12}{7}}$$

$$n = \pm\frac{2\sqrt{21}}{7} \qquad \sqrt{\frac{12}{7}} = \frac{\sqrt{12}}{\sqrt{7}} \cdot \frac{\sqrt{7}}{\sqrt{7}} = \frac{\sqrt{84}}{7} = \frac{\sqrt{4}\sqrt{21}}{7} = \frac{2\sqrt{21}}{7}$$

The solution set is $\left\{\pm\frac{2\sqrt{21}}{7}\right\}$. ■

EXAMPLE 7

Solve $(3n + 1)^2 = 25$.

Solution

$$(3n + 1)^2 = 25$$

$$(3n + 1) = \pm\sqrt{25}$$

$$3n + 1 = \pm5$$

$$3n + 1 = 5 \qquad \text{or} \qquad 3n + 1 = -5$$

$$3n = 4 \qquad \text{or} \qquad 3n = -6$$

$$n = \frac{4}{3} \qquad \text{or} \qquad n = -2$$

The solution set is $\left\{-2, \frac{4}{3}\right\}$. ■

EXAMPLE 8

Solve $(x - 3)^2 = -10$.

Solution

$$(x - 3)^2 = -10$$

$$x - 3 = \pm\sqrt{-10}$$

$$x - 3 = \pm i\sqrt{10}$$

$$x = 3 \pm i\sqrt{10}$$

Thus the solution set is $\{3 \pm i\sqrt{10}\}$. ■

Remark: Take another look at the equations in Examples 5 and 8. We should immediately realize that the solution sets will consist only of nonreal complex numbers, because any nonzero real number squared is positive.

Sometimes it may be necessary to change the form before we can apply Property 6.1. Let's consider one example to illustrate this idea.

<table>
<tr><td>**E X A M P L E   9**</td><td>Solve $3(2x - 3)^2 + 8 = 44$.</td></tr>
</table>

Solution

$$3(2x - 3)^2 + 8 = 44$$
$$3(2x - 3)^2 = 36$$
$$(2x - 3)^2 = 12$$
$$2x - 3 = \pm\sqrt{12}$$
$$2x - 3 = \pm 2\sqrt{3}$$
$$2x = 3 \pm 2\sqrt{3}$$
$$x = \frac{3 \pm 2\sqrt{3}}{2}$$

The solution set is $\left\{ \dfrac{3 \pm 2\sqrt{3}}{2} \right\}$. ∎

■ Back to the Pythagorean Theorem

Our work with radicals, Property 6.1, and the Pythagorean theorem form a basis for solving a variety of problems that pertain to right triangles.

E X A M P L E   1 0

A 50-foot rope hangs from the top of a flagpole. When pulled taut to its full length, the rope reaches a point on the ground 18 feet from the base of the pole. Find the height of the pole to the nearest tenth of a foot.

Solution

Let's make a sketch (Figure 6.1) and record the given information.
Use the Pythagorean theorem to solve for p as follows:

$$p^2 + 18^2 = 50^2$$
$$p^2 + 324 = 2500$$
$$p^2 = 2176$$
$$p = \sqrt{2176} = 46.6, \quad \text{to the nearest tenth}$$

The height of the flagpole is approximately 46.6 feet. ∎

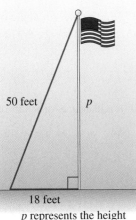

50 feet p

18 feet

p represents the height of the flagpole.

Figure 6.1

There are two special kinds of right triangles that we use extensively in later mathematics courses. The first is the **isosceles right triangle**, which is a right triangle that has both legs of the same length. Let's consider a problem that involves an isosceles right triangle.

EXAMPLE 11

Find the length of each leg of an isosceles right triangle that has a hypotenuse of length 5 meters.

Solution

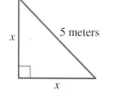

Figure 6.2

Let's sketch an isosceles right triangle and let x represent the length of each leg (Figure 6.2). Then we can apply the Pythagorean theorem.

$$x^2 + x^2 = 5^2$$
$$2x^2 = 25$$
$$x^2 = \frac{25}{2}$$
$$x = \pm\sqrt{\frac{25}{2}} = \pm\frac{5}{\sqrt{2}} = \pm\frac{5\sqrt{2}}{2}$$

Each leg is $\dfrac{5\sqrt{2}}{2}$ meters long. ■

Remark: In Example 10 we made no attempt to express $\sqrt{2176}$ in simplest radical form because the answer was to be given as a rational approximation to the nearest tenth. However, in Example 11 we left the final answer in radical form and therefore expressed it in simplest radical form.

The second special kind of right triangle that we use frequently is one that contains acute angles of 30° and 60°. In such a right triangle, which we refer to as a **30°– 60° right triangle**, the side opposite the 30° angle is equal in length to one-half of the length of the hypotenuse. This relationship, along with the Pythagorean theorem, provides us with another problem-solving technique.

EXAMPLE 12

Suppose that a 20-foot ladder is leaning against a building and makes an angle of 60° with the ground. How far up the building does the top of the ladder reach? Express your answer to the nearest tenth of a foot.

Solution

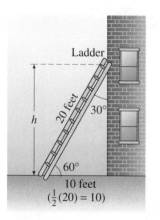

Figure 6.3

Figure 6.3 depicts this situation. The side opposite the 30° angle equals one-half of the hypotenuse, so it is of length $\frac{1}{2}(20) = 10$ feet. Now we can apply the Pythagorean theorem.

$$h^2 + 10^2 = 20^2$$
$$h^2 + 100 = 400$$
$$h^2 = 300$$
$$h = \sqrt{300} = 17.3, \quad \text{to the nearest tenth}$$

The top of the ladder touches the building at a point approximately 17.3 feet from the ground. ■

Problem Set 6.2

For Problems 1–20, solve each of the quadratic equations by factoring and applying the property, $ab = 0$ if and only if $a = 0$ or $b = 0$. If necessary, return to Chapter 3 and review the factoring techniques presented there.

1. $x^2 - 9x = 0$

2. $x^2 + 5x = 0$

3. $x^2 = -3x$

4. $x^2 = 15x$

5. $3y^2 + 12y = 0$

6. $6y^2 - 24y = 0$

7. $5n^2 - 9n = 0$

8. $4n^2 + 13n = 0$

9. $x^2 + x - 30 = 0$

10. $x^2 - 8x - 48 = 0$

11. $x^2 - 19x + 84 = 0$

12. $x^2 - 21x + 104 = 0$

13. $2x^2 + 19x + 24 = 0$

14. $4x^2 + 29x + 30 = 0$

15. $15x^2 + 29x - 14 = 0$

16. $24x^2 + x - 10 = 0$

17. $25x^2 - 30x + 9 = 0$

18. $16x^2 - 8x + 1 = 0$

19. $6x^2 - 5x - 21 = 0$

20. $12x^2 - 4x - 5 = 0$

For Problems 21–26, solve each radical equation. Don't forget, you *must* check potential solutions.

21. $3\sqrt{x} = x + 2$

22. $3\sqrt{2x} = x + 4$

23. $\sqrt{2x} = x - 4$

24. $\sqrt{x} = x - 2$

25. $\sqrt{3x + 6} = x$

26. $\sqrt{5x + 10} = x$

For Problems 27–34, solve each equation for x by factoring and applying the property, $ab = 0$ if and only if $a = 0$ or $b = 0$.

27. $x^2 - 5kx = 0$

28. $x^2 + 7kx = 0$

29. $x^2 = 16k^2x$

30. $x^2 = 25k^2x$

31. $x^2 - 12kx + 35k^2 = 0$

32. $x^2 - 3kx - 18k^2 = 0$

33. $2x^2 + 5kx - 3k^2 = 0$

34. $3x^2 - 20kx - 7k^2 = 0$

For Problems 35–70, use Property 6.1 to help solve each quadratic equation.

35. $x^2 = 1$

36. $x^2 = 81$

37. $x^2 = -36$

38. $x^2 = -49$

39. $x^2 = 14$

40. $x^2 = 22$

41. $n^2 - 28 = 0$

42. $n^2 - 54 = 0$

43. $3t^2 = 54$

44. $4t^2 = 108$

45. $2t^2 = 7$

46. $3t^2 = 8$

47. $15y^2 = 20$

48. $14y^2 = 80$

49. $10x^2 + 48 = 0$

50. $12x^2 + 50 = 0$

51. $24x^2 = 36$

52. $12x^2 = 49$

53. $(x - 2)^2 = 9$

54. $(x + 1)^2 = 16$

55. $(x + 3)^2 = 25$

56. $(x - 2)^2 = 49$

57. $(x + 6)^2 = -4$

58. $(3x + 1)^2 = 9$

59. $(2x - 3)^2 = 1$

60. $(2x + 5)^2 = -4$

61. $(n - 4)^2 = 5$

62. $(n - 7)^2 = 6$

63. $(t + 5)^2 = 12$

64. $(t - 1)^2 = 18$

65. $(3y - 2)^2 = -27$

66. $(4y + 5)^2 = 80$

67. $3(x + 7)^2 + 4 = 79$

68. $2(x + 6)^2 - 9 = 63$

69. $2(5x - 2)^2 + 5 = 25$

70. $3(4x - 1)^2 + 1 = -17$

For Problems 71–76, a and b represent the lengths of the legs of a right triangle, and c represents the length of the hypotenuse. Express answers in simplest radical form.

71. Find c if $a = 4$ centimeters and $b = 6$ centimeters.

72. Find c if $a = 3$ meters and $b = 7$ meters.

73. Find a if $c = 12$ inches and $b = 8$ inches.

74. Find a if $c = 8$ feet and $b = 6$ feet.

75. Find b if $c = 17$ yards and $a = 15$ yards.

76. Find b if $c = 14$ meters and $a = 12$ meters.

For Problems 77–80, use the isosceles right triangle in Figure 6.4. Express your answers in simplest radical form.

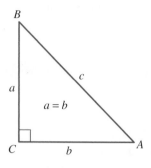

Figure 6.4

77. If $b = 6$ inches, find c.

78. If $a = 7$ centimeters, find c.

79. If $c = 8$ meters, find a and b.

80. If $c = 9$ feet, find a and b.

For Problems 81–86, use the triangle in Figure 6.5. Express your answers in simplest radical form.

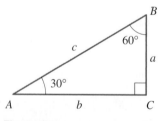

Figure 6.5

81. If $a = 3$ inches, find b and c.

82. If $a = 6$ feet, find b and c.

83. If $c = 14$ centimeters, find a and b.

84. If $c = 9$ centimeters, find a and b.

85. If $b = 10$ feet, find a and c.

86. If $b = 8$ meters, find a and c.

87. A 24-foot ladder resting against a house reaches a windowsill 16 feet above the ground. How far is the foot of the ladder from the foundation of the house? Express your answer to the nearest tenth of a foot.

88. A 62-foot guy-wire makes an angle of 60° with the ground and is attached to a telephone pole (see Figure 6.6). Find the distance from the base of the pole to the point on the pole where the wire is attached. Express your answer to the nearest tenth of a foot.

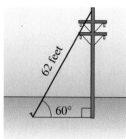

Figure 6.6

89. A rectangular plot measures 16 meters by 34 meters. Find, to the nearest meter, the distance from one corner of the plot to the corner diagonally opposite.

90. Consecutive bases of a square-shaped baseball diamond are 90 feet apart (see Figure 6.7). Find, to the nearest tenth of a foot, the distance from first base diagonally across the diamond to third base.

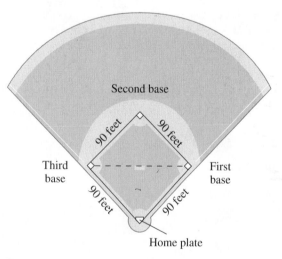

Figure 6.7

91. A diagonal of a square parking lot is 75 meters. Find, to the nearest meter, the length of a side of the lot.

92. Explain why the equation $(x + 2)^2 + 5 = 1$ has no real number solutions.

93. Suppose that your friend solved the equation $(x + 3)^2 = 25$ as follows:

$$(x + 3)^2 = 25$$

$$x^2 + 6x + 9 = 25$$

$$x^2 + 6x - 16 = 0$$

$$(x + 8)(x - 2) = 0$$

$$x + 8 = 0 \quad \text{or} \quad x - 2 = 0$$

$$x = -8 \quad \text{or} \quad x = 2$$

Is this a correct approach to the problem? Would you offer any suggestion about an easier approach to the problem?

94. Suppose that we are given a cube with edges 12 centimeters in length. Find the length of a diagonal from a lower corner to the diagonally opposite upper corner. Express your answer to the nearest tenth of a centimeter.

95. Suppose that we are given a rectangular box with a length of 8 centimeters, a width of 6 centimeters, and a height of 4 centimeters. Find the length of a diagonal from a lower corner to the upper corner diagonally opposite. Express your answer to the nearest tenth of a centimeter.

96. The converse of the Pythagorean theorem is also true. It states, "If the measures a, b, and c of the sides of a triangle are such that $a^2 + b^2 = c^2$, then the triangle is a right triangle with a and b the measures of the legs and c the measure of the hypotenuse." Use the converse of the Pythagorean theorem to determine which of the triangles with sides of the following measures are right triangles.

(**a**) 9, 40, 41   (**b**) 20, 48, 52

(**c**) 19, 21, 26   (**d**) 32, 37, 49

(**e**) 65, 156, 169   (**f**) 21, 72, 75

97. Find the length of the hypotenuse (h) of an isosceles right triangle if each leg is s units long. Then use this relationship to redo Problems 77–80.

98. Suppose that the side opposite the 30° angle in a 30°–60° right triangle is s units long. Express the length of the hypotenuse and the length of the other leg in terms of s. Then use these relationships and redo Problems 81–86.

6.3	**Completing the Square**

Thus far we have solved quadratic equations by factoring and applying the property, $ab = 0$ if and only if $a = 0$ or $b = 0$, or by applying the property, $x^2 = a$ if and only if $x = \pm\sqrt{a}$. In this section we examine another method called **completing the square**, which will give us the power to solve any quadratic equation.

A factoring technique we studied in Chapter 3 relied on recognizing **perfect-square trinomials**. In each of the following, the perfect-square trinomial on the right side is the result of squaring the binomial on the left side.

$$(x + 4)^2 = x^2 + 8x + 16 \qquad (x - 6)^2 = x^2 - 12x + 36$$

$$(x + 7)^2 = x^2 + 14x + 49 \qquad (x - 9)^2 = x^2 - 18x + 81$$

$$(x + a)^2 = x^2 + 2ax + a^2$$

Note that in each of the square trinomials, the constant term is equal to the square of one-half of the coefficient of the x term. This relationship enables us to form a perfect-square trinomial by adding a proper constant term. To find the constant term, take one-half of the coefficient of the x term and then square the result. For example, suppose that we want to form a perfect-square trinomial from $x^2 + 10x$. The coefficient of the x term is 10. Because $\frac{1}{2}(10) = 5$, and $5^2 = 25$, the constant term should be 25. The perfect-square trinomial that can be formed is $x^2 + 10x + 25$. This perfect-square trinomial can be factored and expressed as $(x + 5)^2$. Let's use the previous ideas to help solve some quadratic equations.

E X A M P L E   1

Solve $x^2 + 10x - 2 = 0$.

Solution

$$x^2 + 10x - 2 = 0$$

$$x^2 + 10x = 2 \qquad \text{Isolate the } x^2 \text{ and } x \text{ terms.}$$

$$\frac{1}{2}(10) = 5 \text{ and } 5^2 = 25 \qquad \text{Take } \frac{1}{2} \text{ of the coefficient of the } x \text{ term and then square the result.}$$

$$x^2 + 10x + 25 = 2 + 25 \qquad \text{Add 25 to } both \text{ sides of the equation.}$$

$$(x + 5)^2 = 27 \qquad \text{Factor the perfect-square trinomial.}$$

$$x + 5 = \pm\sqrt{27} \qquad \text{Now solve by applying Property 6.1.}$$

$$x + 5 = \pm 3\sqrt{3}$$

$$x = -5 \pm 3\sqrt{3}$$

The solution set is $\{-5 \pm 3\sqrt{3}\}$. ∎

Note from Example 1 that the method of completing the square to solve a quadratic equation is merely what the name implies. A perfect-square trinomial is formed, then the equation can be changed to the necessary form for applying the property "$x^2 = a$ if and only if $x = \pm\sqrt{a}$." Let's consider another example.

E X A M P L E   2

Solve $x(x + 8) = -23$.

Solution

$$x(x + 8) = -23$$

$$x^2 + 8x = -23 \qquad \text{Apply the distributive property.}$$

$$\frac{1}{2}(8) = 4 \text{ and } 4^2 = 16 \qquad \text{Take } \frac{1}{2} \text{ of the coefficient of the } x \text{ term and then square the result.}$$

$$x^2 + 8x + 16 = -23 + 16 \qquad \text{Add 16 to \textit{both} sides of the equation.}$$

$$(x + 4)^2 = -7 \qquad \text{Factor the perfect-square trinomial.}$$

$$x + 4 = \pm\sqrt{-7} \qquad \text{Now solve by applying Property 6.1.}$$

$$x + 4 = \pm i\sqrt{7}$$

$$x = -4 \pm i\sqrt{7}$$

The solution set is $\{-4 \pm i\sqrt{7}\}$. ∎

EXAMPLE 3

Solve $x^2 - 3x + 1 = 0$.

Solution

$$x^2 - 3x + 1 = 0$$

$$x^2 - 3x = -1$$

$$x^2 - 3x + \frac{9}{4} = -1 + \frac{9}{4} \qquad \frac{1}{2}(3) = \frac{3}{2} \text{ and } \left(\frac{3}{2}\right)^2 = \frac{9}{4}$$

$$\left(x - \frac{3}{2}\right)^2 = \frac{5}{4}$$

$$x - \frac{3}{2} = \pm\sqrt{\frac{5}{4}}$$

$$x - \frac{3}{2} = \pm\frac{\sqrt{5}}{2}$$

$$x = \frac{3}{2} \pm \frac{\sqrt{5}}{2}$$

$$x = \frac{3 \pm \sqrt{5}}{2}$$

The solution set is $\left\{\dfrac{3 \pm \sqrt{5}}{2}\right\}$. ∎

In Example 3 note that because the coefficient of the x term is odd, we are forced into the realm of fractions. Using common fractions rather than decimals enables us to apply our previous work with radicals.

The relationship for a perfect-square trinomial that states that the constant term is equal to the square of one-half of the coefficient of the x term holds only if the coefficient of x^2 is 1. Thus we must make an adjustment when solving quadratic equations that have a coefficient of x^2 other than 1. We will need to apply the multiplication property of equality so that the coefficient of the x^2 term becomes 1. The next example shows how to make this adjustment.

EXAMPLE 4   Solve $2x^2 + 12x - 5 = 0$.

Solution

$$2x^2 + 12x - 5 = 0$$

$$2x^2 + 12x = 5$$

$$x^2 + 6x = \frac{5}{2} \qquad \text{Multiply both sides by } \frac{1}{2}.$$

$$x^2 + 6x + 9 = \frac{5}{2} + 9 \qquad \frac{1}{2}(6) = 3, \text{ and } 3^2 = 9$$

$$x^2 + 6x + 9 = \frac{23}{2}$$

$$(x + 3)^2 = \frac{23}{2}$$

$$x + 3 = \pm\sqrt{\frac{23}{2}}$$

$$x + 3 = \pm\frac{\sqrt{46}}{2} \qquad \sqrt{\frac{23}{2}} = \frac{\sqrt{23}}{\sqrt{2}} \cdot \frac{\sqrt{2}}{\sqrt{2}} = \frac{\sqrt{46}}{2}$$

$$x = -3 \pm \frac{\sqrt{46}}{2}$$

$$x = \frac{-6}{2} \pm \frac{\sqrt{46}}{2} \qquad \text{Common denominator of 2}$$

$$x = \frac{-6 \pm \sqrt{46}}{2}$$

The solution set is $\left\{\dfrac{-6 \pm \sqrt{46}}{2}\right\}$. ∎

As we mentioned earlier, we can use the method of completing the square to solve *any* quadratic equation. To illustrate, let's use it to solve an equation that could also be solved by factoring.

EXAMPLE 5   Solve $x^2 - 2x - 8 = 0$ by completing the square.

Solution

$$x^2 - 2x - 8 = 0$$

$$x^2 - 2x = 8$$

$$x^2 - 2x + 1 = 8 + 1 \qquad \frac{1}{2}(-2) = -1 \text{ and } (-1)^2 = 1$$

$$(x - 1)^2 = 9$$

$$x - 1 = \pm 3$$

$$x - 1 = 3 \quad \text{or} \quad x - 1 = -3$$

$$x = 4 \quad \text{or} \quad x = -2$$

The solution set is $\{-2, 4\}$.   ∎

Solving the equation in Example 5 by factoring would be easier than completing the square. Remember, however, that the method of completing the square will work with any quadratic equation.

Problem Set 6.3

For Problems 1–14, solve each quadratic equation by using (a) the factoring method and (b) the method of completing the square.

1. $x^2 - 4x - 60 = 0$

2. $x^2 + 6x - 16 = 0$

3. $x^2 - 14x = -40$

4. $x^2 - 18x = -72$

5. $x^2 - 5x - 50 = 0$

6. $x^2 + 3x - 18 = 0$

7. $x(x + 7) = 8$

8. $x(x - 1) = 30$

9. $2n^2 - n - 15 = 0$

10. $3n^2 + n - 14 = 0$

11. $3n^2 + 7n - 6 = 0$

12. $2n^2 + 7n - 4 = 0$

13. $n(n + 6) = 160$

14. $n(n - 6) = 216$

For Problems 15–38, use the method of completing the square to solve each quadratic equation.

15. $x^2 + 4x - 2 = 0$

16. $x^2 + 2x - 1 = 0$

17. $x^2 + 6x - 3 = 0$

18. $x^2 + 8x - 4 = 0$

19. $y^2 - 10y = 1$

20. $y^2 - 6y = -10$

21. $n^2 - 8n + 17 = 0$

22. $n^2 - 4n + 2 = 0$

23. $n(n + 12) = -9$

24. $n(n + 14) = -4$

25. $n^2 + 2n + 6 = 0$

26. $n^2 + n - 1 = 0$

27. $x^2 + 3x - 2 = 0$

28. $x^2 + 5x - 3 = 0$

29. $x^2 + 5x + 1 = 0$

30. $x^2 + 7x + 2 = 0$

31. $y^2 - 7y + 3 = 0$

32. $y^2 - 9y + 30 = 0$

33. $2x^2 + 4x - 3 = 0$

34. $2t^2 - 4t + 1 = 0$

35. $3n^2 - 6n + 5 = 0$

36. $3x^2 + 12x - 2 = 0$

37. $3x^2 + 5x - 1 = 0$

38. $2x^2 + 7x - 3 = 0$

For Problems 39–60, solve each quadratic equation using the method that seems most appropriate.

39. $x^2 + 8x - 48 = 0$

40. $x^2 + 5x - 14 = 0$

41. $2n^2 - 8n = -3$

42. $3x^2 + 6x = 1$

43. $(3x - 1)(2x + 9) = 0$

44. $(5x + 2)(x - 4) = 0$

45. $(x + 2)(x - 7) = 10$

46. $(x - 3)(x + 5) = -7$

47. $(x - 3)^2 = 12$

48. $x^2 = 16x$

49. $3n^2 - 6n + 4 = 0$

50. $2n^2 - 2n - 1 = 0$

51. $n(n + 8) = 240$

52. $t(t - 26) = -160$

53. $3x^2 + 5x = -2$

54. $2x^2 - 7x = -5$

55. $4x^2 - 8x + 3 = 0$

56. $9x^2 + 18x + 5 = 0$

57. $x^2 + 12x = 4$

58. $x^2 + 6x = -11$

59. $4(2x + 1)^2 - 1 = 11$

60. $5(x + 2)^2 + 1 = 16$

61. Use the method of completing the square to solve $ax^2 + bx + c = 0$ for x, where a, b, and c are real numbers and $a \neq 0$.

■ ■ ■ THOUGHTS INTO WORDS

62. Explain the process of completing the square to solve a quadratic equation.

63. Give a step-by-step description of how to solve $3x^2 + 9x - 4 = 0$ by completing the square.

■ ■ ■ FURTHER INVESTIGATIONS

Solve Problems 64–67 for the indicated variable. Assume that all letters represent positive numbers.

Solve each of the following equations for x.

64. $\dfrac{x^2}{a^2} - \dfrac{y^2}{b^2} = 1$ for y

65. $\dfrac{x^2}{a^2} + \dfrac{y^2}{b^2} = 1$ for x

66. $s = \dfrac{1}{2}gt^2$ for t

67. $A = \pi r^2$ for r

68. $x^2 + 8ax + 15a^2 = 0$

69. $x^2 - 5ax + 6a^2 = 0$

70. $10x^2 - 31ax - 14a^2 = 0$

71. $6x^2 + ax - 2a^2 = 0$

72. $4x^2 + 4bx + b^2 = 0$

73. $9x^2 - 12bx + 4b^2 = 0$

6.4 Quadratic Formula

As we saw in the last section, the method of completing the square can be used to solve any quadratic equation. Thus if we apply the method of completing the square to the equation $ax^2 + bx + c = 0$, where a, b, and c are real numbers and $a \neq 0$, we can produce a formula for solving quadratic equations. This formula can then be used to solve any quadratic equation. Let's solve $ax^2 + bx + c = 0$ by completing the square.

$$ax^2 + bx + c = 0$$

$$ax^2 + bx = -c \qquad \text{Isolate the } x^2 \text{ and } x \text{ terms.}$$

$$x^2 + \frac{b}{a}x = -\frac{c}{a} \qquad \text{Multiply both sides by } \frac{1}{a}.$$

$$x^2 + \frac{b}{a}x + \frac{b^2}{4a^2} = -\frac{c}{a} + \frac{b^2}{4a^2} \qquad \frac{1}{2}\left(\frac{b}{a}\right) = \frac{b}{2a} \ \text{ and } \ \left(\frac{b}{2a}\right)^2 = \frac{b^2}{4a^2}$$

Complete the square by adding $\dfrac{b^2}{4a^2}$ to both sides.

$$x^2 + \frac{b}{a}x + \frac{b^2}{4a^2} = -\frac{4ac}{4a^2} + \frac{b^2}{4a^2} \qquad \text{Common denominator of } 4a^2 \text{ on right side}$$

$$x^2 + \frac{b}{a}x + \frac{b^2}{4a^2} = \frac{b^2}{4a^2} - \frac{4ac}{4a^2} \qquad \text{Commutative property}$$

$$\left(x + \frac{b}{2a}\right)^2 = \frac{b^2 - 4ac}{4a^2}$$ The right side is combined into a single fraction.

$$x + \frac{b}{2a} = \pm\sqrt{\frac{b^2 - 4ac}{4a^2}}$$

$$x + \frac{b}{2a} = \pm\frac{\sqrt{b^2 - 4ac}}{\sqrt{4a^2}}$$

$$x + \frac{b}{2a} = \pm\frac{\sqrt{b^2 - 4ac}}{2a}$$ $\sqrt{4a^2} = |2a|$ but $2a$ can be used because of the use of $\pm$.

$$x + \frac{b}{2a} = \frac{\sqrt{b^2 - 4ac}}{2a} \qquad \text{or} \qquad x + \frac{b}{2a} = -\frac{\sqrt{b^2 - 4ac}}{2a}$$

$$x = -\frac{b}{2a} + \frac{\sqrt{b^2 - 4ac}}{2a} \qquad \text{or} \qquad x = -\frac{b}{2a} - \frac{\sqrt{b^2 - 4ac}}{2a}$$

$$x = \frac{-b + \sqrt{b^2 - 4ac}}{2a} \qquad \text{or} \qquad x = \frac{-b - \sqrt{b^2 - 4ac}}{2a}$$

The quadratic formula is usually stated as follows:

Quadratic Formula

$$x = \frac{-b \pm \sqrt{b^2 - 4ac}}{2a}, \qquad a \neq 0$$

We can use the quadratic formula to solve *any* quadratic equation by expressing the equation in the standard form $ax^2 + bx + c = 0$ and substituting the values for a, b, and c into the formula. Let's consider some examples.

EXAMPLE 1   Solve $x^2 + 5x + 2 = 0$.

Solution

$$x^2 + 5x + 2 = 0$$

The given equation is in standard form with $a = 1$, $b = 5$, and $c = 2$. Let's substitute these values into the formula and simplify.

$$x = \frac{-b \pm \sqrt{b^2 - 4ac}}{2a}$$

$$x = \frac{-5 \pm \sqrt{5^2 - 4(1)(2)}}{2(1)}$$

$$x = \frac{-5 \pm \sqrt{25 - 8}}{2}$$

$$x = \frac{-5 \pm \sqrt{17}}{2}$$

The solution set is $\left\{\dfrac{-5 \pm \sqrt{17}}{2}\right\}$.

■

EXAMPLE 2   Solve $x^2 - 2x - 4 = 0$.

Solution

$$x^2 - 2x - 4 = 0$$

We need to think of $x^2 - 2x - 4 = 0$ as $x^2 + (-2)x + (-4) = 0$ to determine the values $a = 1$, $b = -2$, and $c = -4$. Let's substitute these values into the quadratic formula and simplify.

$$x = \frac{-b \pm \sqrt{b^2 - 4ac}}{2a}$$

$$x = \frac{-(-2) \pm \sqrt{(-2)^2 - 4(1)(-4)}}{2(1)}$$

$$x = \frac{2 \pm \sqrt{4 + 16}}{2}$$

$$x = \frac{2 \pm \sqrt{20}}{2}$$

$$x = \frac{2 \pm 2\sqrt{5}}{2}$$

$$x = \frac{2(1 \pm \sqrt{5})}{2} = (1 \pm \sqrt{5})$$

The solution set is $\{1 \pm \sqrt{5}\}$.

■

EXAMPLE 3   Solve $x^2 - 2x + 19 = 0$.

Solution

$$x^2 - 2x + 19 = 0$$

We can substitute $a = 1$, $b = -2$, and $c = 19$.

$$x = \frac{-b \pm \sqrt{b^2 - 4ac}}{2a}$$

$$x = \frac{-(-2) \pm \sqrt{(-2)^2 - 4(1)(19)}}{2(1)}$$

$$x = \frac{2 \pm \sqrt{4 - 76}}{2}$$

$$x = \frac{2 \pm \sqrt{-72}}{2}$$

$$x = \frac{2 \pm 6i\sqrt{2}}{2} \qquad \sqrt{-72} = i\sqrt{72} = i\sqrt{36}\sqrt{2} = 6i\sqrt{2}$$

$$x = \frac{2(1 \pm 3i\sqrt{2})}{2} = 1 \pm 3i\sqrt{2}$$

The solution set is $\{1 \pm 3i\sqrt{2}\}$.   ∎

E X A M P L E   4   Solve $2x^2 + 4x - 3 = 0$.

Solution

$$2x^2 + 4x - 3 = 0$$

Here $a = 2$, $b = 4$, and $c = -3$. Solving by using the quadratic formula is unlike solving by completing the square in that there is no need to make the coefficient of x^2 equal to 1.

$$x = \frac{-b \pm \sqrt{b^2 - 4ac}}{2a}$$

$$x = \frac{-4 \pm \sqrt{4^2 - 4(2)(-3)}}{2(2)}$$

$$x = \frac{-4 \pm \sqrt{16 + 24}}{4}$$

$$x = \frac{-4 \pm \sqrt{40}}{4}$$

$$x = \frac{-4 \pm 2\sqrt{10}}{4}$$

$$x = \frac{2(-2 \pm \sqrt{10})}{4}$$

$$x = \frac{-2 \pm \sqrt{10}}{2}$$

The solution set is $\left\{\dfrac{-2 \pm \sqrt{10}}{2}\right\}$.   ∎

EXAMPLE 5

Solve $n(3n - 10) = 25$.

Solution

$$n(3n - 10) = 25$$

First, we need to change the equation to the standard form $an^2 + bn + c = 0$.

$$n(3n - 10) = 25$$

$$3n^2 - 10n = 25$$

$$3n^2 - 10n - 25 = 0$$

Now we can substitute $a = 3$, $b = -10$, and $c = -25$ into the quadratic formula.

$$n = \frac{-b \pm \sqrt{b^2 - 4ac}}{2a}$$

$$n = \frac{-(-10) \pm \sqrt{(-10)^2 - 4(3)(-25)}}{2(3)}$$

$$n = \frac{10 \pm \sqrt{100 + 300}}{2(3)}$$

$$n = \frac{10 \pm \sqrt{400}}{6}$$

$$n = \frac{10 \pm 20}{6}$$

$$n = \frac{10 + 20}{6} \qquad \text{or} \qquad n = \frac{10 - 20}{6}$$

$$n = 5 \qquad \text{or} \qquad n = -\frac{5}{3}$$

The solution set is $\left\{ -\frac{5}{3}, 5 \right\}$. ∎

In Example 5, note that we used the variable n. The quadratic formula is usually stated in terms of x, but it certainly can be applied to quadratic equations in other variables. Also note in Example 5 that the polynomial $3n^2 - 10n - 25$ can be factored as $(3n + 5)(n - 5)$. Therefore, we could also solve the equation $3n^2 - 10n - 25 = 0$ by using the factoring approach. Section 6.5 will offer some guidance in deciding which approach to use for a particular equation.

■ Nature of Roots

The quadratic formula makes it easy to determine the nature of the roots of a quadratic equation without completely solving the equation. The number

$$b^2 - 4ac$$

which appears under the radical sign in the quadratic formula, is called the **discriminant** of the quadratic equation. The discriminant is the indicator of the kind of roots the equation has. For example, suppose that you start to solve the equation $x^2 - 4x + 7 = 0$ as follows:

$$x = \frac{-b \pm \sqrt{b^2 - 4ac}}{2a}$$

$$x = \frac{-(-4) \pm \sqrt{(-4)^2 - 4(1)(7)}}{2(1)}$$

$$x = \frac{4 \pm \sqrt{16 - 28}}{2}$$

$$x = \frac{4 \pm \sqrt{-12}}{2}$$

At this stage you should be able to look ahead and realize that you will obtain two complex solutions for the equation. (Note, by the way, that these solutions are complex conjugates.) In other words, the discriminant, -12, indicates what type of roots you will obtain.

We make the following general statements relative to the roots of a quadratic equation of the form $ax^2 + bx + c = 0$.

1. If $b^2 - 4ac < 0$, then the equation has two nonreal complex solutions.
2. If $b^2 - 4ac = 0$, then the equation has one real solution.
3. If $b^2 - 4ac > 0$, then the equation has two real solutions.

The following examples illustrate each of these situations. (You may want to solve the equations completely to verify the conclusions.)

Equation	Discriminant	Nature of roots
$x^2 - 3x + 7 = 0$	$b^2 - 4ac = (-3)^2 - 4(1)(7)$ $= 9 - 28$ $= -19$	Two nonreal complex solutions
$9x^2 - 12x + 4 = 0$	$b^2 - 4ac = (-12)^2 - 4(9)(4)$ $= 144 - 144$ $= 0$	One real solution
$2x^2 + 5x - 3 = 0$	$b^2 - 4ac = (5)^2 - 4(2)(-3)$ $= 25 + 24$ $= 49$	Two real solutions

There is another very useful relationship that involves the roots of a quadratic equation and the numbers a, b, and c of the general form $ax^2 + bx + c = 0$. Suppose that we let x_1 and x_2 be the two roots generated by the quadratic formula. Thus we have

$$x_1 = \frac{-b + \sqrt{b^2 - 4ac}}{2a} \quad \text{and} \quad x_2 = \frac{-b - \sqrt{b^2 - 4ac}}{2a}$$

Remark: A clarification is called for at this time. Previously, we made the statement that if $b^2 - 4ac = 0$, then the equation has one real solution. Technically, such an equation has two solutions, but they are equal. For example, each factor of $(x - 2)(x - 2) = 0$ produces a solution, but both solutions are the number 2. We sometimes refer to this as one real solution with a *multiplicity of two*. Using the idea of multiplicity of roots, we can say that every quadratic equation has two roots.

Now let's consider the sum and product of the two roots.

$$\textbf{Sum} \quad x_1 + x_2 = \frac{-b + \sqrt{b^2 - 4ac}}{2a} + \frac{-b - \sqrt{b^2 - 4ac}}{2a} = \frac{-2b}{2a} = \boxed{-\frac{b}{a}}$$

$$\textbf{Product} \quad (x_1)(x_2) = \left(\frac{-b + \sqrt{b^2 - 4ac}}{2a}\right)\left(\frac{-b - \sqrt{b^2 - 4ac}}{2a}\right)$$

$$= \frac{b^2 - (b^2 - 4ac)}{4a^2}$$

$$= \frac{b^2 - b^2 + 4ac}{4a^2}$$

$$= \frac{4ac}{4a^2} = \boxed{\frac{c}{a}}$$

These relationships provide another way of checking potential solutions when solving quadratic equations. For instance, back in Example 3 we solved the equation $x^2 - 2x + 19 = 0$ and obtained solutions of $1 + 3i\sqrt{2}$ and $1 - 3i\sqrt{2}$. Let's check these solutions by using the sum and product relationships.

✔ **Check for Example 3**

Sum of roots $\quad (1 + 3i\sqrt{2}) + (1 - 3i\sqrt{2}) = 2 \quad$ and $\quad -\frac{b}{a} = -\frac{-2}{1} = 2$

Product of roots $\quad (1 + 3i\sqrt{2})(1 - 3i\sqrt{2}) = 1 - 18i^2 = 1 + 18 = 19 \quad$ and

$$\frac{c}{a} = \frac{19}{1} = 19$$

Likewise, a check for Example 4 is as follows:

✔ **Check for Example 4**

Sum of roots $\left(\dfrac{-2 + \sqrt{10}}{2}\right) + \left(\dfrac{-2 - \sqrt{10}}{2}\right) = -\dfrac{4}{2} = -2$ and

$$-\dfrac{b}{a} = -\dfrac{4}{2} = -2$$

Product of roots $\left(\dfrac{-2 + \sqrt{10}}{2}\right)\left(\dfrac{-2 - \sqrt{10}}{2}\right) = -\dfrac{6}{4} = -\dfrac{3}{2}$ and

$$\dfrac{c}{a} = \dfrac{-3}{2} = -\dfrac{3}{2}$$

Note that for both Examples 3 and 4, it was much easier to check by using the sum and product relationships than it would have been to check by substituting back into the original equation. Don't forget that the values for a, b, and c come from a quadratic equation of the form $ax^2 + bx + c = 0$. In Example 5, if we are going to check the potential solutions by using the sum and product relationships, we must be certain that we made no errors when changing the given equation $n(3n - 10) = 25$ to the form $3n^2 - 10n - 25 = 0$.

Problem Set 6.4

For each quadratic equation in Problems 1–10, first use the discriminant to determine whether the equation has two nonreal complex solutions, one real solution with a multiplicity of two, or two real solutions. Then solve the equation.

1. $x^2 + 4x - 21 = 0$

2. $x^2 - 3x - 54 = 0$

3. $9x^2 - 6x + 1 = 0$

4. $4x^2 + 20x + 25 = 0$

5. $x^2 - 7x + 13 = 0$

6. $2x^2 - x + 5 = 0$

7. $15x^2 + 17x - 4 = 0$

8. $8x^2 + 18x - 5 = 0$

9. $3x^2 + 4x = 2$

10. $2x^2 - 6x = -1$

For Problems 11–50, use the quadratic formula to solve each of the quadratic equations. Check your solutions by using the sum and product relationships.

11. $x^2 + 2x - 1 = 0$

12. $x^2 + 4x - 1 = 0$

13. $n^2 + 5n - 3 = 0$

14. $n^2 + 3n - 2 = 0$

15. $a^2 - 8a = 4$

16. $a^2 - 6a = 2$

17. $n^2 + 5n + 8 = 0$

18. $2n^2 - 3n + 5 = 0$

19. $x^2 - 18x + 80 = 0$

20. $x^2 + 19x + 70 = 0$

21. $-y^2 = -9y + 5$

22. $-y^2 + 7y = 4$

23. $2x^2 + x - 4 = 0$

24. $2x^2 + 5x - 2 = 0$

25. $4x^2 + 2x + 1 = 0$

26. $3x^2 - 2x + 5 = 0$

27. $3a^2 - 8a + 2 = 0$

28. $2a^2 - 6a + 1 = 0$

29. $-2n^2 + 3n + 5 = 0$

30. $-3n^2 - 11n + 4 = 0$

31. $3x^2 + 19x + 20 = 0$

32. $2x^2 - 17x + 30 = 0$

33. $36n^2 - 60n + 25 = 0$

34. $9n^2 + 42n + 49 = 0$

35. $4x^2 - 2x = 3$

36. $6x^2 - 4x = 3$

37. $5x^2 - 13x = 0$

38. $7x^2 + 12x = 0$

39. $3x^2 = 5$

40. $4x^2 = 3$

41. $6t^2 + t - 3 = 0$

42. $2t^2 + 6t - 3 = 0$

43. $n^2 + 32n + 252 = 0$

44. $n^2 - 4n - 192 = 0$

45. $12x^2 - 73x + 110 = 0$

46. $6x^2 + 11x - 255 = 0$

47. $-2x^2 + 4x - 3 = 0$

48. $-2x^2 + 6x - 5 = 0$

49. $-6x^2 + 2x + 1 = 0$

50. $-2x^2 + 4x + 1 = 0$

■ ■ ■ THOUGHTS INTO WORDS

51. Your friend states that the equation $-2x^2 + 4x - 1 = 0$ must be changed to $2x^2 - 4x + 1 = 0$ (by multiplying both sides by -1) before the quadratic formula can be applied. Is she right about this? If not, how would you convince her she is wrong?

52. Another of your friends claims that the quadratic formula can be used to solve the equation $x^2 - 9 = 0$. How would you react to this claim?

53. Why must we change the equation $3x^2 - 2x = 4$ to $3x^2 - 2x - 4 = 0$ before applying the quadratic formula?

■ ■ ■ FURTHER INVESTIGATIONS

The solution set for $x^2 - 4x - 37 = 0$ is $\{2 \pm \sqrt{41}\}$. With a calculator, we found a rational approximation, to the nearest one-thousandth, for each of these solutions.

$$2 - \sqrt{41} = -4.403 \quad \text{and} \quad 2 + \sqrt{41} = 8.403$$

Thus the solution set is $\{-4.403, 8.403\}$, with the answers rounded to the nearest one-thousandth.

Solve each of the equations in Problems 54–63, expressing solutions to the nearest one-thousandth.

54. $x^2 - 6x - 10 = 0$

55. $x^2 - 16x - 24 = 0$

56. $x^2 + 6x - 44 = 0$

57. $x^2 + 10x - 46 = 0$

58. $x^2 + 8x + 2 = 0$

59. $x^2 + 9x + 3 = 0$

60. $4x^2 - 6x + 1 = 0$

61. $5x^2 - 9x + 1 = 0$

62. $2x^2 - 11x - 5 = 0$

63. $3x^2 - 12x - 10 = 0$

For Problems 64–66, use the discriminant to help solve each problem.

64. Determine k so that the solutions of $x^2 - 2x + k = 0$ are complex but nonreal.

65. Determine k so that $4x^2 - kx + 1 = 0$ has two equal real solutions.

66. Determine k so that $3x^2 - kx - 2 = 0$ has real solutions.

6.5 More Quadratic Equations and Applications

Which method should be used to solve a particular quadratic equation? There is no hard and fast answer to that question; it depends on the type of equation and on your personal preference. In the following examples we will state reasons for choosing a specific technique. However, keep in mind that usually this is a decision you must make as the need arises. That's why you need to be familiar with the strengths and weaknesses of each method.

E X A M P L E   1

Solve $2x^2 - 3x - 1 = 0$.

Solution

Because of the leading coefficient of 2 and the constant term of -1, there are very few factoring possibilities to consider. Therefore, with such problems, first try the factoring approach. Unfortunately, this particular polynomial is not factorable using integers. Let's use the quadratic formula to solve the equation.

$$x = \frac{-b \pm \sqrt{b^2 - 4ac}}{2a}$$

$$x = \frac{-(-3) \pm \sqrt{(-3)^2 - 4(2)(-1)}}{2(2)}$$

$$x = \frac{3 \pm \sqrt{9 + 8}}{4}$$

$$x = \frac{3 \pm \sqrt{17}}{4}$$

Check

We can use the sum-of-roots and the product-of-roots relationships for our checking purposes.

Sum of roots $\dfrac{3 + \sqrt{17}}{4} + \dfrac{3 - \sqrt{17}}{4} = \dfrac{6}{4} = \dfrac{3}{2}$ and $-\dfrac{b}{a} = -\dfrac{-3}{2} = \dfrac{3}{2}$

Product of roots $\left(\dfrac{3 + \sqrt{17}}{4}\right)\left(\dfrac{3 - \sqrt{17}}{4}\right) = \dfrac{9 - 17}{16} = -\dfrac{8}{16} = -\dfrac{1}{2}$ and

$$\frac{c}{a} = \frac{-1}{2} = -\frac{1}{2}$$

The solution set is $\left\{\dfrac{3 \pm \sqrt{17}}{4}\right\}$.

■

E X A M P L E   2

Solve $\dfrac{3}{n} + \dfrac{10}{n + 6} = 1$.

Solution

$$\frac{3}{n} + \frac{10}{n + 6} = 1, \qquad n \neq 0 \text{ and } n \neq -6$$

$$n(n + 6)\left(\frac{3}{n} + \frac{10}{n + 6}\right) = 1(n)(n + 6)$$ Multiply both sides by $n(n + 6)$, which is the LCD.

$$3(n + 6) + 10n = n(n + 6)$$
$$3n + 18 + 10n = n^2 + 6n$$
$$13n + 18 = n^2 + 6n$$
$$0 = n^2 - 7n - 18$$

This equation is an easy one to consider for possible factoring, and it factors as follows:

$$0 = (n - 9)(n + 2)$$
$$n - 9 = 0 \quad \text{or} \quad n + 2 = 0$$
$$n = 9 \quad \text{or} \quad n = -2$$

✔ **Check**

Substituting 9 and -2 back into the original equation, we obtain

$$\frac{3}{n} + \frac{10}{n + 6} = 1 \qquad\qquad \frac{3}{n} + \frac{10}{n + 6} = 1$$

$$\frac{3}{9} + \frac{10}{9 + 6} \overset{?}{=} 1 \qquad\qquad \frac{3}{-2} + \frac{10}{-2 + 6} \overset{?}{=} 1$$

$$\frac{1}{3} + \frac{10}{15} \overset{?}{=} 1 \qquad \text{or} \qquad -\frac{3}{2} + \frac{10}{4} \overset{?}{=} 1$$

$$\frac{1}{3} + \frac{2}{3} \overset{?}{=} 1 \qquad\qquad -\frac{3}{2} + \frac{5}{2} \overset{?}{=} 1$$

$$1 = 1 \qquad\qquad \frac{2}{2} = 1$$

The solution set is $\{-2, 9\}$. ∎

We should make two comments about Example 2. First, note the indication of the initial restrictions $n \neq 0$ and $n \neq -6$. Remember that we need to do this when solving fractional equations. Second, the sum-of-roots and product-of-roots relationships were not used for checking purposes in this problem. Those relationships would check the validity of our work only from the step $0 = n^2 - 7n - 18$ to the finish. In other words, an error made in changing the original equation to quadratic form would not be detected by checking the sum and product of potential roots. With such a problem, the only *absolute check* is to substitute the potential solutions back into the original equation.

E X A M P L E   3

Solve $x^2 + 22x + 112 = 0$.

Solution

The size of the constant term makes the factoring approach a little cumbersome for this problem. Furthermore, because the leading coefficient is 1 and the coefficient of the x term is even, the method of completing the square will work effectively.

$$x^2 + 22x + 112 = 0$$

$$x^2 + 22x = -112$$

$$x^2 + 22x + 121 = -112 + 121$$

$$(x + 11)^2 = 9$$

$$x + 11 = \pm\sqrt{9}$$

$$x + 11 = \pm 3$$

$$x + 11 = 3 \quad \text{or} \quad x + 11 = -3$$

$$x = -8 \quad \text{or} \quad x = -14$$

✔ **Check**

Sum of roots $-8 + (-14) = -22$ and $-\dfrac{b}{a} = -22$

Product of roots $(-8)(-14) = 112$ and $\dfrac{c}{a} = 112$

The solution set is $\{-14, -8\}$.   ■

E X A M P L E   4

Solve $x^4 - 4x^2 - 96 = 0$.

Solution

An equation such as $x^4 - 4x^2 - 96 = 0$ is not a quadratic equation, but we can solve it using the techniques that we use on quadratic equations. That is, we can factor the polynomial and apply the property "$ab = 0$ if and only if $a = 0$ or $b = 0$" as follows:

$$x^4 - 4x^2 - 96 = 0$$

$$(x^2 - 12)(x^2 + 8) = 0$$

$$x^2 - 12 = 0 \quad \text{or} \quad x^2 + 8 = 0$$

$$x^2 = 12 \quad \text{or} \quad x^2 = -8$$

$$x = \pm\sqrt{12} \quad \text{or} \quad x = \pm\sqrt{-8}$$

$$x = \pm 2\sqrt{3} \quad \text{or} \quad x = \pm 2i\sqrt{2}$$

The solution set is $\{\pm 2\sqrt{3}, \pm 2i\sqrt{2}\}$. (We will leave the check for this problem for you to do!)   ■

Remark: Another approach to Example 4 would be to substitute y for x^2 and y^2 for x^4. The equation $x^4 - 4x^2 - 96 = 0$ becomes the quadratic equation

$y^2 - 4y - 96 = 0$. Thus we say that $x^4 - 4x^2 - 96 = 0$ is of *quadratic form*. Then we could solve the quadratic equation $y^2 - 4y - 96 = 0$ and use the equation $y = x^2$ to determine the solutions for x.

■ Applications

Before we conclude this section with some word problems that can be solved using quadratic equations, let's restate the suggestions we made in an earlier chapter for solving word problems.

Suggestions for Solving Word Problems

1. Read the problem carefully, and make certain that you understand the meanings of all the words. Be especially alert for any technical terms used in the statement of the problem.
2. Read the problem a second time (perhaps even a third time) to get an overview of the situation being described and to determine the known facts, as well as what is to be found.
3. Sketch any figure, diagram, or chart that might be helpful in analyzing the problem.
4. Choose a meaningful variable to represent an unknown quantity in the problem (perhaps *l*, if the length of a rectangle is an unknown quantity), and represent any other unknowns in terms of that variable.
5. Look for a guideline that you can use to set up an equation. A guideline might be a formula such as $A = lw$ or a relationship such as "the fractional part of a job done by Bill plus the fractional part of the job done by Mary equals the total job."
6. Form an equation that contains the variable and that translates the conditions of the guideline from English to algebra.
7. Solve the equation and use the solutions to determine all facts requested in the problem.
8. **Check all answers back into the original statement of the problem**.

Keep these suggestions in mind as we now consider some word problems.

PROBLEM 1

A page for a magazine contains 70 square inches of type. The height of a page is twice the width. If the margin around the type is to be 2 inches uniformly, what are the dimensions of a page?

Solution

Let x represent the width of a page. Then $2x$ represents the height of a page. Now let's draw and label a model of a page (Figure 6.8).

Width of   Height of   Area of
typed   typed   typed
material   material   material

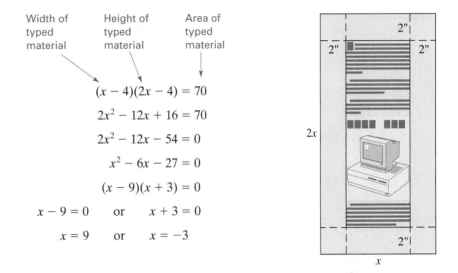

$$(x - 4)(2x - 4) = 70$$

$$2x^2 - 12x + 16 = 70$$

$$2x^2 - 12x - 54 = 0$$

$$x^2 - 6x - 27 = 0$$

$$(x - 9)(x + 3) = 0$$

$$x - 9 = 0 \quad \text{or} \quad x + 3 = 0$$

$$x = 9 \quad \text{or} \quad x = -3$$

Figure 6.8

Disregard the negative solution; the page must be 9 inches wide, and its height is $2(9) = 18$ inches.   ■

Let's use our knowledge of quadratic equations to analyze some applications of the business world. For example, if P dollars is invested at r rate of interest compounded annually for t years, then the amount of money, A, accumulated at the end of t years is given by the formula

$$A = P(1 + r)^t$$

This compound interest formula serves as a guideline for the next problem.

PROBLEM   2

Suppose that $100 is invested at a certain rate of interest compounded annually for 2 years. If the accumulated value at the end of 2 years is $121, find the rate of interest.

Solution

Let r represent the rate of interest. Substitute the known values into the compound interest formula to yield

$$A = P(1 + r)^t$$

$$121 = 100(1 + r)^2$$

Solving this equation, we obtain

$$\frac{121}{100} = (1 + r)^2$$

$$\pm\sqrt{\frac{121}{100}} = (1 + r)$$

$$\pm\frac{11}{10} = 1 + r$$

$$1 + r = \frac{11}{10} \qquad \text{or} \qquad 1 + r = -\frac{11}{10}$$

$$r = -1 + \frac{11}{10} \qquad \text{or} \qquad r = -1 - \frac{11}{10}$$

$$r = \frac{1}{10} \qquad \text{or} \qquad r = -\frac{21}{10}$$

We must disregard the negative solution, so that $r = \dfrac{1}{10}$ is the only solution.

Change $\dfrac{1}{10}$ to a percent, and the rate of interest is 10%. ∎

PROBLEM 3

On a 130-mile trip from Orlando to Sarasota, Roberto encountered a heavy thunderstorm for the last 40 miles of the trip. During the thunderstorm he averaged 20 miles per hour slower than before the storm. The entire trip took $2\dfrac{1}{2}$ hours. How fast did he travel before the storm?

Solution

Let x represent Roberto's rate before the thunderstorm. Then $x - 20$ represents his speed during the thunderstorm. Because $t = \dfrac{d}{r}$, then $\dfrac{90}{x}$ represents the time traveling before the storm, and $\dfrac{40}{x - 20}$ represents the time traveling during the storm. The following guideline sums up the situation.

Time traveling before the storm	Plus	Time traveling after the storm	Equals	Total time
↓		↓		↓
$\dfrac{90}{x}$	$+$	$\dfrac{40}{x - 20}$	$=$	$\dfrac{5}{2}$

Solving this equation, we obtain

$$2x(x - 20)\left(\frac{90}{x} + \frac{40}{x - 20}\right) = 2x(x - 20)\left(\frac{5}{2}\right)$$

$$2x(x - 20)\left(\frac{90}{x}\right) + 2x(x - 20)\left(\frac{40}{x - 20}\right) = 2x(x - 20)\left(\frac{5}{2}\right)$$

$$180(x - 20) + 2x(40) = 5x(x - 20)$$

$$180x - 3600 + 80x = 5x^2 - 100x$$
$$0 = 5x^2 - 360x + 3600$$
$$0 = 5(x^2 - 72x + 720)$$
$$0 = 5(x - 60)(x - 12)$$
$$x - 60 = 0 \quad \text{or} \quad x - 12 = 0$$
$$x = 60 \quad \text{or} \quad x = 12$$

We discard the solution of 12 because it would be impossible to drive 20 miles per hour slower than 12 miles per hour; thus Roberto's rate before the thunderstorm was 60 miles per hour. ∎

P R O B L E M   4

A businesswoman bought a parcel of land on speculation for $120,000. She subdivided the land into lots, and when she had sold all but 18 lots at a profit of $6000 per lot, she had regained the entire cost of the land. How many lots were sold and at what price per lot?

Solution

Let x represent the number of lots sold. Then $x + 18$ represents the total number of lots. Therefore, $\dfrac{120,000}{x}$ represents the selling price per lot, and $\dfrac{120,000}{x + 18}$ represents the cost per lot. The following equation sums up the situation.

Selling price per lot	Equals	Cost per lot	Plus	$6000
↓		↓		↓
$\dfrac{120,000}{x}$	$=$	$\dfrac{120,000}{x + 18}$	$+$	6000

Solving this equation, we obtain

$$x(x + 18)\left(\frac{120,000}{x}\right) = \left(\frac{120,000}{x + 18} + 6000\right)(x)(x + 18)$$
$$120,000(x + 18) = 120,000x + 6000x(x + 18)$$
$$120,000x + 2,160,000 = 120,000x + 6000x^2 + 108,000x$$
$$0 = 6000x^2 + 108,000x - 2,160,000$$
$$0 = x^2 + 18x - 360$$

The method of completing the square works very well with this equation.

$$x^2 + 18x = 360$$
$$x^2 + 18x + 81 = 441$$
$$(x + 9)^2 = 441$$
$$x + 9 = \pm\sqrt{441}$$

$$x + 9 = \pm 21$$

$$x + 9 = 21 \quad \text{or} \quad x + 9 = -21$$

$$x = 12 \quad \text{or} \quad x = -30$$

We discard the negative solution; thus 12 lots were sold at $\dfrac{120,000}{x} = \dfrac{120,000}{12} =$ $10,000 per lot. ∎

PROBLEM 5

Barry bought a number of shares of stock for $600. A week later the value of the stock had increased $3 per share, and he sold all but 10 shares and regained his original investment of $600. How many shares did he sell and at what price per share?

Solution

Let s represent the number of shares Barry sold. Then $s + 10$ represents the number of shares purchased. Therefore, $\dfrac{600}{s}$ represents the selling price per share, and $\dfrac{600}{s + 10}$ represents the cost per share.

$$\underbrace{\frac{600}{s}}_{\substack{\text{Selling price} \\ \text{per share}}} = \underbrace{\frac{600}{s + 10}}_{\substack{\text{Cost per share}}} + 3$$

Solving this equation yields

$$s(s + 10)\left(\frac{600}{s}\right) = \left(\frac{600}{s + 10} + 3\right)(s)(s + 10)$$

$$600(s + 10) = 600s + 3s(s + 10)$$

$$600s + 6000 = 600s + 3s^2 + 30s$$

$$0 = 3s^2 + 30s - 6000$$

$$0 = s^2 + 10s - 2000$$

Use the quadratic formula to obtain

$$s = \frac{-10 \pm \sqrt{10^2 - 4(1)(-2000)}}{2(1)}$$

$$s = \frac{-10 \pm \sqrt{100 + 8000}}{2}$$

$$s = \frac{-10 \pm \sqrt{8100}}{2}$$

$$s = \frac{-10 \pm 90}{2}$$

$$s = \frac{-10 + 90}{2} \quad \text{or} \quad s = \frac{-10 - 90}{2}$$

$$s = 40 \quad \text{or} \quad s = -50$$

We discard the negative solution, and we know that 40 shares were sold at

$$\frac{600}{s} = \frac{600}{40} = \$15 \text{ per share.} \qquad \blacksquare$$

This next problem set contains a large variety of word problems. Not only are there some business applications similar to those we discussed in this section, but there are also more problems of the types we discussed in Chapters 3 and 4. Try to give them your best shot without referring to the examples in earlier chapters.

Problem Set 6.5

For Problems 1–20, solve each quadratic equation using the method that seems most appropriate to you.

1. $x^2 - 4x - 6 = 0$

2. $x^2 - 8x - 4 = 0$

3. $3x^2 + 23x - 36 = 0$

4. $n^2 + 22n + 105 = 0$

5. $x^2 - 18x = 9$

6. $x^2 + 20x = 25$

7. $2x^2 - 3x + 4 = 0$

8. $3y^2 - 2y + 1 = 0$

9. $135 + 24n + n^2 = 0$

10. $28 - x - 2x^2 = 0$

11. $(x - 2)(x + 9) = -10$

12. $(x + 3)(2x + 1) = -3$

13. $2x^2 - 4x + 7 = 0$

14. $3x^2 - 2x + 8 = 0$

15. $x^2 - 18x + 15 = 0$

16. $x^2 - 16x + 14 = 0$

17. $20y^2 + 17y - 10 = 0$

18. $12x^2 + 23x - 9 = 0$

19. $4t^2 + 4t - 1 = 0$

20. $5t^2 + 5t - 1 = 0$

For Problems 21–40, solve each equation.

21. $n + \dfrac{3}{n} = \dfrac{19}{4}$

22. $n - \dfrac{2}{n} = -\dfrac{7}{3}$

23. $\dfrac{3}{x} + \dfrac{7}{x - 1} = 1$

24. $\dfrac{2}{x} + \dfrac{5}{x + 2} = 1$

25. $\dfrac{12}{x - 3} + \dfrac{8}{x} = 14$

26. $\dfrac{16}{x + 5} - \dfrac{12}{x} = -2$

27. $\dfrac{3}{x - 1} - \dfrac{2}{x} = \dfrac{5}{2}$

28. $\dfrac{4}{x + 1} + \dfrac{2}{x} = \dfrac{5}{3}$

29. $\dfrac{6}{x} + \dfrac{40}{x + 5} = 7$

30. $\dfrac{12}{t} + \dfrac{18}{t + 8} = \dfrac{9}{2}$

31. $\dfrac{5}{n - 3} - \dfrac{3}{n + 3} = 1$

32. $\dfrac{3}{t + 2} + \dfrac{4}{t - 2} = 2$

33. $x^4 - 18x^2 + 72 = 0$

34. $x^4 - 21x^2 + 54 = 0$

35. $3x^4 - 35x^2 + 72 = 0$

36. $5x^4 - 32x^2 + 48 = 0$

37. $3x^4 + 17x^2 + 20 = 0$

38. $4x^4 + 11x^2 - 45 = 0$

39. $6x^4 - 29x^2 + 28 = 0$

40. $6x^4 - 31x^2 + 18 = 0$

For Problems 41–70, set up an equation and solve each problem.

41. Find two consecutive whole numbers such that the sum of their squares is 145.

42. Find two consecutive odd whole numbers such that the sum of their squares is 74.

43. Two positive integers differ by 3, and their product is 108. Find the numbers.

44. Suppose that the sum of two numbers is 20, and the sum of their squares is 232. Find the numbers.

45. Find two numbers such that their sum is 10 and their product is 22.

46. Find two numbers such that their sum is 6 and their product is 7.

47. Suppose that the sum of two whole numbers is 9, and the sum of their reciprocals is $\frac{1}{2}$. Find the numbers.

48. The difference between two whole numbers is 8, and the difference between their reciprocals is $\frac{1}{6}$. Find the two numbers.

49. The sum of the lengths of the two legs of a right triangle is 21 inches. If the length of the hypotenuse is 15 inches, find the length of each leg.

50. The length of a rectangular floor is 1 meter less than twice its width. If a diagonal of the rectangle is 17 meters, find the length and width of the floor.

51. A rectangular plot of ground measuring 12 meters by 20 meters is surrounded by a sidewalk of a uniform width (see Figure 6.9). The area of the sidewalk is 68 square meters. Find the width of the walk.

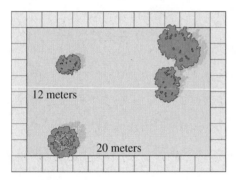

12 meters

20 meters

Figure 6.9

52. A 5-inch by 7-inch picture is surrounded by a frame of uniform width. The area of the picture and frame together is 80 square inches. Find the width of the frame.

53. The perimeter of a rectangle is 44 inches, and its area is 112 square inches. Find the length and width of the rectangle.

54. A rectangular piece of cardboard is 2 units longer than it is wide. From each of its corners a square piece 2 units on a side is cut out. The flaps are then turned up to form an open box that has a volume of 70 cubic units. Find the length and width of the original piece of cardboard.

55. Charlotte's time to travel 250 miles is 1 hour more than Lorraine's time to travel 180 miles. Charlotte drove 5 miles per hour faster than Lorraine. How fast did each one travel?

56. Larry's time to travel 156 miles is 1 hour more than Terrell's time to travel 108 miles. Terrell drove 2 miles per hour faster than Larry. How fast did each one travel?

57. On a 570-mile trip, Andy averaged 5 miles per hour faster for the last 240 miles than he did for the first 330 miles. The entire trip took 10 hours. How fast did he travel for the first 330 miles?

58. On a 135-mile bicycle excursion, Maria averaged 5 miles per hour faster for the first 60 miles than she did for the last 75 miles. The entire trip took 8 hours. Find her rate for the first 60 miles.

59. It takes Terry 2 hours longer to do a certain job than it takes Tom. They worked together for 3 hours; then Tom left and Terry finished the job in 1 hour. How long would it take each of them to do the job alone?

60. Suppose that Arlene can mow the entire lawn in 40 minutes less time with the power mower than she can with the push mower. One day the power mower broke down after she had been mowing for 30 minutes. She finished the lawn with the push mower in 20 minutes. How long does it take Arlene to mow the entire lawn with the power mower?

61. A student did a word processing job for $24. It took him 1 hour longer than he expected, and therefore he earned $4 per hour less than he anticipated. How long did he expect that it would take to do the job?

62. A group of students agreed that each would chip in the same amount to pay for a party that would cost $100. Then they found 5 more students interested in the party and in sharing the expenses. This decreased the amount each had to pay by $1. How many students were involved in the party and how much did each student have to pay?

63. A group of students agreed that each would contribute the same amount to buy their favorite teacher an $80 birthday gift. At the last minute, 2 of the students decided not to chip in. This increased the amount that the remaining students had to pay by $2 per student. How many students actually contributed to the gift?

64. A retailer bought a number of special mugs for $48. She decided to keep two of the mugs for herself but then had to change the price to $3 a mug above the original cost per mug. If she sells the remaining mugs for $70, how many mugs did she buy and at what price per mug did she sell them?

65. Tony bought a number of shares of stock for $720. A month later the value of the stock increased by $8 per share, and he sold all but 20 shares and received $800. How many shares did he sell and at what price per share?

66. The formula $D = \dfrac{n(n-3)}{2}$ yields the number of diagonals, D, in a polygon of n sides. Find the number of sides of a polygon that has 54 diagonals.

67. The formula $S = \dfrac{n(n+1)}{2}$ yields the sum, S, of the first n natural numbers $1, 2, 3, 4, \ldots$. How many consecutive natural numbers starting with 1 will give a sum of 1275?

68. At a point 16 yards from the base of a tower, the distance to the top of the tower is 4 yards more than the height of the tower (see Figure 6.10). Find the height of the tower.

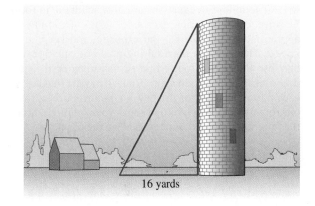

16 yards

Figure 6.10

69. Suppose that $500 is invested at a certain rate of interest compounded annually for 2 years. If the accumulated value at the end of 2 years is $594.05, find the rate of interest.

70. Suppose that $10,000 is invested at a certain rate of interest compounded annually for 2 years. If the accumulated value at the end of 2 years is $12,544, find the rate of interest.

■ ■ ■ THOUGHTS INTO WORDS

71. How would you solve the equation $x^2 - 4x = 252$? Explain your choice of the method that you would use.

72. Explain how you would solve $(x - 2)(x - 7) = 0$ and also how you would solve $(x - 2)(x - 7) = 4$.

73. One of our problem-solving suggestions is to look for a guideline that can be used to help determine an equation. What does this suggestion mean to you?

74. Can a quadratic equation with integral coefficients have exactly one nonreal complex solution? Explain your answer.

■ ■ ■ FURTHER INVESTIGATIONS

For Problems 75–81, solve each equation.

75. $x - 9\sqrt{x} + 18 = 0$ [*Hint:* Let $y = \sqrt{x}$.]

76. $x - 4\sqrt{x} + 3 = 0$

77. $x + \sqrt{x} - 2 = 0$

78. $x^{\frac{2}{3}} + x^{\frac{1}{3}} - 6 = 0$ [*Hint:* Let $y = x^{\frac{1}{3}}$.]

79. $6x^{\frac{2}{3}} - 5x^{\frac{1}{3}} - 6 = 0$

80. $x^{-2} + 4x^{-1} - 12 = 0$

81. $12x^{-2} - 17x^{-1} - 5 = 0$

The following equations are also quadratic in form. To solve, begin by raising each side of the equation to the appropriate power so that the exponent will become an integer. Then, to solve the resulting quadratic equation, you may use the square-root property, factoring, or the quadratic formula, as is most appropriate. Be aware that raising each side of the equation to a power may introduce extraneous roots; therefore, be sure to check your solutions. Study the following example before you begin the problems.

Solve

$$(x + 3)^{\frac{2}{3}} = 1$$

$$\left[(x + 3)^{\frac{2}{3}}\right]^3 = 1^3 \qquad \text{Raise both sides to the third power.}$$

$$(x + 3)^2 = 1$$

$$x^2 + 6x + 9 = 1$$

$$x^2 + 6x + 8 = 0$$

$$(x + 4)(x + 2) = 0$$

$$x + 4 = 0 \quad \text{or} \quad x + 2 = 0$$

$$x = -4 \quad \text{or} \quad x = -2$$

Both solutions do check. The solution set is $\{-4, -2\}$.

For problems 82–90, solve each equation.

82. $(5x + 6)^{\frac{1}{2}} = x$

83. $(3x + 4)^{\frac{1}{2}} = x$

84. $x^{\frac{2}{3}} = 2$

85. $x^{\frac{2}{5}} = 2$

86. $(2x + 6)^{\frac{1}{2}} = x$

87. $(2x - 4)^{\frac{2}{3}} = 1$

88. $(4x + 5)^{\frac{2}{3}} = 2$

89. $(6x + 7)^{\frac{1}{2}} = x + 2$

90. $(5x + 21)^{\frac{1}{2}} = x + 3$

6.6   Quadratic and Other Nonlinear Inequalities

We refer to the equation $ax^2 + bx + c = 0$ as the standard form of a quadratic equation in one variable. Similarly, the following forms express **quadratic inequalities** in one variable.

$$ax^2 + bx + c > 0 \qquad\qquad ax^2 + bx + c < 0$$
$$ax^2 + bx + c \geq 0 \qquad\qquad ax^2 + bx + c \leq 0$$

We can use the number line very effectively to help solve quadratic inequalities where the quadratic polynomial is factorable. Let's consider some examples to illustrate the procedure.

EXAMPLE 1

Solve and graph the solutions for $x^2 + 2x - 8 > 0$.

Solution

First, let's factor the polynomial.

$$x^2 + 2x - 8 > 0$$

$$(x + 4)(x - 2) > 0$$

On a number line (Figure 6.11), we indicate that at $x = 2$ and $x = -4$, the product $(x + 4)(x - 2)$ equals zero. The numbers -4 and 2 divide the number line into three intervals: (1) the numbers less than -4, (2) the numbers between -4 and 2, and (3) the numbers greater than 2. We can choose a **test number** from each of these intervals and see how it affects the signs of the factors $x + 4$ and $x - 2$ and,

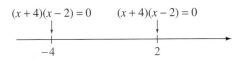

Figure 6.11

consequently, the sign of the product of these factors. For example, if $x < -4$ (try $x = -5$), then $x + 4$ is negative and $x - 2$ is negative, so their product is positive. If $-4 < x < 2$ (try $x = 0$), then $x + 4$ is positive and $x - 2$ is negative, so their product is negative. If $x > 2$ (try $x = 3$), then $x + 4$ is positive and $x - 2$ is positive, so their product is positive. This information can be conveniently arranged using a number line, as shown in Figure 6.12. Note the open circles at -4 and 2 to indicate that they are not included in the solution set.

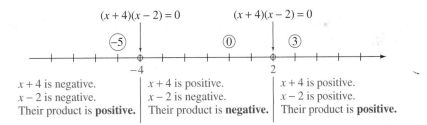

Figure 6.12

Thus the given inequality, $x^2 + 2x - 8 > 0$, is satisfied by numbers less than -4 along with numbers greater than 2. Using interval notation, the solution set is $(-\infty, -4) \cup (2, \infty)$. These solutions can be shown on a number line (Figure 6.13).

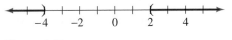

Figure 6.13

We refer to numbers such as -4 and 2 in the preceding example (where the given polynomial or algebraic expression equals zero or is undefined) as **critical numbers**. Let's consider some additional examples that make use of critical numbers and test numbers.

EXAMPLE 2

Solve and graph the solutions for $x^2 + 2x - 3 \leq 0$.

Solution

First, factor the polynomial.

$$x^2 + 2x - 3 \leq 0$$

$$(x + 3)(x - 1) \leq 0$$

Second, locate the values for which $(x + 3)(x - 1)$ equals zero. We put dots at -3 and 1 to remind ourselves that these two numbers are to be included in the solution set because the given statement includes equality. Now let's choose a test number from each of the three intervals, and record the sign behavior of the factors $(x + 3)$ and $(x - 1)$ (Figure 6.14).

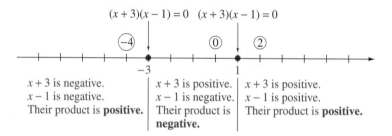

Figure 6.14

Therefore, the solution set is $[-3, 1]$, and it can be graphed as in Figure 6.15.

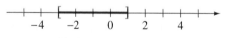

Figure 6.15 ∎

Examples 1 and 2 have indicated a systematic approach for solving quadratic inequalities where the polynomial is factorable. This same type of number line analysis can also be used to solve indicated quotients such as $\dfrac{x + 1}{x - 5} > 0$.

EXAMPLE 3

Solve and graph the solutions for $\dfrac{x + 1}{x - 5} > 0$.

Solution

First, indicate that at $x = -1$ the given quotient equals zero, and at $x = 5$ the quotient is undefined. Second, choose test numbers from each of the three intervals, and record the sign behavior of $(x + 1)$ and $(x - 5)$ as in Figure 6.16.

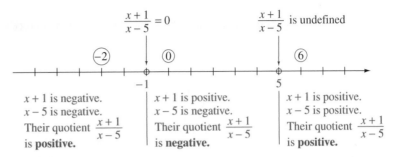

Figure 6.16

Therefore, the solution set is $(-\infty, -1) \cup (5, \infty)$, and its graph is shown in Figure 6.17.

Figure 6.17

EXAMPLE 4

Solve $\dfrac{x + 2}{x + 4} \leq 0$.

Solution

The indicated quotient equals zero at $x = -2$ and is undefined at $x = -4$. (Note that -2 is to be included in the solution set, but -4 is not to be included.) Now let's choose some test numbers and record the sign behavior of $(x + 2)$ and $(x + 4)$ as in Figure 6.18.

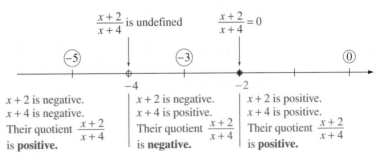

Figure 6.18

Therefore, the solution set is $(-4, -2]$.

The final example illustrates that sometimes we need to change the form of the given inequality before we use the number line analysis.

Solve $\dfrac{x}{x+2} \geq 3$.

Solution

First, let's change the form of the given inequality as follows:

$$\frac{x}{x+2} \geq 3$$

$$\frac{x}{x+2} - 3 \geq 0 \qquad \text{Add } -3 \text{ to both sides.}$$

$$\frac{x - 3(x+2)}{x+2} \geq 0 \qquad \text{Express the left side over a common denominator.}$$

$$\frac{x - 3x - 6}{x+2} \geq 0$$

$$\frac{-2x - 6}{x+2} \geq 0$$

Now we can proceed as we did with the previous examples. If $x = -3$, then $\dfrac{-2x-6}{x+2}$ equals zero; and if $x = -2$, then $\dfrac{-2x-6}{x+2}$ is undefined. Then, choosing test numbers, we can record the sign behavior of $(-2x - 6)$ and $(x + 2)$ as in Figure 6.19.

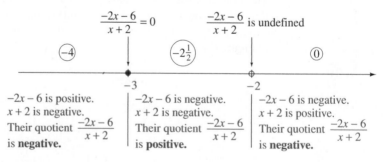

Figure 6.19

Therefore, the solution set is $[-3, -2)$. Perhaps you should check a few numbers from this solution set back into the original inequality! ■

Problem Set 6.6

For Problems 1–20, solve each inequality and graph its solution set on a number line.

1. $(x + 2)(x - 1) > 0$

2. $(x - 2)(x + 3) > 0$

3. $(x + 1)(x + 4) < 0$

4. $(x - 3)(x - 1) < 0$

5. $(2x - 1)(3x + 7) \geq 0$

6. $(3x + 2)(2x - 3) \geq 0$

7. $(x + 2)(4x - 3) \leq 0$

8. $(x - 1)(2x - 7) \leq 0$

9. $(x + 1)(x - 1)(x - 3) > 0$

10. $(x + 2)(x + 1)(x - 2) > 0$

11. $x(x + 2)(x - 4) \leq 0$

12. $x(x + 3)(x - 3) \leq 0$

13. $\dfrac{x + 1}{x - 2} > 0$

14. $\dfrac{x - 1}{x + 2} > 0$

15. $\dfrac{x - 3}{x + 2} < 0$

16. $\dfrac{x + 2}{x - 4} < 0$

17. $\dfrac{2x - 1}{x} \geq 0$

18. $\dfrac{x}{3x + 7} \geq 0$

19. $\dfrac{-x + 2}{x - 1} \leq 0$

20. $\dfrac{3 - x}{x + 4} \leq 0$

For Problems 21–56, solve each inequality.

21. $x^2 + 2x - 35 < 0$

22. $x^2 + 3x - 54 < 0$

23. $x^2 - 11x + 28 > 0$

24. $x^2 + 11x + 18 > 0$

25. $3x^2 + 13x - 10 \leq 0$

26. $4x^2 - x - 14 \leq 0$

27. $8x^2 + 22x + 5 \geq 0$

28. $12x^2 - 20x + 3 \geq 0$

29. $x(5x - 36) > 32$

30. $x(7x + 40) < 12$

31. $x^2 - 14x + 49 \geq 0$

32. $(x + 9)^2 \geq 0$

33. $4x^2 + 20x + 25 \leq 0$

34. $9x^2 - 6x + 1 \leq 0$

35. $(x + 1)(x - 3)^2 > 0$

36. $(x - 4)^2(x - 1) \leq 0$

37. $4 - x^2 < 0$

38. $2x^2 - 18 \geq 0$

39. $4(x^2 - 36) < 0$

40. $-4(x^2 - 36) \geq 0$

41. $5x^2 + 20 > 0$

42. $-3x^2 - 27 \geq 0$

43. $x^2 - 2x \geq 0$

44. $2x^2 + 6x < 0$

45. $3x^3 + 12x^2 > 0$

46. $2x^3 + 4x^2 \leq 0$

47. $\dfrac{2x}{x + 3} > 4$

48. $\dfrac{x}{x - 1} > 2$

49. $\dfrac{x - 1}{x - 5} \leq 2$

50. $\dfrac{x + 2}{x + 4} \leq 3$

51. $\dfrac{x + 2}{x - 3} > -2$

52. $\dfrac{x - 1}{x - 2} < -1$

53. $\dfrac{3x + 2}{x + 4} \leq 2$

54. $\dfrac{2x - 1}{x + 2} \geq -1$

55. $\dfrac{x + 1}{x - 2} < 1$

56. $\dfrac{x + 3}{x - 4} \geq 1$

■ ■ ■ **THOUGHTS INTO WORDS**

57. Explain how to solve the inequality $(x + 1)(x - 2)(x - 3) > 0$.

58. Explain how to solve the inequality $(x - 2)^2 > 0$ by inspection.

59. Your friend looks at the inequality $1 + \dfrac{1}{x} > 2$ and without any computation states that the solution set is all real numbers between 0 and 1. How can she do that?

60. Why is the solution set for $(x - 2)^2 \geq 0$ the set of all real numbers?

61. Why is the solution set for $(x - 2)^2 \leq 0$ the set $\{2\}$?

■ ■ ■ FURTHER INVESTIGATIONS

62. The product $(x - 2)(x + 3)$ is positive if both factors are negative *or* if both factors are positive. Therefore, we can solve $(x - 2)(x + 3) > 0$ as follows:

$$(x - 2 < 0 \text{ and } x + 3 < 0) \text{ or } (x - 2 > 0 \text{ and } x + 3 > 0)$$

$$(x < 2 \text{ and } x < -3) \text{ or } (x > 2 \text{ and } x > -3)$$

$$x < -3 \text{ or } x > 2$$

The solution set is $(-\infty, -3) \cup (2, \infty)$. Use this type of analysis to solve each of the following.

(a) $(x - 2)(x + 7) > 0$

(b) $(x - 3)(x + 9) \geq 0$

(c) $(x + 1)(x - 6) \leq 0$

(d) $(x + 4)(x - 8) < 0$

(e) $\dfrac{x + 4}{x - 7} > 0$

(f) $\dfrac{x - 5}{x + 8} \leq 0$

(6.1) A number of the form $a + bi$, where a and b are real numbers, and i is the imaginary unit defined by $i = \sqrt{-1}$, is a **complex number**.

Two complex numbers $a + bi$ and $c + di$ are said to be equal if and only if $a = c$ and $b = d$.

We describe addition and subtraction of complex numbers as follows:

$$(a + bi) + (c + di) = (a + c) + (b + d)i$$

$$(a + bi) - (c + di) = (a - c) + (b - d)i$$

We can represent a square root of any negative real number as the product of a real number and the imaginary unit i. That is,

$$\sqrt{-b} = i\sqrt{b}, \quad \text{where } b \text{ is a positive real number}$$

The product of two complex numbers conforms with the product of two binomials. The **conjugate** of $a + bi$ is $a - bi$. The product of a complex number and its conjugate is a real number. Therefore, conjugates are used to simplify expressions such as $\dfrac{4 + 3i}{5 - 2i}$, which indicate the quotient of two complex numbers.

(6.2) The **standard form for a quadratic equation** in one variable is

$$ax^2 + bx + c = 0$$

where a, b, and c are real numbers and $a \neq 0$.

Some quadratic equations can be solved by factoring and applying the property, $ab = 0$ if and only if $a = 0$ or $b = 0$.

Don't forget that applying the property, if $a = b$, then $a^n = b^n$ might produce extraneous solutions. Therefore, we *must* check all potential solutions.

We can solve some quadratic equations by applying the property, $x^2 = a$ if and only if $x = \pm\sqrt{a}$.

(6.3) To solve a quadratic equation of the form $x^2 + bx = k$ by **completing the square**, we (1) add $\left(\dfrac{b}{2}\right)^2$ to

both sides, (2) factor the left side, and (3) apply the property, $x^2 = a$ if and only if $x = \pm\sqrt{a}$.

(6.4) We can solve any quadratic equation of the form $ax^2 + bx + c = 0$ by the **quadratic formula**, which we usually state as

$$x = \frac{-b \pm \sqrt{b^2 - 4ac}}{2a}$$

The **discriminant**, $b^2 - 4ac$, can be used to determine the nature of the roots of a quadratic equation as follows:

1. If $b^2 - 4ac < 0$, then the equation has two nonreal complex solutions.

2. If $b^2 - 4ac = 0$, then the equation has two equal real solutions.

3. If $b^2 - 4ac > 0$, then the equation has two unequal real solutions.

If x_1 and x_2 are roots of a quadratic equation, then the following relationships exist.

$$x_1 + x_2 = -\frac{b}{a} \quad \text{and} \quad (x_1)(x_2) = \frac{c}{a}$$

These **sum-of-roots** and **product-of-roots relationships** can be used to check potential solutions of quadratic equations.

(6.5) To review the strengths and weaknesses of the three basic methods for solving a quadratic equation (factoring, completing the square, and the quadratic formula), go back over the examples in this section.

Keep the following suggestions in mind as you solve word problems.

1. Read the problem carefully.

2. Sketch any figure, diagram, or chart that might help you organize and analyze the problem.

3. Choose a meaningful variable.

4. Look for a guideline that can be used to set up an equation.

5. Form an equation that translates the guideline from English to algebra.

6. Solve the equation and use the solutions to determine all facts requested in the problem.

7. Check all answers back into the original statement of the problem.

(6.6) The number line, along with **critical numbers** and **test numbers**, provides a good basis for solving **quadratic inequalities** where the polynomial is factorable. We can use this same basic approach to solve inequalities, such as $\dfrac{3x+1}{x-4} > 0$, that indicate quotients.

Chapter 6   Review Problem Set

For Problems 1–8, perform the indicated operations and express the answers in the standard form of a complex number.

1. $(-7+3i)+(9-5i)$

2. $(4-10i)-(7-9i)$

3. $5i(3-6i)$

4. $(5-7i)(6+8i)$

5. $(-2-3i)(4-8i)$

6. $(4-3i)(4+3i)$

7. $\dfrac{4+3i}{6-2i}$

8. $\dfrac{-1-i}{-2+5i}$

For Problems 9–12, find the discriminant of each equation and determine whether the equation has (1) two nonreal complex solutions, (2) one real solution with a multiplicity of two, or (3) two real solutions. Do not solve the equations.

9. $4x^2-20x+25=0$

10. $5x^2-7x+31=0$

11. $7x^2-2x-14=0$

12. $5x^2-2x=4$

For Problems 13–31, solve each equation.

13. $x^2-17x=0$

14. $(x-2)^2=36$

15. $(2x-1)^2=-64$

16. $x^2-4x-21=0$

17. $x^2+2x-9=0$

18. $x^2-6x=-34$

19. $4\sqrt{x}=x-5$

20. $3n^2+10n-8=0$

21. $n^2-10n=200$

22. $3a^2+a-5=0$

23. $x^2-x+3=0$

24. $2x^2-5x+6=0$

25. $2a^2+4a-5=0$

26. $t(t+5)=36$

27. $x^2+4x+9=0$

28. $(x-4)(x-2)=80$

29. $\dfrac{3}{x}+\dfrac{2}{x+3}=1$

30. $2x^4-23x^2+56=0$

31. $\dfrac{3}{n-2}=\dfrac{n+5}{4}$

For Problems 32–35, solve each inequality and indicate the solution set on a number line graph.

32. $x^2+3x-10>0$

33. $2x^2+x-21\le 0$

34. $\dfrac{x-4}{x+6}\ge 0$

35. $\dfrac{2x-1}{x+1}>4$

For Problems 36–43, set up an equation and solve each problem.

36. Find two numbers whose sum is 6 and whose product is 2.

37. Sherry bought a number of shares of stock for $250. Six months later the value of the stock had increased by $5 per share, and she sold all but 5 shares and regained her original investment plus a profit of $50. How many shares did she sell and at what price per share?

38. Andre traveled 270 miles in 1 hour more time than it took Sandy to travel 260 miles. Sandy drove 7 miles per hour faster than Andre. How fast did each one travel?

39. The area of a square is numerically equal to twice its perimeter. Find the length of a side of the square.

40. Find two consecutive even whole numbers such that the sum of their squares is 164.

41. The perimeter of a rectangle is 38 inches, and its area is 84 square inches. Find the length and width of the rectangle.

42. It takes Billy 2 hours longer to do a certain job than it takes Reena. They worked together for 2 hours; then Reena left, and Billy finished the job in 1 hour. How long would it take each of them to do the job alone?

43. A company has a rectangular parking lot 40 meters wide and 60 meters long. The company plans to increase the area of the lot by 1100 square meters by adding a strip of equal width to one side and one end. Find the width of the strip to be added.

1. Find the product $(3 - 4i)(5 + 6i)$ and express the result in the standard form of a complex number.

2. Find the quotient $\dfrac{2 - 3i}{3 + 4i}$ and express the result in the standard form of a complex number.

For Problems 3–15, solve each equation.

3. $x^2 = 7x$

4. $(x - 3)^2 = 16$

5. $x^2 + 3x - 18 = 0$

6. $x^2 - 2x - 1 = 0$

7. $5x^2 - 2x + 1 = 0$

8. $x^2 + 30x = -224$

9. $(3x - 1)^2 + 36 = 0$

10. $(5x - 6)(4x + 7) = 0$

11. $(2x + 1)(3x - 2) = 55$

12. $n(3n - 2) = 40$

13. $x^4 + 12x^2 - 64 = 0$

14. $\dfrac{3}{x} + \dfrac{2}{x + 1} = 4$

15. $3x^2 - 2x - 3 = 0$

16. Does the equation $4x^2 + 20x + 25 = 0$ have (a) two nonreal complex solutions, (b) two equal real solutions, or (c) two unequal real solutions?

17. Does the equation $4x^2 - 3x = -5$ have (a) two nonreal complex solutions, (b) two equal real solutions, or (c) two unequal real solutions?

For Problems 18–20, solve each inequality and express the solution set using interval notation.

18. $x^2 - 3x - 54 \leq 0$

19. $\dfrac{3x - 1}{x + 2} > 0$

20. $\dfrac{x - 2}{x + 6} \geq 3$

For Problems 21–25, set up an equation and solve each problem.

21. A 24-foot ladder leans against a building and makes an angle of 60° with the ground. How far up on the building does the top of the ladder reach? Express your answer to the nearest tenth of a foot.

22. A rectangular plot of ground measures 16 meters by 24 meters. Find, to the nearest meter, the distance from one corner of the plot to the diagonally opposite corner.

23. Dana bought a number of shares of stock for a total of $3000. Three months later the stock had increased in value by $5 per share, and she sold all but 50 shares and regained her original investment of $3000. How many shares did she sell?

24. The perimeter of a rectangle is 41 inches and its area is 91 square inches. Find the length of its shortest side.

25. The sum of two numbers is 6 and their product is 4. Find the larger of the two numbers.

For Problems 1–5, evaluate each algebraic expression for the given values of the variables.

1. $\dfrac{4a^2b^3}{12a^3b}$ for $a = 5$ and $b = -8$

2. $\dfrac{\dfrac{1}{x} + \dfrac{1}{y}}{\dfrac{1}{x} - \dfrac{1}{y}}$ for $x = 4$ and $y = 7$

3. $\dfrac{3}{n} + \dfrac{5}{2n} - \dfrac{4}{3n}$ for $n = 25$

4. $\dfrac{4}{x - 1} - \dfrac{2}{x + 2}$ for $x = \dfrac{1}{2}$

5. $2\sqrt{2x + y} - 5\sqrt{3x - y}$ for $x = 5$ and $y = 6$

For Problems 6–17, perform the indicated operations and express the answers in simplified form.

6. $(3a^2b)(-2ab)(4ab^3)$

7. $(x + 3)(2x^2 - x - 4)$

8. $\dfrac{6xy^2}{14y} \cdot \dfrac{7x^2y}{8x}$

9. $\dfrac{a^2 + 6a - 40}{a^2 - 4a} \div \dfrac{2a^2 + 19a - 10}{a^3 + a^2}$

10. $\dfrac{3x + 4}{6} - \dfrac{5x - 1}{9}$

11. $\dfrac{4}{x^2 + 3x} + \dfrac{5}{x}$

12. $\dfrac{3n^2 + n}{n^2 + 10n + 16} \cdot \dfrac{2n^2 - 8}{3n^3 - 5n^2 - 2n}$

13. $\dfrac{3}{5x^2 + 3x - 2} - \dfrac{2}{5x^2 - 22x + 8}$

14. $\dfrac{y^3 - 7y^2 + 16y - 12}{y - 2}$

15. $(4x^3 - 17x^2 + 7x + 10) \div (4x - 5)$

16. $(3\sqrt{2} + 2\sqrt{5})(5\sqrt{2} - \sqrt{5})$

17. $(\sqrt{x} - 3\sqrt{y})(2\sqrt{x} + 4\sqrt{y})$

For Problems 18–25, evaluate each of the numerical expressions.

18. $-\sqrt{\dfrac{9}{64}}$

19. $\sqrt[3]{-\dfrac{8}{27}}$

20. $\sqrt[3]{0.008}$

21. $32^{-\frac{1}{5}}$

22. $3^0 + 3^{-1} + 3^{-2}$

23. $-9^{\frac{3}{2}}$

24. $\left(\dfrac{3}{4}\right)^{-2}$

25. $\dfrac{1}{\left(\dfrac{2}{3}\right)^{-3}}$

For Problems 26–31, factor each of the algebraic expressions completely.

26. $3x^4 + 81x$

27. $6x^2 + 19x - 20$

28. $12 + 13x - 14x^2$

29. $9x^4 + 68x^2 - 32$

30. $2ax - ay - 2bx + by$

31. $27x^3 - 8y^3$

For Problems 32–55, solve each of the equations.

32. $3(x - 2) - 2(3x + 5) = 4(x - 1)$

33. $0.06n + 0.08(n + 50) = 25$

34. $4\sqrt{x} + 5 = x$

35. $\sqrt[3]{n^2 - 1} = -1$

36. $6x^2 - 24 = 0$

37. $a^2 + 14a + 49 = 0$

38. $3n^2 + 14n - 24 = 0$

39. $\dfrac{2}{5x - 2} = \dfrac{4}{6x + 1}$

40. $\sqrt{2x - 1} - \sqrt{x + 2} = 0$

41. $5x - 4 = \sqrt{5x - 4}$

42. $|3x - 1| = 11$

43. $(3x - 2)(4x - 1) = 0$

44. $(2x + 1)(x - 2) = 7$

45. $\dfrac{5}{6x} - \dfrac{2}{3} = \dfrac{7}{10x}$

46. $\dfrac{3}{y + 4} + \dfrac{2y - 1}{y^2 - 16} = \dfrac{-2}{y - 4}$

47. $6x^4 - 23x^2 - 4 = 0$

48. $3n^3 + 3n = 0$

49. $n^2 - 13n - 114 = 0$

50. $12x^2 + x - 6 = 0$

51. $x^2 - 2x + 26 = 0$

52. $(x + 2)(x - 6) = -15$

53. $(3x - 1)(x + 4) = 0$

54. $x^2 + 4x + 20 = 0$

55. $2x^2 - x - 4 = 0$

For Problems 56 – 65, solve each inequality and express the solution set using interval notation.

56. $6 - 2x \geq 10$

57. $4(2x - 1) < 3(x + 5)$

58. $\dfrac{n + 1}{4} + \dfrac{n - 2}{12} > \dfrac{1}{6}$

59. $|2x - 1| < 5$

60. $|3x + 2| > 11$

61. $\dfrac{1}{2}(3x - 1) - \dfrac{2}{3}(x + 4) \leq \dfrac{3}{4}(x - 1)$

62. $x^2 - 2x - 8 \leq 0$

63. $3x^2 + 14x - 5 > 0$

64. $\dfrac{x + 2}{x - 7} \geq 0$

65. $\dfrac{2x - 1}{x + 3} < 1$

For Problems 66 –74, solve each problem by setting up and solving an appropriate equation.

66. How many liters of a 60%-acid solution must be added to 14 liters of a 10%-acid solution to produce a 25%-acid solution?

67. A sum of $2250 is to be divided between two people in the ratio of 2 to 3. How much does each person receive?

68. The length of a picture without its border is 7 inches less than twice its width. If the border is 1 inch wide and its area is 62 square inches, what are the dimensions of the picture alone?

69. Working together, Lolita and Doug can paint a shed in 3 hours and 20 minutes. If Doug can paint the shed by himself in 10 hours, how long would it take Lolita to paint the shed by herself?

70. Angie bought some golf balls for $14. If each ball had cost $0.25 less, she could have purchased one more ball for the same amount of money. How many golf balls did Angie buy?

71. A jogger who can run an 8-minute mile starts half a mile ahead of a jogger who can run a 6-minute mile. How long will it take the faster jogger to catch the slower jogger?

72. Suppose that $100 is invested at a certain rate of interest compounded annually for 2 years. If the accumulated value at the end of 2 years is $114.49, find the rate of interest.

73. A room contains 120 chairs arranged in rows. The number of chairs per row is one less than twice the number of rows. Find the number of chairs per row.

74. Bjorn bought a number of shares of stock for $2800. A month later the value of the stock had increased $6 per share, and he sold all but 60 shares and regained his original investment of $2800. How many shares did he sell?

Linear Equations and Inequalities in Two Variables

7.1 Rectangular Coordinate System and Linear Equations

7.2 Graphing Nonlinear Equations

7.3 Linear Inequalities in Two Variables

7.4 Distance and Slope

7.5 Determining the Equation of a Line

René Descartes, a philosopher and mathematician, developed a system for locating a point on a plane. This system is our current rectangular coordinate grid used for graphing; it is named the Cartesian coordinate system.

© Leonard de Selva/CORBIS

René Descartes, a French mathematician of the 17th century, was able to transform geometric problems into an algebraic setting so that he could use the tools of algebra to solve the problems. This connecting of algebraic and geometric ideas is the foundation of a branch of mathematics called **analytic geometry**, today more commonly called **coordinate geometry**. Basically, there are two kinds of problems in coordinate geometry: Given an algebraic equation, find its geometric graph; and given a set of conditions pertaining to a geometric graph, find its algebraic equation. We discuss problems of both types in this chapter.

7.1 Rectangular Coordinate System and Linear Equations

Consider two number lines, one vertical and one horizontal, perpendicular to each other at the point we associate with zero on both lines (Figure 7.1). We refer to these number lines as the **horizontal and vertical axes** or, together, as the **coordinate axes**. They partition the plane into four regions called **quadrants**. The quadrants are numbered counterclockwise from I through IV as indicated in Figure 7.1. The point of intersection of the two axes is called the **origin**.

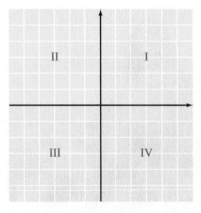

Figure 7.1

It is now possible to set up a one-to-one correspondence between **ordered pairs** of real numbers and the points in a plane. To each ordered pair of real numbers there corresponds a unique point in the plane, and to each point in the plane there corresponds a unique ordered pair of real numbers. A part of this correspondence is illustrated in Figure 7.2. The ordered pair (3, 2) means that the point A is located

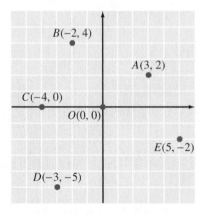

Figure 7.2

three units to the right of, and two units up from, the origin. (The ordered pair $(0, 0)$ is associated with the origin O.) The ordered pair $(-3, -5)$ means that the point D is located three units to the left and five units down from the origin.

Remark: The notation $(-2, 4)$ was used earlier in this text to indicate an interval of the real number line. Now we are using the same notation to indicate an ordered pair of real numbers. This double meaning should not be confusing because the context of the material will always indicate which meaning of the notation is being used. Throughout this chapter, we will be using the ordered-pair interpretation.

In general we refer to the real numbers a and b in an ordered pair (a, b) associated with a point as the **coordinates of the point**. The first number, a, called the **abscissa**, is the directed distance of the point from the vertical axis measured parallel to the horizontal axis. The second number, b, called the **ordinate**, is the directed distance of the point from the horizontal axis measured parallel to the vertical axis (Figure 7.3a). Thus in the first quadrant all points have a positive abscissa and a positive ordinate. In the second quadrant all points have a negative abscissa and a positive ordinate. We have indicated the sign situations for all four quadrants in Figure 7.3(b). This system of associating points in a plane with pairs of real numbers is called the **rectangular coordinate system** or the **Cartesian coordinate system**.

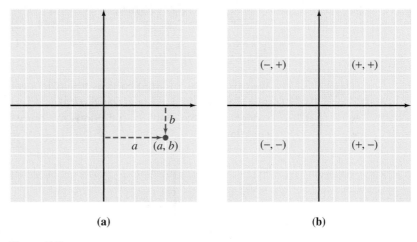

| (a) | (b) |

Figure 7.3

Historically, the rectangular coordinate system provided the basis for the development of the branch of mathematics called **analytic geometry**, or what we presently refer to as **coordinate geometry**. In this discipline, René Descartes, a French 17th-century mathematician, was able to transform geometric problems into an algebraic setting and then use the tools of algebra to solve the problems. Basically, there are two kinds of problems to solve in coordinate geometry:

1. Given an algebraic equation, find its geometric graph.

2. Given a set of conditions pertaining to a geometric figure, find its algebraic equation.

In this chapter we will discuss problems of both types. Let's begin by considering the solutions for the equation $y = x + 2$. A **solution** of an equation in two variables is an ordered pair of real numbers that satisfies the equation. When using the variables x and y, we agree that the first number of an ordered pair is a value of x, and the second number is a value of y. We see that $(1, 3)$ is a solution for $y = x + 2$ because if x is replaced by 1 and y by 3, the true numerical statement $3 = 1 + 2$ is obtained. Likewise, $(-2, 0)$ is a solution because $0 = -2 + 2$ is a true statement. We can find infinitely many pairs of real numbers that satisfy $y = x + 2$ by arbitrarily choosing values for x, and for each value of x we choose, we can determine a corresponding value for y. Let's use a table to record some of the solutions for $y = x + 2$.

Choose x	Determine y from $y = x + 2$	Solutions for $y = x + 2$
0	2	$(0, 2)$
1	3	$(1, 3)$
3	5	$(3, 5)$
5	7	$(5, 7)$
-2	0	$(-2, 0)$
-4	-2	$(-4, -2)$
-6	-4	$(-6, -4)$

We can plot the ordered pairs as points in a coordinate plane and use the horizontal axis as the x axis and the vertical axis as the y axis, as in Figure 7.4(a). The straight line that contains the points in Figure 7.4(b) is called the **graph of the equation** $y = x + 2$.

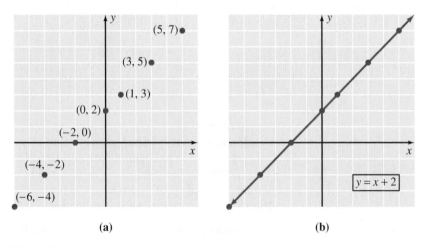

(a)　　　　　　　　　　(b)

Figure 7.4

Remark: It is important to recognize that all points on the x axis have ordered pairs of the form $(a, 0)$ associated with them. That is, the second number in the ordered pair is 0. Likewise, all points on the y axis have ordered pairs of the form $(0, b)$ associated with them.

E X A M P L E   1

Graph $2x + 3y = 6$.

Solution

First, let's find the points of this graph that fall on the coordinate axes. Let $x = 0$; then

$$2(0) + 3y = 6$$
$$3y = 6$$
$$y = 2$$

Thus $(0, 2)$ is a solution and locates a point of the graph on the y axis. Let $y = 0$; then

$$2x + 3(0) = 6$$
$$2x = 6$$
$$x = 3$$

Thus $(3, 0)$ is a solution and locates a point of the graph on the x axis.

Second, let's change the form of the equation to make it easier to find some additional solutions. We can either solve for x in terms of y, or solve for y in terms of x. Let's solve for y in terms of x.

$$2x + 3y = 6$$
$$3y = 6 - 2x$$
$$y = \frac{6 - 2x}{3}$$

Third, a table of values can be formed that includes the two points we found previously.

x	y
0	2
3	0
6	-2
-3	4
-6	6

Plotting these points, we see that they lie in a straight line, and we obtain Figure 7.5.

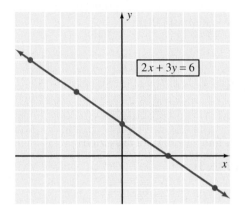

Figure 7.5   ■

Remark: Look again at the table of values in Example 1. Note that values of x were chosen such that integers were obtained for y. That is not necessary, but it does make things easier from a computational standpoint.

The points $(3, 0)$ and $(0, 2)$ in Figure 7.5 are special points. They are the points of the graph that are on the coordinate axes. That is, they yield the x intercept and the y intercept of the graph. Let's define in general the *intercepts* of a graph.

> The x coordinates of the points that a graph has in common with the x axis are called the **x intercepts** of the graph. (To compute the x intercepts, let $y = 0$ and solve for x.)
>
> The y coordinates of the points that a graph has in common with the y axis are called the **y intercepts** of the graph. (To compute the y intercepts, let $x = 0$ and solve for y.)

It is advantageous to be able to recognize the kind of graph that a certain type of equation produces. For example, if we recognize that the graph of $3x + 2y = 12$ is a straight line, then it becomes a simple matter to find two points and sketch the line. Let's pursue the graphing of straight lines in a little more detail.

In general, any equation of the form $Ax + By = C$, where A, B, and C are constants (A and B not both zero) and x and y are variables, is a **linear equation**, and its graph is a straight line. Two points of clarification about this description of a linear equation should be made. First, the choice of x and y for variables is arbitrary. Any two letters could be used to represent the variables. For example, an equation such as $3r + 2s = 9$ can be considered a linear equation in two variables. So that we are not constantly changing the labeling of the coordinate axes when graphing equations, however, it is much easier to use the same two variables in all equations. Thus we will go along with convention and use x and y as variables. Second, the phrase "any equation of the form $Ax + By = C$" technically means "any equation of the form $Ax + By = C$ or equivalent to that form." For example, the equation $y = 2x - 1$ is equivalent to $-2x + y = -1$ and thus is linear and produces a straight-line graph.

The knowledge that any equation of the form $Ax + By = C$ produces a straight-line graph, along with the fact that two points determine a straight line, makes graphing linear equations a simple process. We merely find two solutions (such as the intercepts), plot the corresponding points, and connect the points with a straight line. It is usually wise to find a third point as a check point. Let's consider an example.

EXAMPLE 2

Graph $3x - 2y = 12$.

Solution

First, let's find the intercepts. Let $x = 0$; then

$$3(0) - 2y = 12$$
$$-2y = 12$$
$$y = -6$$

Thus $(0, -6)$ is a solution. Let $y = 0$; then

$$3x - 2(0) = 12$$
$$3x = 12$$
$$x = 4$$

Thus $(4, 0)$ is a solution. Now let's find a third point to serve as a check point. Let $x = 2$; then

$$3(2) - 2y = 12$$
$$6 - 2y = 12$$
$$-2y = 6$$
$$y = -3$$

Thus $(2, -3)$ is a solution. Plot the points associated with these three solutions and connect them with a straight line to produce the graph of $3x - 2y = 12$ in Figure 7.6.

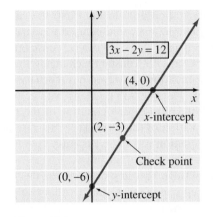

Figure 7.6 ■

Let's review our approach to Example 2. Note that we did not solve the equation for y in terms of x or for x in terms of y. Because we know the graph is a straight line, there is no need for any extensive table of values; thus there is no need to change the form of the original equation. Furthermore, the solution $(2, -3)$ served as a check point. If it had not been on the line determined by the two intercepts, then we would have known that an error had been made.

EXAMPLE 3

Graph $2x + 3y = 7$.

Solution

Without showing all of our work, the following table indicates the intercepts and a check point.

x	y	
0	$\dfrac{7}{3}$	
$\dfrac{7}{2}$	0	Intercepts
2	1	Check point

The points from the table are plotted, and the graph of $2x + 3y = 7$ is shown in Figure 7.7.

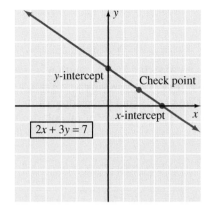

Figure 7.7   ■

It is helpful to recognize some *special* straight lines. For example, the graph of any equation of the form $Ax + By = C$, where $C = 0$ (the constant term is zero), is a straight line that contains the origin. Let's consider an example.

E X A M P L E   4   Graph $y = 2x$.

Solution

Obviously $(0, 0)$ is a solution. (Also, notice that $y = 2x$ is equivalent to $-2x + y = 0$; thus it fits the condition $Ax + By = C$, where $C = 0$.) Because both the x intercept and the y intercept are determined by the point $(0, 0)$, another point is necessary to determine the line. Then a third point should be found as a check point. The graph of $y = 2x$ is shown in Figure 7.8.

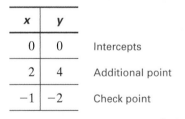

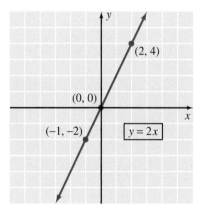

x	y	
0	0	Intercepts
2	4	Additional point
-1	-2	Check point

Figure 7.8   ■

E X A M P L E   5

Graph $x = 2$.

Solution

Because we are considering linear equations in *two variables*, the equation $x = 2$ is equivalent to $x + 0(y) = 2$. Now we can see that any value of y can be used, but the x value must always be 2. Therefore, some of the solutions are $(2, 0)$, $(2, 1)$, $(2, 2)$, $(2, -1)$, and $(2, -2)$. The graph of all solutions of $x = 2$ is the vertical line in Figure 7.9.

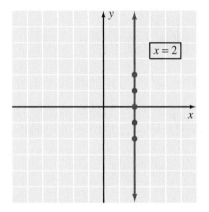

Figure 7.9

E X A M P L E   6

Graph $y = -3$.

Solution

The equation $y = -3$ is equivalent to $0(x) + y = -3$. Thus any value of x can be used, but the value of y must be -3. Some solutions are $(0, -3)$, $(1, -3)$, $(2, -3)$, $(-1, -3)$, and $(-2, -3)$. The graph of $y = -3$ is the horizontal line in Figure 7.10.

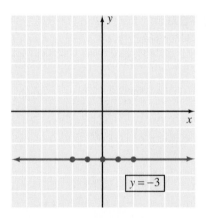

Figure 7.10

In general, the graph of any equation of the form $Ax + By = C$, where $A = 0$ or $B = 0$ (not both), is a line parallel to one of the axes. More specifically, any equation of the form $x = a$, where a is a constant, is a line parallel to the y axis that has an x intercept of a. Any equation of the form $y = b$, where b is a constant, is a line parallel to the x axis that has a y intercept of b.

■ Linear Relationships

There are numerous applications of linear relationships. For example, suppose that a retailer has a number of items that she wants to sell at a profit of 30% of the cost of each item. If we let s represent the selling price and c the cost of each item, then the equation

$$s = c + 0.3c = 1.3c$$

can be used to determine the selling price of each item based on the cost of the item. In other words, if the cost of an item is \$4.50, then it should be sold for $s = (1.3)(4.5) = \$5.85$.

The equation $s = 1.3c$ can be used to determine the following table of values. Reading from the table, we see that if the cost of an item is \$15, then it should be sold for \$19.50 in order to yield a profit of 30% of the cost. Furthermore, because this is a linear relationship, we can obtain exact values between values given in the table.

c	1	5	10	15	20
s	1.3	6.5	13	19.5	26

For example, a c value of 12.5 is halfway between c values of 10 and 15, so the corresponding s value is halfway between the s values of 13 and 19.5. Therefore, a c value of 12.5 produces an s value of

$$s = 13 + \frac{1}{2}(19.5 - 13) = 16.25$$

Thus, if the cost of an item is \$12.50, it should be sold for \$16.25.

Now let's graph this linear relationship. We can label the horizontal axis c, label the vertical axis s, and use the origin along with one ordered pair from the table to produce the straight-line graph in Figure 7.11. (Because of the type of application, we use only nonnegative values for c and s.)

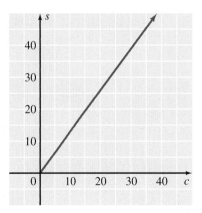

Figure 7.11

From the graph we can approximate s values on the basis of given c values. For example, if $c = 30$, then by reading up from 30 on the c axis to the line and then across to the s axis, we see that s is a little less than 40. (An exact s value of 39 is obtained by using the equation $s = 1.3c$.)

Many formulas that are used in various applications are linear equations in two variables. For example, the formula $C = \dfrac{5}{9}(F - 32)$, which is used to convert temperatures from the Fahrenheit scale to the Celsius scale, is a linear relationship. Using this equation, we can determine that 14°F is equivalent to $C = \dfrac{5}{9}(14 - 32) = \dfrac{5}{9}(-18) = -10$°C. Let's use the equation $C = \dfrac{5}{9}(F - 32)$ to complete the following table.

F	−22	−13	5	32	50	68	86
C	−30	−25	−15	0	10	20	30

Reading from the table, we see, for example, that −13°F = −25°C and 68°F = 20°C.

To graph the equation $C = \dfrac{5}{9}(F - 32)$ we can label the horizontal axis F, label the vertical axis C, and plot two ordered pairs (F, C) from the table. Figure 7.12 shows the graph of the equation.

From the graph we can approximate C values on the basis of given F values. For example, if F = 80°, then by reading up from 80 on the F axis to the line and then across to the C axis, we see that C is approximately 25°. Likewise, we can obtain approximate F values on the basis of given C values. For example, if C = −25°, then by reading across from −25 on the C axis to the line and then up to the F axis, we see that F is approximately −15°.

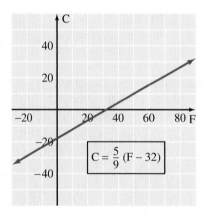

Figure 7.12

■ Graphing Utilities

The term **graphing utility** is used in current literature to refer to either a graphing calculator (see Figure 7.13) or a computer with a graphing software package. (We will frequently use the phrase *use a graphing calculator* to mean "use a graphing calculator or a computer with the appropriate software.")

These devices have a large range of capabilities that enable the user not only to obtain a quick sketch of a graph but also to study various characteristics of it, such as the *x* intercepts, *y* intercepts, and turning points of a curve. We will introduce some of these features of graphing utilities as we need them in the text. Because there are so many different types of graphing utilities available, we will use mostly generic terminology and let you consult your user's manual for specific keypunching instructions. We urge you to study the graphing utility examples in this text even if you do not have access to a graphing calculator or a computer. The examples were chosen to reinforce concepts under discussion.

Courtesy Texas Instruments

Figure 7.13

E X A M P L E   7

Use a graphing utility to obtain a graph of the line $2.1x + 5.3y = 7.9$.

Solution

First, let's solve the equation for *y* in terms of *x*.

$$2.1x + 5.3y = 7.9$$

$$5.3y = 7.9 - 2.1x$$

$$y = \frac{7.9 - 2.1x}{5.3}$$

Now we can enter the expression $\dfrac{7.9 - 2.1x}{5.3}$ for Y_1 and obtain the graph as shown in Figure 7.14.

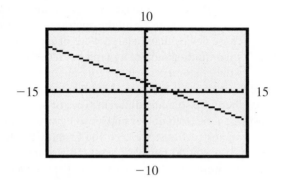

Figure 7.14

Problem Set 7.1

For Problems 1–33, graph each of the linear equations.

1. $x + 2y = 4$

2. $2x + y = 6$

3. $2x - y = 2$

4. $3x - y = 3$

5. $3x + 2y = 6$

6. $2x + 3y = 6$

7. $5x - 4y = 20$

8. $4x - 3y = -12$

9. $x + 4y = -6$

10. $5x + y = -2$

11. $-x - 2y = 3$

12. $-3x - 2y = 12$

13. $y = x + 3$

14. $y = x - 1$

15. $y = -2x - 1$

16. $y = 4x + 3$

17. $y = \dfrac{1}{2}x + \dfrac{2}{3}$

18. $y = \dfrac{2}{3}x - \dfrac{3}{4}$

19. $y = -x$

20. $y = x$

21. $y = 3x$

22. $y = -4x$

23. $x = 2y - 1$

24. $x = -3y + 2$

25. $y = -\dfrac{1}{4}x + \dfrac{1}{6}$

26. $y = -\dfrac{1}{2}x - \dfrac{1}{2}$

27. $2x - 3y = 0$

28. $3x + 4y = 0$

29. $x = 0$

30. $y = 0$

31. $y = 2$

32. $x = -3$

33. $-3y = -x + 3$

34. Suppose that the daily profit from an ice cream stand is given by the equation $p = 2n - 4$, where n represents the number of gallons of ice cream mix used in a day, and p represents the number of dollars of profit. Label the horizontal axis n and the vertical axis p, and graph the equation $p = 2n - 4$ for nonnegative values of n.

35. The cost (c) of playing an online computer game for a time (t) in hours is given by the equation $c = 3t + 5$. Label the horizontal axis t and the vertical axis c, and graph the equation for nonnegative values of t.

36. The area of a sidewalk whose width is fixed at 3 feet can be given by the equation $A = 3l$, where A represents the area in square feet, and l represents the length in feet. Label the horizontal axis l and the vertical axis A, and graph the equation $A = 3l$ for nonnegative values of l.

37. An online grocery store charges for delivery based on the equation $C = 0.30p$, where C represents the cost in dollars, and p represents the weight of the groceries in pounds. Label the horizontal axis p and the vertical axis C, and graph the equation $C = 0.30p$ for non-negative values of p.

38. **(a)** The equation $F = \dfrac{9}{5}C + 32$ can be used to convert from degrees Celsius to degrees Fahrenheit. Complete the following table.

C	0	5	10	15	20	−5	−10	−15	−20	−25
F										

(b) Graph the equation $F = \dfrac{9}{5}C + 32$.

(c) Use your graph from part (b) to approximate values for F when C = 25°, 30°, −30°, and −40°.

(d) Check the accuracy of your readings from the graph in part (c) by using the equation $F = \dfrac{9}{5}C + 32$.

39. **(a)** Digital Solutions charges for help-desk services according to the equation $c = 0.25m + 10$, where c represents the cost in dollars, and m represents the minutes of service. Complete the following table.

m	5	10	15	20	30	60
c						

(b) Label the horizontal axis m and the vertical axis c, and graph the equation $c = 0.25m + 10$ for non-negative values of m.

(c) Use the graph from part (b) to approximate values for c when m = 25, 40, and 45.

(d) Check the accuracy of your readings from the graph in part (c) by using the equation $c = 0.25m + 10$.

■■■ THOUGHTS INTO WORDS

40. How do we know that the graph of $y = -3x$ is a straight line that contains the origin?

41. How do we know that the graphs of $2x - 3y = 6$ and $-2x + 3y = -6$ are the same line?

42. What is the graph of the conjunction $x = 2$ and $y = 4$? What is the graph of the disjunction $x = 2$ or $y = 4$? Explain your answers.

43. Your friend claims that the graph of the equation $x = 2$ is the point $(2, 0)$. How do you react to this claim?

■■■ FURTHER INVESTIGATIONS

From our work with absolute value, we know that $|x + y| = 1$ is equivalent to $x + y = 1$ or $x + y = -1$. Therefore, the graph of $|x + y| = 1$ consists of the two lines $x + y = 1$ and $x + y = -1$. Graph each of the following.

44. $|x + y| = 1$

45. $|x - y| = 4$

46. $|2x - y| = 4$

47. $|3x + 2y| = 6$

GRAPHING CALCULATOR ACTIVITIES

This is the first of many appearances of a group of problems called graphing calculator activities. These problems are specifically designed for those of you who have access to a graphing calculator or a computer with an appropriate software package. Within the framework of these problems, you will be given the opportunity to reinforce concepts we discussed in the text; lay groundwork for concepts we will introduce later in the text; predict shapes and locations of graphs on the basis of your previous graphing experiences; solve problems that are unreasonable or perhaps impossible to solve without a graphing utility; and in general become familiar with the capabilities and limitations of your graphing utility.

48. **(a)** Graph $y = 3x + 4$, $y = 2x + 4$, $y = -4x + 4$, and $y = -2x + 4$ on the same set of axes.

(b) Graph $y = \frac{1}{2}x - 3$, $y = 5x - 3$, $y = 0.1x - 3$, and $y = -7x - 3$ on the same set of axes.

(c) What characteristic do all lines of the form $y = ax + 2$ (where a is any real number) share?

49. (a) Graph $y = 2x - 3$, $y = 2x + 3$, $y = 2x - 6$, and $y = 2x + 5$ on the same set of axes.

(b) Graph $y = -3x + 1$, $y = -3x + 4$, $y = -3x - 2$, and $y = -3x - 5$ on the same set of axes.

(c) Graph $y = \frac{1}{2}x + 3$, $y = \frac{1}{2}x - 4$, $y = \frac{1}{2}x + 5$, and $y = \frac{1}{2}x - 2$ on the same set of axes.

(d) What relationship exists among all lines of the form $y = 3x + b$, where b is any real number?

50. (a) Graph $2x + 3y = 4$, $2x + 3y = -6$, $4x - 6y = 7$, and $8x + 12y = -1$ on the same set of axes.

(b) Graph $5x - 2y = 4$, $5x - 2y = -3$, $10x - 4y = 3$, and $15x - 6y = 30$ on the same set of axes.

(c) Graph $x + 4y = 8$, $2x + 8y = 3$, $x - 4y = 6$, and $3x + 12y = 10$ on the same set of axes.

(d) Graph $3x - 4y = 6$, $3x + 4y = 10$, $6x - 8y = 20$, and $6x - 8y = 24$ on the same set of axes.

(e) For each of the following pairs of lines, (a) predict whether they are parallel lines, and (b) graph each pair of lines to check your prediction.

(1) $5x - 2y = 10$ and $5x - 2y = -4$
(2) $x + y = 6$ and $x - y = 4$
(3) $2x + y = 8$ and $4x + 2y = 2$
(4) $y = 0.2x + 1$ and $y = 0.2x - 4$
(5) $3x - 2y = 4$ and $3x + 2y = 4$
(6) $4x - 3y = 8$ and $8x - 6y = 3$
(7) $2x - y = 10$ and $6x - 3y = 6$
(8) $x + 2y = 6$ and $3x - 6y = 6$

51. Now let's use a graphing calculator to get a graph of $C = \frac{5}{9}(F - 32)$. By letting F = x and C = y, we obtain Figure 7.15.

Pay special attention to the boundaries on x. These values were chosen so that the fraction

$$\frac{\text{(Maximum value of } x) \text{ minus (Minimum value of } x)}{95}$$

would be equal to 1. The viewing window of the graphing calculator used to produce Figure 7.15 is 95 pixels (dots) wide. Therefore, we use 95 as the denominator

of the fraction. We chose the boundaries for y to make sure that the cursor would be visible on the screen when we looked for certain values.

Now let's use the TRACE feature of the graphing calculator to complete the following table. Note that the cursor moves in increments of 1 as we trace along the graph.

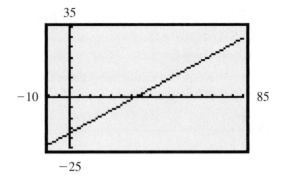

35

−10   85

−25

Figure 7.15

F	−5	5	9	11	12	20	30	45	60
C									

(This was accomplished by setting the aforementioned fraction equal to 1.) By moving the cursor to each of the F values, we can complete the table as follows.

F	−5	5	9	11	12	20	30	45	60
C	−21	−15	−13	−12	−11	−7	−1	7	16

The C values are expressed to the nearest degree. Use your calculator and check the values in the table by using the equation $C = \frac{5}{9}(F - 32)$.

52. (a) Use your graphing calculator to graph $F = \frac{9}{5}C + 32$. Be sure to set boundaries on the horizontal axis so that when you are using the trace feature, the cursor will move in increments of 1.

(b) Use the TRACE feature and check your answers for part (a) of Problem 38.

7.2   Graphing Nonlinear Equations

Equations such as $y = x^2 - 4$, $x = y^2$, $y = \dfrac{1}{x}$, $x^2y = -2$, and $x = y^3$ are all examples of nonlinear equations. The graphs of these equations are figures other than straight lines that can be determined by plotting a sufficient number of points. Let's plot the points and observe some characteristics of these graphs that we then can use to supplement the point-plotting process.

EXAMPLE 1

Graph $y = x^2 - 4$

Solution

Let's begin by finding the intercepts. If $x = 0$, then

$$y = 0^2 - 4 = -4$$

The point $(0, -4)$ is on the graph. If $y = 0$, then

$$0 = x^2 - 4$$

$$0 = (x + 2)(x - 2)$$

$$x + 2 = 0 \quad \text{or} \quad x - 2 = 0$$

$$x = -2 \quad \text{or} \quad x = 2$$

The points $(-2, 0)$ and $(2, 0)$ are on the graph. The given equation is in a convenient form for setting up a table of values.

Plotting these points and connecting them with a smooth curve produces Figure 7.16.

x	y	
0	−4	
−2	0	Intercepts
2	0	
1	−3	
−1	−3	
3	5	Other points
−3	5	

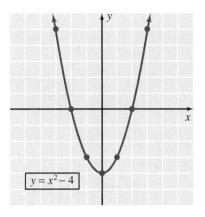

Figure 7.16

The curve in Figure 7.16 is called a parabola; we will study parabolas in more detail in a later chapter. However, at this time we want to emphasize that the parabola in Figure 7.16 is said to be *symmetric with respect to the y axis*. In other words, the y axis is a line of symmetry. Each half of the curve is a mirror image of the other half through the y axis. Note, in the table of values, that for each ordered pair (x, y), the ordered pair $(-x, y)$ is also a solution. A general test for y axis symmetry can be stated as follows:

---

y Axis Symmetry

The graph of an equation is symmetric with respect to the y axis if replacing x with $-x$ results in an equivalent equation.

The equation $y = x^2 - 4$ exhibits symmetry with respect to the y axis because replacing x with $-x$ produces $y = (-x)^2 - 4 = x^2 - 4$. Let's test some equations for such symmetry. We will replace x with $-x$ and check for an equivalent equation.

Equation	Test for symmetry with respect to the *y* axis	Equivalent equation	Symmetric with respect to the *y* axis
$y = -x^2 + 2$	$y = -(-x)^2 + 2 = -x^2 + 2$	Yes	Yes
$y = 2x^2 + 5$	$y = 2(-x)^2 + 5 = 2x^2 + 5$	Yes	Yes
$y = x^4 + x^2$	$y = (-x)^4 + (-x)^2$ $= x^4 + x^2$	Yes	Yes
$y = x^3 + x^2$	$y = (-x)^3 + (-x)^2$ $= -x^3 + x^2$	No	No
$y = x^2 + 4x + 2$	$y = (-x)^2 + 4(-x) + 2$ $= x^2 - 4x + 2$	No	No

Some equations yield graphs that have x axis symmetry. In the next example we will see the graph of a parabola that is symmetric with respect to the x axis.

E X A M P L E   2

Graph $x = y^2$.

Solution

First, we see that $(0, 0)$ is on the graph and determines both intercepts. Second, the given equation is in a convenient form for setting up a table of values.

Plotting these points and connecting them with a smooth curve produces Figure 7.17.

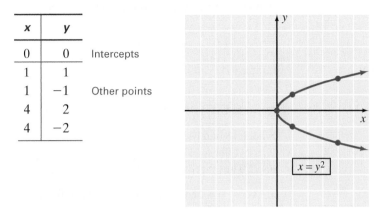

x	y	
0	0	Intercepts
1	1	
1	−1	Other points
4	2	
4	−2	

Figure 7.17   ■

The parabola in Figure 7.17 is said to be *symmetric with respect to the x axis*. Each half of the curve is a mirror image of the other half through the *x* axis. Also note, in the table of values, that for each ordered pair (x, y), the ordered pair $(x, -y)$ is a solution. A general test for *x* axis symmetry can be stated as follows:

x Axis Symmetry

> The graph of an equation is symmetric with respect to the *x* axis if replacing *y* with $-y$ results in an equivalent equation.

The equation $x = y^2$ exhibits *x* axis symmetry because replacing *x* with $-y$ produces $y = (-y)^2 = y^2$. Let's test some equations for *x* axis symmetry. We will replace *y* with $-y$ and check for an equivalent equation.

Equation	Test for symmetry with respect to the *x* axis	Equivalent equation	Symmetric with respect to the *x* axis
$x = y^2 + 5$	$x = (-y)^2 + 5 = y^2 + 5$	Yes	Yes
$x = -3y^2$	$x = -3(-y)^2 = -3y^2$	Yes	Yes
$x = y^3 + 2$	$x = (-y)^3 + 2 = -y^3 + 2$	No	No
$x = y^2 - 5y + 6$	$x = (-y)^2 - 5(-y) + 6$		
	$= y^2 + 5y + 6$	No	No

In addition to y axis and x axis symmetry, some equations yield graphs that have symmetry with respect to the origin. In the next example we will see a graph that is symmetric with respect to the origin.

EXAMPLE 3

Graph $y = \dfrac{1}{x}$.

Solution

First, let's find the intercepts. Let $x = 0$; then $y = \dfrac{1}{x}$ becomes $y = \dfrac{1}{0}$, and $\dfrac{1}{0}$ is undefined. Thus there is no y intercept. Let $y = 0$; then $y = \dfrac{1}{x}$ becomes $0 = \dfrac{1}{x}$, and there are no values of x that will satisfy this equation. In other words, this graph has no points on either the x axis or the y axis. Second, let's set up a table of values and keep in mind that neither x nor y can equal zero.

In Figure 7.18(a) we plotted the points associated with the solutions from the table. Because the graph does not intersect either axis, it must consist of two branches. Thus connecting the points in the first quadrant with a smooth curve and then connecting the points in the third quadrant with a smooth curve, we obtain the graph shown in Figure 7.18(b).

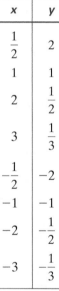

x	y
$\dfrac{1}{2}$	2
1	1
2	$\dfrac{1}{2}$
3	$\dfrac{1}{3}$
$-\dfrac{1}{2}$	-2
-1	-1
-2	$-\dfrac{1}{2}$
-3	$-\dfrac{1}{3}$

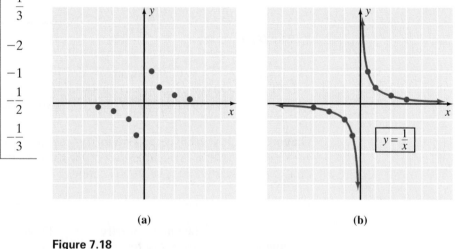

(a)   (b)

Figure 7.18

The curve in Figure 7.18 is said to be *symmetric with respect to the origin*. Each half of the curve is a mirror image of the other half through the origin. Note, in the table of values, that for each ordered pair (x, y), the ordered pair $(-x, -y)$ is also a solution. A general test for origin symmetry can be stated as follows:

Origin Symmetry

> The graph of an equation is symmetric with respect to the origin if replacing x with $-x$ and y with $-y$ results in an equivalent equation.

The equation $y = \dfrac{1}{x}$ exhibits symmetry with respect to the origin because replacing y with $-y$ and x with $-x$ produces $-y = \dfrac{1}{-x}$, which is equivalent to $y = \dfrac{1}{x}$. Let's test some equations for symmetry with respect to the origin. We will replace y with $-y$, replace x with $-x$, and then check for an equivalent equation.

Equation	Test for symmetry with respect to the origin	Equivalent equation	Symmetric with respect to the origin
$y = x^3$	$(-y) = (-x)^3$ $-y = -x^3$ $y = x^3$	Yes	Yes
$x^2 + y^2 = 4$	$(-x)^2 + (-y)^2 = 4$ $x^2 + y^2 = 4$	Yes	Yes
$y = x^2 - 3x + 4$	$(-y) = (-x)^2 - 3(-x) + 4$ $-y = x^2 + 3x + 4$ $y = -x^2 - 3x - 4$	No	No

Let's pause for a moment and pull together the graphing techniques that we have introduced thus far. Following is a list of graphing suggestions. The order of the suggestions indicates the order in which we usually attack a new graphing problem.

1. Determine what type of symmetry the equation exhibits.

2. Find the intercepts.

3. Solve the equation for y in terms of x or for x in terms of y if it is not already in such a form.

4. Set up a table of ordered pairs that satisfy the equation. The type of symmetry will affect your choice of values in the table. (We will illustrate this in a moment.)

5. Plot the points associated with the ordered pairs from the table, and connect them with a smooth curve. Then, if appropriate, reflect this part of the curve according to the symmetry shown by the equation.

EXAMPLE 4

Graph $x^2y = -2$.

Solution

Because replacing x with $-x$ produces $(-x)^2y = -2$ or, equivalently, $x^2y = -2$, the equation exhibits y axis symmetry. There are no intercepts because neither x nor y can equal 0. Solving the equation for y produces $y = \dfrac{-2}{x^2}$. The equation exhibits y axis symmetry, so let's use only positive values for x and then reflect the curve across the y axis.

x	y
1	-2
2	$-\dfrac{1}{2}$
3	$-\dfrac{2}{9}$
4	$-\dfrac{1}{8}$
$\dfrac{1}{2}$	-8

Let's plot the points determined by the table, connect them with a smooth curve, and reflect this portion of the curve across the y axis. Figure 7.19 is the result of this process.

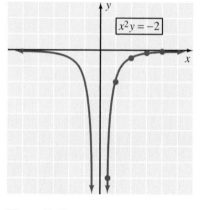

Figure 7.19

EXAMPLE 5

Graph $x = y^3$.

Solution

Because replacing x with $-x$ and y with $-y$ produces $-x = (-y)^3 = -y^3$, which is equivalent to $x = y^3$, the given equation exhibits origin symmetry. If $x = 0$, then $y = 0$, so the origin is a point of the graph. The given equation is in an easy form for deriving a table of values.

x	y
0	0
8	2
$\dfrac{1}{8}$	$\dfrac{1}{2}$
$\dfrac{27}{64}$	$\dfrac{3}{4}$

Let's plot the points determined by the table, connect them with a smooth curve, and reflect this portion of the curve through the origin to produce Figure 7.20.

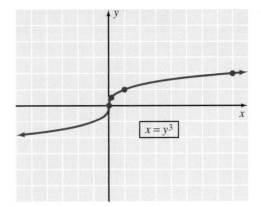

Figure 7.20   ■

EXAMPLE 6

Use a graphing utility to obtain a graph of the equation $x = y^3$.

Solution

First, we may need to solve the equation for y in terms of x. (We say we "may need to" because some graphing utilities are capable of graphing two-variable equations without solving for y in terms of x.)

$$y = \sqrt[3]{x} = x^{1/3}$$

Now we can enter the expression $x^{1/3}$ for Y_1 and obtain the graph shown in Figure 7.21.

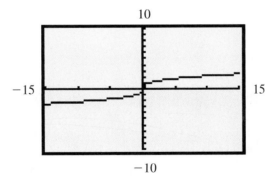

Figure 7.21   ■

As indicated in Figure 7.21, the **viewing rectangle** of a graphing utility is a portion of the xy plane shown on the display of the utility. In this display, the boundaries were set so that $-15 \leq x \leq 15$ and $-10 \leq y \leq 10$. These boundaries were set automatically; however, boundaries can be reassigned as necessary, which is an important feature of graphing utilities.

Problem Set 7.2

For each of the points in Problems 1–5, determine the points that are symmetric with respect to (a) the x axis, (b) the y axis, and (c) the origin.

1. $(-3, 1)$

2. $(-2, -4)$

3. $(7, -2)$

4. $(0, -4)$

5. $(5, 0)$

For Problems 6–25, determine the type(s) of symmetry (symmetry with respect to the x axis, y axis, and/or origin) exhibited by the graph of each of the following equations. Do not sketch the graph.

6. $x^2 + 2y = 4$

7. $-3x + 2y^2 = -4$

8. $x = -y^2 + 5$

9. $y = 4x^2 + 13$

10. $xy = -6$

11. $2x^2y^2 = 5$

12. $2x^2 + 3y^2 = 9$

13. $x^2 - 2x - y^2 = 4$

14. $y = x^2 - 6x - 4$

15. $y = 2x^2 - 7x - 3$

16. $y = x$

17. $y = 2x$

18. $y = x^4 + 4$

19. $y = x^4 - x^2 + 2$

20. $x^2 + y^2 = 13$

21. $x^2 - y^2 = -6$

22. $y = -4x^2 - 2$

23. $x = -y^2 + 9$

24. $x^2 + y^2 - 4x - 12 = 0$

25. $2x^2 + 3y^2 + 8y + 2 = 0$

For Problems 26–59, graph each of the equations.

26. $y = x + 1$

27. $y = x - 4$

28. $y = 3x - 6$

29. $y = 2x + 4$

30. $y = -2x + 1$

31. $y = -3x - 1$

32. $y = \dfrac{2}{3}x - 1$

33. $y = -\dfrac{1}{3}x + 2$

34. $y = \dfrac{1}{3}x$

35. $y = \dfrac{1}{2}x$

36. $2x + y = 6$

37. $2x - y = 4$

38. $x + 3y = -3$

39. $x - 2y = 2$

40. $y = x^2 - 1$

41. $y = x^2 + 2$

42. $y = -x^3$

43. $y = x^3$

44. $y = \dfrac{2}{x^2}$

45. $y = \dfrac{-1}{x^2}$

46. $y = 2x^2$

47. $y = -3x^2$

48. $xy = -3$

49. $xy = 2$

50. $x^2y = 4$

51. $xy^2 = -4$

52. $y^3 = x^2$

53. $y^2 = x^3$

54. $y = \dfrac{-2}{x^2 + 1}$

55. $y = \dfrac{4}{x^2 + 1}$

56. $x = -y^3$

57. $y = x^4$

58. $y = -x^4$

59. $x = -y^3 + 2$

■ ■ ■ THOUGHTS INTO WORDS

60. How would you convince someone that there are infinitely many ordered pairs of real numbers that satisfy $x + y = 7$?

61. What is the graph of $x = 0$? What is the graph of $y = 0$? Explain your answers.

62. Is a graph symmetric with respect to the origin if it is symmetric with respect to both axes? Defend your answer.

63. Is a graph symmetric with respect to both axes if it is symmetric with respect to the origin? Defend your answer.

GRAPHING CALCULATOR ACTIVITIES

This set of activities is designed to help you get started with your graphing utility by setting different boundaries for the viewing rectangle; you will notice the effect on the graphs produced. These boundaries are usually set by using a menu displayed by a key marked either WINDOW or RANGE. You may need to consult the user's manual for specific key-punching instructions.

64. Graph the equation $y = \dfrac{1}{x}$ (Example 4) using the following boundaries.
  (a) $-15 \le x \le 15$ and $-10 \le y \le 10$
  (b) $-10 \le x \le 10$ and $-10 \le y \le 10$
  (c) $-5 \le x \le 5$ and $-5 \le y \le 5$

65. Graph the equation $y = \dfrac{-2}{x^2}$ (Example 5), using the following boundaries.
  (a) $-15 \le x \le 15$ and $-10 \le y \le 10$
  (b) $-5 \le x \le 5$ and $-10 \le y \le 10$
  (c) $-5 \le x \le 5$ and $-10 \le y \le 1$

66. Graph the two equations $y = \pm\sqrt{x}$ (Example 3) on the same set of axes, using the following boundaries.

(Let $Y_1 = \sqrt{x}$ and $Y_2 = -\sqrt{x}$)
  (a) $-15 \le x \le 15$ and $-10 \le y \le 10$
  (b) $-1 \le x \le 15$ and $-10 \le y \le 10$
  (c) $-1 \le x \le 15$ and $-5 \le y \le 5$

67. Graph $y = \dfrac{1}{x}$, $y = \dfrac{5}{x}$, $y = \dfrac{10}{x}$, and $y = \dfrac{20}{x}$ on the same set of axes. (Choose your own boundaries.) What effect does increasing the constant seem to have on the graph?

68. Graph $y = \dfrac{10}{x}$ and $y = \dfrac{-10}{x}$ on the same set of axes. What relationship exists between the two graphs?

69. Graph $y = \dfrac{10}{x^2}$ and $y = \dfrac{-10}{x^2}$ on the same set of axes. What relationship exists between the two graphs?

7.3   Linear Inequalities in Two Variables

Linear inequalities in two variables are of the form $Ax + By > C$ or $Ax + By < C$, where A, B, and C are real numbers. (Combined linear equality and inequality statements are of the form $Ax + By \ge C$ or $Ax + By \le C$.)

Graphing linear inequalities is almost as easy as graphing linear equations. The following discussion leads into a simple, step-by-step process. Let's consider the following equation and related inequalities.

$$x + y = 2 \qquad x + y > 2 \qquad x + y < 2$$

The graph of $x + y = 2$ is shown in Figure 7.22. The line divides the plane into two half planes, one above the line and one below the line. In Figure 7.23(a) we

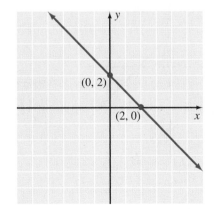

Figure 7.22

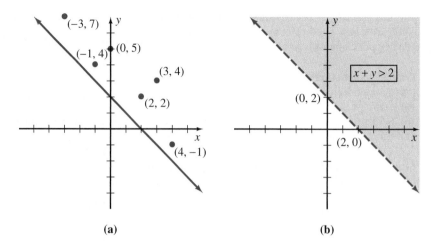

(a)   **(b)**

Figure 7.23

indicated several points in the half plane above the line. Note that for each point, the ordered pair of real numbers satisfies the inequality $x + y > 2$. This is true for *all points* in the half plane above the line. Therefore, the graph of $x + y > 2$ is the half plane above the line, as indicated by the shaded portion in Figure 7.23(b). We use a dashed line to indicate that points on the line do *not* satisfy $x + y > 2$. We would use a solid line if we were graphing $x + y \geq 2$.

In Figure 7.24(a) several points were indicated in the half plane below the line, $x + y = 2$. Note that for each point, the ordered pair of real numbers satisfies the inequality $x + y < 2$. This is true for *all points* in the half plane below the line. Thus the graph of $x + y < 2$ is the half plane below the line, as indicated in Figure 7.24(b).

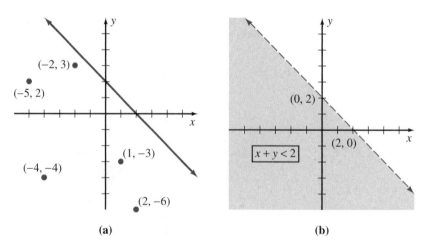

(a)   **(b)**

Figure 7.24

To graph a linear inequality, we suggest the following steps.

1. First, graph the corresponding equality. Use a solid line if equality is included in the original statement. Use a dashed line if equality is not included.

2. Choose a "test point" not on the line and substitute its coordinates into the inequality. (The origin is a convenient point to use if it is not on the line.)

3. The graph of the original inequality is
   (a) the half plane that contains the test point if the inequality is satisfied by that point, or
   (b) the half plane that does not contain the test point if the inequality is not satisfied by the point.

Let's apply these steps to some examples.

E X A M P L E   1

Graph $x - 2y > 4$.

Solution

Step 1   Graph $x - 2y = 4$ as a dashed line because equality is not included in $x - 2y > 4$ (Figure 7.25).

Step 2   Choose the origin as a test point and substitute its coordinates into the inequality.

$$x - 2y > 4 \qquad \text{becomes } 0 - 2(0) > 4, \text{ which is false.}$$

Step 3   Because the test point did not satisfy the given inequality, the graph is the half plane that does not contain the test point. Thus the graph of $x - 2y > 4$ is the half plane below the line, as indicated in Figure 7.25.

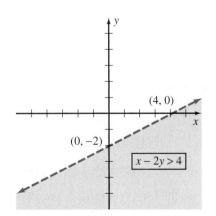

Figure 7.25

EXAMPLE 2   Graph $3x + 2y \leq 6$.

Solution

Step 1   Graph $3x + 2y = 6$ as a solid line because equality is included in $3x + 2y \leq 6$ (Figure 7.26).

Step 2   Choose the origin as a test point and substitute its coordinates into the given statement.

$$3x + 2y \leq 6 \qquad \text{becomes } 3(0) + 2(0) \leq 6, \text{which is true.}$$

Step 3   Because the test point satisfies the given statement, all points in the same half plane as the test point satisfy the statement. Thus the graph of $3x + 2y \leq 6$ consists of the line and the half plane below the line (Figure 7.26).

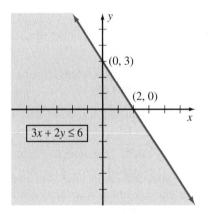

Figure 7.26   ■

EXAMPLE 3   Graph $y \leq 3x$.

Solution

Step 1   Graph $y = 3x$ as a solid line because equality is included in the statement $y \leq 3x$ (Figure 7.27).

Step 2   The origin is on the line, so we must choose some other point as a test point. Let's try $(2, 1)$.

$$y \leq 3x \qquad \text{becomes } 1 \leq 3(2), \text{which is a true statement.}$$

Step 3   Because the test point satisfies the given inequality, the graph is the half plane that contains the test point. Thus the graph of $y \leq 3x$ consists of the line and the half plane below the line, as indicated in Figure 7.27.

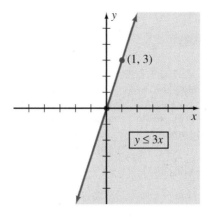

Figure 7.27

Problem Set 7.3

For Problems 1–24, graph each of the inequalities.

1. $x - y > 2$

2. $x + y > 4$

3. $x + 3y < 3$

4. $2x - y > 6$

5. $2x + 5y \geq 10$

6. $3x + 2y \leq 4$

7. $y \leq -x + 2$

8. $y \geq -2x - 1$

9. $y > -x$

10. $y < x$

11. $2x - y \geq 0$

12. $x + 2y \geq 0$

13. $-x + 4y - 4 \leq 0$

14. $-2x + y - 3 \leq 0$

15. $y > -\dfrac{3}{2}x - 3$

16. $2x + 5y > -4$

17. $y < -\dfrac{1}{2}x + 2$

18. $y < -\dfrac{1}{3}x + 1$

19. $x \leq 3$

20. $y \geq -2$

21. $x > 1$ and $y < 3$

22. $x > -2$ and $y > -1$

23. $x \leq -1$ and $y < 1$

24. $x < 2$ and $y \geq -2$

■ ■ ■ THOUGHTS INTO WORDS

25. Why is the point $(-4, 1)$ not a good test point to use when graphing $5x - 2y > -22$?

26. Explain how you would graph the inequality $-3 > x - 3y$.

■ ■ ■ FURTHER INVESTIGATIONS

27. Graph $|x| < 2$. [*Hint:* Remember that $|x| < 2$ is equivalent to $-2 < x < 2$.]

28. Graph $|y| > 1$.

29. Graph $|x + y| < 1$.

30. Graph $|x - y| > 2$.

GRAPHING CALCULATOR ACTIVITIES

31. This is a good time for you to become acquainted with the DRAW features of your graphing calculator. Again, you may need to consult your user's manual for specific key-punching instructions. Return to Examples 1, 2, and 3 of this section, and use your graphing calculator to graph the inequalities.

32. Use a graphing calculator to check your graphs for Problems 1–24.

33. Use the DRAW feature of your graphing calculator to draw each of the following.
   (a) A line segment between $(-2, -4)$ and $(-2, 5)$
   (b) A line segment between $(2, 2)$ and $(5, 2)$
   (c) A line segment between $(2, 3)$ and $(5, 7)$
   (d) A triangle with vertices at $(1, -2)$, $(3, 4)$, and $(-3, 6)$

7.4 Distance and Slope

As we work with the rectangular coordinate system, it is sometimes necessary to express the length of certain line segments. In other words, we need to be able to find the distance between two points. Let's first consider two specific examples and then develop the general distance formula.

EXAMPLE 1

Find the distance between the points $A(2, 2)$ and $B(5, 2)$ and also between the points $C(-2, 5)$ and $D(-2, -4)$.

Solution

Let's plot the points and draw $\overline{AB}$ as in Figure 7.28. Because $\overline{AB}$ is parallel to the x axis, its length can be expressed as $|5 - 2|$ or $|2 - 5|$. (The absolute-value symbol is used to ensure a nonnegative value.) Thus the length of $\overline{AB}$ is 3 units. Likewise, the length of $\overline{CD}$ is $|5 - (-4)| = |-4 - 5| = 9$ units.

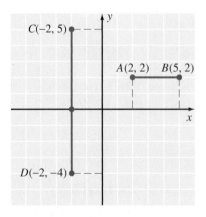

Figure 7.28

EXAMPLE 2

Find the distance between the points $A(2, 3)$ and $B(5, 7)$.

Solution

Let's plot the points and form a right triangle as indicated in Figure 7.29. Note that the coordinates of point C are $(5, 3)$. Because $\overline{AC}$ is parallel to the horizontal axis, its length is easily determined to be 3 units. Likewise, $\overline{CB}$ is parallel to the vertical axis and its length is 4 units. Let d represent the length of $\overline{AB}$, and apply the Pythagorean theorem to obtain

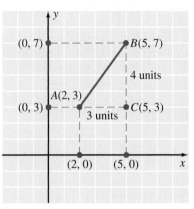

$$d^2 = 3^2 + 4^2$$

$$d^2 = 9 + 16$$

$$d^2 = 25$$

$$d = \pm\sqrt{25} = \pm 5$$

"Distance between" is a nonnegative value, so the length of $\overline{AB}$ is 5 units.

Figure 7.29   ◼

We can use the approach we used in Example 2 to develop a general distance formula for finding the distance between any two points in a coordinate plane. The development proceeds as follows:

1. Let $P_1(x_1, y_1)$ and $P_2(x_2, y_2)$ represent any two points in a coordinate plane.

2. Form a right triangle as indicated in Figure 7.30. The coordinates of the vertex of the right angle, point R, are (x_2, y_1).

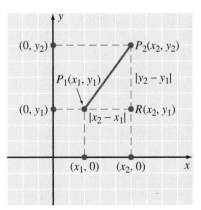

Figure 7.30

The length of $\overline{P_1R}$ is $|x_2 - x_1|$ and the length of $\overline{RP_2}$ is $|y_2 - y_1|$. (The absolute-value symbol is used to ensure a nonnegative value.) Let d represent the length of P_1P_2 and apply the Pythagorean theorem to obtain

$$d^2 = |x_2 - x_1|^2 + |y_2 - y_1|^2$$

Because $|a|^2 = a^2$, the **distance formula** can be stated as

$$d = \sqrt{(x_2 - x_1)^2 + (y_2 - y_1)^2}$$

It makes no difference which point you call P_1 or P_2 when using the distance formula. If you forget the formula, don't panic. Just form a right triangle and apply the Pythagorean theorem as we did in Example 2. Let's consider an example that demonstrates the use of the distance formula.

E X A M P L E   3

Find the distance between $(-1, 4)$ and $(1, 2)$.

Solution

Let $(-1, 4)$ be P_1 and $(1, 2)$ be P_2. Using the distance formula, we obtain

$$\begin{aligned}
d &= \sqrt{[(1 - (-1))]^2 + (2 - 4)^2} \\
&= \sqrt{2^2 + (-2)^2} \\
&= \sqrt{4 + 4} \\
&= \sqrt{8} = 2\sqrt{2} \qquad \text{Express the answer in simplest radical form.}
\end{aligned}$$

The distance between the two points is $2\sqrt{2}$ units.   ∎

In Example 3, we did not sketch a figure because of the simplicity of the problem. However, sometimes it is helpful to use a figure to organize the given information and aid in the analysis of the problem, as we see in the next example.

E X A M P L E   4

Verify that the points $(-3, 6)$, $(3, 4)$, and $(1, -2)$ are vertices of an isosceles triangle. (An isosceles triangle has two sides of the same length.)

Solution

Let's plot the points and draw the triangle (Figure 7.31). Use the distance formula to find the lengths d_1, d_2, and d_3, as follows:

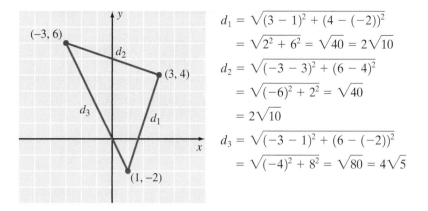

$$d_1 = \sqrt{(3-1)^2 + (4-(-2))^2}$$
$$= \sqrt{2^2 + 6^2} = \sqrt{40} = 2\sqrt{10}$$
$$d_2 = \sqrt{(-3-3)^2 + (6-4)^2}$$
$$= \sqrt{(-6)^2 + 2^2} = \sqrt{40}$$
$$= 2\sqrt{10}$$
$$d_3 = \sqrt{(-3-1)^2 + (6-(-2))^2}$$
$$= \sqrt{(-4)^2 + 8^2} = \sqrt{80} = 4\sqrt{5}$$

Figure 7.31

Because $d_1 = d_2$, we know that it is an isosceles triangle. ∎

■ Slope of a Line

In coordinate geometry, the concept of **slope** is used to describe the "steepness" of lines. The slope of a line is the ratio of the vertical change to the horizontal change as we move from one point on a line to another point. This is illustrated in Figure 7.32 with points P_1 and P_2.

A precise definition for slope can be given by considering the coordinates of the points P_1, P_2, and R as indicated in Figure 7.33. The horizontal change as we move from P_1 to P_2 is $x_2 - x_1$ and the vertical change is $y_2 - y_1$. Thus the following definition for slope is given.

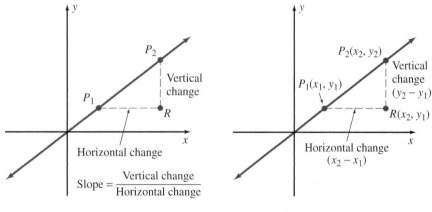

Figure 7.32

Figure 7.33

Definition 7.1

If points P_1 and P_2 with coordinates (x_1, y_1) and (x_2, y_2), respectively, are any two different points on a line, then the slope of the line (denoted by m) is

$$m = \frac{y_2 - y_1}{x_2 - x_1}, \qquad x_2 \neq x_1$$

Because $\dfrac{y_2 - y_1}{x_2 - x_1} = \dfrac{y_1 - y_2}{x_1 - x_2}$, how we designate P_1 and P_2 is not important. Let's use Definition 7.1 to find the slopes of some lines.

E X A M P L E   5

Find the slope of the line determined by each of the following pairs of points, and graph the lines.

(a) $(-1, 1)$ and $(3, 2)$ **(b)** $(4, -2)$ and $(-1, 5)$

(c) $(2, -3)$ and $(-3, -3)$

Solution

(a) Let $(-1, 1)$ be P_1 and $(3, 2)$ be P_2 (Figure 7.34).

$$m = \frac{y_2 - y_1}{x_2 - x_1} = \frac{2 - 1}{3 - (-1)} = \frac{1}{4}$$

(b) Let $(4, -2)$ be P_1 and $(-1, 5)$ be P_2 (Figure 7.35).

$$m = \frac{y_2 - y_1}{x_2 - x_1} = \frac{5 - (-2)}{-1 - 4} = \frac{7}{-5} = -\frac{7}{5}$$

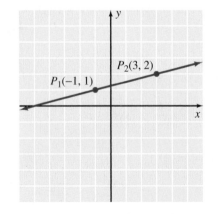

Figure 7.34

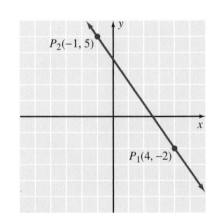

Figure 7.35

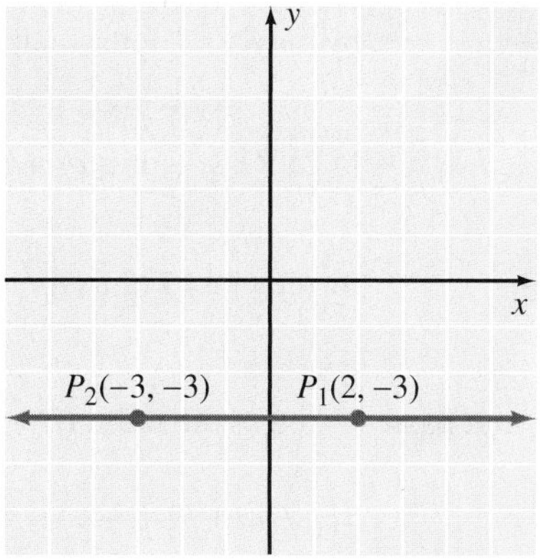

Figure 7.36 ■

(c) Let $(2, -3)$ be P_1 and $(-3, -3)$ be P_2 (Figure 7.36).

$$m = \frac{y_2 - y_1}{x_2 - x_1}$$

$$= \frac{-3 - (-3)}{-3 - 2}$$

$$= \frac{0}{-5} = 0$$

The three parts of Example 5 represent the three basic possibilities for slope; that is, the slope of a line can be positive, negative, or zero. A line that has a positive slope rises as we move from left to right, as in Figure 7.34. A line that has a negative slope falls as we move from left to right, as in Figure 7.35. A horizontal line, as in Figure 7.36, has a slope of zero. Finally, we need to realize that *the concept of slope is undefined for vertical lines*. This is due to the fact that for any vertical line, the horizontal change as we move from one point on the line to another is zero. Thus the ratio $\dfrac{y_2 - y_1}{x_2 - x_1}$ will have a denominator of zero and be undefined. Accordingly, the restriction $x_2 \neq x_1$ is imposed in Definition 7.1.

One final idea pertaining to the concept of slope needs to be emphasized. The slope of a line is a **ratio**, the ratio of vertical change to horizontal change. A slope of $\dfrac{2}{3}$ means that for every 2 units of vertical change there must be a corresponding 3 units of horizontal change. Thus, starting at some point on a line that has a slope of $\dfrac{2}{3}$, we could locate other points on the line as follows:

$$\frac{2}{3} = \frac{4}{6} \qquad \longrightarrow \text{ by moving 4 units } up \text{ and 6 units to the } right$$

$$\frac{2}{3} = \frac{8}{12} \qquad \longrightarrow \text{ by moving 8 units } up \text{ and 12 units to the } right$$

$$\frac{2}{3} = \frac{-2}{-3} \qquad \longrightarrow \text{ by moving 2 units } down \text{ and 3 units to the } left$$

Likewise, if a line has a slope of $-\dfrac{3}{4}$, then by starting at some point on the line we could locate other points on the line as follows:

$$-\frac{3}{4} = \frac{-3}{4} \qquad \longrightarrow \text{ by moving 3 units } down \text{ and 4 units to the } right$$

$$-\frac{3}{4} = \frac{3}{-4} \qquad \longrightarrow \text{ by moving 3 units } up \text{ and 4 units to the } left$$

$$-\frac{3}{4} = \frac{-9}{12} \qquad \longrightarrow \text{ by moving 9 units } down \text{ and 12 units to the } right$$

$$-\frac{3}{4} = \frac{15}{-20} \qquad \longrightarrow \text{ by moving 15 units } up \text{ and 20 units to the } left$$

EXAMPLE 6

Graph the line that passes through the point $(0, -2)$ and has a slope of $\frac{1}{3}$.

Solution

To graph, plot the point $(0, -2)$. Furthermore, because the slope $= \frac{\text{vertical change}}{\text{horizontal change}} = \frac{1}{3}$, we can locate another point on the line by starting from the point $(0, -2)$ and moving 1 unit up and 3 units to the right to obtain the point $(3, -1)$. Because two points determine a line, we can draw the line (Figure 7.37).

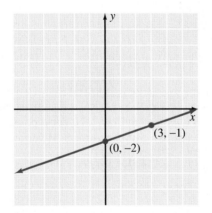

Figure 7.37

Remark: Because $m = \frac{1}{3} = \frac{-1}{-3}$, we can locate another point by moving 1 unit down and 3 units to the left from the point $(0, -2)$. ∎

EXAMPLE 7

Graph the line that passes through the point $(1, 3)$ and has a slope of -2.

Solution

To graph the line, plot the point $(1, 3)$. We know that $m = -2 = \frac{-2}{1}$. Furthermore, because the slope $= \frac{\text{vertical change}}{\text{horizontal change}} = \frac{-2}{1}$, we can locate another point on the line by starting from the point (1, 3) and moving 2 units down and 1 unit to the right to obtain the point (2, 1). Because two points determine a line, we can draw the line (Figure 7.38).

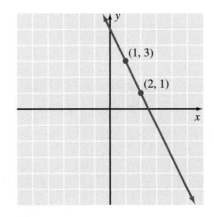

Figure 7.38

Remark: Because $m = -2 = \dfrac{-2}{1} = \dfrac{2}{-1}$ we can locate another point by moving 2 units up and 1 unit to the left from the point (1, 3). ∎

■ Applications of Slope

The concept of slope has many real-world applications even though the word *slope* is often not used. The concept of slope is used in most situations where an incline is involved. Hospital beds are hinged in the middle so that both the head end and the foot end can be raised or lowered; that is, the slope of either end of the bed can be changed. Likewise, treadmills are designed so that the incline (slope) of the platform can be adjusted. A roofer, when making an estimate to replace a roof, is concerned not only about the total area to be covered but also about the pitch of the roof. (Contractors do not define *pitch* as identical with the mathematical definition of slope, but both concepts refer to "steepness.") In Figure 7.39, the two roofs might require the same amount of shingles, but the roof on the left will take longer to complete because the pitch is so great that scaffolding will be required.

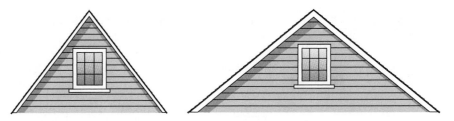

Figure 7.39

The concept of slope is also used in the construction of flights of stairs (Figure 7.40). The terms *rise* and *run* are commonly used, and the steepness (slope) of the stairs can be expressed as the ratio of rise to run. In Figure 7.40, the stairs on the left, where the ratio of rise to run is $\dfrac{10}{11}$, are steeper than the stairs on the right, which have a ratio of $\dfrac{7}{11}$.

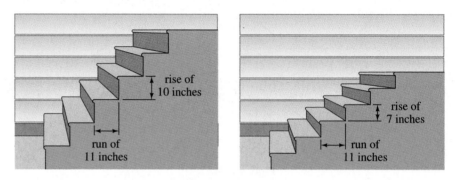

Figure 7.40

In highway construction, the word *grade* is used for the concept of slope. For example, in Figure 7.41 the highway is said to have a grade of 17%. This means that for every horizontal distance of 100 feet, the highway rises or drops 17 feet. In other words, the slope of the highway is $\dfrac{17}{100}$.

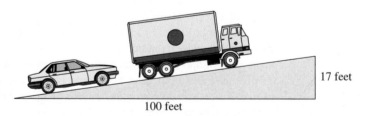

Figure 7.41

EXAMPLE 8

A certain highway has a 3% grade. How many feet does it rise in a horizontal distance of 1 mile?

Solution

A 3% grade means a slope of $\dfrac{3}{100}$. Therefore, if we let *y* represent the unknown vertical distance and use the fact that 1 mile = 5280 feet, we can set up and solve the following proportion.

$$\frac{3}{100} = \frac{y}{5280}$$

$$100y = 3(5280) = 15{,}840$$

$$y = 158.4$$

The highway rises 158.4 feet in a horizontal distance of 1 mile. ∎

Problem Set 7.4

For Problems 1–12, find the distance between each of the pairs of points. Express answers in simplest radical form.

1. $(-2, -1), (7, 11)$ **2.** $(2, 1), (10, 7)$

3. $(1, -1), (3, -4)$ **4.** $(-1, 3), (2, -2)$

5. $(6, -4), (9, -7)$ **6.** $(-5, 2), (-1, 6)$

7. $(-3, 3), (0, -3)$ **8.** $(-2, -4), (4, 0)$

9. $(1, -6), (-5, -6)$ **10.** $(-2, 3), (-2, -7)$

11. $(1, 7), (4, -2)$ **12.** $(6, 4), (-4, -8)$

13. Verify that the points $(-3, 1)$, $(5, 7)$, and $(8, 3)$ are vertices of a right triangle. [*Hint:* If $a^2 + b^2 = c^2$, then it is a right triangle with the right angle opposite side c.]

14. Verify that the points $(0, 3)$, $(2, -3)$, and $(-4, -5)$ are vertices of an isosceles triangle.   2 sides are equal

15. Verify that the points $(7, 12)$ and $(11, 18)$ divide the line segment joining $(3, 6)$ and $(15, 24)$ into three segments of equal length.

16. Verify that $(3, 1)$ is the midpoint of the line segment joining $(-2, 6)$ and $(8, -4)$.

For Problems 17–28, graph the line determined by the two points and find the slope of the line.

17. $(1, 2), (4, 6)$ **18.** $(3, 1), (-2, -2)$

19. $(-4, 5), (-1, -2)$ **20.** $(-2, 5), (3, -1)$

21. $(2, 6), (6, -2)$ **22.** $(-2, -1), (2, -5)$

23. $(-6, 1), (-1, 4)$ **24.** $(-3, 3), (2, 3)$

25. $(-2, -4), (2, -4)$ **26.** $(1, -5), (4, -1)$

27. $(0, -2), (4, 0)$ **28.** $(-4, 0), (0, -6)$

29. Find x if the line through $(-2, 4)$ and $(x, 6)$ has a slope of $\dfrac{2}{9}$.

30. Find y if the line through $(1, y)$ and $(4, 2)$ has a slope of $\dfrac{5}{3}$.

31. Find x if the line through $(x, 4)$ and $(2, -5)$ has a slope of $-\dfrac{9}{4}$.

32. Find y if the line through $(5, 2)$ and $(-3, y)$ has a slope of $-\dfrac{7}{8}$.

For Problems 33–40, you are given one point on a line and the slope of the line. Find the coordinates of three other points on the line.

33. $(2, 5)$, $m = \dfrac{1}{2}$ **34.** $(3, 4)$, $m = \dfrac{5}{6}$

35. $(-3, 4)$, $m = 3$ **36.** $(-3, -6)$, $m = 1$

37. $(5, -2)$, $m = -\dfrac{2}{3}$ **38.** $(4, -1)$, $m = -\dfrac{3}{4}$

39. $(-2, -4)$, $m = -2$ **40.** $(-5, 3)$, $m = -3$

For Problems 41–48, graph the line that passes through the given point and has the given slope.

41. $(3, 1)$ $m = \dfrac{2}{3}$

42. $(-1, 0)$ $m = \dfrac{3}{4}$

43. $(-2, 3)$ $m = -1$

44. $(1, -4)$ $m = -3$

45. $(0, 5)$ $m = \dfrac{-1}{4}$

46. $(-3, 4)$ $m = \dfrac{-3}{2}$

47. $(2, -2)$ $m = \dfrac{3}{2}$

48. $(3, -4)$ $m = \dfrac{5}{2}$

For Problems 49–58, find the coordinates of two points on the given line, and then use those coordinates to find the slope of the line.

49. $2x + 3y = 6$

50. $4x + 5y = 20$

51. $x - 2y = 4$

52. $3x - y = 12$

53. $4x - 7y = 12$

54. $2x + 7y = 11$

55. $y = 4$

56. $x = 3$

57. $y = -5x$

58. $y - 6x = 0$

59. A certain highway has a 2% grade. How many feet does it rise in a horizontal distance of 1 mile? (1 mile = 5280 feet)

60. The grade of a highway up a hill is 30%. How much change in horizontal distance is there if the vertical height of the hill is 75 feet?

61. Suppose that a highway rises a distance of 215 feet in a horizontal distance of 2640 feet. Express the grade of the highway to the nearest tenth of a percent.

62. If the ratio of rise to run is to be $\dfrac{3}{5}$ for some steps and the rise is 19 centimeters, find the run to the nearest centimeter.

63. If the ratio of rise to run is to be $\dfrac{2}{3}$ for some steps, and the run is 28 centimeters, find the rise to the nearest centimeter.

64. Suppose that a county ordinance requires a $2\dfrac{1}{4}\%$ "fall" for a sewage pipe from the house to the main pipe at the street. How much vertical drop must there be for a horizontal distance of 45 feet? Express the answer to the nearest tenth of a foot.

■ ■ ■ THOUGHTS INTO WORDS

65. How would you explain the concept of slope to someone who was absent from class the day it was discussed?

66. If one line has a slope of $\dfrac{2}{5}$, and another line has a slope of $\dfrac{3}{7}$, which line is steeper? Explain your answer.

67. Suppose that a line has a slope of $\dfrac{2}{3}$ and contains the point $(4, 7)$. Are the points $(7, 9)$ and $(1, 3)$ also on the line? Explain your answer.

■ ■ ■ FURTHER INVESTIGATIONS

68. Sometimes it is necessary to find the coordinate of a point on a number line that is located somewhere between two given points. For example, suppose that we want to find the coordinate (x) of the point located two-thirds of the distance from 2 to 8. Because the total distance from 2 to 8 is $8 - 2 = 6$ units, we can start at 2 and move $\dfrac{2}{3}(6) = 4$ units toward 8. Thus $x = 2 + \dfrac{2}{3}(6) = 2 + 4 = 6$.

For each of the following, find the coordinate of the indicated point on a number line.
(a) Two-thirds of the distance from 1 to 10
(b) Three-fourths of the distance from -2 to 14
(c) One-third of the distance from -3 to 7
(d) Two-fifths of the distance from -5 to 6
(e) Three-fifths of the distance from -1 to -11
(f) Five-sixths of the distance from 3 to -7

69. Now suppose that we want to find the coordinates of point P, which is located two-thirds of the distance from $A(1, 2)$ to $B(7, 5)$ in a coordinate plane. We have plotted the given points A and B in Figure 7.42 to help with the analysis of this problem. Point D is two-thirds of the distance from A to C because parallel lines cut off proportional segments on every transversal that intersects the lines. Thus $\overline{AC}$ can be treated as a segment of a number line, as shown in Figure 7.43.

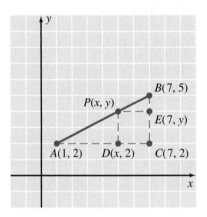

Figure 7.42

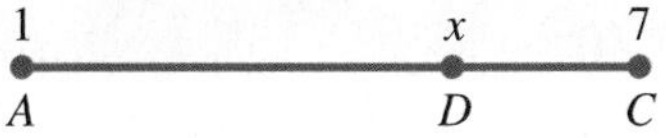

Figure 7.43

Therefore,

$$x = 1 + \frac{2}{3}(7 - 1) = 1 + \frac{2}{3}(6) = 5$$

Similarly, $\overline{CB}$ can be treated as a segment of a number line, as shown in Figure 7.44. Therefore,

$$y = 2 + \frac{2}{3}(5 - 2) = 2 + \frac{2}{3}(3) = 4$$

The coordinates of point P are $(5, 4)$.

Figure 7.44

For each of the following, find the coordinates of the indicated point in the xy plane.
(a) One-third of the distance from $(2, 3)$ to $(5, 9)$
(b) Two-thirds of the distance from $(1, 4)$ to $(7, 13)$
(c) Two-fifths of the distance from $(-2, 1)$ to $(8, 11)$
(d) Three-fifths of the distance from $(2, -3)$ to $(-3, 8)$
(e) Five-eighths of the distance from $(-1, -2)$ to $(4, -10)$
(f) Seven-eighths of the distance from $(-2, 3)$ to $(-1, -9)$

70. Suppose we want to find the coordinates of the midpoint of a line segment. Let $P(x, y)$ represent the midpoint of the line segment from $A(x_1, y_1)$ to $B(x_2, y_2)$. Using the method in Problem 68, the formula for the x coordinate of the midpoint is $x = x_1 + \frac{1}{2}(x_2 - x_1)$. This formula can be simplified algebraically to produce a simpler formula.

$$x = x_1 + \frac{1}{2}(x_2 - x_1)$$

$$x = x_1 + \frac{1}{2}x_2 - \frac{1}{2}x_1$$

$$x = \frac{1}{2}x_1 + \frac{1}{2}x_2$$

$$x = \frac{x_1 + x_2}{2}$$

Hence the x coordinate of the midpoint can be interpreted as the average of the x coordinates of the endpoints of the line segment. A similar argument for the y coordinate of the midpoint gives the following formula.

$$y = \frac{y_1 + y_2}{2}$$

For each of the pairs of points, use the formula to find the midpoint of the line segment between the points.
(a) $(3, 1)$ and $(7, 5)$
(b) $(-2, 8)$ and $(6, 4)$
(c) $(-3, 2)$ and $(5, 8)$
(d) $(4, 10)$ and $(9, 25)$
(e) $(-4, -1)$ and $(-10, 5)$
(f) $(5, 8)$ and $(-1, 7)$

GRAPHING CALCULATOR ACTIVITIES

71. Remember that we did some work with parallel lines back in the graphing calculator activities in Problem Set 7.1. Now let's do some work with perpendicular lines. Be sure to set your boundaries so that the distance between tick marks is the same on both axes.

(a) Graph $y = 4x$ and $y = -\frac{1}{4}x$ on the same set of axes. Do they appear to be perpendicular lines?

(b) Graph $y = 3x$ and $y = \frac{1}{3}x$ on the same set of axes. Do they appear to be perpendicular lines?

(c) Graph $y = \frac{2}{5}x - 1$ and $y = -\frac{5}{2}x + 2$ on the same set of axes. Do they appear to be perpendicular lines?

(d) Graph $y = \frac{3}{4}x - 3$, $y = \frac{4}{3}x + 2$, and $y = -\frac{4}{3}x + 2$ on the same set of axes. Does there appear to be a pair of perpendicular lines?

(e) On the basis of your results in parts (a) through (d), make a statement about how we can recognize perpendicular lines from their equations.

72. For each of the following pairs of equations, (1) predict whether they represent parallel lines, perpendicular lines, or lines that intersect but are not perpendicular, and (2) graph each pair of lines to check your prediction.

(a) $5.2x + 3.3y = 9.4$ and $5.2x + 3.3y = 12.6$
(b) $1.3x - 4.7y = 3.4$ and $1.3x - 4.7y = 11.6$
(c) $2.7x + 3.9y = 1.4$ and $2.7x - 3.9y = 8.2$
(d) $5x - 7y = 17$ and $7x + 5y = 19$
(e) $9x + 2y = 14$ and $2x + 9y = 17$
(f) $2.1x + 3.4y = 11.7$ and $3.4x - 2.1y = 17.3$

7.5   Determining the Equation of a Line

To review, there are basically two types of problems to solve in coordinate geometry:

1. Given an algebraic equation, find its geometric graph.

2. Given a set of conditions pertaining to a geometric figure, find its algebraic equation.

Problems of type 1 have been our primary concern thus far in this chapter. Now let's analyze some problems of type 2 that deal specifically with straight lines. Given certain facts about a line, we need to be able to determine its algebraic equation. Let's consider some examples.

EXAMPLE 1

Find the equation of the line that has a slope of $\frac{2}{3}$ and contains the point $(1, 2)$.

Solution

First, let's draw the line and record the given information. Then choose a point (x, y) that represents any point on the line other than the given point $(1, 2)$. (See Figure 7.45.)

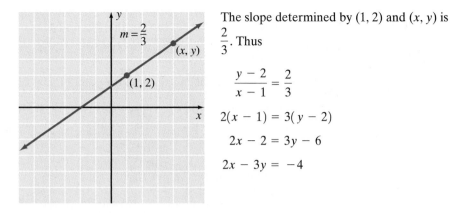

The slope determined by $(1, 2)$ and (x, y) is $\frac{2}{3}$. Thus

$$\frac{y - 2}{x - 1} = \frac{2}{3}$$

$$2(x - 1) = 3(y - 2)$$

$$2x - 2 = 3y - 6$$

$$2x - 3y = -4$$

Figure 7.45   ∎

E X A M P L E   2

Find the equation of the line that contains $(3, 2)$ and $(-2, 5)$.

Solution

First, let's draw the line determined by the given points (Figure 7.46); if we know two points, we can find the slope.

$$m = \frac{y_2 - y_1}{x_2 - x_1} = \frac{3}{-5} = -\frac{3}{5}$$

Now we can use the same approach as in Example 1.

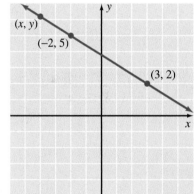

Figure 7.46

Form an equation using a variable point (x, y), one of the two given points, and the slope of $-\frac{3}{5}$.

$$\frac{y - 5}{x + 2} = \frac{3}{-5} \qquad \left(-\frac{3}{5} = \frac{3}{-5} \right)$$

$$3(x + 2) = -5(y - 5)$$

$$3x + 6 = -5y + 25$$

$$3x + 5y = 19$$ ∎

E X A M P L E   3

Find the equation of the line that has a slope of $\dfrac{1}{4}$ and a y intercept of 2.

Solution

A y intercept of 2 means that the point $(0, 2)$ is on the line (Figure 7.47).

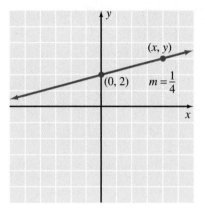

Figure 7.47

Choose a variable point (x, y) and proceed as in the previous examples.

$$\frac{y - 2}{x - 0} = \frac{1}{4}$$

$$1(x - 0) = 4(y - 2)$$

$$x = 4y - 8$$

$$x - 4y = -8$$ ■

Perhaps it would be helpful to pause a moment and look back over Examples 1, 2, and 3. Note that we used the same basic approach in all three situations. We chose a variable point (x, y) and used it to determine the equation that satisfies the conditions given in the problem. The approach we took in the previous examples can be generalized to produce some special forms of equations of straight lines.

■ Point-Slope Form

E X A M P L E   4

Find the equation of the line that has a slope of m and contains the point (x_1, y_1).

Solution

Choose (x, y) to represent any other point on the line (Figure 7.48), and the slope of the line is therefore given by

$$m = \frac{y - y_1}{x - x_1}, \qquad x \neq x_1$$

from which

$$y - y_1 = m(x - x_1)$$

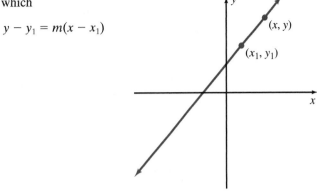

Figure 7.48   ■

We refer to the equation

$$y - y_1 = m(x - x_1)$$

as the **point-slope form** of the equation of a straight line. Instead of the approach we used in Example 1, we could use the point-slope form to write the equation of a line with a given slope that contains a given point. For example, we can determine the equation of the line that has a slope of $\dfrac{3}{5}$ and contains the point $(2, 4)$ as follows:

$$y - y_1 = m(x - x_1)$$

Substitute $(2, 4)$ for (x_1, y_1) and $\dfrac{3}{5}$ for m.

$$y - 4 = \frac{3}{5}(x - 2)$$
$$5(y - 4) = 3(x - 2)$$
$$5y - 20 = 3x - 6$$
$$-14 = 3x - 5y$$

■ Slope-Intercept Form

EXAMPLE 5   Find the equation of the line that has a slope of m and a y intercept of b.

Solution

A y intercept of b means that the line contains the point $(0, b)$, as in Figure 7.49. Therefore, we can use the point-slope form as follows:

$$y - y_1 = m(x - x_1)$$
$$y - b = m(x - 0)$$
$$y - b = mx$$
$$y = mx + b$$

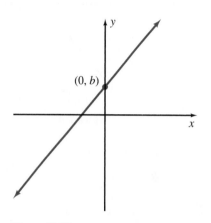

Figure 7.49   ■

We refer to the equation

$$y = mx + b$$

as the **slope-intercept form** of the equation of a straight line. We use it for three primary purposes, as the next three examples illustrate.

E X A M P L E   6   Find the equation of the line that has a slope of $\dfrac{1}{4}$ and a y intercept of 2.

Solution

This is a restatement of Example 3, but this time we will use the slope-intercept form ($y = mx + b$) of a line to write its equation. Because $m = \dfrac{1}{4}$ and $b = 2$, we can substitute these values into $y = mx + b$.

$$y = mx + b$$

$$y = \frac{1}{4}x + 2$$

$$4y = x + 8 \qquad \text{Multiply both sides by 4.}$$

$$x - 4y = -8 \qquad \text{Same result as in Example 3.} \qquad ■$$

EXAMPLE   7

Find the slope of the line when the equation is $3x + 2y = 6$.

Solution

We can solve the equation for y in terms of x and then compare it to the slope-intercept form to determine its slope. Thus

$$3x + 2y = 6$$

$$2y = -3x + 6$$

$$y = -\frac{3}{2}x + 3$$

$$y = -\frac{3}{2}x + 3 \qquad y = mx + b$$

The slope of the line is $-\frac{3}{2}$. Furthermore, the y intercept is 3.   ∎

EXAMPLE   8

Graph the line determined by the equation $y = \frac{2}{3}x - 1$.

Solution

Comparing the given equation to the general slope-intercept form, we see that the slope of the line is $\frac{2}{3}$ and the y intercept is -1. Because the y intercept is -1, we can plot the point $(0, -1)$. Then, because the slope is $\frac{2}{3}$, let's move 3 units to the right and 2 units up from $(0, -1)$ to locate the point $(3, 1)$. The two points $(0, -1)$ and $(3, 1)$ determine the line in Figure 7.50. (Again, you should determine a third point as a check point.)

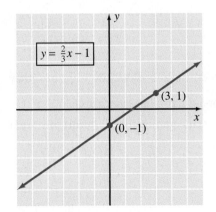

Figure 7.50   ∎

> In general, if the equation of a nonvertical line is written in slope-intercept form ($y = mx + b$), the coefficient of x is the slope of the line, and the constant term is the y intercept. (Remember that the concept of slope is not defined for a vertical line.)

■ Parallel and Perpendicular Lines

We can use two important relationships between lines and their slopes to solve certain kinds of problems. It can be shown that nonvertical parallel lines have the same slope and that two nonvertical lines are perpendicular if the product of their slopes is -1. (Details for verifying these facts are left to another course.) In other words, if two lines have slopes m_1 and m_2, respectively, then

1. The two lines are parallel if and only if $m_1 = m_2$.

2. The two lines are perpendicular if and only if $(m_1)(m_2) = -1$.

The following examples demonstrate the use of these properties.

(a) Verify that the graphs of $2x + 3y = 7$ and $4x + 6y = 11$ are parallel lines.

(b) Verify that the graphs of $8x - 12y = 3$ and $3x + 2y = 2$ are perpendicular lines.

Solution

(a) Let's change each equation to slope-intercept form.

$$2x + 3y = 7 \quad \longrightarrow \quad 3y = -2x + 7$$

$$y = -\frac{2}{3}x + \frac{7}{3}$$

$$4x + 6y = 11 \quad \longrightarrow \quad 6y = -4x + 11$$

$$y = -\frac{4}{6}x + \frac{11}{6}$$

$$y = -\frac{2}{3}x + \frac{11}{6}$$

Both lines have a slope of $-\dfrac{2}{3}$, but they have different y intercepts. Therefore, the two lines are parallel.

(b) Solving each equation for y in terms of x, we obtain

$$8x - 12y = 3 \quad \longrightarrow \quad -12y = -8x + 3$$

$$y = \frac{8}{12}x - \frac{3}{12}$$

$$y = \frac{2}{3}x - \frac{1}{4}$$

$$3x + 2y = 2 \quad \longrightarrow \quad 2y = -3x + 2$$

$$y = -\frac{3}{2}x + 1$$

Because $\left(\dfrac{2}{3}\right)\left(-\dfrac{3}{2}\right) = -1$ (the product of the two slopes is -1), the lines are perpendicular.   ■

Remark: The statement "the product of two slopes is -1" is the same as saying that the two slopes are negative reciprocals of each other; that is, $m_1 = -\dfrac{1}{m_2}$.

EXAMPLE   10

Find the equation of the line that contains the point $(1, 4)$ and is parallel to the line determined by $x + 2y = 5$.

Solution

First, let's draw a figure to help in our analysis of the problem (Figure 7.51). Because the line through $(1, 4)$ is to be parallel to the line determined by $x + 2y = 5$, it must have the same slope. Let's find the slope by changing $x + 2y = 5$ to the slope-intercept form.

$$x + 2y = 5$$

$$2y = -x + 5$$

$$y = -\frac{1}{2}x + \frac{5}{2}$$

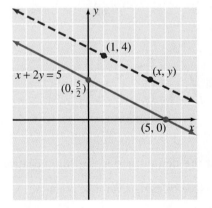

Figure 7.51

The slope of both lines is $-\dfrac{1}{2}$. Now we can choose a variable point (x, y) on the line through $(1, 4)$ and proceed as we did in earlier examples.

$$\frac{y - 4}{x - 1} = \frac{1}{-2}$$

$$1(x - 1) = -2(y - 4)$$

$$x - 1 = -2y + 8$$

$$x + 2y = 9$$ ■

E X A M P L E   1 1

Find the equation of the line that contains the point $(-1, -2)$ and is perpendicular to the line determined by $2x - y = 6$.

Solution

First, let's draw a figure to help in our analysis of the problem (Figure 7.52). Because the line through $(-1, -2)$ is to be perpendicular to the line determined by $2x - y = 6$, its slope must be the negative reciprocal of the slope of $2x - y = 6$. Let's find the slope of $2x - y = 6$ by changing it to the slope-intercept form.

$$2x - y = 6$$

$$-y = -2x + 6$$

$$y = 2x - 6 \qquad \text{The slope is 2.}$$

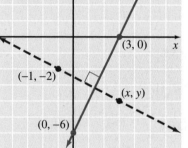

Figure 7.52

The slope of the desired line is $-\dfrac{1}{2}$ (the negative reciprocal of 2), and we can proceed as before by using a variable point (x, y).

$$\frac{y + 2}{x + 1} = \frac{1}{-2}$$

$$1(x + 1) = -2(y + 2)$$

$$x + 1 = -2y - 4$$

$$x + 2y = -5$$ ■

We use two forms of equations of straight lines extensively. They are the **standard form** and the **slope-intercept form**, and we describe them as follows.

Standard Form $Ax + By = C$, where B and C are integers, and A is a nonnegative integer (A and B not both zero).

Slope-Intercept Form $y = mx + b$, where m is a real number representing the slope, and b is a real number representing the y intercept.

Problem Set 7.5

For Problems 1–8, write the equation of the line that has the indicated slope and contains the indicated point. Express final equations in standard form.

1. $m = \dfrac{1}{2}$, $(3, 5)$

2. $m = \dfrac{1}{3}$, $(2, 3)$

3. $m = 3$, $(-2, 4)$

4. $m = -2$, $(-1, 6)$

5. $m = -\dfrac{3}{4}$, $(-1, -3)$

6. $m = -\dfrac{3}{5}$, $(-2, -4)$

7. $m = \dfrac{5}{4}$, $(4, -2)$

8. $m = \dfrac{3}{2}$, $(8, -2)$

For Problems 9–18, write the equation of the line that contains the indicated pair of points. Express final equations in standard form.

9. $(2, 1), (6, 5)$

10. $(-1, 2), (2, 5)$

11. $(-2, -3), (2, 7)$

12. $(-3, -4), (1, 2)$

13. $(-3, 2), (4, 1)$

14. $(-2, 5), (3, -3)$

15. $(-1, -4), (3, -6)$

16. $(3, 8), (7, 2)$

17. $(0, 0), (5, 7)$

18. $(0, 0), (-5, 9)$

For Problems 19–26, write the equation of the line that has the indicated slope (m) and y intercept (b). Express final equations in slope-intercept form.

19. $m = \dfrac{3}{7}$, $b = 4$

20. $m = \dfrac{2}{9}$, $b = 6$

21. $m = 2$, $b = -3$

22. $m = -3$, $b = -1$

23. $m = -\dfrac{2}{5}$, $b = 1$

24. $m = -\dfrac{3}{7}$, $b = 4$

25. $m = 0$, $b = -4$

26. $m = \dfrac{1}{5}$, $b = 0$

For Problems 27–42, write the equation of the line that satisfies the given conditions. Express final equations in standard form.

27. x intercept of 2 and y intercept of -4

28. x intercept of -1 and y intercept of -3

29. x intercept of -3 and slope of $-\dfrac{5}{8}$

30. x intercept of 5 and slope of $-\dfrac{3}{10}$

31. Contains the point $(2, -4)$ and is parallel to the y axis

32. Contains the point $(-3, -7)$ and is parallel to the x axis

33. Contains the point $(5, 6)$ and is perpendicular to the y axis

34. Contains the point $(-4, 7)$ and is perpendicular to the x axis

35. Contains the point $(1, 3)$ and is parallel to the line $x + 5y = 9$

36. Contains the point $(-1, 4)$ and is parallel to the line $x - 2y = 6$

37. Contains the origin and is parallel to the line $4x - 7y = 3$

38. Contains the origin and is parallel to the line $-2x - 9y = 4$

39. Contains the point $(-1, 3)$ and is perpendicular to the line $2x - y = 4$

40. Contains the point $(-2, -3)$ and is perpendicular to the line $x + 4y = 6$

41. Contains the origin and is perpendicular to the line $-2x + 3y = 8$

42. Contains the origin and is perpendicular to the line $y = -5x$

For Problems 43–48, change the equation to slope-intercept form and determine the slope and y intercept of the line.

43. $3x + y = 7$ **44.** $5x - y = 9$

45. $3x + 2y = 9$ **46.** $x - 4y = 3$

47. $x = 5y + 12$ **48.** $-4x - 7y = 14$

For Problems 49–56, use the slope-intercept form to graph the following lines.

49. $y = \dfrac{2}{3}x - 4$ **50.** $y = \dfrac{1}{4}x + 2$

51. $y = 2x + 1$ **52.** $y = 3x - 1$

53. $y = -\dfrac{3}{2}x + 4$ **54.** $y = -\dfrac{5}{3}x + 3$

55. $y = -x + 2$ **56.** $y = -2x + 4$

For Problems 57–66, graph the following lines using the technique that seems most appropriate.

57. $y = -\dfrac{2}{5}x - 1$ **58.** $y = -\dfrac{1}{2}x + 3$

59. $x + 2y = 5$ **60.** $2x - y = 7$

61. $-y = -4x + 7$ **62.** $3x = 2y$

63. $7y = -2x$ **64.** $y = -3$

65. $x = 2$ **66.** $y = -x$

For Problems 67–70, the situations can be described by the use of linear equations in two variables. If two pairs of values are known, then we can determine the equation by using the approach we used in Example 2 of this section. For each of the following, assume that the relationship can be expressed as a linear equation in two variables, and use the given information to determine the equation. Express the equation in slope-intercept form.

67. A company uses 7 pounds of fertilizer for a lawn that measures 5000 square feet and 12 pounds for a lawn that measures 10,000 square feet. Let y represent the pounds of fertilizer and x the square footage of the lawn.

68. A new diet fad claims that a person weighing 140 pounds should consume 1490 daily calories and that a 200-pound person should consume 1700 calories. Let y represent the calories and x the weight of the person in pounds.

69. Two banks on opposite corners of a town square had signs that displayed the current temperature. One bank displayed the temperature in degrees Celsius and the other in degrees Fahrenheit. A temperature of $10°C$ was displayed at the same time as a temperature of $50°F$. On another day, a temperature of $-5°C$ was displayed at the same time as a temperature of $23°F$. Let y represent the temperature in degrees Fahrenheit and x the temperature in degrees Celsius.

70. An accountant has a schedule of depreciation for some business equipment. The schedule shows that after 12 months the equipment is worth \$7600 and that after 20 months it is worth \$6000. Let y represent the worth and x represent the time in months.

■ ■ ■ **THOUGHTS INTO WORDS**

71. What does it mean to say that two points determine a line?

72. How would you help a friend determine the equation of the line that is perpendicular to $x - 5y = 7$ and contains the point $(5, 4)$?

73. Explain how you would find the slope of the line $y = 4$.

■ ■ ■ **FURTHER INVESTIGATIONS**

74. The equation of a line that contains the two points (x_1, y_1) and (x_2, y_2) is $\dfrac{y - y_1}{x - x_1} = \dfrac{y_2 - y_1}{x_2 - x_1}$. We often refer to this as the **two-point form** of the equation of a straight line. Use the two-point form and write the equation of the line that contains each of the indicated pairs of points. Express final equations in standard form.
(a) $(1, 1)$ and $(5, 2)$
(b) $(2, 4)$ and $(-2, -1)$
(c) $(-3, 5)$ and $(3, 1)$
(d) $(-5, 1)$ and $(2, -7)$

75. Let $Ax + By = C$ and $A'x + B'y = C'$ represent two lines. Change both of these equations to slope-intercept form, and then verify each of the following properties.
(a) If $\dfrac{A}{A'} = \dfrac{B}{B'} \neq \dfrac{C}{C'}$, then the lines are parallel.
(b) If $AA' = -BB'$, then the lines are perpendicular.

76. The properties in Problem 75 provide us with another way to write the equation of a line parallel or perpendicular to a given line that contains a given point not on the line. For example, suppose that we want the equation of the line perpendicular to $3x + 4y = 6$ that contains the point $(1, 2)$. The form $4x - 3y = k$, where k is a constant, represents a family of lines perpendicular to $3x + 4y = 6$ because we have satisfied the condition $AA' = -BB'$. Therefore, to find what specific line of the family contains $(1, 2)$, we substitute 1 for x and 2 for y to determine k.

$$4x - 3y = k$$
$$4(1) - 3(2) = k$$
$$-2 = k$$

Thus the equation of the desired line is $4x - 3y = -2$.

Use the properties from Problem 75 to help write the equation of each of the following lines.
(a) Contains $(1, 8)$ and is parallel to $2x + 3y = 6$
(b) Contains $(-1, 4)$ and is parallel to $x - 2y = 4$
(c) Contains $(2, -7)$ and is perpendicular to $3x - 5y = 10$
(d) Contains $(-1, -4)$ and is perpendicular to $2x + 5y = 12$

77. The problem of finding the perpendicular bisector of a line segment presents itself often in the study of analytic geometry. As with any problem of writing the equation of a line, you must determine the slope of the line and a point that the line passes through. A perpendicular bisector passes through the midpoint of the line segment and has a slope that is the negative reciprocal of the slope of the line segment. The problem can be solved as follows:

Find the perpendicular bisector of the line segment between the points $(1, -2)$ and $(7, 8)$.

The midpoint of the line segment is $\left(\dfrac{1 + 7}{2}, \dfrac{-2 + 8}{2} \right)$ $= (4, 3)$.

The slope of the line segment is $m = \dfrac{8 - (-2)}{7 - 1}$ $= \dfrac{10}{6} = \dfrac{5}{3}$.

Hence the perpendicular bisector will pass through the point $(4, 3)$ and have a slope of $m = -\dfrac{3}{5}$.

$$y - 3 = -\frac{3}{5}(x - 4)$$
$$5(y - 3) = -3(x - 4)$$
$$5y - 15 = -3x + 12$$
$$3x + 5y = 27$$

Thus the equation of the perpendicular bisector of the line segment between the points $(1, -2)$ and $(7, 8)$ is $3x + 5y = 27$.

Find the perpendicular bisector of the line segment between the points for the following. Write the equation in standard form.
(a) $(-1, 2)$ and $(3, 0)$
(b) $(6, -10)$ and $(-4, 2)$
(c) $(-7, -3)$ and $(5, 9)$
(d) $(0, 4)$ and $(12, -4)$

GRAPHING CALCULATOR ACTIVITIES

78. Predict whether each of the following pairs of equations represents parallel lines, perpendicular lines, or lines that intersect but are not perpendicular. Then graph each pair of lines to check your predictions. (The properties presented in Problem 75 should be very helpful.)

(a) $5.2x + 3.3y = 9.4$ and $5.2x + 3.3y = 12.6$

(b) $1.3x - 4.7y = 3.4$ and $1.3x - 4.7y = 11.6$

(c) $2.7x + 3.9y = 1.4$ and $2.7x - 3.9y = 8.2$

(d) $5x - 7y = 17$ and $7x + 5y = 19$

(e) $9x + 2y = 14$ and $2x + 9y = 17$

(f) $2.1x + 3.4y = 11.7$ and $3.4x - 2.1y = 17.3$

(g) $7.1x - 2.3y = 6.2$ and $2.3x + 7.1y = 9.9$

(h) $-3x + 9y = 12$ and $9x - 3y = 14$

(i) $2.6x - 5.3y = 3.4$ and $5.2x - 10.6y = 19.2$

(j) $4.8x - 5.6y = 3.4$ and $6.1x + 7.6y = 12.3$

Chapter 7 Summary

(7.1) The **Cartesian** (or **rectangular**) **coordinate system** is used to graph ordered pairs of real numbers. The first number, a, of the ordered pair (a, b) is called the **abscissa**, and the second number, b, is called the **ordinate**; together they are referred to as the **coordinates** of a point.

Two basic kinds of problems exist in coordinate geometry:

1. Given an algebraic equation, find its geometric graph.
2. Given a set of conditions that pertains to a geometric figure, find its algebraic equation.

A **solution** of an equation in two variables is an ordered pair of real numbers that satisfies the equation.

Any equation of the form $Ax + By = C$, where A, B, and C are constants (A and B not both zero) and x and y are variables, is a **linear equation**, and its graph is a **straight line**.

Any equation of the form $Ax + By = C$, where $C = 0$, is a straight line that contains the origin.

Any equation of the form $x = a$, where a is a constant, is a line parallel to the y axis that has an x intercept of a.

Any equation of the form $y = b$, where b is a constant, is a line parallel to the x axis that has a y intercept of b.

(7.2) The following suggestions are offered for **graphing an equation** in two variables.

1. Determine what type of symmetry the equation exhibits.
2. Find the intercepts.
3. Solve the equation for y in terms of x or for x in terms of y if it is not already in such a form.
4. Set up a table of ordered pairs that satisfy the equation. The type of symmetry will affect your choice of values in the table.
5. Plot the points associated with the ordered pairs from the table, and connect them with a smooth curve. Then, if appropriate, reflect this part of the curve according to the symmetry shown by the equation.

(7.3) **Linear inequalities** in two variables are of the form $Ax + By > C$ or $Ax + By < C$. To **graph a linear inequality**, we suggest the following steps.

1. First, graph the corresponding equality. Use a solid line if equality is included in the original statement. Use a dashed line if equality is not included.
2. Choose a test point not on the line and substitute its coordinates into the inequality.
3. The graph of the original inequality is
   (a) the half plane that contains the test point if the inequality is satisfied by that point, or
   (b) the half plane that does not contain the test point if the inequality is not satisfied by the point.

(7.4) The distance between any two points (x_1, y_1) and (x_2, y_2) is given by the **distance formula**,

$$d = \sqrt{(x_2 - x_1)^2 + (y_2 - y_1)^2}$$

The **slope** (denoted by m) of a line determined by the points (x_1, y_1) and (x_2, y_2) is given by the slope formula,

$$m = \frac{y_2 - y_1}{x_2 - x_1}, \qquad x_2 \neq x_1$$

(7.5) The equation $y = mx + b$ is referred to as the **slope-intercept form** of the equation of a straight line. If the equation of a nonvertical line is written in this y form, the coefficient of x is the slope of the line and the constant term is the y intercept.

If two lines have slopes m_1 and m_2, respectively, then

1. The two lines are parallel if and only if $m_1 = m_2$.
2. The two lines are perpendicular if and only if $(m_1)(m_2) = -1$.

To determine the equation of a straight line given a set of conditions, we can use the point-slope form, $y - y_1 = m(x - x_1)$, or $\dfrac{y - y_1}{x - x_1} = m$. The conditions generally fall into one of the following four categories.

1. Given the slope and a point contained in the line
2. Given two points contained in the line
3. Given a point contained in the line and that the line is parallel to another line
4. Given a point contained in the line and that the line is perpendicular to another line

The result can then be expressed in standard form or slope-intercept form.

Chapter 7   Review Problem Set

1. Find the slope of the line determined by each pair of points.
   (a) $(3, 4), (-2, -2)$ (b) $(-2, 3), (4, -1)$

2. Find y if the line through $(-4, 3)$ and $(12, y)$ has a slope of $\frac{1}{8}$.

3. Find x if the line through $(x, 5)$ and $(3, -1)$ has a slope of $-\frac{3}{2}$.

4. Find the slope of each of the following lines.
   (a) $4x + y = 7$ (b) $2x - 7y = 3$

5. Find the lengths of the sides of a triangle whose vertices are at $(2, 3)$, $(5, -1)$, and $(-4, -5)$.

6. Find the distance between each of the pairs of points.
   (a) $(-1, 4), (1, -2)$ (b) $(5, 0), (2, 7)$

7. Verify that $(1, 6)$ is the midpoint of the line segment joining $(3, 2)$ and $(-1, 10)$.

For Problems 8–15, write the equation of the line that satisfies the stated conditions. Express final equations in standard form.

8. Containing the points $(-1, 2)$ and $(3, -5)$

9. Having a slope of $-\frac{3}{7}$ and a y intercept of 4

10. Containing the point $(-1, -6)$ and having a slope of $\frac{2}{3}$

11. Containing the point $(2, 5)$ and parallel to the line $x - 2y = 4$

12. Containing the point $(-2, -6)$ and perpendicular to the line $3x + 2y = 12$

13. Containing the points $(0, 4)$ and $(2, 6)$

14. Containing the point $(3, -5)$ and having a slope of -1

15. Containing the point $(-8, 3)$ and parallel to the line $4x + y = 7$

For Problems 16–35, graph each equation.

16. $2x - y = 6$

17. $y = 2x - 5$

18. $y = -2x - 1$

19. $y = -4x$

20. $-3x - 2y = 6$

21. $x = 2y + 4$

22. $5x - y = -5$

23. $y = -\frac{1}{2}x + 3$

24. $y = \dfrac{3x - 4}{2}$

25. $y = 4$

26. $2x + 3y = 0$

27. $y = \frac{3}{5}x - 4$

28. $x = 1$

29. $x = -3$

30. $y = -2$

31. $2x - 3y = 3$

32. $y = x^3 + 2$

33. $y = -x^3$

34. $y = x^2 + 3$

35. $y = -2x^2 - 1$

For Problems 36–41, graph each inequality.

36. $-x + 3y < -6$

37. $x + 2y \geq 4$

38. $2x - 3y \leq 6$

39. $y > -\frac{1}{2}x + 3$

40. $y < 2x - 5$

41. $y \geq \frac{2}{3}x$

42. A certain highway has a 6% grade. How many feet does it rise in a horizontal distance of 1 mile?

43. If the ratio of rise to run is to be $\frac{2}{3}$ for the steps of a staircase, and the run is 12 inches, find the rise.

44. Find the slope of any line that is perpendicular to the line $-3x + 5y = 7$.

45. Find the slope of any line that is parallel to the line $4x + 5y = 10$.

46. The taxes for a primary residence can be described by a linear relationship. Find the equation for the rela-

tionship if the taxes for a home valued at $200,000 are $2400, and the taxes are $3150 when the home is valued at $250,000. Let y be the taxes and x the value of the home. Write the equation in slope-intercept form.

47. The freight charged by a trucking firm for a parcel under 200 pounds depends on the miles it is being shipped. To ship a 150-pound parcel 300 miles, it costs $40. If the same parcel is shipped 1000 miles, the cost is $180. Assume the relationship between the cost and miles is linear. Find the equation for the relationship. Let y be the cost and x be the miles. Write the equation in slope-intercept form.

48. On a final exam in math class, the number of points earned has a linear relationship with the number of correct answers. John got 96 points when he answered 12 questions correctly. Kimberly got 144 points when she answered 18 questions correctly. Find the equation for the relationship. Let y be the number of points and x be the number of correct answers. Write the equation in slope-intercept form.

49. The time needed to install computer cables has a linear relationship with the number of feet of cable being installed. It takes $1\frac{1}{2}$ hours to install 300 feet, and 1050 feet can be installed in 4 hours. Find the equation for the relationship. Let y be the feet of cable installed and x be the time in hours. Write the equation in slope-intercept form.

50. Determine the type(s) of symmetry (symmetry with respect to the x axis, y axis, and/or origin) exhibited by the graph of each of the following equations. Do not sketch the graph.
    (a) $y = x^2 + 4$ **(b)** $xy = -4$
    (c) $y = -x^3$ **(d)** $x = y^4 + 2y^2$

1. Find the slope of the line determined by the points $(-2, 4)$ and $(3, -2)$.

2. Find the slope of the line determined by the equation $3x - 7y = 12$.

3. Find the length of the line segment whose endpoints are $(4, 2)$ and $(-3, -1)$. Express the answer in simplest radical form.

4. Find the equation of the line that has a slope of $-\dfrac{3}{2}$ and contains the point $(4, -5)$. Express the equation in standard form.

5. Find the equation of the line that contains the points $(-4, 2)$ and $(2, 1)$. Express the equation in slope-intercept form.

6. Find the equation of the line that is parallel to the line $5x + 2y = 7$ and contains the point $(-2, -4)$. Express the equation in standard form.

7. Find the equation of the line that is perpendicular to the line $x - 6y = 9$ and contains the point $(4, 7)$. Express the equation in standard form.

8. What kind(s) of symmetry does the graph of $y = 9x$ exhibit?

9. What kind(s) of symmetry does the graph of $y^2 = x^2 + 6$ exhibit?

10. What kind(s) of symmetry does the graph of $x^2 + 6x + 2y^2 - 8 = 0$ exhibit?

11. What is the slope of all lines that are parallel to the line $7x - 2y = 9$?

12. What is the slope of all lines that are perpendicular to the line $4x + 9y = -6$?

13. Find the x intercept of the line $y = \dfrac{3}{5}x - \dfrac{2}{3}$.

14. Find the y intercept of the line $\dfrac{3}{4}x - \dfrac{2}{5}y = \dfrac{1}{4}$.

15. The grade of a highway up a hill is 25%. How much change in horizontal distance is there if the vertical height of the hill is 120 feet?

16. Suppose that a highway rises 200 feet in a horizontal distance of 3000 feet. Express the grade of the highway to the nearest tenth of a percent.

17. If the ratio of rise to run is to be $\dfrac{3}{4}$ for the steps of a staircase, and the rise is 32 centimeters, find the run to the nearest centimeter.

For Problems 18–23, graph each equation.

18. $y = -x^2 - 3$

19. $y = -x - 3$

20. $-3x + y = 5$

21. $3y = 2x$

22. $\dfrac{1}{3}x + \dfrac{1}{2}y = 2$

23. $y = \dfrac{-x - 1}{4}$

For Problems 24 and 25, graph each inequality.

24. $2x - y < 4$

25. $3x + 2y \geq 6$

Functions

The price of goods may be decided by using a function to describe the relationship between the price and the demand. Such a function gives us a means of studying the demand when the price is varied.

© Bill Aron /PhotoEdit

A golf pro-shop operator finds that she can sell 30 sets of golf clubs in a year at $500 per set. Furthermore, she predicts that for each $25 decrease in price, she could sell three extra sets of golf clubs. At what price should she sell the clubs to maximize gross income? We can use the quadratic function $f(x) = (30 + 3x)(500 - 25x)$ to determine that the clubs should be sold at $375 per set.

One of the fundamental concepts of mathematics is that of a function. Functions unify different areas of mathematics, and they also serve as a meaningful way of applying mathematics to many problems. They provide a means of studying quantities that vary with one another; that is, a change in one produces a corresponding change in another. In this chapter, we will (1) introduce the basic ideas pertaining to functions, (2) use the idea of a function to show how some concepts from previous chapters are related, and (3) discuss some applications in which functions are used.

8.1   Concept of a Function

The notion of correspondence is used in everyday situations and is central to the concept of a function. Consider the following correspondences.

1. To each person in a class, there corresponds an assigned seat.

2. To each day of a year, there corresponds an assigned integer that represents the average temperature for that day in a certain geographic location.

3. To each book in a library, there corresponds a whole number that represents the number of pages in the book.

Such correspondences can be depicted as in Figure 8.1. To each member in set A, there corresponds *one and only one* member in set B. For example, in the first correspondence, set A would consist of the students in a class, and set B would be the assigned seats. In the second example, set A would consist of the days of a year and set B would be a set of integers. Furthermore, the same integer might be assigned to more than one day of the year. (Different days might have the same average temperature.) The key idea is that *one and only one* integer is assigned to *each* day of the year. Likewise, in the third example, more than one book may have the same number of pages, but to each book, there is assigned one and only one number of pages.

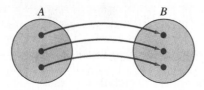

Figure 8.1

Mathematically, the general concept of a function can be defined as follows:

Definition 8.1

> A **function** f is a correspondence between two sets X and Y that assigns to each element x of set X one and only one element y of set Y. The element y being assigned is called the **image** of x. The set X is called the **domain** of the function, and the set of all images is called the **range** of the function.

In Definition 8.1, the image y is usually denoted by $f(x)$. Thus the symbol $f(x)$, which is read "f of x" or "the **value** of f at x," represents the element in the range associated with the element x from the domain. Figure 8.2 depicts this situation. Again we emphasize that each member of the domain has precisely one image in the range; however, different members in the domain, such as a and b in Figure 8.2, may have the same image.

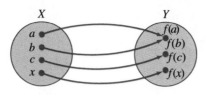

Figure 8.2

In Definition 8.1, we named the function f. It is common to name a function with a single letter, and the letters f, g, and h are often used. We suggest more meaningful choices when functions are used in real-world situations. For example, if a problem involves a profit function, then naming the function p or even P seems natural. Be careful not to confuse f and $f(x)$. Remember that f is used to name a function, whereas $f(x)$ is an element of the range — namely, the element assigned to x by f.

The assignments made by a function are often expressed as ordered pairs. For example, the assignments in Figure 8.2 could be expressed as $(a, f(a))$, $(b, f(b))$, $(c, f(c))$, and $(x, f(x))$, where the first components are from the domain, and the second components are from the range. Thus a function can also be thought of as a set of ordered pairs where no two of the ordered pairs have the same first component.

Remark: In some texts, the concept of a **relation** is introduced first, and then functions are defined as special kinds of relations. A relation is defined as a set of ordered pairs, and a function is defined as a relation in which no two ordered pairs have the same first element.

The ordered pairs that represent a function can be generated by various means, such as a graph or a chart. However, one of the most common ways of generating ordered pairs is by using equations. For example, the equation $f(x) = 2x + 3$ indicates that to each value of x in the domain, we assign $2x + 3$ from the range. For example,

$$f(1) = 2(1) + 3 = 5 \qquad \text{produces the ordered pair } (1, 5)$$
$$f(4) = 2(4) + 3 = 11 \qquad \text{produces the ordered pair } (4, 11)$$
$$f(-2) = 2(-2) + 3 = -1 \quad \text{produces the ordered pair } (-2, -1)$$

It may be helpful for you to picture the concept of a function in terms of a function machine, as illustrated in Figure 8.3. Each time a value of x is put into the machine, the equation $f(x) = 2x + 3$ is used to generate one and only one value for $f(x)$ to be ejected from the machine.

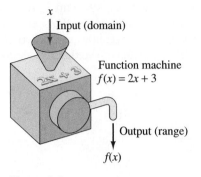

x

Input (domain)

Function machine
$f(x) = 2x + 3$

Output (range)

$f(x)$

Figure 8.3

Using the ordered-pair interpretation of a function, we can define the **graph** of a function f to be the set of all points in a plane of the form $(x, f(x))$, where x is from the domain of f. In other words, the graph of f is the same as the graph of the equation $y = f(x)$. Furthermore, because $f(x)$, or y, takes on only one value for each value of x, we can easily tell whether a given graph represents a function. For example, in Figure 8.4(a), for any choice of x there is only one value for y. Geometrically this means that no vertical line intersects the curve in more than one point. On the other hand, Figure 8.4(b) does not represent the graph of a function because certain values of x (all positive values) produce more than one value for y. In other words, some vertical lines intersect the curve in more than one point, as illustrated in Figure 8.4(b). A **vertical-line test** for functions can be stated as follows.

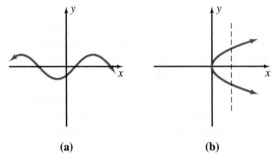

(a)   (b)

Figure 8.4

Vertical-Line Test

If each vertical line intersects a graph in no more than one point, then the graph represents a function.

Let's consider some examples to illustrate these ideas about functions.

E X A M P L E   1

If $f(x) = x^2 - x + 4$ and $g(x) = x^3 - x^2$, find $f(3)$, $f(-1)$, $f(a)$, $f(2a)$, $g(4)$, $g(-3)$, $g(m^2)$, and $g(-m)$.

Solution

$$f(3) = (3)^2 - (3) + 4 = 9 - 3 + 4 = 10 \qquad g(4) = 4^3 - 4^2 = 64 - 16 = 48$$

$$\begin{aligned} f(-1) &= (-1)^2 - (-1) + 4 \\ &= 1 + 1 + 4 = 6 \end{aligned} \qquad \begin{aligned} g(-3) &= (-3)^3 - (-3)^2 \\ &= -27 - 9 = -36 \end{aligned}$$

$$f(a) = (a)^2 - (a) + 4 = a^2 - a + 4 \qquad \begin{aligned} g(m^2) &= (m^2)^3 - (m^2)^2 \\ &= m^6 - m^4 \end{aligned}$$

$$f(2a) = (2a)^2 - (2a) + 4 = 4a^2 - 2a + 4 \qquad \begin{aligned} g(-m) &= (-m)^3 - (-m)^2 \\ &= -m^3 - m^2 \end{aligned}$$

∎

Note that in Example 1 we were working with two different functions in the same problem. That is why we used two different names, f and g. Sometimes the rule of assignment for a function consists of more than one part. Different rules are assigned depending on x, the element in the domain. An everyday example of this concept is that the price of admission to a theme park depends on whether you are a child, an adult, or a senior citizen. In mathematics we often refer to such functions as *piecewise-defined functions*. Let's consider an example of such a function.

E X A M P L E   2

If $f(x) = \begin{cases} 2x + 1 & for\ x \geq 0 \\ 3x - 1 & for\ x < 0 \end{cases}$, find $f(2)$, $f(4)$, $f(-1)$, and $f(-3)$.

Solution

For $x \geq 0$, we use the assignment $f(x) = 2x + 1$.

$$f(2) = 2(2) + 1 = 5$$

$$f(4) = 2(4) + 1 = 9$$

For $x < 0$, we use the assignment $f(x) = 3x - 1$.

$$f(-1) = 3(-1) - 1 = -4$$

$$f(-3) = 3(-3) - 1 = -10$$

∎

The quotient $\dfrac{f(a + h) - f(a)}{h}$ is often called a **difference quotient**. We use it extensively with functions when we study the limit concept in calculus. The next examples illustrate finding the difference quotient for specific functions.

E X A M P L E   3

Find $\dfrac{f(a + h) - f(a)}{h}$ for each of the following functions.

(a) $f(x) = x^2 + 6$ **(b)** $f(x) = 2x^2 + 3x - 4$ **(c)** $f(x) = \dfrac{1}{x}$

Solutions

(a) $f(a) = a^2 + 6$

$f(a + h) = (a + h)^2 + 6 = a^2 + 2ah + h^2 + 6$

Therefore

$$f(a + h) - f(a) = (a^2 + 2ah + h^2 + 6) - (a^2 + 6)$$
$$= a^2 + 2ah + h^2 + 6 - a^2 - 6$$
$$= 2ah + h^2$$

and

$$\frac{f(a + h) - f(a)}{h} = \frac{2ah + h^2}{h} = \frac{\cancel{h}(2a + h)}{\cancel{h}} = 2a + h$$

(b) $f(a) = 2a^2 + 3a - 4$

$f(a + h) = 2(a + h)^2 + 3(a + h) - 4$

$= 2(a^2 + 2ha + h^2) + 3a + 3h - 4$

$= 2a^2 + 4ha + 2h^2 + 3a + 3h - 4$

Therefore

$$f(a + h) - f(a) = (2a^2 + 4ha + 2h^2 + 3a + 3h - 4) - (2a^2 + 3a - 4)$$
$$= 2a^2 + 4ha + 2h^2 + 3a + 3h - 4 - 2a^2 - 3a + 4$$
$$= 4ha + 2h^2 + 3h$$

and

$$\frac{f(a + h) - f(a)}{h} = \frac{4ha + 2h^2 + 3h}{h}$$

$$= \frac{\cancel{h}(4a + 2h + 3)}{\cancel{h}}$$

$$= 4a + 2h + 3$$

(c) $f(a) = \dfrac{1}{a}$

$f(a + h) = \dfrac{1}{a + h}$

Therefore

$$f(a + h) - f(a) = \frac{1}{a + h} - \frac{1}{a}$$

$$= \frac{a}{a(a + h)} - \frac{a + h}{a(a + h)} \qquad \text{Common denominator of } a(a + h)$$

$$= \frac{a - (a + h)}{a(a + h)}$$

$$= \frac{a - a - h}{a(a + h)}$$

$$= \frac{-h}{a(a + h)} \quad \text{or} \quad -\frac{h}{a(a + h)}$$

and

$$\frac{f(a + h) - f(a)}{h} = \frac{-\dfrac{h}{a(a + h)}}{h}$$

$$= -\frac{h}{a(a + h)} \cdot \frac{1}{h}$$

$$= -\frac{1}{a(a + h)}$$

■

For our purposes in this text, if the domain of a function is not specifically indicated or determined by a real-world application, then we will assume the domain is *all real number* replacements for the variable, provided that they represent elements in the domain and produce real number functional values.

<hr>

E X A M P L E   4

For the function $f(x) = \sqrt{x - 1}$, (a) specify the domain, (b) determine the range, and (c) evaluate $f(5)$, $f(50)$, and $f(25)$.

Solutions

(a) The radicand must be nonnegative, so $x - 1 \geq 0$ and thus $x \geq 1$. Therefore the domain (D) is

$$D = \{x | x \geq 1\}$$

(b) The symbol $\sqrt{}$ indicates the nonnegative square root; thus the range (R) is

$$R = \{f(x) | f(x) \geq 0\}$$

(c) $f(5) = \sqrt{4} = 2$

$f(50) = \sqrt{49} = 7$

$f(25) = \sqrt{24} = 2\sqrt{6}$

■

As we will see later, the range of a function is often easier to determine after we have graphed the function. However, our equation- and inequality-solving processes are frequently sufficient to determine the domain of a function. Let's consider some examples.

E X A M P L E   5

Determine the domain for each of the following functions:

(a) $f(x) = \dfrac{3}{2x - 5}$ **(b)** $g(x) = \dfrac{1}{x^2 - 9}$ **(c)** $f(x) = \sqrt{x^2 + 4x - 12}$

Solutions

(a) We need to eliminate any values of x that will make the denominator zero. Therefore let's solve the equation $2x - 5 = 0$:

$$2x - 5 = 0$$

$$2x = 5$$

$$x = \frac{5}{2}$$

We can replace x with any real number except $\dfrac{5}{2}$ because $\dfrac{5}{2}$ makes the denominator zero. Thus the domain is

$$D = \left\{ x \middle| x \neq \frac{5}{2} \right\}$$

(b) We need to eliminate any values of x that will make the denominator zero. Let's solve the equation $x^2 - 9 = 0$:

$$x^2 - 9 = 0$$

$$x^2 = 9$$

$$x = \pm 3$$

The domain is thus the set

$$D = \{x \mid x \neq 3 \text{ and } x \neq -3\}$$

(c) The radicand, $x^2 + 4x - 12$, must be nonnegative. Let's use a number line approach, as we did in Chapter 6, to solve the inequality $x^2 + 4x - 12 \geq 0$ (see Figure 8.5):

$$x^2 + 4x - 12 \geq 0$$

$$(x + 6)(x - 2) \geq 0$$

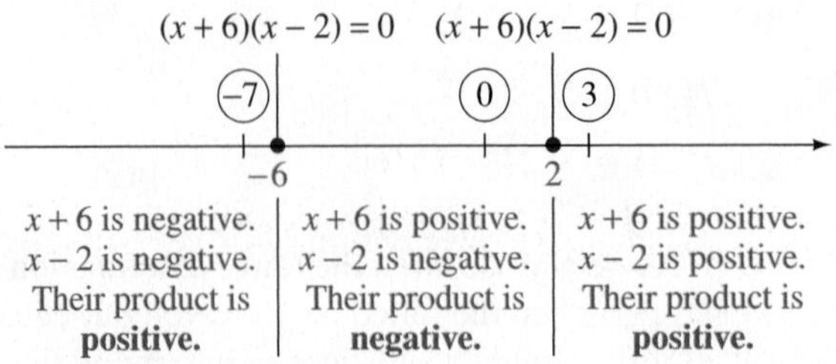

$(x+6)(x-2) = 0$ $(x+6)(x-2) = 0$

ⓧ(-7)   ⓪   ③

―――――――――●――――――●―――――→
         -6   2

$x + 6$ is negative.	$x + 6$ is positive.	$x + 6$ is positive.
$x - 2$ is negative.	$x - 2$ is negative.	$x - 2$ is positive.
Their product is **positive.**	Their product is **negative.**	Their product is **positive.**

Figure 8.5

The product $(x + 6)(x - 2)$ is nonnegative if $x \leq -6$ or $x \geq 2$. Using interval notation, we can express the domain as $(-\infty, -6] \cup [2, \infty)$.   ■

Functions and function notation provide the basis for describing many real-world relationships. The next example illustrates this point.

E X A M P L E   6

Suppose a factory determines that the overhead for producing a quantity of a certain item is \$500 and that the cost for each item is \$25. Express the total expenses as a function of the number of items produced, and compute the expenses for producing 12, 25, 50, 75, and 100 items.

Solution

Let n represent the number of items produced. Then $25n + 500$ represents the total expenses. Using E to represent the expense function, we have

$$E(n) = 25n + 500, \quad \text{where } n \text{ is a whole number}$$

We obtain

$$E(12) = 25(12) + 500 = 800$$

$$E(25) = 25(25) + 500 = 1125$$

$$E(50) = 25(50) + 500 = 1750$$

$$E(75) = 25(75) + 500 = 2375$$

$$E(100) = 25(100) + 500 = 3000$$

Thus the total expenses for producing 12, 25, 50, 75, and 100 items are \$800, \$1125, \$1750, \$2375, and \$3000, respectively.   ■

As we stated before, an equation such as $f(x) = 5x - 7$ that is used to determine a function can also be written $y = 5x - 7$. In either form, we refer to x as the **independent variable** and to y, or $f(x)$, as the **dependent variable**. Many formulas in mathematics and other related areas also determine functions. For example, the area formula for a circular region, $A = \pi r^2$, assigns to each positive real value for r a unique value for A. This formula determines a function f, where $f(r) = \pi r^2$. The variable r is the independent variable, and A, or $f(r)$, is the dependent variable.

Problem Set 8.1

1. If $f(x) = -2x + 5$, find $f(3)$, $f(5)$, and $f(-2)$.

2. If $f(x) = x^2 - 3x - 4$, find $f(2)$, $f(4)$, and $f(-3)$.

3. If $g(x) = -2x^2 + x - 5$, find $g(3)$, $g(-1)$, and $g(2a)$.

4. If $g(x) = -x^2 - 4x + 6$, find $g(0)$, $g(5)$, and $g(-a)$.

5. If $h(x) = \dfrac{2}{3}x - \dfrac{3}{4}$, find $h(3)$, $h(4)$, and $h\left(-\dfrac{1}{2}\right)$.

6. If $h(x) = -\dfrac{1}{2}x + \dfrac{2}{3}$, find $h(-2)$, $h(6)$, and $h\left(-\dfrac{2}{3}\right)$.

7. If $f(x) = \sqrt{2x - 1}$, find $f(5)$, $f\left(\dfrac{1}{2}\right)$, and $f(23)$.

8. If $f(x) = \sqrt{3x + 2}$, find $f\left(\dfrac{14}{3}\right)$, $f(10)$, and $f\left(-\dfrac{1}{3}\right)$.

9. If $f(x) = -2x + 7$, find $f(a)$, $f(a + 2)$, and $f(a + h)$.

10. If $f(x) = x^2 - 7x$, find $f(a)$, $f(a - 3)$, and $f(a + h)$.

11. If $f(x) = x^2 - 4x + 10$, find $f(-a)$, $f(a - 4)$, and $f(a + h)$.

12. If $f(x) = 2x^2 - x - 1$, find $f(-a)$, $f(a + 1)$, and $f(a + h)$.

13. If $f(x) = -x^2 + 3x + 5$, find $f(-a)$, $f(a + 6)$, and $f(-a + 1)$.

14. If $f(x) = -x^2 - 2x - 7$, find $f(-a)$, $f(-a - 2)$, and $f(a + 7)$.

15. If $f(x) = \begin{cases} x & \text{for } x \ge 0 \\ x^2 & \text{for } x < 0 \end{cases}$, find $f(4)$, $f(10)$, $f(-3)$, and $f(-5)$.

16. If $f(x) = \begin{cases} 3x + 2 & \text{for } x \ge 0 \\ 5x - 1 & \text{for } x < 0 \end{cases}$, find $f(2)$, $f(6)$, $f(-1)$, and $f(-4)$.

17. If $f(x) = \begin{cases} 2x & \text{for } x \ge 0 \\ -2x & \text{for } x < 0 \end{cases}$, find $f(3)$, $f(5)$, $f(-3)$, and $f(-5)$.

18. If $f(x) = \begin{cases} 2 & \text{for } x < 0 \\ x^2 + 1 & \text{for } 0 \le x \le 4 \\ -1 & \text{for } x > 4 \end{cases}$, find $f(3)$, $f(6)$, $f(0)$, and $f(-3)$.

19. If $f(x) = \begin{cases} 1 & \text{for } x > 0 \\ 0 & \text{for } -1 < x \le 0 \\ -1 & \text{for } x \le -1 \end{cases}$, find $f(2)$, $f(0)$, $f\left(-\dfrac{1}{2}\right)$, and $f(-4)$.

For Problems 20–31, find $\dfrac{f(a + h) - f(a)}{h}$.

20. $f(x) = 4x + 5$

21. $f(x) = -7x - 2$

22. $f(x) = x^2 - 3x$

23. $f(x) = -x^2 + 4x - 2$

24. $f(x) = 2x^2 + 7x - 4$

25. $f(x) = 3x^2 - x - 4$

26. $f(x) = x^3$

27. $f(x) = x^3 - x^2 + 2x - 1$

28. $f(x) = \dfrac{1}{x + 1}$

29. $f(x) = \dfrac{2}{x - 1}$

30. $f(x) = \dfrac{x}{x + 1}$

31. $f(x) = \dfrac{1}{x^2}$

For Problems 32–39 (Figures 8.6 through 8.13), determine whether the indicated graph represents a function of x.

32.

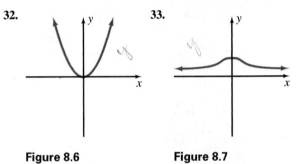

Figure 8.6

33.

Figure 8.7

34.

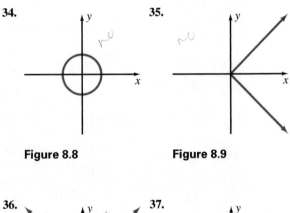

Figure 8.8

35.

Figure 8.9

36.

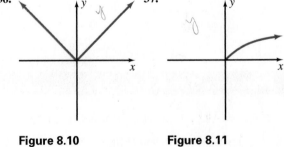

Figure 8.10

37.

Figure 8.11

38.   **39.**

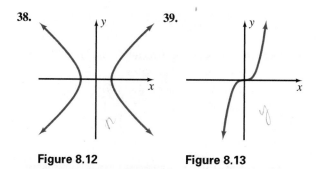

Figure 8.12   Figure 8.13

For Problems 40–47, determine the domain and the range of the given function.

40. $f(x) = \sqrt{x}$

41. $f(x) = \sqrt{3x - 4}$

42. $f(x) = x^2 + 1$

43. $f(x) = x^2 - 2$

44. $f(x) = x^3$

45. $f(x) = |x|$

46. $f(x) = x^4$

47. $f(x) = -\sqrt{x}$

For Problems 48–57, determine the domain of the given function.

48. $f(x) = \dfrac{3}{x - 4}$

49. $f(x) = \dfrac{-4}{x + 2}$

50. $f(x) = \dfrac{2x}{(x - 2)(x + 3)}$

51. $f(x) = \dfrac{5}{(2x - 1)(x + 4)}$

52. $f(x) = \sqrt{5x + 1}$

53. $f(x) = \dfrac{1}{x^2 - 4}$

54. $g(x) = \dfrac{3}{x^2 + 5x + 6}$

55. $f(x) = \dfrac{4x}{x^2 - x - 12}$

56. $g(x) = \dfrac{5}{x^2 + 4x}$

57. $g(x) = \dfrac{x}{6x^2 + 13x - 5}$

For Problems 58–67, express the domain of the given function using interval notation.

58. $f(x) = \sqrt{x^2 - 1}$

59. $f(x) = \sqrt{x^2 - 16}$

60. $f(x) = \sqrt{x^2 + 4}$

61. $f(x) = \sqrt{x^2 + 1} - 4$

62. $f(x) = \sqrt{x^2 - 2x - 24}$

63. $f(x) = \sqrt{x^2 - 3x - 40}$

64. $f(x) = \sqrt{12x^2 + x - 6}$

65. $f(x) = -\sqrt{8x^2 + 6x - 35}$

66. $f(x) = \sqrt{16 - x^2}$

67. $f(x) = \sqrt{1 - x^2}$

For Problems 68–75, solve each problem.

68. Suppose that the profit function for selling n items is given by

$$P(n) = -n^2 + 500n - 61,500$$

Evaluate $P(200)$, $P(230)$, $P(250)$, and $P(260)$.

69. The equation $A(r) = \pi r^2$ expresses the area of a circular region as a function of the length of a radius (r). Compute $A(2)$, $A(3)$, $A(12)$, and $A(17)$ and express your answers to the nearest hundredth.

70. In a physics experiment, it is found that the equation $V(t) = 1667t - 6940t^2$ expresses the velocity of an object as a function of time (t). Compute $V(0.1)$, $V(0.15)$, and $V(0.2)$.

71. The height of a projectile fired vertically into the air (neglecting air resistance) at an initial velocity of 64 feet per second is a function of the time (t) and is given by the equation $h(t) = 64t - 16t^2$. Compute $h(1)$, $h(2)$, $h(3)$, and $h(4)$.

72. A car rental agency charges \$50 per day plus \$0.32 a mile. Therefore the daily charge for renting a car is a function of the number of miles traveled (m) and can be expressed as $C(m) = 50 + 0.32m$. Compute $C(75)$, $C(150)$, $C(225)$, and $C(650)$.

73. The equation $I(r) = 500r$ expresses the amount of simple interest earned by an investment of \$500 for 1 year as a function of the rate of interest (r). Compute $I(0.11)$, $I(0.12)$, $I(0.135)$, and $I(0.15)$.

74. Suppose the height of a semielliptical archway is given by the function $h(x) = \sqrt{64 - 4x^2}$, where x is the distance from the center line of the arch. Compute $h(0)$, $h(2)$, and $h(4)$.

75. The equation $A(r) = 2\pi r^2 + 16\pi r$ expresses the total surface area of a right circular cylinder of height 8 centimeters as a function of the length of a radius (r). Compute $A(2)$, $A(4)$, and $A(8)$ and express your answers to the nearest hundredth.

▪▪▪ THOUGHTS INTO WORDS

76. Expand Definition 8.1 to include a definition for the concept of a relation.

77. What does it mean to say that the domain of a function may be restricted if the function represents a real-world situation? Give three examples of such functions.

78. Does $f(a + b) = f(a) + f(b)$ for all functions? Defend your answer.

79. Are there any functions for which $f(a + b) = f(a) + f(b)$? Defend your answer.

8.2 Linear Functions and Applications

As we use the function concept in our study of mathematics, it is helpful to classify certain types of functions and become familiar with their equations, characteristics, and graphs. This will enhance our problem-solving capabilities.

Any function that can be written in the form

$$f(x) = ax + b$$

where a and b are real numbers, is called a **linear function**. The following equations are examples of linear functions.

$$f(x) = -2x + 4 \qquad f(x) = 3x - 6 \qquad f(x) = \frac{2}{3}x + \frac{5}{6}$$

The equation $f(x) = ax + b$ can also be written as $y = ax + b$. From our work in Section 7.5, we know that $y = ax + b$ is the equation of a straight line that has a slope of a and a y intercept of b. This information can be used to graph linear functions, as illustrated by the following example.

EXAMPLE 1

Graph $f(x) = -2x + 4$.

Solution

Because the y intercept is 4, the point $(0, 4)$ is on the line. Furthermore, because the slope is -2, we can move two units down and one unit to the right of $(0, 4)$ to determine the point $(1, 2)$. The line determined by $(0, 4)$ and $(1, 2)$ is drawn in Figure 8.14.

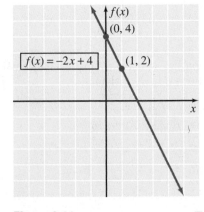

Figure 8.14

Note that in Figure 8.14, we labeled the vertical axis $f(x)$. We could also label it y because $y = f(x)$. We will use the $f(x)$ labeling for most of our work with functions; however, we will continue to refer to y axis symmetry instead of $f(x)$ axis symmetry.

Recall from Section 7.2 that we can also graph linear equations by finding the two intercepts. This same approach can be used with linear functions, as illustrated by the next two examples.

EXAMPLE 2

Graph $f(x) = 3x - 6$.

Solution

First, we see that $f(0) = -6$; thus the point $(0, -6)$ is on the graph. Second, by setting $3x - 6$ equal to zero and solving for x, we obtain

$$3x - 6 = 0$$

$$3x = 6$$

$$x = 2$$

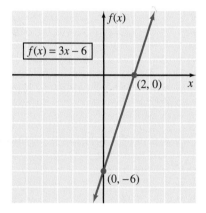

Therefore $f(2) = 3(2) - 6 = 0$, and the point $(2, 0)$ is on the graph. The line determined by $(0, -6)$ and $(2, 0)$ is drawn in Figure 8.15.

Figure 8.15 ■

EXAMPLE 3

Graph the function $f(x) = \dfrac{2}{3}x + \dfrac{5}{6}$.

Solution

Because $f(0) = \dfrac{5}{6}$, the point $\left(0, \dfrac{5}{6}\right)$ is on the graph. By setting $\dfrac{2}{3}x + \dfrac{5}{6}$ equal to zero and solving for x, we obtain

$$\frac{2}{3}x + \frac{5}{6} = 0$$

$$\frac{2}{3}x = -\frac{5}{6}$$

$$x = -\frac{5}{4}$$

Therefore $f\left(-\dfrac{5}{4}\right) = 0$, and the point $\left(-\dfrac{5}{4}, 0\right)$ is on the graph. The line determined by the two points $\left(0, \dfrac{5}{6}\right)$ and $\left(-\dfrac{5}{4}, 0\right)$ is shown in Figure 8.16.

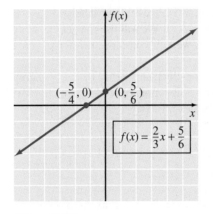

Figure 8.16 ■

As you graph functions using function notation, it is often helpful to think of the ordinate of every point on the graph as the value of the function at a specific value of x. Geometrically the functional value is the directed distance of the point from the x axis. This idea is illustrated in Figure 8.17 for the function $f(x) = x$ and in Figure 8.18 for the function $f(x) = 2$. The linear function $f(x) = x$ is often called the **identity function**. Any linear function of the form $f(x) = ax + b$, where $a = 0$, is called a **constant function**.

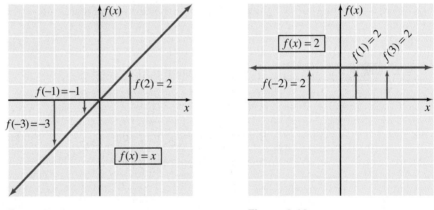

Figure 8.17   **Figure 8.18**

From our previous work with linear equations, we know that parallel lines have equal slopes and that two perpendicular lines have slopes that are negative reciprocals of each other. Thus when we work with linear functions of the form $f(x) = ax + b$, it is easy to recognize parallel and perpendicular lines. For example, the lines determined by $f(x) = 0.21x + 4$ and $g(x) = 0.21x - 3$ are parallel lines because both lines have a slope of 0.21 and different y intercepts. Let's use a graphing calculator to graph these two functions along with $h(x) = 0.21x + 2$ and $p(x) = 0.21x - 7$ (Figure 8.19).

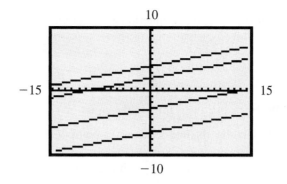

Figure 8.19

The graphs of the functions $f(x) = \frac{2}{5}x + 8$ and $g(x) = -\frac{5}{2}x - 4$ are perpendicular lines because the slopes $\left(\frac{2}{5} \text{ and } -\frac{5}{2}\right)$ of the two lines are negative reciprocals of each other. Again using our graphing calculator, let's graph these two functions along with $h(x) = -\frac{5}{2}x + 2$ and $p(x) = -\frac{5}{2}x - 6$ (Figure 8.20). If the lines do not appear to be perpendicular, you may want to change the window with a zoom square option.

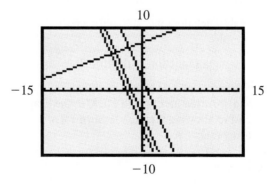

Figure 8.20

Remark: A property of plane geometry states that if two or more lines are perpendicular to the same line, then they are parallel lines. Figure 8.20 is a good illustration of that property.

The function notation can also be used to determine linear functions that satisfy certain conditions. Let's see how this works.

EXAMPLE 4

Determine the linear function whose graph is a line with a slope of $\frac{1}{4}$ that contains the point $(2, 5)$.

Solution

We can substitute $\frac{1}{4}$ for a in the equation $f(x) = ax + b$ to obtain $f(x) = \frac{1}{4}x + b$. The fact that the line contains the point $(2, 5)$ means that $f(2) = 5$. Therefore

$$f(2) = \frac{1}{4}(2) + b = 5$$

$$b = \frac{9}{2}$$

and the function is $f(x) = \frac{1}{4}x + \frac{9}{2}$. ∎

■ Applications of Linear Functions

We worked with some applications of linear equations in Section 7.2. Now let's consider some additional applications that use the concept of a linear function to connect mathematics to the real world.

EXAMPLE 5

The cost for burning a 60-watt light bulb is given by the function $c(h) = 0.0036h$, where h represents the number of hours that the bulb is burning.

(a) How much does it cost to burn a 60-watt bulb for 3 hours per night for a 30-day month?

(b) Graph the function $c(h) = 0.0036h$.

(c) Suppose that a 60-watt light bulb is left burning in a closet for a week before it is discovered and turned off. Use the graph from part (b) to approximate the cost of allowing the bulb to burn for a week. Then use the function to find the exact cost.

Solutions

(a) $c(90) = 0.0036(90) = 0.324$ The cost, to the nearest cent, is $0.32.

(b) Because $c(0) = 0$ and $c(100) = 0.36$, we can use the points $(0, 0)$ and $(100, 0.36)$ to graph the linear function $c(h) = 0.0036h$ (Figure 8.21).

(c) If the bulb burns for 24 hours per day for a week, it burns for $24(7) = 168$ hours. Reading from the graph, we can approximate 168 on the horizontal axis, read up to the line, and then read across to the vertical axis. It looks as though it will cost approximately 60 cents. Using $c(h) = 0.0036h$, we obtain exactly $c(168) = 0.0036(168) = 0.6048$.

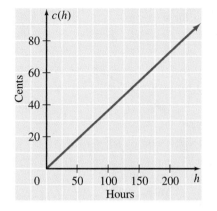

Figure 8.21   ■

EXAMPLE 6

The EZ Car Rental charges a fixed amount per day plus an amount per mile for renting a car. For two different day trips, Ed has rented a car from EZ. He paid $70 for 100 miles on one day and $120 for 350 miles on another day. Determine the linear function that the EZ Car Rental uses to determine its daily rental charges.

Solution

The linear function $f(x) = ax + b$, where x represents the number of miles, models this situation. Ed's two day trips can be represented by the ordered pairs (100, 70) and (350, 120). From these two ordered pairs, we can determine a, which is the slope of the line.

$$a = \frac{120 - 70}{350 - 100} = \frac{50}{250} = \frac{1}{5} = 0.2$$

Thus $f(x) = ax + b$ becomes $f(x) = 0.2x + b$. Now either ordered pair can be used to determine the value of b. Using (100, 70), we have $f(100) = 70$, so

$$f(100) = 0.2(100) + b = 70$$

$$b = 50$$

The linear function is $f(x) = 0.2x + 50$. In other words, the EZ Car Rental charges a daily fee of $50 plus $0.20 per mile.   ■

EXAMPLE 7

Suppose that Ed (Example 6) also has access to the A-OK Car Rental agency, which charges a daily fee of $25 plus $0.30 per mile. Should Ed use EZ Car Rental from Example 6 or A-OK Car Rental?

Solution

The linear function $g(x) = 0.3x + 25$, where x represents the number of miles, can be used to determine the daily charges of A-OK Car Rental. Let's graph this function and $f(x) = 0.2x + 50$ from Example 6 on the same set of axes (Figure 8.22).

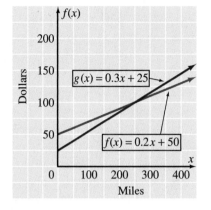

Figure 8.22

Now we see that the two functions have equal values at the point of intersection of the two lines. To find the coordinates of this point, we can set $0.3x + 25$ equal to $0.2x + 50$ and solve for x.

$$0.3x + 25 = 0.2x + 50$$

$$0.1x = 25$$

$$x = 250$$

If $x = 250$, then $0.3(250) + 25 = 100$ and the point of intersection is $(250, 100)$. Again looking at the lines in Figure 8.22, Ed should use A-OK Car Rental for daily trips less than 250 miles, but he should use EZ Car Rental for trips more than 250 miles. ■

Problem Set 8.2

For Problems 1–16, graph each of the linear functions.

1. $f(x) = 2x - 4$

2. $f(x) = 3x + 3$

3. $f(x) = -x + 3$

4. $f(x) = -2x + 6$

5. $f(x) = 3x + 9$

6. $f(x) = 2x - 6$

7. $f(x) = -4x - 4$

8. $f(x) = -x - 5$

9. $f(x) = -3x$

10. $f(x) = -4x$

11. $f(x) = -3$

12. $f(x) = -1$

13. $f(x) = \frac{1}{2}x + 3$

14. $f(x) = \frac{2}{3}x + 4$

15. $f(x) = -\frac{3}{4}x - 6$

16. $f(x) = -\frac{1}{2}x - 1$

17. Determine the linear function whose graph is a line with a slope of $\frac{2}{3}$ and contains the point $(-1, 3)$.

18. Determine the linear function whose graph is a line with a slope of $-\frac{3}{5}$ and contains the point $(4, -5)$.

19. Determine the linear function whose graph is a line that contains the points $(-3, -1)$ and $(2, -6)$.

20. Determine the linear function whose graph is a line that contains the points $(-2, -3)$ and $(4, 3)$.

21. Determine the linear function whose graph is a line that is perpendicular to the line $g(x) = 5x - 2$ and contains the point $(6, 3)$.

22. Determine the linear function whose graph is a line that is parallel to the line $g(x) = -3x - 4$ and contains the point $(2, 7)$.

23. The cost for burning a 75-watt bulb is given by the function $c(h) = 0.0045h$, where h represents the number of hours that the bulb burns.
   (a) How much does it cost to burn a 75-watt bulb for 3 hours per night for a 31-day month? Express your answer to the nearest cent.
   (b) Graph the function $c(h) = 0.0045h$.
   (c) Use the graph in part (b) to approximate the cost of burning a 75-watt bulb for 225 hours.
   (d) Use $c(h) = 0.0045h$ to find the exact cost, to the nearest cent, of burning a 75-watt bulb for 225 hours.

24. The Rent-Me Car Rental charges $15 per day plus $0.22 per mile to rent a car. Determine a linear function that can be used to calculate daily car rentals. Then use that function to determine the cost of renting a car for a day and driving 175 miles; 220 miles; 300 miles; 460 miles.

25. The ABC Car Rental uses the function $f(x) = 26$ for any daily use of a car up to and including 200 miles. For driving more than 200 miles per day, it uses the function $g(x) = 26 + 0.15(x - 200)$ to determine the charges. How much would the company charge for daily driving of 150 miles? of 230 miles? of 360 miles? of 430 miles?

26. Suppose that a car rental agency charges a fixed amount per day plus an amount per mile for renting a car. Heidi rented a car one day and paid $80 for 200 miles. On another day she rented a car from the same agency and paid $117.50 for 350 miles. Determine the linear function that the agency could use to determine its daily rental charges.

27. A retailer has a number of items that she wants to sell and make a profit of 40% of the cost of each item. The function $s(c) = c + 0.4c = 1.4c$, where c represents the cost of an item, can be used to determine the selling price. Find the selling price of items that cost $1.50, $3.25, $14.80, $21, and $24.20.

28. Zack wants to sell five items that cost him $1.20, $2.30, $6.50, $12, and $15.60. He wants to make a profit of 60% of the cost. Create a function that you can use to determine the selling price of each item, and then use the function to calculate each selling price.

29. "All Items 20% Off Marked Price" is a sign at a local golf course. Create a function and then use it to determine how much one has to pay for each of the following marked items: a $9.50 hat, a $15 umbrella, a $75 pair of golf shoes, a $12.50 golf glove, a $750 set of golf clubs.

30. The linear depreciation method assumes that an item depreciates the same amount each year. Suppose a new piece of machinery costs $32,500 and it depreciates $1950 each year for t years.
   (a) Set up a linear function that yields the value of the machinery after t years.
   (b) Find the value of the machinery after 5 years.
   (c) Find the value of the machinery after 8 years.
   (d) Graph the function from part (a).
   (e) Use the graph from part (d) to approximate how many years it takes for the value of the machinery to become zero.
   (f) Use the function to determine how long it takes for the value of the machinery to become zero.

■ ■ ■ THOUGHTS INTO WORDS

31. Is $f(x) = (3x - 2) - (2x + 1)$ a linear function? Explain your answer.

32. Suppose that Bianca walks at a constant rate of 3 miles per hour. Explain what it means that the distance Bianca walks is a linear function of the time that she walks.

■ ■ ■ **FURTHER INVESTIGATIONS**

For Problems 33–37, graph each of the functions.

33. $f(x) = |x|$

34. $f(x) = x + |x|$

35. $f(x) = x - |x|$

36. $f(x) = |x| - x$

37. $f(x) = \dfrac{x}{|x|}$

GRAPHING CALCULATOR ACTIVITIES

38. Use a graphing calculator to check your graphs for Problems 1–16.

39. Use a graphing calculator to do parts (b) and (c) of Example 5.

40. Use a graphing calculator to check our solution for Example 7.

41. Use a graphing calculator to do parts (b) and (c) of Problem 23.

42. Use a graphing calculator to do parts (d) and (e) of Problem 30.

43. Use a graphing calculator to check your graphs for Problems 33–37.

44. (a) Graph $f(x) = |x|$, $f(x) = 2|x|$, $f(x) = 4|x|$, and $f(x) = \dfrac{1}{2}|x|$ on the same set of axes.

(b) Graph $f(x) = |x|$, $f(x) = -|x|$, $f(x) = -3|x|$, and $f(x) = -\dfrac{1}{2}|x|$ on the same set of axes.

(c) Use your results from parts (a) and (b) to make a conjecture about the graphs of $f(x) = a|x|$, where a is a nonzero real number.

(d) Graph $f(x) = |x|$, $f(x) = |x| + 3$, $f(x) = |x| - 4$, and $f(x) = |x| + 1$ on the same set of axes. Make a conjecture about the graphs of $f(x) = |x| + k$, where k is a nonzero real number.

(e) Graph $f(x) = |x|$, $f(x) = |x - 3|$, $f(x) = |x - 1|$, and $f(x) = |x + 4|$ on the same set of axes. Make a conjecture about the graphs of $f(x) = |x - h|$, where h is a nonzero real number.

(f) On the basis of your results from parts (a) through (e), sketch each of the following graphs. Then use a graphing calculator to check your sketches.

(1) $f(x) = |x - 2| + 3$
(2) $f(x) = |x + 1| - 4$
(3) $f(x) = 2|x - 4| - 1$
(4) $f(x) = -3|x + 2| + 4$
(5) $f(x) = -\dfrac{1}{2}|x - 3| - 2$

8.3 Quadratic Functions

Any function that can be written in the form

$$f(x) = ax^2 + bx + c$$

where a, b, and c are real numbers with $a \neq 0$, is called a **quadratic function**. The graph of any quadratic function is a **parabola**. As we work with parabolas, we will use the vocabulary indicated in Figure 8.23.

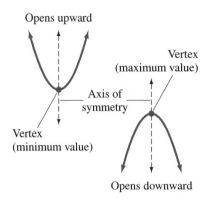

Figure 8.23

Graphing a parabola relies on finding the vertex, determining whether the parabola opens upward or downward, and locating two points on opposite sides of the axis of symmetry. We are also interested in comparing parabolas produced by equations such as $f(x) = x^2 + k$, $f(x) = ax^2$, $f(x) = (x - h)^2$, and $f(x) = a(x - h)^2 + k$ to the basic parabola produced by the equation $f(x) = x^2$. The graph of $f(x) = x^2$ is shown in Figure 8.24. Note that the vertex of the parabola is at the origin, $(0, 0)$, and the graph is symmetric to the y, or $f(x)$, axis. Remember that an equation exhibits y axis symmetry if replacing x with $-x$ produces an equivalent equation. Therefore, because $f(-x) = (-x)^2 = x^2$, the equation exhibits y axis symmetry.

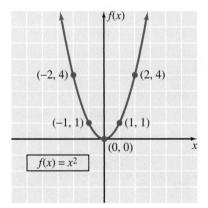

Figure 8.24

Now let's consider an equation of the form $f(x) = x^2 + k$, where k is a constant. (Keep in mind that all such equations exhibit y axis symmetry.)

E X A M P L E   1   Graph $f(x) = x^2 - 2$.

Solution

Let's set up a table to make some comparisons of function values. Because the graph exhibits y axis symmetry, we will calculate only positive values and then reflect the points across the y axis.

x	$f(x) = x^2$	$f(x) = x^2 - 2$
0	0	-2
1	1	-1
2	4	2
3	9	7

It should be observed that the functional values for $f(x) = x^2 - 2$ are 2 less than the corresponding functional values for $f(x) = x^2$. Thus the graph of $f(x) = x^2 - 2$ is the same as the parabola of $f(x) = x^2$ except that it is moved down two units (Figure 8.25).

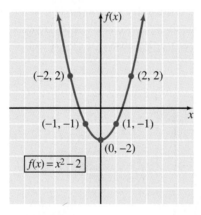

Figure 8.25   ■

> In general, the graph of a quadratic function of the form $f(x) = x^2 + k$ is the same as the graph of $f(x) = x^2$ except that it is moved up or down $|k|$ units, depending on whether k is positive or negative. We say that the graph of $f(x) = x^2 + k$ is a **vertical translation** of the graph of $f(x) = x^2$.

Now let's consider some quadratic functions of the form $f(x) = ax^2$, where a is a nonzero constant. (The graphs of these equations also have y axis symmetry.)

EXAMPLE 2   Graph $f(x) = 2x^2$.

Solution

Let's set up a table to make some comparisons of functional values. Note that in the table, the functional values for $f(x) = 2x^2$ are *twice* the corresponding functional values for $f(x) = x^2$. Thus the parabola associated with $f(x) = 2x^2$ has the same vertex (the origin) as the graph of $f(x) = x^2$, but it is *narrower*, as shown in Figure 8.26.

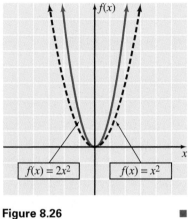

x	$f(x) = x^2$	$f(x) = 2x^2$
0	0	0
1	1	2
2	4	8
3	9	18

Figure 8.26   ■

EXAMPLE 3   Graph $f(x) = \dfrac{1}{2}x^2$.

Solution

As we see from the table, the functional values for $f(x) = \dfrac{1}{2}x^2$ are *one-half* of the corresponding functional values for $f(x) = x^2$. Therefore the parabola associated with $f(x) = \dfrac{1}{2}x^2$ is *wider* than the basic parabola, as shown in Figure 8.27.

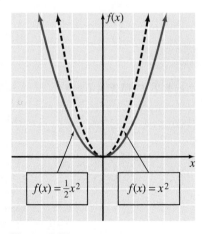

x	$f(x) = x^2$	$f(x) = \frac{1}{2}x^2$
0	0	0
1	1	$\dfrac{1}{2}$
2	4	2
3	9	$\dfrac{9}{2}$
4	16	8

Figure 8.27   ■

| EXAMPLE 4 | Graph $f(x) = -x^2$. |

Solution

It should be evident that the functional values for $f(x) = -x^2$ are the *opposites* of the corresponding functional values for $f(x) = x^2$. Therefore the graph of $f(x) = -x^2$ is a reflection across the x axis of the basic parabola (Figure 8.28).

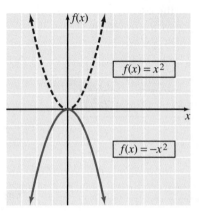

Figure 8.28   ∎

> In general, the graph of a quadratic function of the form $f(x) = ax^2$ has its vertex at the origin and opens upward if a is positive and downward if a is negative. The parabola is narrower than the basic parabola if $|a| > 1$ and wider if $|a| < 1$.

Let's continue our investigation of quadratic functions by considering those of the form $f(x) = (x - h)^2$, where h is a nonzero constant.

| EXAMPLE 5 | Graph $f(x) = (x - 3)^2$. |

Solution

A fairly extensive table of values illustrates a pattern. Note that $f(x) = (x - 3)^2$ and $f(x) = x^2$ take on the same functional values but for different values of x. More specifically, if $f(x) = x^2$ achieves a certain functional value at a specific value of x, then $f(x) = (x - 3)^2$ achieves that same functional value at x *plus three*. In other words, the graph of $f(x) = (x - 3)^2$ is the graph of $f(x) = x^2$ *moved three units to the right* (Figure 8.29).

x	$f(x) = x^2$	$f(x) = (x - 3)^2$
-1	1	16
0	0	9
1	1	4
2	4	1
3	9	0
4	16	1
5	25	4
6	36	9
7	49	16

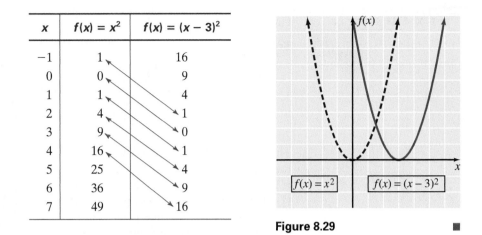

Figure 8.29

In general, the graph of a quadratic function of the form $f(x) = (x - h)^2$ is the same as the graph of $f(x) = x^2$ except that it is moved to the right h units if h is positive or moved to the left $|h|$ units if h is negative. We say that the graph of $f(x) = (x - h)^2$ is a **horizontal translation** of the graph of $f(x) = x^2$.

The following diagram summarizes our work thus far for graphing quadratic functions.

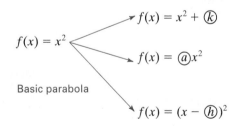

$f(x) = x^2 + ⓚ$ Moves the parabola up or down

$f(x) = x^2$

Basic parabola

$f(x) = ⓐx^2$ Affects the width and the way the parabola opens

$f(x) = (x - ⓗ)^2$ Moves the parabola right or left

We have studied, separately, the effects a, h, and k have on the graph of a quadratic function. However, we need to consider the general form of a quadratic function when all of these effects are present.

In general, the graph of a quadratic function of the form $f(x) = a(x - h)^2 + k$ has its vertex at (h, k) and opens upward if a is positive and downward if a is negative. The parabola is narrower than the basic parabola if $|a| > 1$ and wider if $|a| < 1$.

EXAMPLE 6

Graph $f(x) = 3(x - 2)^2 + 1$.

Solution

$$f(x) = 3(x - 2)^2 + 1$$

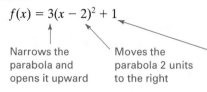

| Narrows the parabola and opens it upward | Moves the parabola 2 units to the right | Moves the parabola 1 unit up |

The vertex is $(2, 1)$ and the line $x = 2$ is the axis of symmetry. If $x = 1$, then $f(1) = 3(1 - 2)^2 + 1 = 4$. Thus the point $(1, 4)$ is on the graph, and so is its reflection, $(3, 4)$, across the line of symmetry. The parabola is shown in Figure 8.30.

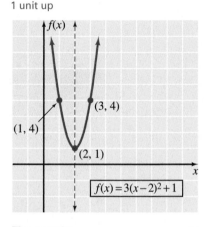

$f(x) = 3(x-2)^2 + 1$

Figure 8.30 ∎

EXAMPLE 7

Graph $f(x) = -\dfrac{1}{2}(x + 1)^2 - 3$.

Solution

$$f(x) = -\frac{1}{2}[x - (-1)]^2 - 3$$

| Widens the parabola and opens it downward | Moves the parabola 1 unit to the left | Moves the parabola 3 units down |

The vertex is at $(-1, -3)$, and the line $x = -1$ is the axis of symmetry. If $x = 0$, then $f(0) = -\dfrac{1}{2}(0 + 1)^2 - 3 = -\dfrac{7}{2}$. Thus the point $\left(0, -\dfrac{7}{2}\right)$ is on the graph, and so is its reflection, $\left(-2, -\dfrac{7}{2}\right)$, across the line of symmetry. The parabola is shown in Figure 8.31.

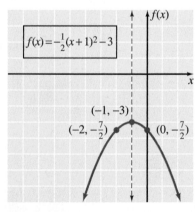

$f(x) = -\frac{1}{2}(x+1)^2 - 3$

$(-1, -3)$

$\left(-2, -\frac{7}{2}\right)$ $\left(0, -\frac{7}{2}\right)$

Figure 8.31 ∎

■ Quadratic Functions of the Form $f(x) = ax^2 + bx + c$

We are now ready to graph quadratic functions of the form $f(x) = ax^2 + bx + c$. The general approach is to change from the form $f(x) = ax^2 + bx + c$ to the form $f(x) = a(x - h)^2 + k$ and then proceed as we did in Examples 6 and 7. The process of *completing the square* serves as the basis for making the change in form. Let's consider two examples to illustrate the details.

EXAMPLE 8

Graph $f(x) = x^2 - 4x + 3$.

Solution

$$f(x) = x^2 - 4x + 3$$
$$= (x^2 - 4x) + 3 \qquad \text{Add 4, which is the square of one-half of the coefficient of } x.$$
$$= (x^2 - 4x + 4) + 3 - 4 \longleftarrow \text{Subtract 4 to compensate for the 4 that was added.}$$
$$= (x - 2)^2 - 1$$

The graph of $f(x) = (x - 2)^2 - 1$ is the basic parabola moved two units to the right and one unit down (Figure 8.32).

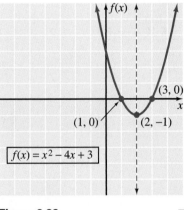

$$f(x) = x^2 - 4x + 3$$

Figure 8.32 ■

EXAMPLE 9

Graph $f(x) = -2x^2 - 4x + 1$.

Solution

$$f(x) = -2x^2 - 4x + 1$$
$$= -2(x^2 + 2x) + 1 \qquad \text{Factor } -2 \text{ from the first two terms.}$$
$$= -2(x^2 + 2x + 1) - (-2)(1) + 1 \qquad \text{Add 1 inside the parentheses to complete the square.}$$
$$\text{Subtract 1, but it must also be multiplied by a factor of } -2.$$
$$= -2(x^2 + 2x + 1) + 2 + 1$$
$$= -2(x + 1)^2 + 3$$

The graph of $f(x) = -2(x + 1)^2 + 3$ is shown in Figure 8.33.

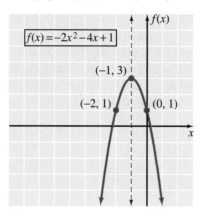

Figure 8.33   ■

Now let's graph a piecewise-defined function that involves both linear and quadratic rules of assignment.

Graph $f(x) = \begin{cases} 2x & \text{for } x \geq 0 \\ x^2 + 1 & \text{for } x < 0 \end{cases}$.

Solution

If $x \geq 0$, then $f(x) = 2x$. Thus for nonnegative values of x, we graph the linear function $f(x) = 2x$. If $x < 0$, then $f(x) = x^2 + 1$. Thus for negative values of x, we graph the quadratic function $f(x) = x^2 + 1$. The complete graph is shown in Figure 8.34.

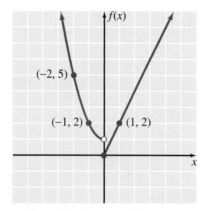

Figure 8.34   ■

What we know about parabolas and the process of completing the square can be helpful when we are using a graphing utility to graph a quadratic function. Consider the following example.

EXAMPLE 11   Use a graphing utility to obtain the graph of the quadratic function

$$f(x) = -x^2 + 37x - 311$$

Solution

First, we know that the parabola opens downward, and its width is the same as that of the basic parabola $f(x) = x^2$. Then we can start the process of completing the square to determine an approximate location of the vertex:

$$f(x) = -x^2 + 37x - 311$$

$$= -(x^2 - 37x) - 311$$

$$= -\left(x^2 - 37x + \left(\frac{37}{2}\right)^2\right) - 311 + \left(\frac{37}{2}\right)^2$$

$$= -(x^2 - 37x + (18.5)^2) - 311 + 342.25$$

Thus the vertex is near $x = 18$ and $y = 31$. Setting the boundaries of the viewing rectangle so that $-2 \le x \le 25$ and $-10 \le y \le 35$, we obtain the graph shown in Figure 8.35.

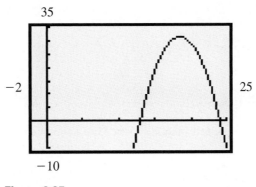

Figure 8.35   ■

Remark: The graph in Figure 8.35 is sufficient for most purposes because it shows the vertex and the x intercepts of the parabola. Certainly we could use other boundaries that would also give this information.

Problem Set 8.3

For Problems 1–26, graph each quadratic function.

1. $f(x) = x^2 + 1$

2. $f(x) = x^2 - 3$

3. $f(x) = 3x^2$

4. $f(x) = -2x^2$

5. $f(x) = -x^2 + 2$

6. $f(x) = -3x^2 - 1$

7. $f(x) = (x + 2)^2$

8. $f(x) = (x - 1)^2$

9. $f(x) = -2(x + 1)^2$

10. $f(x) = 3(x - 2)^2$

11. $f(x) = (x - 1)^2 + 2$

12. $f(x) = -(x + 2)^2 + 3$

13. $f(x) = \dfrac{1}{2}(x - 2)^2 - 3$

14. $f(x) = 2(x - 3)^2 - 1$

15. $f(x) = x^2 + 2x + 4$

16. $f(x) = x^2 - 4x + 2$

17. $f(x) = x^2 - 3x + 1$

18. $f(x) = x^2 + 5x + 5$

19. $f(x) = 2x^2 + 12x + 17$

20. $f(x) = 3x^2 - 6x$

21. $f(x) = -x^2 - 2x + 1$

22. $f(x) = -2x^2 + 12x - 16$

23. $f(x) = 2x^2 - 2x + 3$

24. $f(x) = 2x^2 + 3x - 1$

25. $f(x) = -2x^2 - 5x + 1$

26. $f(x) = -3x^2 + x - 2$

For Problems 27–34, graph each function.

27. $f(x) = \begin{cases} x & \text{for } x \ge 0 \\ 3x & \text{for } x < 0 \end{cases}$

28. $f(x) = \begin{cases} -x & \text{for } x \ge 0 \\ 4x & \text{for } x < 0 \end{cases}$

29. $f(x) = \begin{cases} 2x + 1 & \text{for } x \ge 0 \\ x^2 & \text{for } x < 0 \end{cases}$

30. $f(x) = \begin{cases} -x^2 & \text{for } x \ge 0 \\ 2x^2 & \text{for } x < 0 \end{cases}$

31. $f(x) = \begin{cases} 2 & \text{for } x \ge 0 \\ -1 & \text{for } x < 0 \end{cases}$

32. $f(x) = \begin{cases} 2 & \text{for } x > 2 \\ 1 & \text{for } 0 < x \le 2 \\ -1 & \text{for } x \le 0 \end{cases}$

33. $f(x) = \begin{cases} 1 & \text{for } 0 \le x < 1 \\ 2 & \text{for } 1 \le x < 2 \\ 3 & \text{for } 2 \le x < 3 \\ 4 & \text{for } 3 \le x < 4 \end{cases}$

34. $f(x) = \begin{cases} 2x + 3 & \text{for } x < 0 \\ x^2 & \text{for } 0 \le x < 2 \\ 1 & \text{for } x \ge 2 \end{cases}$

35. The **greatest integer function** is defined by the equation $f(x) = [x]$, where $[x]$ refers to the largest integer less than or equal to x. For example, $[2.6] = 2$, $[\sqrt{2}] = 1$, $[4] = 4$, and $[-1.4] = -2$. Graph $f(x) = [x]$ for $-4 \le x < 4$.

■ ■ ■ **THOUGHTS INTO WORDS**

36. Explain the concept of a piecewise-defined function.

37. Is $f(x) = (3x^2 - 2) - (2x + 1)$ a quadratic function? Explain your answer.

38. Give a step-by-step description of how you would use the ideas presented in this section to graph $f(x) = 5x^2 + 10x + 4$.

GRAPHING CALCULATOR ACTIVITIES

39. This problem is designed to reinforce ideas presented in this section. For each part, first predict the shapes and locations of the parabolas, and then use your graphing calculator to graph them on the same set of axes.

(a) $f(x) = x^2$, $f(x) = x^2 - 4$, $f(x) = x^2 + 1$, $f(x) = x^2 + 5$

(b) $f(x) = x^2$, $f(x) = (x - 5)^2$, $f(x) = (x + 5)^2$, $f(x) = (x - 3)^2$

(c) $f(x) = x^2$, $f(x) = 5x^2$, $f(x) = \dfrac{1}{3}x^2$, $f(x) = -2x^2$

(d) $f(x) = x^2$, $f(x) = (x - 7)^2 - 3$, $f(x) = -(x + 8)^2 + 4$, $f(x) = -3x^2 - 4$

(e) $f(x) = x^2 - 4x - 2$, $f(x) = -x^2 + 4x + 2$, $f(x) = -x^2 - 16x - 58$, $f(x) = x^2 + 16x + 58$

40. (a) Graph both $f(x) = x^2 - 14x + 51$ and $f(x) = x^2 + 14x + 51$ on the same set of axes. What relationship seems to exist between the two graphs?

(b) Graph both $f(x) = x^2 + 12x + 34$ and $f(x) = x^2 - 12x + 34$ on the same set of axes. What relationship seems to exist between the two graphs?

(c) Graph both $f(x) = -x^2 + 8x - 20$ and $f(x) = -x^2 - 8x - 20$ on the same set of axes. What relationship seems to exist between the two graphs?

(d) Make a statement that generalizes your findings in parts (a) through (c).

41. Use your graphing calculator to graph the piecewise-defined functions in Problems 27–34. You may need to consult your user's manual for instructions on graphing these functions.

8.4 More Quadratic Functions and Applications

In the previous section, we used the process of completing the square to change a quadratic function such as $f(x) = x^2 - 4x + 3$ to the form $f(x) = (x - 2)^2 - 1$. From the form $f(x) = (x - 2)^2 - 1$, it is easy to identify the vertex $(2, -1)$ and the axis of symmetry $x = 2$ of the parabola. In general, if we complete the square on

$$f(x) = ax^2 + bx + c$$

we obtain

$$f(x) = a\left(x^2 + \frac{b}{a}x \right) + c$$

$$= a\left(x^2 + \frac{b}{a}x + \frac{b^2}{4a^2} \right) + c - \frac{b^2}{4a}$$

$$= a\left(x + \frac{b}{2a} \right)^2 + \frac{4ac - b^2}{4a}$$

Therefore the parabola associated with the function $f(x) = ax^2 + bx + c$ has its vertex at

$$\left(-\frac{b}{2a}, \frac{4ac - b^2}{4a} \right)$$

and the equation of its axis of symmetry is $x = -b/2a$. These facts are illustrated in Figure 8.36.

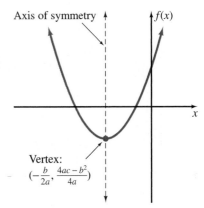

Axis of symmetry

$f(x)$

x

Vertex:
$\left(-\dfrac{b}{2a}, \dfrac{4ac - b^2}{4a}\right)$

Figure 8.36

By using the information from Figure 8.36, we now have another way of graphing quadratic functions of the form $f(x) = ax^2 + bx + c$, as indicated by the following steps:

1. Determine whether the parabola opens upward (if $a > 0$) or downward (if $a < 0$).

2. Find $-b/2a$, which is the x coordinate of the vertex.

3. Find $f(-b/2a)$, which is the y coordinate of the vertex, or find the y coordinate by evaluating

$$\frac{4ac - b^2}{4a}$$

4. Locate another point on the parabola, and also locate its image across the axis of symmetry, which is the line with equation $x = -b/2a$.

The three points found in steps 2, 3, and 4 should determine the general shape of the parabola. Let's illustrate this procedure with two examples.

E X A M P L E   1   Graph $f(x) = 3x^2 - 6x + 5$.

Solution

Step 1   Because $a > 0$, the parabola opens upward.

Step 2 $-\dfrac{b}{2a} = -\dfrac{(-6)}{2(3)} = -\dfrac{(-6)}{6} = 1$

Step 3 $f\left(-\dfrac{b}{2a}\right) = f(1) = 3(1)^2 - 6(1) + 5 = 2$. Thus the vertex is at $(1, 2)$.

Step 4   Letting $x = 2$, we obtain $f(2) = 12 - 12 + 5 = 5$. Thus $(2, 5)$ is on the graph, and so is its reflection, $(0, 5)$, across the line of symmetry, $x = 1$.

The three points $(1, 2), (2, 5)$, and $(0, 5)$ are used to graph the parabola in Figure 8.37.

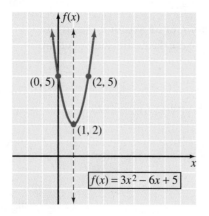

Figure 8.37

EXAMPLE 2

Graph $f(x) = -x^2 - 4x - 7$.

Solution

Step 1   Because $a < 0$, the parabola opens downward.

Step 2 $-\dfrac{b}{2a} = -\dfrac{(-4)}{2(-1)} = -\dfrac{(-4)}{(-2)} = -2$

Step 3 $f\left(-\dfrac{b}{2a}\right) = f(-2) = -(-2)^2 - 4(-2) - 7 = -3$. Thus the vertex is at $(-2, -3)$.

Step 4   Letting $x = 0$, we obtain $f(0) = -7$. Thus $(0, -7)$ is on the graph, and so is its reflection, $(-4, -7)$, across the line of symmetry, $x = -2$.

The three points $(-2, -3)$, $(0, -7)$, and $(-4, -7)$ are used to draw the parabola in Figure 8.38.

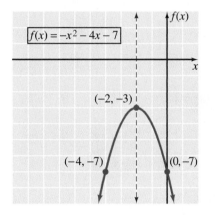

Figure 8.38   ■

In summary, we have two methods to graph a quadratic function:

1. We can express the function in the form $f(x) = a(x - h)^2 + k$ and use the values of a, h, and k to determine the parabola.

2. We can express the function in the form $f(x) = ax^2 + bx + c$ and use the approach demonstrated in Examples 1 and 2.

Parabolas possess various properties that make them very useful. For example, if a parabola is rotated about its axis, a parabolic surface is formed, and such surfaces are used for light and sound reflectors. A projectile fired into the air follows the curvature of a parabola. The trend line of profit and cost functions sometimes follows a parabolic curve. In most applications of the parabola, we are primarily interested in the x intercepts and the vertex. Let's consider some examples of finding the x intercepts and the vertex.

EXAMPLE 3   Find the x intercepts and the vertex for each of the following parabolas.

(a) $f(x) = -x^2 + 11x - 18$ **(b)** $f(x) = x^2 - 8x - 3$ **(c)** $f(x) = 2x^2 - 12x + 23$

Solutions

(a) To find the x intercepts, let $f(x) = 0$ and solve the resulting equation:

$$-x^2 + 11x - 18 = 0$$

$$x^2 - 11x + 18 = 0$$

$$(x - 2)(x - 9) = 0$$

$$x - 2 = 0 \quad \text{or} \quad x - 9 = 0$$

$$x = 2 \qquad\qquad x = 9$$

Therefore the x intercepts are 2 and 9. To find the vertex, let's determine the point $\left(-\dfrac{b}{2a},\ f\left(-\dfrac{b}{2a}\right)\right)$:

$$f(x) = -x^2 + 11x - 18$$

$$-\frac{b}{2a} = -\frac{11}{2(-1)} = -\frac{11}{-2} = \frac{11}{2}$$

$$f\left(\frac{11}{2}\right) = -\left(\frac{11}{2}\right)^2 + 11\left(\frac{11}{2}\right) - 18$$

$$= -\frac{121}{4} + \frac{121}{2} - 18$$

$$= \frac{-121 + 242 - 72}{4}$$

$$= \frac{49}{4}$$

Therefore the vertex is at $\left(\dfrac{11}{2}, \dfrac{49}{4}\right)$.

(b) To find the x intercepts, let $f(x) = 0$, and solve the resulting equation:

$$x^2 - 8x - 3 = 0$$

$$x = \frac{-(-8) \pm \sqrt{(-8)^2 - 4(1)(-3)}}{2(1)}$$

$$= \frac{8 \pm \sqrt{76}}{2}$$

$$= \frac{8 \pm 2\sqrt{19}}{2}$$

$$= 4 \pm \sqrt{19}$$

Therefore the x intercepts are $4 + \sqrt{19}$ and $4 - \sqrt{19}$. This time, to find the vertex, let's complete the square on x:

$$f(x) = x^2 - 8x - 3$$

$$= x^2 - 8x + 16 - 3 - 16$$

$$= (x - 4)^2 - 19$$

Therefore the vertex is at $(4, -19)$.

(c) To find the x intercepts, let $f(x) = 0$ and solve the resulting equation:

$$2x^2 - 12x + 23 = 0$$

$$x = \frac{-(-12) \pm \sqrt{(-12)^2 - 4(2)(23)}}{2(2)}$$

$$= \frac{12 \pm \sqrt{-40}}{4}$$

Because these solutions are nonreal complex numbers, there are no x intercepts. To find the vertex, let's determine the point $\left(-\dfrac{b}{2a}, f\left(-\dfrac{b}{2a}\right)\right)$.

$$f(x) = 2x^2 - 12x + 23$$

$$-\frac{b}{2a} = -\frac{-12}{2(2)}$$

$$= 3$$

$$f(3) = 2(3)^2 - 12(3) + 23$$

$$= 18 - 36 + 23$$

$$= 5$$

Therefore the vertex is at $(3, 5)$.   ■

Remark: Note that in parts (a) and (c), we used the general point

$$\left(-\frac{b}{2a}, f\left(-\frac{b}{2a}\right)\right)$$

to find the vertices. In part (b), however, we completed the square and used that form to determine the vertex. Which approach you use is up to you. We chose to complete the square in part (b) because the algebra involved was quite easy.

In part (a) of Example 3, we solved the equation $-x^2 + 11x - 18 = 0$ to determine that 2 and 9 are the x intercepts of the graph of the function $f(x) = -x^2 + 11x - 18$. The numbers 2 and 9 are also called the **real number zeros** of the function. That is to say, $f(2) = 0$ and $f(9) = 0$. In part (b) of Example 3, the real numbers $4 + \sqrt{19}$ and $4 - \sqrt{19}$ are the x intercepts of the graph of the function $f(x) = x^2 - 8x - 3$ and are the real number zeros of the function. Again, this means that $f(4 + \sqrt{19}) = 0$ and $f(4 - \sqrt{19}) = 0$. In part (c) of Example 3, the nonreal complex numbers $\dfrac{12 \pm \sqrt{-40}}{4}$, which simplify to $\dfrac{6 \pm i\sqrt{10}}{2}$, indicate that the graph of the function $f(x) = 2x^2 - 12x + 23$ has no points on the x axis. The complex numbers are zeros of the function, but they have no physical significance for the graph other than indicating that the graph has no points on the x axis.

Figure 8.39 shows the result we got when we used a graphing calculator to graph the three functions of Example 3 on the same set of axes. This gives us a visual interpretation of the conclusions drawn regarding the x intercepts and vertices.

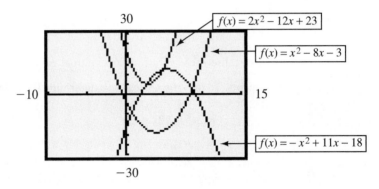

Figure 8.39

■ Back to Problem Solving

As we have seen, the vertex of the graph of a quadratic function is either the lowest or the highest point on the graph. Thus we often speak of the **minimum value** or **maximum value** of a function in applications of the parabola. The x value of the vertex indicates where the minimum or maximum occurs, and $f(x)$ yields the minimum or maximum value of the function. Let's consider some examples that illustrate these ideas.

P R O B L E M   1

A farmer has 120 rods of fencing and wants to enclose a rectangular plot of land that requires fencing on only three sides because it is bounded on one side by a river. Find the length and width of the plot that will maximize the area.

Solution

Let x represent the width; then $120 - 2x$ represents the length, as indicated in Figure 8.40.

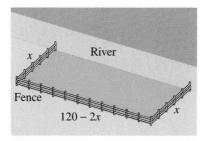

Figure 8.40

The function $A(x) = x(120 - 2x)$ represents the area of the plot in terms of the width x. Because

$$A(x) = x(120 - 2x)$$
$$= 120x - 2x^2$$
$$= -2x^2 + 120x$$

we have a quadratic function with $a = -2$, $b = 120$, and $c = 0$. Therefore the *maximum* value ($a < 0$ so the parabola opens downward) of the function is obtained where the x value is

$$-\frac{b}{2a} = -\frac{120}{2(-2)} = 30$$

If $x = 30$, then $120 - 2x = 120 - 2(30) = 60$. Thus the farmer should make the plot 30 rods wide and 60 rods long to maximize the area at $(30)(60) = 1800$ square rods.   ■

PROBLEM   2

Find two numbers whose sum is 30, such that the sum of their squares is a minimum.

Solution

Let x represent one of the numbers; then $30 - x$ represents the other number. By expressing the sum of their squares as a function of x, we obtain

$$f(x) = x^2 + (30 - x)^2$$

which can be simplified to

$$f(x) = x^2 + 900 - 60x + x^2$$
$$= 2x^2 - 60x + 900$$

This is a quadratic function with $a = 2$, $b = -60$, and $c = 900$. Therefore the x value where the *minimum* occurs is

$$-\frac{b}{2a} = -\frac{-60}{4}$$
$$= 15$$

If $x = 15$, then $30 - x = 30 - 15 = 15$. Thus the two numbers should both be 15.   ■

PROBLEM 3

A golf pro-shop operator finds that she can sell 30 sets of golf clubs at $500 per set in a year. Furthermore, she predicts that for each $25 decrease in price, she could sell three extra sets of golf clubs. At what price should she sell the clubs to maximize gross income?

Solution

In analyzing such a problem, it sometimes helps to start by setting up a table. We use the fact that three additional sets can be sold for each $25 decrease in price.

Number of sets	×	Price per set	=	Income
30	×	$500	=	$15,000
33	×	$475	=	$15,675
36	×	$450	=	$16,200

Let x represent the number of $25 decreases in price. Then the income can be expressed as a function of x.

$$f(x) = (30 + 3x)(500 - 25x)$$

Number of sets ↑   Price per set ↑

Simplifying this, we obtain

$$f(x) = 15,000 - 750x + 1500x - 75x^2$$
$$= -75x^2 + 750x + 15,000$$

We complete the square in order to analyze the parabola.

$$f(x) = -75x^2 + 750x + 15,000$$
$$= -75(x^2 - 10x) + 15,000$$
$$= -75(x^2 - 10x + 25) + 15,000 + 1875$$
$$= -75(x - 5)^2 + 16,875$$

From this form, we know that the vertex of the parabola is at (5, 16,875), and because $a = -75$, we know that a *maximum* occurs at the vertex. Thus five decreases of $25 — that is, a $125 reduction in price — will give a maximum income of $16,875. The golf clubs should be sold at $375 per set.   ∎

We have determined that the vertex of a parabola associated with $f(x) = ax^2 + bx + c$ is located at $\left(-\dfrac{b}{2a}, f\left(-\dfrac{b}{2a}\right)\right)$ and that the x intercepts of the graph can be found by solving the quadratic equation $ax^2 + bx + c = 0$. Therefore a graphing utility does not provide us with much extra power when we are work-ing with quadratic functions. However, as functions become more complex, a graphing utility becomes more helpful. Let's build our confidence in the use of a graphing utility at this time, while we have a way of checking our results.

E X A M P L E   4   Use a graphing utility to graph $f(x) = x^2 - 8x - 3$ and find the x intercepts of the graph. [This is the parabola from part (b) of Example 3.]

Solution

A graph of the parabola is shown in Figure 8.41.

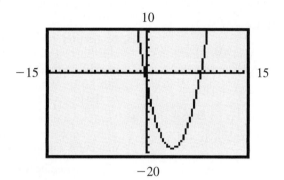

Figure 8.41

One x intercept appears to be between 0 and -1 and the other between 8 and 9. Let's zoom in on the x intercept between 8 and 9. This produces a graph like Figure 8.42.

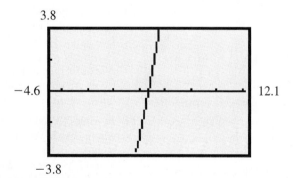

Figure 8.42

Now we can use the TRACE function to determine that this x intercept is at ap-proximately 8.4. (This agrees with the answer of $4 + \sqrt{19}$ that we got in Example 3.) In a similar fashion, we can determine that the other x intercept is at approxi-mately -0.4. ∎

Problem Set 8.4

For Problems 1–12, use the approach of Examples 1 and 2 of this section to graph each quadratic function.

1. $f(x) = x^2 - 8x + 15$
2. $f(x) = x^2 + 6x + 11$

3. $f(x) = 2x^2 + 20x + 52$
4. $f(x) = 3x^2 - 6x - 1$

5. $f(x) = -x^2 + 4x - 7$
6. $f(x) = -x^2 - 6x - 5$

7. $f(x) = -3x^2 + 6x - 5$
8. $f(x) = -2x^2 - 4x + 2$

9. $f(x) = x^2 + 3x - 1$
10. $f(x) = x^2 + 5x + 2$

11. $f(x) = -2x^2 + 5x + 1$
12. $f(x) = -3x^2 + 2x - 1$

For Problems 13–20, use the approach that you think is the most appropriate to graph each quadratic function.

13. $f(x) = -x^2 + 3$
14. $f(x) = (x + 1)^2 + 1$

15. $f(x) = x^2 + x - 1$
16. $f(x) = -x^2 + 3x - 4$

17. $f(x) = -2x^2 + 4x + 1$
18. $f(x) = 4x^2 - 8x + 5$

19. $f(x) = -\left(x + \dfrac{5}{2}\right)^2 + \dfrac{3}{2}$
20. $f(x) = x^2 - 4x$

For Problems 21–36, find the x intercepts and the vertex of each parabola.

21. $f(x) = 3x^2 - 12$

22. $f(x) = 6x^2 - 4$

23. $f(x) = 5x^2 - 10x$

24. $f(x) = 3x^2 + 9x$

25. $f(x) = x^2 - 8x + 15$
26. $f(x) = x^2 - 16x + 63$

27. $f(x) = 2x^2 - 28x + 96$
28. $f(x) = 3x^2 - 60x + 297$

29. $f(x) = -x^2 + 10x - 24$
30. $f(x) = -2x^2 + 36x - 160$

31. $f(x) = x^2 - 14x + 44$
32. $f(x) = x^2 - 18x + 68$

33. $f(x) = -x^2 + 9x - 21$
34. $f(x) = 2x^2 + 3x + 3$

35. $f(x) = -4x^2 + 4x + 4$
36. $f(x) = -2x^2 + 3x + 7$

For Problems 37–42, find the zeros of each function.

37. $f(x) = x^2 + 3x - 88$
38. $f(x) = 6x^2 - 5x - 4$

39. $f(x) = 4x^2 - 48x + 108$
40. $f(x) = x^2 - 6x - 6$

41. $f(x) = x^2 - 4x + 11$
42. $f(x) = x^2 - 23x + 126$

For Problems 43–52, solve each problem.

43. Suppose that the equation $p(x) = -2x^2 + 280x - 1000$, where x represents the number of items sold, describes the profit function for a certain business. How many items should be sold to maximize the profit?

44. Suppose that the cost function for the production of a particular item is given by the equation $C(x) = 2x^2 - 320x + 12,920$, where x represents the number of items. How many items should be produced to minimize the cost?

45. Neglecting air resistance, the height of a projectile fired vertically into the air at an initial velocity of 96 feet per second is a function of time x and is given by the equation $f(x) = 96x - 16x^2$. Find the highest point reached by the projectile.

46. Find two numbers whose sum is 30, such that the sum of the square of one number plus ten times the other number is a minimum.

47. Find two numbers whose sum is 50 and whose product is a maximum.

48. Find two numbers whose difference is 40 and whose product is a minimum.

49. Two hundred and forty meters of fencing is available to enclose a rectangular playground. What should be the dimensions of the playground to maximize the area?

50. Motel managers advertise that they will provide dinner, dancing, and drinks for $50 per couple for a New Year's Eve party. They must have a guarantee of 30 couples. Furthermore, they will agree that for each couple in excess of 30, they will reduce the price per couple by $0.50 for all attending. How many couples will it take to maximize the motel's revenue?

51. A cable TV company has 1000 subscribers, each of whom pays $15 per month. On the basis of a survey, the company believes that for each decrease of $0.25 in the monthly rate, it could obtain 20 additional subscribers. At what rate will the maximum revenue be obtained, and how many subscribers will there be at that rate?

52. A manufacturer finds that for the first 500 units of its product that are produced and sold, the profit is $50 per unit. The profit on each of the units beyond 500 is decreased by $0.10 times the number of additional units sold. What level of output will maximize profit?

■ ■ ■ THOUGHTS INTO WORDS

53. Suppose your friend was absent the day this section was discussed. How would you explain to her the ideas pertaining to x intercepts of the graph of a function, zeros of the function, and solutions of the equation $f(x) = 0$?

54. Give a step-by-step explanation of how to find the x intercepts of the graph of the function $f(x) = 2x^2 + 7x - 4$.

55. Give a step-by-step explanation of how to find the vertex of the parabola determined by the equation $f(x) = -x^2 - 6x - 5$.

GRAPHING CALCULATOR ACTIVITIES

56. Suppose that the viewing window on your graphing calculator is set so that $-15 \leq x \leq 15$ and $-10 \leq y \leq 10$. Now try to graph the function $f(x) = x^2 - 8x + 28$. Nothing appears on the screen, so the parabola must be outside the viewing window. We could arbitrarily expand the window until the parabola appeared. However, let's be a little more systematic and use $\left(-\dfrac{b}{2a}, f\left(-\dfrac{b}{2a}\right)\right)$ to find the vertex. We find the vertex is at $(4, 12)$, so let's change the y values of the window so that $0 \leq y \leq 25$. Now we get a good picture of the parabola.

Graph each of the following parabolas, and keep in mind that you may need to change the dimensions of the viewing window to obtain a good picture.

(a) $f(x) = x^2 - 2x + 12$
(b) $f(x) = -x^2 - 4x - 16$
(c) $f(x) = x^2 + 12x + 44$
(d) $f(x) = x^2 - 30x + 229$
(e) $f(x) = -2x^2 + 8x - 19$

57. Use a graphing calculator to graph each of the following parabolas, and then use the TRACE function to help estimate the x intercepts and the vertex. Finally, use the approach of Example 3 to find the x intercepts and the vertex.
(a) $f(x) = x^2 - 6x + 3$
(b) $f(x) = x^2 - 18x + 66$
(c) $f(x) = -x^2 + 8x - 3$
(d) $f(x) = -x^2 + 24x - 129$
(e) $f(x) = 14x^2 - 7x + 1$
(f) $f(x) = -\dfrac{1}{2}x^2 + 5x - \dfrac{17}{2}$

58. In Problems 21–36, you were asked to find the x intercepts and the vertex of some parabolas. Now use a graphing calculator to graph each parabola and visually justify your answers.

59. For each of the following quadratic functions, use the discriminant to determine the number of real-number zeros, and then graph the function with a graphing calculator to check your answer.
(a) $f(x) = 3x^2 - 15x - 42$
(b) $f(x) = 2x^2 - 36x + 162$
(c) $f(x) = -4x^2 - 48x - 144$
(d) $f(x) = 2x^2 + 2x + 5$
(e) $f(x) = 4x^2 - 4x - 120$
(f) $f(x) = 5x^2 - x + 4$

8.5 Transformations of Some Basic Curves

From our work in Section 8.3, we know that the graph of $f(x) = (x - 5)^2$ is the basic parabola $f(x) = x^2$ translated five units to the right. Likewise, we know that the graph of $f(x) = -x^2 - 2$ is the basic parabola reflected across the x axis and translated downward two units. Translations and reflections apply not only to parabolas but also to curves in general. Therefore, if we know the shapes of a few basic curves, then it is easy to sketch numerous variations of these curves by using the concepts of translation and reflection.

Let's begin this section by establishing the graphs of four basic curves and then apply some transformations to these curves. First, let's restate, in terms of function vocabulary, the graphing suggestions offered in Chapter 7. Pay special attention to suggestions 2 and 3, where we restate the concepts of intercepts and symmetry using function notation.

1. Determine the domain of the function.

2. Find the y intercept [we are labeling the y axis with $f(x)$] by evaluating $f(0)$. Find the x intercept by finding the value(s) of x such that $f(x) = 0$.

3. Determine any types of symmetry that the equation possesses. If $f(-x) = f(x)$, then the function exhibits y axis symmetry. If $f(-x) = -f(x)$, then the function exhibits origin symmetry. (Note that the definition of a function rules out the possibility that the graph of a function has x axis symmetry.)

4. Set up a table of ordered pairs that satisfy the equation. The type of symmetry and the domain will affect your choice of values of x in the table.

5. Plot the points associated with the ordered pairs and connect them with a smooth curve. Then, if appropriate, reflect this part of the curve according to any symmetries possessed by the graph.

E X A M P L E   1

Graph $f(x) = x^3$.

Solution

The domain is the set of real numbers. Because $f(0) = 0$, the origin is on the graph. Because $f(-x) = (-x)^3 = -x^3 = -f(x)$, the graph is symmetric with respect to the origin. Therefore, we can concentrate our table on the positive values of x. By connecting the points associated with the ordered pairs from the table with a smooth curve and then reflecting it through the origin, we get the graph in Figure 8.43.

x	$f(x) = x^3$
0	0
1	1
2	8
$\dfrac{1}{2}$	$\dfrac{1}{8}$

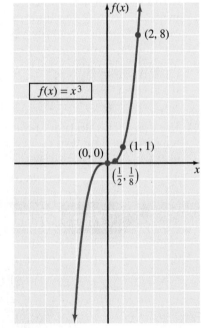

Figure 8.43

EXAMPLE 2   Graph $f(x) = x^4$.

Solution

The domain is the set of real numbers. Because $f(0) = 0$, the origin is on the graph. Because $f(-x) = (-x)^4 = x^4 = f(x)$, the graph has y axis symmetry, and we can concentrate our table of values on the positive values of x. If we connect the points associated with the ordered pairs from the table with a smooth curve and then reflect across the vertical axis, we get the graph in Figure 8.44.

x	$f(x) = x^4$
0	0
1	1
2	16
$\dfrac{1}{2}$	$\dfrac{1}{16}$

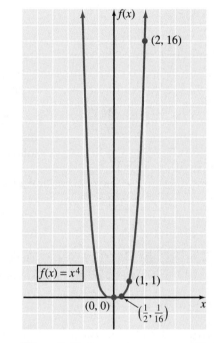

Figure 8.44   ■

Remark: The curve in Figure 8.44 is not a parabola, even though it resembles one; this curve is flatter at the bottom and steeper.

EXAMPLE 3   Graph $f(x) = \sqrt{x}$.

Solution

The domain of the function is the set of nonnegative real numbers. Because $f(0) = 0$, the origin is on the graph. Because $f(-x) \neq f(x)$ and $f(-x) \neq -f(x)$, there is no symmetry, so let's set up a table of values using nonnegative values for x. Plotting the points determined by the table and connecting them with a smooth curve produces Figure 8.45.

x	$f(x) = \sqrt{x}$
0	0
1	1
4	2
9	3

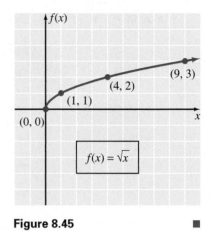

Figure 8.45 ∎

Sometimes a new function is defined in terms of old functions. In such cases, the definition plays an important role in the study of the new function. Consider the following example.

E X A M P L E   4

Graph $f(x) = |x|$.

Solution

The concept of absolute value is defined for all real numbers by

$$|x| = x \quad \text{if } x \geq 0$$
$$|x| = -x \quad \text{if } x < 0$$

Therefore the absolute value function can be expressed as

$$f(x) = |x| = \begin{cases} x & \text{if } x \geq 0 \\ -x & \text{if } x < 0 \end{cases}$$

The graph of $f(x) = x$ for $x \geq 0$ is the ray in the first quadrant, and the graph of $f(x) = -x$ for $x < 0$ is the half line (not including the origin) in the second quadrant, as indicated in Figure 8.46. Note that the graph has y axis symmetry.

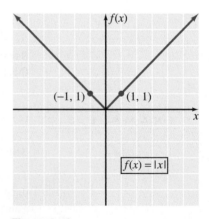

Figure 8.46 ∎

■ Translations of the Basic Curves

From our work in Section 8.3, we know that

1. The graph of $f(x) = x^2 + 3$ is the graph of $f(x) = x^2$ moved up three units.

2. The graph of $f(x) = x^2 - 2$ is the graph of $f(x) = x^2$ moved down two units.

Now let's describe in general the concept of a vertical translation.

Vertical Translation

The graph of $y = f(x) + k$ is the graph of $y = f(x)$ shifted k units upward if $k > 0$ or shifted $|k|$ units downward if $k < 0$.

In Figure 8.47, the graph of $f(x) = |x| + 2$ is obtained by shifting the graph of $f(x) = |x|$ upward two units, and the graph of $f(x) = |x| - 3$ is obtained by shifting the graph of $f(x) = |x|$ downward three units. [Remember that $f(x) = |x| - 3$ can be written as $f(x) = |x| + (-3)$.]

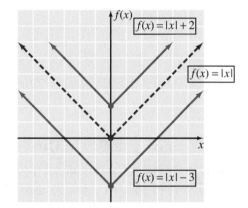

Figure 8.47

We also graphed horizontal translations of the basic parabola in Section 8.3. For example:

1. The graph of $f(x) = (x - 4)^2$ is the graph of $f(x) = x^2$ shifted four units to the right.

2. The graph of $f(x) = (x + 5)^2$ is the graph of $f(x) = x^2$ shifted five units to the left.

The general concept of a horizontal translation can be described as follows.

Horizontal Translation

> The graph of $y = f(x - h)$ is the graph of $y = f(x)$ shifted h units to the right if $h > 0$ or shifted $|h|$ units to the left if $h < 0$.

In Figure 8.48, the graph of $f(x) = (x - 3)^3$ is obtained by shifting the graph of $f(x) = x^3$ three units to the right. Likewise, the graph of $f(x) = (x + 2)^3$ is obtained by shifting the graph of $f(x) = x^3$ two units to the left.

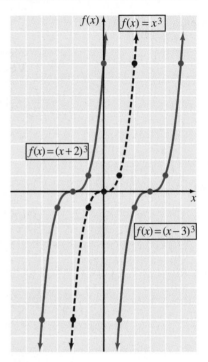

Figure 8.48

■ Reflections of the Basic Curves

From our work in Section 8.3, we know that the graph of $f(x) = -x^2$ is the graph of $f(x) = x^2$ reflected through the x axis. The general concept of an x axis reflection can be described as follows:

x Axis Reflection

> The graph of $y = -f(x)$ is the graph of $y = f(x)$ reflected through the x axis.

In Figure 8.49, the graph of $f(x) = -\sqrt{x}$ is obtained by reflecting the graph of $f(x) = \sqrt{x}$ through the x axis. Reflections are sometimes referred to as **mirror images**. Thus if we think of the x axis in Figure 8.49 as a mirror, then the graphs of $f(x) = \sqrt{x}$ and $f(x) = -\sqrt{x}$ are mirror images of each other.

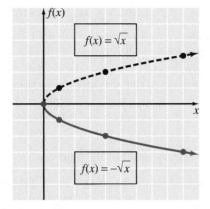

Figure 8.49

In Section 8.3, we did not consider a y axis reflection of the basic parabola $f(x) = x^2$ because it is symmetric with respect to the y axis. In other words, a y axis reflection of $f(x) = x^2$ produces the same figure. However, at this time, let's describe the general concept of a y axis reflection.

y Axis Reflection

The graph of $y = f(-x)$ is the graph of $y = f(x)$ reflected through the y axis.

Now suppose that we want to do a y axis reflection of $f(x) = \sqrt{x}$. Because $f(x) = \sqrt{x}$ is defined for $x \geq 0$, the y axis reflection $f(x) = \sqrt{-x}$ is defined for $-x \geq 0$, which is equivalent to $x \leq 0$. Figure 8.50 shows the y axis reflection of $f(x) = \sqrt{x}$.

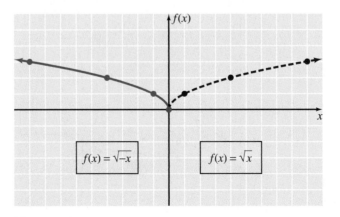

Figure 8.50

■ Vertical Stretching and Shrinking

Translations and reflections are called **rigid transformations** because the basic shape of the curve being transformed is not changed. In other words, only the positions of the graphs are changed. Now we want to consider some transformations that distort the shape of the original figure somewhat.

In Section 8.3, we graphed the function $f(x) = 2x^2$ by doubling the $f(x)$ values of the ordered pairs that satisfy the function $f(x) = x^2$. We obtained a parabola with its vertex at the origin, symmetric to the y axis, but *narrower* than the basic parabola. Likewise, we graphed the function $f(x) = \frac{1}{2}x^2$ by halving the $f(x)$ values of the ordered pairs that satisfy $f(x) = x^2$. In this case, we obtained a parabola with its vertex at the origin, symmetric to the y axis, but *wider* than the basic parabola.

The concepts of *narrower* and *wider* can be used to describe parabolas, but they cannot be used to describe some other curves accurately. Instead, we use the more general concepts of vertical stretching and shrinking.

Vertical Stretching and Shrinking

> The graph of $y = cf(x)$ is obtained from the graph of $y = f(x)$ by multiplying the y coordinates for $y = f(x)$ by c. If $|c| > 1$, the graph is said to be *stretched* by a factor of $|c|$, and if $0 < |c| < 1$, the graph is said to be *shrunk* by a factor of $|c|$.

In Figure 8.51, the graph of $f(x) = 2\sqrt{x}$ is obtained by doubling the y coordinates of points on the graph of $f(x) = \sqrt{x}$. Likewise, the graph of $f(x) = \frac{1}{2}\sqrt{x}$ is obtained by halving the y coordinates of points on the graph of $f(x) = \sqrt{x}$.

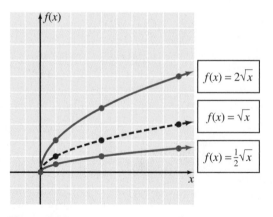

Figure 8.51

■ Successive Transformations

Some curves are the result of performing more than one transformation on a basic curve. Let's consider the graph of a function that involves a stretching, a reflection, a horizontal translation, and a vertical translation of the basic absolute-value function.

EXAMPLE 5

Graph $f(x) = -2|x - 3| + 1$.

Solution

This is the basic absolute-value curve stretched by a factor of 2, reflected through the x axis, shifted three units to the right, and shifted one unit upward. To sketch the graph, we locate the point $(3, 1)$ and then determine a point on each of the rays. The graph is shown in Figure 8.52.

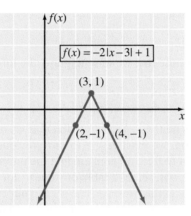

Figure 8.52

Remark: Note that in Example 5, we did not sketch the original basic curve $f(x) = |x|$ or any of the intermediate transformations. However, it is helpful to picture each transformation mentally. This locates the point $(3, 1)$ and establishes the fact that the two rays point downward. Then a point on each ray determines the final graph.

We do need to realize that changing the order of doing the transformations may produce an incorrect graph. In Example 5, performing the translations first, and then performing the stretching and x axis reflection, would locate the vertex of the graph at $(3, -1)$ instead of $(3, 1)$. **Unless parentheses indicate otherwise, stretchings, shrinkings, and reflections should be performed before translations.**

Suppose that you need to graph the function $f(x) = \sqrt{-3 - x}$. Furthermore, suppose that you are not certain which transformations of the basic square root function will produce this function. By plotting a few points and using your knowledge of the general shape of a square root curve, you should be able to sketch the curve as shown in Figure 8.53.

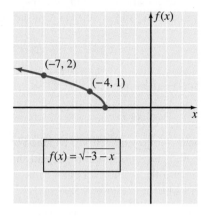

Figure 8.53

Now suppose that we want to graph the following function.

$$f(x) = \frac{2x^2}{x^2 + 4}$$

Because this is neither a basic function that we recognize nor a transformation of a basic function, we must revert to our previous graphing experiences. In other words, we need to find the domain, find the intercepts, check for symmetry, check for any restrictions, set up a table of values, plot the points, and sketch the curve. (If you want to do this now, you can check your result on page 503.) Furthermore, if the new function is defined in terms of an old function, we may be able to apply the definition of the old function and thereby simplify the new function for graphing purposes. Suppose you are asked to graph the function $f(x) = |x| + x$. This function can be simplified by applying the definition of absolute value. We will leave this for you to do in the next problem set.

Finally, let's use a graphing utility to give another illustration of the concept of stretching and shrinking a curve.

E X A M P L E   6

If $f(x) = \sqrt{25 - x^2}$, sketch a graph of $y = 2(f(x))$ and $y = \frac{1}{2}(f(x))$.

Solution

If $y = f(x) = \sqrt{25 - x^2}$, then

$$y = 2(f(x)) = 2\sqrt{25 - x^2} \qquad \text{and} \qquad y = \frac{1}{2}(f(x)) = \frac{1}{2}\sqrt{25 - x^2}$$

Graphing all three of these functions on the same set of axes produces Figure 8.54.

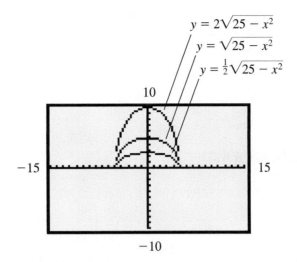

Figure 8.54

Problem Set 8.5

For Problems 1–30, graph each function.

1. $f(x) = x^4 + 2$

2. $f(x) = -x^4 - 1$

3. $f(x) = (x - 2)^4$

4. $f(x) = (x + 3)^4 + 1$

5. $f(x) = -x^3$

6. $f(x) = x^3 - 2$

7. $f(x) = (x + 2)^3$

8. $f(x) = (x - 3)^3 - 1$

9. $f(x) = |x - 1| + 2$

10. $f(x) = -|x + 2|$

11. $f(x) = |x + 1| - 3$

12. $f(x) = 2|x|$

13. $f(x) = x + |x|$

14. $f(x) = \dfrac{|x|}{x}$

15. $f(x) = -|x - 2| - 1$

16. $f(x) = 2|x + 1| - 4$

17. $f(x) = x - |x|$

18. $f(x) = |x| - x$

19. $f(x) = -2\sqrt{x}$

20. $f(x) = 2\sqrt{x - 1}$

21. $f(x) = \sqrt{x + 2} - 3$

22. $f(x) = -\sqrt{x + 2} + 2$

23. $f(x) = \sqrt{2 - x}$

24. $f(x) = \sqrt{-1 - x}$

25. $f(x) = -2x^4 + 1$

26. $f(x) = 2(x - 2)^4 - 4$

27. $f(x) = -2x^3$

28. $f(x) = 2x^3 + 3$

29. $f(x) = 3(x - 2)^3 - 1$

30. $f(x) = -2(x + 1)^3 + 2$

31. Suppose that the graph of $y = f(x)$ with a domain of $-2 \le x \le 2$ is shown in Figure 8.55.

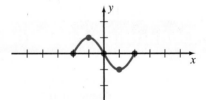

Figure 8.55

Sketch the graph of each of the following transformations of $y = f(x)$.

(a) $y = f(x) + 3$

(b) $y = f(x - 2)$

(c) $y = -f(x)$

(d) $y = f(x + 3) - 4$

■ ■ ■ **THOUGHTS INTO WORDS**

32. Are the graphs of the two functions $f(x) = \sqrt{x - 2}$ and $g(x) = \sqrt{2 - x}$ y axis reflections of each other? Defend your answer.

33. Are the graphs of $f(x) = 2\sqrt{x}$ and $g(x) = \sqrt{2x}$ identical? Defend your answer.

34. Are the graphs of $f(x) = \sqrt{x + 4}$ and $g(x) = \sqrt{-x + 4}$ y axis reflections of each other? Defend your answer.

GRAPHING CALCULATOR ACTIVITIES

35. Use your graphing calculator to check your graphs for Problems 13–30.

36. Graph $f(x) = \sqrt{x^2 + 8}$, $f(x) = \sqrt{x^2 + 4}$, and $f(x) = \sqrt{x^2 + 1}$ on the same set of axes. Look at these graphs and predict the graph of $f(x) = \sqrt{x^2 - 4}$. Now graph it with the calculator to test your prediction.

37. For each of the following, predict the general shape and location of the graph, and then use your calculator to graph the function to check your prediction.
(a) $f(x) = \sqrt{x^2}$ **(b)** $f(x) = \sqrt{x^3}$
(c) $f(x) = |x^2|$ **(d)** $f(x) = |x^3|$

38. Graph $f(x) = x^4 + x^3$. Now predict the graph for each of the following, and check each prediction with your graphing calculator.

(a) $f(x) = x^4 + x^3 - 4$
(b) $f(x) = (x - 3)^4 + (x - 3)^3$
(c) $f(x) = -x^4 - x^3$
(d) $f(x) = x^4 - x^3$

39. Graph $f(x) = \sqrt[3]{x}$. Now predict the graph for each of the following, and check each prediction with your graphing calculator.
(a) $f(x) = 5 + \sqrt[3]{x}$ **(b)** $f(x) = \sqrt[3]{x + 4}$
(c) $f(x) = -\sqrt[3]{x}$ **(d)** $f(x) = \sqrt[3]{x - 3} - 5$
(e) $f(x) = \sqrt[3]{-x}$

8.6   Combining Functions

In subsequent mathematics courses, it is common to encounter functions that are defined in terms of sums, differences, products, and quotients of simpler functions. For example, if $h(x) = x^2 + \sqrt{x - 1}$, then we may consider the function h as the sum of f and g, where $f(x) = x^2$ and $g(x) = \sqrt{x - 1}$. In general, if f and g are functions and D is the intersection of their domains, then the following definitions can be made:

Sum	$(f + g)(x) = f(x) + g(x)$
Difference	$(f - g)(x) = f(x) - g(x)$
Product	$(f \cdot g)(x) = f(x) \cdot g(x)$
Quotient	$\left(\dfrac{f}{g}\right)(x) = \dfrac{f(x)}{g(x)}, \quad g(x) \neq 0$

EXAMPLE 1

If $f(x) = 3x - 1$ and $g(x) = x^2 - x - 2$, find (a) $(f + g)(x)$; (b) $(f - g)(x)$; (c) $(f \cdot g)(x)$; and (d) $(f/g)(x)$. Determine the domain of each.

Solutions

(a) $(f + g)(x) = f(x) + g(x) = (3x - 1) + (x^2 - x - 2) = x^2 + 2x - 3$

(b) $(f - g)(x) = f(x) - g(x)$

$$= (3x - 1) - (x^2 - x - 2)$$

$$= 3x - 1 - x^2 + x + 2$$

$$= -x^2 + 4x + 1$$

(c) $(f \cdot g)(x) = f(x) \cdot g(x)$

$$= (3x - 1)(x^2 - x - 2)$$

$$= 3x^3 - 3x^2 - 6x - x^2 + x + 2$$

$$= 3x^3 - 4x^2 - 5x + 2$$

(d) $\left(\dfrac{f}{g}\right)(x) = \dfrac{f(x)}{g(x)} = \dfrac{3x - 1}{x^2 - x - 2}$

The domain of both f and g is the set of all real numbers. Therefore the domain of $f + g$, $f - g$, and $f \cdot g$ is the set of all real numbers. For f/g, the denominator $x^2 - x - 2$ cannot equal zero. Solving $x^2 - x - 2 = 0$ produces

$$(x - 2)(x + 1) = 0$$

$$x - 2 = 0 \quad \text{or} \quad x + 1 = 0$$

$$x = 2 \qquad\qquad x = -1$$

Therefore the domain for f/g is the set of all real numbers except 2 and -1. ■

Graphs of functions can help us visually sort out our thought processes. For example, suppose that $f(x) = 0.46x - 4$ and $g(x) = 3$. If we think in terms of ordinate values, it seems reasonable that the graph of $f + g$ is the graph of f moved up three units. Likewise, the graph of $f - g$ should be the graph of f moved down three units. Let's use a graphing calculator to support these conclusions. Letting $Y_1 = 0.46x - 4$, $Y_2 = 3$, $Y_3 = Y_1 + Y_2$, and $Y_4 = Y_1 - Y_2$, we obtain Figure 8.56.

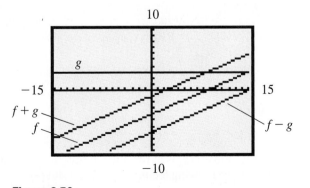

Figure 8.56

Certainly this figure supports our conclusions. This type of graphical analysis becomes more important as the functions become more complex.

■ Composition of Functions

Besides adding, subtracting, multiplying, and dividing functions, there is another important operation called *composition*. The composition of two functions can be defined as follows:

Definition 8.2

The **composition** of functions f and g is defined by

$$(f \circ g)(x) = f(g(x))$$

for all x in the domain of g such that $g(x)$ is in the domain of f.

The left side, $(f \circ g)(x)$, of the equation in Definition 8.2 is read "the composition of f and g," and the right side is read "f of g of x." It may also be helpful for you to have a mental picture of Definition 8.2 as two function machines hooked together to produce another function (called the **composite function**), as illustrated in Figure 8.57. Note that what comes out of the g function is substituted into the f function. Thus composition is sometimes called the **substitution of functions**.

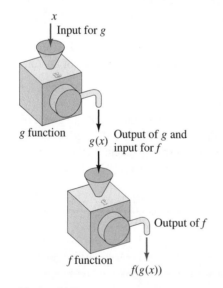

Figure 8.57

Figure 8.57 also illustrates the fact that $f \circ g$ is defined for all x in the domain of g such that $g(x)$ is in the domain of f. In other words, what comes out of g must be capable of being fed into f. Let's consider some examples.

EXAMPLE　2

If $f(x) = x^2$ and $g(x) = 3x - 4$, find $(f \circ g)(x)$ and determine its domain.

Solution

Apply Definition 8.2 to obtain

$$
\begin{aligned}
(f \circ g)(x) &= f(g(x)) \\
&= f(3x - 4) \\
&= (3x - 4)^2 \\
&= 9x^2 - 24x + 16
\end{aligned}
$$

Because g and f are both defined for all real numbers, so is $f \circ g$. ■

Definition 8.2, with f and g interchanged, defines the composition of g and f as $(g \circ f)(x) = g(f(x))$.

EXAMPLE　3

If $f(x) = x^2$ and $g(x) = 3x - 4$, find $(g \circ f)(x)$ and determine its domain.

Solution

$$
\begin{aligned}
(g \circ f)(x) &= g(f(x)) \\
&= g(x^2) \\
&= 3x^2 - 4
\end{aligned}
$$

Because f and g are defined for all real numbers, so is $g \circ f$. ■

The results of Examples 2 and 3 demonstrate an important idea: **The composition of functions is not a commutative operation**. In other words, $f \circ g \neq g \circ f$ for all functions f and g. However, as we will see in Section 10.3, there is a special class of functions for which $f \circ g = g \circ f$.

EXAMPLE　4

If $f(x) = \sqrt{x}$ and $g(x) = 2x - 1$, find $(f \circ g)(x)$ and $(g \circ f)(x)$. Also determine the domain of each composite function.

Solution

$$
\begin{aligned}
(f \circ g)(x) &= f(g(x)) \\
&= f(2x - 1) \\
&= \sqrt{2x - 1}
\end{aligned}
$$

The domain and range of g are the set of all real numbers, but the domain of f is all *nonnegative* real numbers. Therefore $g(x)$, which is $2x - 1$, must be nonnegative.

$$2x - 1 \geq 0$$

$$2x \geq 1$$

$$x \geq \frac{1}{2}$$

Thus the domain of $f \circ g$ is $D = \left\{ x \middle| x \geq \frac{1}{2} \right\}$.

$$(g \circ f)(x) = g(f(x))$$

$$= g(\sqrt{x})$$

$$= 2\sqrt{x} - 1$$

The domain and range of f are the set of nonnegative real numbers. The domain of g is the set of all real numbers. Therefore the domain of $g \circ f$ is $D = \{x|x \geq 0\}$.   ∎

E X A M P L E   5   If $f(x) = \dfrac{3}{x - 1}$ and $g(x) = \dfrac{1}{2x}$, find $(f \circ g)(x)$ and $(g \circ f)(x)$. Determine the domain for each composite function.

Solution

$$(f \circ g)(x) = f(g(x))$$

$$= f\left(\frac{1}{2x}\right)$$

$$= \frac{3}{\dfrac{1}{2x} - 1} = \frac{3}{\dfrac{1}{2x} - \dfrac{2x}{2x}} = \frac{3}{\dfrac{1 - 2x}{2x}}$$

$$= \frac{6x}{1 - 2x}$$

The domain of g is all real numbers except 0, and the domain of f is all real numbers except 1. Therefore $g(x) \neq 1$. So we need to solve $g(x) = 1$ to find the values of x that will make $g(x) = 1$.

$$g(x) = 1$$

$$\frac{1}{2x} = 1$$

$$1 = 2x$$

$$\frac{1}{2} = x$$

Therefore $x \neq \dfrac{1}{2}$, so the domain of $f \circ g$ is $D = \left\{ x \middle| x \neq 0 \text{ and } x \neq \dfrac{1}{2} \right\}$.

$$(g \circ f)(x) = g(f(x))$$

$$= g\left(\frac{3}{x - 1}\right)$$

$$= \frac{1}{2\left(\dfrac{3}{x - 1}\right)} = \frac{1}{\dfrac{6}{x - 1}}$$

$$= \frac{x - 1}{6}$$

The domain of f is all real numbers except 1, and the domain of g is all real numbers except 0. Because $f(x)$, which is $3/(x - 1)$, will never equal 0, the domain of $g \circ f$ is $D = \{x | x \neq 1\}$.   ∎

A graphing utility can be used to find the graph of a composite function without actually forming the function algebraically. Let's see how this works.

EXAMPLE 6

If $f(x) = x^3$ and $g(x) = x - 4$, use a graphing utility to obtain the graphs of $y = (f \circ g)(x)$ and of $y = (g \circ f)(x)$.

Solution

To find the graph of $y = (f \circ g)(x)$, we can make the following assignments:

$$Y_1 = x - 4$$
$$Y_2 = (Y_1)^3$$

[Note that we have substituted Y_1 for x in $f(x)$ and assigned this expression to Y_2, much the same way as we would do it algebraically.] The graph of $y = (f \circ g)(x)$ is shown in Figure 8.58.

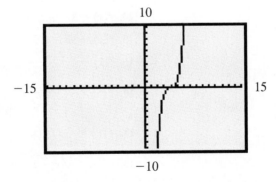

Figure 8.58

To find the graph of $y = (g \circ f)(x)$, we can make the following assignments.

$$Y_1 = x^3$$
$$Y_2 = Y_1 - 4$$

The graph of $y = (g \circ f)(x)$ is shown in Figure 8.59.

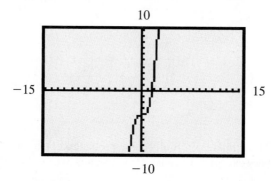

Figure 8.59

      Take another look at Figures 8.58 and 8.59. Note that in Figure 8.58, the graph of $y = (f \circ g)(x)$ is the basic cubic curve $f(x) = x^3$ translated four units to the right. Likewise, in Figure 8.59, the graph of $y = (g \circ f)(x)$ is the basic cubic curve translated four units downward. These are examples of a more general concept of using composite functions to represent various geometric transformations.

Problem Set 8.6

For Problems 1–8, find $f + g$, $f - g$, $f \cdot g$, and f/g. Also specify the domain for each.

1. $f(x) = 3x - 4$, $g(x) = 5x + 2$

2. $f(x) = -6x - 1$, $g(x) = -8x + 7$

3. $f(x) = x^2 - 6x + 4$, $g(x) = -x - 1$

4. $f(x) = 2x^2 - 3x + 5$, $g(x) = x^2 - 4$

5. $f(x) = x^2 - x - 1$, $g(x) = x^2 + 4x - 5$

6. $f(x) = x^2 - 2x - 24$, $g(x) = x^2 - x - 30$

7. $f(x) = \sqrt{x - 1}$, $g(x) = \sqrt{x}$

8. $f(x) = \sqrt{x + 2}$, $g(x) = \sqrt{3x - 1}$

For Problems 9–26, find $(f \circ g)(x)$ and $(g \circ f)(x)$. Also specify the domain for each.

9. $f(x) = 2x$, $g(x) = 3x - 1$

10. $f(x) = 4x + 1$, $g(x) = 3x$

11. $f(x) = 5x - 3$, $g(x) = 2x + 1$

12. $f(x) = 3 - 2x$, $g(x) = -4x$

13. $f(x) = 3x + 4$, $g(x) = x^2 + 1$

14. $f(x) = 3$, $g(x) = -3x^2 - 1$

15. $f(x) = 3x - 4$, $g(x) = x^2 + 3x - 4$

16. $f(x) = 2x^2 - x - 1$, $g(x) = x + 4$

17. $f(x) = \dfrac{1}{x}$, $g(x) = 2x + 7$

18. $f(x) = \dfrac{1}{x^2}$, $g(x) = x$

19. $f(x) = \sqrt{x - 2}$, $g(x) = 3x - 1$

20. $f(x) = \dfrac{1}{x}$, $g(x) = \dfrac{1}{x^2}$

21. $f(x) = \dfrac{1}{x - 1}$, $g(x) = \dfrac{2}{x}$

22. $f(x) = \dfrac{4}{x + 2}, \quad g(x) = \dfrac{3}{2x}$

23. $f(x) = 2x + 1, \quad g(x) = \sqrt{x - 1}$

24. $f(x) = \sqrt{x + 1}, \quad g(x) = 5x - 2$

25. $f(x) = \dfrac{1}{x - 1}, \quad g(x) = \dfrac{x + 1}{x}$

26. $f(x) = \dfrac{x - 1}{x + 2}, \quad g(x) = \dfrac{1}{x}$

For Problems 27–32, solve each problem.

27. If $f(x) = 3x - 2$ and $g(x) = x^2 + 1$, find $(f \circ g)(-1)$ and $(g \circ f)(3)$.

28. If $f(x) = x^2 - 2$ and $g(x) = x + 4$, find $(f \circ g)(2)$ and $(g \circ f)(-4)$.

29. If $f(x) = 2x - 3$ and $g(x) = x^2 - 3x - 4$, find $(f \circ g)(-2)$ and $(g \circ f)(1)$.

30. If $f(x) = 1/x$ and $g(x) = 2x + 1$, find $(f \circ g)(1)$ and $(g \circ f)(2)$.

31. If $f(x) = \sqrt{x}$ and $g(x) = 3x - 1$, find $(f \circ g)(4)$ and $(g \circ f)(4)$.

32. If $f(x) = x + 5$ and $g(x) = |x|$, find $(f \circ g)(-4)$ and $(g \circ f)(-4)$.

For Problems 33–38, show that $(f \circ g)(x) = x$ and that $(g \circ f)(x) = x$.

33. $f(x) = 2x, \quad g(x) = \dfrac{1}{2}x$

34. $f(x) = \dfrac{3}{4}x, \quad g(x) = \dfrac{4}{3}x$

35. $f(x) = x - 2, \quad g(x) = x + 2$

36. $f(x) = 2x + 1, \quad g(x) = \dfrac{x - 1}{2}$

37. $f(x) = 3x + 4 \quad g(x) = \dfrac{x - 4}{3}$

38. $f(x) = 4x - 3, \quad g(x) = \dfrac{x + 3}{4}$

■■■ **THOUGHTS INTO WORDS**

39. Discuss whether addition, subtraction, multiplication, and division of functions are commutative operations.

40. Explain why the composition of two functions is not a commutative operation.

41. Explain how to find the domain of

$$\left(\dfrac{f}{g}\right)(x) \text{ if } f(x) = \dfrac{x - 1}{x + 2} \text{ and } g(x) = \dfrac{x + 3}{x - 5}.$$

■■■ **FURTHER INVESTIGATIONS**

42. If $f(x) = 3x - 4$ and $g(x) = ax + b$, find conditions on a and b that will guarantee that $f \circ g = g \circ f$.

43. If $f(x) = x^2$ and $g(x) = \sqrt{x}$, with both having a domain of the set of nonnegative real numbers, then show that $(f \circ g)(x) = x$ and $(g \circ f)(x) = x$.

44. If $f(x) = 3x^2 - 2x - 1$ and $g(x) = x$, find $f \circ g$ and $g \circ f$. (Recall that we have previously named $g(x) = x$ the "identity function.")

GRAPHING CALCULATOR ACTIVITIES

45. For each of the following, predict the general shape and location of the graph, and then use your calculator to graph the function to check your prediction. (Your knowledge of the graphs of the basic functions that are being added or subtracted should be helpful when you are making your predictions.)

(a) $f(x) = x^4 + x^2$ **(b)** $f(x) = x^3 + x^2$
(c) $f(x) = x^4 - x^2$ **(d)** $f(x) = x^2 - x^4$
(e) $f(x) = x^2 - x^3$ **(f)** $f(x) = x^3 - x^2$
(g) $f(x) = |x| + \sqrt{x}$ **(h)** $f(x) = |x| - \sqrt{x}$

46. For each of the following, find the graph of $y = (f \circ g)(x)$ and of $y = (g \circ f)(x)$.

(a) $f(x) = x^2$ and $g(x) = x + 5$
(b) $f(x) = x^3$ and $g(x) = x + 3$
(c) $f(x) = x - 6$ and $g(x) = -x^3$
(d) $f(x) = x^2 - 4$ and $g(x) = \sqrt{x}$
(e) $f(x) = \sqrt{x}$ and $g(x) = x^2 + 4$
(f) $f(x) = \sqrt[3]{x}$ and $g(x) = x^3 - 5$

8.7   Direct and Inverse Variation

The amount of simple interest earned by a fixed amount of money invested at a certain rate *varies directly* as the time.

At a constant temperature, the volume of an enclosed gas *varies inversely* as the pressure.

Such statements illustrate two basic types of functional relationships, **direct variation** and **inverse variation**, that are widely used, especially in the physical sciences. These relationships can be expressed by equations that determine functions. The purpose of this section is to investigate these special functions.

■ Direct Variation

The statement "*y* varies directly as *x*" means

$$y = kx$$

where k is a nonzero constant called the **constant of variation**. The phrase "*y* is directly proportional to *x*" is also used to indicate direct variation; k is then referred to as the **constant of proportionality**.

Remark: Note that the equation $y = kx$ defines a function and can be written $f(x) = kx$. However, in this section, it is more convenient not to use function notation but instead to use variables that are meaningful in terms of the physical entities involved in the particular problem.

Statements that indicate direct variation may also involve powers of a variable. For example, "y varies directly as the square of x" can be written $y = kx^2$. In general, y varies directly as the nth power of x $(n > 0)$ means

$$y = kx^n$$

There are three basic types of problems in which we deal with direct variation:

1. Translating an English statement into an equation expressing the direct variation;

2. Finding the constant of variation from given values of the variables; and

3. Finding additional values of the variables once the constant of variation has been determined.

Let's consider an example of each type of problem.

E X A M P L E   1

Translate the statement "The tension on a spring varies directly as the distance it is stretched" into an equation, using k as the constant of variation.

Solution

Let t represent the tension and d the distance; the equation is

$$t = kd$$
∎

E X A M P L E   2

If A varies directly as the square of e, and if $A = 96$ when $e = 4$, find the constant of variation.

Solution

Because A varies directly as the square of e, we have

$$A = ke^2$$

Substitute 96 for A and 4 for e to obtain

$$96 = k(4)^2$$

$$96 = 16k$$

$$6 = k$$

The constant of variation is 6.
∎

EXAMPLE 3   If y is directly proportional to x, and if $y = 6$ when $x = 8$, find the value of y when $x = 24$.

Solution

The statement "y is directly proportional to x" translates into

$$y = kx$$

Let $y = 6$ and $x = 8$; the constant of variation becomes

$$6 = k(8)$$

$$\frac{6}{8} = k$$

$$\frac{3}{4} = k$$

Thus the specific equation is

$$y = \frac{3}{4}x$$

Now let $x = 24$ to obtain

$$y = \frac{3}{4}(24) = 18$$

∎

■ Inverse Variation

The second basic type of variation is *inverse variation*. The statement "y varies inversely as x" means

$$y = \frac{k}{x}$$

where k is a nonzero constant, which is again referred to as the constant of variation. The phrase "y is inversely proportional to x" is also used to express inverse variation. As with direct variation, statements indicating inverse variation may involve powers of x. For example, "y varies inversely as the square of x" can be written $y = k/x^2$. In general, y varies inversely as the nth power of x ($n > 0$) means

$$y = \frac{k}{x^n}$$

The following examples illustrate the three basic kinds of problems that involve inverse variation.

EXAMPLE 4

Translate the statement "The length of a rectangle of fixed area varies inversely as the width" into an equation, using k as the constant of variation.

Solution

Let l represent the length and w the width; the equation is

$$l = \frac{k}{w}$$

■

EXAMPLE 5

If y is inversely proportional to x, and if $y = 14$ when $x = 4$, find the constant of variation.

Solution

Because y is inversely proportional to x, we have

$$y = \frac{k}{x}$$

Substitute 4 for x and 14 for y to obtain

$$14 = \frac{k}{4}$$

Solving this equation yields

$$k = 56$$

The constant of variation is 56.

■

EXAMPLE 6

The time required for a car to travel a certain distance varies inversely as the rate at which it travels. If it takes 4 hours at 50 miles per hour to travel the distance, how long will it take at 40 miles per hour?

Solution

Let t represent time and r rate. The phrase "time required . . . varies inversely as the rate" translates into

$$t = \frac{k}{r}$$

Substitute 4 for t and 50 for r to find the constant of variation.

$$4 = \frac{k}{50}$$

$$k = 200$$

Thus the specific equation is

$$t = \frac{200}{r}$$

Now substitute 40 for r to produce

$$t = \frac{200}{40}$$

$$= 5$$

It will take 5 hours at 40 miles per hour. ■

The terms *direct* and *inverse*, as applied to variation, refer to the relative behavior of the variables involved in the equation. That is, in *direct variation* ($y = kx$), an assignment of **increasing absolute values for x** produces **increasing absolute values for y**. However, in *inverse variation* ($y = k/x$), an assignment of **increasing absolute values for x** produces **decreasing absolute values for y**.

■ Joint Variation

Variation may involve more than two variables. The following table illustrates some different types of variation statements and their equivalent algebraic equations that use k as the constant of variation. Statements 1, 2, and 3 illustrate the concept of **joint variation**. Statements 4 and 5 show that both direct and inverse variation may occur in the same problem. Statement 6 combines joint variation with inverse variation.

Variation Statement	Algebraic Equation
1. y varies jointly as x and z.	$y = kxz$
2. y varies jointly as x, z, and w.	$y = kxzw$
3. V varies jointly as h and the square of r.	$V = khr^2$
4. h varies directly as V and inversely as w.	$h = \dfrac{kV}{w}$
5. y is directly proportional to x and inversely proportional to the square of z.	$y = \dfrac{kx}{z^2}$
6. y varies jointly as w and z and inversely as x.	$y = \dfrac{kwz}{x}$

The final two examples of this section illustrate different kinds of problems involving some of these variation situations.

E X A M P L E   7

The volume of a pyramid varies jointly as its altitude and the area of its base. If a pyramid with an altitude of 9 feet and a base with an area of 17 square feet has a volume of 51 cubic feet, find the volume of a pyramid with an altitude of 14 feet and a base with an area of 45 square feet.

Solution

Let's use the following variables:

$$V = \text{volume} \qquad h = \text{altitude}$$

$$B = \text{area of base} \qquad k = \text{constant of variation}$$

The fact that the volume varies jointly as the altitude and the area of the base can be represented by the equation

$$V = kBh$$

Substitute 51 for V, 17 for B, and 9 for h to obtain

$$51 = k(17)(9)$$

$$51 = 153k$$

$$\frac{51}{153} = k$$

$$\frac{1}{3} = k$$

Therefore the specific equation is $V = \dfrac{1}{3}Bh$. Now substitute 45 for B and 14 for h to obtain

$$V = \frac{1}{3}(45)(14) = (15)(14) = 210$$

The volume is 210 cubic feet.   ■

E X A M P L E   8

Suppose that y varies jointly as x and z and inversely as w. If $y = 154$ when $x = 6$, $z = 11$, and $w = 3$, find y when $x = 8$, $z = 9$, and $w = 6$.

Solution

The statement "y varies jointly as x and z and inversely as w" translates into the equation

$$y = \frac{kxz}{w}$$

Substitute 154 for y, 6 for x, 11 for z, and 3 for w to produce

$$154 = \frac{(k)(6)(11)}{3}$$

$$154 = 22k$$

$$7 = k$$

Thus the specific equation is

$$y = \frac{7xz}{w}$$

Now substitute 8 for x, 9 for z, and 6 for w to obtain

$$y = \frac{7(8)(9)}{6} = 84$$

■

Problem Set 8.7

For Problems 1–8, translate each statement of variation into an equation; use k as the constant of variation.

1. y varies directly as the cube of x.

2. a varies inversely as the square of b.

3. A varies jointly as l and w.

4. s varies jointly as g and the square of t.

5. At a constant temperature, the volume (V) of a gas varies inversely as the pressure (P).

6. y varies directly as the square of x and inversely as the cube of w.

7. The volume (V) of a cone varies jointly as its height (h) and the square of a radius (r).

8. l is directly proportional to r and t.

For Problems 9–18, find the constant of variation for each stated condition.

9. y varies directly as x, and $y = 72$ when $x = 3$.

10. y varies inversely as the square of x, and $y = 4$ when $x = 2$.

11. A varies directly as the square of r, and $A = 154$ when $r = 7$.

12. V varies jointly as B and h, and $V = 104$ when $B = 24$ and $h = 13$.

13. A varies jointly as b and h, and $A = 81$ when $b = 9$ and $h = 18$.

14. s varies jointly as g and the square of t, and $s = -108$ when $g = 24$ and $t = 3$.

15. y varies jointly as x and z and inversely as w, and $y = 154$ when $x = 6$, $z = 11$, and $w = 3$.

16. V varies jointly as h and the square of r, and $V = 1100$ when $h = 14$ and $r = 5$.

17. y is directly proportional to the square of x and inversely proportional to the cube of w, and $y = 18$ when $x = 9$ and $w = 3$.

18. y is directly proportional to x and inversely proportional to the square root of w, and $y = \frac{1}{5}$ when $x = 9$ and $w = 10$.

For Problems 19–32, solve each problem.

19. If y is directly proportional to x, and $y = 5$ when $x = -15$, find the value of y when $x = -24$.

20. If y is inversely proportional to the square of x, and $y = \frac{1}{8}$ when $x = 4$, find y when $x = 8$.

21. If V varies jointly as B and h, and $V = 96$ when $B = 36$ and $h = 8$, find V when $B = 48$ and $h = 6$.

22. If A varies directly as the square of e, and $A = 150$ when $e = 5$, find A when $e = 10$.

23. The time required for a car to travel a certain distance varies inversely as the rate at which it travels. If it takes 3 hours to travel the distance at 50 miles per hour, how long will it take at 30 miles per hour?

24. The distance that a freely falling body falls varies directly as the square of the time it falls. If a body falls 144 feet in 3 seconds, how far will it fall in 5 seconds?

25. The period (the time required for one complete oscillation) of a simple pendulum varies directly as the square root of its length. If a pendulum 12 feet long has a period of 4 seconds, find the period of a pendulum of length 3 feet.

26. Suppose the number of days it takes to complete a construction job varies inversely as the number of people assigned to the job. If it takes 7 people 8 days to do the job, how long will it take 10 people to complete the job?

27. The number of days needed to assemble some machines varies directly as the number of machines and inversely as the number of people working. If it takes 4 people 32 days to assemble 16 machines, how many days will it take 8 people to assemble 24 machines?

28. The volume of a gas at a constant temperature varies inversely as the pressure. What is the volume of a gas under a pressure of 25 pounds if the gas occupies 15 cubic centimeters under a pressure of 20 pounds?

29. The volume (V) of a gas varies directly as the temperature (T) and inversely as the pressure (P). If $V = 48$ when $T = 320$ and $P = 20$, find V when $T = 280$ and $P = 30$.

30. The volume of a cylinder varies jointly as its altitude and the square of the radius of its base. If the volume of a cylinder is 1386 cubic centimeters when the radius of the base is 7 centimeters, and its altitude is 9 centimeters, find the volume of a cylinder that has a base of radius 14 centimeters if the altitude of the cylinder is 5 centimeters.

31. The cost of labor varies jointly as the number of workers and the number of days that they work. If it costs \$900 to have 15 people work for 5 days, how much will it cost to have 20 people work for 10 days?

32. The cost of publishing pamphlets varies directly as the number of pamphlets produced. If it costs \$96 to publish 600 pamphlets, how much does it cost to publish 800 pamphlets?

■ ■ ■ **THOUGHTS INTO WORDS**

33. How would you explain the difference between direct variation and inverse variation?

34. Suppose that y varies directly as the square of x. Does doubling the value of x also double the value of y? Explain your answer.

35. Suppose that y varies inversely as x. Does doubling the value of x also double the value of y? Explain your answer.

■ ■ ■ **FURTHER INVESTIGATIONS**

[C] In the previous problems, we chose numbers to make computations reasonable without the use of a calculator. However, variation-type problems often involve messy computations, and the calculator becomes a very useful tool. Use your calculator to help solve the following problems.

36. The simple interest earned by a certain amount of money varies jointly as the rate of interest and the time (in years) that the money is invested.
   (a) If some money invested at 11% for 2 years earns $385, how much would the same amount earn at 12% for 1 year?
   (b) If some money invested at 12% for 3 years earns $819, how much would the same amount earn at 14% for 2 years?
   (c) If some money invested at 14% for 4 years earns $1960, how much would the same amount earn at 15% for 2 years?

37. The period (the time required for one complete oscillation) of a simple pendulum varies directly as the square root of its length. If a pendulum 9 inches long has a period of 2.4 seconds, find the period of a pendulum of length 12 inches. Express the answer to the nearest tenth of a second.

38. The volume of a cylinder varies jointly as its altitude and the square of the radius of its base. If the volume of a cylinder is 549.5 cubic meters when the radius of the base is 5 meters and its altitude is 7 meters, find the volume of a cylinder that has a base of radius 9 meters and an altitude of 14 meters.

39. If y is directly proportional to x and inversely proportional to the square of z, and if $y = 0.336$ when $x = 6$ and $z = 5$, find the constant of variation.

40. If y is inversely proportional to the square root of x, and $y = 0.08$ when $x = 225$, find y when $x = 625$.

(8.1) A **function** f is a correspondence between two sets X and Y that assigns to each element x of set X one and only one element y of set Y. We call the element y being assigned the image of x. We call the set X the **domain** of the function, and we call the set of all images the **range** of the function.

A function can also be thought of as a set of ordered pairs, no two of which have the same first element.

Vertical-Line Test If each vertical line intersects a graph in no more than one point, then the graph represents a function.

Single letters such as f, g, and h are commonly used as symbols to name functions. The symbol $f(x)$ represents the element in the range associated with x from the domain. Thus if $f(x) = 3x + 7$, then $f(1) = 3(1) + 7 = 10$.

(8.2) Any function that can be written in the form

$$f(x) = ax + b$$

where a and b are real numbers, is a **linear function**. The graph of a linear function is a straight line.

The linear function $f(x) = x$ is called the **identity function**.

Any linear function of the form $f(x) = ax + b$, where $a = 0$, is called a **constant function**.

Linear functions provide a natural connection between mathematics and the real world.

(8.3) and **(8.4)** Any function that can be written in the form

$$f(x) = ax^2 + bx + c$$

where a, b, and c are real numbers and $a \neq 0$, is a **quadratic function**. The graph of any quadratic function is a **parabola**, which can be drawn using either one of the following methods.

1. Express the function in the form $f(x) = a(x - h)^2 + k$, and use the values of a, h, and k to determine the parabola.

2. Express the function in the form $f(x) = ax^2 + bx + c$, and use the fact that the vertex is at

$$\left(-\frac{b}{2a}, f\left(-\frac{b}{2a} \right) \right)$$

and the axis of symmetry is

$$x = -\frac{b}{2a}$$

Quadratic functions produce parabolas that have either a **minimum** or a **maximum** value. Therefore a real-world minimum- or maximum-value problem that can be described by a quadratic function can be solved using the techniques of this chapter.

(8.5) Another important skill in graphing is to be able to recognize equations of the transformations of basic curves. We worked with the following transformations in this chapter:

Vertical Translation The graph of $y = f(x) + k$ is the graph of $y = f(x)$ shifted k units upward if $k > 0$ or shifted $|k|$ units downward if $k < 0$.

Horizontal Translation The graph of $y = f(x - h)$ is the graph of $y = f(x)$ shifted h units to the right if $h > 0$ or shifted $|h|$ units to the left if $h < 0$.

x Axis Reflection The graph of $y = -f(x)$ is the graph of $y = f(x)$ reflected through the x axis.

y Axis Reflection The graph of $y = f(-x)$ is the graph of $y = f(x)$ reflected through the y axis.

Vertical Stretching and Shrinking The graph of $y = cf(x)$ is obtained from the graph of $y = f(x)$ by multiplying the y coordinates of $y = f(x)$ by c. If $|c| > 1$, the graph is said to be **stretched** by a factor of $|c|$, and if $0 < |c| < 1$, the graph is said to be **shrunk** by a factor of $|c|$.

The following suggestions are helpful for graphing functions that are unfamiliar.

1. Determine the domain of the function.

2. Find the intercepts.

3. Determine what type of symmetry the equation exhibits.

4. Set up a table of values that satisfy the equation. The type of symmetry and the domain will affect your choice of values for x in the table.

5. Plot the points associated with the ordered pairs and connect them with a smooth curve. Then, if appropriate, reflect this part of the curve according to the symmetry the graph exhibits.

(8.6) Functions can be added, subtracted, multiplied, and divided according to the following definition: If f and g are functions, and x is in the domain of both functions, then

1. $(f + g)(x) = f(x) + g(x)$

2. $(f - g)(x) = f(x) - g(x)$

3. $(f \cdot g)(x) = f(x) \cdot g(x)$

4. $\left(\dfrac{f}{g}\right)(x) = \dfrac{f(x)}{g(x)}, \quad g(x) \neq 0$

To find the domain of a sum, difference, product, or quotient of two functions, we can proceed as follows.

1. Find the domain of each function individually.

2. Find the set of values common to each domain. This set of values is the domain of the sum, difference, and product of the functions. The domain of the quotient is this set of values common to both domains, except for any values that would lead to division by zero.

The **composition** of functions f and g is defined by

$$(f \circ g)(x) = f(g(x))$$

for all x in the domain of g such that $g(x)$ is in the domain of f.

Remember that the composition of functions is *not* a commutative operation.

(8.7) Relationships that involve **direct** and **inverse variation** can be expressed by equations that determine functions. The statement "y varies directly as x" means

$$y = kx$$

where k is the **constant of variation**. The statement "y varies directly as the nth power of x" $(n > 0)$ means

$$y = kx^n$$

The statement "y varies inversely as x" means $y = \dfrac{k}{x}$.

The statement "y varies inversely as the nth power of x" $(n > 0)$ means $y = \dfrac{k}{x^n}$.

The statement "y varies jointly as x and w" means $y = kxw$.

Chapter 8   Review Problem Set

1. If $f(x) = 3x^2 - 2x - 1$, find $f(2)$, $f(-1)$, and $f(-3)$.

2. For each of the following functions, find
$$\dfrac{f(a + h) - f(a)}{h}.$$
(a) $f(x) = -5x + 4$ **(b)** $f(x) = 2x^2 - x + 4$
(c) $f(x) = -3x^2 + 2x - 5$

3. Determine the domain and range of the function $f(x) = x^2 + 5$.

4. Determine the domain of the function
$$f(x) = \dfrac{2}{2x^2 + 7x - 4}.$$

5. Express the domain of $f(x) = \sqrt{x^2 - 7x + 10}$ using interval notation.

For Problems 6–23, graph each function.

6. $f(x) = -2x + 2$ **7.** $f(x) = 2x^2 - 1$

8. $f(x) = -\sqrt{x - 2} + 1$ **9.** $f(x) = x^2 - 8x + 17$

10. $f(x) = -x^3 + 2$ **11.** $f(x) = 2|x - 1| + 3$

12. $f(x) = -2x^2 - 12x - 19$ **13.** $f(x) = -\dfrac{1}{3}x + 1$

14. $f(x) = -\dfrac{2}{x^2}$ **15.** $f(x) = 2|x| - x$

16. $f(x) = (x - 2)^2$

17. $f(x) = \sqrt{-x + 4}$

18. $f(x) = -(x + 1)^2 - 3$

19. $f(x) = \sqrt{x + 3} - 2$

20. $f(x) = -|x| + 4$

21. $f(x) = (x - 2)^3$

22. $f(x) = \begin{cases} x^2 - 1 & \text{for } x < 0 \\ 3x - 1 & \text{for } x \geq 0 \end{cases}$

23. $f(x) = \begin{cases} 3 & \text{for } x \leq -3 \\ |x| & \text{for } -3 < x < 3 \\ 2x - 3 & \text{for } x \geq 3 \end{cases}$

24. If $f(x) = 2x + 3$ and $g(x) = x^2 - 4x - 3$, find $f + g$, $f - g$, $f \cdot g$, and f/g.

For Problems 25–30, find $(f \circ g)(x)$ and $(g \circ f)(x)$. Also specify the domain for each.

25. $f(x) = 3x - 9$ and $g(x) = -2x + 7$

26. $f(x) = x^2 - 5$ and $g(x) = 5x - 4$

27. $f(x) = \sqrt{x - 5}$ and $g(x) = x + 2$

28. $f(x) = \dfrac{1}{x}$ and $g(x) = x^2 - x - 6$

29. $f(x) = x^2$ and $g(x) = \sqrt{x - 1}$

30. $f(x) = \dfrac{1}{x - 3}$ and $g(x) = \dfrac{1}{x + 2}$

31. If $f(x) = \begin{cases} x^2 - 2 & \text{for } x \geq 0 \\ -3x + 4 & \text{for } x < 0 \end{cases}$

find $f(5)$, $f(0)$, and $f(-3)$.

32. If $f(x) = -x^2 - x + 4$ and $g(x) = \sqrt{x - 2}$, find $f(g(6))$ and $g(f(-2))$.

33. If $f(x) = |x|$ and $g(x) = x^2 - x - 1$, find $(f \circ g)(1)$ and $(g \circ f)(-3)$.

34. Determine the linear function whose graph is a line that is parallel to the line determined by $g(x) = \dfrac{2}{3}x + 4$ and contains the point $(5, -2)$.

35. Determine the linear function whose graph is a line that is perpendicular to the line determined by $g(x) = -\dfrac{1}{2}x - 6$ and contains the point $(-6, 3)$.

36. The cost for burning a 100-watt light bulb is given by the function $c(h) = 0.006h$, where h represents the number of hours that the bulb burns. How much, to the nearest cent, does it cost to burn a 100-watt bulb for 4 hours per night for a 30-day month?

37. "All Items 30% Off Marked Price" is a sign in a local department store. Form a function and then use it to determine how much one has to pay for each of the following marked items: a $65 pair of shoes, a $48 pair of slacks, a $15.50 belt.

For Problems 38–40, find the x intercepts and the vertex for each parabola.

38. $f(x) = 3x^2 + 6x - 24$

39. $f(x) = x^2 - 6x - 5$

40. $f(x) = 2x^2 - 28x + 101$

41. Find two numbers whose sum is 10, such that the sum of the square of one number plus four times the other number is a minimum.

42. A group of students is arranging a chartered flight to Europe. The charge per person is $496 if 100 students go on the flight. If more than 100 students go, the charge per student is reduced by an amount equal to $4 times the number of students above 100. How many students should the airline try to get in order to maximize its revenue?

43. If y varies directly as x and inversely as w, and if $y = 27$ when $x = 18$ and $w = 6$, find the constant of variation.

44. If y varies jointly as x and the square root of w, and if $y = 140$ when $x = 5$ and $w = 16$, find y when $x = 9$ and $w = 49$.

45. The weight of a body above the surface of the earth varies inversely as the square of its distance from the center of the earth. Assuming the radius of the earth to be 4000 miles, determine how much a man would weigh 1000 miles above the earth's surface if he weighs 200 pounds on the surface.

46. The number of hours needed to assemble some furniture varies directly as the number of pieces of furniture and inversely as the number of people working. If it takes 3 people 10 hours to assemble 20 pieces of furniture, how many hours will it take 4 people to assemble 40 pieces of furniture?

1. If $f(x) = -\dfrac{1}{2}x + \dfrac{1}{3}$, find $f(-3)$.

2. If $f(x) = -x^2 - 6x + 3$, find $f(-2)$.

3. If $f(x) = 3x^2 + 2x - 5$, find $\dfrac{f(a + h) - f(a)}{h}$.

4. Determine the domain of the function $f(x) = \dfrac{-3}{2x^2 + 7x - 4}$.

5. Determine the domain of the function $f(x) = \sqrt{5 - 3x}$.

6. If $f(x) = 3x - 1$ and $g(x) = 2x^2 - x - 5$, find $f + g$, $f - g$, and $f \cdot g$.

7. If $f(x) = -3x + 4$ and $g(x) = 7x + 2$, find $(f \circ g)(x)$.

8. If $f(x) = 2x + 5$ and $g(x) = 2x^2 - x + 3$, find $(g \circ f)(x)$.

9. If $f(x) = \dfrac{3}{x - 2}$ and $g(x) = \dfrac{2}{x}$, find $(f \circ g)(x)$.

10. If $f(x) = x^2 - 2x - 3$ and $g(x) = |x - 3|$, find $f(g(-2))$ and $g(f(1))$.

11. Determine the linear function whose graph is a line that has a slope of $-\dfrac{5}{6}$ and contains the point $(4, -8)$.

12. If $f(x) = \dfrac{3}{x}$ and $g(x) = \dfrac{2}{x - 1}$, determine the domain of $\left(\dfrac{f}{g}\right)(x)$.

13. If $f(x) = 2x^2 - x + 1$ and $g(x) = x^2 + 3$, find $(f + g)(-2)$, $(f - g)(4)$, and $(g - f)(-1)$.

14. If $f(x) = x^2 + 5x - 6$ and $g(x) = x - 1$, find $(f \cdot g)(x)$ and $\left(\dfrac{f}{g}\right)(x)$.

15. Find two numbers whose sum is 60, such that the sum of the square of one number plus 12 times the other number is a minimum.

16. If y varies jointly as x and z, and if $y = 18$ when $x = 8$ and $z = 9$, find y when $x = 5$ and $z = 12$.

17. If y varies inversely as x, and if $y = \dfrac{1}{2}$ when $x = -8$, find the constant of variation.

18. The simple interest earned by a certain amount of money varies jointly as the rate of interest and the time (in years) that the money is invested. If \$140 is earned for the money invested at 7% for 5 years, how much is earned if the same amount is invested at 8% for 3 years?

19. A retailer has a number of items that he wants to sell at a profit of 35% of the cost. What linear function can be used to determine selling prices of the items? What price should he charge for a tie that cost him \$13?

20. Find the x intercepts and the vertex of the parabola $f(x) = 4x^2 - 16x - 48$.

For Problems 21–25, graph each function.

21. $f(x) = (x - 2)^3 - 3$

22. $f(x) = -2x^2 - 12x - 14$

23. $f(x) = 3|x - 2| - 1$

24. $f(x) = \sqrt{-x + 2}$

25. $f(x) = -x - 1$

Polynomial and Rational Functions

© Michele Westmoreland/CORBIS

The graphs of polynomial functions are smooth curves that can be used to describe the path of objects such as a roller coaster.

Earlier in this text we solved linear and quadratic equations and graphed linear and quadratic functions. In this chapter we will expand our equation-solving processes and graphing techniques to include more general polynomial equations and functions. Then our knowledge of polynomial functions will allow us to work with rational functions. The function concept will again serve as a unifying thread throughout the chapter. To facilitate our study in this chapter, we will first review the concept of dividing polynomials, and we will introduce theorems about division.

9.1 Synthetic Division

In Section 4.5 we discussed the process of dividing polynomials and the simplified process of synthetic division when the divisor is of the form $x - c$. Because polynomial division is central to the study of polynomial functions, we want to review the division process and state the algorithms and theorems for the division of polynomials.

Earlier we discussed the process of dividing polynomials by using the following format:

$$
\begin{array}{r}
x^2 - 2x + 4 \\
3x + 1\overline{)3x^3 - 5x^2 + 10x + 1} \\
\underline{3x^3 + x^2} \\
-6x^2 + 10x + 1 \\
\underline{-6x^2 - 2x} \\
12x + 1 \\
\underline{12x + 4} \\
-3
\end{array}
$$

We also suggested writing the final result as

$$
\frac{3x^3 - 5x^2 + 10x + 1}{3x + 1} = x^2 - 2x + 4 + \frac{-3}{3x + 1}
$$

Multiplying both sides of this equation by $3x + 1$ produces

$$
3x^3 - 5x^2 + 10x + 1 = (3x + 1)(x^2 - 2x + 4) + (-3)
$$

which is of the familiar form

Dividend = (Divisor)(Quotient) + Remainder

This result is commonly called the **division algorithm for polynomials**, and it can be stated in general terms as follows:

Division Algorithm for Polynomials

If $f(x)$ and $d(x)$ are polynomials and $d(x) \neq 0$, then there exist unique polynomials $q(x)$ and $r(x)$ such that

$$
f(x) = d(x)q(x) + r(x)
$$

Dividend   Divisor   Quotient   Remainder

where $r(x) = 0$ or the degree of $r(x)$ is less than the degree of $d(x)$.

If the divisor is of the form $x - c$, where c is a constant, then the typical long-division algorithm can be conveniently simplified into a process called **synthetic division**. First, let's consider an example using the usual algorithm. Then, in a step-by-step fashion, we will list some shortcuts to use that will lead us into the synthetic-division procedure. Consider the division problem $(3x^4 + x^3 - 15x^2 + 6x - 8) \div (x - 2)$:

$$
\begin{array}{r}
3x^3 + 7x^2 - x + 4 \\
x - 2 \overline{)3x^4 + x^3 - 15x^2 + 6x - 8} \\
\underline{3x^4 - 6x^3} \\
7x^3 - 15x^2 \\
\underline{7x^3 - 14x^2} \\
-x^2 + 6x \\
\underline{-x^2 + 2x} \\
4x - 8 \\
\underline{4x - 8}
\end{array}
$$

Note that because the dividend $(3x^4 + x^3 - 15x^2 + 6x - 8)$ is written in descending powers of x, the quotient $(3x^3 + 7x^2 - x + 4)$ is also in descending powers of x. In other words, the numerical coefficients are the key, so let's rewrite this problem in terms of its coefficients.

$$
\begin{array}{r}
3 \quad 7 \quad -1 \quad 4 \\
1 - 2 \overline{)3 \quad 1 \quad -15 \quad 6 \quad -8} \\
③ -6 \\
7 \quad ㊟{-15} \\
⑦ -14 \\
-1 \quad ⑥ \\
㊟{-1} \quad 2 \\
4 \quad ㊟{-8} \\
④ \quad -8
\end{array}
$$

Now observe that the numbers circled are simply repetitions of the numbers directly above them in the format. Thus the circled numbers could be omitted and the format would be as follows. (Disregard the arrows for the moment.)

$$
\begin{array}{r}
3 \quad 7 \quad -1 \quad 4 \\
1 - 2 \overline{)3 \quad 1 \quad -15 \quad 6 \quad -8} \\
\underline{-6} \\
7 \\
-14 \\
\underline{-1} \\
2 \\
\underline{4} \\
-8
\end{array}
$$

Next, move some numbers up as indicated by the arrows, and omit writing 1 as the coefficient of x in the divisor to yield the following more compact form:

$$
\begin{array}{cccc}
3 & 7 & -1 & 4 \\
\end{array}
\tag{1}
$$

$$
-2\overline{)\begin{array}{ccccc} 3 & 1 & -15 & 6 & -8 \end{array}}
\tag{2}
$$

$$
\begin{array}{cccc}
-6 & -14 & 2 & -8 \\
\end{array}
\tag{3}
$$

$$
\begin{array}{ccc}
7 & -1 & 4 \\
\end{array}
\tag{4}
$$

Note that line (4) reveals all of the coefficients of the quotient [line (1)] except for the first coefficient, 3. Thus we can omit line (1), begin line (4) with the first coefficient, and then use the following form:

$$
-2\overline{)\begin{array}{ccccc} 3 & 1 & -15 & 6 & -8 \end{array}}
\tag{5}
$$

$$
\begin{array}{ccccc}
& -6 & -14 & 2 & -8 \\
\end{array}
\tag{6}
$$

$$
\begin{array}{ccccc}
3 & 7 & -1 & 4 & 0 \\
\end{array}
\tag{7}
$$

Line (7) contains the coefficients of the quotient; the 0 indicates the remainder. Finally, changing the constant in the divisor to 2 (instead of -2), which will change the signs of the numbers in line (6), allows us to add the corresponding entries in lines (5) and (6) rather than subtract them. Thus the final synthetic-division form for this problem is

$$
2\overline{)\begin{array}{ccccc} 3 & 1 & -15 & 6 & -8 \end{array}}
$$

$$
\begin{array}{ccccc}
& 6 & 14 & -2 & 8 \\
\hline
3 & 7 & -1 & 4 & 0
\end{array}
$$

Now we will consider another problem and follow a step-by-step procedure for setting up and carrying out the synthetic division. Suppose that we want to do the following division problem.

$$
x + 4\overline{)2x^3 + 5x^2 - 13x - 2}
$$

1. Write the coefficients of the dividend as follows.

$$
\overline{)\begin{array}{cccc} 2 & 5 & -13 & -2 \end{array}}
$$

2. In the divisor, use -4 instead of 4 so that later we can add rather than subtract.

$$
-4\overline{)\begin{array}{cccc} 2 & 5 & -13 & -2 \end{array}}
$$

3. Bring down the first coefficient of the dividend.

$$
-4\overline{)\begin{array}{cccc} 2 & 5 & -13 & -2 \end{array}}
$$

$$
\begin{array}{c}
\hline
2
\end{array}
$$

4. Multiply that first coefficient by the divisor, which yields $2(-4) = -8$. This result is added to the second coefficient of the dividend.

$$
-4\overline{)\begin{array}{cccc} 2 & 5 & -13 & -2 \end{array}}
$$

$$
\begin{array}{cc}
& -8 \\
\hline
2 & -3
\end{array}
$$

5. Multiply $(-3)(-4)$, which yields 12; this result is added to the third coefficient of the dividend.

$$
\begin{array}{r}
-4\,\overline{)\,2 \quad\ \ 5 \quad -13 \quad -2} \\
-8 \quad\ \ 12 \\
\hline
2 \quad -3 \quad -\ 1
\end{array}
$$

6. Multiply $(-1)(-4)$, which yields 4; this result is added to the last term of the dividend.

$$
\begin{array}{r}
-4\,\overline{)\,2 \quad\ \ 5 \quad -13 \quad -2} \\
-8 \quad\ \ 12 \quad\ \ 4 \\
\hline
2 \quad -3 \quad -\ 1 \quad\ \ 2
\end{array}
$$

The last row indicates a quotient of $2x^2 - 3x - 1$ and a remainder of 2.

Let's consider three more examples, showing only the final compact form for synthetic division.

E X A M P L E   1

Find the quotient and remainder for $(2x^3 - 5x^2 + 6x + 4) \div (x - 2)$.

Solution

$$
\begin{array}{r}
2\,\overline{)\,2 \quad -5 \quad\ \ 6 \quad\ \ 4} \\
4 \quad -2 \quad\ \ 8 \\
\hline
2 \quad -1 \quad\ \ 4 \quad 12
\end{array}
$$

Therefore the quotient is $2x^2 - x + 4$, and the remainder is 12. ■

E X A M P L E   2

Find the quotient and remainder for $(4x^4 - 2x^3 + 6x - 1) \div (x - 1)$.

Solution

$$
\begin{array}{r}
1\,\overline{)\,4 \quad -2 \quad 0 \quad\ \ 6 \quad -1} \\
4 \quad\ \ 2 \quad 2 \quad\ \ 8 \\
\hline
4 \quad\ \ 2 \quad 2 \quad\ \ 8 \quad\ \ 7
\end{array}
$$

Note that a 0 has been inserted as the coefficient of the missing x^2 term.

Thus the quotient is $4x^3 + 2x^2 + 2x + 8$, and the remainder is 7. ■

E X A M P L E   3

Find the quotient and remainder for $(x^3 + 8x^2 + 13x - 6) \div (x + 3)$.

Solution

$$
\begin{array}{r}
-3\,\overline{)\,1 \quad\ \ 8 \quad\ \ 13 \quad -6} \\
-3 \quad -15 \quad\ \ 6 \\
\hline
1 \quad\ \ 5 \quad -\ 2 \quad\ \ 0
\end{array}
$$

Thus the quotient is $x^2 + 5x - 2$, and the remainder is 0. ■

In Example 3, because the remainder is 0, we can say that $x + 3$ is a factor of $x^3 + 8x^2 + 13x - 6$. We will use this idea a bit later when we solve polynomial equations.

Problem Set 9.1

Use synthetic division to determine the quotient and remainder for each problem.

1. $(4x^2 - 5x - 6) \div (x - 2)$

2. $(5x^2 - 9x + 4) \div (x - 1)$

3. $(2x^2 - x - 21) \div (x + 3)$

4. $(3x^2 + 8x + 4) \div (x + 2)$

5. $(3x^2 - 16x + 17) \div (x - 4)$

6. $(6x^2 - 29x - 8) \div (x - 5)$

7. $(4x^2 + 19x - 32) \div (x + 6)$

8. $(7x^2 + 26x - 2) \div (x + 4)$

9. $(x^3 + 2x^2 - 7x + 4) \div (x - 1)$

10. $(2x^3 - 7x^2 + 2x + 3) \div (x - 3)$

11. $(3x^3 + 8x^2 - 8) \div (x + 2)$

12. $(4x^3 + 17x^2 + 75) \div (x + 5)$

13. $(5x^3 - 9x^2 - 3x - 2) \div (x - 2)$

14. $(x^3 - 6x^2 + 5x + 14) \div (x - 4)$

15. $(x^3 + 6x^2 - 8x + 1) \div (x + 7)$

16. $(2x^3 + 11x^2 - 5x + 1) \div (x + 6)$

17. $(-x^3 + 7x^2 - 14x + 6) \div (x - 3)$

18. $(-2x^3 - 3x^2 + 4x + 5) \div (x + 1)$

19. $(-3x^3 + x^2 + 2x + 2) \div (x + 1)$

20. $(-x^3 + 4x^2 + 31x + 2) \div (x - 8)$

21. $(3x^3 - 2x - 5) \div (x - 2)$

22. $(2x^3 - x - 4) \div (x + 3)$

23. $(2x^4 + x^3 + 3x^2 + 2x - 2) \div (x + 1)$

24. $(x^4 - 3x^3 - 6x^2 + 11x - 12) \div (x - 4)$

25. $(x^4 + 4x^3 - 7x - 1) \div (x - 3)$

26. $(3x^4 - x^3 + 2x^2 - 7x - 1) \div (x + 1)$

27. $(x^4 + 5x^3 - x^2 + 25) \div (x + 5)$

28. $(2x^4 + 3x^2 + 3) \div (x + 2)$

29. $(x^4 - 16) \div (x - 2)$

30. $(x^4 - 16) \div (x + 2)$

31. $(x^5 - 1) \div (x + 1)$

32. $(x^5 - 1) \div (x - 1)$

33. $(x^5 + 1) \div (x + 1)$

34. $(x^5 + 1) \div (x - 1)$

35. $(x^5 + 3x^4 - 5x^3 - 3x^2 + 3x - 4) \div (x + 4)$

36. $(2x^5 + 3x^4 - 4x^3 - x^2 + 5x - 2) \div (x + 2)$

37. $(4x^5 - 6x^4 + 2x^3 + 2x^2 - 5x + 2) \div (x - 1)$

38. $(3x^5 - 8x^4 + 5x^3 + 2x^2 - 9x + 4) \div (x - 2)$

39. $(9x^3 - 6x^2 + 3x - 4) \div \left(x - \dfrac{1}{3}\right)$

40. $(2x^3 + 3x^2 - 2x + 3) \div \left(x + \dfrac{1}{2}\right)$

41. $(3x^4 - 2x^3 + 5x^2 - x - 1) \div \left(x + \dfrac{1}{3}\right)$

42. $(4x^4 - 5x^2 + 1) \div \left(x - \dfrac{1}{2}\right)$

■ ■ ■ THOUGHTS INTO WORDS

43. How would you give a general description of what is accomplished with synthetic division to someone who had just completed an elementary algebra course?

44. Why is synthetic division restricted to situations where the divisor is of the form $x - c$?

9.2 Remainder and Factor Theorems

Let's consider the division algorithm (stated in the previous section) when the dividend, $f(x)$, is divided by a linear polynomial of the form $x - c$. Then the division algorithm

$$f(x) = d(x)q(x) + r(x)$$

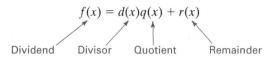

Dividend   Divisor   Quotient   Remainder

becomes

$$f(x) = (x - c)q(x) + r(x)$$

Because the degree of the remainder, $r(x)$, must be less than the degree of the divisor, $x - c$, the remainder is a constant. Therefore, letting R represent the remainder, we have

$$f(x) = (x - c)q(x) + R$$

If the functional value at c is found, we obtain

$$f(c) = (c - c)q(c) + R$$

$$= 0 \cdot q(c) + R$$

$$= R$$

In other words, if a polynomial is divided by a linear polynomial of the form $x - c$, then the remainder is given by the value of the polynomial at c. Let's state this result more formally as the remainder theorem.

Property 9.1 Remainder Theorem

If the polynomial $f(x)$ is divided by $x - c$, then the remainder is equal to $f(c)$.

EXAMPLE 1

If $f(x) = x^3 + 2x^2 - 5x - 1$, find $f(2)$ by (a) using synthetic division and the remainder theorem, and (b) evaluating $f(2)$ directly.

Solution

(a)
$$
\begin{array}{r|rrrr}
2) & 1 & 2 & -5 & -1 \\
  &   & 2 & 8 & 6 \\
\hline
  & 1 & 4 & 3 & ⑤
\end{array}
\quad \longleftarrow \quad R = f(2)
$$

(b) $f(2) = 2^3 + 2(2)^2 - 5(2) - 1 = 8 + 8 - 10 - 1 = 5$

∎

E X A M P L E   2

If $f(x) = x^4 + 7x^3 + 8x^2 + 11x + 5$, find $f(-6)$ by (a) using synthetic division and the remainder theorem and (b) evaluating $f(-6)$ directly.

Solution

(a)
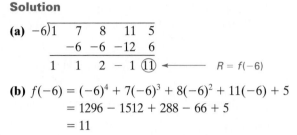

$$-6 \overline{\big)\, 1 \quad 7 \quad 8 \quad 11 \quad 5}$$
$$ -6 \quad -6 \quad -12 \quad 6$$
$$\overline{\,1 \quad 1 \quad 2 \quad -1 \quad \textcircled{11}} \longleftarrow \quad R = f(-6)$$

(b) $f(-6) = (-6)^4 + 7(-6)^3 + 8(-6)^2 + 11(-6) + 5$
$$= 1296 - 1512 + 288 - 66 + 5$$
$$= 11$$

■

In Example 2, note that the computations involved in finding $f(-6)$ by using synthetic division and the remainder theorem are much easier than those required to evaluate $f(-6)$ directly. This is not always the case, but using synthetic division is often easier than evaluating $f(c)$ directly.

E X A M P L E   3

Find the remainder when $x^3 + 3x^2 - 13x - 15$ is divided by $x + 1$.

Solution

Let $f(x) = x^3 + 3x^2 - 13x - 15$, write $x + 1$ as $x - (-1)$, and apply the remainder theorem:

$$f(-1) = (-1)^3 + 3(-1)^2 - 13(-1) - 15 = 0$$

Thus the remainder is 0.

■

Example 3 illustrates an important aspect of the remainder theorem — the situation in which the remainder is *zero*. Thus we can say that $x + 1$ is a factor of $x^3 + 3x^2 - 13x - 15$.

■ Factor Theorem

A general factor theorem can be formulated by considering the equation

$$f(x) = (x - c)q(x) + R$$

If $x - c$ is a factor of $f(x)$, then the remainder R, which is also $f(c)$, must be zero. Conversely, if $R = f(c) = 0$, then $f(x) = (x - c)q(x)$; in other words, $x - c$ is a factor of $f(x)$. The factor theorem can be stated as follows:

Property 9.2 Factor Theorem

A polynomial $f(x)$ has a factor $x - c$ if and only if $f(c) = 0$.

EXAMPLE 4

Is $x - 1$ a factor of $x^3 + 5x^2 + 2x - 8$?

Solution

Let $f(x) = x^3 + 5x^2 + 2x - 8$ and compute $f(1)$ to obtain

$$f(1) = 1^3 + 5(1)^2 + 2(1) - 8 = 0$$

By the factor theorem, therefore, $x - 1$ is a factor of $f(x)$. ∎

EXAMPLE 5

Is $x + 3$ a factor of $2x^3 + 5x^2 - 6x - 7$?

Solution

Use synthetic division to obtain the following:

$$
\begin{array}{r}
-3\overline{)\begin{array}{rrrr} 2 & 5 & -6 & -7 \end{array}} \\
\begin{array}{rrrr} & -6 & 3 & 9 \end{array} \\
\hline
\begin{array}{rrrr} 2 & -1 & -3 & \;\;②\end{array}
\end{array} \longleftarrow R = f(-3)
$$

Because $R \neq 0$, we know that $x + 3$ is not a factor of the given polynomial. ∎

In Examples 4 and 5, we were concerned only with determining whether a linear polynomial of the form $x - c$ was a factor of another polynomial. For such problems, it is reasonable to compute $f(c)$ either directly or by synthetic division, whichever way seems easier for a particular problem. However, if more information is required, such as the complete factorization of the given polynomial, then the use of synthetic division is appropriate, as the next two examples illustrate.

EXAMPLE 6

Show that $x - 1$ is a factor of $x^3 - 2x^2 - 11x + 12$, and find the other linear factors of the polynomial.

Solution

Let's use synthetic division to divide $x^3 - 2x^2 - 11x + 12$ by $x - 1$.

$$
\begin{array}{r}
1\overline{)\begin{array}{rrrr} 1 & -2 & -11 & 12 \end{array}} \\
\begin{array}{rrrr} & 1 & -1 & -12 \end{array} \\
\hline
\begin{array}{rrrr} 1 & -1 & -12 & 0 \end{array}
\end{array}
$$

The last line indicates a quotient of $x^2 - x - 12$ and a remainder of 0. The remainder of 0 means that $x - 1$ is a factor. Furthermore, we can write

$$x^3 - 2x^2 - 11x + 12 = (x - 1)(x^2 - x - 12)$$

The quadratic polynomial $x^2 - x - 12$ can be factored as $(x - 4)(x + 3)$ using our conventional factoring techniques. Thus we obtain

$$x^3 - 2x^2 - 11x + 12 = (x - 1)(x - 4)(x + 3)$$ ∎

E X A M P L E   7

Show that $x + 4$ is a factor of $f(x) = x^3 - 5x^2 - 22x + 56$, and complete the factorization of $f(x)$.

Solution

Use synthetic division to divide $x^3 - 5x^2 - 22x + 56$ by $x + 4$.

$$
\begin{array}{r|rrrr}
-4 & 1 & -5 & -22 & 56 \\
   &   & -4 & 36 & -56 \\
\hline
   & 1 & -9 & 14 & 0
\end{array}
$$

The last line indicates a quotient of $x^2 - 9x + 14$ and a remainder of 0. The remainder of 0 means that $x + 4$ is a factor. Furthermore, we can write

$$x^3 - 5x^2 - 22x + 56 = (x + 4)(x^2 - 9x + 14)$$

and then complete the factoring to obtain

$$x^3 - 5x^2 - 22x + 56 = (x + 4)(x - 7)(x - 2)$$ ■

The factor theorem also plays a significant role in determining some general factorization ideas, as the last example of this section demonstrates.

E X A M P L E   8

Verify that $x + 1$ is a factor of $x^n + 1$ for all odd positive integral values of n.

Solution

Let $f(x) = x^n + 1$ and compute $f(-1)$.

$$f(-1) = (-1)^n + 1$$

$$= -1 + 1 \qquad \text{Any odd power of } -1 \text{ is } -1.$$

$$= 0$$

Because $f(-1) = 0$, we know that $x + 1$ is a factor of $f(x)$. ■

Problem Set 9.2

For Problems 1–10, find $f(c)$ by (a) evaluating $f(c)$ directly, and (b) using synthetic division and the remainder theorem.

1. $f(x) = x^2 + 2x - 6$ and $c = 3$

2. $f(x) = x^2 - 7x + 4$ and $c = 2$

3. $f(x) = x^3 - 2x^2 + 3x - 1$ and $c = -1$

4. $f(x) = x^3 + 3x^2 - 4x - 7$ and $c = -2$

5. $f(x) = 2x^4 - x^3 - 3x^2 + 4x - 1$ and $c = 2$

6. $f(x) = 3x^4 - 4x^3 + 5x^2 - 7x + 6$ and $c = 1$

7. $f(n) = 6n^3 - 35n^2 + 8n - 10$ and $c = 6$

8. $f(n) = 8n^3 - 39n^2 - 7n - 1$ and $c = 5$

9. $f(n) = 2n^5 - 1$ and $c = -2$

10. $f(n) = 3n^4 - 2n^3 + 4n - 1$ and $c = 3$

For Problems 11–20, find $f(c)$ *either* by using synthetic division and the remainder theorem *or* by evaluating $f(c)$ directly.

11. $f(x) = 6x^5 - 3x^3 + 2$ and $c = -1$

12. $f(x) = -4x^4 + x^3 - 2x^2 - 5$ and $c = 2$

13. $f(x) = 2x^4 - 15x^3 - 9x^2 - 2x - 3$ and $c = 8$

14. $f(x) = x^4 - 8x^3 + 9x^2 - 15x + 2$ and $c = 7$

15. $f(n) = 4n^7 + 3$ and $c = 3$

16. $f(n) = -3n^6 - 2$ and $c = -3$

17. $f(n) = 3n^5 + 17n^4 - 4n^3 + 10n^2 - 15n + 13$ and $c = -6$

18. $f(n) = -2n^5 - 9n^4 + 7n^3 + 14n^2 + 19n - 38$ and $c = -5$

19. $f(x) = -4x^4 - 6x^2 + 7$ and $c = 4$

20. $f(x) = 3x^5 - 7x^3 - 6$ and $c = 5$

For Problems 21–34, use the factor theorem to help answer some questions about factors.

21. Is $x - 2$ a factor of $5x^2 - 17x + 14$?

22. Is $x + 1$ a factor of $3x^2 - 5x - 8$?

23. Is $x + 3$ a factor of $6x^2 + 13x - 14$?

24. Is $x - 5$ a factor of $8x^2 - 47x + 32$?

25. Is $x - 1$ a factor of $4x^3 - 13x^2 + 21x - 12$?

26. Is $x - 4$ a factor of $2x^3 - 11x^2 + 10x + 8$?

27. Is $x + 2$ a factor of $x^3 + 7x^2 + x - 18$?

28. Is $x + 3$ a factor of $x^3 + x^2 - 14x - 24$?

29. Is $x - 3$ a factor of $3x^3 - 5x^2 - 17x + 17$?

30. Is $x + 4$ a factor of $2x^3 + 9x^2 - 5x - 39$?

31. Is $x + 2$ a factor of $x^3 + 8$?

32. Is $x - 2$ a factor of $x^3 - 8$?

33. Is $x - 3$ a factor of $x^4 - 81$?

34. Is $x + 3$ a factor of $x^4 - 81$?

For Problems 35–44, use synthetic division to show that $g(x)$ is a factor of $f(x)$, and complete the factorization of $f(x)$.

35. $g(x) = x - 2$, $f(x) = x^3 - 6x^2 - 13x + 42$

36. $g(x) = x + 1$, $f(x) = x^3 + 6x^2 - 31x - 36$

37. $g(x) = x + 2$, $f(x) = 12x^3 + 29x^2 + 8x - 4$

38. $g(x) = x - 3$, $f(x) = 6x^3 - 17x^2 - 5x + 6$

39. $g(x) = x + 1$, $f(x) = x^3 - 2x^2 - 7x - 4$

40. $g(x) = x - 5$, $f(x) = 2x^3 + x^2 - 61x + 30$

41. $g(x) = x - 6$, $f(x) = x^5 - 6x^4 - 16x + 96$

42. $g(x) = x + 3$, $f(x) = x^5 + 3x^4 - x - 3$

43. $g(x) = x + 5$, $f(x) = 9x^3 + 21x^2 - 104x + 80$

44. $g(x) = x + 4$, $f(x) = 4x^3 + 4x^2 - 39x + 36$

For Problems 45–48, find the value(s) of k that makes the second polynomial a factor of the first.

45. $k^2x^4 + 3kx^2 - 4; x - 1$

46. $x^3 - kx^2 + 5x + k; x - 2$

47. $kx^3 + 19x^2 + x - 6; x + 3$

48. $x^3 + 4x^2 - 11x + k; x + 2$

49. Argue that $f(x) = 3x^4 + 2x^2 + 5$ has no factor of the form $x - c$, where c is a real number.

50. Show that $x + 2$ is a factor of $x^{12} - 4096$.

51. Verify that $x + 1$ is a factor of $x^n - 1$ for all even positive integral values of n.

52. Verify that $x - 1$ is a factor of $x^n - 1$ for all positive integral values of n.

53. **(a)** Verify that $x - y$ is a factor of $x^n - y^n$ for all positive integral values of n.
  (b) Verify that $x + y$ is a factor of $x^n - y^n$ for all even positive integral values of n.
  (c) Verify that $x + y$ is a factor of $x^n + y^n$ for all odd positive integral values of n.

■ ■ ■ **THOUGHTS INTO WORDS**

54. State the remainder theorem in your own words.

55. Discuss some of the uses of the factor theorem.

The remainder and factor theorems are true for any complex value of c. Therefore, for Problems 56–58, find $f(c)$ by (a) using synthetic division and the remainder theorem, and (b) evaluating $f(c)$ directly.

56. $f(x) = x^3 - 5x^2 + 2x + 1$ and $c = i$

57. $f(x) = x^2 + 4x - 2$ and $c = 1 + i$

58. $f(x) = x^3 + 2x^2 + x - 2$ and $c = 2 - 3i$

59. Show that $x - 2i$ is a factor of $f(x) = x^4 + 6x^2 + 8$.

60. Show that $x + 3i$ is a factor of $f(x) = x^4 + 14x^2 + 45$.

61. Consider changing the form of the polynomial $f(x) = x^3 + 4x^2 - 3x + 2$ as follows:

$$f(x) = x^3 + 4x^2 - 3x + 2$$

$$= x(x^2 + 4x - 3) + 2$$

$$= x[x(x + 4) - 3] + 2$$

The final form $f(x) = x[x(x + 4) - 3] + 2$ is called the **nested form** of the polynomial. It is particularly well suited for evaluating functional values of f either by hand or with a calculator. For each of the following, find the indicated functional values using the nested form of the given polynomial.

(a) $f(4)$, $f(-5)$, and $f(7)$ for $f(x) = x^3 + 5x^2 - 2x + 1$

(b) $f(3)$, $f(6)$, and $f(-7)$ for $f(x) = 2x^3 - 4x^2 - 3x + 2$

(c) $f(4)$, $f(5)$, and $f(-3)$ for $f(x) = -2x^3 + 5x^2 - 6x - 7$

(d) $f(5)$, $f(6)$, and $f(-3)$ for $f(x) = x^4 + 3x^3 - 2x^2 + 5x - 1$

9.3 Polynomial Equations

We have solved a large variety of linear equations of the form $ax + b = 0$ and quadratic equations of the form $ax^2 + bx + c = 0$. Linear and quadratic equations are special cases of a general class of equations we refer to as **polynomial equations**. The equation

$$a_n x^n + a_{n-1} x^{n-1} + \cdots + a_1 x + a_0 = 0$$

where the coefficients $a_0, a_1, \ldots, a_n$ are real numbers and n is a positive integer, is called a **polynomial equation of degree n**. The following are examples of polynomial equations:

$$\sqrt{2}x - 6 = 0 \qquad \text{Degree 1}$$

$$\frac{3}{4}x^2 - \frac{2}{3}x + 5 = 0 \qquad \text{Degree 2}$$

$$4x^3 - 3x^2 - 7x - 9 = 0 \qquad \text{Degree 3}$$

$$5x^4 - x + 6 = 0 \qquad \text{Degree 4}$$

Remark: The most general polynomial equation would allow complex numbers as coefficients. However, for our purposes in this text, we will restrict the coefficients to real numbers. We often refer to such equations as **polynomial equations over the reals**.

In general, solving polynomial equations of degree greater than 2 can be very difficult and often requires mathematics beyond the scope of this text. However, there are some general properties pertaining to the solving of polynomial equations that you should be familiar with; furthermore, there are certain types of polynomial equations that we can solve using the techniques available to us at this time. We can also use a graphical approach to approximate solutions, which, in some cases, is shorter than using an algebraic approach.

Let's begin by listing some polynomial equations and corresponding solution sets that we have already encountered in this text.

Equation	Solution set
$3x + 4 = 7$	$\{1\}$
$x^2 + x - 6 = 0$	$\{-3, 2\}$
$2x^3 - 3x^2 - 2x + 3 = 0$	$\left\{-1, 1, \dfrac{3}{2}\right\}$
$x^4 - 16 = 0$	$\{-2, 2, -2i, 2i\}$

Note that in each of these examples, the number of solutions corresponds to the degree of the equation. The first-degree equation has one solution, the second-degree equation has two solutions, the third-degree equation has three solutions, and the fourth-degree equation has four solutions. Now consider the equation

$$(x - 4)^2(x + 5)^3 = 0$$

It can be written as

$$(x - 4)(x - 4)(x + 5)(x + 5)(x + 5) = 0$$

which implies that

$$x - 4 = 0 \quad \text{or} \quad x - 4 = 0 \quad \text{or} \quad x + 5 = 0 \quad \text{or}$$

$$x + 5 = 0 \quad \text{or} \quad x + 5 = 0$$

Therefore

$$x = 4 \quad \text{or} \quad x = 4 \quad \text{or} \quad x = -5 \quad \text{or}$$

$$x = -5 \quad \text{or} \quad x = -5$$

We state that the solution set of the original equation is $\{-5, 4\}$, but we also say that the equation has a solution of 4 with a **multiplicity of two** and a solution of -5 with a **multiplicity of three**. Furthermore, note that the sum of the multiplicities is 5, which agrees with the degree of the equation. The following general property can be stated:

Property 9.3

> A polynomial equation of degree n has n solutions, where any solution of multiplicity p is counted p times.

■ Finding Rational Solutions

Although solving polynomial equations of degree greater than 2 can, in general, be very difficult, **rational solutions of polynomial equations with integral coefficients** can be found using techniques presented in this chapter. The following property restricts the potential rational solutions of such equations:

Property 9.4 Rational Root Theorem

> Consider the polynomial equation
>
> $$a_n x^n + a_{n-1} x^{n-1} + \cdots + a_1 x + a_0 = 0$$
>
> where the coefficients $a_0, a_1, \ldots, a_n$ are *integers*. If the rational number $\dfrac{c}{d}$, reduced to lowest terms, is a solution of the equation, then c is a factor of the constant term a_0 and d is a factor of the leading coefficient a_n.

The "why" behind the rational root theorem is based on some simple factoring ideas, as indicated by the following outline of a proof for the theorem.

Outline of Proof   If $\dfrac{c}{d}$ is to be a solution, then

$$a_n\left(\frac{c}{d}\right)^n + a_{n-1}\left(\frac{c}{d}\right)^{n-1} + \cdots + a_1\left(\frac{c}{d}\right) + a_0 = 0$$

Multiply both sides of this equation by d^n and add $-a_0 d^n$ to both sides to yield

$$a_n c^n + a_{n-1} c^{n-1} d + \cdots + a_1 c d^{n-1} = -a_0 d^n$$

Because c is a factor of the left side of this equation, c must also be a factor of $-a_0 d^n$. Furthermore, because $\dfrac{c}{d}$ is in reduced form, c and d have no common factors other than -1 or 1. Thus c is a factor of a_0. In the same way, from the equation

$$a_{n-1} c^{n-1} d + \cdots + a_1 c d^{n-1} + a_0 d^n = -a_n c^n$$

we can conclude that d is a factor of the left side, and therefore d is also a factor of a_n.

The rational root theorem, a graph, synthetic division, the factor theorem, and some previous knowledge pertaining to solving linear and quadratic equations form a basis for finding rational solutions. Let's consider some examples.

EXAMPLE 1

Find all rational solutions of $3x^3 + 8x^2 - 15x + 4 = 0$.

Solution

If $\dfrac{c}{d}$ is a rational solution, then c must be a factor of 4, and d must be a factor of 3.

Therefore, the possible values for c and d are as follows:

> For c: $\pm 1, \pm 2, \pm 4$
> For d: $\pm 1, \pm 3$

Thus the possible values for $\dfrac{c}{d}$ are

$$\pm 1, \pm\frac{1}{3}, \pm 2, \pm\frac{2}{3}, \pm 4, \pm\frac{4}{3}$$

Now let's use a graph of $y = 3x^3 + 8x^2 - 15x + 4$ to shorten the list of possible rational solutions (see Figure 9.1).

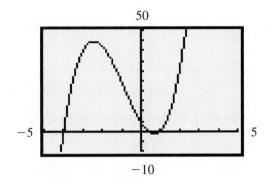

Figure 9.1

The x intercepts appear to be at -4, at 1, and between 0 and 1. Using synthetic division,

$$
\begin{array}{r|rrrr}
1 & 3 & 8 & -15 & 4 \\
  &   & 3 & 11 & -4 \\
\hline
  & 3 & 11 & -4 & 0 \\
\end{array}
$$

we can show that $x - 1$ is a factor of the given polynomial, and therefore 1 is a rational solution of the equation. Furthermore, the result of the synthetic division also indicates that we can factor the given polynomial as follows:

$$3x^3 + 8x^2 - 15x + 4 = 0$$

$$(x - 1)(3x^2 + 11x - 4) = 0$$

The quadratic factor can be factored further using our previous techniques; we can proceed as follows:

$$(x - 1)(3x^2 + 11x - 4) = 0$$

$$(x - 1)(3x - 1)(x + 4) = 0$$

$$x - 1 = 0 \qquad \text{or} \qquad 3x - 1 = 0 \qquad \text{or} \qquad x + 4 = 0$$

$$x = 1 \qquad \text{or} \qquad x = \frac{1}{3} \qquad \text{or} \qquad x = -4$$

Thus the entire solution set consists of rational numbers, which can be listed as $\left\{-4, \frac{1}{3}, 1\right\}$

∎

Remark: The graphs used in this section are done with a graphing utility. In the next section, we will discuss some special situations for which freehand sketches of the graphs are easily obtained.

In Example 1 we used a graph to help shorten the list of possible rational solutions determined by the rational root theorem. Without using a graph, one needs to conduct an organized search of the list of possible rational solutions, as the next example demonstrates.

E X A M P L E   2   Find all rational solutions of $3x^3 + 7x^2 - 22x - 8 = 0$.

Solution

If $\dfrac{c}{d}$ is a rational solution, then c must be a factor of -8, and d must be a factor of 3.

Therefore, the possible values for c and d are as follows:

For c: $\pm 1, \pm 2, \pm 4, \pm 8$
For d: $\pm 1, \pm 3$

Thus the possible values for $\dfrac{c}{d}$ are

$$\pm 1, \pm \frac{1}{3}, \pm 2, \pm \frac{2}{3}, \pm 4, \pm \frac{4}{3}, \pm 8, \pm \frac{8}{3}$$

Let's begin our search for rational solutions; we will try the integers first.

```
1)3    7   -22   - 8
       3    10   -12
   ─────────────────
   3   10   -12   ⊝20  ←──── This remainder indicates that x − 1 is not
                               a factor, and thus 1 is not a solution.
```

```
-1)3    7   -22   - 8
       - 3  - 4    26
   ─────────────────
   3    4   -26   ⑱ ←──── This remainder indicates that −1 is not
                           a solution.
```

```
2)3    7   -22   - 8
       6    26     8
   ────────────────
   3   13    4     0
```

Now we know that $x - 2$ is a factor; we can proceed as follows:

$$3x^3 + 7x^2 - 22x - 8 = 0$$

$$(x - 2)(3x^2 + 13x + 4) = 0$$

$$(x - 2)(3x + 1)(x + 4) = 0$$

$$x - 2 = 0 \quad \text{or} \quad 3x + 1 = 0 \quad \text{or} \quad x + 4 = 0$$

$$x = 2 \quad \text{or} \quad 3x = -1 \quad \text{or} \quad x = -4$$

$$x = 2 \quad \text{or} \quad x = -\frac{1}{3} \quad \text{or} \quad x = -4$$

The solution set is $\left\{ -4, -\dfrac{1}{3}, 2 \right\}$ ∎

In Examples 1 and 2, we were solving third-degree equations. Therefore, after finding one linear factor by synthetic division, we were able to factor the remaining quadratic factor in the usual way. However, if the given equation is of degree 4 or more, we may need to find more than one linear factor by synthetic division, as the next example illustrates.

EXAMPLE 3

Solve $x^4 - 6x^3 + 22x^2 - 30x + 13 = 0$.

Solution

The possible values for $\dfrac{c}{d}$ are as follows:

For $\dfrac{c}{d}$: $\pm 1, \pm 13$

By synthetic division, we find that

$$
\begin{array}{r|rrrr}
1) & 1 & -6 & 22 & -30 & 13 \\
  &   & 1 & -5 & 17 & -13 \\
\hline
  & 1 & -5 & 17 & -13 & 0
\end{array}
$$

which indicates that $x - 1$ is a factor of the given polynomial. The bottom line of the synthetic division indicates that the given polynomial can be factored as follows:

$$x^4 - 6x^3 + 22x^2 - 30x + 13 = 0$$

$$(x - 1)(x^3 - 5x^2 + 17x - 13) = 0$$

Therefore

$$x - 1 = 0 \quad \text{or} \quad x^3 - 5x^2 + 17x - 13 = 0$$

Now we can use the same approach to look for rational solutions of the expression $x^3 - 5x^2 + 17x - 13 = 0$. The possible values for $\dfrac{c}{d}$ are as follows:

For $\dfrac{c}{d}$: $\pm 1, \pm 13$

By synthetic division, we find that

$$1\,\overline{)1 \quad -5 \quad 17 \quad -13}$$
$$\;\;\;\;\;1 \;\; -\;4 \quad\;\; 13$$
$$\overline{1 \quad -4 \quad\;\; 13 \quad\;\;\; 0}$$

which indicates that $x - 1$ is a factor of $x^3 - 5x^2 + 17x - 13$ and that the other factor is $x^2 - 4x + 13$.

Now we can solve the original equation as follows:

$$x^4 - 6x^3 + 22x^2 - 30x + 13 = 0$$
$$(x - 1)(x^3 - 5x^2 + 17x - 13) = 0$$
$$(x - 1)(x - 1)(x^2 - 4x + 13) = 0$$

$x - 1 = 0$	or	$x - 1 = 0$	or	$x^2 - 4x + 13 = 0$
$x = 1$	or	$x = 1$	or	$x^2 - 4x + 13 = 0$

Use the quadratic formula on $x^2 - 4x + 13 = 0$:

$$x = \frac{4 \pm \sqrt{16 - 52}}{2} = \frac{4 \pm \sqrt{-36}}{2}$$
$$= \frac{4 \pm 6i}{2} = 2 \pm 3i$$

Thus the original equation has a rational solution of 1 with a multiplicity of two and two complex solutions, $2 + 3i$ and $2 - 3i$. The solution set is listed as $\{1, 2 \pm 3i\}$. ∎

Let's graph the equation $y = x^4 - 6x^3 + 22x^2 - 30x + 13$ to give some visual support for our work in Example 3. The graph in Figure 9.2 indicates only an x intercept at 1. This is consistent with the solution set of $\{1, 2 \pm 3i\}$.

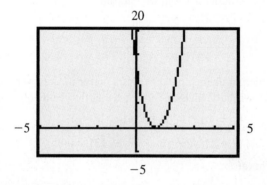

Figure 9.2

Example 3 illustrates two general properties. First, note that the coefficient of x^4 is 1, and thus the possible rational solutions must be integers. In general, the possible rational solutions of $x^n + a_{n-1}x^{n-1} + \cdots + a_1x + a_0 = 0$ are the integral factors of a_0. Second, note that the complex solutions of Example 3 are conjugates of each other. The following general property can be stated:

Property 9.5

> Nonreal complex solutions of polynomial equations with real coefficients, if they exist, must occur in conjugate pairs.

Each of Properties 9.3, 9.4, and 9.5 yields some information about the solutions of a polynomial equation. Before we state the final property of this section, which will give us some additional information, we need to consider two ideas.

First, in a polynomial that is arranged in descending powers of x, if two successive terms differ in sign, then there is said to be a **variation in sign**. (We disregard terms with zero coefficients when sign variations are counted.) For example, the polynomial

$$3x^3 - 2x^2 + 4x + 7$$

has *two* sign variations, whereas the polynomial

$$x^5 - 4x^3 + x - 5$$

has *three* variations.

Second, the solutions of

$$a_n(-x)^n + a_{n-1}(-x)^{n-1} + \cdots + a_1(-x) + a_0 = 0$$

are the opposites of the solutions of

$$a_n x^n + a_{n-1} x^{n-1} + \cdots + a_1 x + a_0 = 0$$

In other words, if a new equation is formed by replacing x with $-x$ in a given equation, then the solutions of the newly formed equation are the opposites of the solutions of the given equation. For example, the solution set of $x^2 + 7x + 12 = 0$ is $\{-4, -3\}$, and the solution set of $(-x)^2 + 7(-x) + 12 = 0$, which simplifies to $x^2 - 7x + 12 = 0$, is $\{3, 4\}$.

Now we can state a property that can help us to determine the nature of the solutions of a polynomial equation without actually solving the equation.

Property 9.6 Descartes' Rule of Signs

> Let $a_n x^n + a_{n-1} x^{n-1} + \cdots + a_1 x + a_0 = 0$ be a polynomial equation with real coefficients.
>
> **1.** The number of *positive real solutions* of the given equation either is equal to the number of variations in sign of the polynomial or is less than the number of variations by a positive even integer.
>
> **2.** The number of *negative real solutions* of the given equation either is equal to the number of variations in sign of the polynomial $a_n(-x)^n + a_{n-1}(-x)^{n-1} + \cdots + a_1(-x) + a_0$ or is less than the number of variations by a positive even integer.

Along with Properties 9.3 and 9.5, Property 9.6 allows us to acquire some information about the solutions of a polynomial equation without actually solving the equation. Let's consider some equations and see how much we know about their solutions without solving them.

1. $x^3 + 3x^2 + 5x + 4 = 0$
   (a) No variations of sign in $x^3 + 3x^2 + 5x + 4$ means that there are *no positive solutions*.
   (b) Replacing x with $-x$ in the given polynomial produces $(-x)^3 + 3(-x)^2 + 5(-x) + 4$, which simplifies to $-x^3 + 3x^2 - 5x + 4$ and contains three variations of sign; thus there are *three or one negative solutions*.

Conclusion The given equation has three negative real solutions or else one negative real solution and two nonreal complex solutions.

2. $2x^4 + 3x^2 - x - 1 = 0$
   (a) There is one variation of sign; thus the equation has *one positive solution*.
   (b) Replacing x with $-x$ produces $2(-x)^4 + 3(-x)^2 - (-x) - 1$, which simplifies to $2x^4 + 3x^2 + x - 1$ and contains one variation of sign. Thus the equation has *one negative solution*.

Conclusion The given equation has one positive, one negative, and two nonreal complex solutions.

3. $3x^4 + 2x^2 + 5 = 0$
   (a) No variations of sign in the given polynomial means that there are *no positive solutions*.
   (b) Replacing x with $-x$ produces $3(-x)^4 + 2(-x)^2 + 5$, which simplifies to $3x^4 + 2x^2 + 5$ and contains no variations of sign. Thus there are *no negative solutions*.

Conclusion The given equation contains four nonreal complex solutions. These solutions will appear in conjugate pairs.

4. $2x^5 - 4x^3 + 2x - 5 = 0$
   (a) The fact that there are three variations of sign in the given polynomial implies that there are *three or one positive solutions*.
   (b) Replacing x with $-x$ produces $2(-x)^5 - 4(-x)^3 + 2(-x) - 5$, which simplifies to $-2x^5 + 4x^3 - 2x - 5$ and contains two variations of sign. Thus there are *two or zero negative solutions*.

Conclusion The given equation has either three positive and two negative solutions; three positive and two nonreal complex solutions; one positive, two negative, and two nonreal complex solutions; or one positive and four nonreal complex solutions.

It should be evident from the previous discussions that sometimes we can truly pinpoint the nature of the solutions of a polynomial equation. However, for some equations (such as the last example), the best we can do with the properties discussed in this section is to restrict the possibilities for the nature of the solutions. It might be helpful for you to review Examples 1, 2, and 3 of this section and show that the solution sets do satisfy Properties 9.3, 9.5, and 9.6.

Finally, let's consider a situation for which the graphing calculator becomes a very useful tool.

EXAMPLE 4

Find the real number solutions of the equation $x^4 - 2x^3 - 5 = 0$.

Solution

First, let's use a graphing calculator to get a graph of $y = x^4 - 2x^3 - 5$, as shown in Figure 9.3. Obviously, there are two x intercepts, one between -2 and -1 and another between 2 and 3. From the rational root theorem, we know that the only possible rational roots of the given equation are ± 1 and ± 5. Therefore these x intercepts must be irrational numbers. We can use the ZOOM and TRACE features of the graphing calculator to approximate these values at -1.2 and 2.4, to the nearest tenth. Thus the real number solutions of $x^4 - 2x^3 - 5 = 0$ are approximately -1.2 and 2.4. The other two solutions must be conjugate complex numbers.

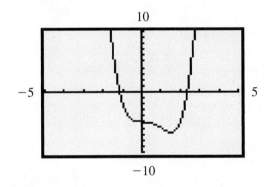

Figure 9.3   ∎

Problem Set 9.3

For Problems 1–20, use the rational root theorem and the factor theorem to help solve each equation. Be sure that the number of solutions for each equation agrees with Property 9.3, taking into account multiplicity of solutions.

1. $x^3 - 2x^2 - 11x + 12 = 0$

2. $x^3 + x^2 - 4x - 4 = 0$

3. $15x^3 + 14x^2 - 3x - 2 = 0$

4. $3x^3 + 13x^2 - 52x + 28 = 0$

5. $8x^3 - 2x^2 - 41x - 10 = 0$

6. $6x^3 + x^2 - 10x + 3 = 0$

7. $x^3 - x^2 - 8x + 12 = 0$

8. $x^3 - 2x^2 - 7x - 4 = 0$

9. $x^3 - 4x^2 + 8 = 0$

10. $x^3 - 10x - 12 = 0$

11. $x^4 + 4x^3 - x^2 - 16x - 12 = 0$

12. $x^4 - 4x^3 - 7x^2 + 34x - 24 = 0$

13. $x^4 + x^3 - 3x^2 - 17x - 30 = 0$

14. $x^4 - 3x^3 + 2x^2 + 2x - 4 = 0$

15. $x^3 - x^2 + x - 1 = 0$

16. $6x^4 - 13x^3 - 19x^2 + 12x = 0$

17. $2x^4 + 3x^3 - 11x^2 - 9x + 15 = 0$

18. $3x^4 - x^3 - 8x^2 + 2x + 4 = 0$

19. $4x^4 + 12x^3 + x^2 - 12x + 4 = 0$

20. $2x^5 - 5x^4 + x^3 + x^2 - x + 6 = 0$

For Problems 21–26, verify that the equations do not have any rational number solutions.

21. $x^4 + 3x - 2 = 0$

22. $x^4 - x^3 - 8x^2 - 3x + 1 = 0$

23. $3x^4 - 4x^3 - 10x^2 + 3x - 4 = 0$

24. $2x^4 - 3x^3 + 6x^2 - 24x + 5 = 0$

25. $x^5 + 2x^4 - 2x^3 + 5x^2 - 2x - 3 = 0$

26. $x^5 - 2x^4 + 3x^3 + 4x^2 + 7x - 1 = 0$

For Problems 27–30, solve each equation by first applying the multiplication property of equality to produce an equivalent equation with integral coefficients.

27. $\dfrac{1}{10}x^3 + \dfrac{1}{5}x^2 - \dfrac{1}{2}x - \dfrac{3}{5} = 0$

28. $\dfrac{1}{10}x^3 + \dfrac{1}{2}x^2 + \dfrac{1}{5}x - \dfrac{4}{5} = 0$

29. $x^3 - \dfrac{5}{6}x^2 - \dfrac{22}{3}x + \dfrac{5}{2} = 0$

30. $x^3 + \dfrac{9}{2}x^2 - x - 12 = 0$

For Problems 31–40, use Descartes' rule of signs (Property 9.6) to help list the possibilities for the nature of the solutions for each equation. *Do not* solve the equations.

31. $6x^2 + 7x - 20 = 0$

32. $8x^2 - 14x + 3 = 0$

33. $2x^3 + x - 3 = 0$

34. $4x^3 + 3x + 7 = 0$

35. $3x^3 - 2x^2 + 6x + 5 = 0$

36. $4x^3 + 5x^2 - 6x - 2 = 0$

37. $x^5 - 3x^4 + 5x^3 - x^2 + 2x - 1 = 0$

38. $2x^5 + 3x^3 - x + 1 = 0$

39. $x^5 + 32 = 0$

40. $2x^6 + 3x^4 - 2x^2 - 1 = 0$

■ ■ ■ THOUGHTS INTO WORDS

41. Explain what it means to say that the equation $(x + 3)^2 = 0$ has a solution of -3 with a multiplicity of two.

42. Describe how to use the rational root theorem to show that the equation $x^2 - 3 = 0$ has no rational solutions.

■ ■ ■ FURTHER INVESTIGATIONS

43. Use the rational root theorem to argue that $\sqrt{2}$ is not a rational number. [*Hint*: The solutions of $x^2 - 2 = 0$ are $\pm\sqrt{2}$.]

44. Use the rational root theorem to argue that $\sqrt{12}$ is not a rational number.

45. Defend this statement: "Every polynomial equation of odd degree with real coefficients has at least one real number solution."

46. The following synthetic division shows that 2 is a solution of $x^4 + x^3 + x^2 - 9x - 10 = 0$:

$$
\begin{array}{r|rrrrr}
2) & 1 & 1 & 1 & -9 & -10 \\
  &   & 2 & 6 & 14 & 10 \\
\hline
  & 1 & 3 & 7 & 5 & 0 \longleftarrow
\end{array}
$$

Note that the new quotient row (indicated by the arrow) consists entirely of nonnegative numbers. This indicates that searching for solutions greater than 2 would be a waste of time because larger divisors would continue to increase each of the numbers (except the one on the far left) in the new quotient row. (Try 3 as a divisor!) Thus we say that 2 is an *upper bound* for the real number solutions of the given equation.

Now consider the following synthetic division, which shows that -1 is also a solution of $x^4 + x^3 + x^2 - 9x - 10 = 0$:

$$
\begin{array}{r|rrrrr}
-1) & 1 & 1 & 1 & -9 & -10 \\
   &   & -1 & 0 & -1 & 10 \\
\hline
   & 1 & 0 & 1 & -10 & 0 \longleftarrow
\end{array}
$$

The new quotient row (indicated by the arrow) shows that there is no need to look for solutions less than -1 because any divisor less than -1 would increase the absolute value of each number (except the one on the far left) in the new quotient row. (Try -2 as a divisor!) Thus we say that -1 is a *lower bound* for the real number solutions of the given equation.

The following general property can be stated:

If $a_n x^n + a_{n-1} x^{n-1} + \cdots + a_1 x + a_0 = 0$ is a polynomial equation with real coefficients, where $a_n > 0$, and if the polynomial is divided synthetically by $x - c$, then

1. If $c > 0$ and all numbers in the new quotient row of the synthetic division are nonnegative, then c is an upper bound of the solutions of the given equation.

2. If $c < 0$ and the numbers in the new quotient row alternate in sign (with 0 considered either positive or negative, as needed), then c is a lower bound of the solutions of the given equation.

Find the smallest positive integer and the largest negative integer that are upper and lower bounds, respectively, for the real number solutions of each of the following equations. Keep in mind that the integers that serve as bounds do not necessarily have to be solutions of the equation.

(a) $x^3 - 3x^2 + 25x - 75 = 0$

(b) $x^3 + x^2 - 4x - 4 = 0$

(c) $x^4 + 4x^3 - 7x^2 - 22x + 24 = 0$

(d) $3x^3 + 7x^2 - 22x - 8 = 0$

(e) $x^4 - 2x^3 - 9x^2 + 2x + 8 = 0$

GRAPHING CALCULATOR ACTIVITIES

47. Solve each of the following equations, using a graphing calculator whenever it seems to be helpful. Express all irrational solutions in lowest radical form.

(a) $x^3 + 2x^2 - 14x - 40 = 0$

(b) $x^3 + x^2 - 7x + 65 = 0$

(c) $x^4 - 6x^3 - 6x^2 + 32x + 24 = 0$

(d) $x^4 + 3x^3 - 39x^2 + 11x + 24 = 0$

(e) $x^3 - 14x^2 + 26x - 24 = 0$

(f) $x^4 + 2x^3 - 3x^2 - 4x + 4 = 0$

48. Find approximations, to the nearest hundredth, of the real number solutions of each of the following equations:

(a) $x^2 - 4x + 1 = 0$

(b) $3x^3 - 2x^2 + 12x - 8 = 0$

(c) $x^4 - 8x^3 + 14x^2 - 8x + 13 = 0$

(d) $x^4 + 6x^3 - 10x^2 - 22x + 161 = 0$

(e) $7x^5 - 5x^4 + 35x^3 - 25x^2 + 28x - 20 = 0$

9.4   Graphing Polynomial Functions

The terms with which we classify functions are analogous to those with which we describe the linear equations, quadratic equations, and polynomial equations. In Chapter 8 we defined a linear function in terms of the equation

$$f(x) = ax + b$$

and a quadratic function in terms of the equation

$$f(x) = ax^2 + bx + c$$

Both are special cases of a general class of functions called polynomial functions. Any function of the form

$$f(x) = a_n x^n + a_{n-1} x^{n-1} + \cdots + a_1 x + a_0$$

is called a **polynomial function of degree n**, where a_n is a nonzero real number, $a_{n-1}, \ldots, a_1, a_0$ are real numbers, and n is a nonnegative integer. The following are examples of polynomial functions:

$$f(x) = 5x^3 - 2x^2 + x - 4 \qquad \text{Degree 3}$$

$$f(x) = -2x^4 - 5x^3 + 3x^2 + 4x - 1 \qquad \text{Degree 4}$$

$$f(x) = 3x^5 + 2x^2 - 3 \qquad \text{Degree 5}$$

Remark: Our previous work with polynomial equations is sometimes presented as "finding zeros of polynomial functions." The **solutions**, or **roots**, of a polynomial equation are also called the **zeros** of the polynomial function. For example, -2 and 2 are solutions of $x^2 - 4 = 0$, and they are zeros of $f(x) = x^2 - 4$. That is, $f(-2) = 0$ and $f(2) = 0$.

For a complete discussion of graphing polynomial functions, we would need some tools from calculus. However, the graphing techniques that we have discussed in this text will allow us to graph certain kinds of polynomial functions. For example, polynomial functions of the form

$$f(x) = ax^n$$

are quite easy to graph. We know from our previous work that if $n = 1$, then functions such as $f(x) = 2x, f(x) = -3x$, and $f(x) = \frac{1}{2}x$ are lines through the origin that have slopes of 2, -3, and $\frac{1}{2}$, respectively.

Furthermore, if $n = 2$, we know that the graphs of functions of the form $f(x) = ax^2$ are parabolas that are symmetric with respect to the y axis and have their vertices at the origin.

We have also previously graphed the special case of $f(x) = ax^n$, where $a = 1$ and $n = 3$—namely, the function $f(x) = x^3$. This graph is shown in Figure 9.4.

The graphs of functions of the form $f(x) = ax^3$, where $a \neq 1$, are slight variations of $f(x) = x^3$ and can be determined easily by plotting a few points. The graphs of $f(x) = \frac{1}{2}x^3$ and $f(x) = -x^3$ appear in Figure 9.5.

Two general patterns emerge from studying functions of the form $f(x) = x^n$. If n is odd and greater than 3, the graphs closely resemble Figure 9.4. The graph of $f(x) = x^5$ is shown in Figure 9.6. Note that the curve "flattens out" a little more around the origin than it does in the graph of $f(x) = x^3$; it increases and decreases more rapidly because of the larger exponent. If n is even and greater than 2, the graphs of $f(x) = x^n$ are not parabolas. They resemble the basic parabola, but they are flatter at the bottom and steeper on the sides. Figure 9.7 shows the graph of $f(x) = x^4$.

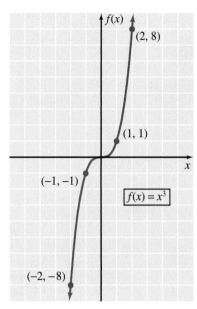

Figure 9.4

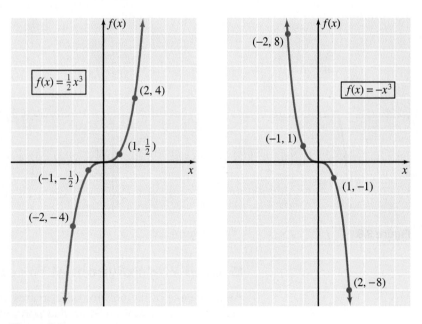

Figure 9.5

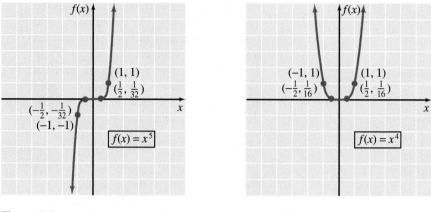

Figure 9.6

Figure 9.7

Graphs of functions of the form $f(x) = ax^n$, where n is an integer greater than 2 and $a \neq 1$, are variations of those shown in Figures 9.4 and 9.7. If n is odd, the curve is symmetric about the origin. If n is even, the graph is symmetric about the y axis.

Remember from our work in Chapter 8 that transformations of basic curves are easy to sketch. For example, in Figure 9.8, we translated the graph of $f(x) = x^3$ upward two units to produce the graph of $f(x) = x^3 + 2$. Figure 9.9 shows the graph of $f(x) = (x - 1)^5$, obtained by translating the graph of $f(x) = x^5$ one unit to the right. In Figure 9.10, we sketched the graph of $f(x) = -x^4$ as the x axis reflection of $f(x) = x^4$.

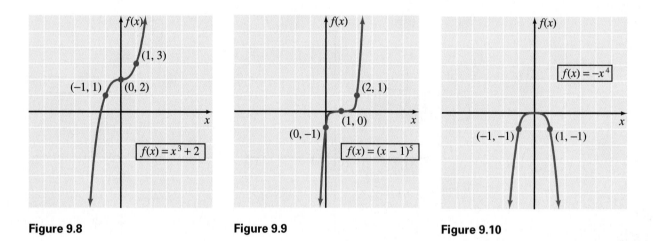

Figure 9.8

Figure 9.9

Figure 9.10

■ Graphing Polynomial Functions in Factored Form

As the degree of the polynomial increases, the graphs often become more complicated. We do know, however, that polynomial functions produce smooth continuous curves with a number of turning points, as illustrated in Figures 9.11 and 9.12. Some typical graphs of polynomial functions of odd degree are shown in

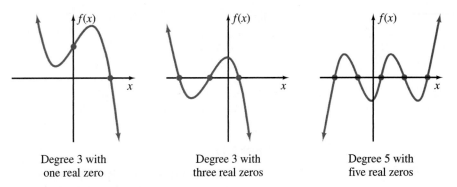

Degree 3 with
one real zero

Degree 3 with
three real zeros

Degree 5 with
five real zeros

Figure 9.11

Figure 9.11. As the graphs suggest, every polynomial function of odd degree has at least one *real zero*— that is, at least one real number c such that $f(c) = 0$. Geometrically, the zeros of the function are the x intercepts of the graph. Figure 9.12 illustrates some possible graphs of polynomial functions of even degree.

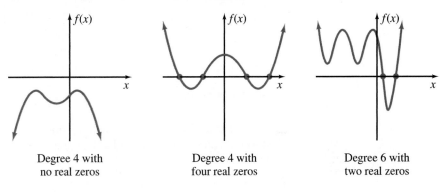

Degree 4 with
no real zeros

Degree 4 with
four real zeros

Degree 6 with
two real zeros

Figure 9.12

The **turning points** are the places where the function changes either from increasing to decreasing or from decreasing to increasing. Using calculus, we are able to verify that a polynomial function of degree n has at most $n - 1$ turning points. Now let's illustrate how we can use this information, along with some other techniques, to graph polynomial functions that are expressed in factored form.

EXAMPLE 1

Graph $f(x) = (x + 2)(x - 1)(x - 3)$.

Solution

First, let's find the x intercepts (zeros of the function) by setting each factor equal to zero and solving for x:

$$x + 2 = 0 \quad \text{or} \quad x - 1 = 0 \quad \text{or} \quad x - 3 = 0$$

$$x = -2 \qquad\qquad x = 1 \qquad\qquad x = 3$$

Thus the points $(-2, 0)$, $(1, 0)$, and $(3, 0)$ are on the graph. Second, the points associated with the x intercepts divide the x axis into four intervals as shown in Figure 9.13.

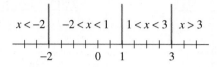

Figure 9.13

In each of these intervals, $f(x)$ is either always positive or always negative. That is to say, the graph is either above or below the x axis. Selecting a test value for x in each of the intervals will determine whether x is positive or negative. Any additional points that are easily obtained improve the accuracy of the graph. The table summarizes these results.

Interval	Test value	Sign of $f(x)$	Location of graph
$x < -2$	$f(-3) = -24$	Negative	Below x axis
$-2 < x < 1$	$f(0) = 6$	Positive	Above x axis
$1 < x < 3$	$f(2) = -4$	Negative	Below x axis
$x > 3$	$f(4) = 18$	Positive	Above x axis

Additional values: $f(-1) = 8$

Making use of the x intercepts and the information in the table, we can sketch the graph in Figure 9.14. The points $(-3, -24)$ and $(4, 18)$ are not shown,

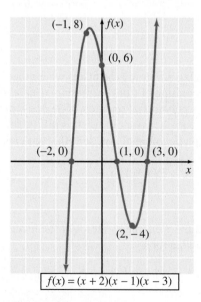

$$f(x) = (x + 2)(x - 1)(x - 3)$$

Figure 9.14

but they are used to indicate a rapid decrease and increase of the curve in those regions. ∎

Remark: In Figure 9.14, the approximate turning points of the graph are indicated at $(2, -4)$ and $(-1, 8)$. Keep in mind that these are only integral approximations. Using the ZOOM and TRACE features of a graphing calculator, we found that the points $(-0.8, 8.2)$ and $(2.1, -4.1)$ are approximations to the nearest tenth. Again, the tools of calculus are needed to find the exact turning points.

E X A M P L E   2

Graph $f(x) = -x^4 + 3x^3 - 2x^2$.

Solution

The polynomial can be factored as follows:

$$f(x) = -x^4 + 3x^3 - 2x^2$$
$$= -x^2(x^2 - 3x + 2)$$
$$= -x^2(x - 1)(x - 2)$$

Now we can find the x intercepts.

$$-x^2 = 0 \quad \text{or} \quad x - 1 = 0 \quad \text{or} \quad x - 2 = 0$$
$$x = 0 \quad \text{or} \quad x = 1 \quad \text{or} \quad x = 2$$

The points $(0, 0)$, $(1, 0)$, and $(2, 0)$ are on the graph and divide the x axis into four intervals as shown in Figure 9.15.

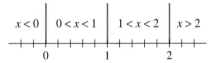

Figure 9.15

In the following table, we determine some points and summarize the sign behavior of $f(x)$.

Interval	Test value	Sign of $f(x)$	Location of graph
$x < 0$	$f(-1) = -6$	Negative	Below x axis
$0 < x < 1$	$f\left(\dfrac{1}{2}\right) = -\dfrac{3}{16}$	Negative	Below x axis
$1 < x < 2$	$f\left(\dfrac{3}{2}\right) = -\dfrac{9}{16}$	Positive	Above x axis
$x > 2$	$f(3) = -18$	Negative	Below x axis

Making use of the table and the x intercepts, we can draw the graph, as illustrated in Figure 9.16.

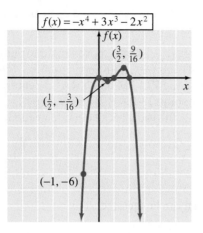

Figure 9.16   ∎

E X A M P L E   3

Graph $f(x) = x^3 + 3x^2 - 4$.

Solution

Use the rational root theorem, synthetic division, and the factor theorem to factor the given polynomial as follows.

$$f(x) = x^3 + 3x^2 - 4$$
$$= (x - 1)(x^2 + 4x + 4)$$
$$= (x - 1)(x + 2)^2$$

Now we can find the x intercepts.

$$x - 1 = 0 \quad \text{or} \quad (x + 2)^2 = 0$$
$$x = 1 \quad \text{or} \quad x = -2$$

Figure 9.17

The points $(-2, 0)$ and $(1, 0)$ are on the graph and divide the x axis into three intervals as shown in Figure 9.17.

In the following table, we determine some points and summarize the sign behavior of $f(x)$.

Interval	Test value	Sign of $f(x)$	Location of graph
$x < -2$	$f(-3) = -4$	Negative	Below x axis
$-2 < x < 1$	$f(0) = -4$	Negative	Below x axis
$x > 1$	$f(2) = 16$	Positive	Above x axis

Additional values: $f(-1) = -2$
$$f(-4) = -20$$

As a result of the table and the x intercepts, we can sketch the graph as shown in Figure 9.18.

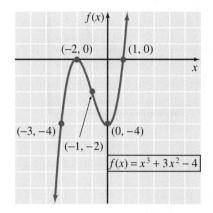

Figure 9.18   ■

Finally, let's use a graphical approach to solve a problem involving a polynomial function.

E X A M P L E   4

Suppose that we have a rectangular piece of cardboard that measures 20 inches by 14 inches. From each corner, a square piece is cut out, and then the flaps are turned up to form an open box (see Figure 9.19). Determine the length of a side of the square pieces to be cut out so that the volume of the box is as large as possible.

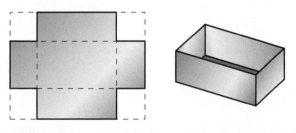

Figure 9.19

Solution

Let x represent the length of a side of the squares to be cut from each corner. Then $20 - 2x$ represents the length of the open box, and $14 - 2x$ represents the width. The volume of a rectangular box is given by the formula $V = lwh$, so the volume of this box can be represented by $V = x(20 - 2x)(14 - 2x)$. Now let $y = V$, and graph the function $y = x(20 - 2x)(14 - 2x)$ as shown in Figure 9.20. For this problem, we are interested only in the part of the graph between $x = 0$ and $x = 7$ because the length of a side of the squares has to be less than 7 inches for a box to be

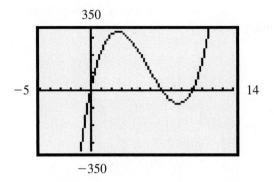

Figure 9.20

formed. Figure 9.21 gives us a view of that part of the graph. Now we can use the ZOOM and TRACE features to determine that when x equals approximately 2.7, the value of y is a maximum of approximately 339.0. Thus square pieces of length approximately 2.7 inches on a side should be cut from each corner of the rectangular piece of cardboard. The open box formed will have a volume of approximately 339.0 cubic inches.

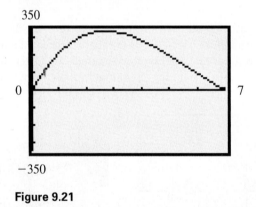

Figure 9.21   ■

Problem Set 9.4

For Problems 1–22, graph each of the polynomial functions.

1. $f(x) = -(x - 3)^3$

2. $f(x) = (x - 2)^3 + 1$

3. $f(x) = (x + 1)^3$

4. $f(x) = x^3 - 3$

5. $f(x) = (x + 3)^4$

6. $f(x) = x^4 - 2$

7. $f(x) = -(x - 2)^4$

8. $f(x) = (x - 1)^5 + 2$

9. $f(x) = (x + 1)^4 + 3$

10. $f(x) = -x^5$

11. $f(x) = (x - 2)(x + 1)(x + 3)$

12. $f(x) = (x - 1)(x + 1)(x - 3)$

13. $f(x) = x(x + 2)(2 - x)$

14. $f(x) = (x + 4)(x + 1)(1 - x)$

15. $f(x) = -x^2(x - 1)(x + 1)$

16. $f(x) = -x(x + 3)(x - 2)$

17. $f(x) = (2x - 1)(x - 2)(x - 3)$

18. $f(x) = x(x - 2)^2(x - 1)$

19. $f(x) = (x - 2)(x - 1)(x + 1)(x + 2)$

20. $f(x) = (x - 1)^2(x + 2)$

21. $f(x) = x(x - 2)^2(x + 1)$

22. $f(x) = (x + 1)^2(x - 1)^2$

For Problems 23–34, graph each polynomial function by first factoring the given polynomial. You may need to use some factoring techniques from Chapter 3 as well as the rational root theorem and the factor theorem.

23. $f(x) = -x^3 - x^2 + 6x$

24. $f(x) = x^3 + x^2 - 2x$

25. $f(x) = x^4 - 5x^3 + 6x^2$

26. $f(x) = -x^4 - 3x^3 - 2x^2$

27. $f(x) = x^3 + 2x^2 - x - 2$

28. $f(x) = x^3 - x^2 - 4x + 4$

29. $f(x) = x^3 - 8x^2 + 19x - 12$

30. $f(x) = x^3 + 6x^2 + 11x + 6$

31. $f(x) = 2x^3 - 3x^2 - 3x + 2$

32. $f(x) = x^3 + 2x^2 - x - 2$

33. $f(x) = x^4 - 5x^2 + 4$

34. $f(x) = -x^4 + 5x^2 - 4$

For Problems 35–42, (a) find the y intercepts, (b) find the x intercepts, and (c) find the intervals of x where $f(x) > 0$ and those where $f(x) < 0$. *Do not* sketch the graphs.

35. $f(x) = (x + 3)(x - 6)(8 - x)$

36. $f(x) = (x - 5)(x + 4)(x - 3)$

37. $f(x) = (x + 3)^4(x - 1)^3$

38. $f(x) = (x - 4)^2(x + 3)^3$

39. $f(x) = x(x - 6)^2(x + 4)$

40. $f(x) = (x + 2)^2(x - 1)^3(x - 2)$

41. $f(x) = x^2(2 - x)(x + 3)$

42. $f(x) = (x + 2)^5(x - 4)^2$

■ ■ ■ THOUGHTS INTO WORDS

43. How would you defend the statement that the equation $2x^4 + 3x^3 + x^2 + 5 = 0$ has no positive solutions? Does it have any negative solutions? Defend your answer.

44. How do you know by inspection that the graph of $f(x) = (x + 1)^2(x - 2)^2$ in Figure 9.22 is incorrect?

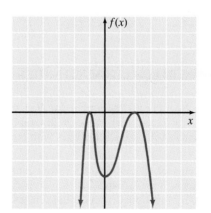

Figure 9.22

■ ■ ■ FURTHER INVESTIGATIONS

45. A polynomial function with real coefficients is continuous everywhere; that is, its graph has no holes or breaks. This is the basis for the following property: If $f(x)$ is a polynomial with real coefficients, and if $f(a)$ and $f(b)$ are of opposite sign, then there is at least one real zero between a and b. This property, along with our knowledge of polynomial functions, provides the basis for locating and approximating irrational solutions of a polynomial equation.

Consider the equation $x^3 + 2x - 4 = 0$. Applying Descartes' rule of signs, we can determine that this equation has one positive real solution and two nonreal complex solutions. (You may want to confirm this!) The rational root theorem indicates that the only possible rational solutions are 1, 2, and 4. Using a little more compact format for synthetic division, we obtain the following results when testing for 1 and 2 as possible solutions:

	1	0	2	−4
1	1	1	3	−1
2	1	2	6	8

Because $f(1) = -1$ and $f(2) = 8$, there must be an irrational solution between 1 and 2. Furthermore, -1 is closer to 0 than is 8, so our guess is that the solution is closer to 1 than to 2. Let's start looking at 1.0, 1.1, 1.2, and so on, until we can place the solution between two numbers.

	1	0	2	−4
1.0	1	1	3	−1
1.1	1	1.1	3.21	−0.469
1.2	1	1.2	3.44	0.128

A calculator is very helpful at this time.

Because $f(1.1) = -0.469$ and $f(1.2) = 0.128$, the irrational solution must be between 1.1 and 1.2. Furthermore, because 0.128 is closer to 0 than is -0.469, our guess is that the solution is closer to 1.2 than to 1.1. Let's start looking at 1.15, 1.16, and so on.

	1	0	2	−4
1.15	1	1.15	3.3225	−0.179
1.16	1	1.16	3.3456	−0.119
1.17	1	1.17	3.3689	−0.058
1.18	1	1.18	3.3924	0.003

Because $f(1.17) = -0.058$ and $f(1.18) = 0.003$, the irrational solution must be between 1.17 and 1.18. Therefore we can use 1.2 as a rational approximation to the nearest tenth.

For each of the following equations, (a) verify that the equation has exactly one irrational solution, and (b) find an approximation, to the nearest tenth, of that solution.
(a) $x^3 + x - 6 = 0$
(b) $x^3 - 6x - 6 = 0$
(c) $x^3 - 27x - 60 = 0$
(d) $x^3 - x^2 - x - 1 = 0$
(e) $x^3 - 2x - 10 = 0$
(f) $x^3 - 5x^2 - 1 = 0$

■ GRAPHING CALCULATOR ACTIVITIES

46. Graph $f(x) = x^3$. Now predict the graphs for $f(x) = x^3 + 2$, $f(x) = -x^3 + 2$, and $f(x) = -x^3 - 2$. Graph these three functions on the same set of axes with the graph of $f(x) = x^3$.

47. Draw a rough sketch of the graphs of the functions $f(x) = x^3 - x^2$, $f(x) = -x^3 + x^2$, and $f(x) = -x^3 - x^2$. Now graph these three functions to check your sketches.

48. Graph $f(x) = x^4 + x^3 + x^2$. What should the graphs of $f(x) = x^4 - x^3 + x^2$ and $f(x) = -x^4 - x^3 - x^2$ look like? Graph them to see if you were right.

49. How should the graphs of $f(x) = x^3$, $f(x) = x^5$, and $f(x) = x^7$ compare? Graph these three functions on the same set of axes.

50. How should the graphs of $f(x) = x^2$, $f(x) = x^4$, and $f(x) = x^6$ compare? Graph these three functions on the same set of axes.

51. For each of the following functions, find the x intercepts, and find the intervals of x where $f(x) > 0$ and those where $f(x) < 0$.

(a) $f(x) = x^3 - 3x^2 - 6x + 8$
(b) $f(x) = x^3 - 8x^2 - x + 8$
(c) $f(x) = x^3 - 7x^2 + 16x - 12$
(d) $f(x) = x^3 - 19x^2 + 90x - 72$
(e) $f(x) = x^4 + 3x^3 - 3x^2 - 11x - 6$
(f) $f(x) = x^4 + 12x^2 - 64$

52. Find the coordinates of the turning points of each of the following graphs. Express x and y values to the nearest integer.
(a) $f(x) = 2x^3 - 3x^2 - 12x + 40$
(b) $f(x) = 2x^3 - 33x^2 + 60x + 1050$
(c) $f(x) = -2x^3 - 9x^2 + 24x + 100$
(d) $f(x) = x^4 - 4x^3 - 2x^2 + 12x + 3$
(e) $f(x) = x^3 - 30x^2 + 288x - 900$
(f) $f(x) = x^5 - 2x^4 - 3x^3 - 2x^2 + x - 1$

53. For each of the following functions, find the x intercepts and find the turning points. Express your answers to the nearest tenth.
(a) $f(x) = x^3 + 2x^2 - 3x + 4$
(b) $f(x) = 42x^3 - x^2 - 246x - 35$
(c) $f(x) = x^4 - 4x^2 - 4$

54. A rectangular piece of cardboard is 13 inches long and 9 inches wide. From each corner, a square piece is cut out, and then the flaps are turned up to form an open box. Determine the length of a side of the square pieces so that the volume of the box is as large as possible.

55. A company determines that its weekly profit from manufacturing and selling x units of a certain item is given by $P(x) = -x^3 + 3x^2 + 2880x - 500$. What weekly production rate will maximize the profit?

9.5 Graphing Rational Functions

Let's begin this section by using a graphing calculator to graph the function $f(x) = \dfrac{x^2}{x^2 - x - 2}$ twice using different boundaries, as indicated in Figures 9.23 and 9.24. It should be evident from the two figures that we really cannot tell what the graph of the function looks like. This happens frequently in graphing rational functions with a graphing calculator. Thus we need to do a careful analysis of rational functions, emphasizing the use of hand-drawn graphs. (By the way, if you are interested in seeing the complete graph of this function, turn to the first example of the next section.)

A function of the form

$$f(x) = \frac{p(x)}{q(x)}, \quad q(x) \neq 0$$

where $p(x)$ and $q(x)$ are polynomials, is called a **rational function**.

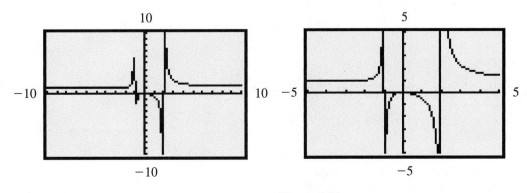

Figure 9.23   **Figure 9.24**

The following are examples of rational functions:

$$f(x) = \frac{2}{x-1} \qquad f(x) = \frac{x}{x-2}$$

$$f(x) = \frac{x^2}{x^2-x-6} \qquad f(x) = \frac{x^3-8}{x+4}$$

In each of these examples, the domain of the rational function is the set of all real numbers except those that make the denominator zero. For example, the domain of $f(x) = \dfrac{2}{x-1}$ is the set of all real numbers except 1. As we will soon see, these exclusions from the domain are important numbers from a graphing standpoint; they represent breaks in an otherwise continuous curve.

Let's set the stage for graphing rational functions by considering in detail the function $f(x) = \dfrac{1}{x}$. First, note that at $x = 0$ the function is undefined. Second, let's consider a rather extensive table of values to find some number trends and to build a basis for defining the concept of an asymptote.

x	$f(x) = \dfrac{1}{x}$	
1	1	
2	0.5	These values indicate that the value of $f(x)$ is positive and approaches zero from above as x gets larger and larger.
10	0.1	
100	0.01	
1000	0.001	
0.5	2	
0.1	10	These values indicate that $f(x)$ is positive and is getting larger and larger as x approaches zero from the right.
0.01	100	
0.001	1000	
0.0001	10,000	
−0.5	−2	
−0.1	−10	These values indicate that $f(x)$ is negative and is getting smaller and smaller as x approaches zero from the left.
−0.01	−100	
−0.001	−1000	
−0.0001	−10,000	
−1	−1	
−2	−0.5	These values indicate that $f(x)$ is negative and approaches zero from below as x gets smaller and smaller without bound.
−10	−0.1	
−100	−0.01	
−1000	−0.001	

Figure 9.25 shows a sketch of $f(x) = \dfrac{1}{x}$, which is drawn using a few points from this table and the patterns discussed. Note that the graph approaches but does not touch either axis. We say that the y axis [or the $f(x)$ axis] is a **vertical asymptote** and that the x axis is a **horizontal asymptote**.

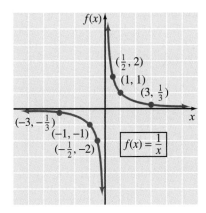

Figure 9.25

Remark: We know that the equation $f(x) = \dfrac{1}{x}$ exhibits origin symmetry because $f(-x) = -f(x)$. Thus the graph in Figure 9.25 could have been drawn by first determining the part of the curve in the first quadrant and then reflecting that curve through the origin.

Now let's define the concepts of vertical and horizontal asymptotes.

Vertical Asymptote

A line $x = a$ is a vertical asymptote for the graph of a function f if:

1. $f(x)$ either increases or decreases without bound as x approaches a from the left, as in Figure 9.26,

or

Figure 9.26

2. $f(x)$ either increases or decreases without bound as x approaches a from the right, as in Figure 9.27.

Figure 9.27

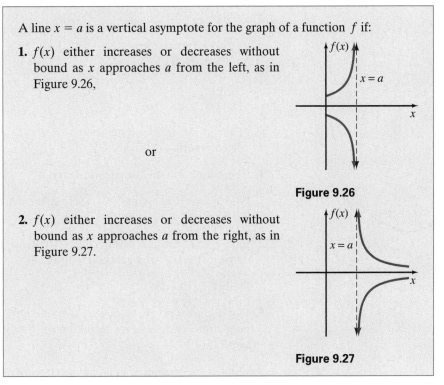

Horizontal Asymptote

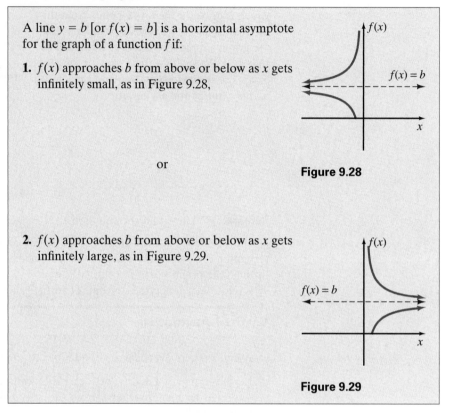

A line $y = b$ [or $f(x) = b$] is a horizontal asymptote for the graph of a function f if:

1. $f(x)$ approaches b from above or below as x gets infinitely small, as in Figure 9.28,

or

Figure 9.28

2. $f(x)$ approaches b from above or below as x gets infinitely large, as in Figure 9.29.

Figure 9.29

Following are some suggestions for graphing rational functions of the type we are considering in this section.

1. Check for y axis and origin symmetry.

2. Find any vertical asymptote by setting the denominator equal to zero and solving for x.

3. Find any horizontal asymptote by studying the behavior of $f(x)$ as x gets infinitely large or as x gets infinitely small.

4. Study the behavior of the graph when it is close to the asymptotes.

5. Plot as many points as necessary to determine the shape of the graph. The number needed may be affected by whether or not the graph has any kind of symmetry.

Keep these suggestions in mind as you study the following examples.

EXAMPLE 1   Graph $f(x) = \dfrac{-2}{x - 1}$.

Solution

Because $x = 1$ makes the denominator zero, the line $x = 1$ is a vertical asymptote. We have indicated this with a dashed line in Figure 9.30. Now let's look for a horizontal asymptote by checking some large and some small values of x.

x	$f(x)$
10	$-\dfrac{2}{9}$
100	$-\dfrac{2}{99}$
1000	$-\dfrac{2}{999}$

This portion of the table shows that as x gets very large, the value of $f(x)$ approaches zero from below.

x	$f(x)$
-10	$\dfrac{2}{11}$
-100	$\dfrac{2}{101}$
-1000	$\dfrac{2}{1001}$

This portion shows that as x gets very small, the value of $f(x)$ approaches zero from above.

Therefore the x axis is a horizontal asymptote.

Finally, let's check the behavior of the graph near the vertical asymptote.

x	$f(x)$
2	-2
1.5	-4
1.1	-20
1.01	-200
1.001	-2000
0	2
0.5	4
0.9	20
0.99	200
0.999	2000

As x approaches 1 from the right side, the value of $f(x)$ gets smaller and smaller.

As x approaches 1 from the left side, the value of $f(x)$ gets larger and larger.

The graph of $f(x) = \dfrac{-2}{x - 1}$ is shown in Figure 9.30.

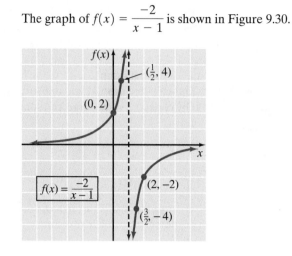

Figure 9.30

E X A M P L E   2

Graph $f(x) = \dfrac{x}{x + 2}$.

Solution

Because $x = -2$ makes the denominator zero, the line $x = -2$ is a vertical asymptote. To study the behavior of $f(x)$ as x gets very large or very small, let's change the form of the rational expression by dividing numerator and denominator by x:

$$f(x) = \frac{x}{x + 2} = \frac{\dfrac{x}{x}}{\dfrac{x + 2}{x}} = \frac{1}{\dfrac{x}{x} + \dfrac{2}{x}} = \frac{1}{1 + \dfrac{2}{x}}$$

Now we can see that as x gets larger and larger, the value of $f(x)$ approaches 1 from below; as x gets smaller and smaller, the value of $f(x)$ approaches 1 from above. (Perhaps you should check these claims by plugging in some values for x.) Thus the line $f(x) = 1$ is a horizontal asymptote. Drawing the asymptotes (dashed lines) and plotting a few points, we complete the graph in Figure 9.31.

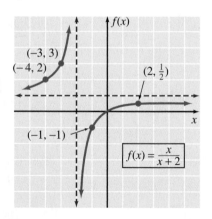

Figure 9.31

In the next two examples, pay special attention to the role of symmetry. It will allow us to direct our efforts toward quadrants I and IV and then to reflect those portions of the curve across the vertical axis to complete the graph.

EXAMPLE 3

Graph $f(x) = \dfrac{2x^2}{x^2 + 4}$.

Solution

First, note that $f(-x) = f(x)$; therefore this graph is symmetric with respect to the vertical axis. Second, the denominator $x^2 + 4$ cannot equal zero for any real number value of x; thus there is no vertical asymptote. Third, dividing both numerator and denominator of the rational expression by x^2 produces

$$f(x) = \frac{2x^2}{x^2 + 4} = \frac{\dfrac{2x^2}{x^2}}{\dfrac{x^2 + 4}{x^2}} = \frac{2}{\dfrac{x^2}{x^2} + \dfrac{4}{x^2}}$$

$$= \frac{2}{1 + \dfrac{4}{x^2}}$$

Now we can see that as x gets larger and larger, the value of $f(x)$ approaches 2 from below. Therefore the line $f(x) = 2$ is a horizontal asymptote. We can plot a few points using positive values for x, sketch this part of the curve, and then reflect across the $f(x)$ axis to obtain the complete graph, as shown in Figure 9.32.

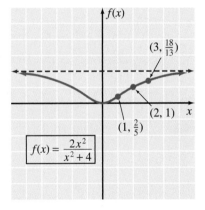

$\left(3, \frac{18}{13}\right)$

$(2, 1)$

$\left(1, \frac{2}{5}\right)$

$f(x) = \dfrac{2x^2}{x^2 + 4}$

Figure 9.32 ■

EXAMPLE 4

Graph $f(x) = \dfrac{3}{x^2 - 4}$.

Solution

First, note that $f(-x) = f(x)$; therefore this graph is symmetric about the y axis. Thus, by setting the denominator equal to zero and solving for x, we obtain

$$x^2 - 4 = 0$$

$$x^2 = 4$$

$$x = \pm 2$$

The lines $x = 2$ and $x = -2$ are vertical asymptotes. Next, we can see that $\dfrac{3}{x^2 - 4}$ approaches zero from above as x gets larger and larger. Finally, we can plot a few points using positive values for x (other than 2), sketch this part of the curve, and then reflect it across the $f(x)$ axis to obtain the complete graph shown in Figure 9.33.

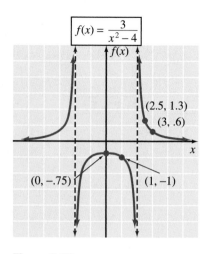

Figure 9.33 ∎

Now suppose that we are going to use a graphing utility to obtain a graph of the function $f(x) = \dfrac{4x^2}{x^4 - 16}$. Before we enter this function into a graphing utility, let's analyze what we know about the graph.

1. Because $f(0) = 0$, the origin is a point on the graph.

2. Because $f(-x) = f(x)$, the graph is symmetric with respect to the y axis.

3. By setting the denominator equal to zero and solving for x, we can determine the vertical asymptotes.

$$x^4 - 16 = 0$$
$$(x^2 + 4)(x^2 - 4) = 0$$
$$x^2 + 4 = 0 \quad \text{or} \quad x^2 - 4 = 0$$
$$x^2 = -4 \qquad\qquad x^2 = 4$$
$$x = \pm 2i \qquad\qquad x = \pm 2$$

Remember that we are working with ordered pairs of real numbers. Thus the lines $x = -2$ and $x = 2$ are vertical asymptotes.

4. Divide both the numerator and the denominator of the rational expression by x^4 to produce

$$\frac{4x^2}{x^4 - 16} = \frac{\dfrac{4x^2}{x^4}}{\dfrac{x^4 - 16}{x^4}} = \frac{\dfrac{4}{x^2}}{1 - \dfrac{16}{x^4}}$$

From the last expression, we see that as $|x|$ gets larger and larger, the value of $f(x)$ approaches zero from above. Therefore the x axis is a horizontal asymptote.

Now let's enter the function into a graphing calculator and set the boundaries so that we show the behavior of the function close to the asymptotes. Note that the graph shown in Figure 9.34 is consistent with all of the information that we determined before using the graphing calculator. In other words, our knowledge of graphing techniques enhances our use of a graphing utility.

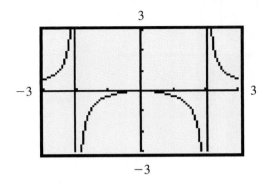

Figure 9.34

Back in Section 2.4 we solved problems of the following type: How much pure alcohol should be added to 6 liters of a 40% alcohol solution to raise it to a 60% alcohol solution? The answer of 3 liters can be found by solving the following equation, where x represents the amount of pure alcohol to be added:

Pure alcohol to start with	+	Pure alcohol added	=	Pure alcohol in final solution
↓		↓		↓
$0.40(6)$	+	x	=	$0.60(6 + x)$

Now let's consider this problem in a more general setting by writing a function where x represents the amount of pure alcohol to be added and y represents the concentration of pure alcohol in the final solution.

$$0.40(6) + x = y(6 + x)$$

$$2.4 + x = y(6 + x)$$

$$\frac{2.4 + x}{6 + x} = y$$

We can graph the rational function $y = \dfrac{2.4 + x}{6 + x}$ as shown in Figure 9.35. For this particular problem, x is nonnegative, so we are interested only in the part of the graph that is in the first quadrant. We change the boundaries of the viewing rectangle so that $0 \le x \le 15$ and $0 \le y \le 2$ to obtain Figure 9.36. Now we are ready to answer questions about this situation.

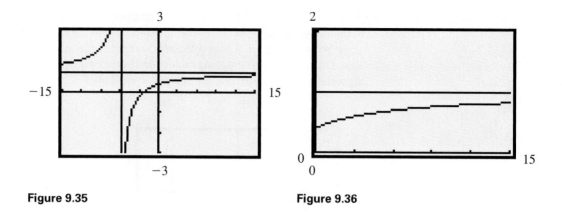

Figure 9.35   **Figure 9.36**

1. How much pure alcohol needs to be added to raise the 40% solution to a 60% solution? [*Hint:* We are looking for the value of x when y is 0.60. (*Answer:* Using the TRACE feature of the graphing utility, we find that when $y = 0.60$, $x = 3$. Therefore 3 liters of pure alcohol need to be added.)]

2. How much pure alcohol needs to be added to raise the 40% solution to a 70% solution? (*Answer:* Using the TRACE feature of the graphing utility, we find that when $y = 0.70$, $x = 6$. Therefore 6 liters of pure alcohol need to be added.)

3. What concentration in percent of alcohol do we obtain if we add 9 liters of pure alcohol to 6 liters of a 40% solution? (*Answer:* Using the TRACE feature of the graphing utility, we find that when $x = 9$, $y = 0.76$. Therefore adding 9 liters of pure alcohol will give us a 76% alcohol solution.)

Problem Set 9.5

Graph each of the following rational functions:

1. $f(x) = \dfrac{1}{x^2}$

2. $f(x) = \dfrac{-1}{x}$

3. $f(x) = \dfrac{-1}{x - 3}$

4. $f(x) = \dfrac{3}{x + 1}$

5. $f(x) = \dfrac{-3}{(x + 2)^2}$

6. $f(x) = \dfrac{2}{(x - 1)^2}$

7. $f(x) = \dfrac{2x}{x - 1}$

8. $f(x) = \dfrac{x}{x - 3}$

9. $f(x) = \dfrac{-x}{x + 1}$

10. $f(x) = \dfrac{-3x}{x + 2}$

11. $f(x) = \dfrac{-2}{x^2 - 4}$

12. $f(x) = \dfrac{1}{x^2 - 1}$

13. $f(x) = \dfrac{3}{(x + 2)(x - 4)}$

14. $f(x) = \dfrac{-2}{(x + 1)(x - 2)}$

15. $f(x) = \dfrac{-1}{x^2 + x - 6}$

16. $f(x) = \dfrac{2}{x^2 + x - 2}$

17. $f(x) = \dfrac{2x - 1}{x}$

18. $f(x) = \dfrac{x + 2}{x}$

19. $f(x) = \dfrac{4x^2}{x^2 + 1}$

20. $f(x) = \dfrac{4}{x^2 + 2}$

21. $f(x) = \dfrac{x^2 - 4}{x^2}$

22. $f(x) = \dfrac{2x^4}{x^4 + 1}$

■ ■ ■ THOUGHTS INTO WORDS

23. How would you explain the concept of an asymptote to an elementary algebra student?

24. Give a step-by-step description of how you would go about graphing $f(x) = \dfrac{-2}{x^2 - 9}$.

■ ■ ■ FURTHER INVESTIGATIONS

25. The rational function $f(x) = \dfrac{(x - 2)(x + 3)}{x - 2}$ has a domain of all real numbers except 2 and can be simplified to $f(x) = x + 3$. Thus its graph is a straight line with a hole at $(2, 5)$. Graph each of the following functions.

(a) $f(x) = \dfrac{(x + 4)(x - 1)}{x + 4}$

(b) $f(x) = \dfrac{x^2 - 5x + 6}{x - 2}$

(c) $f(x) = \dfrac{x - 1}{x^2 - 1}$

(d) $f(x) = \dfrac{x + 2}{x^2 + 6x + 8}$

26. Graph the function $f(x) = x + 2 + \dfrac{3}{x - 2}$. It may be necessary to plot a rather large number of points. Also, defend the statement that $f(x) = x + 2$ is an **oblique asymptote**.

▲ GRAPHING CALCULATOR ACTIVITIES

27. Use a graphing calculator to check your graphs for Problem 25. What feature of the graphs does not show up on the calculator?

28. Each of the following graphs is a transformation of $f(x) = \dfrac{1}{x}$. First predict the general shape and location of the graph, and then check your prediction with a graphing calculator.

(a) $f(x) = \dfrac{1}{x} - 2$

(b) $f(x) = \dfrac{1}{x + 3}$

(c) $f(x) = -\dfrac{1}{x}$

(d) $f(x) = \dfrac{1}{x - 2} + 3$

(e) $f(x) = \dfrac{2x + 1}{x}$

29. Graph $f(x) = \dfrac{1}{x^2}$. How should the graph of $f(x) = \dfrac{1}{(x - 4)^2}$, $f(x) = \dfrac{1 + 3x^2}{x^2}$, and $f(x) = \dfrac{1}{x^2}$ com-

pare to the graph of $f(x) = \dfrac{1}{x^2}$? Graph the three functions on the same set of axes with the graph of $f(x) = \dfrac{1}{x^2}$.

30. Graph $f(x) = \dfrac{1}{x^3}$. How should the graphs of $f(x) = \dfrac{2x^3 + 1}{x^3}$, $f(x) = \dfrac{1}{(x + 2)^3}$, and $f(x) = \dfrac{-1}{x^3}$ com-

pare to the graph of $f(x) = \dfrac{1}{x^3}$? Graph the three functions on the same set of axes with the graph of $f(x) = \dfrac{1}{x^3}$.

31. Use a graphing calculator to check your graphs for Problems 19–22.

32. Suppose that x ounces of pure acid have been added to 14 ounces of a 15% acid solution.

(a) Set up the rational expression that represents the concentration of pure acid in the final solution.

(b) Graph the rational function that displays the concentration.

(c) How many ounces of pure acid need to be added to the 14 ounces of a 15% solution to raise it to a 40.5% solution? Check your answer.

(d) How many ounces of pure acid need to be added to the 14 ounces of a 15% solution to raise it to a 50% solution? Check your answer.

(e) What concentration of acid do we obtain if we add 12 ounces of pure acid to the 14 ounces of a 15% solution? Check your answer.

33. Solve the following problem both algebraically and graphically: One solution contains 50% alcohol, and another solution contains 80% alcohol. How many liters of each solution should be mixed to produce 10.5 liters of a 70% alcohol solution? Check your answer.

34. Graph each of the following functions. Be sure that you get a complete graph for each one. Sketch each graph on a sheet of paper, and keep them all handy as you study the next section.

(a) $f(x) = \dfrac{x^2}{x^2 - x - 2}$ **(b)** $f(x) = \dfrac{x}{x^2 - 4}$

(c) $f(x) = \dfrac{3x}{x^2 + 1}$ **(d)** $f(x) = \dfrac{x^2 - 1}{x - 2}$

9.6 More on Graphing Rational Functions

The rational functions that we studied in the previous section "behaved rather well." In fact, once we established the vertical and horizontal asymptotes, a little bit of point plotting usually determined the graph fairly easily. Such is not always the case with rational functions. In this section, we want to investigate some rational functions that behave a little differently.

Vertical asymptotes occur at values of x where the denominator is zero, so no points of a graph can be on a vertical asymptote. However, recall that horizontal asymptotes are created by the behavior of $f(x)$ as x gets infinitely large or infinitely small. This does not restrict the possibility that for some values of x, points of the graph will be on the horizontal asymptote. Let's consider some examples.

EXAMPLE 1

Graph $f(x) = \dfrac{x^2}{x^2 - x - 2}$.

Solution

First, let's identify the vertical asymptotes by setting the denominator equal to zero and solving for x:

$$x^2 - x - 2 = 0$$

$$(x - 2)(x + 1) = 0$$

$$x - 2 = 0 \quad \text{or} \quad x + 1 = 0$$

$$x = 2 \qquad\qquad x = -1$$

Thus the lines $x = 2$ and $x = -1$ are vertical asymptotes. Next, we can divide both the numerator and the denominator of the rational expression by x^2.

$$f(x) = \frac{x^2}{x^2 - x - 2} = \frac{\dfrac{x^2}{x^2}}{\dfrac{x^2 - x - 2}{x^2}} = \frac{1}{1 - \dfrac{1}{x} - \dfrac{2}{x^2}}$$

Now we can see that as x gets larger and larger, the value of $f(x)$ approaches 1 from above. Thus the line $f(x) = 1$ is a horizontal asymptote. To determine whether any points of the graph are *on* the horizontal asymptote, we can see whether the equation

$$\frac{x^2}{x^2 - x - 2} = 1$$

has any solutions.

$$\frac{x^2}{x^2 - x - 2} = 1$$

$$x^2 = x^2 - x - 2$$

$$0 = -x - 2$$

$$x = -2$$

Therefore the point $(-2, 1)$ is on the graph. Now, by drawing the asymptotes, plotting a few points [including $(-2, 1)$], and studying the behavior of the function close to the asymptotes, we can sketch the curve shown in Figure 9.37.

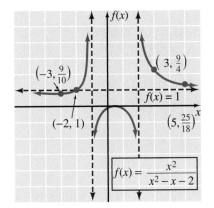

Figure 9.37

E X A M P L E   2   Graph $f(x) = \dfrac{x}{x^2 - 4}$.

Solution

First, note that $f(-x) = -f(x)$; therefore this graph is symmetric with respect to the origin. Second, let's identify the vertical asymptotes:

$$x^2 - 4 = 0$$

$$x^2 = 4$$

$$x = \pm 2$$

Thus the lines $x = -2$ and $x = 2$ are vertical asymptotes. Next, by dividing the numerator and the denominator of the rational expression by x^2, we obtain

$$f(x) = \frac{x}{x^2 - 4} = \frac{\dfrac{x}{x^2}}{\dfrac{x^2 - 4}{x^2}} = \frac{\dfrac{1}{x}}{1 - \dfrac{4}{x^2}}$$

From this form, we can see that as x gets larger and larger, the value of $f(x)$ approaches zero from above. Therefore the x axis is a horizontal asymptote. Because $f(0) = 0$, we know that the origin is a point of the graph. Finally, by concentrating our point plotting on positive values of x, we can sketch the portion of the curve to the right of the vertical axis, and then use the fact that the graph is symmetric with respect to the origin to complete the graph. Figure 9.38 shows the completed graph.

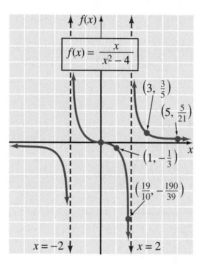

Figure 9.38 ∎

E X A M P L E   3   Graph $f(x) = \dfrac{3x}{x^2 + 1}$.

Solution

First, observe that $f(-x) = -f(x)$; therefore this graph is symmetric with respect to the origin. Second, because $x^2 + 1$ is a positive number for all real number values of x, there are no vertical asymptotes for this graph. Next, by dividing the numerator and the denominator of the rational expression by x^2, we obtain

$$f(x) = \frac{3x}{x^2 + 1} = \frac{\dfrac{3x}{x^2}}{\dfrac{x^2 + 1}{x^2}} = \frac{\dfrac{3}{x}}{1 + \dfrac{1}{x^2}}$$

From this form, we see that as x gets larger and larger, the value of $f(x)$ approaches zero from above. Thus the x axis is a horizontal asymptote. Because $f(0) = 0$, the origin is a point of the graph. Finally, by concentrating our point plotting on positive values of x, we can sketch the portion of the curve to the right of the vertical axis, and then use origin symmetry to complete the graph, as shown in Figure 9.39.

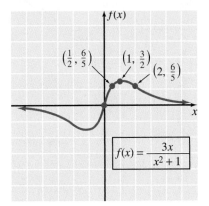

Figure 9.39

■ Oblique Asymptotes

Thus far we have restricted our study of rational functions to those where the degree of the numerator is less than or equal to the degree of the denominator. As our final examples of graphing rational functions, we will consider functions where the degree of the numerator is one greater than the degree of the denominator.

E X A M P L E   4   Graph $f(x) = \dfrac{x^2 - 1}{x - 2}$.

Solution

First, let's observe that $x = 2$ is a vertical asymptote. Second, because the degree of the numerator is greater than the degree of the denominator, we can change the form of the rational expression by division. We use synthetic division.

$$
\begin{array}{r}
2\overline{)\,1 \quad 0 \quad -1} \\
\underline{\quad 2 \quad 4} \\
1 \quad 2 \quad 3
\end{array}
$$

Therefore the original function can be rewritten as

$$f(x) = \frac{x^2 - 1}{x - 2} = x + 2 + \frac{3}{x - 2}$$

Now, for very large values of $|x|$, the fraction $\dfrac{3}{x-2}$ is close to zero. Therefore, as $|x|$ gets larger and larger, the graph of $f(x) = x + 2 + \dfrac{3}{x-2}$ gets closer and closer to the line $f(x) = x + 2$. We call this line an **oblique asymptote** and indicate it with a dashed line in Figure 9.40. Finally, because this is a new situation, it may be necessary to plot a large number of points on both sides of the vertical asymptote, so let's make an extensive table of values. The graph of the function is shown in Figure 9.40.

x	$f(x) = \dfrac{x^2 - 1}{x - 2}$	
2.1	34.1	
2.5	10.5	
3	8	
4	7.5	These values indicate the behavior of $f(x)$ to the right of the vertical asymptote $x = 2$.
5	8	
6	8.75	
10	12.375	
1.9	−26.1	
1.5	−2.5	
1	0	
0	0.5	These values indicate the behavior of $f(x)$ to the left of the vertical asymptote $x = 2$.
−1	0	
−3	−1.6	
−5	−3.4	
−10	−8.25	

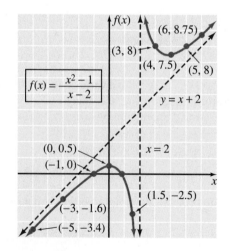

Figure 9.40

If the degree of the numerator of a rational function is *exactly one more* than the degree of its denominator, then the graph of the function has an oblique asymptote. [If the graph is a line, as is the case with $f(x) = \dfrac{(x-2)(x+1)}{x-2}$, then we consider it to be its own asymptote.] As in Example 4, we find the equation of the oblique asymptote by changing the form of the function using long division. Let's consider another example.

EXAMPLE   5

Graph $f(x) = \dfrac{x^2 - x - 2}{x - 1}$.

Solution

From the given form of the function, we see that $x = 1$ is a vertical asymptote. Then, by factoring the numerator, we can change the form to

$$f(x) = \frac{(x-2)(x+1)}{x-1}$$

which indicates x intercepts of 2 and -1. Then, by long division, we can change the original form of the function to

$$f(x) = x - \frac{2}{x-1}$$

which indicates an oblique asymptote $f(x) = x$. Finally, by plotting a few additional points, we can determine the graph as shown in Figure 9.41.

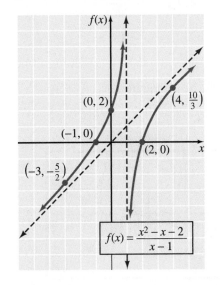

Figure 9.41

Finally, let's combine our knowledge of rational functions with the use of a graphing utility to obtain the graph of a fairly complex rational function.

E X A M P L E   6

Graph the rational function $f(x) = \dfrac{x^3 - 2x^2 - x - 1}{x^2 - 36}$.

Solution

Before entering this function into a graphing utility, let's analyze what we know about the graph.

1. Because $f(0) = \dfrac{1}{36}$, the point $\left(0, \dfrac{1}{36}\right)$ is on the graph.

2. Because $f(-x) \neq f(x)$ and $f(-x) \neq -f(x)$, there is no symmetry with respect to the origin or the y axis.

3. The denominator is zero at $x = \pm 6$. Thus the lines $x = 6$ and $x = -6$ are vertical asymptotes.

4. Let's change the form of the rational expression by division.

$$
\begin{array}{r}
x - 2 \\
x^2 - 36\overline{)x^3 - 2x^2 - x - 1} \\
\underline{x^3 - 36x } \\
-2x^2 + 35x - 1 \\
\underline{-2x^2 + 72} \\
35x - 73
\end{array}
$$

Thus the original function can be rewritten as

$$f(x) = x - 2 + \frac{35x - 73}{x^2 - 36}$$

Therefore the line $y = x - 2$ is an oblique asymptote. Now let $Y_1 = x - 2$ and $Y_2 = \dfrac{x^3 - 2x^2 - x - 1}{x^2 - 36}$ and use a viewing rectangle where $-15 \leq x \leq 15$ and $-30 \leq y \leq 30$. We get Figure 9.42.

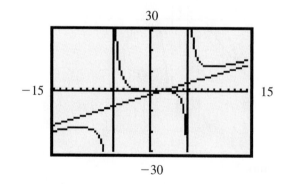

Figure 9.42

Note that the graph in Figure 9.42 is consistent with the information we had before we used the graphing calculator. Keep in mind that the oblique line and the two vertical lines are asymptotes and not part of the graph. Furthermore, the graph may appear to be symmetric about the origin, but remember that the test for origin symmetry failed. For example, the point $\left(0, \dfrac{1}{36}\right)$ is on the graph but the point $\left(0, -\dfrac{1}{36}\right)$ is not on the graph. Also note that the curve does intersect the oblique asymptote. We can use the ZOOM and TRACE features of the graphing calculator to approximate this point of intersection, or we can use an algebraic approach as follows: Because $y = \dfrac{x^3 - 2x^2 - x - 1}{x^2 - 36}$ and $y = x - 2$, we can equate the two expressions for y and solve the resulting equation for x.

$$\frac{x^3 - 2x^2 - x - 1}{x^2 - 36} = x - 2$$

$$x^3 - 2x^2 - x - 1 = (x - 2)(x^2 - 36)$$

$$x^3 - 2x^2 - x - 1 = x^3 - 2x^2 - 36x + 72$$

$$35x = 73$$

$$x = \frac{73}{35}$$

If $x = \dfrac{73}{35}$, then $y = x - 2 = \dfrac{73}{35} - 2 = \dfrac{3}{35}$. The point of intersection of the curve and the oblique asymptote is $\left(\dfrac{73}{35}, \dfrac{3}{35}\right)$.

Problem Set 9.6

For Problems 1–20, graph each rational function. Check first for symmetry, and identify the asymptotes.

1. $f(x) = \dfrac{x^2}{x^2 + x - 2}$

2. $f(x) = \dfrac{x^2}{x^2 + 2x - 3}$

3. $f(x) = \dfrac{2x^2}{x^2 - 2x - 8}$

4. $f(x) = \dfrac{-x^2}{x^2 + 3x - 4}$

5. $f(x) = \dfrac{-x}{x^2 - 1}$

6. $f(x) = \dfrac{2x}{x^2 - 9}$

7. $f(x) = \dfrac{x}{x^2 + x - 6}$

8. $f(x) = \dfrac{-x}{x^2 - 2x - 8}$

9. $f(x) = \dfrac{x^2}{x^2 - 4x + 3}$

10. $f(x) = \dfrac{1}{x^3 + x^2 - 6x}$

11. $f(x) = \dfrac{x}{x^2 + 2}$

12. $f(x) = \dfrac{6x}{x^2 + 1}$

13. $f(x) = \dfrac{-4x}{x^2 + 1}$

14. $f(x) = \dfrac{-5x}{x^2 + 2}$

15. $f(x) = \dfrac{x^2 + 2}{x - 1}$

16. $f(x) = \dfrac{x^2 - 3}{x + 1}$

17. $f(x) = \dfrac{x^2 - x - 6}{x + 1}$

18. $f(x) = \dfrac{x^2 + 4}{x + 2}$

19. $f(x) = \dfrac{x^2 + 1}{1 - x}$

20. $f(x) = \dfrac{x^3 + 8}{x^2}$

■ ■ ■ **THOUGHTS INTO WORDS**

21. Explain the concept of an oblique asymptote.

22. Explain why it is possible for curves to intersect horizontal and oblique asymptotes but not to intersect vertical asymptotes.

23. Give a step-by-step description of how you would go about graphing $f(x) = \dfrac{x^2 - x - 12}{x - 2}$.

24. Your friend is having difficulty finding the point of intersection of a curve and the oblique asymptote. How would you help?

GRAPHING CALCULATOR ACTIVITIES

25. First check for symmetry and identify the asymptotes for the graphs of the following rational functions. Then use your graphing utility to graph each function.

(a) $f(x) = \dfrac{4x^2}{x^2 + x - 2}$

(b) $f(x) = \dfrac{-2x}{x^2 - 5x - 6}$

(c) $f(x) = \dfrac{x^2}{x^2 - 9}$

(d) $f(x) = \dfrac{x^2 - 4}{x^2 - 9}$

(e) $f(x) = \dfrac{x^2 - 9}{x^2 - 4}$

(f) $f(x) = \dfrac{x^2 + 2x + 1}{x^2 - 5x + 6}$

(c) $f(x) = \dfrac{2x^2 + x + 1}{x + 1}$

(d) $f(x) = \dfrac{x^2 + 4}{x - 3}$

(e) $f(x) = \dfrac{3x^2 - x - 2}{x - 2}$

(f) $f(x) = \dfrac{4x^2 + x + 1}{x + 1}$

(g) $f(x) = \dfrac{x^3 + x^2 - x - 1}{x^2 + 2x + 3}$

(h) $f(x) = \dfrac{x^3 + 2x^2 + x - 3}{x^2 - 4}$

26. For each of the following rational functions, first determine and graph any oblique asymptotes. Then, on the same set of axes, graph the function.

(a) $f(x) = \dfrac{x^2 - 1}{x - 2}$

(b) $f(x) = \dfrac{x^2 + 1}{x + 2}$

Chapter 9   Summary

(9.1) If the divisor is of the form $x - c$, where c is a constant, then the typical long-division format for dividing polynomials can be simplified to a process called **synthetic division**. Review this process by studying the examples in this section.

The division algorithm for polynomials states that if $f(x)$ and $d(x)$ are polynomials and $d(x) \neq 0$, then there exist unique polynomials $q(x)$ and $r(x)$ such that

$$f(x) = d(x)q(x) + r(x)$$

where $r(x) = 0$ or the degree of $r(x)$ is less than the degree of $d(x)$.

(9.2) The remainder theorem states that if a polynomial $f(x)$ is divided by $x - c$, then the remainder is equal to $f(c)$. Thus a polynomial can be evaluated for a given number either by direct substitution or by using synthetic division.

The factor theorem states that a polynomial $f(x)$ has a factor $x - c$ if and only if $f(c) = 0$.

(9.3) The following concepts and properties provide a basis for solving polynomial equations.

1. Synthetic division.

2. The factor theorem.

3. Property 9.3: A polynomial equation of degree n has n solutions, where any solution of multiplicity p is counted p times.

4. The rational root theorem: Consider the polynomial equation

$$a_n x^n + a_{n-1} x^{n-1} + \cdots + a_1 x + a_0 = 0$$

where the coefficients are *integers*. If the rational number $\dfrac{c}{d}$, reduced to lowest terms, is a solution of the equation, then c is a factor of the constant term a_0 and d is a factor of the leading coefficient a_n.

5. Property 9.5: Nonreal complex solutions of polynomial equations with real coefficients, if they exist, must occur in conjugate pairs.

6. Descartes' rule of signs: Let $a_n x^n + a_{n-1} x^{n-1} + \cdots + a_1 x + a_0 = 0$ be a polynomial equation with real coefficients.

(a) The number of *positive real solutions* either is equal to the number of sign variations or is less than the number of sign variations by a positive even integer.

(b) The number of *negative real solutions* either is equal to the number of sign variations in

$$a_n(-x)^n + a_{n-1}(-x)^{n-1} + \cdots + a_1(-x) + a_0$$

or is less than the number of sign variations by a positive even integer.

(9.4) The following steps may be used to graph a polynomial function that is expressed in factored form:

1. Find the x intercepts, which are also called the **zeros** of the polynomial function.

2. Use a test value in each of the intervals determined by the x intercepts to find out whether the function is positive or negative over that interval.

3. Plot any additional points that are needed to determine the graph.

(9.5) and **(9.6)** To graph a rational function, the following steps are useful.

1. Check for symmetry with respect to the vertical axis and with respect to the origin.

2. Find any vertical asymptotes by setting the denominator equal to zero and solving it for x.

3. Find any horizontal asymptotes by studying the behavior of $f(x)$ as x gets very large or very small. This may require changing the form of the original rational expression.

4. If the degree of the numerator is one greater than the degree of the denominator, determine the equation of the oblique asymptote.

5. Study the behavior of the graph when it is close to the asymptotic lines.

6. Plot as many points as necessary to determine the graph. This may be affected by whether the graph has any symmetries.

Chapter 9   Review Problem Set

For Problems 1–4, use synthetic division to determine the quotient and the remainder.

1. $(3x^3 - 4x^2 + 6x - 2) \div (x - 1)$

2. $(5x^3 + 7x^2 - 9x + 10) \div (x + 2)$

3. $(-2x^4 + x^3 - 2x^2 - x - 1) \div (x + 4)$

4. $(-3x^4 - 5x^2 + 9) \div (x - 3)$

For Problems 5–8, find $f(c)$ either by using synthetic division and the remainder theorem or by evaluating $f(c)$ directly.

5. $f(x) = 4x^5 - 3x^3 + x^2 - 1$ and $c = 1$

6. $f(x) = 4x^3 - 7x^2 + 6x - 8$ and $c = -3$

7. $f(x) = -x^4 + 9x^2 - x - 2$ and $c = -2$

8. $f(x) = x^4 - 9x^3 + 9x^2 - 10x + 16$ and $c = 8$

For Problems 9–12, use the factor theorem to help answer some questions about factors.

9. Is $x + 2$ a factor of $2x^3 + x^2 - 7x - 2$?

10. Is $x - 3$ a factor of $x^4 + 5x^3 - 7x^2 - x + 3$?

11. Is $x - 4$ a factor of $x^5 - 1024$?

12. Is $x + 1$ a factor of $x^5 + 1$?

For Problems 13–16, use the rational root theorem and the factor theorem to help solve each of the equations.

13. $x^3 - 3x^2 - 13x + 15 = 0$

14. $8x^3 + 26x^2 - 17x - 35 = 0$

15. $x^4 - 5x^3 + 34x^2 - 82x + 52 = 0$

16. $x^3 - 4x^2 - 10x + 4 = 0$

For Problems 17 and 18, use Descartes' rule of signs (Property 9.6) to help list the possibilities for the nature of the solutions. *Do not solve the equations.*

17. $4x^4 - 3x^3 + 2x^2 + x + 4 = 0$

18. $x^5 + 3x^3 + x + 7 = 0$

For Problems 19–22, graph each of the polynomial functions.

19. $f(x) = -(x - 2)^3 + 3$

20. $f(x) = (x + 3)(x - 1)(3 - x)$

21. $f(x) = x^4 - 4x^2$

22. $f(x) = x^3 - 4x^2 + x + 6$

For Problems 23–26, graph each of the rational functions. Be sure to identify the asymptotes.

23. $f(x) = \dfrac{2x}{x - 3}$

24. $f(x) = \dfrac{-3}{x^2 + 1}$

25. $f(x) = \dfrac{-x^2}{x^2 - x - 6}$

26. $f(x) = \dfrac{x^2 + 3}{x + 1}$

1. Find the quotient and remainder when $3x^3 + 5x^2 - 14x - 6$ is divided by $x + 3$.

2. Find the quotient and remainder when $4x^4 - 7x^2 - x + 4$ is divided by $x - 2$.

3. If $f(x) = x^5 - 8x^4 + 9x^3 - 13x^2 - 9x - 10$, find $f(7)$.

4. If $f(x) = 3x^4 + 20x^3 - 6x^2 + 9x + 19$, find $f(-7)$.

5. If $f(x) = x^5 - 35x^3 - 32x + 15$, find $f(6)$.

6. Is $x - 5$ a factor of $3x^3 - 11x^2 - 22x - 20$?

7. Is $x + 2$ a factor of $5x^3 + 9x^2 - 9x - 17$?

8. Is $x + 3$ a factor of $x^4 - 16x^2 - 17x + 12$?

9. Is $x - 6$ a factor of $x^4 - 2x^2 + 3x - 12$?

For Problems 10–14, solve each equation.

10. $x^3 - 13x + 12 = 0$

11. $2x^3 + 5x^2 - 13x - 4 = 0$

12. $x^4 - 4x^3 - 5x^2 + 38x - 30 = 0$

13. $2x^3 + 3x^2 - 17x + 12 = 0$

14. $3x^3 - 7x^2 - 8x + 20 = 0$

15. Use Descartes' rule of signs to determine the nature of the roots of $5x^4 + 3x^3 - x^2 - 9 = 0$.

16. Find the x intercepts of the graph of the function $f(x) = 3x^3 + 19x^2 - 14x$.

17. Find the equation of the vertical asymptote for the graph of the function $f(x) = \dfrac{5x}{x + 3}$.

18. Find the equation of the horizontal asymptote for the graph of the function $f(x) = \dfrac{5x^2}{x^2 - 4}$.

19. What type of symmetry does the graph of the equation $f(x) = \dfrac{x^2}{x^2 + 2}$ exhibit?

20. What type of symmetry does the graph of the equation $f(x) = \dfrac{-3x}{x^2 + 1}$ exhibit?

For Problems 21–25, graph each of the functions.

21. $f(x) = (2 - x)(x - 1)(x + 1)$

22. $f(x) = -x(x - 3)(x + 2)$

23. $f(x) = \dfrac{-x}{x - 3}$

24. $f(x) = \dfrac{-2}{x^2 - 4}$

25. $f(x) = \dfrac{4x^2 + x + 1}{x + 1}$

10

Exponential and Logarithmic Functions

10.1 Exponents and Exponential Functions

10.2 Applications of Exponential Functions

10.3 Inverse Functions

10.4 Logarithms

10.5 Logarithmic Functions

10.6 Exponential Equations, Logarithmic Equations, and Problem Solving

Because Richter numbers for reporting the intensity of an earthquake are calculated from logarithms, they are referred to as being on a logarithmic scale. Logarithmic scales are commonly used in science and mathematics to transform very large numbers to a smaller scale.

© AP/Wide World Photos

How long will it take \$100 to triple if it is invested at 8% interest compounded continuously? We can use the formula $A = Pe^{rt}$ to generate the equation $300 = 100e^{0.08t}$, which can be solved for t using logarithms. It will take approximately 13.7 years for the money to triple.

In this chapter, we will (1) extend our understanding of exponents, (2) work with some exponential functions, (3) consider the concept of a logarithm, (4) work with some logarithmic functions, and (5) use the concepts of exponents and logarithms to expand our problem-solving skills. Your calculator will be a valuable tool throughout this chapter.

10.1   **Exponents and Exponential Functions**

In Chapter 1 the expression b^n was defined to mean n factors of b, where n is any positive integer and b is any real number. For example,

$$2^3 = 2 \cdot 2 \cdot 2 = 8 \qquad \left(\frac{1}{3}\right)^4 = \left(\frac{1}{3}\right)\left(\frac{1}{3}\right)\left(\frac{1}{3}\right)\left(\frac{1}{3}\right) = \frac{1}{81}$$

$$(-4)^2 = (-4)(-4) = 16 \qquad -(0.5)^3 = -[(0.5)(0.5)(0.5)] = -0.125$$

Then in Chapter 5, by defining $b^0 = 1$ and $b^{-n} = \dfrac{1}{b^n}$, where n is any positive integer and b is any nonzero real number, we extended the concept of an exponent to include all integers. Examples include

$$(0.76)^0 = 1 \qquad\qquad 2^{-3} = \frac{1}{2^3} = \frac{1}{8}$$

$$\left(\frac{2}{3}\right)^{-2} = \frac{1}{\left(\frac{2}{3}\right)^2} = \frac{1}{\frac{4}{9}} = \frac{9}{4} \qquad (0.4)^{-1} = \frac{1}{(0.4)^1} = \frac{1}{0.4} = 2.5$$

In Chapter 5 we also provided for the use of all rational numbers as exponents by defining

$$b^{m/n} = \sqrt[n]{b^m} = (\sqrt[n]{b})^m$$

where n is a positive integer greater than 1, and b is a real number such that $\sqrt[n]{b}$ exists. Some examples are

$$27^{2/3} = (\sqrt[3]{27})^2 = 9 \qquad 16^{1/4} = \sqrt[4]{16^1} = 2$$

$$\left(\frac{1}{9}\right)^{1/2} = \sqrt{\frac{1}{9}} = \frac{1}{3} \qquad 32^{-1/5} = \frac{1}{32^{1/5}} = \frac{1}{\sqrt[5]{32}} = \frac{1}{2}$$

Formally extending the concept of an exponent to include the use of irrational numbers requires some ideas from calculus and is therefore beyond the scope of this text. However, we can take a brief glimpse at the general idea involved. Consider the number $2^{\sqrt{3}}$. By using the nonterminating and nonrepeating decimal representation $1.73205\ldots$ for $\sqrt{3}$, we can form the sequence of numbers $2^1, 2^{1.7}, 2^{1.73}, 2^{1.732}, 2^{1.7320}, 2^{1.73205}, \ldots$. It seems reasonable that each successive power gets closer to $2^{\sqrt{3}}$. This is precisely what happens if b^n, where n is irrational, is properly defined using the concept of a limit. Furthermore, this will ensure that an expression such as 2^x will yield exactly one value for each value of x.

From now on, then, we can use any real number as an exponent, and we can extend the basic properties stated in Chapter 5 to include all real numbers as exponents. Let's restate those properties with the restriction that the bases a and b

must be positive numbers so that we avoid expressions such as $(-4)^{1/2}$, which do not represent real numbers.

Property 10.1

If a and b are positive real numbers, and m and n are any real numbers, then

1. $b^n \cdot b^m = b^{n+m}$ Product of two powers

2. $(b^n)^m = b^{mn}$ Power of a power

3. $(ab)^n = a^n b^n$ Power of a product

4. $\left(\dfrac{a}{b}\right)^n = \dfrac{a^n}{b^n}$ Power of a quotient

5. $\dfrac{b^n}{b^m} = b^{n-m}$ Quotient of two powers

Another property that we can use to solve certain types of equations that involve exponents can be stated as follows:

Property 10.2

If $b > 0$, $b \neq 1$, and m and n are real numbers, then $b^n = b^m$ if and only if $n = m$.

The following examples illustrate the use of Property 10.2. To use the property to solve equations, we will want both sides of the equation to have the same base number.

EXAMPLE 1   Solve $2^x = 32$.

Solution

$$2^x = 32$$

$$2^x = 2^5 \qquad 32 = 2^5$$

$$x = 5 \qquad \text{Property 10.2}$$

The solution set is $\{5\}$.   ■

E X A M P L E   2

Solve $3^{2x} = \dfrac{1}{9}$.

Solution

$$3^{2x} = \frac{1}{9} = \frac{1}{3^2}$$

$$3^{2x} = 3^{-2}$$

$$2x = -2 \qquad \text{Property 10.2}$$

$$x = -1$$

The solution set is $\{-1\}$.

E X A M P L E   3

Solve $\left(\dfrac{1}{5}\right)^{x-4} = \dfrac{1}{125}$.

Solution

$$\left(\frac{1}{5}\right)^{x-4} = \frac{1}{125}$$

$$\left(\frac{1}{5}\right)^{x-4} = \left(\frac{1}{5}\right)^{3}$$

$$x - 4 = 3 \qquad \text{Property 10.2}$$

$$x = 7$$

The solution set is $\{7\}$.

E X A M P L E   4

Solve $8^x = 32$.

Solution

$$8^x = 32$$

$$(2^3)^x = 2^5 \qquad 8 = 2^3$$

$$2^{3x} = 2^5$$

$$3x = 5 \qquad \text{Property 10.2}$$

$$x = \frac{5}{3}$$

The solution set is $\left\{\dfrac{5}{3}\right\}$.

E X A M P L E   5

Solve $(3^{x+1})(9^{x-2}) = 27$.

Solution

$$(3^{x+1})(9^{x-2}) = 27$$

$$(3^{x+1})(3^2)^{x-2} = 3^3$$

$$(3^{x+1})(3^{2x-4}) = 3^3$$
$$3^{3x-3} = 3^3$$
$$3x - 3 = 3 \qquad \text{Property 10.2}$$
$$3x = 6$$
$$x = 2$$

The solution set is {2}. ■

■ Exponential Functions

If b is any positive number, then the expression b^x designates exactly one real number for every real value of x. Therefore the equation $f(x) = b^x$ defines a function whose domain is the set of real numbers. Furthermore, if we include the additional restriction $b \neq 1$, then any equation of the form $f(x) = b^x$ describes what we will call later a one-to-one function and is known as an **exponential function**. This leads to the following definition:

Definition 10.1

If $b > 0$ and $b \neq 1$, then the function f defined by

$$f(x) = b^x$$

where x is any real number, is called the **exponential function with base b**.

Now let's consider graphing some exponential functions.

EXAMPLE 6

Graph the function $f(x) = 2^x$.

Solution

Let's set up a table of values; keep in mind that the domain is the set of real numbers and that the equation $f(x) = 2^x$ exhibits no symmetry. Plot these points and connect them with a smooth curve to produce Figure 10.1.

x	2^x
-2	$\dfrac{1}{4}$
-1	$\dfrac{1}{2}$
0	1
1	2
2	4
3	8

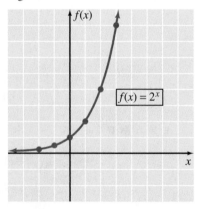

Figure 10.1 ■

In the table for Example 6, we chose integral values for x to keep the computation simple. However, with the use of a calculator, we could easily acquire functional values by using nonintegral exponents. Consider the following additional values for $f(x) = 2^x$:

$$f(0.5) \approx 1.41 \quad f(1.7) \approx 3.25$$

$$f(-0.5) \approx 0.71 \quad f(-2.6) \approx 0.16$$

Use your calculator to check these results. Also note that the points generated by these values do fit the graph in Figure 10.1.

Graph $f(x) = \left(\dfrac{1}{2}\right)^x$.

Solution

Again, let's set up a table of values, plot the points, and connect them with a smooth curve. The graph is shown in Figure 10.2.

x	$\left(\dfrac{1}{2}\right)^x$
-3	8
-2	4
-1	2
0	1
1	$\dfrac{1}{2}$
2	$\dfrac{1}{4}$
3	$\dfrac{1}{8}$

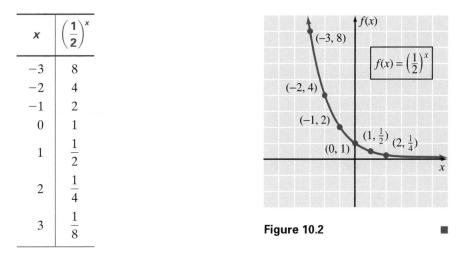

Figure 10.2

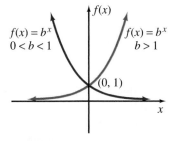

Figure 10.3

Remark: Because $\left(\dfrac{1}{2}\right)^x = \dfrac{1}{2^x} = 2^{-x}$, the graphs of $f(x) = 2^x$ and $f(x) = \left(\dfrac{1}{2}\right)^x$ are reflections of each other across the y axis. Therefore Figure 10.2 could have been drawn by reflecting Figure 10.1 across the y axis.

The graphs in Figures 10.1 and 10.2 illustrate a general behavior pattern of exponential functions. That is to say, if $b > 1$, then the graph of $f(x) = b^x$ *goes up to the right*, and the function is called an **increasing function**. If $0 < b < 1$, then the graph of $f(x) = b^x$ *goes down to the right*, and the function is called a **decreasing function**. These facts are illustrated in Figure 10.3. Note that $b^0 = 1$ for any $b > 0$; thus all graphs of $f(x) = b^x$ contain the point $(0, 1)$.

As you graph exponential functions, don't forget your previous graphing experiences.

1. The graph of $f(x) = 2^x - 4$ is the graph of $f(x) = 2^x$ *moved down four units.*

2. The graph of $f(x) = 2^{x+3}$ is the graph of $f(x) = 2^x$ *moved three units to the left.*

3. The graph of $f(x) = -2^x$ is the graph of $f(x) = 2^x$ *reflected across the x axis.*

We used a graphing calculator to graph these four functions on the same set of axes, as shown in Figure 10.4.

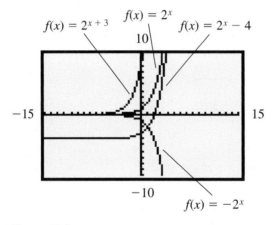

Figure 10.4

If you are faced with an exponential function that is not of the basic form $f(x) = b^x$ or a variation thereof, don't forget the graphing suggestions offered in earlier chapters. Let's consider one such example.

E X A M P L E   8   Graph $f(x) = 2^{-x^2}$.

Solution

Because $f(-x) = 2^{-(-x)^2} = 2^{-x^2} = f(x)$, we know that this curve is symmetric with respect to the y axis. Therefore let's set up a table of values using non-negative values for x. Plot these points, connect them with a smooth curve, and reflect this portion of the curve across the y axis to produce the graph in Figure 10.5.

x	2^{x^2}
0	1
$\dfrac{1}{2}$	0.84
1	0.5
$\dfrac{3}{2}$	0.21
2	0.06

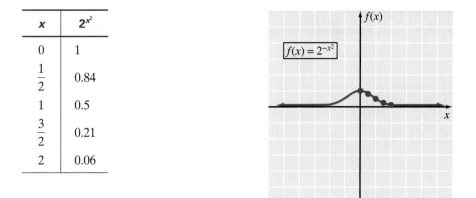

Figure 10.5

Finally, let's consider a problem in which a graphing utility gives us an approximate solution.

E X A M P L E   9

Use a graphing utility to obtain a graph of $f(x) = 50(2^x)$ and find an approximate value for x when $f(x) = 15,000$.

Solution

First, we must find an appropriate viewing rectangle. Because $50(2^{10}) = 51,200$, let's set the boundaries so that $0 \leq x \leq 10$ and $0 \leq y \leq 50,000$ with a scale of 10,000 on the y axis. (Certainly other boundaries could be used, but these will give us a graph that we can work with for this problem.) The graph of $f(x) = 50(2^x)$ is shown in Figure 10.6. Now we can use the TRACE and ZOOM features of the graphing utility to find that $x \approx 8.2$ at $y = 15,000$.

50,000

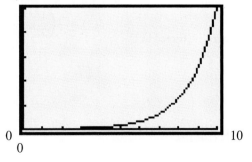

Figure 10.6

Remark: In Example 9 we used a graphical approach to solve the equation $50(2^x) = 15,000$. In Section 10.6 we will use an algebraic approach for solving that same kind of equation.

Problem Set 10.1

For Problems 1–26, solve each of the equations.

1. $2^x = 64$

2. $3^x = 81$

3. $3^{2x} = 27$

4. $2^{2x} = 16$

5. $\left(\dfrac{1}{2}\right)^x = \dfrac{1}{128}$

6. $\left(\dfrac{1}{4}\right)^x = \dfrac{1}{256}$

7. $3^{-x} = \dfrac{1}{243}$

8. $3^{x+1} = 9$

9. $6^{3x-1} = 36$

10. $2^{2x+3} = 32$

11. $\left(\dfrac{3}{4}\right)^n = \dfrac{64}{27}$

12. $\left(\dfrac{2}{3}\right)^n = \dfrac{9}{4}$

13. $16^x = 64$

14. $4^x = 8$

15. $27^{4x} = 9^{x+1}$

16. $32^x = 16^{1-x}$

17. $9^{4x-2} = \dfrac{1}{81}$

18. $8^{3x+2} = \dfrac{1}{16}$

19. $10^x = 0.1$

20. $10^x = 0.0001$

21. $(2^{x+1})(2^x) = 64$

22. $(2^{2x-1})(2^{x+2}) = 32$

23. $(27)(3^x) = 9^x$

24. $(3^x)(3^{5x}) = 81$

25. $(4^x)(16^{3x-1}) = 8$

26. $(8^{2x})(4^{2x-1}) = 16$

For Problems 27–46, graph each of the exponential functions.

27. $f(x) = 3^x$

28. $f(x) = 4^x$

29. $f(x) = \left(\dfrac{1}{3}\right)^x$

30. $f(x) = \left(\dfrac{1}{4}\right)^x$

31. $f(x) = \left(\dfrac{3}{2}\right)^x$

32. $f(x) = \left(\dfrac{2}{3}\right)^x$

33. $f(x) = 2^x - 3$

34. $f(x) = 2^x + 1$

35. $f(x) = 2^{x+2}$

36. $f(x) = 2^{x-1}$

37. $f(x) = -2^x$

38. $f(x) = -3^x$

39. $f(x) = 2^{-x-2}$

40. $f(x) = 2^{-x+1}$

41. $f(x) = 2^{x^2}$

42. $f(x) = 2^x + 2^{-x}$

43. $f(x) = 2^{|x|}$

44. $f(x) = 3^{1-x^2}$

45. $f(x) = 2^x - 2^{-x}$

46. $f(x) = 2^{-|x|}$

■ ■ ■ THOUGHTS INTO WORDS

47. Explain how you would solve the equation $(2^{x+1})(8^{2x-3}) = 64$.

48. Why is the base of an exponential function restricted to positive numbers not including 1?

49. Explain how you would graph the function
$$f(x) = -\left(\dfrac{1}{3}\right)^x.$$

GRAPHING CALCULATOR ACTIVITIES

50. Use a graphing calculator to check your graphs for Problems 27–46.

51. Graph $f(x) = 2^x$. Where should the graphs of $f(x) = 2^{x-5}$, $f(x) = 2^{x-7}$, and $f(x) = 2^{x+5}$ be located? Graph all three functions on the same set of axes with $f(x) = 2^x$.

52. Graph $f(x) = 3^x$. Where should the graphs of $f(x) = 3^x + 2$, $f(x) = 3^x - 3$, and $f(x) = 3^x - 7$ be located? Graph all three functions on the same set of axes with $f(x) = 3^x$.

53. Graph $f(x) = \left(\dfrac{1}{2}\right)^x$. Where should the graphs of

$f(x) = -\left(\dfrac{1}{2}\right)^x$, $f(x) = \left(\dfrac{1}{2}\right)^{-x}$, and $f(x) = -\left(\dfrac{1}{2}\right)^{-x}$ be

located? Graph all three functions on the same set of

axes with $f(x) = \left(\dfrac{1}{2}\right)^x$.

54. Graph $f(x) = (1.5)^x$, $f(x) = (5.5)^x$, $f(x) = (0.3)^x$, and $f(x) = (0.7)^x$ on the same set of axes. Are these graphs consistent with Figure 10.3?

55. What is the solution for $3^x = 5$? Do you agree that it is between 1 and 2 because $3^1 = 3$ and $3^2 = 9$? Now graph

$f(x) = 3^x - 5$ and use the ZOOM and TRACE features of your graphing calculator to find an approximation, to the nearest hundredth, for the x intercept. You should get an answer of 1.46. Do you see that this is an approximation for the solution of $3^x = 5$? Try it; raise 3 to the 1.46 power.

Find an approximate solution, to the nearest hundredth, for each of the following equations by graphing the appropriate function and finding the x intercept.

(a) $2^x = 19$ **(b)** $3^x = 50$ **(c)** $4^x = 47$
(d) $5^x = 120$ **(e)** $2^x = 1500$ **(f)** $3^{x-1} = 34$

10.2 Applications of Exponential Functions

We can represent many real-world situations exhibiting growth or decay with equations that describe exponential functions. For example, suppose an economist predicts an annual inflation rate of 5% per year for the next 10 years. This means that an item that presently costs $8 will cost $8(105\%) = 8(1.05) = \$8.40$ a year from now. The same item will cost $[8(105\%)](105\%) = 8(1.05)^2 = \$8.82$ in 2 years. In general, the equation

$$\tilde{P} = P_0(1.05)^t$$

yields the predicted price P of an item in t years if the present cost is P_0 and the annual inflation rate is 5%. Using this equation, we can look at some future prices based on the prediction of a 5% inflation rate.

A $0.79 jar of mustard will cost $0.79(1.05)^3 = \$0.91$ in 3 years.
A $2.69 bag of potato chips will cost $2.69(1.05)^5 = \$3.43$ in 5 years.
A $6.69 can of coffee will cost $6.69(1.05)^7 = \$9.41$ in 7 years.

■ Compound Interest

Compound interest provides another illustration of exponential growth. Suppose that $500, called the **principal**, is invested at an interest rate of 8% compounded annually. The interest earned the first year is $500(0.08) = \$40$, and this amount is added to the original $500 to form a new principal of $540 for the second year. The interest earned during the second year is $540(0.08) = \$43.20$, and this amount is added to $540 to form a new principal of $583.20 for the third year. Each year a new principal is formed by reinvesting the interest earned during that year.

In general, suppose that a sum of money P (the principal) is invested at an interest rate of r percent compounded annually. The interest earned the first year is Pr, and the new principal for the second year is $P + Pr$, or $P(1 + r)$. Note that the new principal for the second year can be found by multiplying the original principal P by $(1 + r)$. In like fashion, the new principal for the third year can be found by multiplying the previous principal $P(1 + r)$ by $1 + r$, thus obtaining $P(1 + r)^2$. If this process is continued, after t years the total amount of money accumulated, (A), is given by

$$A = P(1 + r)^t$$

Consider the following examples of investments made at a certain rate of interest compounded annually:

1. $750 invested for 5 years at 9% compounded annually produces

$$A = \$750(1.09)^5 = \$1153.97$$

2. $1000 invested for 10 years at 7% compounded annually produces

$$A = \$1000(1.07)^{10} = \$1967.15$$

3. $5000 invested for 20 years at 6% compounded annually produces

$$A = \$5000(1.06)^{20} = \$16,035.68$$

We can use the compound interest formula to determine what rate of interest is needed to accumulate a certain amount of money based on a given initial investment. The next example illustrates this idea.

E X A M P L E   1

What rate of interest is needed for an investment of $1000 to yield $4000 in 10 years if the interest is compounded annually?

Solution

Let's substitute $1000 for P, $4000 for A, and 10 years for t in the compound interest formula and solve for r.

$$A = P(1 + r)^t$$

$$4000 = 1000(1 + r)^{10}$$

$$4 = (1 + r)^{10}$$

$$4^{0.1} = [(1 + r)^{10}]^{0.1} \qquad \text{Raise both sides to the 0.1 power.}$$

$$1.148698355 \approx 1 + r$$

$$0.148698355 \approx r$$

$$r = 14.9\% \quad \text{to the nearest tenth of a percent}$$

Therefore a rate of interest of approximately 14.9% is needed. (Perhaps you should check this answer.) ∎

If money invested at a certain rate of interest is compounded more than once a year, then the basic formula $A = P(1 + r)^t$ can be adjusted according to the number of compounding periods in a year. For example, for *semiannual compounding*, the formula becomes $A = P\left(1 + \dfrac{r}{2}\right)^{2t}$; for *quarterly compounding*, the formula

becomes $A = P\left(1 + \dfrac{r}{4}\right)^{4t}$. In general, if n represents the number of compounding periods in a year, the formula becomes

$$A = P\left(1 + \frac{r}{n}\right)^{nt}$$

The following examples illustrate the use of the formula:

1. $750 invested for 5 years at 9% compounded semiannually produces

$$A = \$750\left(1 + \frac{0.09}{2}\right)^{2(5)} = \$750(1.045)^{10} = \$1164.73$$

2. $1000 invested for 10 years at 7% compounded quarterly produces

$$A = \$1000\left(1 + \frac{0.07}{4}\right)^{4(10)} = \$1000(1.0175)^{40} = \$2001.60$$

3. $5000 invested for 20 years at 6% compounded monthly produces

$$A = \$5000\left(1 + \frac{0.06}{12}\right)^{12(20)} = \$5000(1.005)^{240} = \$16{,}551.02$$

You may find it interesting to compare these results with those we obtained earlier for annual compounding.

■ Exponential Decay

Suppose that the value of a car depreciates 15% per year for the first 5 years. Therefore a car that costs $9500 will be worth $9500(100% − 15%) = $9500(85%) = $9500(0.85) = $8075 in 1 year. In 2 years the value of the car will have depreciated $9500(0.85)^2 = $6864 (to the nearest dollar). The equation

$$V = V_0(0.85)^t$$

yields the value V of a car in t years if the initial cost is V_0, and the value depreciates 15% per year. Therefore we can estimate some car values to the nearest dollar as follows:

A $13,000 car will be worth $13,000(0.85)^3 = $7984 in 3 years.

A $17,000 car will be worth $17,000(0.85)^5 = $7543 in 5 years.

A $25,000 car will be worth $25,000(0.85)^4 = $13,050 in 4 years.

Another example of exponential decay is associated with radioactive substances. The rate of decay can be described exponentially and is based on the half-life of a substance. The **half-life** of a radioactive substance is the amount of time that it takes for one-half of an initial amount of the substance to disappear as the result of decay. For example, suppose that we have 200 grams of a certain substance that has a half-life of 5 days. After 5 days, $200\left(\dfrac{1}{2}\right) = 100$ grams remain. After 10 days, $200\left(\dfrac{1}{2}\right)^2 = 50$ grams remain. After 15 days, $200\left(\dfrac{1}{2}\right)^3 = 25$ grams remain. In general, after t days, $200\left(\dfrac{1}{2}\right)^{\frac{t}{5}}$ grams remain.

The previous discussion leads to the following half-life formula. Suppose there is an initial amount (Q_0) of a radioactive substance with a half-life of h. The amount of substance remaining (Q) after a time period of t is given by the formula

$$Q = Q_0\left(\frac{1}{2}\right)^{\frac{t}{h}}$$

The units of measure for t and h must be the same.

EXAMPLE 2

Barium-140 has a half-life of 13 days. If there are 500 milligrams of barium initially, how many milligrams remain after 26 days? After 100 days?

Solution

When we use $Q_0 = 500$ and $h = 13$, the half-life formula becomes

$$Q = 500\left(\frac{1}{2}\right)^{\frac{t}{13}}$$

If $t = 26$, then

$$Q = 500\left(\frac{1}{2}\right)^{\frac{26}{13}}$$

$$= 500\left(\frac{1}{2}\right)^2$$

$$= 500\left(\frac{1}{4}\right)$$

$$= 125$$

Thus 125 milligrams remain after 26 days. If $t = 100$, then

$$Q = 500\left(\frac{1}{2}\right)^{\frac{100}{13}}$$

$$= 500(0.5)^{\frac{100}{13}}$$

$$= 2.4 \quad \text{to the nearest tenth of a milligram}$$

Approximately 2.4 milligrams remain after 100 days.   ■

Remark: Example 2 clearly illustrates that a calculator is useful at times but unnecessary at other times. We solved the first part of the problem very easily without a calculator, but it certainly was helpful for the second part of the problem.

■ Number *e*

An interesting situation occurs if we consider the compound interest formula for $P = \$1$, $r = 100\%$, and $t = 1$ year. The formula becomes $A = 1\left(1 + \dfrac{1}{n}\right)^n$. The following table shows some values, rounded to eight decimal places, of $\left(1 + \dfrac{1}{n}\right)^n$ for different values of n.

n	$\left(1 + \dfrac{1}{n}\right)^n$
1	2.00000000
10	2.59374246
100	2.70481383
1000	2.71692393
10,000	2.71814593
100,000	2.71826824
1,000,000	2.71828047
10,000,000	2.71828169
100,000,000	2.71828181
1,000,000,000	2.71828183

The table suggests that as n increases, the value of $\left(1 + \dfrac{1}{n}\right)^n$ gets closer and closer to some fixed number. This does happen, and the fixed number is called *e*. To five decimal places, $e = 2.71828$.

The function defined by the equation $f(x) = e^x$ is the **natural exponential function**. It has a great many real-world applications, some of which we will look at in a moment. First, however, let's get a picture of the natural exponential function. Because $2 < e < 3$, the graph of $f(x) = e^x$ must fall between the graphs of $f(x) = 2^x$ and $f(x) = 3^x$. To be more specific, let's use our calculator to determine a table of values. Use the $\boxed{e^x}$ key, and round the results to the nearest tenth to obtain the following table. Plot the points determined by this table, and connect them with a smooth curve to produce Figure 10.7.

x	$f(x) = e^x$
0	1.0
1	2.7
2	7.4
-1	0.4
-2	0.1

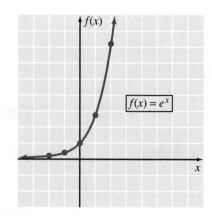

Figure 10.7

■ Back to Compound Interest

Let's return to the concept of compound interest. If the number of compounding periods in a year is increased indefinitely, we arrive at the concept of **compounding continuously**. Mathematically, we can accomplish this by applying the limit concept to the expression $P\left(1 + \dfrac{r}{n}\right)^{nt}$. We will not show the details here, but the following result is obtained. The formula

$$A = Pe^{rt}$$

yields the accumulated value (A) of a sum of money (P) that has been invested for t years at a rate of r percent compounded continuously. The following examples illustrate the use of the formula.

1. $750 invested for 5 years at 9% compounded continuously produces

$$A = \$750e^{(0.09)(5)} = 750e^{0.45} = \$1176.23$$

2. $1000 invested for 10 years at 7% compounded continuously produces

$$A = \$1000e^{(0.07)(10)} = 1000e^{0.7} = \$2013.75$$

3. $5000 invested for 20 years at 6% compounded continuously produces

$$A = \$5000e^{(0.06)(20)} = 5000e^{1.2} = \$16,600.58$$

Again, you may find it interesting to compare these results with those you obtained earlier when you were using a different number of compounding periods.

Is it better to invest at 6% compounded quarterly or at 5.75% compounded continuously? To answer such a question, we can use the concept of **effective yield** (sometimes called effective annual rate of interest). The effective yield of an investment is the simple interest rate that would yield the same amount in 1 year.

Thus, for the investment at 6% compounded quarterly, we can calculate the effective yield as follows:

$$P(1 + r) = P\left(1 + \frac{0.06}{4}\right)^4$$

$$1 + r = \left(1 + \frac{0.06}{4}\right)^4 \qquad \text{Multiply both sides by } \frac{1}{P}.$$

$$1 + r = (1.015)^4$$

$$r = (1.015)^4 - 1$$

$$r \approx 0.0613635506$$

$$r = 6.14\% \quad \text{to the nearest hundredth of a percent}$$

Likewise, for the investment at 5.75% compounded continuously, we can calculate the effective yield as follows:

$$P(1 + r) = Pe^{0.0575}$$

$$1 + r = e^{0.0575}$$

$$r = e^{0.0575} - 1$$

$$r \approx 0.0591852707$$

$$r = 5.92\% \quad \text{to the nearest hundredth of a percent}$$

Therefore, comparing the two effective yields, we see that it is better to invest at 6% compounded quarterly than to invest at 5.75% compounded continuously.

■ Law of Exponential Growth

The ideas behind "compounded continuously" carry over to other growth situations. We use the law of exponential growth,

$$Q(t) = Q_0 e^{kt}$$

as a mathematical model for numerous growth-and-decay applications. In this equation, $Q(t)$ represents the quantity of a given substance at any time t, Q_0 is the initial amount of the substance (when $t = 0$), and k is a constant that depends on the particular application. If $k < 0$, then $Q(t)$ decreases as t increases, and we refer to the model as the **law of decay**.

Let's consider some growth-and-decay applications.

E X A M P L E   3

Suppose that in a certain culture, the equation $Q(t) = 15{,}000e^{0.3t}$ expresses the number of bacteria present as a function of the time t, where t is expressed in hours. Find (a) the initial number of bacteria and (b) the number of bacteria after 3 hours.

Solution

(a) The initial number of bacteria is produced when $t = 0$.

$$Q(0) = 15{,}000e^{0.3(0)}$$

$$= 15{,}000e^0$$

$$= 15{,}000 \qquad e^0 = 1$$

(b) $Q(3) = 15{,}000e^{0.3(3)}$

$$= 15{,}000e^{0.9}$$

$$= 36{,}894 \quad \text{to the nearest whole number}$$

Therefore approximately 36,894 bacteria should be present after 3 hours. ∎

EXAMPLE 4   Suppose the number of bacteria present in a certain culture after t minutes is given by the equation $Q(t) = Q_0 e^{0.05t}$, where Q_0 represents the initial number of bacteria. If 5000 bacteria are present after 20 minutes, how many bacteria were present initially?

Solution

If 5000 bacteria are present after 20 minutes, then $Q(20) = 5000$.

$$5000 = Q_0 e^{0.05(20)}$$

$$5000 = Q_0 e^1$$

$$\frac{5000}{e} = Q_0$$

$$1839 = Q_0 \quad \text{to the nearest whole number}$$

Therefore, approximately 1839 bacteria were present initially. ∎

EXAMPLE 5   The number of grams of a certain radioactive substance present after t seconds is given by the equation $Q(t) = 200e^{-0.3t}$. How many grams remain after 7 seconds?

Solution

Use $Q(t) = 200e^{-0.3t}$ to obtain

$$Q(7) = 200e^{(-0.3)(7)}$$

$$= 200e^{-2.1}$$

$$= 24.5 \quad \text{to the nearest tenth}$$

Thus approximately 24.5 grams remain after 7 seconds. ∎

Finally, let's consider two examples where we use a graphing utility to produce the graph.

EXAMPLE 6

Suppose that $1000 was invested at 6.5% interest compounded continuously. How long would it take for the money to double?

Solution

Substitute $1000 for P and 0.065 for r in the formula $A = Pe^{rt}$ to produce $A = 1000e^{0.065t}$. If we let $y = A$ and $x = t$, we can graph the equation $y = 1000e^{0.065x}$. By letting $x = 20$, we obtain $y = 1000e^{0.065(20)} = 1000e^{1.3} \approx 3670$. Therefore let's set the boundaries of the viewing rectangle so that $0 \le x \le 20$ and $0 \le y \le 3700$ with a y scale of 1000. Then we obtain the graph in Figure 10.8. Now we want to find the value of x so that $y = 2000$. (The money is to double.) Using the ZOOM and TRACE features of the graphing utility, we can determine that an x value of approximately 10.7 will produce a y value of 2000. Thus it will take approximately 10.7 years for the $1000 investment to double.

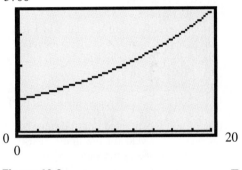

Figure 10.8   ∎

EXAMPLE 7

Graph the function $y = \dfrac{1}{\sqrt{2\pi}} e^{-x^2/2}$ and find its maximum value.

Solution

If $x = 0$, then $y = \dfrac{1}{\sqrt{2\pi}} e^0 = \dfrac{1}{\sqrt{2\pi}} \approx 0.4$, so let's set the boundaries of the viewing rectangle so that $-5 \le x \le 5$ and $0 \le y \le 1$ with a y scale of 0.1; the graph of the function is shown in Figure 10.9. From the graph, we see that the maximum value of the function occurs at $x = 0$, which we have already determined to be approximately 0.4.

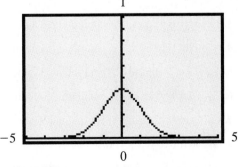

Figure 10.9   ∎

Remark: The curve in Figure 10.9 is called a **normal distribution curve**. You may want to ask your instructor to explain what it means to assign grades on the basis of the normal distribution curve.

Problem Set 10.2

1. Assuming that the rate of inflation is 4% per year, the equation $P = P_0(1.04)^t$ yields the predicted price (P) of an item in t years that presently costs P_0. Find the predicted price of each of the following items for the indicated years ahead:
   (a) $0.77 can of soup in 3 years
   (b) $3.43 container of cocoa mix in 5 years
   (c) $1.99 jar of coffee creamer in 4 years
   (d) $1.05 can of beans and bacon in 10 years
   (e) $18,000 car in 5 years (nearest dollar)
   (f) $120,000 house in 8 years (nearest dollar)
   (g) $500 TV set in 7 years (nearest dollar)

2. Suppose it is estimated that the value of a car depreciates 30% per year for the first 5 years. The equation $A = P_0(0.7)^t$ yields the value (A) of a car after t years if the original price is P_0. Find the value (to the nearest dollar) of each of the following cars after the indicated time:
   (a) $16,500 car after 4 years
   (b) $22,000 car after 2 years
   (c) $27,000 car after 5 years
   (d) $40,000 car after 3 years

For Problems 3–14, use the formula $A = P\left(1 + \dfrac{r}{n}\right)^{nt}$ to find the total amount of money accumulated at the end of the indicated time period for each of the following investments:

3. $200 for 6 years at 6% compounded annually

4. $250 for 5 years at 7% compounded annually

5. $500 for 7 years at 8% compounded semiannually

6. $750 for 8 years at 8% compounded semiannually

7. $800 for 9 years at 9% compounded quarterly

8. $1200 for 10 years at 10% compounded quarterly

9. $1500 for 5 years at 12% compounded monthly

10. $2000 for 10 years at 9% compounded monthly

11. $5000 for 15 years at 8.5% compounded annually

12. $7500 for 20 years at 9.5% compounded semiannually

13. $8000 for 10 years at 10.5% compounded quarterly

14. $10,000 for 25 years at 9.25% compounded monthly

For Problems 15–23, use the formula $A = Pe^{rt}$ to find the total amount of money accumulated at the end of the indicated time period by compounding continuously.

15. $400 for 5 years at 7%

16. $500 for 7 years at 6%

17. $750 for 8 years at 8%

18. $1000 for 10 years at 9%

19. $2000 for 15 years at 10%

20. $5000 for 20 years at 11%

21. $7500 for 10 years at 8.5%

22. $10,000 for 25 years at 9.25%

23. $15,000 for 10 years at 7.75%

24. What rate of interest, to the nearest tenth of a percent, compounded annually is needed for an investment of $200 to grow to $350 in 5 years?

25. What rate of interest, to the nearest tenth of a percent, compounded quarterly is needed for an investment of $1500 to grow to $2700 in 10 years?

26. Find the effective yield, to the nearest tenth of a percent, of an investment at 7.5% compounded monthly.

27. Find the effective yield, to the nearest hundredth of a percent, of an investment at 7.75% compounded continuously.

28. What investment yields the greater return: 7% compounded monthly or 6.85% compounded continuously?

29. What investment yields the greater return: 8.25% compounded quarterly or 8.3% compounded semiannually?

30. Suppose that a certain radioactive substance has a half-life of 20 years. If there are presently 2500 milligrams of the substance, how much, to the nearest milligram, will remain after 40 years? After 50 years?

31. Strontium-90 has a half-life of 29 years. If there are 400 grams of strontium-90 initially, how much, to the nearest gram, will remain after 87 years? After 100 years?

32. The half-life of radium is approximately 1600 years. If the present amount of radium in a certain location is 500 grams, how much will remain after 800 years? Express your answer to the nearest gram.

33. Suppose that in a certain culture, the equation $Q(t) = 1000e^{0.4t}$ expresses the number of bacteria present as a function of the time t, where t is expressed in hours. How many bacteria are present at the end of 2 hours? 3 hours? 5 hours?

34. The number of bacteria present at a given time under certain conditions is given by the equation $Q = 5000e^{0.05t}$, where t is expressed in minutes. How many bacteria are present at the end of 10 minutes? 30 minutes? 1 hour?

35. The number of bacteria present in a certain culture after t hours is given by the equation $Q = Q_0e^{0.3t}$, where Q_0 represents the initial number of bacteria. If 6640 bacteria are present after 4 hours, how many bacteria were present initially?

36. The number of grams Q of a certain radioactive substance present after t seconds is given by the equation $Q = 1500e^{-0.4t}$. How many grams remain after 5 seconds? 10 seconds? 20 seconds?

37. The atmospheric pressure, measured in pounds per square inch, is a function of the altitude above sea level. The equation $P(a) = 14.7e^{-0.21a}$, where a is the altitude measured in miles, can be used to approximate atmospheric pressure. Find the atmospheric pressure at each of the following locations:
   (a) Mount McKinley in Alaska: altitude of 3.85 miles
   (b) Denver, Colorado: the "mile-high" city
   (c) Asheville, North Carolina: altitude of 1985 feet
   (d) Phoenix, Arizona: altitude of 1090 feet

38. Suppose that the present population of a city is 75,000. Using the equation $P(t) = 75,000e^{0.01t}$ to estimate future growth, estimate the population (a) 10 years from now, (b) 15 years from now, and (c) 25 years from now.

For Problems 39–44, graph each of the exponential functions.

39. $f(x) = e^x + 1$

40. $f(x) = e^x - 2$

41. $f(x) = 2e^x$

42. $f(x) = -e^x$

43. $f(x) = e^{2x}$

44. $f(x) = e^{-x}$

■ ■ ■ THOUGHTS INTO WORDS

45. Explain the difference between simple interest and compound interest.

46. Would it be better to invest $5000 at 6.25% interest compounded annually for 5 years or to invest $5000 at 6.25% interest compounded continuously for 5 years? Explain your answer.

47. How would you explain the concept of effective yield to someone who missed class when it was discussed?

48. How would you explain the half-life formula to someone who missed class when it was discussed?

■ ■ ■ **FURTHER INVESTIGATIONS**

49. Complete the following chart, which illustrates what happens to $1000 invested at various rates of interest for different lengths of time but always compounded continuously. Round your answers to the nearest dollar.

$1000 Compounded continuously

	8%	10%	12%	14%
5 years				
10 years				
15 years				
20 years				
25 years				

50. Complete the following chart, which illustrates what happens to $1000 invested at 12% for different lengths of time and different numbers of compounding periods. Round all of your answers to the nearest dollar.

$1000 at 12%

	1 year	5 years	10 years	20 years
Compounded annually				
Compounded semiannually				
Compounded quarterly				
Compounded monthly				
Compounded continuously				

51. Complete the following chart, which illustrates what happens to $1000 in 10 years based on different rates of interest and different numbers of compounding periods. Round your answers to the nearest dollar.

$1000 for 10 years

	8%	10%	12%	14%
Compounded annually				
Compounded semiannually				
Compounded quarterly				
Compounded monthly				
Compounded continuously				

For Problems 52–56, graph each of the functions.

52. $f(x) = x(2^x)$

53. $f(x) = \dfrac{e^x + e^{-x}}{2}$

54. $f(x) = \dfrac{2}{e^x + e^{-x}}$

55. $f(x) = \dfrac{e^x - e^{-x}}{2}$

56. $f(x) = \dfrac{2}{e^x - e^{-x}}$

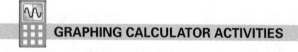 **GRAPHING CALCULATOR ACTIVITIES**

57. Use a graphing calculator to check your graphs for Problems 52–56.

58. Graph $f(x) = 2^x$, $f(x) = e^x$, and $f(x) = 3^x$ on the same set of axes. Are these graphs consistent with the discussion prior to Figure 10.7?

59. Graph $f(x) = e^x$. Where should the graphs of $f(x) = e^{x-4}$, $f(x) = e^{x-6}$, and $f(x) = e^{x+5}$ be located? Graph all three functions on the same set of axes with $f(x) = e^x$.

60. Graph $f(x) = e^x$. Now predict the graphs for $f(x) = -e^x$, $f(x) = e^{-x}$, and $f(x) = -e^{-x}$. Graph all three functions on the same set of axes with $f(x) = e^x$.

61. How do you think the graphs of $f(x) = e^x$, $f(x) = e^{2x}$, and $f(x) = 2e^x$ will compare? Graph them on the same set of axes to see if you were correct.

62. Find an approximate solution, to the nearest hundredth, for each of the following equations by graphing the appropriate function and finding the x intercept.
 (a) $e^x = 7$ **(b)** $e^x = 21$ **(c)** $e^x = 53$
 (d) $2e^x = 60$ **(e)** $e^{x+1} = 150$ **(f)** $e^{x-2} = 300$

63. Use a graphing approach to argue that it is better to invest money at 6% compounded quarterly than at 5.75% compounded continuously.

64. How long will it take $500 to be worth $1500 if it is invested at 7.5% interest compounded semiannually?

65. How long will it take $5000 to triple if it is invested at 6.75% interest compounded quarterly?

10.3 Inverse Functions

Recall the vertical-line test: If each vertical line intersects a graph in no more than one point, then the graph represents a function. There is also a useful distinction between two basic types of functions. Consider the graphs of the two functions in Figure 10.10: $f(x) = 2x - 1$ and $g(x) = x^2$. In Figure 10.10(a), any *horizontal line* will intersect the graph in no more than one point. Therefore every value of $f(x)$ has only one value of x associated with it. Any function that has this property of having exactly one value of x associated with each value of $f(x)$ is called a **one-to-one function**. Thus $g(x) = x^2$ is not a one-to-one function because the horizontal line in Figure 10.10(b) intersects the parabola in two points.

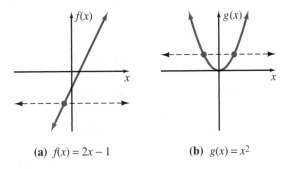

(a) $f(x) = 2x - 1$ **(b)** $g(x) = x^2$

Figure 10.10

The statement that for a function f to be a one-to-one function, every value of $f(x)$ has only one value of x associated with it can be equivalently stated: If $f(x_1) = f(x_2)$ for x_1 and x_2 in the domain of f, then $x_1 = x_2$. Let's use this last if-then statement to verify that $f(x) = 2x - 1$ is a one-to-one function. We start with the assumption that $f(x_1) = f(x_2)$:

$$2x_1 - 1 = 2x_2 - 1$$

$$2x_1 = 2x_2$$

$$x_1 = x_2$$

Thus $f(x) = 2x - 1$ is a one-to-one function.

To show that $g(x) = x^2$ is not a one-to-one function, we simply need to find two distinct real numbers in the domain of f that produce the same functional value. For example, $g(-2) = (-2)^2 = 4$ and $g(2) = 2^2 = 4$. Thus $g(x) = x^2$ is not a one-to-one function.

Now let's consider a one-to-one function f that assigns to each x in its domain D the value $f(x)$ in its range R (Figure 10.11(a)). We can define a new function g that goes from R to D; it assigns $f(x)$ in R back to x in D, as indicated in Figure 10.11(b). The functions f and g are called **inverse functions** of each other. The following definition precisely states this concept.

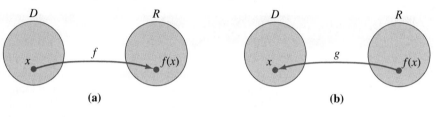

(a)   (b)

Figure 10.11

Definition 10.2

Let f be a one-to-one function with a domain of X and a range of Y. A function g with a domain of Y and a range of X is called the **inverse function** of f if

$$(f \circ g)(x) = x \quad \text{for every } x \text{ in } Y$$

and

$$(g \circ f)(x) = x \quad \text{for every } x \text{ in } X$$

In Definition 10.2, note that for f and g to be inverses of each other, the domain of f must equal the range of g, and the range of f must equal the domain of g. Furthermore, g must reverse the correspondences given by f, and f must reverse the correspondences given by g. In other words, inverse functions *undo* each other. Let's use Definition 10.2 to verify that two specific functions are inverses of each other.

E X A M P L E 1

Verify that $f(x) = 4x - 5$ and $g(x) = \dfrac{x + 5}{4}$ are inverse functions.

Solution

Because the set of real numbers is the domain and range of both functions, we know that the domain of f equals the range of g and that the range of f equals the domain of g. Furthermore,

$$(f \circ g)(x) = f(g(x))$$

$$= f\left(\frac{x + 5}{4}\right)$$

$$= 4\left(\frac{x + 5}{4}\right) - 5 = x$$

and

$$(g \circ f)(x) = g(f(x))$$
$$= g(4x - 5)$$
$$= \frac{4x - 5 + 5}{4} = x$$

Therefore f and g are inverses of each other. ∎

EXAMPLE 2

Verify that $f(x) = x^2 + 1$ for $x \geq 0$ and $g(x) = \sqrt{x - 1}$ for $x \geq 1$ are inverse functions.

Solution

First, note that the domain of f equals the range of g—namely, the set of nonnegative real numbers. Also, the range of f equals the domain of g—namely, the set of real numbers greater than or equal to 1. Furthermore,

$$(f \circ g)(x) = f(g(x))$$
$$= f(\sqrt{x - 1})$$
$$= (\sqrt{x - 1})^2 + 1$$
$$= x - 1 + 1 = x$$

and

$$(g \circ f)(x) = g(f(x))$$
$$= g(x^2 + 1)$$
$$= \sqrt{x^2 + 1 - 1} = \sqrt{x^2} = x \qquad \sqrt{x^2} = x \text{ because } x \geq 1$$

Therefore f and g are inverses of each other. ∎

The inverse of a function f is commonly denoted by f^{-1} (read "f inverse" or "the inverse of f"). Do not confuse the -1 in f^{-1} with a negative exponent. The symbol f^{-1} does *not* mean $1/f^1$ but rather refers to the inverse function of function f.

Remember that a function can also be thought of as a set of ordered pairs no two of which have the same first element. Along those lines, a one-to-one function further requires that no two of the ordered pairs have the same second element. Then, if the components of each ordered pair of a given one-to-one function are interchanged, the resulting function and the given function are inverses of each other. Thus, if

$$f = \{(1, 4), (2, 7), (5, 9)\}$$

then

$$f^{-1} = \{(4, 1), (7, 2), (9, 5)\}$$

Graphically, two functions that are inverses of each other are **mirror images with reference to the line $y = x$**. This is because ordered pairs (a, b) and (b, a) are reflections of each other with respect to the line $y = x$, as illustrated in Figure 10.12. (You will verify this in the next set of exercises.) Therefore, if the graph of a function f is known, as in Figure 10.13(a), then the graph of f^{-1} can be determined by reflecting f across the line $y = x$, as in Figure 10.13(b).

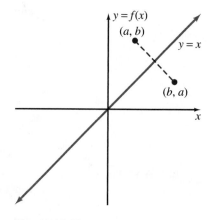

Figure 10.12

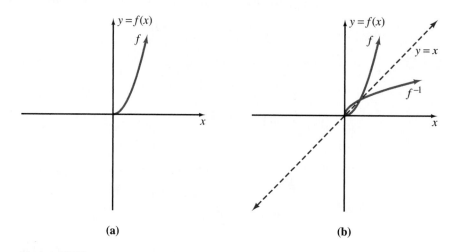

(a)   (b)

Figure 10.13

■ Finding Inverse Functions

The idea of inverse functions *undoing each other* provides the basis for an informal approach to finding the inverse of a function. Consider the function

$$f(x) = 2x + 1$$

To each x, this function assigns twice x plus 1. To undo this function, we can subtract 1 and divide by 2. Hence the inverse is

$$f^{-1}(x) = \frac{x - 1}{2}$$

Now let's verify that f and f^{-1} are indeed inverses of each other:

$$(f \circ f^{-1})(x) = f(f^{-1}(x)) \qquad\qquad (f^{-1} \circ f)(x) = f^{-1}(f(x))$$

$$= f\left(\frac{x-1}{2}\right) \qquad\qquad = f^{-1}(2x+1)$$

$$= 2\left(\frac{x-1}{2}\right) + 1 \qquad\qquad = \frac{2x+1-1}{2}$$

$$= x - 1 + 1 = x \qquad\qquad = \frac{2x}{2} = x$$

Thus the inverse of $f(x) = 2x + 1$ is $f^{-1}(x) = \dfrac{x-1}{2}$.

This informal approach may not work very well with more complex functions, but it does emphasize how inverse functions are related to each other. A more formal and systematic technique for finding the inverse of a function can be described as follows:

1. Replace the symbol $f(x)$ with y.

2. Interchange x and y.

3. Solve the equation for y in terms of x.

4. Replace y with the symbol $f^{-1}(x)$.

The following examples illustrate this technique.

E X A M P L E 3

Find the inverse of $f(x) = \dfrac{2}{3}x + \dfrac{3}{5}$.

Solution

When we replace $f(x)$ with y, the equation becomes $y = \dfrac{2}{3}x + \dfrac{3}{5}$. Interchanging x and y produces $x = \dfrac{2}{3}y + \dfrac{3}{5}$. Now, solving for y, we obtain

$$x = \frac{2}{3}y + \frac{3}{5}$$

$$15(x) = 15\left(\frac{2}{3}y + \frac{3}{5}\right)$$

$$15x = 10y + 9$$

$$15x - 9 = 10y$$

$$\frac{15x - 9}{10} = y$$

Finally, by replacing y with $f^{-1}(x)$, we can express the inverse function as

$$f^{-1}(x) = \frac{15x - 9}{10}$$

The domain of f is equal to the range of f^{-1} (both are the set of real numbers), and the range of f equals the domain of f^{-1} (both are the set of real numbers). Furthermore, we could show that $(f \circ f^{-1})(x) = x$ and $(f^{-1} \circ f)(x) = x$. We leave this for you to complete.   ∎

Does the function $f(x) = x^2 - 2$ have an inverse? Sometimes a graph of the function helps answer such a question. In Figure 10.14(a), it should be evident that f is not a one-to-one function and therefore cannot have an inverse. However, it should also be apparent from the graph that if we restrict the domain of f to the nonnegative real numbers, Figure 10.14(b), then it is a one-to-one function and should have an inverse function. The next example illustrates how to find the inverse function.

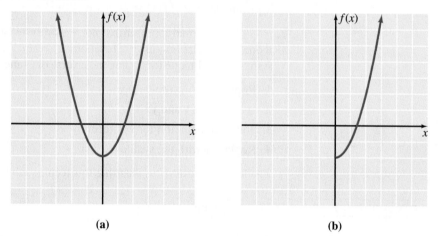

(a)   (b)

Figure 10.14

EXAMPLE   4

Find the inverse of $f(x) = x^2 - 2$, where $x \geq 0$.

Solution

When we replace $f(x)$ with y, the equation becomes

$$y = x^2 - 2, \qquad x \geq 0$$

Interchanging x and y produces

$$x = y^2 - 2, \qquad y \geq 0$$

Now let's solve for y; keep in mind that y is to be nonnegative.

$$x = y^2 - 2$$
$$x + 2 = y^2$$
$$\sqrt{x + 2} = y, \qquad x \geq -2$$

Finally, by replacing y with $f^{-1}(x)$, we can express the inverse function as

$$f^{-1}(x) = \sqrt{x + 2}, \qquad x \geq -2$$

The domain of f equals the range of f^{-1} (both are the nonnegative real numbers), and the range of f equals the domain of f^{-1} (both are the real numbers greater than or equal to -2). It can also be shown that $(f \circ f^{-1})(x) = x$ and $(f^{-1} \circ f)(x) = x$. Again, we leave this for you to complete. ∎

■ Increasing and Decreasing Functions

In Section 10.1, we used exponential functions as examples of increasing and decreasing functions. In reality, one function can be both increasing and decreasing over certain intervals. For example, in Figure 10.15, the function f is said to be *increasing* on the intervals $(-\infty, x_1]$ and $[x_2, \infty)$, and f is said to be *decreasing* on the interval $[x_1, x_2]$. More specifically, increasing and decreasing functions are defined as follows:

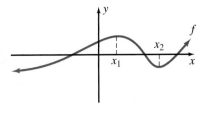

Figure 10.15

Definition 10.3

Let f be a function, with the interval I a subset of the domain of f. Let x_1 and x_2 be in I. Then:

1. f is *increasing* on I if $f(x_1) < f(x_2)$ whenever $x_1 < x_2$.

2. f is *decreasing* on I if $f(x_1) > f(x_2)$ whenever $x_1 < x_2$.

3. f is *constant* on I if $f(x_1) = f(x_2)$ for every x_1 and x_2.

Apply Definition 10.3, and you will see that the quadratic function $f(x) = x^2$ shown in Figure 10.16 is decreasing on $(-\infty, 0]$ and increasing on $[0, \infty)$. Likewise, the linear function $f(x) = 2x$ in Figure 10.17 is increasing throughout its domain of

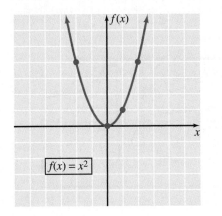

Figure 10.16

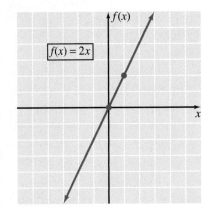

Figure 10.17

real numbers, so we say that it is increasing on $(-\infty, \infty)$. The function $f(x) = -2x$ in Figure 10.18 is decreasing on $(-\infty, \infty)$. For our purposes in this text, we will rely on our knowledge of the graphs of the functions to determine where functions are increasing and decreasing. More formal techniques for determining where functions increase and decrease will be developed in calculus.

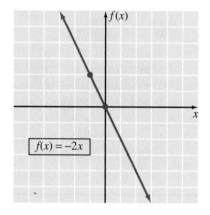

Figure 10.18

A function that is always increasing (or is always decreasing) over its entire domain is a one-to-one function and so has an inverse function. Furthermore, as illustrated by Example 4, even if a function is not one-to-one over its entire domain, it may be so over some subset of the domain. It then has an inverse function over this restricted domain. As functions become more complex, a graphing utility can be used to help with problems like those we have discussed in this section. For example, suppose that we want to know whether the function $f(x) = \dfrac{3x + 1}{x - 4}$ is a one-to-one function and therefore has an inverse function. Using a graphing utility, we can quickly get a sketch of the graph (Figure 10.19). Then, by applying the horizontal-line test to the graph, we can be fairly certain that the function is one-to-one.

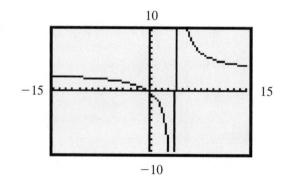

Figure 10.19

A graphing utility can also be used to help determine the intervals on which a function is increasing or decreasing. For example, to determine such intervals for the function $f(x) = \sqrt{x^2 + 4}$, let's use a graphing utility to get a sketch of the curve (Figure 10.20). From this graph, we see that the function is decreasing on the interval $(-\infty, 0]$ and is increasing on the interval $[0, \infty)$.

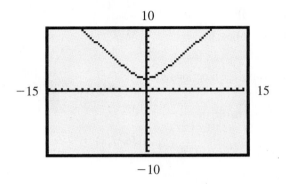

Figure 10.20

Problem Set 10.3

For Problems 1–6, determine whether the graph represents a one-to-one function.

1.

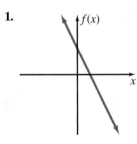

Figure 10.21

2.

Figure 10.22

5.

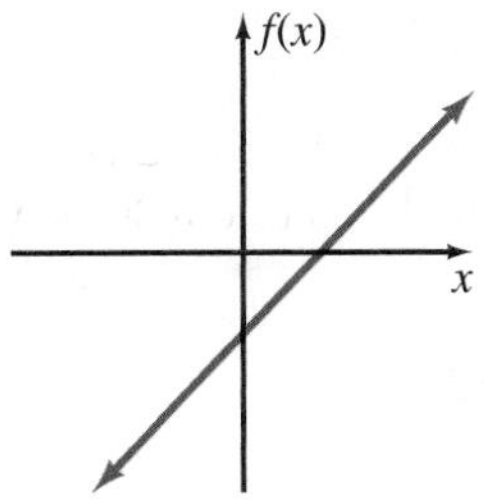

Figure 10.25

6.

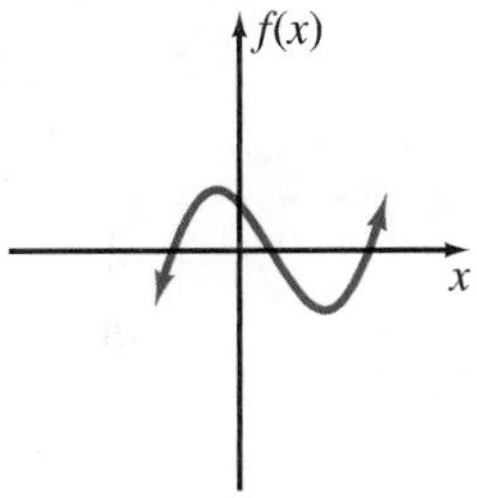

Figure 10.26

3.

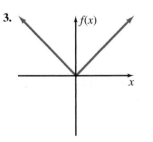

Figure 10.23

4.

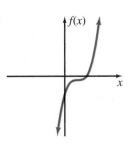

Figure 10.24

For Problems 7–14, determine whether the function f is one-to-one.

7. $f(x) = 5x + 4$

8. $f(x) = -3x + 4$

9. $f(x) = x^3$

10. $f(x) = x^5 + 1$

11. $f(x) = |x| + 1$

12. $f(x) = -|x| - 2$

13. $f(x) = -x^4$

14. $f(x) = x^4 + 1$

For Problems 15–18, (a) list the domain and range of the function, (b) form the inverse function f^{-1}, and (c) list the domain and range of f^{-1}.

15. $f = \{(1, 5), (2, 9), (5, 21)\}$

16. $f = \{(1, 1), (4, 2), (9, 3), (16, 4)\}$

17. $f = \{(0, 0), (2, 8), (-1, -1), (-2, -8)\}$

18. $f = \{(-1, 1), (-2, 4), (-3, 9), (-4, 16)\}$

For Problems 19–26, verify that the two given functions are inverses of each other.

19. $f(x) = 5x - 9$ and $g(x) = \dfrac{x + 9}{5}$

20. $f(x) = -3x + 4$ and $g(x) = \dfrac{4 - x}{3}$

21. $f(x) = -\dfrac{1}{2}x + \dfrac{5}{6}$ and $g(x) = -2x + \dfrac{5}{3}$

22. $f(x) = x^3 + 1$ and $g(x) = \sqrt[3]{x - 1}$

23. $f(x) = \dfrac{1}{x - 1}$ for $x > 1$ and

$g(x) = \dfrac{x + 1}{x}$ for $x > 0$

24. $f(x) = x^2 + 2$ for $x \geq 0$ and

$g(x) = \sqrt{x - 2}$ for $x \geq 2$

25. $f(x) = \sqrt{2x - 4}$ for $x \geq 2$ and

$g(x) = \dfrac{x^2 + 4}{2}$ for $x \geq 0$

26. $f(x) = x^2 - 4$ for $x \geq 0$. and

$g(x) = \sqrt{x + 4}$ for $x \geq -4$

For Problems 27–36, determine whether f and g are inverse functions.

27. $f(x) = 3x$ and $g(x) = -\dfrac{1}{3}x$

28. $f(x) = \dfrac{3}{4}x - 2$ and $g(x) = \dfrac{4}{3}x + \dfrac{8}{3}$

29. $f(x) = x^3$ and $g(x) = \sqrt[3]{x}$

30. $f(x) = \dfrac{1}{x + 1}$ and $g(x) = \dfrac{1 - x}{x}$

31. $f(x) = x$ and $g(x) = \dfrac{1}{x}$

32. $f(x) = \dfrac{3}{5}x + \dfrac{1}{3}$ and $g(x) = \dfrac{5}{3}x - 3$

33. $f(x) = x^2 - 3$ for $x \geq 0$ and
$g(x) = \sqrt{x + 3}$ for $x \geq -3$

34. $f(x) = |x - 1|$ for $x \geq 1$ and
$g(x) = |x + 1|$ for $x \geq 0$

35. $f(x) = \sqrt{x + 1}$ and $g(x) = x^2 - 1$ for $x \geq 0$

36. $f(x) = \sqrt{2x - 2}$ and $g(x) = \dfrac{1}{2}x^2 + 1$

For Problems 37–50, (a) find f^{-1} and (b) verify that $(f \circ f^{-1})(x) = x$ and $(f^{-1} \circ f)(x) = x$.

37. $f(x) = x - 4$

38. $f(x) = 2x - 1$

39. $f(x) = -3x - 4$

40. $f(x) = -5x + 6$

41. $f(x) = \dfrac{3}{4}x - \dfrac{5}{6}$

42. $f(x) = \dfrac{2}{3}x - \dfrac{1}{4}$

43. $f(x) = -\dfrac{2}{3}x$

44. $f(x) = \dfrac{4}{3}x$

45. $f(x) = \sqrt{x}$ for $x \geq 0$

46. $f(x) = \dfrac{1}{x}$ for $x \neq 0$

47. $f(x) = x^2 + 4$ for $x \geq 0$

48. $f(x) = x^2 + 1$ for $x \leq 0$

49. $f(x) = 1 + \dfrac{1}{x}$ for $x > 0$

50. $f(x) = \dfrac{x}{x + 1}$ for $x > -1$

For Problems 51–58, (a) find f^{-1} and (b) graph f and f^{-1} on the same set of axes.

51. $f(x) = 3x$

52. $f(x) = -x$

53. $f(x) = 2x + 1$

54. $f(x) = -3x - 3$

55. $f(x) = \dfrac{2}{x-1}$ for $x > 1$

56. $f(x) = \dfrac{-1}{x-2}$ for $x > 2$

57. $f(x) = x^2 - 4$ for $x \geq 0$

58. $f(x) = \sqrt{x-3}$ for $x \geq 3$

61. $f(x) = -3x + 1$

62. $f(x) = (x-3)^2 + 1$

63. $f(x) = -(x+2)^2 - 1$

64. $f(x) = x^2 - 2x + 6$

65. $f(x) = -2x^2 - 16x - 35$

66. $f(x) = x^2 + 3x - 1$

For Problems 59–66, find the intervals on which the given function is increasing and the intervals on which it is decreasing.

59. $f(x) = x^2 + 1$

60. $f(x) = x^3$

■■■ THOUGHTS INTO WORDS

67. Does the function $f(x) = 4$ have an inverse? Explain your answer.

68. Explain why every nonconstant linear function has an inverse.

69. Are the functions $f(x) = x^4$ and $g(x) = \sqrt[4]{x}$ inverses of each other? Explain your answer.

70. What does it mean to say that 2 and -2 are additive inverses of each other? What does it mean to say that 2 and $\dfrac{1}{2}$ are multiplicative inverses of each other? What does it mean to say that the functions $f(x) = x - 2$ and $f(x) = x + 2$ are inverses of each other? Do you think that the concept of "inverse" is being used in a consistent manner? Explain your answer.

■■■ FURTHER INVESTIGATIONS

71. The function notation and the operation of composition can be used to find inverses as follows: To find the inverse of $f(x) = 5x + 3$, we know that $f(f^{-1}(x))$ must produce x. Therefore

$$f(f^{-1}(x)) = 5[f^{-1}(x)] + 3 = x$$

$$5[f^{-1}(x)] = x - 3$$

$$f^{-1}(x) = \frac{x-3}{5}$$

Use this approach to find the inverse of each of the following functions.

(a) $f(x) = 3x - 9$ **(b)** $f(x) = -2x + 6$

(c) $f(x) = -x + 1$ **(d)** $f(x) = 2x$

(e) $f(x) = -5x$ **(f)** $f(x) = x^2 + 6$ for $x \geq 0$

72. If $f(x) = 2x + 3$ and $g(x) = 3x - 5$, find

(a) $(f \circ g)^{-1}(x)$ **(b)** $(f^{-1} \circ g^{-1})(x)$

(c) $(g^{-1} \circ f^{-1})(x)$

GRAPHING CALCULATOR ACTIVITIES

73. For Problems 37–44, graph the given function, the inverse function that you found, and $f(x) = x$ on the same set of axes. In each case, the given function and its inverse should produce graphs that are reflections of each other through the line $f(x) = x$.

74. There is another way we can use the graphing calculator to help show that two functions are inverses of each other. Suppose we want to show that $f(x) = x^2 - 2$ for $x \geq 0$ and $g(x) = \sqrt{x + 2}$ for $x \geq -2$ are inverses of each other. Let's make the following assignments for our graphing calculator.

$$f: \quad Y_1 = x^2 - 2$$

$$g: \quad Y_2 = \sqrt{x + 2}$$

$$f \circ g: \quad Y_3 = (Y_2)^2 - 2$$

$$g \circ f: \quad Y_4 = \sqrt{Y_1 + 2}$$

Now we can proceed as follows:

1. Graph $Y_1 = x^2 - 2$, and note that for $x > 0$, the range is greater than or equal to -2.

2. Graph $Y_2 = \sqrt{x + 2}$, and note that for $x \geq -2$, the range is greater than or equal to 0.
Thus the domain of f equals the range of g, and the range of f equals the domain of g.

3. Graph $Y_3 = (Y_2)^2 - 2$ for $x \geq -2$, and observe the line $y = x$ for $x \geq -2$.

4. Graph $Y_4 = \sqrt{Y_1 + 2}$ for $x \geq 0$, and observe the line $y = x$ for $x \geq 0$.
Thus $(f \circ g)(x) = x$ and $(g \circ f)(x) = x$, and the two functions are inverses of each other.

Use this approach to check your answers for Problems 45–50.

75. Use the technique demonstrated in Problem 74 to show that

$$f(x) = \frac{x}{\sqrt{x^2 + 1}}$$

and

$$g(x) = \frac{x}{\sqrt{1 - x^2}} \qquad \text{for } -1 < x < 1$$

are inverses of each other.

10.4   Logarithms

In Sections 10.1 and 10.2 we discussed exponential expressions of the form b^n, where b is any positive real number and n is any real number; we used exponential expressions of the form b^n to define exponential functions; and we used exponential functions to help solve problems. In the next three sections, we will follow the same basic pattern with respect to a new concept—logarithms. Let's begin with the following definition:

Definition 10.4

If r is any positive real number, then the unique exponent t such that $b^t = r$ is called the **logarithm of r with base b** and is denoted by $\log_b r$.

According to Definition 10.4, the logarithm of 16 base 2 is the exponent t such that $2^t = 16$; thus we can write $\log_2 16 = 4$. Likewise, we can write $\log_{10} 1000 = 3$ because $10^3 = 1000$. In general, we can remember Definition 10.4 by the statement

$$\log_b r = t \quad \text{is equivalent to} \quad b^t = r$$

Therefore we can easily switch back and forth between exponential and logarithmic forms of equations, as the next examples illustrate.

$$\log_2 8 = 3 \quad \text{is equivalent to} \quad 2^3 = 8$$

$$\log_{10} 100 = 2 \quad \text{is equivalent to} \quad 10^2 = 100$$

$$\log_3 81 = 4 \quad \text{is equivalent to} \quad 3^4 = 81$$

$$\log_{10} 0.001 = -3 \quad \text{is equivalent to} \quad 10^{-3} = 0.001$$

$$\log_m n = p \quad \text{is equivalent to} \quad m^p = n$$

$$2^7 = 128 \quad \text{is equivalent to} \quad \log_2 128 = 7$$

$$5^3 = 125 \quad \text{is equivalent to} \quad \log_5 125 = 3$$

$$\left(\frac{1}{2}\right)^4 = \frac{1}{16} \quad \text{is equivalent to} \quad \log_{1/2}\left(\frac{1}{16}\right) = 4$$

$$10^{-2} = 0.01 \quad \text{is equivalent to} \quad \log_{10} 0.01 = -2$$

$$a^b = c \quad \text{is equivalent to} \quad \log_a c = b$$

Some logarithms can be determined by changing to exponential form and using the properties of exponents, as the next two examples illustrate.

EXAMPLE   1

Evaluate $\log_{10} 0.0001$.

Solution

Let $\log_{10} 0.0001 = x$. Then, by changing to exponential form, we have $10^x = 0.0001$, which can be solved as follows:

$$10^x = 0.0001$$

$$10^x = 10^{-4} \qquad 0.0001 = \frac{1}{10{,}000} = \frac{1}{10^4} = 10^{-4}$$

$$x = -4$$

Thus we have $\log_{10} 0.0001 = -4$.   ∎

EXAMPLE 2

Evaluate $\log_9\left(\dfrac{\sqrt[5]{27}}{3}\right)$.

Solution

Let $\log_9\left(\dfrac{\sqrt[5]{27}}{3}\right) = x$. Then, by changing to exponential form, we have $9^x = \dfrac{\sqrt[5]{27}}{3}$, which can be solved as follows:

$$9^x = \frac{(27)^{1/5}}{3}$$

$$(3^2)^x = \frac{(3^3)^{1/5}}{3}$$

$$3^{2x} = \frac{3^{3/5}}{3}$$

$$3^{2x} = 3^{-2/5}$$

$$2x = -\frac{2}{5}$$

$$x = -\frac{1}{5}$$

Therefore we have $\log_9\left(\dfrac{\sqrt[5]{27}}{3}\right) = -\dfrac{1}{5}$. ∎

Some equations that involve logarithms can also be solved by changing to exponential form and using our knowledge of exponents.

EXAMPLE 3

Solve $\log_8 x = \dfrac{2}{3}$.

Solution

Changing $\log_8 x = \dfrac{2}{3}$ to exponential form, we obtain

$$8^{2/3} = x$$

Therefore

$$x = (\sqrt[3]{8})^2$$

$$= 2^2$$

$$= 4$$

The solution set is {4}. ∎

EXAMPLE 4

Solve $\log_b\left(\dfrac{27}{64}\right) = 3$.

Solution

Change $\log_b\left(\dfrac{27}{64}\right) = 3$ to exponential form to obtain

$$b^3 = \frac{27}{64}$$

Therefore

$$b = \sqrt[3]{\frac{27}{64}}$$

$$= \frac{3}{4}$$

The solution set is $\left\{\dfrac{3}{4}\right\}$.   ∎

■ Properties of Logarithms

There are some properties of logarithms that are a direct consequence of Definition 10.2 and the properties of exponents. For example, the following property is obtained by writing the exponential equations $b^1 = b$ and $b^0 = 1$ in logarithmic form.

Property 10.3

For $b > 0$ and $b \neq 1$,

$$\log_b b = 1 \qquad \text{and} \qquad \log_b 1 = 0$$

Therefore according to Property 10.3, we can write

$$\log_{10} 10 = 1 \qquad \log_4 4 = 1$$

$$\log_{10} 1 \ = 0 \qquad \log_5 1 = 0$$

Also, from Definition 10.2, we know that $\log_b r$ is the exponent t such that $b^t = r$. Therefore, raising b to the $\log_b r$ power must produce r. This fact is stated in Property 10.4.

Property 10.4

For $b > 0$, $b \neq 1$, and $r > 0$,

$$b^{\log_b r} = r$$

Therefore according to Property 10.4, we can write

$$10^{\log_{10} 72} = 72 \qquad 3^{\log_3 85} = 85 \qquad e^{\log_e 7} = 7$$

Because a logarithm is by definition an exponent, it seems reasonable to predict that some properties of logarithms correspond to the basic exponential properties. This is an accurate prediction; these properties provide a basis for computational work with logarithms. Let's state the first of these properties and show how we can use our knowledge of exponents to verify it.

Property 10.5

> For positive numbers b, r, and s, where $b \neq 1$,
>
> $$\log_b rs = \log_b r + \log_b s$$

To verify Property 10.5, we can proceed as follows. Let $m = \log_b r$ and $n = \log_b s$. Change each of these equations to exponential form:

$$m = \log_b r \quad \text{becomes} \quad r = b^m$$

$$n = \log_b s \quad \text{becomes} \quad s = b^n$$

Thus the product rs becomes

$$rs = b^m \cdot b^n = b^{m+n}$$

Now, by changing $rs = b^{m+n}$ back to logarithmic form, we obtain

$$\log_b rs = m + n$$

Replace m with $\log_b r$ and replace n with $\log_b s$ to yield

$$\log_b rs = \log_b r + \log_b s$$

The following two examples illustrate the use of Property 10.5.

E X A M P L E   5

If $\log_2 5 = 2.3222$ and $\log_2 3 = 1.5850$, evaluate $\log_2 15$.

Solution

Because $15 = 5 \cdot 3$, we can apply Property 10.5 as follows:

$$\log_2 15 = \log_2(5 \cdot 3)$$
$$= \log_2 5 + \log_2 3$$
$$= 2.3222 + 1.5850$$
$$= 3.9072 \qquad \blacksquare$$

EXAMPLE 6   Given that $\log_{10} 178 = 2.2504$ and $\log_{10} 89 = 1.9494$, evaluate $\log_{10}(178 \cdot 89)$.

Solution

$$\log_{10}(178 \cdot 89) = \log_{10} 178 + \log_{10} 89$$
$$= 2.2504 + 1.9494$$
$$= 4.1998 \qquad ■$$

Because $\dfrac{b^m}{b^n} = b^{m-n}$, we would expect a corresponding property that pertains to logarithms. Property 10.6 is that property. We can verify it by using an approach similar to the one we used to verify Property 10.5. This verification is left for you to do as an exercise in the next problem set.

Property 10.6

For positive numbers b, r, and s, where $b \neq 1$,

$$\log_b\left(\frac{r}{s}\right) = \log_b r - \log_b s$$

We can use Property 10.6 to change a division problem into an equivalent subtraction problem, as the next two examples illustrate.

EXAMPLE 7   If $\log_5 36 = 2.2265$ and $\log_5 4 = 0.8614$, evaluate $\log_5 9$.

Solution

Because $9 = \dfrac{36}{4}$, we can use Property 10.6 as follows:

$$\log_5 9 = \log_5\left(\frac{36}{4}\right)$$
$$= \log_5 36 - \log_5 4$$
$$= 2.2265 - 0.8614$$
$$= 1.3651 \qquad ■$$

EXAMPLE 8   Evaluate $\log_{10}\left(\dfrac{379}{86}\right)$ given that $\log_{10} 379 = 2.5786$ and $\log_{10} 86 = 1.9345$.

Solution

$$\log_{10}\left(\frac{379}{86}\right) = \log_{10} 379 - \log_{10} 86$$
$$= 2.5786 - 1.9345$$
$$= 0.6441 \qquad ■$$

Another property of exponents states that $(b^n)^m = b^{mn}$. The corresponding property of logarithms is stated in Property 10.7. Again, we will leave the verification of this property as an exercise for you to do in the next set of problems.

Property 10.7

> If r is a positive real number, b is a positive real number other than 1, and p is any real number, then
>
> $$\log_b r^p = p(\log_b r)$$

We will use Property 10.7 in the next two examples.

EXAMPLE 9

Evaluate $\log_2 22^{1/3}$ given that $\log_2 22 = 4.4598$.

Solution

$$\log_2 22^{1/3} = \frac{1}{3}\log_2 22 \qquad \text{Property 10.7}$$

$$= \frac{1}{3}(4.4598)$$

$$= 1.4866 \qquad\qquad\blacksquare$$

EXAMPLE 10

Evaluate $\log_{10}(8540)^{3/5}$ given that $\log_{10} 8540 = 3.9315$.

Solution

$$\log_{10}(8540)^{3/5} = \frac{3}{5}\log_{10} 8540$$

$$= \frac{3}{5}(3.9315)$$

$$= 2.3589 \qquad\qquad\blacksquare$$

Used together, the properties of logarithms allow us to change the forms of various logarithmic expressions. For example, we can rewrite an expression such as $\log_b \sqrt{\dfrac{xy}{z}}$ in terms of sums and differences of simpler logarithmic quantities as follows:

$$\log_b \sqrt{\frac{xy}{z}} = \log_b \left(\frac{xy}{z}\right)^{1/2}$$

$$= \frac{1}{2}\log_b \left(\frac{xy}{z}\right) \qquad\qquad \text{Property 10.7}$$

$$= \frac{1}{2}(\log_b xy - \log_b z) \qquad\qquad \text{Property 10.6}$$

$$= \frac{1}{2}(\log_b x + \log_b y - \log_b z) \qquad\qquad \text{Property 10.5}$$

Sometimes we need to change from an indicated sum or difference of logarithmic quantities to an indicated product or quotient. This is especially helpful when solving certain kinds of equations that involve logarithms. Note in these next two examples how we can use the properties, along with the process of changing from logarithmic form to exponential form, to solve some equations.

EXAMPLE   11

Solve $\log_{10} x + \log_{10}(x + 9) = 1$.

Solution

$$\log_{10} x + \log_{10}(x + 9) = 1$$

$$\log_{10}[x(x + 9)] = 1 \qquad \text{Property 10.5}$$

$$10^1 = x(x + 9) \qquad \text{Change to exponential form.}$$

$$10 = x^2 + 9x$$

$$0 = x^2 + 9x - 10$$

$$0 = (x + 10)(x - 1)$$

$$x + 10 = 0 \qquad \text{or} \qquad x - 1 = 0$$

$$x = -10 \qquad \text{or} \qquad x = 1$$

Logarithms are defined only for positive numbers, so x and $x + 9$ have to be positive. Therefore the solution of -10 must be discarded. The solution set is $\{1\}$.   ■

EXAMPLE   12

Solve $\log_5(x + 4) - \log_5 x = 2$.

Solution

$$\log_5(x + 4) - \log_5 x = 2$$

$$\log_5\left(\frac{x + 4}{x}\right) = 2 \qquad \text{Property 10.6}$$

$$5^2 = \frac{x + 4}{x} \qquad \text{Change to exponential form.}$$

$$25 = \frac{x + 4}{x}$$

$$25x = x + 4$$

$$24x = 4$$

$$x = \frac{4}{24} = \frac{1}{6}$$

The solution set is $\left\{\dfrac{1}{6}\right\}$.   ■

Because logarithms are defined only for positive numbers, we should realize that some logarithmic equations may not have any solutions. (In those cases, the solution set is the null set.) It is also possible for a logarithmic equation to have a negative solution as the next example illustrates.

E X A M P L E   1 3

Solve $\log_2 3 + \log_2(x + 4) = 3$.

Solution

$$\log_2 3 + \log_2(x + 4) = 3$$

$$\log_2 3(x + 4) = 3 \qquad \text{Property 10.5}$$

$$3(x + 4) = 2^3 \qquad \text{Change to exponential form.}$$

$$3x + 12 = 8$$

$$3x = -4$$

$$x = -\frac{4}{3}$$

The only restriction is that $x + 4 > 0$ or $x > -4$. Therefore, the solution set is $\left\{-\dfrac{4}{3}\right\}$.

Perhaps you should check this answer.   ∎

Problem Set 10.4

For Problems 1–10, write each exponential statement in logarithmic form. For example, $2^5 = 32$ becomes $\log_2 32 = 5$ in logarithmic form.

1. $2^7 = 128$

2. $3^3 = 27$

3. $5^3 = 125$

4. $2^6 = 64$

5. $10^3 = 1000$

6. $10^1 = 10$

7. $2^{-2} = \dfrac{1}{4}$

8. $3^{-4} = \dfrac{1}{81}$

9. $10^{-1} = 0.1$

10. $10^{-2} = 0.01$

For Problems 11–20, write each logarithmic statement in exponential form. For example, $\log_2 8 = 3$ becomes $2^3 = 8$ in exponential form.

11. $\log_3 81 = 4$

12. $\log_2 256 = 8$

13. $\log_4 64 = 3$

14. $\log_5 25 = 2$

15. $\log_{10} 10,000 = 4$

16. $\log_{10} 100,000 = 5$

17. $\log_2\left(\dfrac{1}{16}\right) = -4$

18. $\log_5\left(\dfrac{1}{125}\right) = -3$

19. $\log_{10} 0.001 = -3$

20. $\log_{10} 0.000001 = -6$

For Problems 21–40, evaluate each logarithmic expression.

21. $\log_2 16$ = 4

22. $\log_3 9$ = 2

23. $\log_3 81$ = 4

24. $\log_2 512$ = 9

25. $\log_6 216$ = 3

26. $\log_4 256$ = 4

27. $\log_7 \sqrt{7}$ = $\frac{1}{2}$

28. $\log_2 \sqrt[3]{2}$ = $\frac{1}{3}$

29. $\log_{10} 1$ = 0

30. $\log_{10} 10$ = 1

31. $\log_{10} 0.1$ = −1

32. $\log_{10} 0.0001$ = −4

33. $10^{\log_{10} 5}$ = 5

34. $10^{\log_{10} 14}$ = 14

35. $\log_2\left(\dfrac{1}{32}\right)$ **36.** $\log_5\left(\dfrac{1}{25}\right)$

37. $\log_5(\log_2 32)$ **38.** $\log_2(\log_4 16)$

39. $\log_{10}(\log_7 7)$ **40.** $\log_2(\log_5 5)$

For Problems 41–50, solve each equation.

41. $\log_7 x = 2$ **42.** $\log_2 x = 5$

43. $\log_8 x = \dfrac{4}{3}$ **44.** $\log_{16} x = \dfrac{3}{2}$

45. $\log_9 x = \dfrac{3}{2}$ **46.** $\log_8 x = -\dfrac{2}{3}$

47. $\log_4 x = -\dfrac{3}{2}$ **48.** $\log_9 x = -\dfrac{5}{2}$

49. $\log_x 2 = \dfrac{1}{2}$ **50.** $\log_x 3 = \dfrac{1}{2}$

For Problems 51–59, given that $\log_2 5 = 2.3219$ and $\log_2 7 = 2.8074$, evaluate each expression by using Properties 10.5–10.7.

51. $\log_2 35$ **52.** $\log_2\left(\dfrac{7}{5}\right)$

53. $\log_2 125$ **54.** $\log_2 49$

55. $\log_2 \sqrt{7}$ **56.** $\log_2 \sqrt[3]{5}$

57. $\log_2 175$ **58.** $\log_2 56$

59. $\log_2 80$

For Problems 60–68, given that $\log_8 5 = 0.7740$ and $\log_8 11 = 1.1531$, evaluate each expression using Properties 10.5–10.7.

60. $\log_8 55$ **61.** $\log_8\left(\dfrac{5}{11}\right)$

62. $\log_8 25$ **63.** $\log_8 \sqrt{11}$

64. $\log_8 (5)^{2/3}$ **65.** $\log_8 88$

66. $\log_8 320$ **67.** $\log_8\left(\dfrac{25}{11}\right)$

68. $\log_8\left(\dfrac{121}{25}\right)$

For Problems 69–80, express each of the following as the sum or difference of simpler logarithmic quantities. Assume that all variables represent positive real numbers. For example,

$$\log_b \frac{x^3}{y^2} = \log_b x^3 - \log_b y^2$$

$$= 3\log_b x - 2\log_b y$$

69. $\log_b xyz$ **70.** $\log_b 5x$

71. $\log_b\left(\dfrac{y}{z}\right)$ **72.** $\log_b\left(\dfrac{x^2}{y}\right)$

73. $\log_b y^3 z^4$ **74.** $\log_b x^2 y^3$

75. $\log_b\left(\dfrac{x^{1/2} y^{1/3}}{z^4}\right)$ **76.** $\log_b x^{2/3} y^{3/4}$

77. $\log_b \sqrt[3]{x^2 z}$ **78.** $\log_b \sqrt{xy}$

79. $\log_b\left(x\sqrt{\dfrac{x}{y}}\right)$ **80.** $\log_b \sqrt{\dfrac{x}{y}}$

For Problems 81–88, express each of the following as a single logarithm. (Assume that all variables represent positive real numbers.) For example,

$$3\log_b x + 5\log_b y = \log_b x^3 y^5$$

81. $2\log_b x - 4\log_b y$

82. $\log_b x + \log_b y - \log_b z$

83. $\log_b x - (\log_b y - \log_b z)$

84. $(\log_b x - \log_b y) - \log_b z$

85. $2\log_b x + 4\log_b y - 3\log_b z$

86. $\log_b x + \dfrac{1}{2}\log_b y$

87. $\dfrac{1}{2}\log_b x - \log_b x + 4\log_b y$

88. $2\log_b x + \dfrac{1}{2}\log_b(x - 1) - 4\log_b(2x + 5)$

For Problems 89–106, solve each equation.

89. $\log_3 x + \log_3 4 = 2$

90. $\log_7 5 + \log_7 x = 1$

91. $\log_{10} x + \log_{10}(x - 21) = 2$

92. $\log_{10} x + \log_{10}(x - 3) = 1$

93. $\log_2 x + \log_2(x - 3) = 2$

94. $\log_3 x + \log_3(x - 2) = 1$

95. $\log_3(x + 3) + \log_3(x + 5) = 1$

96. $\log_2(x + 2) = 1 - \log_2(x + 3)$

97. $\log_2 3 + \log_2(x + 4) = 3$

98. $\log_4 7 + \log_4(x + 3) = 2$

99. $\log_{10}(2x - 1) - \log_{10}(x - 2) = 1$

100. $\log_{10}(9x - 2) = 1 + \log_{10}(x - 4)$

101. $\log_5(3x - 2) = 1 + \log_5(x - 4)$

102. $\log_6 x + \log_6(x + 5) = 2$

103. $\log_2(x - 1) - \log_2(x + 3) = 2$

104. $\log_5 x = \log_5(x + 2) + 1$

105. $\log_8(x + 7) + \log_8 x = 1$

106. $\log_6(x + 1) + \log_6(x - 4) = 2$

107. Verify Property 10.6.

108. Verify Property 10.7.

■ ■ ■ THOUGHTS INTO WORDS

109. Explain, without using Property 10.4, why $4^{\log_4 9}$ equals 9.

110. How would you explain the concept of a logarithm to someone who had just completed an elementary algebra course?

111. In the next section, we will show that the logarithmic function $f(x) = \log_2 x$ is the inverse of the exponential function $f(x) = 2^x$. From that information, how could you sketch a graph of $f(x) = \log_2 x$?

10.5 Logarithmic Functions

We can now use the concept of a logarithm to define a logarithmic function as follows:

Definition 10.5

If $b > 0$ and $b \neq 1$, then the function defined by

$$f(x) = \log_b x$$

where x is any positive real number, is called the **logarithmic function with base b**.

We can obtain the graph of a specific logarithmic function in various ways. For example, the equation $y = \log_2 x$ can be changed to the exponential equation $2^y = x$, where we can determine a table of values. The next set of exercises asks you to use this approach to graph some logarithmic functions. We can also set up a table of values directly from the logarithmic equation and sketch the graph from the table. Example 1 illustrates this approach.

EXAMPLE 1

Graph $f(x) = \log_2 x$.

Solution

Let's choose some values for x where we can easily determine the corresponding values for $\log_2 x$. (Remember that logarithms are defined only for the positive real numbers.)

x	$f(x)$
$\dfrac{1}{8}$	-3
$\dfrac{1}{4}$	-2
$\dfrac{1}{2}$	-1
1	0
2	1
4	2
8	3

$\text{Log}_2 \dfrac{1}{8} = -3$ because $2^{-3} = \dfrac{1}{2^3} = \dfrac{1}{8}$.

$\text{Log}_2 1 = 0$ because $2^0 = 1$.

Plot these points and connect them with a smooth curve to produce Figure 10.27.

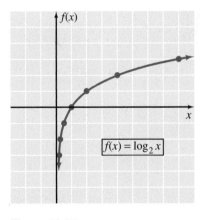

$$f(x) = \log_2 x$$

Figure 10.27 ∎

Now suppose that we consider two functions f and g as follows:

$f(x) = b^x$ Domain: all real numbers
                      Range: positive real numbers

$g(x) = \log_b x$ Domain: positive real numbers
                      Range: all real numbers

Furthermore, suppose that we consider the composition of f and g and the composition of g and f.

$$(f \circ g)(x) = f(g(x)) = f(\log_b x) = b^{\log_b x} = x$$

$$(g \circ f)(x) = g(f(x)) = g(b^x) = \log_b b^x = x \log_b b = x(1) = x$$

Because the domain of f is the range of g, the range of f is the domain of g, $f(g(x)) = x$, and $g(f(x)) = x$, the two functions f and g are *inverses* of each other.

Remember that the graph of a function and the graph of its inverse are reflections of each other through the line $y = x$. Thus we can determine the graph of a logarithmic function by reflecting the graph of its inverse exponential function through the line $y = x$. We demonstrate this idea in Figure 10.28, where the graph of $y = 2^x$ has been reflected across the line $y = x$ to produce the graph of $y = \log_2 x$.

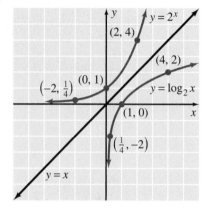

The general behavior patterns of exponential functions were illustrated back in Figure 10.3. We can now reflect each of these graphs through the line $y = x$ and observe the general behavior patterns of logarithmic functions, shown in Figure 10.29.

Figure 10.28

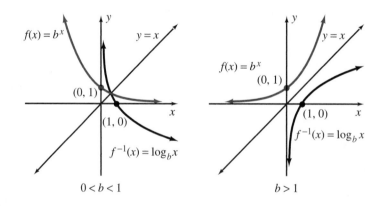

Figure 10.29

As you graph logarithmic functions, don't forget about transformations of basic curves.

1. The graph of $f(x) = 3 + \log_2 x$ is the graph of $f(x) = \log_2 x$ moved up three units. (Because $\log_2 x + 3$ is apt to be confused with $\log_2(x + 3)$, we commonly write $3 + \log_2 x$.)

2. The graph of $f(x) = \log_2(x - 4)$ is the graph of $f(x) = \log_2 x$ moved four units to the right.

3. The graph of $f(x) = -\log_2 x$ is the graph of $f(x) = \log_2 x$ reflected across the x axis.

■ Common Logarithms—Base 10

The properties of logarithms we discussed in Section 10.4 are true for any valid base. However, because the Hindu-Arabic numeration system that we use is a base-10 system, logarithms to base 10 have historically been used for computational purposes. Base-10 logarithms are called **common logarithms**.

Originally, common logarithms were developed to aid in complicated numerical calculations that involve products, quotients, and powers of real numbers. Today they are seldom used for that purpose because the calculator and computer can much more effectively handle the messy computational problems. However, common logarithms do still occur in applications, so they deserve our attention.

As we know from earlier work, the definition of a logarithm provides the basis for evaluating $\log_{10} x$ for values of x that are integral powers of 10. Consider the following examples:

$$\log_{10} 1000 = 3 \quad \text{because } 10^3 = 1000$$

$$\log_{10} 100 = 2 \quad \text{because } 10^2 = 100$$

$$\log_{10} 10 = 1 \quad \text{because } 10^1 = 10$$

$$\log_{10} 1 = 0 \quad \text{because } 10^0 = 1$$

$$\log_{10} 0.1 = -1 \quad \text{because } 10^{-1} = \frac{1}{10} = 0.1$$

$$\log_{10} 0.01 = -2 \quad \text{because } 10^{-2} = \frac{1}{10^2} = 0.01$$

$$\log_{10} 0.001 = -3 \quad \text{because } 10^{-3} = \frac{1}{10^3} = 0.001$$

When working exclusively with base-10 logarithms, it is customary to omit writing the numeral 10 to designate the base. Thus the expression $\log_{10} x$ is written as $\log x$, and a statement such as $\log_{10} 1000 = 3$ becomes $\log 1000 = 3$. We will follow this practice from now on in this chapter, but don't forget that the base is understood to be 10.

$$\log_{10} x = \log x$$

To find the common logarithm of a positive number that is not an integral power of 10, we can use an appropriately equipped calculator. A calculator that has a common logarithm function (ordinarily, a key labeled $\boxed{\text{log}}$ is used) gives us the following results rounded to four decimal places:

$\log 1.75 = 0.2430$

$\log 23.8 = 1.3766$ Be sure that you can use a calculator and obtain these results.

$\log 134 = 2.1271$

$\log 0.192 = -0.7167$

$\log 0.0246 = -1.6091$

In order to use logarithms to solve problems, we sometimes need to be able to determine a number when the logarithm of the number is known. That is to say, we may need to determine x if $\log x$ is known. Let's consider an example.

EXAMPLE 2   Find x if $\log x = 0.2430$.

Solution

If $\log x = 0.2430$, then by changing to exponential form we have $10^{0.2430} = x$. Use the $\boxed{10^x}$ key to find x:

$$x = 10^{0.2430} \approx 1.749846689$$

Therefore $x = 1.7498$ rounded to five significant digits.   ■

Be sure that you can use your calculator and obtain the following results. We rounded the values for x to five significant digits.

If $\log x = 0.7629$, then $x = 10^{0.7629} = 5.7930$.
If $\log x = 1.4825$, then $x = 10^{1.4825} = 30.374$.
If $\log x = 4.0214$, then $x = 10^{4.0214} = 10,505$.
If $\log x = -1.5162$, then $x = 10^{-1.5162} = 0.030465$.
If $\log x = -3.8921$, then $x = 10^{-3.8921} = 0.00012820$.

The **common logarithmic function** is defined by the equation $f(x) = \log x$. It should now be a simple matter to set up a table of values and sketch the function. You will do this in the next set of exercises. Remember that $f(x) = 10^x$ and $g(x) = \log x$ are inverses of each other. Therefore we could also get the graph of $g(x) = \log x$ by reflecting the exponential curve $f(x) = 10^x$ across the line $y = x$.

■ Natural Logarithms—Base *e*

In many practical applications of logarithms, the number e (remember that $e \approx 2.71828$) is used as a base. Logarithms with a base of e are called **natural logarithms**, and the symbol $\ln x$ is commonly used instead of $\log_e x$.

$$\log_e x = \ln x$$

Natural logarithms can be found with an appropriately equipped calculator. A calculator that has a natural logarithm function (ordinarily, a key labeled $\boxed{\ln x}$) gives us the following results rounded to four decimal places:

$$\ln 3.21 = 1.1663$$

$$\ln 47.28 = 3.8561$$

$$\ln 842 = 6.7358$$

$$\ln 0.21 = -1.5606$$

$$\ln 0.0046 = -5.3817$$

$$\ln 10 = 2.3026$$

Be sure that you can use your calculator to obtain these results. Keep in mind the significance of a statement such as $\ln 3.21 = 1.1663$. By changing to exponential form, we are claiming that e raised to the 1.1663 power is approximately 3.21. Using a calculator, we obtain $e^{1.1663} = 3.210093293$.

Let's do a few more problems to find x when given $\ln x$. Be sure that you agree with these results.

If $\ln x = 2.4156$, then $x = e^{2.4156} = 11.196$.

If $\ln x = 0.9847$, then $x = e^{0.9847} = 2.6770$.

If $\ln x = 4.1482$, then $x = e^{4.1482} = 63.320$.

If $\ln x = -1.7654$, then $x = e^{-1.7654} = 0.17112$.

The **natural logarithmic function** is defined by the equation $f(x) = \ln x$. It is the inverse of the natural exponential function $f(x) = e^x$. Thus one way to graph $f(x) = \ln x$ is to reflect the graph of $f(x) = e^x$ across the line $y = x$. We will ask you to do this in the next set of problems.

In Figure 10.30 we used a graphing utility to sketch the graph of $f(x) = e^x$. Now, on the basis of our previous work with transformations, we should be able to make the following statements:

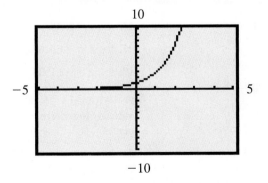

Figure 10.30

1. The graph of $f(x) = -e^x$ is the graph of $f(x) = e^x$ reflected through the x axis.

2. The graph of $f(x) = e^{-x}$ is the graph of $f(x) = e^x$ reflected through the y axis.

3. The graph of $f(x) = e^x + 4$ is the graph of $f(x) = e^x$ shifted upward four units.

4. The graph of $f(x) = e^{x+2}$ is the graph of $f(x) = e^x$ shifted two units to the left.

These statements are verified in Figure 10.31, which shows the result of graphing these four functions on the same set of axes by using a graphing utility.

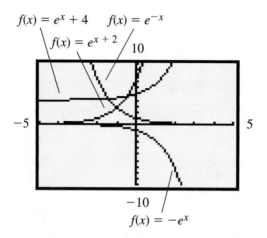

Figure 10.31

Remark: So far, we have used a graphing utility to graph only common logarithmic and natural logarithmic functions. In the next section, we will see how logarithms with bases other than 10 or e are related to common and natural logarithms. This will provide a way of using a graphing utility to graph a logarithmic function with any valid base.

Problem Set 10.5

For Problems 1–10, use a calculator to find each **common logarithm**. Express answers to four decimal places.

1. log 7.24

2. log 2.05

3. log 52.23

4. log 825.8

5. log 3214.1

6. log 14,189

7. log 0.729

8. log 0.04376

9. log 0.00034

10. log 0.000069

For Problems 11–20, use your calculator to find x when given log x. Express answers to five significant digits.

11. log $x = 2.6143$

12. log $x = 1.5263$

13. log $x = 4.9547$

14. log $x = 3.9335$

15. log $x = 1.9006$

16. log $x = 0.5517$

17. log $x = -1.3148$

18. log $x = -0.1452$

19. log $x = -2.1928$

20. log $x = -2.6542$

For Problems 21–30, use your calculator to find each **natural logarithm**. Express answers to four decimal places.

21. ln 5

22. ln 18

23. ln 32.6

24. ln 79.5

25. ln 430

26. ln 371.8

27. ln 0.46

28. ln 0.524

29. ln 0.0314

30. ln 0.008142

For Problems 31–40, use your calculator to find x when given $\ln x$. Express answers to five significant digits.

31. $\ln x = 0.4721$

32. $\ln x = 0.9413$

33. $\ln x = 1.1425$

34. $\ln x = 2.7619$

35. $\ln x = 4.6873$

36. $\ln x = 3.0259$

37. $\ln x = -0.7284$

38. $\ln x = -1.6246$

39. $\ln x = -3.3244$

40. $\ln x = -2.3745$

41. (a) Complete the following table, and then graph $f(x) = \log x$. (Express the values for $\log x$ to the nearest tenth.)

x	0.1	0.5	1	2	4	8	10
$\log x$							

(b) Complete the following table, expressing values for 10^x to the nearest tenth.

x	-1	-0.3	0	0.3	0.6	0.9	1
10^x							

Then graph $f(x) = 10^x$, and reflect it across the line $y = x$ to produce the graph for $f(x) = \log x$.

42. (a) Complete the following table, and then graph $f(x) = \ln x$. (Express the values for $\ln x$ to the nearest tenth.)

x	0.1	0.5	1	2	4	8	10
$\ln x$							

(b) Complete the following table, expressing values for e^x to the nearest tenth.

x	-2.3	-0.7	0	0.7	1.4	2.1	2.3
e^x							

Then graph $f(x) = e^x$, and reflect it across the line $y = x$ to produce the graph for $f(x) = \ln x$.

43. Graph $y = \log_{\frac{1}{2}} x$ by graphing $\left(\dfrac{1}{2}\right)^y = x$.

44. Graph $y = \log_2 x$ by graphing $2^y = x$.

45. Graph $f(x) = \log_3 x$ by reflecting the graph of $g(x) = 3^x$ across the line $y = x$.

46. Graph $f(x) = \log_4 x$ by reflecting the graph of $g(x) = 4^x$ across the line $y = x$.

For Problems 47–53, graph each of the functions. Remember that the graph of $f(x) = \log_2 x$ is given in Figure 10.27.

47. $f(x) = 3 + \log_2 x$

48. $f(x) = -2 + \log_2 x$

49. $f(x) = \log_2(x + 3)$

50. $f(x) = \log_2(x - 2)$

51. $f(x) = \log_2 2x$

52. $f(x) = -\log_2 x$

53. $f(x) = 2\log_2 x$

For Problems 54–61, perform the following calculations and express answers to the nearest hundredth. (These calculations are in preparation for our work in the next section.)

54. $\dfrac{\log 7}{\log 3}$

55. $\dfrac{\ln 2}{\ln 7}$

56. $\dfrac{2\ln 3}{\ln 8}$

57. $\dfrac{\ln 5}{2\ln 3}$

58. $\dfrac{\ln 3}{0.04}$

59. $\dfrac{\ln 2}{0.03}$

60. $\dfrac{\log 2}{5\log 1.02}$

61. $\dfrac{\log 5}{3\log 1.07}$

■ ■ ■ THOUGHTS INTO WORDS

62. Why is the number 1 excluded from being a base of a logarithm?

63. How do we know that $\log_2 6$ is between 2 and 3?

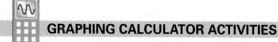

GRAPHING CALCULATOR ACTIVITIES

64. Graph $f(x) = x$, $f(x) = e^x$, and $f(x) = \ln x$ on the same set of axes.

65. Graph $f(x) = x$, $f(x) = 10^x$, and $f(x) = \log x$ on the same set of axes.

66. Graph $f(x) = \ln x$. How should the graphs of $f(x) = 2 \ln x$, $f(x) = 4 \ln x$, and $f(x) = 6 \ln x$ compare to the graph of $f(x) = \ln x$? Graph the three functions on the same set of axes with $f(x) = \ln x$.

67. Graph $f(x) = \log x$. Now predict the graphs for $f(x) = 2 + \log x$, $f(x) = -2 + \log x$, and $f(x) = -6 + \log x$. Graph the three functions on the same set of axes with $f(x) = \log x$.

68. Graph $\ln x$. Now predict the graphs for $f(x) = \ln(x - 2)$, $f(x) = \ln(x - 6)$, and $f(x) = \ln(x + 4)$. Graph the three functions on the same set of axes with $f(x) = \ln x$.

69. For each of the following, (a) predict the general shape and location of the graph, and (b) use your graphing calculator to graph the function to check your prediction.

(a) $f(x) = \log x + \ln x$ **(b)** $f(x) = \log x - \ln x$

(c) $f(x) = \ln x - \log x$ **(d)** $f(x) = \ln x^2$

10.6   Exponential Equations, Logarithmic Equations, and Problem Solving

In Section 10.1 we solved exponential equations such as $3^x = 81$ by expressing both sides of the equation as a power of 3 and then applying the property: If $b^n = b^m$, then $n = m$. However, if we try this same approach with an equation such as $3^x = 5$, we face the difficulty of expressing 5 as a power of 3. We can solve this type of problem by using the properties of logarithms and the following property of equality:

Property 10.8

> If $x > 0$, $y > 0$, $b > 0$, and $b \neq 1$, then $x = y$ if and only if $\log_b x = \log_b y$.

Property 10.8 is stated in terms of any valid base b; however, for most applications, we use either common logarithms or natural logarithms. Let's consider some examples.

EXAMPLE 1

Solve $3^x = 5$ to the nearest hundredth.

Solution

By using common logarithms, we can proceed as follows:

$$3^x = 5$$

$$\log 3^x = \log 5 \qquad \text{Property 10.8}$$

$$x \log 3 = \log 5 \qquad \log r^p = p \log r$$

$$x = \frac{\log 5}{\log 3}$$

$$x = 1.46 \quad \text{to the nearest hundredth}$$

✔ **Check**

Because $3^{1.46} \approx 4.972754647$, we say that, to the nearest hundredth, the solution set for $3^x = 5$ is $\{1.46\}$. ■

E X A M P L E   2

Solve $e^{x+1} = 5$ to the nearest hundredth.

Solution

Because base e is used in the exponential expression, let's use natural logarithms to help solve this equation.

$$e^{x+1} = 5$$

$$\ln e^{x+1} = \ln 5 \qquad \text{Property 10.8}$$

$$(x+1) \ln e = \ln 5 \qquad \ln r^p = p \ln r$$

$$(x+1)(1) = \ln 5 \qquad \ln e = 1$$

$$x = \ln 5 - 1$$

$$x = 0.61 \quad \text{to the nearest hundredth}$$

The solution set is $\{0.61\}$. Check it! ■

E X A M P L E   3

Solve $2^{3x-2} = 3^{2x+1}$ to the nearest hundredth.

Solution

$$2^{3x-2} = 3^{2x+1}$$

$$\log 2^{3x-2} = \log 3^{2x+1}$$

$$(3x-2)\log 2 = (2x+1)\log 3$$

$$3x \log 2 - 2 \log 2 = 2x \log 3 + \log 3$$

$$3x \log 2 - 2x \log 3 = \log 3 + 2 \log 2$$

$$x(3 \log 2 - 2 \log 3) = \log 3 + 2 \log 2$$

$$x = \frac{\log 3 + 2 \log 2}{3 \log 2 - 2 \log 3}$$

$$x = -21.10 \quad \text{to the nearest hundredth}$$

The solution set is $\{-21.10\}$. Check it! ■

■ Logarithmic Equations

In Example 11 of Section 10.4, we solved the logarithmic equation

$$\log_{10} x + \log_{10}(x + 9) = 1$$

by simplifying the left side of the equation to $\log_{10}[x(x + 9)]$ and then changing the equation to exponential form to complete the solution. Now, using Property 10.8, we can solve such a logarithmic equation another way and also expand our equation-solving capabilities. Let's consider some examples.

E X A M P L E   4

Solve $\log x + \log(x - 15) = 2$.

Solution

Because $\log 100 = 2$, the given equation becomes

$$\log x + \log(x - 15) = \log 100$$

Now simplify the left side, apply Property 10.8, and proceed as follows.

$$\log(x)(x - 15) = \log 100$$

$$x(x - 15) = 100$$

$$x^2 - 15x - 100 = 0$$

$$(x - 20)(x + 5) = 0$$

$$x - 20 = 0 \quad \text{or} \quad x + 5 = 0$$

$$x = 20 \quad \text{or} \quad x = -5$$

The domain of a logarithmic function must contain only positive numbers, so x and $x - 15$ must be positive in this problem. Therefore we discard the solution -5; the solution set is $\{20\}$. ■

E X A M P L E   5

Solve $\ln(x + 2) = \ln(x - 4) + \ln 3$.

Solution

$$\ln(x + 2) = \ln(x - 4) + \ln 3$$

$$\ln(x + 2) = \ln[3(x - 4)]$$

$$x + 2 = 3(x - 4)$$

$$x + 2 = 3x - 12$$

$$14 = 2x$$

$$7 = x$$

The solution set is $\{7\}$. ■

EXAMPLE 6

Solve $\log_b(x + 2) + \log_b(2x - 1) = \log_b x$.

Solution

$$\log_b(x + 2) + \log_b(2x - 1) = \log_b x$$

$$\log_b[(x + 2)(2x - 1)] = \log_b x$$

$$(x + 2)(2x - 1) = x$$

$$2x^2 + 3x - 2 = x$$

$$2x^2 + 2x - 2 = 0$$

$$x^2 + x - 1 = 0$$

Using the quadratic formula, we obtain

$$x = \frac{-1 \pm \sqrt{1 + 4}}{2}$$

$$= \frac{-1 \pm \sqrt{5}}{2}$$

Because $x + 2$, $2x - 1$, and x have to be positive, we must discard the solution $\dfrac{-1 - \sqrt{5}}{2}$; the solution set is $\left\{\dfrac{-1 + \sqrt{5}}{2}\right\}$.   ∎

■ Problem Solving

In Section 10.2 we used the compound interest formula

$$A = P\left(1 + \frac{r}{n}\right)^{nt}$$

to determine the amount of money (A) accumulated at the end of t years if P dollars is invested at rate of interest r compounded n times per year. Now let's use this formula to solve other types of problems that deal with compound interest.

EXAMPLE 7

How long will it take for $500 to double if it is invested at 12% interest compounded quarterly?

Solution

"To double" means that the $500 must grow into $1000. Thus

$$1000 = 500\left(1 + \frac{0.12}{4}\right)^{4t}$$

$$= 500(1 + 0.03)^{4t}$$

$$= 500(1.03)^{4t}$$

Multiplying both sides of $1000 = 500(1.03)^{4t}$ by $\dfrac{1}{500}$ yields

$$2 = (1.03)^{4t}$$

Therefore

$$\log 2 = \log(1.03)^{4t} \qquad \text{Property 10.8}$$
$$\qquad = 4t \log 1.03 \qquad \log r^p = p \log r$$

Now let's solve for t.

$$4t \log 1.03 = \log 2$$

$$t = \frac{\log 2}{4 \log 1.03}$$

$$t = 5.9 \quad \text{to the nearest tenth}$$

Therefore we are claiming that $500 invested at 12% interest compounded quarterly will double in approximately 5.9 years.

✔ **Check**

$500 invested at 12% interest compounded quarterly for 5.9 years will produce

$$A = \$500\left(1 + \frac{0.12}{4}\right)^{4(5.9)}$$

$$= \$500(1.03)^{23.6}$$

$$= \$1004.45$$

■

E X A M P L E   8

Suppose that the number of bacteria present in a certain culture after t minutes is given by the equation $Q(t) = Q_0 e^{0.04t}$, where Q_0 represents the initial number of bacteria. How long will it take for the bacteria count to grow from 500 to 2000?

Solution

Substituting into $Q(t) = Q_0 e^{0.04t}$ and solving for t, we obtain the following.

$$2000 = 500e^{0.04t}$$

$$4 = e^{0.04t}$$

$$\ln 4 = \ln e^{0.04t}$$

$$\ln 4 = 0.04t \ln e$$

$$\ln 4 = 0.04t \qquad \ln e = 1$$

$$\frac{\ln 4}{0.04} = t$$

$$34.7 = t \quad \text{to the nearest tenth}$$

It should take approximately 34.7 minutes.

■

■ Richter Numbers

Seismologists use the Richter scale to measure and report the magnitude of earthquakes. The equation

$$R = \log \frac{I}{I_0}$$ *R* is called a Richter number.

compares the intensity I of an earthquake to a minimal or reference intensity I_0. The reference intensity is the smallest earth movement that can be recorded on a seismograph. Suppose that the intensity of an earthquake was determined to be 50,000 times the reference intensity. In this case, $I = 50,000\ I_0$, and the Richter number is calculated as follows:

$$R = \log \frac{50,000\ I_0}{I_0}$$

$$= \log 50,000$$

$$\approx 4.698970004$$

Thus a Richter number of 4.7 would be reported. Let's consider two more examples that involve Richter numbers.

EXAMPLE 9

An earthquake in the San Francisco area in 1989 was reported to have a Richter number of 6.9. How did its intensity compare to the reference intensity?

Solution

$$6.9 = \log \frac{I}{I_0}$$

$$10^{6.9} = \frac{I}{I_0}$$

$$I = (10^{6.9})(I_0)$$

$$I \approx 7,943,282\ I_0$$

Its intensity was a little less than 8 million times the reference intensity.   ■

EXAMPLE 10

An earthquake in Iran in 1990 had a Richter number of 7.7. Compare the intensity of this earthquake to that of the one in San Francisco referred to in Example 9.

Solution

From Example 9 we have $I = (10^{6.9})(I_0)$ for the earthquake in San Francisco. Then, using a Richter number of 7.7, we obtain $I = (10^{7.7})(I_0)$ for the earthquake in Iran. Therefore, by comparison,

$$\frac{(10^{7.7})(I_0)}{(10^{6.9})(I_0)} = 10^{7.7-6.9} = 10^{0.8} \approx 6.3$$

The earthquake in Iran was about 6 times as intense as the one in San Francisco.   ■

■ Logarithms with Base Other Than 10 or *e*

The basic approach whereby we apply Property 10.8 and use either common or natural logarithms can also be used to evaluate a logarithm to some base other than 10 or *e*. Consider the following example.

E X A M P L E   1 1

Evaluate $\log_3 41$.

Solution

Let $x = \log_3 41$. Change to exponential form to obtain

$$3^x = 41$$

Now we can apply Property 10.8 and proceed as follows.

$$\log 3^x = \log 41$$

$$x \log 3 = \log 41$$

$$x = \frac{\log 41}{\log 3}$$

$$x = 3.3802 \quad \text{rounded to four decimal places}$$

Therefore we are claiming that 3 raised to the 3.3802 power will produce approximately 41. Check it! ■

The method of Example 11 to evaluate $\log_a r$ produces the following formula, which we often refer to as the **change-of-base formula for logarithms**.

Property 10.9

If a, b, and r are positive numbers, with $a \neq 1$ and $b \neq 1$, then

$$\log_a r = \frac{\log_b r}{\log_b a}$$

By using Property 10.9, we can easily determine a relationship between logarithms of different bases. For example, suppose that in Property 10.9 we let $a = 10$ and $b = e$. Then

$$\log_a r = \frac{\log_b r}{\log_b a}$$

becomes

$$\log_{10} r = \frac{\log_e r}{\log_e 10}$$

$$\log_e r = (\log_e 10)(\log_{10} r)$$

$$\log_e r = (2.3026)(\log_{10} r)$$

Thus the natural logarithm of any positive number is approximately equal to the common logarithm of the number times 2.3026.

Now we can use a graphing utility to graph logarithmic functions such as $f(x) = \log_2 x$. Using the change-of-base formula, we can express this function as $f(x) = \dfrac{\log x}{\log 2}$ or as $f(x) = \dfrac{\ln x}{\ln 2}$. The graph of $f(x) = \log_2 x$ is shown in Figure 10.32.

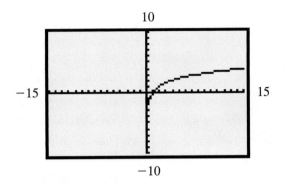

Figure 10.32

Finally, let's use a graphical approach to solve an equation that is cumbersome to solve with an algebraic approach.

EXAMPLE 12

Solve the equation $(5^x - 5^{-x})/2 = 3$.

Solution

First, we need to recognize that the solutions for the equation $(5^x - 5^{-x})/2 = 3$ are the x intercepts of the graph of the equation $y = (5^x - 5^{-x})/2 - 3$. We can use a graphing utility to obtain the graph of this equation as shown in Figure 10.33. Use the ZOOM and TRACE features to determine that the graph crosses the x axis at approximately 1.13. Thus the solution set of the original expression is {1.13}.

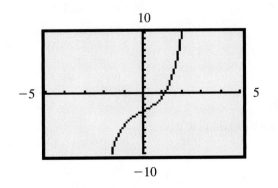

Figure 10.33

Problem Set 10.6

For Problems 1–20, solve each exponential equation and express approximate solutions to the nearest hundredth.

1. $3^x = 13$

2. $2^x = 21$

3. $4^n = 35$

4. $5^n = 75$

5. $2^x + 7 = 50$

6. $3^x - 6 = 25$

7. $3^{x-2} = 11$

8. $2^{x+1} = 7$

9. $5^{3t+1} = 9$

10. $7^{2t-1} = 35$

11. $e^x = 27$

12. $e^x = 86$

13. $e^{x-2} = 13.1$

14. $e^{x-1} = 8.2$

15. $3e^x - 1 = 17$

16. $2e^x = 12.4$

17. $5^{2x+1} = 7^{x+3}$

18. $3^{x-1} = 2^{x+3}$

19. $3^{2x+1} = 2^{3x+2}$

20. $5^{x-1} = 2^{2x+1}$

For Problems 21–32, solve each logarithmic equation and express irrational solutions in lowest radical form.

21. $\log x + \log(x + 21) = 2$

22. $\log x + \log(x + 3) = 1$

23. $\log(3x - 1) = 1 + \log(5x - 2)$

24. $\log(2x - 1) - \log(x - 3) = 1$

25. $\log(x + 1) = \log 3 - \log(2x - 1)$

26. $\log(x - 2) = 1 - \log(x + 3)$

27. $\log(x + 2) - \log(2x + 1) = \log x$

28. $\log(x + 1) - \log(x + 2) = \log \dfrac{1}{x}$

29. $\ln(2t + 5) = \ln 3 + \ln(t - 1)$

30. $\ln(3t - 4) - \ln(t + 1) = \ln 2$

31. $\log \sqrt{x} = \sqrt{\log x}$

32. $\log x^2 = (\log x)^2$

For Problems 33–42, approximate each logarithm to three decimal places. (Example 11 and/or Property 10.9 should be of some help.)

33. $\log_2 40$

34. $\log_2 93$

35. $\log_3 16$

36. $\log_3 37$

37. $\log_4 1.6$

38. $\log_4 3.2$

39. $\log_5 0.26$

40. $\log_5 0.047$

41. $\log_7 500$

42. $\log_8 750$

For Problems 43–55, solve each problem and express answers to the nearest tenth unless stated otherwise.

43. How long will it take $750 to be worth $1000 if it is invested at 12% interest compounded quarterly?

44. How long will it take $1000 to double if it is invested at 9% interest compounded semiannually?

45. How long will it take $2000 to double if it is invested at 13% interest compounded continuously?

46. How long will it take $500 to triple if it is invested at 9% interest compounded continuously?

47. What rate of interest compounded continuously is needed for an investment of $500 to grow to $900 in 10 years?

48. What rate of interest compounded continuously is needed for an investment of $2500 to grow to $10,000 in 20 years?

49. For a certain strain of bacteria, the number of bacteria present after t hours is given by the equation $Q = Q_0 e^{0.34t}$, where Q_0 represents the initial number of bacteria. How long will it take 400 bacteria to increase to 4000 bacteria?

50. A piece of machinery valued at $30,000 depreciates at a rate of 10% yearly. How long will it take for it to reach a value of $15,000?

51. The equation $P(a) = 14.7e^{-0.21a}$, where a is the altitude above sea level measured in miles, yields the atmospheric pressure in pounds per square inch. If the atmospheric pressure at Cheyenne, Wyoming, is approximately 11.53 pounds per square inch, find that city's altitude above sea level. Express your answer to the nearest hundred feet.

52. The number of grams of a certain radioactive substance present after t hours is given by the equation $Q = Q_0 e^{-0.45t}$, where Q_0 represents the initial number of grams. How long will it take 2500 grams to be reduced to 1250 grams?

53. For a certain culture, the equation $Q(t) = Q_0 e^{0.4t}$, where Q_0 is an initial number of bacteria and t is time measured in hours, yields the number of bacteria as a function of time. How long will it take 500 bacteria to increase to 2000?

54. Suppose that the equation $P(t) = P_0 e^{0.02t}$, where P_0 represents an initial population and t is the time in years, is used to predict population growth. How long will it take a city of 50,000 to double its population?

55. An earthquake in Los Angeles in 1971 had an intensity of approximately 5 million times the reference inten-

sity. What was the Richter number associated with that earthquake?

56. An earthquake in San Francisco in 1906 was reported to have a Richter number of 8.3. How did its intensity compare to the reference intensity?

57. Calculate how many times more intense an earthquake with a Richter number of 7.3 is than an earthquake with a Richter number of 6.4.

58. Calculate how many times more intense an earthquake is with a Richter number of 8.9 than an earthquake with a Richter number of 6.2.

■ ■ ■ THOUGHTS INTO WORDS

59. Explain how to determine $\log_4 76$ without using Property 10.9.

60. Explain the concept of a Richter number.

61. Explain how you would solve the equation $2^x = 64$ and also how you would solve the equation $2^x = 53$.

62. How do logarithms with a base of 9 compare to logarithms with a base of 3? Explain how you reached this conclusion.

■ ■ ■ FURTHER INVESTIGATIONS

63. Use the approach of Example 11 to develop Property 10.9.

64. Let $r = b$ in Property 10.9, and verify that $\log_a b = \dfrac{1}{\log_b a}$.

65. Solve the equation $\dfrac{5^x - 5^{-x}}{2} = 3$. Express your answer to the nearest hundredth.

66. Solve the equation $y = \dfrac{10^x + 10^{-x}}{2}$ for x in terms of y.

67. Solve the equation $y = \dfrac{e^x - e^{-x}}{2}$ for x in terms of y.

GRAPHING CALCULATOR ACTIVITIES

68. Check your answers for Problems 17–20 by graphing the appropriate function and finding the x intercept.

69. Graph $f(x) = x$, $f(x) = 2^x$, and $f(x) = \log_2 x$ on the same set of axes.

70. Graph $f(x) = x$, $f(x) = (0.5)^x$, and $f(x) = \log_{0.5} x$ on the same set of axes.

71. Graph $f(x) = \log_2 x$. Now predict the graphs for $f(x) = \log_3 x$, $f(x) = \log_4 x$, and $f(x) = \log_8 x$. Graph

these three functions on the same set of axes with $f(x) = \log_2 x$.

72. Graph $f(x) = \log_5 x$. Now predict the graphs for $f(x) = 2 \log_5 x$, $f(x) = -4 \log_5 x$, and $f(x) = \log_5(x + 4)$. Graph these three functions on the same set of axes with $f(x) = \log_5 x$.

73. Use both a graphical and an algebraic approach to solve the equation $\dfrac{2^x - 2^{-x}}{3} = 4$.

Chapter 10 Summary

(10.1) If a and b are positive real numbers, and m and n are any real numbers, then

1. $b^n \cdot b^m = b^{n+m}$ Product of two powers

2. $(b^n)^m = b^{mn}$ Power of a power

3. $(ab)^n = a^n b^n$ Power of a product

4. $\left(\dfrac{a}{b}\right)^n = \dfrac{a^n}{b^n}$ Power of a quotient

5. $\dfrac{b^n}{b^m} = b^{n-m}$ Quotient of two powers

If $b > 0$, $b \neq 1$, and m and n are real numbers, then $b^n = b^m$ if and only if $n = m$.

A function defined by an equation of the form

$$f(x) = b^x, \qquad b > 0 \text{ and } b \neq 1$$

is called an **exponential function**.

(10.2) A general formula for any principal P compounded n times per year for any number t, at a rate r, is

$$A = P\left(1 + \frac{r}{n}\right)^{nt}$$

where A represents the total amount of money accumulated at the end of the t years. The value of $\left(1 + \dfrac{1}{n}\right)^{n}$, as n gets infinitely large, approaches the number e, where e equals 2.71828 to five decimal places.

The formula

$$A = Pe^{rt}$$

yields the accumulated value A of a sum of money P that has been invested for t years at a rate of r percent **compounded continuously**.

The formula

$$Q = Q_0 \left(\frac{1}{2}\right)^{\frac{t}{h}}$$

is referred to as the **half-life** formula.

The equation

$$Q(t) = Q_0 e^{kt}$$

is used as a mathematical model for many growth-and-decay applications.

(10.3) A function f is said to be a **one-to-one function** if every value of $f(x)$ has only one value of x associated with it.

In terms of ordered pairs, a one-to-one function is a function such that no two ordered pairs have the same second component.

Let f be a one-to-one function with a domain of X and a range of Y. A function g, with a domain of Y and a range of X is called the **inverse function** of f if $(f \circ g)(x) = x$ for every x in Y and $(g \circ f)(x) = x$ for every x in X.

If the components of each ordered pair of a given one-to-one function are interchanged, the resulting function and the given function are **inverses** of each other.

The inverse of a function f is denoted by f^{-1}. Graphically, two functions that are inverses of each other are **mirror images** with reference to the line $y = x$.

A systematic technique for finding the inverse of a function can be described as follows:

1. Let $y = f(x)$.

2. Interchange x and y.

3. Solve the equation for y in terms of x.

4. The inverse function $f^{-1}(x)$ is determined by the equation in step 3.

Don't forget that the domain of f must equal the range of f^{-1}, and the domain of f^{-1} must equal the range of f.

Let f be a function, with the interval I a subset of the domain of f. Let x_1 and x_2 be in I.

1. f is increasing on I if $f(x_1) < f(x_2)$ whenever $x_1 < x_2$.
2. f is decreasing on I if $f(x_1) > f(x_2)$ whenever $x_1 < x_2$.
3. f is constant on I if $f(x_1) = f(x_2)$ for every x_1 and x_2.

(10.4) If r is any positive real number, then the unique exponent t such that $b^t = r$ is called the **logarithm of r with base b** and is denoted by $\log_b r$.

1. $\log_b b = 1$

2. $\log_b 1 = 0$ $\quad$ for $b \geq 0, b \neq 1, r > 0$

3. $r = b^{\log_b r}$

The following properties of logarithms are derived from the definition of a logarithm and the properties of exponents. For positive real numbers b, r, and s, where $b \neq 1$,

1. $\log_b rs = \log_b r + \log_b s$

2. $\log_b\left(\dfrac{r}{s}\right) = \log_b r - \log_b s$

3. $\log_b r^p = p \log_b r, \qquad$ where p is any real number

(10.5) A function defined by an equation of the form

$$f(x) = \log_b x, \qquad b > 0 \text{ and } b \neq 1$$

is called a **logarithmic function**. The equation $y = \log_b x$ is equivalent to $x = b^y$. The two functions $f(x) = b^x$ and $g(x) = \log_b x$ are inverses of each other.

Logarithms with a base of 10 are called **common logarithms**. The expression $\log_{10} x$ is commonly written as $\log x$.

Many calculators are equipped with a common logarithm function. Often a key labeled $\boxed{\log}$ is used to find common logarithms.

Natural logarithms are logarithms that have a base of e, where e is an irrational number whose decimal approximation to eight digits is 2.7182818. Natural logarithms are denoted by $\log_e x$ or $\ln x$.

Many calculators are also equipped with a natural logarithm function. Often a key labeled $\boxed{\ln x}$ is used for this purpose.

(10.6) The properties of equality and the properties of exponents and logarithms merge to help us solve a variety of exponential and logarithmic equations. These properties also help us solve problems that deal with various applications, including compound interest and growth problems.

The formula

$$R = \log \frac{I}{I_0}$$

yields the Richter number associated with an earthquake.

The formula

$$\log_a r = \frac{\log_b r}{\log_b a}$$

is often called the **change-of-base formula**.

Chapter 10   Review Problem Set

For Problems 1–10, evaluate each of the following:

1. $8^{5/3}$

2. $-25^{3/2}$

3. $(-27)^{4/3}$

4. $\log_6 216$

5. $\log_7\left(\dfrac{1}{49}\right)$

6. $\log_2 \sqrt[3]{2}$

7. $\log_2\left(\dfrac{\sqrt[4]{32}}{2}\right)$

8. $\log_{10} 0.00001$

9. $\ln e$

10. $7^{\log_7 12}$

For Problems 11–24, solve each equation. Express approximate solutions to the nearest hundredth.

11. $\log_{10} 2 + \log_{10} x = 1$

12. $\log_3 x = -2$

13. $4^x = 128$

14. $3^t = 42$

15. $\log_2 x = 3$

16. $\left(\dfrac{1}{27}\right)^{3x} = 3^{2x-1}$

17. $2e^x = 14$

18. $2^{2x+1} = 3^{x+1}$

19. $\ln(x + 4) - \ln(x + 2) = \ln x$

20. $\log x + \log(x - 15) = 2$

21. $\log(\log x) = 2$

22. $\log(7x - 4) - \log(x - 1) = 1$

23. $\ln(2t - 1) = \ln 4 + \ln(t - 3)$

24. $64^{2t+1} = 8^{-t+2}$

For Problems 25–28, if $\log 3 = 0.4771$ and $\log 7 = 0.8451$, evaluate each of the following:

25. $\log\left(\dfrac{7}{3}\right)$ **26.** $\log 21$

27. $\log 27$ **28.** $\log 7^{2/3}$

29. Express each of the following as the sum or difference of simpler logarithmic quantities. Assume that all variables represent positive real numbers.

    (a) $\log_b\left(\dfrac{x}{y^2}\right)$ **(b)** $\log_b \sqrt[4]{xy^2}$

    (c) $\log_b\left(\dfrac{\sqrt{x}}{y^3}\right)$

30. Express each of the following as a single logarithm. Assume that all variables represent positive real numbers.

    (a) $3 \log_b x + 2 \log_b y$

    (b) $\dfrac{1}{2} \log_b y - 4 \log_b x$

    (c) $\dfrac{1}{2}(\log_b x + \log_b y) - 2 \log_b z$

For Problems 31–34, approximate each of the logarithms to three decimal places.

31. $\log_2 3$ **32.** $\log_3 2$

33. $\log_4 191$ **34.** $\log_2 0.23$

For Problems 35–42, graph each of the functions.

35. **(a)** $f(x) = \left(\dfrac{3}{4}\right)^x$

    (b) $f(x) = \left(\dfrac{3}{4}\right)^x + 2$

    (c) $f(x) = \left(\dfrac{3}{4}\right)^{-x}$

36. **(a)** $f(x) = 2^x$

    (b) $f(x) = 2^{x+2}$

    (c) $f(x) = -2^x$

37. **(a)** $f(x) = e^{x-1}$

    (b) $f(x) = e^x - 1$

    (c) $f(x) = e^{-x+1}$

38. **(a)** $f(x) = -1 + \log x$

    (b) $f(x) = \log(x - 1)$

    (c) $f(x) = -1 - \log x$

39. $f(x) = 3^x - 3^{-x}$ **40.** $f(x) = e^{-x^2/2}$

41. $f(x) = \log_2(x - 3)$ **42.** $f(x) = 3 \log_3 x$

For Problems 43–45, use the compound interest formula $A = P\left(1 + \dfrac{r}{n}\right)^{nt}$ to find the total amount of money accumulated at the end of the indicated time period for each of the investments.

43. \$750 for 10 years at 11% compounded quarterly

44. \$1250 for 15 years at 9% compounded monthly

45. \$2500 for 20 years at 9.5% compounded semiannually

For Problems 46–49, determine whether f and g are inverse functions.

46. $f(x) = 7x - 1$ and $g(x) = \dfrac{x + 1}{7}$

47. $f(x) = -\dfrac{2}{3}x$ and $g(x) = \dfrac{3}{2}x$

48. $f(x) = x^2 - 6$ for $x \geq 0$ and $g(x) = \sqrt{x + 6}$ for $x \geq -6$

49. $f(x) = 2 - x^2$ for $x \geq 0$ and $g(x) = \sqrt{2 - x}$ for $x \leq 2$

For Problems 50–53, (a) find f^{-1}, and (b) verify that $(f \circ f^{-1})(x) = x$ and $(f^{-1} \circ f)(x) = x$.

50. $f(x) = 4x + 5$ **51.** $f(x) = -3x - 7$

52. $f(x) = \dfrac{5}{6}x - \dfrac{1}{3}$

53. $f(x) = -2 - x^2$ for $x \geq 0$

For Problems 54 and 55, find the intervals on which the function is increasing and the intervals on which it is decreasing.

54. $f(x) = -2x^2 + 16x - 35$

55. $f(x) = 2\sqrt{x - 3}$

56. How long will it take $100 to double if it is invested at 14% interest compounded annually?

57. How long will it take $1000 to be worth $3500 if it is invested at 10.5% interest compounded quarterly?

58. What rate of interest (to the nearest tenth of a percent) compounded continuously is needed for an investment of $500 to grow to $1000 in 8 years?

59. Suppose that the present population of a city is 50,000. Use the equation $P(t) = P_0 e^{0.02t}$ (where P_0 represents an initial population) to estimate future populations, and estimate the population of that city in 10 years, 15 years, and 20 years.

60. The number of bacteria present in a certain culture after t hours is given by the equation $Q = Q_0 e^{0.29t}$, where Q_0 represents the initial number of bacteria. How long will it take 500 bacteria to increase to 2000 bacteria?

61. Suppose that a certain radioactive substance has a half-life of 40 days. If there are presently 750 grams of the substance, how much, to the nearest gram, will remain after 100 days?

62. An earthquake occurred in Mexico City in 1985 that had an intensity level about 125,000,000 times the reference intensity. Find the Richter number for that earthquake.

For Problems 1–4, evaluate each expression.

1. $\log_3 \sqrt{3}$

2. $\log_2(\log_2 4)$

3. $-2 + \ln e^3$

4. $\log_2(0.5)$

For Problems 5–10, solve each equation.

5. $4^x = \dfrac{1}{64}$

6. $9^x = \dfrac{1}{27}$

7. $2^{3x-1} = 128$

8. $\log_9 x = \dfrac{5}{2}$

9. $\log x + \log(x + 48) = 2$

10. $\ln x = \ln 2 + \ln(3x - 1)$

For Problems 11–13, given that $\log_3 4 = 1.2619$ and $\log_3 5 = 1.4650$, evaluate each of the following.

11. $\log_3 100$

12. $\log_3 \dfrac{5}{4}$

13. $\log_3 \sqrt{5}$

14. Find the inverse of the function $f(x) = -3x - 6$.

15. Solve $e^x = 176$ to the nearest hundredth.

16. Solve $2^{x-2} = 314$ to the nearest hundredth.

17. Determine $\log_5 632$ to four decimal places.

18. Find the inverse of the function $f(x) = \dfrac{2}{3}x - \dfrac{3}{5}$.

19. If $3500 is invested at 7.5% interest compounded quarterly, how much money has accumulated at the end of 8 years?

20. How long will it take $5000 to be worth $12,500 if it is invested at 7% compounded annually? Express your answer to the nearest tenth of a year.

21. The number of bacteria present in a certain culture after t hours is given by $Q(t) = Q_0 e^{0.23t}$, where Q_0 represents the initial number of bacteria. How long will it take 400 bacteria to increase to 2400 bacteria? Express your answer to the nearest tenth of an hour.

22. Suppose that a certain radioactive substance has a half-life of 50 years. If there are presently 7500 grams of the substance, how much will remain after 32 years? Express your answer to the nearest gram.

For Problems 23–25, graph each of the functions.

23. $f(x) = e^x - 2$

24. $f(x) = -3^{-x}$

25. $f(x) = \log_2(x - 2)$

For Problems 1–5, evaluate each algebraic expression for the given values of the variables.

1. $-5(x - 1) - 3(2x + 4) + 3(3x - 1)$ for $x = -2$

2. $\dfrac{14a^3b^2}{7a^2b}$ for $a = -1$ and $b = 4$

3. $\dfrac{2}{n} - \dfrac{3}{2n} + \dfrac{5}{3n}$ for $n = 4$

4. $4\sqrt{2x - y} + 5\sqrt{3x + y}$ for $x = 16$ and $y = 16$

5. $\dfrac{3}{x - 2} - \dfrac{5}{x + 3}$ for $x = 3$

For Problems 6–15, perform the indicated operations and express answers in simplified form.

6. $(-5\sqrt{6})(3\sqrt{12})$

7. $(2\sqrt{x} - 3)(\sqrt{x} + 4)$

8. $(3\sqrt{2} - \sqrt{6})(\sqrt{2} + 4\sqrt{6})$

9. $(2x - 1)(x^2 + 6x - 4)$

10. $\dfrac{x^2 - x}{x + 5} \cdot \dfrac{x^2 + 5x + 4}{x^4 - x^2}$

11. $\dfrac{16x^2y}{24xy^3} \div \dfrac{9xy}{8x^2y^2}$

12. $\dfrac{x + 3}{10} + \dfrac{2x + 1}{15} - \dfrac{x - 2}{18}$

13. $\dfrac{7}{12ab} - \dfrac{11}{15a^2}$

14. $\dfrac{8}{x^2 - 4x} + \dfrac{2}{x}$

15. $(8x^3 - 6x^2 - 15x + 4) \div (4x - 1)$

For Problems 16–19, simplify each of the complex fractions.

16. $\dfrac{\dfrac{5}{x^2} - \dfrac{3}{x}}{\dfrac{1}{y} + \dfrac{2}{y^2}}$

17. $\dfrac{\dfrac{2}{x} - 3}{\dfrac{3}{y} + 4}$

18. $\dfrac{2 - \dfrac{1}{n - 2}}{3 + \dfrac{4}{n + 3}}$

19. $\dfrac{3a}{2 - \dfrac{1}{a}} - 1$

For Problems 20–25, factor each of the algebraic expressions completely.

20. $20x^2 + 7x - 6$ **21.** $16x^3 + 54$

22. $4x^4 - 25x^2 + 36$ **23.** $12x^3 - 52x^2 - 40x$

24. $xy - 6x + 3y - 18$ **25.** $10 + 9x - 9x^2$

For Problems 26–35, evaluate each of the numerical expressions.

26. $\left(\dfrac{2}{3}\right)^{-4}$ **27.** $\dfrac{3}{\left(\dfrac{4}{3}\right)^{-1}}$

28. $\sqrt[3]{-\dfrac{27}{64}}$ **29.** $-\sqrt{0.09}$

30. $(27)^{-4/3}$ **31.** $4^0 + 4^{-1} + 4^{-2}$

32. $\left(\dfrac{3^{-1}}{2^{-3}}\right)^{-2}$ **33.** $(2^{-3} - 3^{-2})^{-1}$

34. $\log_2 64$ **35.** $\log_3\left(\dfrac{1}{9}\right)$

For Problems 36–38, find the indicated products and quotients; express final answers with positive integral exponents only.

36. $(-3x^{-1}y^2)(4x^{-2}y^{-3})$

37. $\dfrac{48x^{-4}y^2}{6xy}$ **38.** $\left(\dfrac{27a^{-4}b^{-3}}{-3a^{-1}b^{-4}}\right)^{-1}$

For Problems 39–46, express each radical expression in simplest radical form.

39. $\sqrt{80}$ **40.** $-2\sqrt{54}$

41. $\sqrt{\dfrac{75}{81}}$ **42.** $\dfrac{4\sqrt{6}}{3\sqrt{8}}$

43. $\sqrt[3]{56}$

44. $\dfrac{\sqrt[3]{3}}{\sqrt[3]{4}}$

45. $4\sqrt{52x^3y^2}$

46. $\sqrt{\dfrac{2x}{3y}}$

For Problems 47–49, use the distributive property to help simplify each of the following:

47. $-3\sqrt{24} + 6\sqrt{54} - \sqrt{6}$

48. $\dfrac{\sqrt{8}}{3} - \dfrac{3\sqrt{18}}{4} - \dfrac{5\sqrt{50}}{2}$

49. $8\sqrt[3]{3} - 6\sqrt[3]{24} - 4\sqrt[3]{81}$

For Problems 50 and 51, rationalize the denominator and simplify.

50. $\dfrac{\sqrt{3}}{\sqrt{6} - 2\sqrt{2}}$

51. $\dfrac{3\sqrt{5} - \sqrt{3}}{2\sqrt{3} + \sqrt{7}}$

For Problems 52–54, use scientific notation to help perform the indicated operations.

52. $\dfrac{(0.00016)(300)(0.028)}{0.064}$

53. $\dfrac{0.00072}{0.0000024}$

54. $\sqrt{0.00000009}$

For Problems 55–58, find each of the indicated products or quotients, and express answers in standard form.

55. $(5 - 2i)(4 + 6i)$

56. $(-3 - i)(5 - 2i)$

57. $\dfrac{5}{4i}$

58. $\dfrac{-1 + 6i}{7 - 2i}$

59. Find the slope of the line determined by the points $(2, -3)$ and $(-1, 7)$.

60. Find the slope of the line determined by the equation $4x - 7y = 9$.

61. Find the length of the line segment whose endpoints are $(4, 5)$ and $(-2, 1)$.

62. Write the equation of the line that contains the points $(3, -1)$ and $(7, 4)$.

63. Write the equation of the line that is perpendicular to the line $3x - 4y = 6$ and contains the point $(-3, -2)$.

64. Find the center and the length of a radius of the circle $x^2 + 4x + y^2 - 12y + 31 = 0$.

65. Find the coordinates of the vertex of the parabola $y = x^2 + 10x + 21$.

66. Find the length of the major axis of the ellipse $x^2 + 4y^2 = 16$.

For Problems 67–76, graph each of the functions.

67. $f(x) = -2x - 4$

68. $f(x) = -2x^2 - 2$

69. $f(x) = x^2 - 2x - 2$

70. $f(x) = \sqrt{x + 1} + 2$

71. $f(x) = 2x^2 + 8x + 9$

72. $f(x) = -|x - 2| + 1$

73. $f(x) = 2^x + 2$

74. $f(x) = \log_2(x - 2)$

75. $f(x) = -x(x + 1)(x - 2)$

76. $f(x) = \dfrac{-x}{x + 2}$

77. If $f(x) = x - 3$ and $g(x) = 2x^2 - x - 1$, find $(g \circ f)(x)$ and $(f \circ g)(x)$.

78. Find the inverse (f^{-1}) of $f(x) = 3x - 7$.

79. Find the inverse of $f(x) = -\dfrac{1}{2}x + \dfrac{2}{3}$.

80. Find the constant of variation if y varies directly as x, and $y = 2$ when $x = -\dfrac{2}{3}$.

81. If y is inversely proportional to the square of x, and $y = 4$ when $x = 3$, find y when $x = 6$.

82. The volume of gas at a constant temperature varies inversely as the pressure. What is the volume of a gas under a pressure of 25 pounds if the gas occupies 15 cubic centimeters under a pressure of 20 pounds?

For Problems 83–110, solve each equation.

83. $3(2x - 1) - 2(5x + 1) = 4(3x + 4)$

84. $n + \dfrac{3n - 1}{9} - 4 = \dfrac{3n + 1}{3}$

85. $0.92 + 0.9(x - 0.3) = 2x - 5.95$

86. $|4x - 1| = 11$

87. $3x^2 = 7x$

88. $x^3 - 36x = 0$

89. $30x^2 + 13x - 10 = 0$

90. $8x^3 + 12x^2 - 36x = 0$

91. $x^4 + 8x^2 - 9 = 0$

92. $(n + 4)(n - 6) = 11$

93. $2 - \dfrac{3x}{x - 4} = \dfrac{14}{x + 7}$

94. $\dfrac{2n}{6n^2 + 7n - 3} - \dfrac{n - 3}{3n^2 + 11n - 4} = \dfrac{5}{2n^2 + 11n + 12}$

95. $\sqrt{3y} - y = -6$

96. $\sqrt{x + 19} - \sqrt{x + 28} = -1$

97. $(3x - 1)^2 = 45$

98. $(2x + 5)^2 = -32$

99. $2x^2 - 3x + 4 = 0$

100. $3n^2 - 6n + 2 = 0$

101. $\dfrac{5}{n - 3} - \dfrac{3}{n + 3} = 1$

102. $12x^4 - 19x^2 + 5 = 0$

103. $2x^2 + 5x + 5 = 0$

104. $x^3 - 4x^2 - 25x + 28 = 0$

105. $6x^3 - 19x^2 + 9x + 10 = 0$

106. $16^x = 64$

107. $\log_3 x = 4$

108. $\log_{10} x + \log_{10} 25 = 2$

109. $\ln(3x - 4) - \ln(x + 1) = \ln 2$

110. $27^{4x} = 9^{x+1}$

For Problems 111–120, solve each inequality.

111. $-5(y - 1) + 3 > 3y - 4 - 4y$

112. $0.06x + 0.08(250 - x) \geq 19$

113. $|5x - 2| > 13$

114. $|6x + 2| < 8$

115. $\dfrac{x - 2}{5} - \dfrac{3x - 1}{4} \leq \dfrac{3}{10}$

116. $(x - 2)(x + 4) \leq 0$

117. $(3x - 1)(x - 4) > 0$

118. $x(x + 5) < 24$

119. $\dfrac{x - 3}{x - 7} \geq 0$

120. $\dfrac{2x}{x + 3} > 4$

For Problems 121–135, set up an equation or an inequality to help solve each problem.

121. Find three consecutive odd integers whose sum is 57.

122. Eric has a collection of 63 coins consisting of nickels, dimes, and quarters. The number of dimes is 6 more than the number of nickels, and the number of quarters is 1 more than twice the number of nickels. How many coins of each kind are in the collection?

123. One of two supplementary angles is 4° more than one-third of the other angle. Find the measure of each of the angles.

124. If a ring costs a jeweler $300, at what price should it be sold to make a profit of 50% on the selling price?

125. Beth invested a certain amount of money at 8% and $300 more than that amount at 9%. Her total yearly interest was $316. How much did she invest at each rate?

126. Two trains leave the same depot at the same time, one traveling east and the other traveling west. At the end of $4\dfrac{1}{2}$ hours, they are 639 miles apart. If the rate of the train traveling east is 10 miles per hour faster than the other train, find their rates.

127. A 10-quart radiator contains a 50% solution of antifreeze. How much needs to be drained out and replaced with pure antifreeze to obtain a 70% antifreeze solution?

128. Sam shot rounds of 70, 73, and 76 on the first 3 days of a golf tournament. What must he shoot on the fourth day of the tournament to average 72 or less for the 4 days?

129. The cube of a number equals nine times the same number. Find the number.

130. A strip of uniform width is to be cut off both sides and both ends of a sheet of paper that is 8 inches by

587

14 inches to reduce the size of the paper to an area of 72 square inches. Find the width of the strip.

131. A sum of $2450 is to be divided between two people in the ratio of 3 to 4. How much does each person receive?

132. Working together, Sue and Dean can complete a task in $1\frac{1}{5}$ hours. Dean can do the task by himself in 2 hours. How long would it take Sue to complete the task by herself?

133. Dudley bought a number of shares of stock for $300. A month later he sold all but 10 shares at a profit of $5 per share and regained his original investment of $300. How many shares did he originally buy, and at what price per share?

134. The units digit of a two-digit number is 1 more than twice the tens digit. The sum of the digits is 10. Find the number.

135. The sum of the two smallest angles of a triangle is 40° less than the other angle. The sum of the smallest and largest angles is twice the other angle. Find the measures of the three angles of the triangle.

Systems of Equations

When mixing different solutions, a chemist could use a system of equations to determine how much of each solution is needed to produce a specific concentration.

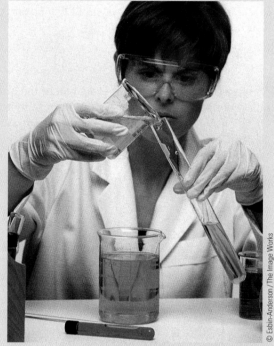

© Esbin-Anderson / The Image Works

A 10% salt solution is to be mixed with a 20% salt solution to produce 20 gallons of a 17.5% salt solution. How many gallons of the 10% solution and how many gallons of the 20% solution should be mixed? The two equations $x + y = 20$ and $0.10x + 0.20y = 0.175(20)$ algebraically represent the conditions of the problem; x represents the number of gallons of the 10% solution, and y represents the number of gallons of the 20% solution. The two equations considered together form a system of linear equations, and the problem can be solved by solving the system of equations.

Throughout most of this chapter, we consider systems of linear equations and their applications. We will discuss various techniques for solving systems of linear equations.

11.1 Systems of Two Linear Equations in Two Variables

In Chapter 7 we stated that any equation of the form $Ax + By = C$, where A, B, and C are real numbers (A and B not both zero), is a **linear equation** in the two variables x and y, and its graph is a straight line. Two linear equations in two variables considered together form a **system of two linear equations in two variables**, as illustrated by the following examples:

$$\begin{pmatrix} x + y = 6 \\ x - y = 2 \end{pmatrix} \qquad \begin{pmatrix} 3x + 2y = 1 \\ 5x - 2y = 23 \end{pmatrix} \qquad \begin{pmatrix} 4x - 5y = 21 \\ -3x + y = -7 \end{pmatrix}$$

To **solve** such a system means to find all of the ordered pairs that simultaneously satisfy both equations in the system. For example, if we graph the two equations $x + y = 6$ and $x - y = 2$ on the same set of axes, as in Figure 11.1, then the ordered pair associated with the point of intersection of the two lines is the **solution of the system**. Thus we say that $\{(4, 2)\}$ is the solution set of the system

$$\begin{pmatrix} x + y = 6 \\ x - y = 2 \end{pmatrix}$$

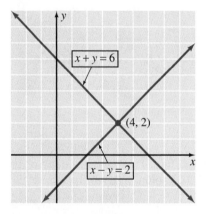

Figure 11.1

To check the solution, we substitute 4 for x and 2 for y in the two equations.

$x + y = 6$ becomes $4 + 2 = 6$, a true statement

$x - y = 2$ becomes $4 - 2 = 2$, a true statement

Because the graph of a linear equation in two variables is a straight line, three possible situations can occur when we are solving a system of two linear equations in two variables. These situations are shown in Figure 11.2.

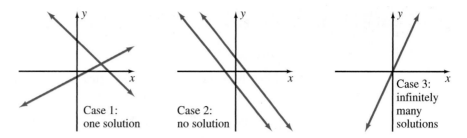

Figure 11.2

Case 1   The graphs of the two equations are two lines intersecting in one point. There is exactly one solution, and the system is called a **consistent system**.

Case 2   The graphs of the two equations are parallel lines. There is *no solution*, and the system is called an **inconsistent system**.

Case 3   The graphs of the two equations are the same line, and there are *infinitely many solutions* of the system. Any pair of real numbers that satisfies one of the equations also satisfies the other equation, and we say that the equations are dependent.

Thus, as we solve a system of two linear equations in two variables, we can expect one of three outcomes: The system will have *no* solutions, *one* ordered pair as a solution, or *infinitely many* ordered pairs as solutions.

■ The Substitution Method

Solving specific systems of equations by graphing requires accurate graphs. However, unless the solutions are integers, it is difficult to obtain exact solutions from a graph. Therefore we will consider some other techniques for solving systems of equations.

The **substitution method**, which works especially well with systems of two equations in two unknowns, can be described as follows.

Step 1   Solve one of the equations for one variable in terms of the other. (If possible, make a choice that will avoid fractions.)

Step 2   Substitute the expression obtained in step 1 into the other equation, producing an equation in one variable.

Step 3   Solve the equation obtained in step 2.

Step 4   Use the solution obtained in step 3, along with the expression obtained in step 1, to determine the solution of the system.

E X A M P L E   1

Solve the system $\left(\begin{array}{l} x - 3y = -25 \\ 4x + 5y = 19 \end{array} \right)$.

Solution

Solve the first equation for x in terms of y to produce

$$x = 3y - 25$$

Substitute $3y - 25$ for x in the second equation and solve for y.

$$4x + 5y = 19$$
$$4(3y - 25) + 5y = 19$$
$$12y - 100 + 5y = 19$$
$$17y = 119$$
$$y = 7$$

Next, substitute 7 for y in the equation $x = 3y - 25$ to obtain

$$x = 3(7) - 25 = -4$$

The solution set of the given system is $\{(-4, 7)\}$. (You should check this solution in both of the original equations.) ∎

E X A M P L E   2

Solve the system $\left(\begin{array}{l} 5x + 9y = -2 \\ 2x + 4y = -1 \end{array} \right)$.

Solution

A glance at the system should tell us that solving either equation for either variable will produce a fractional form, so let's just use the first equation and solve for x in terms of y.

$$5x + 9y = -2$$
$$5x = -9y - 2$$
$$x = \frac{-9y - 2}{5}$$

Now we can substitute this value for x into the second equation and solve for y.

$$2x + 4y = -1$$
$$2\left(\frac{-9y - 2}{5}\right) + 4y = -1$$
$$2(-9y - 2) + 20y = -5 \qquad \text{Multiplied both sides by 5.}$$
$$-18y - 4 + 20y = -5$$
$$2y - 4 = -5$$
$$2y = -1$$
$$y = -\frac{1}{2}$$

Now we can substitute $-\dfrac{1}{2}$ for y in $x = \dfrac{-9y - 2}{5}$.

$$x = \frac{-9\left(-\dfrac{1}{2}\right) - 2}{5} = \frac{\dfrac{9}{2} - 2}{5} = \frac{1}{2}$$

The solution set is $\left\{\left(\dfrac{1}{2}, -\dfrac{1}{2}\right)\right\}$.   ■

E X A M P L E   3   Solve the system

$$\left(\begin{array}{l} 6x - 4y = 18 \\ y = \dfrac{3}{2}x - \dfrac{9}{2} \end{array}\right)$$

Solution

The second equation is given in appropriate form for us to begin the substitution process. Substitute $\dfrac{3}{2}x - \dfrac{9}{2}$ for y in the first equation to yield

$$6x - 4y = 18$$

$$6x - 4\left(\frac{3}{2}x - \frac{9}{2}\right) = 18$$

$$6x - 6x + 18 = 18$$

$$18 = 18$$

Our obtaining a true numerical statement ($18 = 18$) indicates that the system has infinitely many solutions. Any ordered pair that satisfies one of the equations will also satisfy the other equation. Thus in the second equation of the original system, if we let $x = k$, then $y = \dfrac{3}{2}k - \dfrac{9}{2}$. Therefore the solution set can be expressed as $\left\{\left(k, \dfrac{3}{2}k - \dfrac{9}{2}\right) \mid k \text{ is a real number}\right\}$. If some specific solutions are needed, they can be generated by the ordered pair $\left(k, \dfrac{3}{2}k - \dfrac{9}{2}\right)$. For example, if we let $k = 1$, then we get $\dfrac{3}{2}(1) - \dfrac{9}{2} = -\dfrac{6}{2} = -3$. Thus the ordered pair $(1, -3)$ is a member of the solution set of the given system.   ■

■ The Elimination-by-Addition Method

Now let's consider the **elimination-by-addition method** for solving a system of equations. This is a very important method because it is the basis for developing other techniques for solving systems that contain many equations and variables.

The method involves replacing systems of equations with *simpler equivalent systems* until we obtain a system where the solutions are obvious. **Equivalent systems of equations** are systems that have exactly the same solution set. The following operations or transformations can be applied to a system of equations to produce an equivalent system:

1. Any two equations of the system can be interchanged.

2. Both sides of any equation of the system can be multiplied by any nonzero real number.

3. Any equation of the system can be replaced by the sum of that equation and a nonzero multiple of another equation.

EXAMPLE 4

Solve the system $\left(\begin{array}{l} 3x + 5y = -9 \\ 2x - 3y = 13 \end{array} \right)$.

$$(1)$$
$$(2)$$

Solution

We can replace the given system with an equivalent system by multiplying equation (2) by -3.

$$\left(\begin{array}{r} 3x + 5y = -9 \\ -6x + 9y = -39 \end{array} \right)$$

$$(3)$$
$$(4)$$

Now let's replace equation (4) with an equation formed by multiplying equation (3) by 2 and adding this result to equation (4).

$$\left(\begin{array}{r} 3x + 5y = -9 \\ 19y = -57 \end{array} \right)$$

$$(5)$$
$$(6)$$

From equation (6), we can easily determine that $y = -3$. Then, substituting -3 for y in equation (5) produces

$$3x + 5(-3) = -9$$
$$3x - 15 = -9$$
$$3x = 6$$
$$x = 2$$

The solution set for the given system is $\{(2, -3)\}$.   ■

Remark: We are using a format for the elimination-by-addition method that highlights the use of equivalent systems. In Section 11.3 this format will lead naturally to an approach using matrices. Thus it is beneficial to stress the use of equivalent systems at this time.

EXAMPLE 5

Solve the system

$$\left(\begin{array}{l} \dfrac{1}{2}x + \dfrac{2}{3}y = -4 \\[2mm] \dfrac{1}{4}x - \dfrac{3}{2}y = 20 \end{array} \right)$$

$$(7)$$

$$(8)$$

Solution

The given system can be replaced with an equivalent system by multiplying equation (7) by 6 and equation (8) by 4.

$$\begin{pmatrix} 3x + 4y = -24 \\ x - 6y = 80 \end{pmatrix} \qquad \begin{matrix} (9) \\ (10) \end{matrix}$$

Now let's exchange equations (9) and (10).

$$\begin{pmatrix} x - 6y = 80 \\ 3x + 4y = -24 \end{pmatrix} \qquad \begin{matrix} (11) \\ (12) \end{matrix}$$

We can replace equation (12) with an equation formed by multiplying equation (11) by -3 and adding this result to equation (12).

$$\begin{pmatrix} x - 6y = 80 \\ 22y = -264 \end{pmatrix} \qquad \begin{matrix} (13) \\ (14) \end{matrix}$$

From equation (14) we can determine that $y = -12$. Then, substituting -12 for y in equation (13) produces

$$x - 6(-12) = 80$$

$$x + 72 = 80$$

$$x = 8$$

The solution set of the given system is $\{(8, -12)\}$. (Check this!) ∎

EXAMPLE 6

Solve the system $\begin{pmatrix} x - 4y = 9 \\ x - 4y = 3 \end{pmatrix}$. $\begin{matrix} (15) \\ (16) \end{matrix}$

Solution

We can replace equation (16) with an equation formed by multiplying equation (15) by -1 and adding this result to equation (16).

$$\begin{pmatrix} x - 4y = 9 \\ 0 = -6 \end{pmatrix} \qquad \begin{matrix} (17) \\ (18) \end{matrix}$$

The statement $0 = -6$ is a contradiction, and therefore the original system is *inconsistent*; it has no solution. The solution set is $\varnothing$. ∎

Both the elimination-by-addition and the substitution methods can be used to obtain exact solutions for any system of two linear equations in two unknowns. Sometimes it is a matter of deciding which method to use on a particular system. Some systems lend themselves to one or the other of the methods by virtue of the original format of the equations. We will illustrate this idea in a moment when we solve some word problems.

▪ Using Systems to Solve Problems

Many word problems that we solved earlier in this text with one variable and one equation can also be solved by using a system of two linear equations in two variables. In fact, in many of these problems, you may find it more natural to use two variables and two equations.

The two-variable expression, $10t + u$, can be used to represent any two-digit whole number. The t represents the tens digit, and the u represents the units digit. For example, if $t = 4$ and $u = 8$, then $10t + u$ becomes $10(4) + 8 = 48$. Now let's use this general representation for a two-digit number to help solve a problem.

P R O B L E M   1

The units digit of a two-digit number is 1 more than twice the tens digit. The number with the digits reversed is 45 larger than the original number. Find the original number.

Solution

Let u represent the units digit of the original number, and let t represent the tens digit. Then $10t + u$ represents the original number, and $10u + t$ represents the new number with the digits reversed. The problem translates into the following system:

$$\begin{pmatrix} u = 2t + 1 \\ 10u + t = 10t + u + 45 \end{pmatrix}$$

The units digit is 1 more than twice the tens digit.

The number with the digits reversed is 45 larger than the original number.

Simplify the second equation, and the system becomes

$$\begin{pmatrix} u = 2t + 1 \\ u - t = 5 \end{pmatrix}$$

Because of the form of the first equation, this system lends itself to solving by the substitution method. Substitute $2t + 1$ for u in the second equation to produce

$$(2t + 1) - t = 5$$

$$t + 1 = 5$$

$$t = 4$$

Now substitute 4 for t in the equation $u = 2t + 1$ to get

$$u = 2(4) + 1 = 9$$

The tens digit is 4 and the units digit is 9, so the number is 49.   ▪

P R O B L E M   2

Lucinda invested $950, part of it at 11% interest and the remainder at 12%. Her total yearly income from the two investments was $111.50. How much did she invest at each rate?

Solution

Let x represent the amount invested at 11% and y the amount invested at 12%. The problem translates into the following system:

$$\begin{pmatrix} x + y = 950 \\ 0.11x + 0.12y = 111.50 \end{pmatrix}$$

⟵ The two investments total $950.
⟵ The yearly interest from the two investments totals $111.50.

Multiply the second equation by 100 to produce an equivalent system.

$$\begin{pmatrix} x + y = 950 \\ 11x + 12y = 11150 \end{pmatrix}$$

Because neither equation is solved for one variable in terms of the other, let's use the elimination-by-addition method to solve the system. The second equation can be replaced by an equation formed by multiplying the first equation by -11 and adding this result to the second equation.

$$\begin{pmatrix} x + y = 950 \\ y = 700 \end{pmatrix}$$

Now we substitute 700 for y in the equation $x + y = 950$.

$$x + 700 = 950$$

$$x = 250$$

Therefore Lucinda must have invested $250 at 11% and $700 at 12%. ∎

In our final example of this section, we will use a graphing utility to help solve a system of equations.

EXAMPLE 7

Solve the system $\begin{pmatrix} 1.14x + 2.35y = -7.12 \\ 3.26x - 5.05y = 26.72 \end{pmatrix}$.

Solution

First, we need to solve each equation for y in terms of x. Thus the system becomes

$$\begin{pmatrix} y = \dfrac{-7.12 - 1.14x}{2.35} \\ y = \dfrac{3.26x - 26.72}{5.05} \end{pmatrix}$$

Now we can enter both of these equations into a graphing utility and obtain Figure 11.3. From this figure it appears that the point of intersection is at approximately $x = 2$ and $y = -4$. By direct substitution into the given equations, we can verify that the point of intersection is exactly $(2, -4)$.

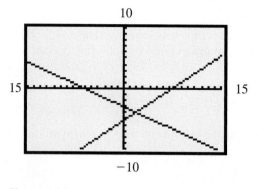

Figure 11.3

Problem Set 11.1

For Problems 1–10, use the graphing approach to determine whether the system is consistent, the system is inconsistent, or the equations are dependent. If the system is consistent, find the solution set from the graph and check it.

1. $\begin{pmatrix} x - y = 1 \\ 2x + y = 8 \end{pmatrix}$

2. $\begin{pmatrix} 3x + y = 0 \\ x - 2y = -7 \end{pmatrix}$

3. $\begin{pmatrix} 4x + 3y = -5 \\ 2x - 3y = -7 \end{pmatrix}$

4. $\begin{pmatrix} 2x - y = 9 \\ 4x - 2y = 11 \end{pmatrix}$

5. $\begin{pmatrix} \dfrac{1}{2}x + \dfrac{1}{4}y = 9 \\ 4x + 2y = 72 \end{pmatrix}$

6. $\begin{pmatrix} 5x + 2y = -9 \\ 4x - 3y = 2 \end{pmatrix}$

7. $\begin{pmatrix} \dfrac{1}{2}x - \dfrac{1}{3}y = 3 \\ x + 4y = -8 \end{pmatrix}$

8. $\begin{pmatrix} 4x - 9y = -60 \\ \dfrac{1}{3}x - \dfrac{3}{4}y = -5 \end{pmatrix}$

9. $\begin{pmatrix} x - \dfrac{y}{2} = -4 \\ 8x - 4y = -1 \end{pmatrix}$

10. $\begin{pmatrix} 3x - 2y = 7 \\ 6x + 5y = -4 \end{pmatrix}$

13. $\begin{pmatrix} x = 3y - 25 \\ 4x + 5y = 19 \end{pmatrix}$

14. $\begin{pmatrix} 3x - 5y = 25 \\ x = y + 7 \end{pmatrix}$

15. $\begin{pmatrix} y = \dfrac{2}{3}x - 1 \\ 5x - 7y = 9 \end{pmatrix}$

16. $\begin{pmatrix} y = \dfrac{3}{4}x + 5 \\ 4x - 3y = -1 \end{pmatrix}$

17. $\begin{pmatrix} a = 4b + 13 \\ 3a + 6b = -33 \end{pmatrix}$

18. $\begin{pmatrix} 9a - 2b = 28 \\ b = -3a + 1 \end{pmatrix}$

19. $\begin{pmatrix} 2x - 3y = 4 \\ y = \dfrac{2}{3}x - \dfrac{4}{3} \end{pmatrix}$

20. $\begin{pmatrix} t + u = 11 \\ t = u + 7 \end{pmatrix}$

21. $\begin{pmatrix} u = t - 2 \\ t + u = 12 \end{pmatrix}$

22. $\begin{pmatrix} y = 5x - 9 \\ 5x - y = 9 \end{pmatrix}$

23. $\begin{pmatrix} 4x + 3y = -7 \\ 3x - 2y = 16 \end{pmatrix}$

24. $\begin{pmatrix} 5x - 3y = -34 \\ 2x + 7y = -30 \end{pmatrix}$

25. $\begin{pmatrix} 5x - y = 4 \\ y = 5x + 9 \end{pmatrix}$

26. $\begin{pmatrix} 2x + 3y = 3 \\ 4x - 9y = -4 \end{pmatrix}$

27. $\begin{pmatrix} 4x - 5y = 3 \\ 8x + 15y = -24 \end{pmatrix}$

28. $\begin{pmatrix} 4x + y = 9 \\ y = 15 - 4x \end{pmatrix}$

For Problems 11–28, solve each system by using the substitution method.

11. $\begin{pmatrix} x + y = 16 \\ y = x + 2 \end{pmatrix}$

12. $\begin{pmatrix} 2x + 3y = -5 \\ y = 2x + 9 \end{pmatrix}$

For Problems 29–44, solve each system by using the elimination-by-addition method.

29. $\begin{pmatrix} 3x + 2y = 1 \\ 5x - 2y = 23 \end{pmatrix}$

30. $\begin{pmatrix} 4x + 3y = -22 \\ 4x - 5y = 26 \end{pmatrix}$

31. $\begin{pmatrix} x - 3y = -22 \\ 2x + 7y = 60 \end{pmatrix}$

32. $\begin{pmatrix} 6x - y = 3 \\ 5x + 3y = -9 \end{pmatrix}$

33. $\begin{pmatrix} 4x - 5y = 21 \\ 3x + 7y = -38 \end{pmatrix}$

34. $\begin{pmatrix} 5x - 3y = -34 \\ 2x + 7y = -30 \end{pmatrix}$

35. $\begin{pmatrix} 5x - 2y = 19 \\ 5x - 2y = 7 \end{pmatrix}$

36. $\begin{pmatrix} 4a + 2b = -4 \\ 6a - 5b = 18 \end{pmatrix}$

37. $\begin{pmatrix} 5a + 6b = 8 \\ 2a - 15b = 9 \end{pmatrix}$

38. $\begin{pmatrix} 7x + 2y = 11 \\ 7x + 2y = -4 \end{pmatrix}$

39. $\begin{pmatrix} \dfrac{2}{3}s + \dfrac{1}{4}t = -1 \\[2mm] \dfrac{1}{2}s - \dfrac{1}{3}t = -7 \end{pmatrix}$

40. $\begin{pmatrix} \dfrac{1}{4}s - \dfrac{2}{3}t = -3 \\[2mm] \dfrac{1}{3}s + \dfrac{1}{3}t = 7 \end{pmatrix}$

41. $\begin{pmatrix} \dfrac{x}{2} - \dfrac{2y}{5} = \dfrac{-23}{60} \\[2mm] \dfrac{2x}{3} + \dfrac{y}{4} = \dfrac{-1}{4} \end{pmatrix}$

42. $\begin{pmatrix} \dfrac{2x}{3} - \dfrac{y}{2} = \dfrac{3}{5} \\[2mm] \dfrac{x}{4} + \dfrac{y}{2} = \dfrac{7}{80} \end{pmatrix}$

43. $\begin{pmatrix} \dfrac{2}{3}x + \dfrac{1}{2}y = \dfrac{1}{6} \\[2mm] 4x + 6y = -1 \end{pmatrix}$

44. $\begin{pmatrix} \dfrac{1}{2}x + \dfrac{2}{3}y = -\dfrac{3}{10} \\[2mm] 5x + 4y = -1 \end{pmatrix}$

For Problems 45–60, solve each system by using either the substitution method or the elimination-by-addition method, whichever seems more appropriate.

45. $\begin{pmatrix} 5x - y = -22 \\ 2x + 3y = -2 \end{pmatrix}$

46. $\begin{pmatrix} 4x + 5y = -41 \\ 3x - 2y = 21 \end{pmatrix}$

47. $\begin{pmatrix} x = 3y - 10 \\ x = -2y + 15 \end{pmatrix}$

48. $\begin{pmatrix} y = 4x - 24 \\ 7x + y = 42 \end{pmatrix}$

49. $\begin{pmatrix} 3x - 5y = 9 \\ 6x - 10y = -1 \end{pmatrix}$

50. $\begin{pmatrix} y = \dfrac{2}{5}x - 3 \\[2mm] 4x - 7y = 33 \end{pmatrix}$

51. $\begin{pmatrix} \dfrac{1}{2}x - \dfrac{2}{3}y = 22 \\[2mm] \dfrac{1}{2}x + \dfrac{1}{4}y = 0 \end{pmatrix}$

52. $\begin{pmatrix} \dfrac{2}{5}x - \dfrac{1}{3}y = -9 \\[2mm] \dfrac{3}{4}x + \dfrac{1}{3}y = -14 \end{pmatrix}$

53. $\begin{pmatrix} t = 2u + 2 \\ 9u - 9t = -45 \end{pmatrix}$

54. $\begin{pmatrix} 9u - 9t = 36 \\ u = 2t + 1 \end{pmatrix}$

55. $\begin{pmatrix} x + y = 1000 \\ 0.12x + 0.14y = 136 \end{pmatrix}$

56. $\begin{pmatrix} x + y = 10 \\ 0.3x + 0.7y = 4 \end{pmatrix}$

57. $\begin{pmatrix} y = 2x \\ 0.09x + 0.12y = 132 \end{pmatrix}$

58. $\begin{pmatrix} y = 3x \\ 0.1x + 0.11y = 64.5 \end{pmatrix}$

59. $\begin{pmatrix} x + y = 10.5 \\ 0.5x + 0.8y = 7.35 \end{pmatrix}$

60. $\begin{pmatrix} 2x + y = 7.75 \\ 3x + 2y = 12.5 \end{pmatrix}$

For Problems 61–80, solve each problem by using a system of equations.

61. The sum of two numbers is 53, and their difference is 19. Find the numbers.

62. The sum of two numbers is −3 and their difference is 25. Find the numbers.

63. The measure of the larger of two complementary angles is 15° more than four times the measure of the smaller angle. Find the measures of both angles.

64. Assume that a plane is flying at a constant speed under unvarying wind conditions. Traveling against a head wind, the plane takes 4 hours to travel 1540 miles. Traveling with a tail wind, the plane flies 1365 miles in 3 hours. Find the speed of the plane and the speed of the wind.

65. The tens digit of a two-digit number is 1 more than three times the units digit. If the sum of the digits is 9, find the number.

66. The units digit of a two-digit number is 1 less than twice the tens digit. The sum of the digits is 8. Find the number.

67. The sum of the digits of a two-digit number is 7. If the digits are reversed, the newly formed number is 9 larger than the original number. Find the original number.

68. The units digit of a two-digit number is 1 less than twice the tens digit. If the digits are reversed, the newly formed number is 27 larger than the original number. Find the original number.

69. A motel rents double rooms at $32 per day and single rooms at $26 per day. If 23 rooms were rented one day for a total of $688, how many rooms of each kind were rented?

70. An apartment complex rents one-bedroom apartments for $325 per month and two-bedroom apartments for $375 per month. One month the number of one-bedroom apartments rented was twice the number of two-bedroom apartments. If the total income for that month was $12,300, how many apartments of each kind were rented?

71. The income from a student production was $10,000. The price of a student ticket was $3, and nonstudent tickets were sold at $5 each. Three thousand tickets were sold. How many tickets of each kind were sold?

72. Michelle can enter a small business as a full partner and receive a salary of $10,000 a year and 15% of the year's profit, or she can be sales manager for a salary of $25,000 plus 5% of the year's profit. What must the year's profit be for her total earnings to be the same whether she is a full partner or a sales manager?

73. Melinda invested three times as much money at 11% yearly interest as she did at 9%. Her total yearly interest from the two investments was $210. How much did she invest at each rate?

74. Sam invested $1950, part of it at 10% and the rest at 12% yearly interest. The yearly income on the 12% investment was $6 less than twice the income from the 10% investment. How much did he invest at each rate?

75. One day last summer, Jim went kayaking on the Little Susitna River in Alaska. Paddling upstream against the current, he traveled 20 miles in 4 hours. Then he turned around and paddled twice as fast downstream and, with the help of the current, traveled 19 miles in 1 hour. Find the rate of the current.

76. One solution contains 30% alcohol and a second solution contains 70% alcohol. How many liters of each solution should be mixed to make 10 liters containing 40% alcohol?

77. Bill bought 4 tennis balls and 3 golf balls for a total of $10.25. Bret went into the same store and bought 2 tennis balls and 5 golf balls for $11.25. What was the price for a tennis ball and the price for a golf ball?

78. Six cans of pop and 2 bags of potato chips cost $5.12. At the same prices, 8 cans of pop and 5 bags of potato chips cost $9.86. Find the price per can of pop and the price per bag of potato chips.

79. A cash drawer contains only five- and ten-dollar bills. There are 12 more five-dollar bills than ten-dollar bills. If the drawer contains $330, find the number of each kind of bill.

80. Brad has a collection of dimes and quarters totaling $47.50. The number of quarters is 10 more than twice the number of dimes. How many coins of each kind does he have?

■ ■ ■ THOUGHTS INTO WORDS

81. Give a general description of how to use the substitution method to solve a system of two linear equations in two variables.

82. Give a general description of how to use the elimination-by-addition method to solve a system of two linear equations in two variables.

83. Which method would you use to solve the system $\begin{pmatrix} 9x + 4y = 7 \\ 3x + 2y = 6 \end{pmatrix}$? Why?

84. Which method would you use to solve the system $\begin{pmatrix} 5x + 3y = 12 \\ 3x - y = 10 \end{pmatrix}$? Why?

■ ■ ■ FURTHER INVESTIGATIONS

A system such as

$$\begin{pmatrix} \dfrac{2}{x} + \dfrac{3}{y} = \dfrac{19}{15} \\ -\dfrac{2}{x} + \dfrac{1}{y} = -\dfrac{7}{15} \end{pmatrix}$$

is not a linear system, but it can be solved using the elimination-by-addition method as follows. Add the first equation to the second to produce the equivalent system

$$\begin{pmatrix} \dfrac{2}{x} + \dfrac{3}{y} = \dfrac{19}{15} \\ \dfrac{4}{y} = \dfrac{12}{15} \end{pmatrix}$$

Now solve $\dfrac{4}{y} = \dfrac{12}{15}$ to produce $y = 5$.

Substitute 5 for y in the first equation and solve for x to produce

$$\frac{2}{x} + \frac{3}{5} = \frac{19}{15}$$

$$\frac{2}{x} = \frac{10}{15}$$

$$10x = 30$$

$$x = 3$$

The solution set of the original system is $\{(3, 5)\}$.

For Problems 85–90, solve each system.

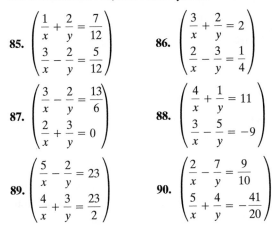

85. $\left(\begin{array}{l} \dfrac{1}{x} + \dfrac{2}{y} = \dfrac{7}{12} \\ \dfrac{3}{x} - \dfrac{2}{y} = \dfrac{5}{12} \end{array} \right)$

86. $\left(\begin{array}{l} \dfrac{3}{x} + \dfrac{2}{y} = 2 \\ \dfrac{2}{x} - \dfrac{3}{y} = \dfrac{1}{4} \end{array} \right)$

87. $\left(\begin{array}{l} \dfrac{3}{x} - \dfrac{2}{y} = \dfrac{13}{6} \\ \dfrac{2}{x} + \dfrac{3}{y} = 0 \end{array} \right)$

88. $\left(\begin{array}{l} \dfrac{4}{x} + \dfrac{1}{y} = 11 \\ \dfrac{3}{x} - \dfrac{5}{y} = -9 \end{array} \right)$

89. $\left(\begin{array}{l} \dfrac{5}{x} - \dfrac{2}{y} = 23 \\ \dfrac{4}{x} + \dfrac{3}{y} = \dfrac{23}{2} \end{array} \right)$

90. $\left(\begin{array}{l} \dfrac{2}{x} - \dfrac{7}{y} = \dfrac{9}{10} \\ \dfrac{5}{x} + \dfrac{4}{y} = -\dfrac{41}{20} \end{array} \right)$

91. Consider the linear system $\left(\begin{array}{l} a_1 x + b_1 y = c_1 \\ a_2 x + b_2 y = c_2 \end{array} \right)$.

   (a) Prove that this system has exactly one solution if and only if $\dfrac{a_1}{a_2} \neq \dfrac{b_1}{b_2}$.

   (b) Prove that this system has no solution if and only if $\dfrac{a_1}{a_2} = \dfrac{b_1}{b_2} \neq \dfrac{c_1}{c_2}$.

   (c) Prove that this system has infinitely many solutions if and only if $\dfrac{a_1}{a_2} = \dfrac{b_1}{b_2} = \dfrac{c_1}{c_2}$.

92. For each of the following systems, use the results from Problem 91 to determine whether the system is consistent or inconsistent or whether the equations are dependent.

   (a) $\left(\begin{array}{l} 5x + y = 9 \\ x - 5y = 4 \end{array} \right)$

   (b) $\left(\begin{array}{l} 3x - 2y = 14 \\ 2x + 3y = 9 \end{array} \right)$

   (c) $\left(\begin{array}{l} x - 7y = 4 \\ x - 7y = 9 \end{array} \right)$

   (d) $\left(\begin{array}{l} 3x - 5y = 10 \\ 6x - 10y = 1 \end{array} \right)$

   (e) $\left(\begin{array}{l} 3x + 6y = 2 \\ \dfrac{3}{5}x + \dfrac{6}{5}y = \dfrac{2}{5} \end{array} \right)$

   (f) $\left(\begin{array}{l} \dfrac{2}{3}x - \dfrac{3}{4}y = 2 \\ \dfrac{1}{2}x + \dfrac{2}{5}y = 9 \end{array} \right)$

   (g) $\left(\begin{array}{l} 7x + 9y = 14 \\ 8x - 3y = 12 \end{array} \right)$

   (h) $\left(\begin{array}{l} 4x - 5y = 3 \\ 12x - 15y = 9 \end{array} \right)$

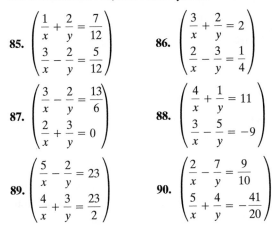 **GRAPHING CALCULATOR ACTIVITIES**

93. For each of the systems of equations in Problem 92, use your graphing calculator to help determine whether the system is consistent or inconsistent or whether the equations are dependent.

94. Use your graphing calculator to help determine the solution set for each of the following systems. Be sure to check your answers.

   (a) $\left(\begin{array}{l} y = 3x - 1 \\ y = 9 - 2x \end{array} \right)$

   (b) $\left(\begin{array}{l} 5x + y = -9 \\ 3x - 2y = 5 \end{array} \right)$

   (c) $\left(\begin{array}{l} 4x - 3y = 18 \\ 5x + 6y = 3 \end{array} \right)$

   (d) $\left(\begin{array}{l} 2x - y = 20 \\ 7x + y = 79 \end{array} \right)$

   (e) $\left(\begin{array}{l} 13x - 12y = 37 \\ 15x + 13y = -11 \end{array} \right)$

   (f) $\left(\begin{array}{l} 1.98x + 2.49y = 13.92 \\ 1.19x + 3.45y = 16.18 \end{array} \right)$

11.2 Systems of Three Linear Equations in Three Variables

Consider a linear equation in three variables x, y, and z, such as $3x - 2y + z = 7$. Any **ordered triple** (x, y, z) that makes the equation a true numerical statement is said to be a *solution* of the equation. For example, the ordered triple $(2, 1, 3)$ is a solution because $3(2) - 2(1) + 3 = 7$. However, the ordered triple $(5, 2, 4)$ is not a solution because $3(5) - 2(2) + 4 \neq 7$. There are infinitely many solutions in the solution set.

Remark: The idea of a linear equation is generalized to include equations of more than two variables. Thus an equation such as $5x - 2y + 9z = 8$ is called a linear equation in three variables, the equation $5x - 7y + 2z - 11w = 1$ is called a linear equation in four variables, and so on.

To *solve* a system of three linear equations in three variables, such as

$$\begin{pmatrix} 3x - y + 2z = 13 \\ 4x + 2y + 5z = 30 \\ 5x - 3y - z = 3 \end{pmatrix}$$

means to find all of the ordered triples that satisfy all three equations. In other words, the solution set of the system is the intersection of the solution sets of all three equations in the system.

The graph of a linear equation in three variables is a *plane*, not a line. In fact, graphing equations in three variables requires the use of a three-dimensional coordinate system. Thus using a graphing approach to solve systems of three linear equations in three variables is not at all practical. However, a simple graphical analysis does provide us with some indication of what we can expect as we begin solving such systems.

In general, because each linear equation in three variables produces a plane, a system of three such equations produces three planes. There are various ways in which three planes can be related. For example, they may be mutually parallel; or two of the planes may be parallel, with the third intersecting the other two. (You may want to analyze all of the other possibilities for the three planes!) However, for our purposes at this time, we need to realize that from a solution set viewpoint, a system of three linear equations in three variables produces one of the following possibilities:

1. There is *one ordered triple* that satisfies all three equations. The three planes have a common *point* of intersection, as indicated in Figure 11.4.

2. There are *infinitely many ordered triples* in the solution set, all of which are coordinates of points on a *line* common to the three planes. This can happen if the three planes have a

Figure 11.4

common line of intersection as in Figure 11.5(a), or if two of the planes coincide and the third plane intersects them as in Figure 11.5(b).

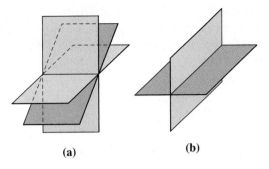

(a)   (b)

Figure 11.5

3. There are *infinitely many ordered triples* in the solution set, all of which are coordinates of points on a
plane. This can happen if the three planes coincide,
as illustrated in Figure 11.6.

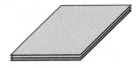

Figure 11.6

4. The solution set is *empty*; thus we write ∅. This can happen in various ways, as
illustrated in Figure 11.7. Note that in each situation there are no points common to all three planes.

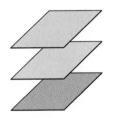

(a) Three parallel planes

(b) Two planes coincide
and the third one is
parallel to the
coinciding planes.

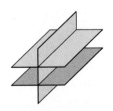

(c) Two planes are
parallel and the third
intersects them in
parallel lines.

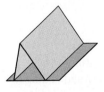

(d) No two planes are
parallel, but two of
them intersect in a
line that is parallel
to the third plane.

Figure 11.7

Now that we know what possibilities exist, let's consider finding the solution sets for some systems. Our approach will be the elimination-by-addition method, whereby systems are replaced with equivalent systems until a system is obtained where we can easily determine the solution set. The details of this approach will become apparent as we work a few examples.

E X A M P L E   1

Solve the system

$$\left(\begin{array}{r} 4x - 3y - 2z = 5 \\ 5y + z = -11 \\ 3z = 12 \end{array}\right)$$

(1)
(2)
(3)

Solution

The form of this system makes it easy to solve. From equation (3), we obtain $z = 4$. Then, substituting 4 for z in equation (2), we get

$$5y + 4 = -11$$

$$5y = -15$$

$$y = -3$$

Finally, substituting 4 for z and -3 for y in equation (1) yields

$$4x - 3(-3) - 2(4) = 5$$

$$4x + 1 = 5$$

$$4x = 4$$

$$x = 1$$

Thus the solution set of the given system is $\{(1, -3, 4)\}$.   ■

E X A M P L E   2

Solve the system

$$\left(\begin{array}{r} x - 2y + 3z = 22 \\ 2x - 3y - z = 5 \\ 3x + y - 5z = -32 \end{array}\right)$$

(4)
(5)
(6)

Solution

Equation (5) can be replaced with the equation formed by multiplying equation (4) by -2 and adding this result to equation (5). Equation (6) can be replaced with the equation formed by multiplying equation (4) by -3 and adding this result to equation (6). The following equivalent system is produced, in which equations (8) and (9) contain only the two variables y and z:

$$\left(\begin{array}{r} x - 2y + 3z = 22 \\ y - 7z = -39 \\ 7y - 14z = -98 \end{array}\right)$$

(7)
(8)
(9)

Equation (9) can be replaced with the equation formed by multiplying equation (8) by -7 and adding this result to equation (9). This produces the following equivalent system:

$$\begin{pmatrix} x - 2y + 3z = 22 \\ y - 7z = -39 \\ 35z = 175 \end{pmatrix} \qquad \begin{array}{l} (10) \\ (11) \\ (12) \end{array}$$

From equation (12), we obtain $z = 5$. Then, substituting 5 for z in equation (11), we obtain

$$y - 7(5) = -39$$

$$y - 35 = -39$$

$$y = -4$$

Finally, substituting -4 for y and 5 for z in equation (10) produces

$$x - 2(-4) + 3(5) = 22$$

$$x + 8 + 15 = 22$$

$$x + 23 = 22$$

$$x = -1$$

The solution set of the original system is $\{(-1, -4, 5)\}$. (Perhaps you should check this ordered triple in all three of the original equations.)   ■

EXAMPLE 3

Solve the system

$$\begin{pmatrix} 3x - y + 2z = 13 \\ 5x - 3y - z = 3 \\ 4x + 2y + 5z = 30 \end{pmatrix} \qquad \begin{array}{l} (13) \\ (14) \\ (15) \end{array}$$

Solution

Equation (14) can be replaced with the equation formed by multiplying equation (13) by -3 and adding this result to equation (14). Equation (15) can be replaced with the equation formed by multiplying equation (13) by 2 and adding this result to equation (15). Thus we produce the following equivalent system, in which equations (17) and (18) contain only the two variables x and z:

$$\begin{pmatrix} 3x - y + 2z = 13 \\ -4x - 7z = -36 \\ 10x + 9z = 56 \end{pmatrix} \qquad \begin{array}{l} (16) \\ (17) \\ (18) \end{array}$$

Now, if we multiply equation (17) by 5 and equation (18) by 2, we get the following equivalent system:

$$\begin{pmatrix} 3x - y + 2z = 13 \\ -20x - 35z = -180 \\ 20x + 18z = 112 \end{pmatrix} \qquad \begin{array}{l} (19) \\ (20) \\ (21) \end{array}$$

Equation (21) can be replaced with the equation formed by adding equation (20) to equation (21).

$$\begin{pmatrix} 3x - y + 2z = 13 \\ -20x \quad\quad - 35z = -180 \\ - 17z = -68 \end{pmatrix}$$

(22)
(23)
(24)

From equation (24), we obtain $z = 4$. Then we can substitute 4 for z in equation (23).

$$-20x - 35(4) = -180$$

$$-20x - 140 = -180$$

$$-20x = -40$$

$$x = 2$$

Now we can substitute 2 for x and 4 for z in equation (22).

$$3(2) - y + 2(4) = 13$$

$$6 - y + 8 = 13$$

$$-y + 14 = 13$$

$$-y = -1$$

$$y = 1$$

The solution set of the original system is $\{(2, 1, 4)\}$. ∎

EXAMPLE 4

Solve the system

$$\begin{pmatrix} 2x + 3y + z = 14 \\ 3x - 4y - 2z = -30 \\ 5x + 7y + 3z = 32 \end{pmatrix}$$

(25)
(26)
(27)

Solution

Equation (26) can be replaced with the equation formed by multiplying equation (25) by 2 and adding this result to equation (26). Equation (27) can be replaced with the equation formed by multiplying equation (25) by -3 and adding this result to equation (27). The following equivalent system is produced, in which equations (29) and (30) contain only the two variables x and y:

$$\begin{pmatrix} 2x + 3y + z = 14 \\ 7x + 2y \quad\quad = -2 \\ -x - 2y \quad\quad = -10 \end{pmatrix}$$

(28)
(29)
(30)

Now, equation (30) can be replaced with the equation formed by adding equation (29) to equation (30).

$$\begin{pmatrix} 2x + 3y + z = 14 \\ 7x + 2y \quad\quad = -2 \\ 6x \quad\quad\quad = -12 \end{pmatrix}$$

(31)
(32)
(33)

From equation (33), we obtain $x = -2$. Then, substituting -2 for x in equation (32), we obtain

$$7(-2) + 2y = -2$$

$$2y = 12$$

$$y = 6$$

Finally, substituting 6 for y and -2 for x in equation (31) yields

$$2(-2) + 3(6) + z = 14$$

$$14 + z = 14$$

$$z = 0$$

The solution set of the original system is $\{(-2, 6, 0)\}$. ∎

The ability to solve systems of three linear equations in three unknowns enhances our problem-solving capabilities. Let's conclude this section with a problem that we can solve using such a system.

PROBLEM 1

A small company that manufactures sporting equipment produces three different styles of golf shirts. Each style of shirt requires the services of three departments, as indicated by the following table:

	Style A	Style B	Style C
Cutting department	0.1 hour	0.1 hour	0.3 hour
Sewing department	0.3 hour	0.2 hour	0.4 hour
Packaging department	0.1 hour	0.2 hour	0.1 hour

The cutting, sewing, and packaging departments have available a maximum of 340, 580, and 255 work-hours per week, respectively. How many of each style of golf shirt should be produced each week so that the company is operating at full capacity?

Solution

Let a represent the number of shirts of style A produced per week, b the number of style B per week, and c the number of style C per week. Then the problem translates into the following system of equations:

$$\begin{pmatrix} 0.1a + 0.1b + 0.3c = 340 \\ 0.3a + 0.2b + 0.4c = 580 \\ 0.1a + 0.2b + 0.1c = 255 \end{pmatrix} \begin{array}{l} \longleftarrow \text{ Cutting department} \\ \longleftarrow \text{ Sewing department} \\ \longleftarrow \text{ Packaging department} \end{array}$$

Solving this system (we will leave the details for you to carry out) produces $a = 500$, $b = 650$, and $c = 750$. Thus the company should produce 500 golf shirts of style A, 650 of style B, and 750 of style C per week. ∎

Problem Set 11.2

For Problems 1–20, solve each system.

1. $\begin{pmatrix} 2x - 3y + 4z = 10 \\ 5y - 2z = -16 \\ 3z = 9 \end{pmatrix}$

2. $\begin{pmatrix} -3x + 2y + z = -9 \\ 4x - 3z = 18 \\ 4z = -8 \end{pmatrix}$

3. $\begin{pmatrix} x + 2y - 3z = 2 \\ 3y - z = 13 \\ 3y + 5z = 25 \end{pmatrix}$

4. $\begin{pmatrix} 2x + 3y - 4z = -10 \\ 2y + 3z = 16 \\ 2y - 5z = -16 \end{pmatrix}$

5. $\begin{pmatrix} 3x + 2y - 2z = 14 \\ x - 6z = 16 \\ 2x + 5z = -2 \end{pmatrix}$

6. $\begin{pmatrix} 3x + 2y - z = -11 \\ 2x - 3y = -1 \\ 4x + 5y = -13 \end{pmatrix}$

7. $\begin{pmatrix} x - 2y + 3z = 7 \\ 2x + y + 5z = 17 \\ 3x - 4y - 2z = 1 \end{pmatrix}$

8. $\begin{pmatrix} x - 2y + z = -4 \\ 2x + 4y - 3z = -1 \\ -3x - 6y + 7z = 4 \end{pmatrix}$

9. $\begin{pmatrix} 2x - y + z = 0 \\ 3x - 2y + 4z = 11 \\ 5x + y - 6z = -32 \end{pmatrix}$

10. $\begin{pmatrix} 2x - y + 3z = -14 \\ 4x + 2y - z = 12 \\ 6x - 3y + 4z = -22 \end{pmatrix}$

11. $\begin{pmatrix} 3x + 2y - z = -11 \\ 2x - 3y + 4z = 11 \\ 5x + y - 2z = -17 \end{pmatrix}$

12. $\begin{pmatrix} 9x + 4y - z = 0 \\ 3x - 2y + 4z = 6 \\ 6x - 8y - 3z = 3 \end{pmatrix}$

13. $\begin{pmatrix} 2x + 3y - 4z = -10 \\ 4x - 5y + 3z = 2 \\ 2y + z = 8 \end{pmatrix}$

14. $\begin{pmatrix} x + 2y - 3z = 2 \\ 3x - z = -8 \\ 2x - 3y + 5z = -9 \end{pmatrix}$

15. $\begin{pmatrix} 3x + 2y - 2z = 14 \\ 2x - 5y + 3z = 7 \\ 4x - 3y + 7z = 5 \end{pmatrix}$

16. $\begin{pmatrix} 4x + 3y - 2z = -11 \\ 3x - 7y + 3z = 10 \\ 9x - 8y + 5z = 9 \end{pmatrix}$

17. $\begin{pmatrix} 2x - 3y + 4z = -12 \\ 4x + 2y - 3z = -13 \\ 6x - 5y + 7z = -31 \end{pmatrix}$

18. $\begin{pmatrix} 3x + 5y - 2z = -27 \\ 5x - 2y + 4z = 27 \\ 7x + 3y - 6z = -55 \end{pmatrix}$

19. $\begin{pmatrix} 5x - 3y - 6z = 22 \\ x - y + z = -3 \\ -3x + 7y - 5z = 23 \end{pmatrix}$

20. $\begin{pmatrix} 4x + 3y - 5z = -29 \\ 3x - 7y - z = -19 \\ 2x + 5y + 2z = -10 \end{pmatrix}$

For Problems 21–30, solve each problem by setting up and solving a system of three linear equations in three variables.

21. A gift store is making a mixture of almonds, pecans, and peanuts, which sells for $3.50 per pound, $4 per pound, and $2 per pound, respectively. The store-keeper wants to make 20 pounds of the mix to sell at $2.70 per pound. The number of pounds of peanuts is to be three times the number of pounds of pecans. Find the number of pounds of each to be used in the mixture.

22. The organizer for a church picnic ordered coleslaw, potato salad, and beans amounting to 50 pounds. There was to be three times as much potato salad as coleslaw. The number of pounds of beans was to be 6 less than the number of pounds of potato salad. Find the number of pounds of each.

23. A box contains $7.15 in nickels, dimes, and quarters. There are 42 coins in all, and the sum of the numbers of nickels and dimes is 2 less than the number of quarters. How many coins of each kind are there?

24. A handful of 65 coins consists of pennies, nickels, and dimes. The number of nickels is 4 less than twice the number of pennies, and there are 13 more dimes than nickels. How many coins of each kind are there?

25. The measure of the largest angle of a triangle is twice the measure of the smallest angle. The sum of the smallest angle and the largest angle is twice the other angle. Find the measure of each angle.

26. The perimeter of a triangle is 45 centimeters. The longest side is 4 centimeters less than twice the shortest side. The sum of the lengths of the shortest and longest sides is 7 centimeters less than three times the length of the remaining side. Find the lengths of all three sides of the triangle.

27. Part of $3000 is invested at 12%, another part at 13%, and the remainder at 14% yearly interest. The total yearly income from the three investments is $400. The sum of the amounts invested at 12% and 13% equals the amount invested at 14%. How much is invested at each rate?

28. Different amounts are invested at 10%, 11%, and 12% yearly interest. The amount invested at 11% is $300 more than what is invested at 10%, and the total yearly income from all three investments is $324. A total of $2900 is invested. Find the amount invested at each rate.

29. A small company makes three different types of bird houses. Each type requires the services of three different departments, as indicated by the following table.

	Type A	Type B	Type C
Cutting department	0.1 hour	0.2 hour	0.1 hour
Finishing department	0.4 hour	0.4 hour	0.3 hour
Assembly department	0.2 hour	0.1 hour	0.3 hour

The cutting, finishing, and assembly departments have available a maximum of 35, 95, and 62.5 work-hours per week, respectively. How many bird houses of each type should be made per week so that the company is operating at full capacity?

30. A certain diet consists of dishes A, B, and C. Each serving of A has 1 gram of fat, 2 grams of carbohydrate, and 4 grams of protein. Each serving of B has 2 grams of fat, 1 gram of carbohydrate, and 3 grams of protein. Each serving of C has 2 grams of fat, 4 grams of carbohydrate, and 3 grams of protein. The diet allows 15 grams of fat, 24 grams of carbohydrate, and 30 grams of protein. How many servings of each dish can be eaten?

■ ■ ■ THOUGHTS INTO WORDS

31. Give a general description of how to solve a system of three linear equations in three variables.

32. Give a step-by-step description of how to solve the system

$$\begin{pmatrix} x - 2y + 3z = -23 \\ 5y - 2z = 32 \\ 4z = -24 \end{pmatrix}$$

33. Give a step-by-step description of how to solve the system

$$\begin{pmatrix} 3x - 2y + 7z = 9 \\ x \quad\quad - 3z = 4 \\ 2x \quad\quad + z = 9 \end{pmatrix}.$$

11.3   Matrix Approach to Solving Linear Systems

In the first two sections of this chapter, we found that the substitution and elimination-by-addition techniques worked effectively with two equations and two unknowns, but they started to get a bit cumbersome with three equations and three unknowns. Therefore we shall now begin to analyze some techniques that lend themselves to use with larger systems of equations. Furthermore, some of these techniques form the basis for using a computer to solve systems. Even though these techniques are primarily designed for large systems of equations, we shall study them in the context of small systems so that we won't get bogged down with the computational aspects of the techniques.

■ Matrices

A **matrix** is an array of numbers arranged in horizontal rows and vertical columns and enclosed in brackets. For example, the matrix

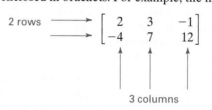

has 2 rows and 3 columns and is called a 2×3 (this is read "two by three") matrix. Each number in a matrix is called an **element** of the matrix. Some additional examples of matrices (*matrices* is the plural of *matrix*) follow:

$$
\begin{matrix}
3 \times 2 & 2 \times 2 & 1 \times 2 & 4 \times 1 \\
\begin{bmatrix} 2 & 1 \\ 1 & -4 \\ \dfrac{1}{2} & \dfrac{2}{3} \end{bmatrix} & \begin{bmatrix} 17 & 18 \\ -14 & 16 \end{bmatrix} & [7 \quad 14] & \begin{bmatrix} 3 \\ -2 \\ 1 \\ 19 \end{bmatrix}
\end{matrix}
$$

In general, a matrix of m rows and n columns is called a matrix of **dimension $m \times n$** or **order $m \times n$**.

With every system of linear equations, we can associate a matrix that consists of the coefficients and constant terms. For example, with the system

$$\begin{pmatrix} a_1x + b_1 y + c_1z = d_1 \\ a_2x + b_2 y + c_2z = d_2 \\ a_3x + b_3 y + c_3z = d_3 \end{pmatrix}$$

we can associate the matrix

$$\begin{bmatrix} a_1 & b_1 & c_1 & \vdots & d_1 \\ a_2 & b_2 & c_2 & \vdots & d_2 \\ a_3 & b_3 & c_3 & \vdots & d_3 \end{bmatrix}$$

which is commonly called the **augmented matrix** of the system of equations. The dashed line simply separates the coefficients from the constant terms and reminds us that we are working with an augmented matrix.

In Section 11.1 we listed the operations or transformations that can be applied to a system of equations to produce an equivalent system. Because augmented matrices are essentially abbreviated forms of systems of linear equations, there are analogous transformations that can be applied to augmented matrices. These transformations are usually referred to as **elementary row operations** and can be stated as follows:

> For any augmented matrix of a system of linear equations, the following elementary row operations will produce a matrix of an equivalent system:
>
> **1.** Any two rows of the matrix can be interchanged.
>
> **2.** Any row of the matrix can be multiplied by a nonzero real number.
>
> **3.** Any row of the matrix can be replaced by the sum of a nonzero multiple of another row plus that row.

Let's illustrate the use of augmented matrices and elementary row operations to solve a system of two linear equations in two variables.

EXAMPLE 1

Solve the system

$$\begin{pmatrix} x - 3y = -17 \\ 2x + 7y = 31 \end{pmatrix}$$

Solution

The augmented matrix of the system is

$$\begin{bmatrix} 1 & -3 & \vdots & -17 \\ 2 & 7 & \vdots & 31 \end{bmatrix}$$

We would like to change this matrix to one of the form

$$\begin{bmatrix} 1 & 0 & \vdots & a \\ 0 & 1 & \vdots & b \end{bmatrix}$$

where we can easily determine that the solution is $x = a$ and $y = b$. Let's begin by adding -2 times row 1 to row 2 to produce a new row 2.

$$\begin{bmatrix} 1 & -3 & \vdots & -17 \\ 0 & 13 & \vdots & 65 \end{bmatrix}$$

Now we can multiply row 2 by $\dfrac{1}{13}$.

$$\begin{bmatrix} 1 & -3 & \vdots & -17 \\ 0 & 1 & \vdots & 5 \end{bmatrix}$$

Finally, we can add 3 times row 2 to row 1 to produce a new row 1.

$$\begin{bmatrix} 1 & 0 & \vdots & -2 \\ 0 & 1 & \vdots & 5 \end{bmatrix}$$

From this last matrix, we see that $x = -2$ and $y = 5$. In other words, the solution set of the original system is $\{(-2, 5)\}$. ∎

It may seem that the matrix approach does not provide us with much extra power for solving systems of two linear equations in two unknowns. However, as the systems get larger, the compactness of the matrix approach becomes more convenient. Let's consider a system of three equations in three variables.

EXAMPLE 2

Solve the system

$$\begin{pmatrix} x + 2y - 3z = 15 \\ -2x - 3y + z = -15 \\ 4x + 9y - 4z = 49 \end{pmatrix}$$

Solution

The augmented matrix of this system is

$$\begin{bmatrix} 1 & 2 & -3 & \vdots & 15 \\ -2 & -3 & 1 & \vdots & -15 \\ 4 & 9 & -4 & \vdots & 49 \end{bmatrix}$$

If the system has a unique solution, then we will be able to change the augmented matrix to the form

$$\begin{bmatrix} 1 & 0 & 0 & \vdots & a \\ 0 & 1 & 0 & \vdots & b \\ 0 & 0 & 1 & \vdots & c \end{bmatrix}$$

where we will be able to read the solution $x = a$, $y = b$, and $z = c$.

Add 2 times row 1 to row 2 to produce a new row 2. Likewise, add -4 times row 1 to row 3 to produce a new row 3.

$$\begin{bmatrix} 1 & 2 & -3 & \vdots & 15 \\ 0 & 1 & -5 & \vdots & 15 \\ 0 & 1 & 8 & \vdots & -11 \end{bmatrix}$$

Now add -2 times row 2 to row 1 to produce a new row 1. Also, add -1 times row 2 to row 3 to produce a new row 3.

$$\begin{bmatrix} 1 & 0 & 7 & \vdots & -15 \\ 0 & 1 & -5 & \vdots & 15 \\ 0 & 0 & 13 & \vdots & -26 \end{bmatrix}$$

Now let's multiply row 3 by $\dfrac{1}{13}$.

$$\begin{bmatrix} 1 & 0 & 7 & \vdots & -15 \\ 0 & 1 & -5 & \vdots & 15 \\ 0 & 0 & 1 & \vdots & -2 \end{bmatrix}$$

Finally, we can add -7 times row 3 to row 1 to produce a new row 1, and we can add 5 times row 3 to row 2 for a new row 2.

$$\left[\begin{array}{ccc|c} 1 & 0 & 0 & -1 \\ 0 & 1 & 0 & 5 \\ 0 & 0 & 1 & -2 \end{array}\right]$$

From this last matrix, we can see that the solution set of the original system is $\{(-1, 5, -2)\}$.   ∎

The final matrices of Examples 1 and 2,

$$\left[\begin{array}{cc|c} 1 & 0 & -2 \\ 0 & 1 & 5 \end{array}\right] \quad \text{and} \quad \left[\begin{array}{ccc|c} 1 & 0 & 0 & -1 \\ 0 & 1 & 0 & 5 \\ 0 & 0 & 1 & -2 \end{array}\right]$$

are said to be in **reduced echelon form**. In general, a matrix is in reduced echelon form if the following conditions are satisfied:

1. As we read from left to right, the first nonzero entry of each row is 1.

2. In the column containing the leftmost 1 of a row, all the other entries are zeros.

3. The leftmost 1 of any row is to the right of the leftmost 1 of the preceding row.

4. Rows containing only zeros are below all the rows containing nonzero entries.

Like the final matrices of Examples 1 and 2, the following are in reduced echelon form:

$$\left[\begin{array}{cc|c} 1 & 2 & -3 \\ 0 & 0 & 0 \end{array}\right] \quad \left[\begin{array}{ccc|c} 1 & 0 & -2 & 5 \\ 0 & 1 & 4 & 7 \\ 0 & 0 & 0 & 0 \end{array}\right] \quad \left[\begin{array}{cccc|c} 1 & 0 & 0 & 0 & 8 \\ 0 & 1 & 0 & 0 & -9 \\ 0 & 0 & 1 & 0 & -2 \\ 0 & 0 & 0 & 1 & 12 \end{array}\right]$$

In contrast, the following matrices are *not* in reduced echelon form for the reason indicated below each matrix:

$$\left[\begin{array}{ccc|c} 1 & 0 & 0 & 11 \\ 0 & 3 & 0 & -1 \\ 0 & 0 & 1 & -2 \end{array}\right] \quad \left[\begin{array}{ccc|c} 1 & 2 & -3 & 5 \\ 0 & 1 & 7 & 9 \\ 0 & 0 & 1 & -6 \end{array}\right]$$

Violates condition 1   Violates condition 2

$$\left[\begin{array}{ccc|c} 1 & 0 & 0 & 7 \\ 0 & 0 & 1 & -8 \\ 0 & 1 & 0 & 14 \end{array}\right] \quad \left[\begin{array}{cccc|c} 1 & 0 & 0 & 0 & -1 \\ 0 & 0 & 0 & 0 & 0 \\ 0 & 0 & 1 & 0 & 7 \\ 0 & 0 & 0 & 0 & 0 \end{array}\right]$$

Violates condition 3   Violates condition 4

Once we have an augmented matrix in reduced echelon form, it is easy to determine the solution set of the system. Furthermore, the procedure for changing a given augmented matrix to reduced echelon form can be described in a very systematic way. For example, if an augmented matrix of a system of three linear equations in three unknowns has a unique solution, then it can be changed to reduced echelon form as follows:

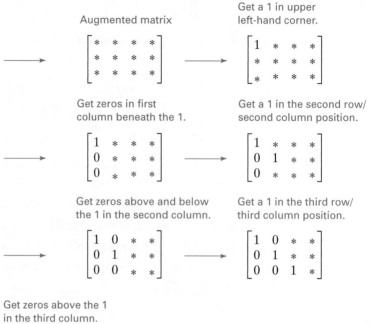

Augmented matrix

Get a 1 in upper left-hand corner.

Get zeros in first column beneath the 1.

Get a 1 in the second row/ second column position.

Get zeros above and below the 1 in the second column.

Get a 1 in the third row/ third column position.

Get zeros above the 1 in the third column.

$$\begin{bmatrix} 1 & 0 & 0 & * \\ 0 & 1 & 0 & * \\ 0 & 0 & 1 & * \end{bmatrix}$$

We can identify inconsistent and dependent systems while we are changing a matrix to reduced echelon form. We will show some examples of such cases in a moment, but first let's consider another example of a system of three linear equations in three unknowns where there is a unique solution.

EXAMPLE 3

Solve the system

$$\begin{pmatrix} 2x + 4y - 5z = 37 \\ x + 3y - 4z = 29 \\ 5x - y + 3z = -20 \end{pmatrix}$$

Solution

The augmented matrix

$$\begin{bmatrix} 2 & 4 & -5 & \vdots & 37 \\ 1 & 3 & -4 & \vdots & 29 \\ 5 & -1 & 3 & \vdots & -20 \end{bmatrix}$$

does not have a 1 in the upper left-hand corner, but this can be remedied by exchanging rows 1 and 2.

$$\begin{bmatrix} 1 & 3 & -4 & \vdots & 29 \\ 2 & 4 & -5 & \vdots & 37 \\ 5 & -1 & 3 & \vdots & -20 \end{bmatrix}$$

Now we can get zeros in the first column beneath the 1 by adding -2 times row 1 to row 2 and by adding -5 times row 1 to row 3.

$$\begin{bmatrix} 1 & 3 & -4 & \vdots & 29 \\ 0 & -2 & 3 & \vdots & -21 \\ 0 & -16 & 23 & \vdots & -165 \end{bmatrix}$$

Next, we can get a 1 for the first nonzero entry of the second row by multiplying the second row by $-\dfrac{1}{2}$.

$$\begin{bmatrix} 1 & 3 & -4 & \vdots & 29 \\ 0 & 1 & -\dfrac{3}{2} & \vdots & \dfrac{21}{2} \\ 0 & -16 & 23 & \vdots & -165 \end{bmatrix}$$

Now we can get zeros above and below the 1 in the second column by adding -3 times row 2 to row 1 and by adding 16 times row 2 to row 3.

$$\begin{bmatrix} 1 & 0 & \dfrac{1}{2} & \vdots & -\dfrac{5}{2} \\ 0 & 1 & -\dfrac{3}{2} & \vdots & \dfrac{21}{2} \\ 0 & 0 & -1 & \vdots & 3 \end{bmatrix}$$

Next, we can get a 1 in the first nonzero entry of the third row by multiplying the third row by -1.

$$\begin{bmatrix} 1 & 0 & \dfrac{1}{2} & \vdots & -\dfrac{5}{2} \\ 0 & 1 & -\dfrac{3}{2} & \vdots & \dfrac{21}{2} \\ 0 & 0 & 1 & \vdots & -3 \end{bmatrix}$$

Finally, we can get zeros above the 1 in the third column by adding $-\dfrac{1}{2}$ times row 3 to row 1 and by adding $\dfrac{3}{2}$ times row 3 to row 2.

$$\begin{bmatrix} 1 & 0 & 0 & \vdots & -1 \\ 0 & 1 & 0 & \vdots & 6 \\ 0 & 0 & 1 & \vdots & -3 \end{bmatrix}$$

From this last matrix, we see that the solution set of the original system is $\{(-1, 6, -3)\}$. ∎

Example 3 illustrates that even though the process of changing to reduced echelon form can be systematically described, it can involve some rather messy calculations. However, with the aid of a computer, such calculations are not troublesome. For our purposes in this text, the examples and problems involve systems that minimize messy calculations. This will allow us to concentrate on the procedures.

We want to call your attention to another issue in the solution of Example 3. Consider the matrix

$$\begin{bmatrix} 1 & 3 & -4 & \vdots & 29 \\ 0 & 1 & -\dfrac{3}{2} & \vdots & \dfrac{21}{2} \\ 0 & -16 & 23 & \vdots & -165 \end{bmatrix}$$

which is obtained about halfway through the solution. At this step, it seems evident that the calculations are getting a little messy. Therefore, instead of continuing toward the reduced echelon form, let's add 16 times row 2 to row 3 to produce a new row 3.

$$\begin{bmatrix} 1 & 3 & -4 & \vdots & 29 \\ 0 & 1 & -\dfrac{3}{2} & \vdots & \dfrac{21}{2} \\ 0 & 0 & -1 & \vdots & 3 \end{bmatrix}$$

The system represented by this matrix is

$$\left(\begin{array}{l} x + 3y - 4z = 29 \\ \quad y - \dfrac{3}{2}z = \dfrac{21}{2} \\ \quad\quad -z = 3 \end{array} \right)$$

and it is said to be in **triangular form**. The last equation determines the value for z; then we can use the process of back-substitution to determine the values for y and x.

Finally, let's consider two examples to illustrate what happens when we use the matrix approach on inconsistent and dependent systems.

EXAMPLE 4

Solve the system

$$\left(\begin{array}{l} x - 2y + 3z = 3 \\ 5x - 9y + 4z = 2 \\ 2x - 4y + 6z = -1 \end{array} \right)$$

Solution

The augmented matrix of the system is

$$\begin{bmatrix} 1 & -2 & 3 & \vdots & 3 \\ 5 & -9 & 4 & \vdots & 2 \\ 2 & -4 & 6 & \vdots & -1 \end{bmatrix}$$

We can get zeros below the 1 in the first column by adding -5 times row 1 to row 2 and by adding -2 times row 1 to row 3.

$$\begin{bmatrix} 1 & -2 & 3 & \vdots & 3 \\ 0 & 1 & -11 & \vdots & -13 \\ 0 & 0 & 0 & \vdots & -7 \end{bmatrix}$$

At this step, we can stop because the bottom row of the matrix represents the statement $0(x) + 0(y) + 0(z) = -7$, which is obviously false for all values of x, y, and z. Thus the original system is inconsistent; its solution set is $\emptyset$. ∎

EXAMPLE 5   Solve the system

$$\begin{pmatrix} x + 2y + 2z = 9 \\ x + 3y - 4z = 5 \\ 2x + 5y - 2z = 14 \end{pmatrix}$$

Solution

The augmented matrix of the system is

$$\begin{bmatrix} 1 & 2 & 2 & \vdots & 9 \\ 1 & 3 & -4 & \vdots & 5 \\ 2 & 5 & -2 & \vdots & 14 \end{bmatrix}$$

We can get zeros in the first column below the 1 in the upper left-hand corner by adding -1 times row 1 to row 2 and adding -2 times row 1 to row 3.

$$\begin{bmatrix} 1 & 2 & 2 & \vdots & 9 \\ 0 & 1 & -6 & \vdots & -4 \\ 0 & 1 & -6 & \vdots & -4 \end{bmatrix}$$

Now we can get zeros in the second column above and below the 1 in the second row by adding -2 times row 2 to row 1 and adding -1 times row 2 to row 3.

$$\begin{bmatrix} 1 & 0 & 14 & \vdots & 17 \\ 0 & 1 & -6 & \vdots & -4 \\ 0 & 0 & 0 & \vdots & 0 \end{bmatrix}$$

The bottom row of zeros represents the statement $0(x) + 0(y) + 0(z) = 0$, which is true for all values of x, y, and z. The second row represents the statement $y - 6z = -4$, which can be rewritten $y = 6z - 4$. The top row represents the statement $x + 14z = 17$, which can be rewritten $x = -14z + 17$. Therefore, if we let $z = k$, where k is any real number, the solution set of infinitely many ordered triples can be represented by $\{(-14k + 17, 6k - 4, k) | k$ is a real number$\}$. Specific solutions can be generated by letting k take on a value. For example, if $k = 2$, then $6k - 4$ becomes $6(2) - 4 = 8$ and $-14k + 17$ becomes $-14(2) + 17 = -11$. Thus the ordered triple $(-11, 8, 2)$ is a member of the solution set. ∎

Problem Set 11.3

For Problems 1–10, indicate whether each matrix is in reduced echelon form.

1. $\begin{bmatrix} 1 & 0 & \vdots & -4 \\ 0 & 1 & \vdots & 14 \end{bmatrix}$

2. $\begin{bmatrix} 1 & 2 & \vdots & 8 \\ 0 & 0 & \vdots & 0 \end{bmatrix}$

3. $\begin{bmatrix} 1 & 0 & 2 & \vdots & 5 \\ 0 & 1 & 3 & \vdots & 7 \\ 0 & 0 & 0 & \vdots & 0 \end{bmatrix}$

4. $\begin{bmatrix} 1 & 0 & 0 & \vdots & 5 \\ 0 & 3 & 0 & \vdots & 8 \\ 0 & 0 & 1 & \vdots & -11 \end{bmatrix}$

5. $\begin{bmatrix} 1 & 0 & 0 & \vdots & 17 \\ 0 & 0 & 0 & \vdots & 0 \\ 0 & 1 & 0 & \vdots & -14 \end{bmatrix}$

6. $\begin{bmatrix} 1 & 0 & 0 & \vdots & -7 \\ 0 & 1 & 0 & \vdots & 0 \\ 0 & 0 & 1 & \vdots & 9 \end{bmatrix}$

7. $\begin{bmatrix} 1 & 1 & 0 & \vdots & -3 \\ 0 & 1 & 2 & \vdots & 5 \\ 0 & 0 & 1 & \vdots & 7 \end{bmatrix}$

8. $\begin{bmatrix} 1 & 0 & 3 & \vdots & 8 \\ 0 & 1 & 2 & \vdots & -6 \\ 0 & 0 & 0 & \vdots & 0 \end{bmatrix}$

9. $\begin{bmatrix} 1 & 0 & 0 & 3 & \vdots & 4 \\ 0 & 1 & 0 & 5 & \vdots & -3 \\ 0 & 0 & 1 & -1 & \vdots & 7 \\ 0 & 0 & 0 & 0 & \vdots & 0 \end{bmatrix}$

10. $\begin{bmatrix} 1 & 0 & 0 & 0 & \vdots & 2 \\ 0 & 0 & 1 & 0 & \vdots & 4 \\ 0 & 1 & 0 & 0 & \vdots & -3 \\ 0 & 0 & 0 & 1 & \vdots & 9 \end{bmatrix}$

For Problems 11–30, use a matrix approach to solve each system.

11. $\begin{pmatrix} x - 3y = 14 \\ 3x + 2y = -13 \end{pmatrix}$

12. $\begin{pmatrix} x + 5y = -18 \\ -2x + 3y = -16 \end{pmatrix}$

13. $\begin{pmatrix} 3x - 4y = 33 \\ x + 7y = -39 \end{pmatrix}$

14. $\begin{pmatrix} 2x + 7y = -55 \\ x - 4y = 25 \end{pmatrix}$

15. $\begin{pmatrix} x - 6y = -2 \\ 2x - 12y = 5 \end{pmatrix}$

fractions

16. $\begin{pmatrix} 2x - 3y = -12 \\ 3x + 2y = 8 \end{pmatrix}$

fractions

17. $\begin{pmatrix} 3x - 5y = 39 \\ 2x + 7y = -67 \end{pmatrix}$

18. $\begin{pmatrix} 3x + 9y = -1 \\ x + 3y = 10 \end{pmatrix}$

no sol

19. $\begin{pmatrix} x - 2y - 3z = -6 \\ 3x - 5y - z = 4 \\ 2x + y + 2z = 2 \end{pmatrix}$

20. $\begin{pmatrix} x + 3y - 4z = 13 \\ 2x + 7y - 3z = 11 \\ -2x - y + 2z = -8 \end{pmatrix}$

21. $\begin{pmatrix} -2x - 5y + 3z = 11 \\ x + 3y - 3z = -12 \\ 3x - 2y + 5z = 31 \end{pmatrix}$

22. $\begin{pmatrix} -3x + 2y + z = 17 \\ x - y + 5z = -2 \\ 4x - 5y - 3z = -36 \end{pmatrix}$

23. $\begin{pmatrix} x - 3y - z = 2 \\ 3x + y - 4z = -18 \\ -2x + 5y + 3z = 2 \end{pmatrix}$

24. $\begin{pmatrix} x - 4y + 3z = 16 \\ 2x + 3y - 4z = -22 \\ -3x + 11y - z = -36 \end{pmatrix}$

25. $\begin{pmatrix} x - y + 2z = 1 \\ -3x + 4y - z = 4 \\ -x + 2y + 3z = 6 \end{pmatrix}$

26. $\begin{pmatrix} x + 2y - 5z = -1 \\ 2x + 3y - 2z = 2 \\ 3x + 5y - 7z = 4 \end{pmatrix}$

27. $\begin{pmatrix} -2x + y + 5z = -5 \\ 3x + 8y - z = -34 \\ x + 2y + z = -12 \end{pmatrix}$

28. $\begin{pmatrix} 4x - 10y + 3z = -19 \\ 2x + 5y - z = -7 \\ x - 3y - 2z = -2 \end{pmatrix}$

29. $\begin{pmatrix} 2x + 3y - z = 7 \\ 3x + 4y + 5z = -2 \\ 5x + y + 3z = 13 \end{pmatrix}$

30. $\begin{pmatrix} 4x + 3y - z = 0 \\ 3x + 2y + 5z = 6 \\ 5x - y - 3z = 3 \end{pmatrix}$

Subscript notation is frequently used for working with larger systems of equations. For Problems 31–34, use a matrix approach to solve each system. Express the solutions as 4-tuples of the form (x_1, x_2, x_3, x_4).

31. $\begin{pmatrix} x_1 - 3x_2 - 2x_3 + x_4 = -3 \\ -2x_1 + 7x_2 + x_3 - 2x_4 = -1 \\ 3x_1 - 7x_2 - 3x_3 + 3x_4 = -5 \\ 5x_1 + x_2 + 4x_3 - 2x_4 = 18 \end{pmatrix}$

32. $\begin{pmatrix} x_1 - 2x_2 + 2x_3 - x_4 = -2 \\ -3x_1 + 5x_2 - x_3 - 3x_4 = 2 \\ 2x_1 + 3x_2 + 3x_3 + 5x_4 = -9 \\ 4x_1 - x_2 - x_3 - 2x_4 = 8 \end{pmatrix}$

33. $\begin{pmatrix} x_1 + 3x_2 - x_3 + 2x_4 = -2 \\ 2x_1 + 7x_2 + 2x_3 - x_4 = 19 \\ -3x_1 - 8x_2 + 3x_3 + x_4 = -7 \\ 4x_1 + 11x_2 - 2x_3 - 3x_4 = 19 \end{pmatrix}$

x yzw

34. $\left(\begin{array}{l} x_1 + 2x_2 - 3x_3 + x_4 = -2 \\ -2x_1 - 3x_2 + x_3 - x_4 = 5 \\ 4x_1 + 9x_2 - 2x_3 - 2x_4 = -28 \\ -5x_1 - 9x_2 + 2x_3 - 3x_4 = 14 \end{array} \right)$

In Problems 35–42, each matrix is the reduced echelon matrix for a system with variables x_1, x_2, x_3, and x_4. Find the solution set of each system.

35. $\left[\begin{array}{cccc|c} 1 & 0 & 0 & 0 & -2 \\ 0 & 1 & 0 & 0 & 4 \\ 0 & 0 & 1 & 0 & -3 \\ 0 & 0 & 0 & 1 & 0 \end{array} \right]$

36. $\left[\begin{array}{cccc|c} 1 & 0 & 0 & 0 & 0 \\ 0 & 1 & 0 & 0 & -5 \\ 0 & 0 & 1 & 0 & 0 \\ 0 & 0 & 0 & 1 & 4 \end{array} \right]$

37. $\left[\begin{array}{cccc|c} 1 & 0 & 0 & 0 & -8 \\ 0 & 1 & 0 & 0 & 5 \\ 0 & 0 & 1 & 0 & -2 \\ 0 & 0 & 0 & 0 & 1 \end{array} \right]$

38. $\left[\begin{array}{cccc|c} 1 & 0 & 0 & 0 & 2 \\ 0 & 1 & 0 & 2 & -3 \\ 0 & 0 & 1 & 3 & 4 \\ 0 & 0 & 0 & 0 & 0 \end{array} \right]$

39. $\left[\begin{array}{cccc|c} 1 & 0 & 0 & 3 & 5 \\ 0 & 1 & 0 & 0 & -1 \\ 0 & 0 & 1 & 4 & 2 \\ 0 & 0 & 0 & 0 & 0 \end{array} \right]$

40. $\left[\begin{array}{cccc|c} 1 & 3 & 0 & 2 & 0 \\ 0 & 0 & 1 & 0 & 0 \\ 0 & 0 & 0 & 0 & 1 \\ 0 & 0 & 0 & 0 & 0 \end{array} \right]$

41. $\left[\begin{array}{cccc|c} 1 & 3 & 0 & 0 & 9 \\ 0 & 0 & 1 & 0 & 2 \\ 0 & 0 & 0 & 1 & -3 \\ 0 & 0 & 0 & 0 & 0 \end{array} \right]$

42. $\left[\begin{array}{cccc|c} 1 & 0 & 0 & 0 & 7 \\ 0 & 1 & 0 & 0 & -3 \\ 0 & 0 & 1 & -2 & 5 \\ 0 & 0 & 0 & 0 & 0 \end{array} \right]$

■ ■ ■ THOUGHTS INTO WORDS

43. What is a matrix? What is an augmented matrix of a system of linear equations?

44. Describe how to use matrices to solve the system $\left(\begin{array}{l} x - 2y = 5 \\ 2x + 7y = 9 \end{array} \right)$.

■ ■ ■ FURTHER INVESTIGATIONS

For Problems 45–50, change each augmented matrix of the system to reduced echelon form and then indicate the solutions of the system.

45. $\left(\begin{array}{l} x - 2y + 3z = 4 \\ 3x - 5y - z = 7 \end{array} \right)$

46. $\left(\begin{array}{l} x + 3y - 2z = -1 \\ 2x - 5y + 7z = 4 \end{array} \right)$

47. $\left(\begin{array}{l} 2x - 4y + 3z = 8 \\ 3x + 5y - z = 7 \end{array} \right)$

48. $\left(\begin{array}{l} 3x + 6y - z = 9 \\ 2x - 3y + 4z = 1 \end{array} \right)$

49. $\left(\begin{array}{l} x - 2y + 4z = 9 \\ 2x - 4y + 8z = 3 \end{array} \right)$

50. $\left(\begin{array}{l} x + y - 2z = -1 \\ 3x + 3y - 6z = -3 \end{array} \right)$

GRAPHING CALCULATOR ACTIVITIES

51. If your graphing calculator has the capability of manipulating matrices, this is a good time to become familiar with those operations. You may need to refer to your user's manual for the key-punching instructions. To begin the familiarization process, load your calculator with the three augmented matrices in Examples 1, 2, and 3. Then, for each one, carry out the row operations as described in the text.

11.4 Determinants

Before we introduce the concept of a determinant, let's agree on some convenient new notation. A **general $m \times n$ matrix** can be represented by

$$A = \begin{bmatrix} a_{11} & a_{12} & a_{13} & \cdots & a_{1n} \\ a_{21} & a_{22} & a_{23} & \cdots & a_{2n} \\ \cdot & \cdot & \cdot & & \cdot \\ \cdot & \cdot & \cdot & & \cdot \\ \cdot & \cdot & \cdot & & \cdot \\ a_{m1} & a_{m2} & a_{m3} & \cdots & a_{mn} \end{bmatrix}$$

where the double subscripts are used to identify the number of the row and the number of the column, in that order. For example, a_{23} is the entry at the intersection of the second row and the third column. In general, the entry at the intersection of row i and column j is denoted by a_{ij}.

A **square matrix** is one that has the same number of rows as columns. Each square matrix A with real number entries can be associated with a real number called the **determinant** of the matrix, denoted by $|A|$. We will first define $|A|$ for a 2×2 matrix.

Definition 11.1

If $A = \begin{bmatrix} a_{11} & a_{12} \\ a_{21} & a_{22} \end{bmatrix}$, then

$$|A| = \begin{vmatrix} a_{11} & a_{12} \\ a_{21} & a_{22} \end{vmatrix} = a_{11}a_{22} - a_{12}a_{21}$$

EXAMPLE 1 If $A = \begin{bmatrix} 3 & -2 \\ 5 & 8 \end{bmatrix}$, find $|A|$.

Solution

Use Definition 11.1 to obtain

$$|A| = \begin{vmatrix} 3 & -2 \\ 5 & 8 \end{vmatrix} = 3(8) - (-2)(5)$$
$$= 24 + 10$$
$$= 34$$ ∎

Finding the determinant of a square matrix is commonly called **evaluating the determinant**, and the matrix notation is often omitted.

EXAMPLE 2

Evaluate $\begin{vmatrix} -3 & 6 \\ 2 & 8 \end{vmatrix}$.

Solution

$$\begin{vmatrix} -3 & 6 \\ 2 & 8 \end{vmatrix} = (-3)(8) - (6)(2)$$
$$= -24 - 12$$
$$= -36$$

To find the determinants of 3×3 and larger square matrices, it is convenient to introduce some additional terminology.

Definition 11.2

If A is a 3×3 matrix, then the **minor** (denoted by M_{ij}) of the a_{ij} element is the determinant of the 2×2 matrix obtained by deleting row i and column j of A.

EXAMPLE 3

If $A = \begin{bmatrix} 2 & 1 & 4 \\ -6 & 3 & -2 \\ 4 & 2 & 5 \end{bmatrix}$, find (a) M_{11} and (b) M_{23}.

Solution

(a) To find M_{11}, we first delete row 1 and column 1 of matrix A.

$$\begin{bmatrix} \cancel{2} & \cancel{1} & \cancel{4} \\ -\cancel{6} & 3 & -2 \\ \cancel{4} & 2 & 5 \end{bmatrix}$$

Thus

$$M_{11} = \begin{vmatrix} 3 & -2 \\ 2 & 5 \end{vmatrix} = 3(5) - (-2)(2) = 19$$

(b) To find M_{23}, we first delete row 2 and column 3 of matrix A.

$$\begin{bmatrix} 2 & 1 & \cancel{4} \\ \cancel{-6} & \cancel{3} & \cancel{-2} \\ 4 & 2 & \cancel{5} \end{bmatrix}$$

Thus

$$M_{23} = \begin{vmatrix} 2 & 1 \\ 4 & 2 \end{vmatrix} = 2(2) - (1)(4) = 0$$

The following definition will also be used.

Definition 11.3

If A is a 3×3 matrix, then the **cofactor** (denoted by C_{ij}) of the element a_{ij} is defined by

$$C_{ij} = (-1)^{i+j} M_{ij}$$

According to Definition 11.3, to find the cofactor of any element a_{ij} of a square matrix A, we find the minor of a_{ij} and multiply it by 1 if $i + j$ is even, or multiply it by -1 if $i + j$ is odd.

EXAMPLE 4   If $A = \begin{bmatrix} 3 & 2 & -4 \\ 1 & 5 & 4 \\ 2 & -3 & 1 \end{bmatrix}$, find C_{32}.

Solution

First, let's find M_{32} by deleting row 3 and column 2 of matrix A.

$$\begin{bmatrix} 3 & 2 & -4 \\ 1 & 5 & 4 \\ 2 & 3 & 1 \end{bmatrix}$$

Thus

$$M_{32} = \begin{vmatrix} 3 & -4 \\ 1 & 4 \end{vmatrix} = 3(4) - (-4)(1) = 16$$

Therefore

$$C_{32} = (-1)^{3+2} M_{32} = (-1)^5(16) = -16 \qquad \blacksquare$$

The concept of a cofactor can be used to define the determinant of a 3×3 matrix as follows:

Definition 11.4

If $A = \begin{bmatrix} a_{11} & a_{12} & a_{13} \\ a_{21} & a_{22} & a_{23} \\ a_{31} & a_{32} & a_{33} \end{bmatrix}$, then

$$|A| = a_{11}C_{11} + a_{21}C_{21} + a_{31}C_{31}$$

Definition 11.4 simply states that the determinant of a 3×3 matrix can be found by multiplying each element of the first column by its corresponding cofactor and then adding the three results. Let's illustrate this procedure.

EXAMPLE 5

Find $|A|$ if $A = \begin{bmatrix} -2 & 1 & 4 \\ 3 & 0 & 5 \\ 1 & -4 & -6 \end{bmatrix}$.

Solution

$$|A| = a_{11}C_{11} + a_{21}C_{21} + a_{31}C_{31}$$

$$= (-2)(-1)^{1+1}\begin{vmatrix} 0 & 5 \\ -4 & -6 \end{vmatrix} + (3)(-1)^{2+1}\begin{vmatrix} 1 & 4 \\ -4 & -6 \end{vmatrix} + (1)(-1)^{3+1}\begin{vmatrix} 1 & 4 \\ 0 & 5 \end{vmatrix}$$

$$= (-2)(1)(20) + (3)(-1)(10) + (1)(1)(5)$$

$$= -40 - 30 + 5$$

$$= -65$$

When we use Definition 11.4, we often say that "the determinant is being expanded about the first column." It can also be shown that **any row or column can be used to expand a determinant**. For example, for matrix A in Example 5, the expansion of the determinant about the *second row* is as follows:

$$\begin{vmatrix} -2 & 1 & 4 \\ 3 & 0 & 5 \\ 1 & -4 & -6 \end{vmatrix} = (3)(-1)^{2+1}\begin{vmatrix} 1 & 4 \\ -4 & -6 \end{vmatrix} + (0)(-1)^{2+2}\begin{vmatrix} -2 & 4 \\ 1 & -6 \end{vmatrix} + (5)(-1)^{2+3}\begin{vmatrix} -2 & 1 \\ 1 & -4 \end{vmatrix}$$

$$= (3)(-1)(10) + (0)(1)(8) + (5)(-1)(7)$$

$$= -30 + 0 - 35$$

$$= -65$$

Note that when we expanded about the second row, the computation was simplified by the presence of a zero. In general, it is helpful to expand about the row or column that contains the most zeros.

The concepts of minor and cofactor have been defined in terms of 3×3 matrices. Analogous definitions can be given for any square matrix (that is, any $n \times n$ matrix with $n \geq 2$), and the determinant can then be expanded about any row or column. Certainly, as the matrices become larger than 3×3, the computations get more tedious. We will concentrate most of our efforts in this text on 2×2 and 3×3 matrices.

■ Properties of Determinants

Determinants have several interesting properties, some of which are important primarily from a theoretical standpoint. But some of the properties are also very useful when evaluating determinants. We will state these properties for square matrices in general, but we will use 2×2 or 3×3 matrices as examples. We can demonstrate some of the proofs of these properties by evaluating the determinants involved, and some of the proofs for 3×3 matrices will be left for you to verify in the next problem set.

Property 11.1

> If any row (or column) of a square matrix A contains only zeros, then $|A| = 0$.

If every element of a row (or column) of a square matrix A is zero, then it should be evident that expanding the determinant about that row (or column) of zeros will produce 0.

Property 11.2

> If square matrix B is obtained from square matrix A by interchanging two rows (or two columns), then $|B| = -|A|$.

Property 11.2 states that interchanging two rows (or columns) changes the sign of the determinant. As an example of this property, suppose that

$$A = \begin{bmatrix} 2 & 5 \\ -1 & 6 \end{bmatrix}$$

and that rows 1 and 2 are interchanged to form

$$B = \begin{bmatrix} -1 & 6 \\ 2 & 5 \end{bmatrix}$$

Calculating $|A|$ and $|B|$ yields

$$|A| = \begin{vmatrix} 2 & 5 \\ -1 & 6 \end{vmatrix} = 2(6) - (5)(-1) = 17$$

and

$$|B| = \begin{vmatrix} -1 & 6 \\ 2 & 5 \end{vmatrix} = (-1)(5) - (6)(2) = -17$$

Property 11.3

> If square matrix B is obtained from square matrix A by multiplying each element of any row (or column) of A by some real number k, then $|B| = k|A|$.

Property 11.3 states that multiplying any row (or column) by a factor of k affects the value of the determinant by a factor of k. As an example of this property, suppose that

$$A = \begin{bmatrix} 1 & -2 & 8 \\ 2 & 1 & 12 \\ 3 & 2 & -16 \end{bmatrix}$$

and that B is formed by multiplying each element of the third column by $\dfrac{1}{4}$:

$$B = \begin{bmatrix} 1 & -2 & 2 \\ 2 & 1 & 3 \\ 3 & 2 & -4 \end{bmatrix}$$

Now let's calculate $|A|$ and $|B|$ by expanding about the third column in each case.

$$|A| = \begin{vmatrix} 1 & -2 & 8 \\ 2 & 1 & 12 \\ 3 & 2 & -16 \end{vmatrix} = (8)(-1)^{1+3} \begin{vmatrix} 2 & 1 \\ 3 & 2 \end{vmatrix} + (12)(-1)^{2+3} \begin{vmatrix} 1 & -2 \\ 3 & 2 \end{vmatrix} + (-16)(-1)^{3+3} \begin{vmatrix} 1 & -2 \\ 2 & 1 \end{vmatrix}$$

$$= (8)(1)(1) + (12)(-1)(8) + (-16)(1)(5)$$

$$= -168$$

$$|B| = \begin{vmatrix} 1 & -2 & 2 \\ 2 & 1 & 3 \\ 3 & 2 & 4 \end{vmatrix} = (2)(-1)^{1+3} \begin{vmatrix} 2 & 1 \\ 3 & 2 \end{vmatrix} + (3)(-1)^{2+3} \begin{vmatrix} 1 & -2 \\ 3 & 2 \end{vmatrix} + (-4)(-1)^{3+3} \begin{vmatrix} 1 & -2 \\ 2 & 1 \end{vmatrix}$$

$$= (2)(1)(1) + (3)(-1)(8) + (-4)(1)(5)$$

$$= -42$$

We see that $|B| = \dfrac{1}{4}|A|$. This example also illustrates the usual computational use of Property 11.3: We can factor out a common factor from a row or column and then adjust the value of the determinant by that factor. For example,

$$\begin{vmatrix} 2 & 6 & 8 \\ -1 & 2 & 7 \\ 5 & 2 & 1 \end{vmatrix} = 2 \begin{vmatrix} 1 & 3 & 4 \\ -1 & 2 & 7 \\ 5 & 2 & 1 \end{vmatrix}$$

Factor a 2 from
the top row.

Property 11.4

If square matrix B is obtained from square matrix A by adding k times a row (or column) of A to another row (or column) of A, then $|B| = |A|$.

Property 11.4 states that adding the product of k times a row (or column) to another row (or column) does not affect the value of the determinant. As an example of this property, suppose that

$$A = \begin{bmatrix} 1 & 2 & 4 \\ 2 & 4 & 7 \\ -1 & 3 & 5 \end{bmatrix}$$

Now let's form B by replacing row 2 with the result of adding -2 times row 1 to row 2.

$$B = \begin{bmatrix} 1 & 2 & 4 \\ 0 & 0 & -1 \\ -1 & 3 & 5 \end{bmatrix}$$

Next, let's evaluate $|A|$ and $|B|$ by expanding about the second row in each case.

$$|A| = \begin{vmatrix} 1 & 2 & 4 \\ 2 & 4 & 7 \\ -1 & 3 & 5 \end{vmatrix} = (2)(-1)^{2+1} \begin{vmatrix} 2 & 4 \\ 3 & 5 \end{vmatrix} + (4)(-1)^{2+2} \begin{vmatrix} 1 & 4 \\ -1 & 5 \end{vmatrix} + (7)(-1)^{2+3} \begin{vmatrix} 1 & 2 \\ -1 & 3 \end{vmatrix}$$

$$= 2(-1)(-2) + (4)(1)(9) + (7)(-1)(5)$$

$$= 5$$

$$|B| = \begin{vmatrix} 1 & 2 & 4 \\ 0 & 0 & -1 \\ -1 & 3 & 5 \end{vmatrix} = (0)(-1)^{2+1} \begin{vmatrix} 2 & 4 \\ 3 & 5 \end{vmatrix} + (0)(-1)^{2+2} \begin{vmatrix} 1 & 4 \\ -1 & 5 \end{vmatrix} + (-1)(-1)^{2+3} \begin{vmatrix} 1 & 2 \\ -1 & 3 \end{vmatrix}$$

$$= 0 + 0 + (-1)(-1)(5)$$

$$= 5$$

Note that $|B| = |A|$. Furthermore, note that because of the zeros in the second row, evaluating $|B|$ is much easier than evaluating $|A|$. Property 11.4 can often be used to obtain some zeros before we evaluate a determinant.

A word of caution is in order at this time. Be careful not to confuse Properties 11.2, 11.3, and 11.4 with the three elementary row transformations of augmented matrices that were used in Section 11.3. The statements of the two sets of properties do resemble each other, but the properties pertain to *two different concepts*, so be sure you understand the distinction between them.

One final property of determinants should be mentioned.

Property 11.5

> If two rows (or columns) of a square matrix A are identical, then $|A| = 0$.

Property 11.5 is a direct consequence of Property 11.2. Suppose that A is a square matrix (any size) with two identical rows. Square matrix B can be formed from A by interchanging the two identical rows. Because identical rows were interchanged, $|B| = |A|$. *But* by Property 11.2, $|B| = -|A|$. For both of these statements to hold, $|A| = 0$.

Let's conclude this section by evaluating a 4 × 4 determinant, using Properties 11.3 and 11.4 to facilitate the computation.

EXAMPLE 6

Evaluate $\begin{vmatrix} 6 & 2 & 1 & -2 \\ 9 & -1 & 4 & 1 \\ 12 & -2 & 3 & -1 \\ 0 & 0 & 9 & 3 \end{vmatrix}$.

Solution

First, let's add -3 times the fourth column to the third column.

$$\begin{vmatrix} 6 & 2 & 7 & -2 \\ 9 & -1 & 1 & 1 \\ 12 & -2 & 6 & -1 \\ 0 & 0 & 0 & 3 \end{vmatrix}$$

Now, if we expand about the fourth row, we get only one nonzero product.

$$(3)(-1)^{4+4}\begin{vmatrix} 6 & 2 & 7 \\ 9 & -1 & 1 \\ 12 & -2 & 6 \end{vmatrix}$$

Factoring a 3 out of the first column of the 3 × 3 determinant yields

$$(3)(-1)^8(3)\begin{vmatrix} 2 & 2 & 7 \\ 3 & -1 & 1 \\ 4 & -2 & 6 \end{vmatrix}$$

Next, working with the 3 × 3 determinant, we can first add column 3 to column 2 and then add -3 times column 3 to column 1.

$$(3)(-1)^8(3)\begin{vmatrix} -19 & 9 & 7 \\ 0 & 0 & 1 \\ -14 & 4 & 6 \end{vmatrix}$$

Finally, by expanding this 3 × 3 determinant about the second row, we obtain

$$(3)(-1)^8(3)(1)(-1)^{2+3}\begin{vmatrix} -19 & 9 \\ -14 & 4 \end{vmatrix}$$

Our final result is

$$(3)(-1)^8(3)(1)(-1)^5(50) = -450 \qquad \blacksquare$$

Problem Set 11.4

For Problems 1–12, evaluate each 2 × 2 determinant by using Definition 11.1.

1. $\begin{vmatrix} 4 & 3 \\ 2 & 7 \end{vmatrix}$

2. $\begin{vmatrix} 3 & 5 \\ 6 & 4 \end{vmatrix}$

3. $\begin{vmatrix} -3 & 2 \\ 7 & 5 \end{vmatrix}$

4. $\begin{vmatrix} 5 & 3 \\ 6 & -1 \end{vmatrix}$

5. $\begin{vmatrix} 2 & -3 \\ 8 & -2 \end{vmatrix}$

6. $\begin{vmatrix} -5 & 5 \\ -6 & 2 \end{vmatrix}$

7. $\begin{vmatrix} -2 & -3 \\ -1 & -4 \end{vmatrix}$

8. $\begin{vmatrix} -4 & -3 \\ -5 & -7 \end{vmatrix}$

9. $\begin{vmatrix} \dfrac{1}{2} & \dfrac{1}{3} \\ -3 & -6 \end{vmatrix}$

10. $\begin{vmatrix} \dfrac{2}{3} & \dfrac{3}{4} \\ 8 & 6 \end{vmatrix}$

11. $\begin{vmatrix} \dfrac{1}{2} & \dfrac{2}{3} \\ \dfrac{3}{4} & -\dfrac{1}{3} \end{vmatrix}$

12. $\begin{vmatrix} \dfrac{2}{3} & \dfrac{1}{5} \\ -\dfrac{1}{4} & \dfrac{3}{2} \end{vmatrix}$

For Problems 13–28, evaluate each 3×3 determinant. Use the properties of determinants to your advantage.

13. $\begin{vmatrix} 1 & 2 & -1 \\ 3 & 1 & 2 \\ 2 & 4 & 3 \end{vmatrix}$

14. $\begin{vmatrix} 1 & -2 & 1 \\ 2 & 1 & -1 \\ 3 & 2 & 4 \end{vmatrix}$

15. $\begin{vmatrix} 1 & -4 & 1 \\ 2 & 5 & -1 \\ 3 & 3 & 4 \end{vmatrix}$

16. $\begin{vmatrix} 3 & -2 & 1 \\ 2 & 1 & 4 \\ -1 & 3 & 5 \end{vmatrix}$

17. $\begin{vmatrix} 6 & 12 & 3 \\ -1 & 5 & 1 \\ -3 & 6 & 2 \end{vmatrix}$

18. $\begin{vmatrix} 2 & 35 & 5 \\ 1 & -5 & 1 \\ -4 & 15 & 2 \end{vmatrix}$

19. $\begin{vmatrix} 2 & -1 & 3 \\ 0 & 3 & 1 \\ 1 & -2 & -1 \end{vmatrix}$

20. $\begin{vmatrix} 2 & -17 & 3 \\ 0 & 5 & 1 \\ 1 & -3 & -1 \end{vmatrix}$

21. $\begin{vmatrix} -3 & -2 & 1 \\ 5 & 0 & 6 \\ 2 & 1 & -4 \end{vmatrix}$

22. $\begin{vmatrix} -5 & 1 & -1 \\ 3 & 4 & 2 \\ 0 & 2 & -3 \end{vmatrix}$

23. $\begin{vmatrix} 3 & -4 & -2 \\ 5 & -2 & 1 \\ 1 & 0 & 0 \end{vmatrix}$

24. $\begin{vmatrix} -6 & 5 & 3 \\ 2 & 0 & -1 \\ 4 & 0 & 7 \end{vmatrix}$

25. $\begin{vmatrix} 24 & -1 & 4 \\ 40 & 2 & 0 \\ -16 & 6 & 0 \end{vmatrix}$

26. $\begin{vmatrix} 2 & -1 & 3 \\ 0 & 3 & 1 \\ 4 & -8 & -4 \end{vmatrix}$

27. $\begin{vmatrix} 2 & 3 & -4 \\ 4 & 6 & -1 \\ -6 & 1 & -2 \end{vmatrix}$

28. $\begin{vmatrix} 1 & 2 & -3 \\ -3 & -1 & 1 \\ 4 & 5 & 4 \end{vmatrix}$

For Problems 29–32, evaluate each 4×4 determinant. Use the properties of determinants to your advantage.

29. $\begin{vmatrix} 1 & -2 & 3 & 2 \\ 2 & -1 & 0 & 4 \\ -3 & 4 & 0 & -2 \\ -1 & 1 & 1 & 5 \end{vmatrix}$

30. $\begin{vmatrix} 1 & 2 & 5 & 7 \\ -6 & 3 & 0 & 9 \\ -3 & 5 & 2 & 7 \\ 2 & 1 & 4 & 3 \end{vmatrix}$

31. $\begin{vmatrix} 3 & -1 & 2 & 3 \\ 1 & 0 & 2 & 1 \\ 2 & 3 & 0 & 1 \\ 5 & 2 & 4 & -5 \end{vmatrix}$

32. $\begin{vmatrix} 1 & 2 & 0 & 0 \\ 3 & -1 & 4 & 5 \\ -2 & 4 & 1 & 6 \\ 2 & -1 & -2 & -3 \end{vmatrix}$

For Problems 33–42, use the appropriate property of determinants from this section to justify each true statement. *Do not* evaluate the determinants.

33. $(-4) \begin{vmatrix} 2 & 1 & -1 \\ 3 & 2 & 1 \\ 2 & 1 & 3 \end{vmatrix} = \begin{vmatrix} 2 & -4 & -1 \\ 3 & -8 & 1 \\ 2 & -4 & 3 \end{vmatrix}$

34. $\begin{vmatrix} 1 & -2 & 3 \\ 4 & -6 & -8 \\ 0 & 2 & 7 \end{vmatrix} = (-2) \begin{vmatrix} 1 & -2 & 3 \\ -2 & 3 & 4 \\ 0 & 2 & 7 \end{vmatrix}$

35. $\begin{vmatrix} 4 & 7 & 9 \\ 6 & -8 & 2 \\ 4 & 3 & -1 \end{vmatrix} = -\begin{vmatrix} 4 & 9 & 7 \\ 6 & 2 & -8 \\ 4 & -1 & 3 \end{vmatrix}$

36. $\begin{vmatrix} 3 & -1 & 4 \\ 5 & 2 & 7 \\ 3 & -1 & 4 \end{vmatrix} = 0$

37. $\begin{vmatrix} 1 & 3 & 4 \\ -2 & 5 & 7 \\ -3 & -1 & 2 \end{vmatrix} = \begin{vmatrix} 1 & 3 & 4 \\ -2 & 5 & 7 \\ 0 & 8 & 14 \end{vmatrix}$

38. $\begin{vmatrix} 3 & 2 & 0 \\ 1 & 4 & 1 \\ -4 & 9 & 2 \end{vmatrix} = \begin{vmatrix} 3 & 2 & -3 \\ 1 & 4 & 0 \\ -4 & 9 & 6 \end{vmatrix}$

39. $\begin{vmatrix} 6 & 2 & 2 \\ 3 & -1 & 4 \\ 9 & -3 & 6 \end{vmatrix} = 6 \begin{vmatrix} 2 & 2 & 1 \\ 1 & -1 & 2 \\ 3 & -3 & 3 \end{vmatrix} = 18 \begin{vmatrix} 2 & 2 & 1 \\ 1 & -1 & 2 \\ 1 & -1 & 1 \end{vmatrix}$

40. $\begin{vmatrix} 2 & 1 & -3 \\ 0 & 2 & -4 \\ -5 & 1 & 3 \end{vmatrix} = -\begin{vmatrix} 2 & 1 & -3 \\ -5 & 1 & 3 \\ 0 & 2 & -4 \end{vmatrix}$

41. $\begin{vmatrix} 2 & -3 & 2 \\ 1 & -4 & 1 \\ 7 & 8 & 7 \end{vmatrix} = 0$

42. $\begin{vmatrix} 3 & 1 & 2 \\ -4 & 5 & -1 \\ 2 & -2 & -4 \end{vmatrix} = \begin{vmatrix} 3 & 1 & 0 \\ -4 & 5 & -11 \\ 2 & -2 & 0 \end{vmatrix}$

■ ■ ■ **THOUGHTS INTO WORDS**

43. Explain the difference between a matrix and a determinant.

44. Explain the concept of a cofactor and how it is used to help expand a determinant.

45. What does it mean to say that any row or column can be used to expand a determinant?

46. Give a step-by-step explanation of how to evaluate the determinant

$$\begin{vmatrix} 3 & 0 & 2 \\ 1 & -2 & 5 \\ 6 & 0 & 9 \end{vmatrix}$$

■ ■ ■ **FURTHER INVESTIGATIONS**

For Problems 47–50, use

$$A = \begin{bmatrix} a_{11} & a_{12} & a_{13} \\ a_{21} & a_{22} & a_{23} \\ a_{31} & a_{32} & a_{33} \end{bmatrix}$$

as a general representation for any 3×3 matrix.

47. Verify Property 11.2 for 3×3 matrices.

48. Verify Property 11.3 for 3×3 matrices.

49. Verify Property 11.4 for 3×3 matrices.

50. Show that $|A| = a_{11}a_{22}a_{33}a_{44}$ if

$$A = \begin{bmatrix} a_{11} & a_{12} & a_{13} & a_{14} \\ 0 & a_{22} & a_{23} & a_{24} \\ 0 & 0 & a_{33} & a_{34} \\ 0 & 0 & 0 & a_{44} \end{bmatrix}$$

■ ■ ■ **GRAPHING CALCULATOR ACTIVITIES**

51. Use a calculator to check your answers for Problems 29–32.

52. Consider the following matrix:

$$A = \begin{bmatrix} 2 & 5 & 7 & 9 \\ -4 & 6 & 2 & 4 \\ 6 & 9 & 12 & 3 \\ 5 & 4 & -2 & 8 \end{bmatrix}$$

Form matrix B by interchanging rows 1 and 3 of matrix A. Now use your calculator to show that $|B| = -|A|$.

53. Consider the following matrix:

$$A = \begin{bmatrix} 2 & 1 & 7 & 6 & 8 \\ 3 & -2 & 4 & 5 & -1 \\ 6 & 7 & 9 & 12 & 13 \\ -4 & -7 & 6 & 2 & 1 \\ 9 & 8 & 12 & 14 & 17 \end{bmatrix}$$

Form matrix B by multiplying each element of the second row of matrix A by 3. Now use your calculator to show that $|B| = 3|A|$.

54. Consider the following matrix:

$$A = \begin{bmatrix} 4 & 3 & 2 & 1 & 5 & -3 \\ 5 & 2 & 7 & 8 & 6 & 3 \\ 0 & 9 & 1 & 4 & 7 & 2 \\ 4 & 3 & 2 & 1 & 5 & -3 \\ -4 & -6 & 7 & 12 & 11 & 9 \\ 5 & 8 & 6 & -3 & 2 & -1 \end{bmatrix}$$

Use your calculator to show that $|A| = 0$.

| **11.5** | **Cramer's Rule** |

Determinants provide the basis for another method of solving linear systems. Consider the following linear system of two equations and two unknowns:

$$\begin{pmatrix} a_1 x + b_1 y = c_1 \\ a_2 x + b_2 y = c_2 \end{pmatrix}$$

The augmented matrix of this system is

$$\begin{bmatrix} a_1 & b_1 & \vdots & c_1 \\ a_2 & b_2 & \vdots & c_2 \end{bmatrix}$$

Using the elementary row transformation of augmented matrices, we can change this matrix to the following reduced echelon form. (The details are left for you to do as an exercise.)

$$\begin{bmatrix} 1 & 0 & \vdots & \dfrac{c_1 b_2 - c_2 b_1}{a_1 b_2 - a_2 b_1} \\ 0 & 1 & \vdots & \dfrac{a_1 c_2 - a_2 c_1}{a_1 b_2 - a_2 b_1} \end{bmatrix}, \qquad a_1 b_2 - a_2 b_1 \neq 0$$

The solution for x and y can be expressed in determinant form as follows:

$$x = \frac{c_1 b_2 - c_2 b_1}{a_1 b_2 - a_2 b_1} = \frac{\begin{vmatrix} c_1 & b_1 \\ c_2 & b_2 \end{vmatrix}}{\begin{vmatrix} a_1 & b_1 \\ a_2 & b_2 \end{vmatrix}} \qquad y = \frac{a_1 c_2 - a_2 c_1}{a_1 b_2 - a_2 b_1} = \frac{\begin{vmatrix} a_1 & c_1 \\ a_2 & c_2 \end{vmatrix}}{\begin{vmatrix} a_1 & b_1 \\ a_2 & b_2 \end{vmatrix}}$$

This method of using determinants to solve a system of two linear equations in two variables is called **Cramer's rule** and can be stated as follows:

Cramer's Rule (2 × 2 case)

Given the system

$$\begin{pmatrix} a_1 x + b_1 y = c_1 \\ a_2 x + b_2 y = c_2 \end{pmatrix}$$

with

$$D = \begin{vmatrix} a_1 & b_1 \\ a_2 & b_2 \end{vmatrix} \neq 0 \qquad D_x = \begin{vmatrix} c_1 & b_1 \\ c_2 & b_2 \end{vmatrix} \qquad \text{and} \qquad D_y = \begin{vmatrix} a_1 & c_1 \\ a_2 & c_2 \end{vmatrix}$$

then the solution for this system is given by

$$x = \frac{D_x}{D} \qquad \text{and} \qquad y = \frac{D_y}{D}$$

Note that the elements of D are the coefficients of the variables in the given system. In D_x, the coefficients of x are replaced by the corresponding constants, and in D_y, the coefficients of y are replaced by the corresponding constants. Let's illustrate the use of Cramer's rule to solve some systems.

E X A M P L E   1

Solve the system $\left(\begin{array}{l} 6x + 3y = 2 \\ 3x + 2y = -4 \end{array}\right)$.

Solution

The system is in the proper form for us to apply Cramer's rule, so let's determine D, D_x, and D_y.

$$D = \begin{vmatrix} 6 & 3 \\ 3 & 2 \end{vmatrix} = 12 - 9 = 3$$

$$D_x = \begin{vmatrix} 2 & 3 \\ -4 & 2 \end{vmatrix} = 4 + 12 = 16$$

$$D_y = \begin{vmatrix} 6 & 2 \\ 3 & -4 \end{vmatrix} = -24 - 6 = -30$$

Therefore

$$x = \frac{D_x}{D} = \frac{16}{3}$$

and

$$y = \frac{D_y}{D} = \frac{-30}{3} = -10$$

The solution set is $\left\{ \left(\dfrac{16}{3}, -10 \right) \right\}$. ∎

E X A M P L E   2

Solve the system $\left(\begin{array}{l} y = -2x - 2 \\ 4x - 5y = 17 \end{array}\right)$.

Solution

To begin, we must change the form of the first equation so that the system fits the form given in Cramer's rule. The equation $y = -2x - 2$ can be rewritten $2x + y = -2$. The system now becomes

$$\left(\begin{array}{l} 2x + y = -2 \\ 4x - 5y = 17 \end{array}\right)$$

and we can proceed to determine D, D_x, and D_y.

$$D = \begin{vmatrix} 2 & 1 \\ 4 & -5 \end{vmatrix} = -10 - 4 = -14$$

$$D_x = \begin{vmatrix} -2 & 1 \\ 17 & -5 \end{vmatrix} = 10 - 17 = -7$$

$$D_y = \begin{vmatrix} 2 & -2 \\ 4 & 17 \end{vmatrix} = 34 - (-8) = 42$$

Thus

$$x = \frac{D_x}{D} = \frac{-7}{-14} = \frac{1}{2} \quad \text{and} \quad y = \frac{D_y}{D} = \frac{42}{-14} = -3$$

The solution set is $\left\{ \left(\frac{1}{2}, -3 \right) \right\}$, which can be verified, as always, by substituting back into the original equations. ■

EXAMPLE 3   Solve the system

$$\begin{pmatrix} \dfrac{1}{2}x + \dfrac{2}{3}y = -4 \\ \dfrac{1}{4}x - \dfrac{3}{2}y = 20 \end{pmatrix}$$

Solution

With such a system, either we can first produce an equivalent system with integral coefficients and then apply Cramer's rule, or we can apply the rule immediately. Let's avoid some work with fractions by multiplying the first equation by 6 and the second equation by 4 to produce the following equivalent system:

$$\begin{pmatrix} 3x + 4y = -24 \\ x - 6y = 80 \end{pmatrix}$$

Now we can proceed as before.

$$D = \begin{vmatrix} 3 & 4 \\ 1 & -6 \end{vmatrix} = -18 - 4 = -22$$

$$D_x = \begin{vmatrix} -24 & 4 \\ 80 & -6 \end{vmatrix} = 144 - 320 = -176$$

$$D_y = \begin{vmatrix} 3 & -24 \\ 1 & 80 \end{vmatrix} = 240 - (-24) = 264$$

Therefore

$$x = \frac{D_x}{D} = \frac{-176}{-22} = 8 \quad \text{and} \quad y = \frac{D_y}{D} = \frac{264}{-22} = -12$$

The solution set is $\{(8, -12)\}$. ∎

In the statement of Cramer's rule, the condition that $D \neq 0$ was imposed. If $D = 0$ and either D_x or D_y (or both) is nonzero, then the system is inconsistent and has no solution. If $D = 0$, $D_x = 0$, and $D_y = 0$, then the equations are dependent and there are infinitely many solutions.

■ Cramer's Rule Extended

Without showing the details, we will simply state that Cramer's rule also applies to solving systems of three linear equations in three variables. It can be stated as follows:

Cramer's Rule (3 × 3 case)

Given the system

$$\begin{pmatrix} a_1x + b_1y + c_1z = d_1 \\ a_2x + b_2y + c_2z = d_2 \\ a_3x + b_3y + c_3z = d_3 \end{pmatrix}$$

with

$$D = \begin{vmatrix} a_1 & b_1 & c_1 \\ a_2 & b_2 & c_2 \\ a_3 & b_3 & c_3 \end{vmatrix} \neq 0 \quad D_x = \begin{vmatrix} d_1 & b_1 & c_1 \\ d_2 & b_2 & c_2 \\ d_3 & b_3 & c_3 \end{vmatrix}$$

$$D_y = \begin{vmatrix} a_1 & d_1 & c_1 \\ a_2 & d_2 & c_2 \\ a_3 & d_3 & c_3 \end{vmatrix} \quad D_z = \begin{vmatrix} a_1 & b_1 & d_1 \\ a_2 & b_2 & d_2 \\ a_3 & b_3 & d_3 \end{vmatrix}$$

then

$$x = \frac{D_x}{D} \quad y = \frac{D_y}{D} \quad \text{and} \quad z = \frac{D_z}{D}$$

Again, note the restriction that $D \neq 0$. If $D = 0$ and at least one of D_x, D_y, and D_z is not zero, then the system is inconsistent. If D, D_x, D_y, and D_z are all zero, then the equations are dependent, and there are infinitely many solutions.

E X A M P L E   4

Solve the system

$$\left(\begin{array}{l} x - 2y + z = -4 \\ 2x + y - z = 5 \\ 3x + 2y + 4z = 3 \end{array} \right)$$

Solution

We will simply indicate the values of D, D_x, D_y, and D_z and leave the computations for you to check.

$$D = \begin{vmatrix} 1 & -2 & 1 \\ 2 & 1 & -1 \\ 3 & 2 & 4 \end{vmatrix} = 29 \qquad D_x = \begin{vmatrix} -4 & -2 & 1 \\ 5 & 1 & -1 \\ 3 & 2 & 4 \end{vmatrix} = 29$$

$$D_y = \begin{vmatrix} 1 & -4 & 1 \\ 2 & 5 & -1 \\ 3 & 3 & 4 \end{vmatrix} = 58 \qquad D_z = \begin{vmatrix} 1 & -2 & -4 \\ 2 & 1 & 5 \\ 3 & 2 & 3 \end{vmatrix} = -29$$

Therefore

$$x = \frac{D_x}{D} = \frac{29}{29} = 1$$

$$y = \frac{D_y}{D} = \frac{58}{29} = 2$$

$$z = \frac{D_z}{D} = \frac{-29}{29} = -1$$

The solution set is $\{(1, 2, -1)\}$. (Be sure to check it!)   ■

E X A M P L E   5

Solve the system

$$\left(\begin{array}{l} x + 3y - z = 4 \\ 3x - 2y + z = 7 \\ 2x + 6y - 2z = 1 \end{array} \right)$$

Solution

$$D = \begin{vmatrix} 1 & 3 & -1 \\ 3 & -2 & 1 \\ 2 & 6 & -2 \end{vmatrix} = 2 \begin{vmatrix} 1 & 3 & -1 \\ 3 & -2 & 1 \\ 1 & 3 & -1 \end{vmatrix} = 2(0) = 0$$

$$D_x = \begin{vmatrix} 4 & 3 & -1 \\ 7 & -2 & 1 \\ 1 & 6 & -2 \end{vmatrix} = -7$$

Therefore, because $D = 0$ and at least one of D_x, D_y, and D_z is not zero, the system is inconsistent. The solution set is $\varnothing$.   ■

Example 5 illustrates why D should be determined first. Once we found that $D = 0$ and $D_x \neq 0$, we knew that the system was inconsistent, and there was no need to find D_y and D_z.

Finally, it should be noted that Cramer's rule can be extended to systems of n linear equations in n variables; however, that method is not considered to be a very efficient way of solving a large system of linear equations.

Problem Set 11.5

For Problems 1–32, use Cramer's rule to find the solution set for each system. If the equations are dependent, simply indicate that there are infinitely many solutions.

1. $\begin{pmatrix} 2x - y = -2 \\ 3x + 2y = 11 \end{pmatrix}$

2. $\begin{pmatrix} 3x + y = -9 \\ 4x - 3y = 1 \end{pmatrix}$

3. $\begin{pmatrix} 5x + 2y = 5 \\ 3x - 4y = 29 \end{pmatrix}$

4. $\begin{pmatrix} 4x - 7y = -23 \\ 2x + 5y = -3 \end{pmatrix}$

5. $\begin{pmatrix} 5x - 4y = 14 \\ -x + 2y = -4 \end{pmatrix}$

6. $\begin{pmatrix} -x + 2y = 10 \\ 3x - y = -10 \end{pmatrix}$

7. $\begin{pmatrix} y = 2x - 4 \\ 6x - 3y = 1 \end{pmatrix}$

8. $\begin{pmatrix} -3x - 4y = 14 \\ -2x + 3y = -19 \end{pmatrix}$

9. $\begin{pmatrix} -4x + 3y = 3 \\ 4x - 6y = -5 \end{pmatrix}$

10. $\begin{pmatrix} x = 4y - 1 \\ 2x - 8y = -2 \end{pmatrix}$

11. $\begin{pmatrix} 9x - y = -2 \\ 8x + y = 4 \end{pmatrix}$

12. $\begin{pmatrix} 6x - 5y = 1 \\ 4x - 7y = 2 \end{pmatrix}$

13. $\begin{pmatrix} -\dfrac{2}{3}x + \dfrac{1}{2}y = -7 \\ \dfrac{1}{3}x - \dfrac{3}{2}y = 6 \end{pmatrix}$

14. $\begin{pmatrix} \dfrac{1}{2}x + \dfrac{2}{3}y = -6 \\ \dfrac{1}{4}x - \dfrac{1}{3}y = -1 \end{pmatrix}$

15. $\begin{pmatrix} 2x + 7y = -1 \\ x = 2 \end{pmatrix}$

16. $\begin{pmatrix} 5x - 3y = 2 \\ y = 4 \end{pmatrix}$

17. $\begin{pmatrix} x - y + 2z = -8 \\ 2x + 3y - 4z = 18 \\ -x + 2y - z = 7 \end{pmatrix}$

18. $\begin{pmatrix} x - 2y + z = 3 \\ 3x + 2y + z = -3 \\ 2x - 3y - 3z = -5 \end{pmatrix}$

19. $\begin{pmatrix} 2x - 3y + z = -7 \\ -3x + y - z = -7 \\ x - 2y - 5z = -45 \end{pmatrix}$

20. $\begin{pmatrix} 3x - y - z = 18 \\ 4x + 3y - 2z = 10 \\ -5x - 2y + 3z = -22 \end{pmatrix}$

21. $\begin{pmatrix} 4x + 5y - 2z = -14 \\ 7x - y + 2z = 42 \\ 3x + y + 4z = 28 \end{pmatrix}$

22. $\begin{pmatrix} -5x + 6y + 4z = -4 \\ -7x - 8y + 2z = -2 \\ 2x + 9y - z = 1 \end{pmatrix}$

23. $\begin{pmatrix} 2x - y + 3z = -17 \\ 3y + z = 5 \\ x - 2y - z = -3 \end{pmatrix}$

24. $\begin{pmatrix} 2x - y + 3z = -5 \\ 3x + 4y - 2z = -25 \\ -x + z = 6 \end{pmatrix}$

25. $\begin{pmatrix} x + 3y - 4z = -1 \\ 2x - y + z = 2 \\ 4x + 5y - 7z = 0 \end{pmatrix}$

26. $\begin{pmatrix} x - 2y + z = 1 \\ 3x + y - z = 2 \\ 2x - 4y + 2z = -1 \end{pmatrix}$

27. $\begin{pmatrix} 3x - 2y - 3z = -5 \\ x + 2y + 3z = -3 \\ -x + 4y - 6z = 8 \end{pmatrix}$

28. $\begin{pmatrix} 3x - 2y + z = 11 \\ 5x + 3y = 17 \\ x + y - 2z = 6 \end{pmatrix}$

29. $\begin{pmatrix} x - 2y + 3z = 1 \\ -2x + 4y - 3z = -3 \\ 5x - 6y + 6z = 10 \end{pmatrix}$

30. $\begin{pmatrix} 2x - y + 2z = -1 \\ 4x + 3y - 4z = 2 \\ x + 5y - z = 9 \end{pmatrix}$

31. $\begin{pmatrix} -x - y + 3z = -2 \\ -2x + y + 7z = 14 \\ 3x + 4y - 5z = 12 \end{pmatrix}$

32. $\begin{pmatrix} -2x + y - 3z = -4 \\ x + 5y - 4z = 13 \\ 7x - 2y - z = 37 \end{pmatrix}$

■ ■ ■ THOUGHTS INTO WORDS

33. Give a step-by-step description of how you would solve the system

$$\begin{pmatrix} 2x - y + 3z = 31 \\ x - 2y - z = 8 \\ 3x + 5y + 8z = 35 \end{pmatrix}$$

34. Give a step-by-step description of how you would find the value of x in the solution for the system

$$\begin{pmatrix} x + 5y - z = -9 \\ 2x - y + z = 11 \\ -3x - 2y + 4z = 20 \end{pmatrix}$$

■ ■ ■ FURTHER INVESTIGATIONS

35. A linear system in which the constant terms are all zero is called a **homogeneous system**.

(a) Verify that for a 3×3 homogeneous system, if $D \neq 0$, then $(0, 0, 0)$ is the only solution for the system.

(b) Verify that for a 3×3 homogeneous system, if $D = 0$, then the equations are dependent.

For Problems 36–39, solve each of the homogeneous systems (see Problem 35). If the equations are dependent, indicate that the system has infinitely many solutions.

36. $\begin{pmatrix} x - 2y + 5z = 0 \\ 3x + y - 2z = 0 \\ 4x - y + 3z = 0 \end{pmatrix}$

37. $\begin{pmatrix} 2x - y + z = 0 \\ 3x + 2y + 5z = 0 \\ 4x - 7y + z = 0 \end{pmatrix}$

cross at origin

38. $\begin{pmatrix} 3x + y - z = 0 \\ x - y + 2z = 0 \\ 4x - 5y - 2z = 0 \end{pmatrix}$

39. $\begin{pmatrix} 2x - y + 2z = 0 \\ x + 2y + z = 0 \\ x - 3y + z = 0 \end{pmatrix}$

〰 ▦ GRAPHING CALCULATOR ACTIVITIES

40. Use determinants and your calculator to solve each of the following systems:

(a) $\begin{pmatrix} 4x - 3y + z = 10 \\ 8x + 5y - 2z = -6 \\ -12x - 2y + 3z = -2 \end{pmatrix}$

(b) $\begin{pmatrix} 2x + y - z + w = -4 \\ x + 2y + 2z - 3w = 6 \\ 3x - y - z + 2w = 0 \\ 2x + 3y + z + 4w = -5 \end{pmatrix}$

(c) $\begin{pmatrix} x - 2y + z - 3w = 4 \\ 2x + 3y - z - 2w = -4 \\ 3x - 4y + 2z - 4w = 12 \\ 2x - y - 3z + 2w = -2 \end{pmatrix}$

(d) $\begin{pmatrix} 1.98x + 2.49y + 3.45z = 80.10 \\ 2.15x + 3.20y + 4.19z = 97.16 \\ 1.49x + 4.49y + 2.79z = 83.92 \end{pmatrix}$

11.6   Partial Fractions (Optional)

In Chapter 4, we reviewed the process of adding rational expressions. For example,

$$\frac{3}{x-2} + \frac{2}{x+3} = \frac{3(x+3) + 2(x-2)}{(x-2)(x+3)} = \frac{3x+9+2x-4}{(x-2)(x+3)} = \frac{5x+5}{(x-2)(x+3)}$$

Now suppose that we want to reverse the process. That is, suppose we are given the rational expression

$$\frac{5x+5}{(x-2)(x+3)}$$

and we want to express it as the sum of two simpler rational expressions called **partial fractions**. This process, called **partial fraction decomposition**, has several applications in calculus and differential equations. The following property provides the basis for partial fraction decomposition.

Property 11.6

Let $f(x)$ and $g(x)$ be polynomials with real coefficients, such that the degree of $f(x)$ is less than the degree of $g(x)$. The indicated quotient $f(x)/g(x)$ can be decomposed into partial fractions as follows.

1. If $g(x)$ has a linear factor of the form $ax + b$, then the partial fraction decomposition will contain a term of the form

$$\frac{A}{ax+b}, \quad \text{where } A \text{ is a constant}$$

2. If $g(x)$ has a linear factor of the form $ax + b$ raised to the kth power, then the partial fraction decomposition will contain terms of the form

$$\frac{A_1}{ax+b} + \frac{A_2}{(ax+b)^2} + \cdots + \frac{A_k}{(ax+b)^k}$$

where $A_1, A_2, \ldots, A_k$ are constants.

3. If $g(x)$ has a quadratic factor of the form $ax^2 + bx + c$, where $b^2 - 4ac < 0$, then the partial fraction decomposition will contain a term of the form

$$\frac{Ax+B}{ax^2+bx+c}, \quad \text{where } A \text{ and } B \text{ are constants.}$$

4. If $g(x)$ has a quadratic factor of the form $ax^2 + bx + c$ raised to the kth power, where $b^2 - 4ac < 0$, then the partial fraction decomposition will contain terms of the form

$$\frac{A_1x+B_1}{ax^2+bx+c} + \frac{A_2x+B_2}{(ax^2+bx+c)^2} + \cdots + \frac{A_kx+B_kx}{(ax^2+bx+c)^k}$$

where $A_1, A_2, \ldots, A_k$, and $B_1, B_2, \ldots, B_k$ are constants.

Note that Property 11.6 applies only to **proper fractions**—that is, fractions in which the degree of the numerator is less than the degree of the denominator. If the numerator is not of lower degree, we can divide and then apply Property 11.6 to the remainder, which will be a proper fraction. For example,

$$\frac{x^3 - 3x^2 - 3x - 5}{x^2 - 4} = x - 3 + \frac{x - 17}{x^2 - 4}$$

and the proper fraction $\dfrac{x - 17}{x^2 - 4}$ can be decomposed into partial fractions by applying Property 11.6. Now let's consider some examples to illustrate the four cases in Property 11.6.

EXAMPLE 1

Find the partial fraction decomposition of $\dfrac{11x + 2}{2x^2 + x - 1}$.

Solution

The denominator can be expressed as $(x + 1)(2x - 1)$. Therefore, according to part 1 of Property 11.6, each of the linear factors produces a partial fraction of the form *constant over linear factor*. In other words, we can write

$$\frac{11x + 2}{(x + 1)(2x - 1)} = \frac{A}{x + 1} + \frac{B}{2x - 1} \tag{1}$$

for some constants A and B. To find A and B, we multiply both sides of equation (1) by the least common denominator $(x + 1)(2x - 1)$:

$$11x + 2 = A(2x - 1) + B(x + 1) \tag{2}$$

Equation (2) is an **identity**: *It is true for all values of x.* Therefore, let's choose some convenient values for x that will determine the values for A and B. If we let $x = -1$, then equation (2) becomes an equation in only A.

$$11(-1) + 2 = A[2(-1) - 1] + B(-1 + 1)$$

$$-9 = -3A$$

$$3 = A$$

If we let $x = \dfrac{1}{2}$, then equation (2) becomes an equation only in B.

$$11\left(\frac{1}{2}\right) + 2 = A\left[2\left(\frac{1}{2}\right) - 1\right] + B\left(\frac{1}{2} + 1\right)$$

$$\frac{15}{2} = \frac{3}{2}B$$

$$5 = B$$

Therefore, the given rational expression can now be written

$$\frac{11x + 2}{2x^2 + x - 1} = \frac{3}{x + 1} + \frac{5}{2x - 1}$$ ∎

The key idea in Example 1 is the statement that equation (2) is true for all values of x. If we had chosen *any* two values for x, we still would have been able to determine the values for A and B. For example, letting $x = 1$ and then $x = 2$ produces the equations $13 = A + 2B$ and $24 = 3A + 3B$. Solving this system of two equations in two unknowns produces $A = 3$ and $B = 5$. In Example 1, our choices of letting $x = -1$ and then $x = \frac{1}{2}$ simply eliminated the need for solving a system of equations to find A and B.

EXAMPLE 2

Find the partial fraction decomposition of

$$\frac{-2x^2 + 7x + 2}{x(x - 1)^2}$$

Solution

Apply part 1 of Property 11.6 to determine that there is a partial fraction of the form A/x corresponding to the factor of x. Next, applying part 2 of Property 11.6 and the squared factor $(x - 1)^2$ gives rise to a sum of partial fractions of the form

$$\frac{B}{x - 1} + \frac{C}{(x - 1)^2}$$

Therefore, the complete partial fraction decomposition is of the form

$$\frac{-2x^2 + 7x + 2}{x(x - 1)^2} = \frac{A}{x} + \frac{B}{x - 1} + \frac{C}{(x - 1)^2} \tag{1}$$

Multiply both sides of equation (1) by $x(x - 1)^2$ to produce

$$-2x^2 + 7x + 2 = A(x - 1)^2 + Bx(x - 1) + Cx \tag{2}$$

which is true for all values of x. If we let $x = 1$, then equation (2) becomes an equation in only C.

$$-2(1)^2 + 7(1) + 2 = A(1 - 1)^2 + B(1)(1 - 1) + C(1)$$

$$7 = C$$

If we let $x = 0$, then equation (2) becomes an equation in just A.

$$-2(0)^2 + 7(0) + 2 = A(0 - 1)^2 + B(0)(0 - 1) + C(0)$$

$$2 = A$$

If we let $x = 2$, then equation (2) becomes an equation in A, B, and C.

$$-2(2)^2 + 7(2) + 2 = A(2 - 1)^2 + B(2)(2 - 1) + C(2)$$
$$8 = A + 2B + 2C$$

But we already know that $A = 2$ and $C = 7$, so we can easily determine B.

$$8 = 2 + 2B + 14$$
$$-8 = 2B$$
$$-4 = B$$

Therefore, the original rational expression can be written

$$\frac{-2x^2 + 7x + 2}{x(x - 1)^2} = \frac{2}{x} - \frac{4}{x - 1} + \frac{7}{(x - 1)^2}$$ ∎

E X A M P L E   3

Find the partial fraction decomposition of

$$\frac{4x^2 + 6x - 10}{(x + 3)(x^2 + x + 2)}$$

Solution

Apply part 1 of Property 11.6 to determine that there is a partial fraction of the form $A/(x + 3)$ that corresponds to the factor $x + 3$. Apply part 3 of Property 11.6 to determine that there is also a partial fraction of the form

$$\frac{Bx + C}{x^2 + x + 2}$$

Thus, the complete partial fraction decomposition is of the form

$$\frac{4x^2 + 6x - 10}{(x + 3)(x^2 + x + 2)} = \frac{A}{x + 3} + \frac{Bx + C}{x^2 + x + 2} \tag{1}$$

Multiply both sides of equation (1) by $(x + 3)(x^2 + x + 2)$ to produce

$$4x^2 + 6x - 10 = A(x^2 + x + 2) + (Bx + C)(x + 3) \tag{2}$$

which is true for all values of x. If we let $x = -3$, then equation (2) becomes an equation in A alone.

$$4(-3)^2 + 6(-3) - 10 = A[(-3)^2 + (-3) + 2] + [B(-3) + C][(-3) + 3]$$
$$8 = 8A$$
$$1 = A$$

If we let $x = 0$, then equation (2) becomes an equation in A and C.

$$4(0)^2 + 6(0) - 10 = A(0^2 + 0 + 2) + [B(0) + C](0 + 3)$$
$$-10 = 2A + 3C$$

Because $A = 1$, we obtain the value of C.

$$-10 = 2 + 3C$$
$$-12 = 3C$$
$$-4 = C$$

If we let $x = 1$, then equation (2) becomes an equation in A, B, and C.

$$4(1)^2 + 6(1) - 10 = A(1^2 + 1 + 2) + [B(1) + C](1 + 3)$$
$$0 = 4A + 4B + 4C$$
$$0 = A + B + C$$

But because $A = 1$ and $C = -4$, we obtain the value of B.

$$0 = A + B + C$$
$$0 = 1 + B + (-4)$$
$$3 = B$$

Therefore, the original rational expression can now be written

$$\frac{4x^2 + 6x - 10}{(x + 3)(x^2 + x + 2)} = \frac{1}{x + 3} + \frac{3x - 4}{x^2 + x + 2}$$ ■

EXAMPLE 4

Find the partial fraction decomposition of

$$\frac{x^3 + x^2 + x + 3}{(x^2 + 1)^2}$$

Solution

Apply part 4 of Property 11.6 to determine that the partial fraction decomposition of this fraction is of the form

$$\frac{x^3 + x^2 + x + 3}{(x^2 + 1)^2} = \frac{Ax + B}{x^2 + 1} + \frac{Cx + D}{(x^2 + 1)^2} \tag{1}$$

Multiply both sides of equation (1) by $(x^2 + 1)^2$ to produce

$$x^3 + x^2 + x + 3 = (Ax + B)(x^2 + 1) + Cx + D \tag{2}$$

which is true for all values of x. Equation (2) is an identity, so we know that the coefficients of similar terms on both sides of the equation must be equal. Therefore, let's collect similar terms on the right side of equation (2).

$$x^3 + x^2 + x + 3 = Ax^3 + Ax + Bx^2 + B + Cx + D$$
$$= Ax^3 + Bx^2 + (A + C)x + B + D$$

Now we can equate coefficients from both sides:

$$1 = A \qquad 1 = B \qquad 1 = A + C \qquad \text{and} \qquad 3 = B + D$$

From these equations, we can determine that $A = 1$, $B = 1$, $C = 0$, and $D = 2$. Therefore, the original rational expression can be written

$$\frac{x^3 + x^2 + x + 3}{(x^2 + 1)^2} = \frac{x + 1}{x^2 + 1} + \frac{2}{(x^2 + 1)^2}$$ ■

Problem Set 11.6

For Problems 1–22, find the partial fraction decomposition for each rational expression.

1. $\dfrac{11x - 10}{(x - 2)(x + 1)}$

2. $\dfrac{11x - 2}{(x + 3)(x - 4)}$

3. $\dfrac{-2x - 8}{x^2 - 1}$

4. $\dfrac{-2x + 32}{x^2 - 4}$

5. $\dfrac{20x - 3}{6x^2 + 7x - 3}$

6. $\dfrac{-2x - 8}{10x^2 - x - 2}$

7. $\dfrac{x^2 - 18x + 5}{(x - 1)(x + 2)(x - 3)}$

8. $\dfrac{-9x^2 + 7x - 4}{x^3 - 3x^2 - 4x}$

9. $\dfrac{-6x^2 + 7x + 1}{x(2x - 1)(4x + 1)}$

10. $\dfrac{15x^2 + 20x + 30}{(x + 3)(3x + 2)(2x + 3)}$

11. $\dfrac{2x + 1}{(x - 2)^2}$

12. $\dfrac{-3x + 1}{(x + 1)^2}$

13. $\dfrac{-6x^2 + 19x + 21}{x^2(x + 3)}$

14. $\dfrac{10x^2 - 73x + 144}{x(x - 4)^2}$

15. $\dfrac{-2x^2 - 3x + 10}{(x^2 + 1)(x - 4)}$

16. $\dfrac{8x^2 + 15x + 12}{(x^2 + 4)(3x - 4)}$

17. $\dfrac{3x^2 + 10x + 9}{(x + 2)^3}$

18. $\dfrac{2x^3 + 8x^2 + 2x + 4}{(x + 1)^2(x^2 + 3)}$

19. $\dfrac{5x^2 + 3x + 6}{x(x^2 - x + 3)}$

20. $\dfrac{x^3 + x^2 + 2}{(x^2 + 2)^2}$

21. $\dfrac{2x^3 + x + 3}{(x^2 + 1)^2}$

22. $\dfrac{4x^2 + 3x + 14}{x^3 - 8}$

■ ■ ■ THOUGHTS INTO WORDS

23. Give a general description of partial fraction decomposition for someone who missed class the day it was discussed.

24. Give a step-by-step explanation of how to find the partial fraction decomposition of $\dfrac{11x + 5}{2x^2 + 5x - 3}$.

Chapter 11 Summary

(11.1 and 11.2) The primary focus of this entire chapter is the development of different techniques for solving systems of linear equations.

■ Substitution Method

With the aid of an example, we can describe the substitution method as follows. Suppose we want to solve the system

$$\left(\begin{array}{c} x - 2y = 22 \\ 3x + 4y = -24 \end{array} \right)$$

Step 1 Solve the first equation for x in terms of y.

$$x - 2y = 22$$
$$x = 2y + 22$$

Step 2 Substitute $2y + 22$ for x in the second equation.

$$3(2y + 22) + 4y = -24$$

Step 3 Solve the equation obtained in step 2.

$$6y + 66 + 4y = -24$$
$$10y + 66 = -24$$
$$10y = -90$$
$$y = -9$$

Step 4 Substitute -9 for y in the equation of step 1.

$$x = 2(-9) + 22 = 4$$

The solution set is $\{(4, -9)\}$.

■ Elimination-by-Addition Method

This method allows us to replace systems of equations with *simpler equivalent systems* until we obtain a system for which we can easily determine the solution. The following operations produce equivalent systems:

1. Any two equations of a system can be interchanged.
2. Both sides of any equation of the system can be multiplied by any nonzero real number.

3. Any equation of the system can be replaced by the sum of a nonzero multiple of another equation plus that equation.

For example, through a sequence of operations, we can transform the system

$$\left(\begin{array}{l} 5x + 3y = -28 \\ \dfrac{1}{2}x - y = -8 \end{array} \right)$$

to the equivalent system

$$\left(\begin{array}{l} x - 2y = -16 \\ 13y = 52 \end{array} \right)$$

for which we can easily determine the solution set $\{(-8, 4)\}$.

■ Matrix Approach

(11.3) We can change the augmented matrix of a system to reduced echelon form by applying the following elementary row operations:

1. Any two rows of the matrix can be interchanged.
2. Any row of the matrix can be multiplied by a nonzero real number.
3. Any row of the matrix can be replaced by the sum of a nonzero multiple of another row plus that row.

For example, the augmented matrix of the system

$$\left(\begin{array}{l} x - 2y + 3z = 4 \\ 2x + y - 4z = 3 \\ -3x + 4y - z = -2 \end{array} \right)$$

is

$$\begin{bmatrix} 1 & -2 & 3 & \vdots & 4 \\ 2 & 1 & -4 & \vdots & 3 \\ -3 & 4 & -1 & \vdots & -2 \end{bmatrix}$$

We can change this matrix to the reduced echelon form

$$\begin{bmatrix} 1 & 0 & 0 & \vdots & 4 \\ 0 & 1 & 0 & \vdots & 3 \\ 0 & 0 & 1 & \vdots & 2 \end{bmatrix}$$

where the solution set $\{(4, 3, 2)\}$ is obvious.

(11.4) A rectangular array of numbers is called a **matrix**. A **square matrix** has the same number of rows as columns. For a 2×2 matrix

$$\begin{bmatrix} a_1 & b_1 \\ a_2 & b_2 \end{bmatrix}$$

the **determinant** of the matrix is written as

$$\begin{vmatrix} a_1 & b_1 \\ a_2 & b_2 \end{vmatrix}$$

and is defined by

$$\begin{vmatrix} a_1 & b_1 \\ a_2 & b_2 \end{vmatrix} = a_1 b_2 - a_2 b_1$$

The determinant of a 3×3 (or larger) square matrix can be evaluated by expansion of minors of the elements of any row or any column. The concepts of minor and cofactor are needed for this purpose; these terms are defined in Definitions 11.2 and 11.3.

The following properties are helpful when evaluating determinants:

1. If any row (or column) of a square matrix A contains only zeros, then $|A| = 0$.

2. If square matrix B is obtained from square matrix A by interchanging two rows (or two columns), then $|B| = -|A|$.

3. If square matrix B is obtained from square matrix A by multiplying each element of any row (or column) of A by some real number k, then $|B| = k|A|$.

4. If square matrix B is obtained from square matrix A by adding k times a row (or column) of A to another row (or column) of A, then $|B| = |A|$.

5. If two rows (or columns) of a square matrix A are identical, then $|A| = 0$.

(11.5) Cramer's rule for solving a system of two linear equations in two variables is stated as follows: Given the system

$$\left(\begin{array}{l} a_1 x + b_1 y = c_1 \\ a_2 x + b_2 y = c_2 \end{array} \right)$$

with

$$D = \begin{vmatrix} a_1 & b_1 \\ a_2 & b_2 \end{vmatrix} \neq 0$$

$$D_x = \begin{vmatrix} c_1 & b_1 \\ c_2 & b_2 \end{vmatrix} \qquad D_y = \begin{vmatrix} a_1 & c_1 \\ a_2 & c_2 \end{vmatrix}$$

then

$$x = \frac{D_x}{D} \qquad \text{and} \qquad y = \frac{D_y}{D}$$

Cramer's rule for solving a system of three linear equations in three variables is stated as follows: Given the system

$$\left(\begin{array}{l} a_1 x + b_1 y + c_1 z = d_1 \\ a_2 x + b_2 y + c_2 z = d_2 \\ a_3 x + b_3 y + c_3 z = d_3 \end{array} \right)$$

with

$$D = \begin{vmatrix} a_1 & b_1 & c_1 \\ a_2 & b_2 & c_2 \\ a_3 & b_3 & c_3 \end{vmatrix} \neq 0 \qquad D_x = \begin{vmatrix} d_1 & b_1 & c_1 \\ d_2 & b_2 & c_2 \\ d_3 & b_3 & c_3 \end{vmatrix}$$

$$D_y = \begin{vmatrix} a_1 & d_1 & c_1 \\ a_2 & d_2 & c_2 \\ a_3 & d_3 & c_3 \end{vmatrix} \qquad D_z = \begin{vmatrix} a_1 & b_1 & d_1 \\ a_2 & b_2 & d_2 \\ a_3 & b_3 & d_3 \end{vmatrix}$$

then

$$x = \frac{D_x}{D}, \qquad y = \frac{D_y}{D}, \qquad \text{and} \qquad z = \frac{D_z}{D}$$

Chapter 11   Review Problem Set

For Problems 1–4, solve each system by using the *substitution* method.

1. $\left(\begin{array}{l} 3x - y = 16 \\ 5x + 7y = -34 \end{array} \right)$

2. $\left(\begin{array}{l} 6x + 5y = -21 \\ x - 4y = 11 \end{array} \right)$

3. $\left(\begin{array}{l} 2x - 3y = 12 \\ 3x + 5y = -20 \end{array} \right)$

4. $\left(\begin{array}{l} 5x + 8y = 1 \\ 4x + 7y = -2 \end{array} \right)$

For Problems 5–8, solve each system by using the *elimination-by-addition* method.

5. $\begin{pmatrix} 4x - 3y = 34 \\ 3x + 2y = 0 \end{pmatrix}$

6. $\begin{pmatrix} \dfrac{1}{2}x - \dfrac{2}{3}y = 1 \\ \dfrac{3}{4}x + \dfrac{1}{6}y = -1 \end{pmatrix}$

7. $\begin{pmatrix} 2x - y + 3z = -19 \\ 3x + 2y - 4z = 21 \\ 5x - 4y - z = -8 \end{pmatrix}$

8. $\begin{pmatrix} 3x + 2y - 4z = 4 \\ 5x + 3y - z = 2 \\ 4x - 2y + 3z = 11 \end{pmatrix}$

For Problems 9–12, solve each system by *changing the augmented matrix to reduced echelon form.*

9. $\begin{pmatrix} x - 3y = 17 \\ -3x + 2y = -23 \end{pmatrix}$

10. $\begin{pmatrix} 2x + 3y = 25 \\ 3x - 5y = -29 \end{pmatrix}$

11. $\begin{pmatrix} x - 2y + z = -7 \\ 2x - 3y + 4z = -14 \\ -3x + y - 2z = 10 \end{pmatrix}$

12. $\begin{pmatrix} -2x - 7y + z = 9 \\ x + 3y - 4z = -11 \\ 4x + 5y - 3z = -11 \end{pmatrix}$

For Problems 13–16, solve each system by using *Cramer's rule.*

13. $\begin{pmatrix} 5x + 3y = -18 \\ 4x - 9y = -3 \end{pmatrix}$

14. $\begin{pmatrix} 0.2x + 0.3y = 2.6 \\ 0.5x - 0.1y = 1.4 \end{pmatrix}$

15. $\begin{pmatrix} 2x - 3y - 3z = 25 \\ 3x + y + 2z = -5 \\ 5x - 2y - 4z = 32 \end{pmatrix}$

16. $\begin{pmatrix} 3x - y + z = -10 \\ 6x - 2y + 5z = -35 \\ 7x + 3y - 4z = 19 \end{pmatrix}$

For Problems 17–24, solve each system by using the method you think is most appropriate.

17. $\begin{pmatrix} 4x + 7y = -15 \\ 3x - 2y = 25 \end{pmatrix}$

18. $\begin{pmatrix} \dfrac{3}{4}x - \dfrac{1}{2}y = -15 \\ \dfrac{2}{3}x + \dfrac{1}{4}y = -5 \end{pmatrix}$

19. $\begin{pmatrix} x + 4y = 3 \\ 3x - 2y = 1 \end{pmatrix}$

20. $\begin{pmatrix} 7x - 3y = -49 \\ y = \dfrac{3}{5}x - 1 \end{pmatrix}$

21. $\begin{pmatrix} x - y - z = 4 \\ -3x + 2y + 5z = -21 \\ 5x - 3y - 7z = 30 \end{pmatrix}$

22. $\begin{pmatrix} 2x - y + z = -7 \\ -5x + 2y - 3z = 17 \\ 3x + y + 7z = -5 \end{pmatrix}$

23. $\begin{pmatrix} 3x - 2y - 5z = 2 \\ -4x + 3y + 11z = 3 \\ 2x - y + z = -1 \end{pmatrix}$

24. $\begin{pmatrix} 7x - y + z = -4 \\ -2x + 9y - 3z = -50 \\ x - 5y + 4z = 42 \end{pmatrix}$

For Problems 25–30, evaluate each determinant.

25. $\begin{vmatrix} -2 & 6 \\ 3 & 8 \end{vmatrix}$

26. $\begin{vmatrix} 5 & -4 \\ 7 & -3 \end{vmatrix}$

27. $\begin{vmatrix} 2 & 3 & -1 \\ 3 & 4 & -5 \\ 6 & 4 & 2 \end{vmatrix}$

28. $\begin{vmatrix} 3 & -2 & 4 \\ 1 & 0 & 6 \\ 3 & -3 & 5 \end{vmatrix}$

29. $\begin{vmatrix} 5 & 4 & 3 \\ 2 & -7 & 0 \\ 3 & -2 & 0 \end{vmatrix}$

30. $\begin{vmatrix} 5 & -4 & 2 & 1 \\ 3 & 7 & 6 & -2 \\ 2 & 1 & -5 & 0 \\ 3 & -2 & 4 & 0 \end{vmatrix}$

For Problems 31–34, solve each problem by setting up and solving a system of linear equations.

31. The sum of the digits of a two-digit number is 9. If the digits are reversed, the newly formed number is 45 less than the original number. Find the original number.

32. Sara invested $2500, part of it at 10% and the rest at 12% yearly interest. The yearly income on the 12% investment was $102 more than the income on the 10% investment. How much money did she invest at each rate?

33. A box contains $17.70 in nickels, dimes, and quarters. The number of dimes is 8 less than twice the number of nickels. The number of quarters is 2 more than the sum of the numbers of nickels and dimes. How many coins of each kind are there in the box?

34. The measure of the largest angle of a triangle is 10° more than four times the smallest angle. The sum of the smallest and largest angles is three times the measure of the other angle. Find the measure of each angle of the triangle.

Chapter 11　Test

For Problems 1–4, refer to the following systems of equations:

I. $\begin{pmatrix} 3x - 2y = 4 \\ 9x - 6y = 12 \end{pmatrix}$　　II. $\begin{pmatrix} 5x - y = 4 \\ 3x + 7y = 9 \end{pmatrix}$

III. $\begin{pmatrix} 2x - y = 4 \\ 2x - y = -6 \end{pmatrix}$

1. For which system are the graphs parallel lines?

2. For which system are the equations dependent?

3. For which system is the solution set $\varnothing$?

4. Which system is consistent?

For Problems 5–8, evaluate each determinant.

5. $\begin{vmatrix} -2 & 4 \\ -5 & 6 \end{vmatrix}$

6. $\begin{vmatrix} \frac{1}{2} & \frac{1}{3} \\ \frac{3}{4} & -\frac{2}{3} \end{vmatrix}$

7. $\begin{vmatrix} -1 & 2 & 1 \\ 3 & 1 & -2 \\ 2 & -1 & 1 \end{vmatrix}$

8. $\begin{vmatrix} 2 & 4 & -5 \\ -4 & 3 & 0 \\ -2 & 6 & 1 \end{vmatrix}$

9. How many ordered pairs of real numbers are in the solution set for the system $\begin{pmatrix} y = 3x - 4 \\ 9x - 3y = 12 \end{pmatrix}$?

10. Solve the system $\begin{pmatrix} 3x - 2y = -14 \\ 7x + 2y = -6 \end{pmatrix}$

11. Solve the system $\begin{pmatrix} 4x - 5y = 17 \\ y = -3x + 8 \end{pmatrix}$

12. Find the value of x in the solution for the system
$$\begin{pmatrix} \frac{3}{4}x - \frac{1}{2}y = -21 \\ \frac{2}{3}x + \frac{1}{6}y = -4 \end{pmatrix}$$

13. Find the value of y in the solution for the system
$$\begin{pmatrix} 4x - y = 7 \\ 3x + 2y = 2 \end{pmatrix}.$$

14. Suppose that the augmented matrix of a system of three linear equations in the three variables x, y, and z can be changed to the matrix
$$\begin{bmatrix} 1 & 1 & -4 & \vdots & 3 \\ 0 & 1 & 4 & \vdots & 5 \\ 0 & 0 & 3 & \vdots & 6 \end{bmatrix}$$

Find the value of x in the solution for the system.

15. Suppose that the augmented matrix of a system of three linear equations in the three variables x, y, and z can be changed to the matrix
$$\begin{bmatrix} 1 & 2 & -3 & \vdots & 4 \\ 0 & 1 & 2 & \vdots & 5 \\ 0 & 0 & 2 & \vdots & -8 \end{bmatrix}$$

Find the value of y in the solution for the system.

16. How many ordered triples are there in the solution set for the following system?
$$\begin{pmatrix} x + 3y - z = 5 \\ 2x - y - z = 7 \\ 5x + 8y - 4z = 22 \end{pmatrix}$$

17. How many ordered triples are there in the solution set for the following system?
$$\begin{pmatrix} 3x - y - 2z = 1 \\ 4x + 2y + z = 5 \\ 6x - 2y - 4z = 9 \end{pmatrix}$$

18. Solve the following system:
$$\begin{pmatrix} 5x - 3y - 2z = -1 \\ 4y + 7z = 3 \\ 4z = -12 \end{pmatrix}$$

19. Solve the following system:
$$\begin{pmatrix} x - 2y + z = 0 \\ y - 3z = -1 \\ 2y + 5z = -2 \end{pmatrix}$$

20. Find the value of x in the solution for the system
$$\begin{pmatrix} x - 4y + z = 12 \\ -2x + 3y - z = -11 \\ 5x - 3y + 2z = 17 \end{pmatrix}$$

21. Find the value of y in the solution for the system

$$\begin{pmatrix} x - 3y + z = -13 \\ 3x + 5y - z = 17 \\ 5x - 2y + 2z = -13 \end{pmatrix}$$

22. One solution is 30% alcohol and another solution is 70% alcohol. Some of each of the two solutions is mixed to produce 8 liters of a 40% solution. How many liters of the 70% solution should be used?

23. A box contain $7.25 in nickels, dimes, and quarters. There are 43 coins, and the number of quarters is 1 more than three times the number of nickels. Find the number of quarters in the box.

24. A catering company makes batches of three different types of pastries to serve at brunches. Each batch requires the services of three different operations, as indicated by the following table:

	Cream Puffs	Eclairs	Danish Rolls
Dough	0.2 hour	0.5 hour	0.4 hour
Baking	0.3 hour	0.1 hour	0.2 hour
Frosting	0.1 hour	0.5 hour	0.3 hour

The dough, baking, and frosting operations have available a maximum of 7.0, 3.9, and 5.5 hours, respectively. How many batches of each type should be made so that the company is operating at full capacity?

25. The measure of the largest angle of a triangle is 20° more than the sum of the measures of the other two angles. The difference in the measures of the largest and smallest angles is 65°. Find the measure of each angle.

Algebra of Matrices

© PhotoDisc/Getty Images

A financial planner might use the techniques of linear programming when developing a plan for clients.

In Section 11.3, we used matrices strictly as a device to help solve systems of linear equations. Our primary objective was the development of techniques for solving systems of equations, not the study of matrices. However, matrices can be studied from an algebraic viewpoint, much as we study the set of real numbers. That is, we can define certain operations on matrices and verify properties of those operations. This algebraic approach to matrices is the focal point of this chapter. In order to get a simplified view of the algebra of matrices, we will begin by studying 2 × 2 matrices, and then later we will enlarge our discussion to include *m* × *n* matrices. As a bonus, another technique for solving systems of equations will emerge from our study. In the final section of this chapter, we expand our problem-solving capabilities by studying systems of linear inequalities.

12.1 **Algebra of 2 × 2 Matrices**

Throughout these next two sections, we will be working primarily with 2×2 matrices; therefore any reference to matrices means 2×2 matrices unless stated otherwise. The following 2×2 matrix notation will be used frequently.

$$A = \begin{bmatrix} a_{11} & a_{12} \\ a_{21} & a_{22} \end{bmatrix} \qquad B = \begin{bmatrix} b_{11} & b_{12} \\ b_{21} & b_{22} \end{bmatrix} \qquad C = \begin{bmatrix} c_{11} & c_{12} \\ c_{21} & c_{22} \end{bmatrix}$$

Two matrices are **equal** if and only if all elements in corresponding positions are equal. Thus $A = B$ if and only if $a_{11} = b_{11}$, $a_{12} = b_{12}$, $a_{21} = b_{21}$, and $a_{22} = b_{22}$.

■ Addition of Matrices

To **add** two matrices, we add the elements that appear in corresponding positions. Therefore the sum of matrix A and matrix B is defined as follows:

Definition 12.1

$$A + B = \begin{bmatrix} a_{11} & a_{12} \\ a_{21} & a_{22} \end{bmatrix} + \begin{bmatrix} b_{11} & b_{12} \\ b_{21} & b_{22} \end{bmatrix}$$

$$= \begin{bmatrix} a_{11} + b_{11} & a_{12} + b_{12} \\ a_{21} + b_{21} & a_{22} + b_{22} \end{bmatrix}$$

For example,

$$\begin{bmatrix} 2 & -1 \\ -3 & 4 \end{bmatrix} + \begin{bmatrix} -5 & 4 \\ -1 & 7 \end{bmatrix} = \begin{bmatrix} -3 & 3 \\ -4 & 11 \end{bmatrix}$$

It is not difficult to show that the **commutative** and **associative properties** are valid for the addition of matrices. Thus we can state that

$$A + B = B + A \qquad \text{and} \qquad (A + B) + C = A + (B + C)$$

Because

$$\begin{bmatrix} a_{11} & a_{12} \\ a_{21} & a_{22} \end{bmatrix} + \begin{bmatrix} 0 & 0 \\ 0 & 0 \end{bmatrix} = \begin{bmatrix} a_{11} & a_{12} \\ a_{21} & a_{22} \end{bmatrix}$$

we see that $\begin{bmatrix} 0 & 0 \\ 0 & 0 \end{bmatrix}$, which is called the **zero matrix**, represented by O, is the **additive identity element**. Thus we can state that

$$A + O = O + A = A$$

Because every real number has an additive inverse, it follows that any matrix A has an **additive inverse**, $-A$, that is formed by taking the additive inverse of each element of A. For example, if

$$A = \begin{bmatrix} 4 & -2 \\ -1 & 0 \end{bmatrix} \quad \text{then} -A = \begin{bmatrix} -4 & 2 \\ 1 & 0 \end{bmatrix}$$

and

$$A + (-A) = \begin{bmatrix} 4 & -2 \\ -1 & 0 \end{bmatrix} + \begin{bmatrix} -4 & 2 \\ 1 & 0 \end{bmatrix} = \begin{bmatrix} 0 & 0 \\ 0 & 0 \end{bmatrix}$$

In general, we can state that every matrix A has an additive inverse $-A$ such that

$$A + (-A) = (-A) + A = O$$

■ Subtraction of Matrices

Again like the algebra of real numbers, **subtraction** of matrices can be defined in terms of *adding the additive inverse*. Therefore we can define subtraction as follows:

Definition 12.2

$$A - B = A + (-B)$$

For example,

$$\begin{bmatrix} 2 & -7 \\ -6 & 5 \end{bmatrix} - \begin{bmatrix} 3 & 4 \\ -2 & -1 \end{bmatrix} = \begin{bmatrix} 2 & -7 \\ -6 & 5 \end{bmatrix} + \begin{bmatrix} -3 & -4 \\ 2 & 1 \end{bmatrix}$$

$$= \begin{bmatrix} -1 & -11 \\ -4 & 6 \end{bmatrix}$$

■ Scalar Multiplication

When we work with matrices, we commonly refer to a single real number as a **scalar** to distinguish it from a matrix. Then, taking the **product** of a scalar and a matrix (often referred to as **scalar multiplication**) can be accomplished by multiplying each element of the matrix by the scalar. For example,

$$3\begin{bmatrix} -4 & -6 \\ 1 & -2 \end{bmatrix} = \begin{bmatrix} 3(-4) & 3(-6) \\ 3(1) & 3(-2) \end{bmatrix} = \begin{bmatrix} -12 & -18 \\ 3 & -6 \end{bmatrix}$$

In general, scalar multiplication can be defined as follows:

Definition 12.3

$$kA = k\begin{bmatrix} a_{11} & a_{12} \\ a_{21} & a_{22} \end{bmatrix} = \begin{bmatrix} ka_{11} & ka_{12} \\ ka_{21} & ka_{22} \end{bmatrix}$$

where k is any real number.

EXAMPLE 1   If $A = \begin{bmatrix} -4 & 3 \\ 2 & -5 \end{bmatrix}$ and $B = \begin{bmatrix} 2 & -3 \\ 7 & -6 \end{bmatrix}$, find

(a) $-2A$ (b) $3A + 2B$ (c) $A - 4B$

Solutions

(a) $-2A = -2\begin{bmatrix} -4 & 3 \\ 2 & -5 \end{bmatrix} = \begin{bmatrix} 8 & -6 \\ -4 & 10 \end{bmatrix}$

(b) $3A + 2B = 3\begin{bmatrix} -4 & 3 \\ 2 & -5 \end{bmatrix} + 2\begin{bmatrix} 2 & -3 \\ 7 & -6 \end{bmatrix}$

$= \begin{bmatrix} -12 & 9 \\ 6 & -15 \end{bmatrix} + \begin{bmatrix} 4 & -6 \\ 14 & -12 \end{bmatrix}$

$= \begin{bmatrix} -8 & 3 \\ 20 & -27 \end{bmatrix}$

(c) $A - 4B = \begin{bmatrix} -4 & 3 \\ 2 & -5 \end{bmatrix} - 4\begin{bmatrix} 2 & -3 \\ 7 & -6 \end{bmatrix}$

$= \begin{bmatrix} -4 & 3 \\ 2 & -5 \end{bmatrix} - \begin{bmatrix} 8 & -12 \\ 28 & -24 \end{bmatrix}$

$= \begin{bmatrix} -4 & 3 \\ 2 & -5 \end{bmatrix} + \begin{bmatrix} -8 & 12 \\ -28 & 24 \end{bmatrix}$

$= \begin{bmatrix} -12 & 15 \\ -26 & 19 \end{bmatrix}$ ∎

The following properties, which are easy to check, pertain to scalar multiplication and matrix addition (where k and l represent any real numbers):

$$k(A + B) = kA + kB$$
$$(k + l)A = kA + lA$$
$$(kl)A = k(lA)$$

■ Multiplication of Matrices

At this time, it probably would seem quite natural to define matrix multiplication by multiplying corresponding elements of two matrices. However, it turns out that such a definition does not have many worthwhile applications. Therefore we use a special type of **matrix multiplication**, sometimes referred to as a "row-by-column multiplication." We will state the definition, paraphrase what it says, and then give some examples.

Definition 12.4

$$AB = \begin{bmatrix} a_{11} & a_{12} \\ a_{21} & a_{22} \end{bmatrix} \begin{bmatrix} b_{11} & b_{12} \\ b_{21} & b_{22} \end{bmatrix}$$

$$= \begin{bmatrix} a_{11}b_{11} + a_{12}b_{21} & a_{11}b_{12} + a_{12}b_{22} \\ a_{21}b_{11} + a_{22}b_{21} & a_{21}b_{12} + a_{22}b_{22} \end{bmatrix}$$

Note the row-by-column pattern of Definition 12.4. We multiply the rows of A times the columns of B in a pairwise entry fashion, adding the results. For example, the element in the first row and second column of the product is obtained by multiplying the elements of the first row of A times the elements of the second column of B and adding the results.

$$\begin{bmatrix} a_{11} & a_{12} \\ a_{21} & a_{22} \end{bmatrix} \begin{bmatrix} b_{11} & b_{12} \\ b_{21} & b_{22} \end{bmatrix} = [\qquad\quad a_{11}b_{12} + a_{12}b_{22}]$$

Now let's look at some specific examples.

EXAMPLE 2

If $A = \begin{bmatrix} -2 & 1 \\ 4 & 5 \end{bmatrix}$ and $B = \begin{bmatrix} 3 & -2 \\ -1 & 7 \end{bmatrix}$, find **(a)** AB and **(b)** BA.

Solutions

(a) $AB = \begin{bmatrix} -2 & 1 \\ 4 & 5 \end{bmatrix} \begin{bmatrix} 3 & -2 \\ -1 & 7 \end{bmatrix}$

$= \begin{bmatrix} (-2)(3) + (1)(-1) & (-2)(-2) + (1)(7) \\ (4)(3) + (5)(-1) & (4)(-2) + (5)(7) \end{bmatrix}$

$= \begin{bmatrix} -7 & 11 \\ 7 & 27 \end{bmatrix}$

(b) $BA = \begin{bmatrix} 3 & -2 \\ -1 & 7 \end{bmatrix} \begin{bmatrix} -2 & 1 \\ 4 & 5 \end{bmatrix}$

$= \begin{bmatrix} (3)(-2) + (-2)(4) & (3)(1) + (-2)(5) \\ (-1)(-2) + (7)(4) & (-1)(1) + (7)(5) \end{bmatrix}$

$= \begin{bmatrix} -14 & -7 \\ 30 & 34 \end{bmatrix}$

■

Example 2 makes it immediately apparent that matrix multiplication is *not* a **commutative** operation.

If $A = \begin{bmatrix} 2 & -6 \\ -3 & 9 \end{bmatrix}$ and $B = \begin{bmatrix} -3 & 6 \\ -1 & 2 \end{bmatrix}$, find AB.

Solution

Once you feel comfortable with Definition 12.4, you can do the addition mentally.

$$AB = \begin{bmatrix} 2 & -6 \\ -3 & 9 \end{bmatrix}\begin{bmatrix} -3 & 6 \\ -1 & 2 \end{bmatrix} = \begin{bmatrix} 0 & 0 \\ 0 & 0 \end{bmatrix}$$ ∎

Example 3 illustrates that the product of two matrices can be the zero matrix even though neither of the two matrices is the zero matrix. This is different from the property of real numbers that states $ab = 0$ if and only if $a = 0$ or $b = 0$.

As we illustrated and stated earlier, matrix multiplication is *not* a commutative operation. However, it is an **associative** operation and it does exhibit two **distributive properties**. These properties can be stated as follows:

$$(AB)C = A(BC)$$
$$A(B + C) = AB + AC$$
$$(B + C)A = BA + CA$$

We will ask you to verify these properties in the next set of problems.

Problem Set 12.1

For Problems 1–12, compute the indicated matrix by using the following matrices:

$A = \begin{bmatrix} 1 & -2 \\ 3 & 4 \end{bmatrix}$ $B = \begin{bmatrix} 2 & -3 \\ 5 & -1 \end{bmatrix}$

$C = \begin{bmatrix} 0 & 6 \\ -4 & 2 \end{bmatrix}$ $D = \begin{bmatrix} -2 & 3 \\ 5 & -4 \end{bmatrix}$

$E = \begin{bmatrix} 2 & 5 \\ 7 & 3 \end{bmatrix}$

1. $A + B$
2. $B - C$
3. $3C + D$
4. $2D - E$
5. $4A - 3B$
6. $2B + 3D$
7. $(A - B) - C$
8. $B - (D - E)$
9. $2D - 4E$
10. $3A - 4E$
11. $B - (D + E)$
12. $A - (B + C)$

For Problems 13–26, compute AB and BA.

13. $A = \begin{bmatrix} 1 & -1 \\ 2 & -2 \end{bmatrix}$, $B = \begin{bmatrix} 3 & -4 \\ -1 & 2 \end{bmatrix}$

14. $A = \begin{bmatrix} -3 & 4 \\ 2 & 1 \end{bmatrix}$, $B = \begin{bmatrix} -2 & 5 \\ 6 & -1 \end{bmatrix}$

15. $A = \begin{bmatrix} 1 & -3 \\ -4 & 6 \end{bmatrix}$, $B = \begin{bmatrix} 7 & -3 \\ 4 & 5 \end{bmatrix}$

16. $A = \begin{bmatrix} 5 & 0 \\ -2 & 3 \end{bmatrix}$, $B = \begin{bmatrix} -3 & 6 \\ 4 & 1 \end{bmatrix}$

17. $A = \begin{bmatrix} 2 & -4 \\ 1 & -2 \end{bmatrix}$, $B = \begin{bmatrix} 1 & -2 \\ -3 & 6 \end{bmatrix}$

18. $A = \begin{bmatrix} 1 & 2 \\ 1 & 2 \end{bmatrix}$, $B = \begin{bmatrix} 2 & 2 \\ -1 & -1 \end{bmatrix}$

19. $A = \begin{bmatrix} -3 & -2 \\ -4 & -1 \end{bmatrix}$, $B = \begin{bmatrix} 2 & -1 \\ 4 & 5 \end{bmatrix}$

20. $A = \begin{bmatrix} -2 & 3 \\ -1 & 7 \end{bmatrix}$, $B = \begin{bmatrix} -1 & -3 \\ -5 & -7 \end{bmatrix}$

21. $A = \begin{bmatrix} 2 & -1 \\ -5 & 3 \end{bmatrix}$, $B = \begin{bmatrix} 3 & 1 \\ 5 & 2 \end{bmatrix}$

22. $A = \begin{bmatrix} -8 & -5 \\ 3 & 2 \end{bmatrix}$, $B = \begin{bmatrix} -2 & -5 \\ 3 & 8 \end{bmatrix}$

23. $A = \begin{bmatrix} \dfrac{1}{2} & -\dfrac{1}{3} \\ \dfrac{1}{3} & \dfrac{1}{4} \end{bmatrix}$, $B = \begin{bmatrix} 4 & -6 \\ 6 & -4 \end{bmatrix}$

24. $A = \begin{bmatrix} \dfrac{1}{3} & -\dfrac{1}{2} \\ \dfrac{3}{2} & -\dfrac{2}{3} \end{bmatrix}$, $B = \begin{bmatrix} -6 & -18 \\ 12 & -12 \end{bmatrix}$

25. $A = \begin{bmatrix} 5 & 6 \\ 2 & 3 \end{bmatrix}$, $B = \begin{bmatrix} 1 & -2 \\ -\dfrac{2}{3} & \dfrac{5}{3} \end{bmatrix}$

26. $A = \begin{bmatrix} -3 & -5 \\ 2 & 4 \end{bmatrix}$, $B = \begin{bmatrix} -2 & -\dfrac{5}{2} \\ 1 & \dfrac{3}{2} \end{bmatrix}$

For Problems 27–30, use the following matrices.

$$A = \begin{bmatrix} -2 & 3 \\ 5 & 4 \end{bmatrix} \quad B = \begin{bmatrix} 0 & 1 \\ 1 & 0 \end{bmatrix}$$

$$C = \begin{bmatrix} 1 & 0 \\ 1 & 0 \end{bmatrix} \quad D = \begin{bmatrix} 1 & 1 \\ 1 & 1 \end{bmatrix}$$

$$I = \begin{bmatrix} 1 & 0 \\ 0 & 1 \end{bmatrix}$$

27. Compute AB and BA.

28. Compute AC and CA.

29. Compute AD and DA.

30. Compute AI and IA.

For Problems 31–34, use the following matrices.

$$A = \begin{bmatrix} 2 & 4 \\ 5 & -3 \end{bmatrix} \quad B = \begin{bmatrix} -2 & 3 \\ -1 & 2 \end{bmatrix}$$

$$C = \begin{bmatrix} 2 & 1 \\ 3 & 7 \end{bmatrix}$$

31. Show that $(AB)C = A(BC)$.

32. Show that $A(B + C) = AB + AC$.

33. Show that $(A + B)C = AC + BC$.

34. Show that $(3 + 2)A = 3A + 2A$.

For Problems 35–43, use the following matrices.

$$A = \begin{bmatrix} a_{11} & a_{12} \\ a_{21} & a_{22} \end{bmatrix} \quad B = \begin{bmatrix} b_{11} & b_{12} \\ b_{21} & b_{22} \end{bmatrix}$$

$$C = \begin{bmatrix} c_{11} & c_{12} \\ c_{21} & c_{22} \end{bmatrix} \quad O = \begin{bmatrix} 0 & 0 \\ 0 & 0 \end{bmatrix}$$

35. Show that $A + B = B + A$.

36. Show that $(A + B) + C = A + (B + C)$.

37. Show that $A + (-A) = O$.

38. Show that $k(A + B) = kA + kB$ for any real number k.

39. Show that $(k + l)A = kA + lA$ for any real numbers k and l.

40. Show that $(kl)A = k(lA)$ for any real numbers k and l.

41. Show that $(AB)C = A(BC)$.

42. Show that $A(B + C) = AB + AC$.

43. Show that $(A + B)C = AC + BC$.

■ ■ ■ THOUGHTS INTO WORDS

44. How would you show that addition of 2×2 matrices is a commutative operation?

45. How would you show that subtraction of 2×2 matrices is not a commutative operation?

46. How would you explain matrix multiplication to someone who missed class the day it was discussed?

47. Your friend says that because multiplication of real numbers is a commutative operation, it seems reasonable that multiplication of matrices should also be a commutative operation. How would you react to that statement?

■ ■ ■ FURTHER INVESTIGATIONS

48. If $A = \begin{bmatrix} 2 & 0 \\ 0 & 3 \end{bmatrix}$, calculate A^2 and A^3, where A^2 means AA, and A^3 means AAA.

49. If $A = \begin{bmatrix} 1 & -1 \\ 2 & 3 \end{bmatrix}$, calculate A^2 and A^3.

50. Does $(A + B)(A - B) = A^2 - B^2$ for all 2×2 matrices? Defend your answer.

📉 GRAPHING CALCULATOR ACTIVITIES

51. Use a calculator to check the answers to all three parts of Example 1.

52. Use a calculator to check your answers for Problems 21–26.

53. Use the following matrices:

$$A = \begin{bmatrix} 7 & -4 \\ 6 & 9 \end{bmatrix} \qquad B = \begin{bmatrix} -3 & 8 \\ -5 & 7 \end{bmatrix}$$

$$C = \begin{bmatrix} 8 & -2 \\ 4 & -7 \end{bmatrix}$$

(a) Show that $(AB)C = A(BC)$.
(b) Show that $A(B + C) = AB + AC$.
(c) Show that $(B + C)A = BA + CA$.

12.2   Multiplicative Inverses

We know that 1 is a multiplicative identity element for the set of real numbers. That is, $a(1) = 1(a) = a$ for any real number a. Is there a multiplicative identity element for 2×2 matrices? Yes. The matrix

$$I = \begin{bmatrix} 1 & 0 \\ 0 & 1 \end{bmatrix}$$

is the **multiplicative identity element** because

$$\begin{bmatrix} 1 & 0 \\ 0 & 1 \end{bmatrix}\begin{bmatrix} a_{11} & a_{12} \\ a_{21} & a_{22} \end{bmatrix} = \begin{bmatrix} a_{11} & a_{12} \\ a_{21} & a_{22} \end{bmatrix}$$

and

$$\begin{bmatrix} a_{11} & a_{12} \\ a_{21} & a_{22} \end{bmatrix}\begin{bmatrix} 1 & 0 \\ 0 & 1 \end{bmatrix} = \begin{bmatrix} a_{11} & a_{12} \\ a_{21} & a_{22} \end{bmatrix}$$

Therefore we can state that

$$AI = IA = A$$

for all 2×2 matrices.

Again, refer to the real numbers, where every nonzero real number a has a multiplicative inverse $1/a$ such that $a(1/a) = (1/a)a = 1$. Does every 2×2 matrix have a multiplicative inverse? To help answer this question, let's think about finding the multiplicative inverse (if one exists) for a specific matrix. This should give us some clues about a general approach.

EXAMPLE 1   Find the multiplicative inverse of $A = \begin{bmatrix} 3 & 5 \\ 2 & 4 \end{bmatrix}$.

Solution

We are looking for a matrix A^{-1} such that $AA^{-1} = A^{-1}A = I$. In other words, we want to solve the following matrix equation:

$$\begin{bmatrix} 3 & 5 \\ 2 & 4 \end{bmatrix}\begin{bmatrix} x & y \\ z & w \end{bmatrix} = \begin{bmatrix} 1 & 0 \\ 0 & 1 \end{bmatrix}$$

We need to multiply the two matrices on the left side of this equation and then set the elements of the product matrix equal to the corresponding elements of the identity matrix. We obtain the following system of equations:

$$\left(\begin{array}{l} 3x + 5z = 1 \\ 3y + 5w = 0 \\ 2x + 4z = 0 \\ 2y + 4w = 1 \end{array}\right)$$

(1)
(2)
(3)
(4)

Solving equations (1) and (3) simultaneously produces values for x and z.

$$x = \frac{\begin{vmatrix} 1 & 5 \\ 0 & 4 \end{vmatrix}}{\begin{vmatrix} 3 & 5 \\ 2 & 4 \end{vmatrix}} = \frac{1(4) - 5(0)}{3(4) - 5(2)} = \frac{4}{2} = 2$$

$$z = \frac{\begin{vmatrix} 3 & 1 \\ 2 & 0 \end{vmatrix}}{\begin{vmatrix} 3 & 5 \\ 2 & 4 \end{vmatrix}} = \frac{3(0) - 1(2)}{3(4) - 5(2)} = \frac{-2}{2} = -1$$

Likewise, solving equations (2) and (4) simultaneously produces values for y and w.

$$y = \frac{\begin{vmatrix} 0 & 5 \\ 1 & 4 \end{vmatrix}}{\begin{vmatrix} 3 & 5 \\ 2 & 4 \end{vmatrix}} = \frac{0(4) - 5(1)}{3(4) - 5(2)} = \frac{-5}{2} = -\frac{5}{2}$$

$$w = \frac{\begin{vmatrix} 3 & 0 \\ 2 & 1 \end{vmatrix}}{\begin{vmatrix} 3 & 5 \\ 2 & 4 \end{vmatrix}} = \frac{3(1) - 0(2)}{3(4) - 5(2)} = \frac{3}{2}$$

Therefore

$$A^{-1} = \begin{bmatrix} x & y \\ z & w \end{bmatrix} = \begin{bmatrix} 2 & -\dfrac{5}{2} \\ -1 & \dfrac{3}{2} \end{bmatrix}$$

To check this, we perform the following multiplication:

$$\begin{bmatrix} 3 & 5 \\ 2 & 4 \end{bmatrix} \begin{bmatrix} 2 & -\dfrac{5}{2} \\ -1 & \dfrac{3}{2} \end{bmatrix} = \begin{bmatrix} 2 & -\dfrac{5}{2} \\ -1 & \dfrac{3}{2} \end{bmatrix} \begin{bmatrix} 3 & 5 \\ 2 & 4 \end{bmatrix} = \begin{bmatrix} 1 & 0 \\ 0 & 1 \end{bmatrix}$$

∎

Now let's use the approach in Example 1 on the general matrix

$$A = \begin{bmatrix} a_{11} & a_{12} \\ a_{21} & a_{22} \end{bmatrix}$$

We want to find

$$A^{-1} = \begin{bmatrix} x & y \\ z & w \end{bmatrix}$$

such that $AA^{-1} = I$. Therefore we need to solve the matrix equation

$$\begin{bmatrix} a_{11} & a_{12} \\ a_{21} & a_{22} \end{bmatrix} \begin{bmatrix} x & y \\ z & w \end{bmatrix} = \begin{bmatrix} 1 & 0 \\ 0 & 1 \end{bmatrix}$$

for x, y, z, and w. Once again, we multiply the two matrices on the left side of the equation and set the elements of this product matrix equal to the corresponding elements of the identity matrix. We then obtain the following system of equations:

$$\begin{pmatrix} a_{11}x + a_{12}z = 1 \\ a_{11}y + a_{12}w = 0 \\ a_{21}x + a_{22}z = 0 \\ a_{21}y + a_{22}w = 1 \end{pmatrix}$$

Solving this system produces

$$x = \frac{a_{22}}{a_{11}a_{22} - a_{12}a_{21}} \qquad y = \frac{-a_{12}}{a_{11}a_{22} - a_{12}a_{21}}$$

$$z = \frac{-a_{21}}{a_{11}a_{22} - a_{12}a_{21}} \qquad w = \frac{a_{11}}{a_{11}a_{22} - a_{12}a_{21}}$$

Note that the number in each denominator, $a_{11}a_{22} - a_{12}a_{21}$, is the determinant of the matrix A. Thus, if $|A| \neq 0$, then

$$A^{-1} = \frac{1}{|A|} \begin{bmatrix} a_{22} & -a_{12} \\ -a_{21} & a_{11} \end{bmatrix}$$

Matrix multiplication will show that $AA^{-1} = A^{-1}A = I$. If $|A| = 0$, then the matrix A has *no* multiplicative inverse.

EXAMPLE 2

Find A^{-1} if $A = \begin{bmatrix} 3 & 5 \\ -2 & -4 \end{bmatrix}$.

Solution

First let's find $|A|$.

$$|A| = (3)(-4) - (5)(-2) = -2$$

Therefore

$$A^{-1} = \frac{1}{-2} \begin{bmatrix} -4 & -5 \\ 2 & 3 \end{bmatrix} = -\frac{1}{2} \begin{bmatrix} -4 & -5 \\ 2 & 3 \end{bmatrix} = \begin{bmatrix} 2 & \frac{5}{2} \\ -1 & -\frac{3}{2} \end{bmatrix}$$

It is easy to check that $AA^{-1} = A^{-1}A = I$. ∎

EXAMPLE 3

Find A^{-1} if $A = \begin{bmatrix} 8 & -2 \\ -12 & 3 \end{bmatrix}$.

Solution

$$|A| = (8)(3) - (-2)(-12) = 0$$

Therefore A has no multiplicative inverse. ∎

■ More About Multiplication of Matrices

Thus far we have found the products of only 2 × 2 matrices. The row-by-column multiplication pattern can be applied to many different kinds of matrices, which we shall see in the next section. For now, let's find the product of a 2 × 2 matrix and a 2 × 1 matrix, with the 2 × 2 matrix on the left, as follows:

$$\begin{bmatrix} a_{11} & a_{12} \\ a_{21} & a_{22} \end{bmatrix} \begin{bmatrix} b_{11} \\ b_{21} \end{bmatrix} = \begin{bmatrix} a_{11}b_{11} + a_{12}b_{21} \\ a_{21}b_{11} + a_{22}b_{21} \end{bmatrix}$$

Note that the product matrix is a 2×1 matrix. The following example illustrates this pattern:

$$\begin{bmatrix} -2 & 3 \\ 1 & -4 \end{bmatrix} \begin{bmatrix} 5 \\ 7 \end{bmatrix} = \begin{bmatrix} (-2)(5) + (3)(7) \\ (1)(5) + (-4)(7) \end{bmatrix} = \begin{bmatrix} 11 \\ -23 \end{bmatrix}$$

■ Back to Solving Systems of Equations

The linear system of equations

$$\begin{pmatrix} a_{11}x + a_{12}y = d_1 \\ a_{21}x + a_{22}y = d_2 \end{pmatrix}$$

can be represented by the matrix equation

$$\begin{bmatrix} a_{11} & a_{12} \\ a_{21} & a_{22} \end{bmatrix} \begin{bmatrix} x \\ y \end{bmatrix} = \begin{bmatrix} d_1 \\ d_2 \end{bmatrix}$$

If we let

$$A = \begin{bmatrix} a_{11} & a_{12} \\ a_{21} & a_{22} \end{bmatrix} \qquad X = \begin{bmatrix} x \\ y \end{bmatrix} \qquad \text{and} \qquad B = \begin{bmatrix} d_1 \\ d_2 \end{bmatrix}$$

then the previous matrix equation can be written $AX = B$.

If A^{-1} exists, then we can multiply both sides of $AX = B$ by A^{-1} (on the left) and simplify as follows:

$$AX = B$$
$$A^{-1}(AX) = A^{-1}(B)$$
$$(A^{-1}A)X = A^{-1}B$$
$$IX = A^{-1}B$$
$$X = A^{-1}B$$

Therefore the product $A^{-1}B$ is the solution of the system.

EXAMPLE 4   Solve the system $\begin{pmatrix} 5x + 4y = 10 \\ 6x + 5y = 13 \end{pmatrix}$.

Solution

If we let

$$A = \begin{bmatrix} 5 & 4 \\ 6 & 5 \end{bmatrix} \qquad X = \begin{bmatrix} x \\ y \end{bmatrix} \qquad \text{and} \qquad B = \begin{bmatrix} 10 \\ 13 \end{bmatrix}$$

then the given system can be represented by the matrix equation $AX = B$. From our previous discussion, we know that the solution of this equation is $X = A^{-1}B$, so we need to find A^{-1} and the product $A^{-1}B$.

$$A^{-1} = \frac{1}{|A|} \begin{bmatrix} 5 & -4 \\ -6 & 5 \end{bmatrix} = \frac{1}{1} \begin{bmatrix} 5 & -4 \\ -6 & 5 \end{bmatrix} = \begin{bmatrix} 5 & -4 \\ -6 & 5 \end{bmatrix}$$

Therefore

$$A^{-1}B = \begin{bmatrix} 5 & -4 \\ -6 & 5 \end{bmatrix}\begin{bmatrix} 10 \\ 13 \end{bmatrix} = \begin{bmatrix} -2 \\ 5 \end{bmatrix}$$

The solution set of the given system is $\{(-2, 5)\}$.

■

E X A M P L E   5

Solve the system $\begin{pmatrix} 3x - 2y = 9 \\ 4x + 7y = -17 \end{pmatrix}$.

Solution

If we let

$$A = \begin{bmatrix} 3 & -2 \\ 4 & 7 \end{bmatrix} \qquad X = \begin{bmatrix} x \\ y \end{bmatrix} \quad \text{and} \quad B = \begin{bmatrix} 9 \\ -17 \end{bmatrix}$$

then the system is represented by $AX = B$, where $X = A^{-1}B$ and

$$A^{-1} = \frac{1}{|A|}\begin{bmatrix} 7 & 2 \\ -4 & 3 \end{bmatrix} = \frac{1}{29}\begin{bmatrix} 7 & 2 \\ -4 & 3 \end{bmatrix} = \begin{bmatrix} \dfrac{7}{29} & \dfrac{2}{29} \\ -\dfrac{4}{29} & \dfrac{3}{29} \end{bmatrix}$$

Therefore

$$A^{-1}B = \begin{bmatrix} \dfrac{7}{29} & \dfrac{2}{29} \\ -\dfrac{4}{29} & \dfrac{3}{29} \end{bmatrix}\begin{bmatrix} 9 \\ -17 \end{bmatrix} = \begin{bmatrix} 1 \\ -3 \end{bmatrix}$$

The solution set of the given system is $\{(1, -3)\}$.

■

This technique of using matrix inverses to solve systems of linear equations is especially useful when there are many systems to be solved that have the same coefficients but different constant terms.

Problem Set 12.2

For Problems 1–18, find the multiplicative inverse (if one exists) of each matrix.

1. $\begin{bmatrix} 5 & 7 \\ 2 & 3 \end{bmatrix}$

2. $\begin{bmatrix} 3 & 4 \\ 2 & 3 \end{bmatrix}$

3. $\begin{bmatrix} 3 & 8 \\ 2 & 5 \end{bmatrix}$

4. $\begin{bmatrix} 2 & 9 \\ 3 & 13 \end{bmatrix}$

5. $\begin{bmatrix} -1 & 2 \\ 3 & 4 \end{bmatrix}$

6. $\begin{bmatrix} 1 & -2 \\ 4 & -3 \end{bmatrix}$

7. $\begin{bmatrix} -2 & -3 \\ 4 & 6 \end{bmatrix}$

8. $\begin{bmatrix} 5 & -1 \\ 3 & 4 \end{bmatrix}$

9. $\begin{bmatrix} -3 & 2 \\ -4 & 5 \end{bmatrix}$

10. $\begin{bmatrix} 3 & -4 \\ 6 & -8 \end{bmatrix}$

11. $\begin{bmatrix} 0 & 1 \\ 5 & 3 \end{bmatrix}$

12. $\begin{bmatrix} -2 & 0 \\ -3 & 5 \end{bmatrix}$

13. $\begin{bmatrix} -2 & -3 \\ -1 & -4 \end{bmatrix}$

14. $\begin{bmatrix} -2 & -5 \\ -3 & -6 \end{bmatrix}$

15. $\begin{bmatrix} -2 & 5 \\ -3 & 6 \end{bmatrix}$

16. $\begin{bmatrix} -3 & 4 \\ 1 & -2 \end{bmatrix}$

17. $\begin{bmatrix} 1 & 1 \\ 1 & -1 \end{bmatrix}$

18. $\begin{bmatrix} 1 & -1 \\ 1 & 1 \end{bmatrix}$

For Problems 19–26, compute AB.

19. $A = \begin{bmatrix} 4 & 3 \\ 2 & 5 \end{bmatrix}$, $B = \begin{bmatrix} 3 \\ 6 \end{bmatrix}$

20. $A = \begin{bmatrix} 5 & -2 \\ 3 & 1 \end{bmatrix}$, $B = \begin{bmatrix} 5 \\ 8 \end{bmatrix}$

21. $A = \begin{bmatrix} -3 & -4 \\ 2 & 1 \end{bmatrix}$, $B = \begin{bmatrix} 4 \\ -3 \end{bmatrix}$

22. $A = \begin{bmatrix} 5 & 2 \\ -1 & -3 \end{bmatrix}$, $B = \begin{bmatrix} 3 \\ -5 \end{bmatrix}$

23. $A = \begin{bmatrix} -4 & 2 \\ 7 & -5 \end{bmatrix}$, $B = \begin{bmatrix} -1 \\ -4 \end{bmatrix}$

24. $A = \begin{bmatrix} 0 & -3 \\ 2 & 9 \end{bmatrix}$, $B = \begin{bmatrix} -3 \\ -6 \end{bmatrix}$

25. $A = \begin{bmatrix} -2 & -3 \\ -5 & -6 \end{bmatrix}$, $B = \begin{bmatrix} 5 \\ -2 \end{bmatrix}$

26. $A = \begin{bmatrix} -3 & -5 \\ 4 & -7 \end{bmatrix}$, $B = \begin{bmatrix} -3 \\ -10 \end{bmatrix}$

For Problems 27–40, use the method of matrix inverses to solve each system.

27. $\begin{pmatrix} 2x + 3y = 13 \\ x + 2y = 8 \end{pmatrix}$

28. $\begin{pmatrix} 3x + 2y = 10 \\ 7x + 5y = 23 \end{pmatrix}$

29. $\begin{pmatrix} 4x - 3y = -23 \\ -3x + 2y = 16 \end{pmatrix}$

30. $\begin{pmatrix} 6x - y = -14 \\ 3x + 2y = -17 \end{pmatrix}$

31. $\begin{pmatrix} x - 7y = 7 \\ 6x + 5y = -5 \end{pmatrix}$

32. $\begin{pmatrix} x + 9y = -5 \\ 4x - 7y = -20 \end{pmatrix}$

33. $\begin{pmatrix} 3x - 5y = 2 \\ 4x - 3y = -1 \end{pmatrix}$

34. $\begin{pmatrix} 5x - 2y = 6 \\ 7x - 3y = 8 \end{pmatrix}$

35. $\begin{pmatrix} y = 19 - 3x \\ 9x - 5y = 1 \end{pmatrix}$

36. $\begin{pmatrix} 4x + 3y = 31 \\ x = 5y + 2 \end{pmatrix}$

37. $\begin{pmatrix} 3x + 2y = 0 \\ 30x - 18y = -19 \end{pmatrix}$

38. $\begin{pmatrix} 12x + 30y = 23 \\ 12x - 24y = -13 \end{pmatrix}$

39. $\begin{pmatrix} \dfrac{1}{3}x + \dfrac{3}{4}y = 12 \\ \dfrac{2}{3}x + \dfrac{1}{5}y = -2 \end{pmatrix}$

40. $\begin{pmatrix} \dfrac{3}{2}x + \dfrac{1}{6}y = 11 \\ \dfrac{2}{3}x - \dfrac{1}{4}y = 1 \end{pmatrix}$

■ ■ ■ THOUGHTS INTO WORDS

41. Describe how to solve the system $\begin{pmatrix} x - 2y = -10 \\ 3x + 5y = 14 \end{pmatrix}$ using each of the following techniques.
   (a) substitution method
   (b) elimination-by-addition method
   (c) reduced echelon form of the augmented matrix
   (d) determinants
   (e) the method of matrix inverses

GRAPHING CALCULATOR ACTIVITIES

42. Use your calculator to find the multiplicative inverse (if one exists) of each of the following matrices. Be sure to check your answers by showing that $A^{-1}A = I$.

(a) $\begin{bmatrix} 7 & 6 \\ 8 & 7 \end{bmatrix}$

(b) $\begin{bmatrix} -12 & 5 \\ -19 & 8 \end{bmatrix}$

(c) $\begin{bmatrix} -7 & 9 \\ 6 & -8 \end{bmatrix}$

(d) $\begin{bmatrix} -6 & -11 \\ -4 & -8 \end{bmatrix}$

(e) $\begin{bmatrix} 13 & 12 \\ 4 & 4 \end{bmatrix}$

(f) $\begin{bmatrix} 15 & -8 \\ -9 & 5 \end{bmatrix}$

(g) $\begin{bmatrix} 9 & 36 \\ 3 & 12 \end{bmatrix}$

(h) $\begin{bmatrix} 1.2 & 1.5 \\ 7.6 & 4.5 \end{bmatrix}$

43. Use your calculator to find the multiplicative inverse of
$\begin{bmatrix} 1 & 2 \\ 2 & 5 \\ 3 & 1 \\ 4 & 4 \end{bmatrix}$ What difficulty did you encounter?

44. Use your calculator and the method of matrix inverses to solve each of the following systems. Be sure to check your solutions.

(a) $\begin{pmatrix} 5x + 7y = 82 \\ 7x + 10y = 116 \end{pmatrix}$ **(b)** $\begin{pmatrix} 9x - 8y = -150 \\ -10x + 9y = 168 \end{pmatrix}$

(c) $\begin{pmatrix} 15x - 8y = -15 \\ -9x + 5y = 12 \end{pmatrix}$ **(d)** $\begin{pmatrix} 1.2x + 1.5y = 5.85 \\ 7.6x + 4.5y = 19.55 \end{pmatrix}$

(e) $\begin{pmatrix} 12x - 7y = -34.5 \\ 8x + 9y = 79.5 \end{pmatrix}$ **(f)** $\begin{pmatrix} \dfrac{3x}{2} + \dfrac{y}{6} = 11 \\ \dfrac{2x}{3} - \dfrac{y}{4} = 1 \end{pmatrix}$

(g) $\begin{pmatrix} 114x + 129y = 2832 \\ 127x + 214y = 4139 \end{pmatrix}$

(h) $\begin{pmatrix} \dfrac{x}{2} + \dfrac{2y}{5} = 14 \\ \dfrac{3x}{4} + \dfrac{y}{4} = 14 \end{pmatrix}$

12.3 $m \times n$ Matrices

Now let's see how much of the algebra of 2×2 matrices extends to $m \times n$ matrices — that is, to matrices of any dimension. In Section 11.4 we represented a general $m \times n$ matrix by

$$A = \begin{bmatrix} a_{11} & a_{12} & a_{13} & \cdots & a_{1n} \\ a_{21} & a_{22} & a_{23} & \cdots & a_{2n} \\ \cdot & \cdot & \cdot & & \cdot \\ \cdot & \cdot & \cdot & & \cdot \\ \cdot & \cdot & \cdot & & \cdot \\ a_{m1} & a_{m2} & a_{m3} & \cdots & a_{mn} \end{bmatrix}$$

We denote the element at the intersection of row i and column j by a_{ij}. It is also customary to denote a matrix A with the abbreviated notation (a_{ij}).

Addition of matrices can be extended to matrices of any dimension by the following definition:

Definition 12.5

Let $A = (a_{ij})$ and $B = (b_{ij})$ be two matrices of the *same dimension*. Then
$$A + B = (a_{ij}) + (b_{ij}) = (a_{ij} + b_{ij})$$

Definition 12.5 states that to add two matrices, we add the elements that appear in corresponding positions in the matrices. For this to work, the matrices must be of the same dimension. An example of the sum of two 3×2 matrices is

$$\begin{bmatrix} 3 & 2 \\ 4 & -1 \\ -3 & 8 \end{bmatrix} + \begin{bmatrix} -2 & 1 \\ -3 & -7 \\ 5 & 9 \end{bmatrix} = \begin{bmatrix} 1 & 3 \\ 1 & -8 \\ 2 & 17 \end{bmatrix}$$

The **commutative** and **associative properties** hold for any matrices that can be added. The $m \times n$ **zero matrix**, denoted by O, is the matrix that contains all zeros. It is the **identity element for addition**. For example,

$$\begin{bmatrix} 2 & 3 & -1 & -5 \\ -7 & 6 & 2 & 8 \end{bmatrix} + \begin{bmatrix} 0 & 0 & 0 & 0 \\ 0 & 0 & 0 & 0 \end{bmatrix} = \begin{bmatrix} 2 & 3 & -1 & -5 \\ -7 & 6 & 2 & 8 \end{bmatrix}$$

Every matrix A has an **additive inverse**, $-A$, that can be found by changing the sign of each element of A. For example, if

$$A = [2 \quad -3 \quad 0 \quad 4 \quad -7]$$

then

$$-A = [-2 \quad 3 \quad 0 \quad -4 \quad 7]$$

Furthermore, $A + (-A) = O$ for all matrices.

The definition we gave earlier for subtraction, $A - B = A + (-B)$, can be extended to any two matrices of the same dimension. For example,

$$[-4 \quad 3 \quad -5] - [7 \quad -4 \quad -1] = [-4 \quad 3 \quad -5] + [-7 \quad 4 \quad 1]$$
$$= [-11 \quad 7 \quad -4]$$

The **scalar product** of any real number k and any $m \times n$ matrix $A = (a_{ij})$ is defined by

$$kA = (ka_{ij})$$

In other words, to find kA, we simply multiply each element of A by k. For example,

$$(-4)\begin{bmatrix} 1 & -1 \\ -2 & 3 \\ 4 & 5 \\ 0 & -8 \end{bmatrix} = \begin{bmatrix} -4 & 4 \\ 8 & -12 \\ -16 & -20 \\ 0 & 32 \end{bmatrix}$$

The properties $k(A + B) = kA + kB$, $(k + l)A = kA + lA$, and $(kl)A = k(lA)$ hold for all matrices. The matrices A and B must be of the same dimension to be added.

The row-by-column definition for multiplying two matrices can be extended, but we must take care. In order for us to define the product AB of two matrices A and B, **the number of columns of A must equal the number of rows of B.** Suppose $A = (a_{ij})$ is $m \times n$, and $B = (b_{ij})$ is $n \times p$. Then

$$AB = \begin{bmatrix} a_{11} & a_{12} & \cdots & a_{1n} \\ & & & \\ & & & \\ a_{i1} & a_{i2} & \cdots & a_{in} \\ & & & \\ & & & \\ a_{m1} & a_{m2} & \cdots & a_{mn} \end{bmatrix} \begin{bmatrix} b_{11} & \cdots & b_{1j} & \cdots & b_{1p} \\ b_{21} & \cdots & b_{2j} & \cdots & b_{2p} \\ & & & & \\ & & & & \\ b_{n1} & \cdots & b_{nj} & \cdots & b_{np} \end{bmatrix} = C$$

The product matrix C is of the dimension $m \times p$, and the general element, c_{ij}, is determined as follows.

$$c_{ij} = a_{i1}b_{1j} + a_{i2}b_{2j} + \cdots + a_{in}b_{nj}$$

A specific element of the product matrix, such as c_{23}, is the result of multiplying the elements in row 2 of matrix A by the elements in column 3 of matrix B and adding the results. Therefore

$$c_{23} = a_{21}b_{13} + a_{22}b_{23} + \cdots + a_{2n}b_{n3}$$

The following example illustrates the product of a 2×3 matrix and a 3×2 matrix:

A
$m \times n$

B
$n \times p$

Number of columns
of A must equal the
number of rows of B.

Dimension of product is $m \times p$.

A B C

$$\begin{bmatrix} 2 & -3 & 1 \\ -4 & 0 & 5 \end{bmatrix} \begin{bmatrix} -1 & -5 \\ 4 & -2 \\ 6 & 1 \end{bmatrix} = \begin{bmatrix} -8 & -3 \\ 34 & 25 \end{bmatrix}$$

$$c_{11} = (2)(-1) + (-3)(4) + (1)(6) = -8$$

$$c_{12} = (2)(-5) + (-3)(-2) + (1)(1) = -3$$

$$c_{21} = (-4)(-1) + (0)(4) + (5)(6) = 34$$

$$c_{22} = (-4)(-5) + (0)(-2) + (5)(1) = 25$$

Recall that matrix multiplication is *not* commutative. In fact, it may be that AB is defined and BA is not defined. For example, if A is a 2×3 matrix and B is a 3×4 matrix, then the product AB is a 2×4 matrix, but the product BA is not defined because the number of columns of B does not equal the number of rows of A.

The **associative property for multiplication** and the two **distributive properties** hold if the matrices have the proper number of rows and columns for the operations to be defined. In that case, we have $(AB)C = A(BC)$, $A(B + C) = AB + AC$, and $(A + B)C = AC + BC$.

■ Square Matrices

Now let's extend some of the algebra of 2×2 matrices to all square matrices (where the number of rows equals the number of columns). For example, the general **multiplicative identity element** for square matrices contains 1s in the main diagonal from the upper left-hand corner to the lower right-hand corner and 0s elsewhere. Therefore, for 3×3 and 4×4 matrices, the multiplicative identity elements are as follows:

$$I_3 = \begin{bmatrix} 1 & 0 & 0 \\ 0 & 1 & 0 \\ 0 & 0 & 1 \end{bmatrix} \qquad I_4 = \begin{bmatrix} 1 & 0 & 0 & 0 \\ 0 & 1 & 0 & 0 \\ 0 & 0 & 1 & 0 \\ 0 & 0 & 0 & 1 \end{bmatrix}$$

We saw in Section 12.2 that some, but not all, 2×2 matrices have multiplicative inverses. In general, some, but not all, square matrices of a particular dimension have multiplicative inverses. If an $n \times n$ square matrix A does have a multiplicative inverse A^{-1}, then

$$AA^{-1} = A^{-1}A = I_n$$

The technique used in Section 12.2 for finding multiplicative inverses of 2×2 matrices does generalize, but it becomes quite complicated. Therefore, we shall now describe another technique that works for all square matrices. Given an $n \times n$ matrix A, we begin by forming the $n \times 2n$ matrix

$$\begin{bmatrix} a_{11} & a_{12} & \cdots & a_{1n} & 1 & 0 & 0 & \cdots & 0 \\ a_{21} & a_{22} & \cdots & a_{2n} & 0 & 1 & 0 & \cdots & 0 \\ \cdot & \cdot & & \cdot & \cdot & \cdot & & & \cdot \\ \cdot & \cdot & & \cdot & \cdot & \cdot & \cdot & & \cdot \\ \cdot & \cdot & & \cdot & \cdot & \cdot & \cdot & & \cdot \\ a_{n1} & a_{n2} & \cdots & a_{nn} & 0 & 0 & 0 & \cdots & 1 \end{bmatrix}$$

where the identity matrix I_n appears to the right of A. Now we apply a succession of elementary row transformations to this double matrix until we obtain a matrix of the form

$$\begin{bmatrix} 1 & 0 & 0 & \cdots & 0 & b_{11} & b_{12} & \cdots & b_{1n} \\ 0 & 1 & 0 & \cdots & 0 & b_{21} & b_{22} & \cdots & b_{2n} \\ \cdot & \cdot & \cdot & & \cdot & \cdot & \cdot & & \cdot \\ \cdot & \cdot & \cdot & & \cdot & \cdot & \cdot & & \cdot \\ \cdot & \cdot & \cdot & & \cdot & \cdot & \cdot & & \cdot \\ 0 & 0 & 0 & \cdots & 1 & b_{n1} & b_{n2} & \cdots & b_{nn} \end{bmatrix}$$

The B matrix in this matrix is the desired inverse A^{-1}. If A does not have an inverse, then it is impossible to change the original matrix to this final form.

E X A M P L E   1

Find A^{-1} if $A = \begin{bmatrix} 2 & 4 \\ 3 & 5 \end{bmatrix}$.

Solution

First form the matrix

$$\left[\begin{array}{cc:cc} 2 & 4 & 1 & 0 \\ 3 & 5 & 0 & 1 \end{array}\right]$$

Now multiply row 1 by $\frac{1}{2}$.

$$\left[\begin{array}{cc:cc} 1 & 2 & \frac{1}{2} & 0 \\ 3 & 5 & 0 & 1 \end{array}\right]$$

Next, add -3 times row 1 to row 2 to form a new row 2.

$$\begin{bmatrix} 1 & 2 & \vdots & \dfrac{1}{2} & 0 \\[2ex] 0 & -1 & \vdots & -\dfrac{3}{2} & 1 \end{bmatrix}$$

Then multiply row 2 by -1.

$$\begin{bmatrix} 1 & 2 & \vdots & \dfrac{1}{2} & 0 \\[2ex] 0 & 1 & \vdots & \dfrac{3}{2} & -1 \end{bmatrix}$$

Finally, add -2 times row 2 to row 1 to form a new row 1.

$$\begin{bmatrix} 1 & 0 & \vdots & -\dfrac{5}{2} & 2 \\[2ex] 0 & 1 & \vdots & \dfrac{3}{2} & -1 \end{bmatrix}$$

The matrix inside the box is A^{-1}; that is,

$$A^{-1} = \begin{bmatrix} -\dfrac{5}{2} & 2 \\[2ex] \dfrac{3}{2} & -1 \end{bmatrix}$$

This can be checked, as always, by showing that $AA^{-1} = A^{-1}A = I_2$. ■

EXAMPLE 2   Find A^{-1} if $A = \begin{bmatrix} 1 & 1 & 2 \\ 2 & 3 & -1 \\ -3 & 1 & -2 \end{bmatrix}$.

Solution

Form the matrix $\begin{bmatrix} 1 & 1 & 2 & \vdots & 1 & 0 & 0 \\ 2 & 3 & -1 & \vdots & 0 & 1 & 0 \\ -3 & 1 & -2 & \vdots & 0 & 0 & 1 \end{bmatrix}$.

Add -2 times row 1 to row 2, and add 3 times row 1 to row 3.

$$\begin{bmatrix} 1 & 1 & 2 & \vdots & 1 & 0 & 0 \\ 0 & 1 & -5 & \vdots & -2 & 1 & 0 \\ 0 & 4 & 4 & \vdots & 3 & 0 & 1 \end{bmatrix}$$

Add -1 times row 2 to row 1, and add -4 times row 2 to row 3.

$$\begin{bmatrix} 1 & 0 & 7 & \vdots & 3 & -1 & 0 \\ 0 & 1 & -5 & \vdots & -2 & 1 & 0 \\ 0 & 0 & 24 & \vdots & 11 & -4 & 1 \end{bmatrix}$$

Multiply row 3 by $\dfrac{1}{24}$.

$$\begin{bmatrix} 1 & 0 & 7 & \vdots & 3 & -1 & 0 \\ 0 & 1 & -5 & \vdots & -2 & 1 & 0 \\ 0 & 0 & 1 & \vdots & \dfrac{11}{24} & -\dfrac{1}{6} & \dfrac{1}{24} \end{bmatrix}$$

Add -7 times row 3 to row 1, and add 5 times row 3 to row 2.

$$\begin{bmatrix} 1 & 0 & 0 & \vdots & -\dfrac{5}{24} & \dfrac{1}{6} & -\dfrac{7}{24} \\ 0 & 1 & 0 & \vdots & \dfrac{7}{24} & \dfrac{1}{6} & \dfrac{5}{24} \\ 0 & 0 & 1 & \vdots & \dfrac{11}{24} & -\dfrac{1}{6} & \dfrac{1}{24} \end{bmatrix}$$

Therefore

$$A^{-1} = \begin{bmatrix} -\dfrac{5}{24} & \dfrac{1}{6} & -\dfrac{7}{24} \\ \dfrac{7}{24} & \dfrac{1}{6} & \dfrac{5}{24} \\ \dfrac{11}{24} & -\dfrac{1}{6} & \dfrac{1}{24} \end{bmatrix}$$ Be sure to check this!

∎

■ Systems of Equations

In Section 12.2 we used the concept of the multiplicative inverse to solve systems of two linear equations in two variables. This same technique can be applied to general systems of *n* linear equations in *n* variables. Let's consider one such example involving three equations in three variables.

E X A M P L E   3   Solve the system

$$\begin{pmatrix} x + y + 2z = -8 \\ 2x + 3y - z = 3 \\ -3x + y - 2z = 4 \end{pmatrix}$$

Solution

If we let

$$A = \begin{bmatrix} 1 & 1 & 2 \\ 2 & 3 & -1 \\ -3 & 1 & -2 \end{bmatrix} \qquad X = \begin{bmatrix} x \\ y \\ z \end{bmatrix} \qquad \text{and} \qquad B = \begin{bmatrix} -8 \\ 3 \\ 4 \end{bmatrix}$$

then the given system can be represented by the matrix equation $AX = B$. Therefore, we know that $X = A^{-1}B$, so we need to find A^{-1} and the product $A^{-1}B$. The matrix A^{-1} was found in Example 2, so let's use that result and find $A^{-1}B$.

$$X = A^{-1}B = \begin{bmatrix} -\dfrac{5}{24} & \dfrac{1}{6} & -\dfrac{7}{24} \\ \dfrac{7}{24} & \dfrac{1}{6} & \dfrac{5}{24} \\ \dfrac{11}{24} & -\dfrac{1}{6} & \dfrac{1}{24} \end{bmatrix} \begin{bmatrix} -8 \\ 3 \\ 4 \end{bmatrix} = \begin{bmatrix} 1 \\ -1 \\ -4 \end{bmatrix}$$

The solution set of the given system is $\{(1, -1, -4)\}$. ∎

Problem Set 12.3

For Problems 1–8, find $A + B$, $A - B$, $2A + 3B$, and $4A - 2B$.

1. $A = \begin{bmatrix} 2 & -1 & 4 \\ -2 & 0 & 5 \end{bmatrix}$, $\quad B = \begin{bmatrix} -1 & 4 & -7 \\ 5 & -6 & 2 \end{bmatrix}$

2. $A = \begin{bmatrix} 3 & -6 \\ 2 & -1 \\ -4 & 5 \end{bmatrix}$, $\quad B = \begin{bmatrix} 1 & 0 \\ 5 & -7 \\ -6 & 9 \end{bmatrix}$

3. $A = [2 \quad -1 \quad 4 \quad 12]$, $\quad B = [-3 \quad -6 \quad 9 \quad -5]$

4. $A = \begin{bmatrix} 3 \\ -9 \\ 7 \end{bmatrix}$, $\quad B = \begin{bmatrix} -6 \\ 12 \\ 9 \end{bmatrix}$

5. $A = \begin{bmatrix} 3 & -2 & 1 \\ -1 & 4 & -7 \\ 0 & 5 & 9 \end{bmatrix}$, $\quad B = \begin{bmatrix} 5 & -1 & -3 \\ 10 & -2 & 4 \\ 7 & 0 & 12 \end{bmatrix}$

6. $A = \begin{bmatrix} 7 & -4 \\ -5 & 9 \\ -1 & 2 \end{bmatrix}$, $\quad B = \begin{bmatrix} 12 & 3 \\ -2 & -4 \\ -6 & 7 \end{bmatrix}$

7. $A = \begin{bmatrix} -1 & 0 \\ 2 & 3 \\ -5 & -4 \\ -7 & 11 \end{bmatrix}$, $\quad B = \begin{bmatrix} 1 & 2 \\ -3 & 7 \\ 6 & -5 \\ 9 & -2 \end{bmatrix}$

8. $A = \begin{bmatrix} 0 & -1 & -2 \\ 3 & -4 & 6 \\ 5 & 4 & -9 \end{bmatrix}$, $\quad B = \begin{bmatrix} 2 & 1 & -7 \\ -6 & 4 & 5 \\ 3 & -2 & -1 \end{bmatrix}$

For Problems 9–20, find AB and BA, whenever they exist.

9. $A = \begin{bmatrix} 2 & -1 \\ 0 & -4 \\ -5 & 3 \end{bmatrix}$, $\quad B = \begin{bmatrix} 5 & -2 & 6 \\ -1 & 4 & -2 \end{bmatrix}$

10. $A = \begin{bmatrix} -2 & 3 & -1 \\ 7 & -4 & 5 \end{bmatrix}$, $\quad B = \begin{bmatrix} 1 & -1 \\ -2 & 3 \\ -5 & -6 \end{bmatrix}$

11. $A = \begin{bmatrix} 2 & -1 & -3 \\ 0 & -4 & 7 \end{bmatrix}$, $\quad B = \begin{bmatrix} 2 & 1 & -1 & 4 \\ 0 & -2 & 3 & 5 \\ -6 & 4 & -2 & 0 \end{bmatrix}$

12. $A = \begin{bmatrix} 3 & -1 & -4 \\ -5 & 2 & 2 \end{bmatrix}$, $\quad B = \begin{bmatrix} 3 & -2 \\ -4 & -1 \end{bmatrix}$

13. $A = \begin{bmatrix} 1 & -1 & 2 \\ 0 & 1 & -2 \\ 3 & 1 & 4 \end{bmatrix}$, $\quad B = \begin{bmatrix} 2 & 3 & -1 \\ 4 & 0 & 2 \\ -5 & 1 & -1 \end{bmatrix}$

14. $A = \begin{bmatrix} 1 & 0 & 1 \\ 0 & 1 & 1 \\ -1 & 2 & 3 \end{bmatrix}$, $\quad B = \begin{bmatrix} -1 & -1 & 1 \\ 0 & 1 & 0 \\ 2 & -3 & 1 \end{bmatrix}$

15. $A = [2 \quad -1 \quad 3 \quad 4]$, $\quad B = \begin{bmatrix} -1 \\ -3 \\ 2 \\ -4 \end{bmatrix}$

16. $A = \begin{bmatrix} -2 \\ 3 \\ -5 \end{bmatrix}$, $\quad B = [3 \quad -4 \quad -5]$

17. $A = \begin{bmatrix} 2 \\ -7 \end{bmatrix}$, $B = \begin{bmatrix} 3 & -2 \\ 1 & 0 \\ -1 & 4 \end{bmatrix}$

18. $A = \begin{bmatrix} 3 & -2 & 2 & -4 \\ 1 & 0 & -1 & 2 \end{bmatrix}$, $B = \begin{bmatrix} 3 & -2 & 1 \\ -3 & 1 & 4 \\ 5 & 2 & 0 \\ -4 & -1 & -2 \end{bmatrix}$

19. $A = \begin{bmatrix} 3 \\ -4 \\ 2 \end{bmatrix}$, $B = \begin{bmatrix} 3 & -4 \end{bmatrix}$

20. $A = \begin{bmatrix} 3 & -7 \end{bmatrix}$, $B = \begin{bmatrix} 8 \\ -9 \end{bmatrix}$

For Problems 21–36, use the technique discussed in this section to find the multiplicative inverse (if one exists) of each matrix.

21. $\begin{bmatrix} 1 & 3 \\ 4 & 2 \end{bmatrix}$ **22.** $\begin{bmatrix} 1 & 2 \\ 2 & -3 \end{bmatrix}$

23. $\begin{bmatrix} 2 & 1 \\ 7 & 4 \end{bmatrix}$ **24.** $\begin{bmatrix} 3 & 7 \\ 2 & 5 \end{bmatrix}$

25. $\begin{bmatrix} -2 & 1 \\ 3 & -4 \end{bmatrix}$ **26.** $\begin{bmatrix} -3 & 1 \\ 3 & -2 \end{bmatrix}$

27. $\begin{bmatrix} 1 & 2 & 3 \\ 1 & 3 & 4 \\ 1 & 4 & 3 \end{bmatrix}$ **28.** $\begin{bmatrix} 1 & 3 & -2 \\ 1 & 4 & -1 \\ -2 & -7 & 5 \end{bmatrix}$

29. $\begin{bmatrix} 1 & -2 & 1 \\ -2 & 5 & 3 \\ 3 & -5 & 7 \end{bmatrix}$ **30.** $\begin{bmatrix} 1 & 4 & -2 \\ -3 & -11 & 1 \\ 2 & 7 & 3 \end{bmatrix}$

31. $\begin{bmatrix} 2 & 3 & -4 \\ 3 & -1 & -2 \\ 1 & -4 & 2 \end{bmatrix}$ **32.** $\begin{bmatrix} -2 & 2 & 3 \\ 1 & -1 & 0 \\ 0 & 1 & 4 \end{bmatrix}$

33. $\begin{bmatrix} 1 & 2 & 3 \\ -3 & -4 & 3 \\ 2 & 4 & -1 \end{bmatrix}$ **34.** $\begin{bmatrix} 1 & -2 & 3 \\ -1 & 3 & -2 \\ -2 & 6 & 1 \end{bmatrix}$

35. $\begin{bmatrix} 2 & 0 & 0 \\ 0 & 4 & 0 \\ 0 & 0 & 10 \end{bmatrix}$ **36.** $\begin{bmatrix} 1 & -3 & 5 \\ 0 & 1 & 2 \\ 0 & 0 & 1 \end{bmatrix}$

For Problems 37–46, use the method of matrix inverses to solve each system. The required multiplicative inverses were found in Problems 21–36.

37. $\begin{pmatrix} 2x + y = -4 \\ 7x + 4y = -13 \end{pmatrix}$ **38.** $\begin{pmatrix} 3x + 7y = -38 \\ 2x + 5y = -27 \end{pmatrix}$

39. $\begin{pmatrix} -2x + y = 1 \\ 3x - 4y = -14 \end{pmatrix}$ **40.** $\begin{pmatrix} -3x + y = -18 \\ 3x - 2y = 15 \end{pmatrix}$

41. $\begin{pmatrix} x + 2y + 3z = -2 \\ x + 3y + 4z = -3 \\ x + 4y + 3z = -6 \end{pmatrix}$

42. $\begin{pmatrix} x + 3y - 2z = 5 \\ x + 4y - z = 3 \\ -2x - 7y + 5z = -12 \end{pmatrix}$

43. $\begin{pmatrix} x - 2y + z = -3 \\ -2x + 5y + 3z = 34 \\ 3x - 5y + 7z = 14 \end{pmatrix}$

44. $\begin{pmatrix} x + 4y - 2z = 2 \\ -3x - 11y + z = -2 \\ 2x + 7y + 3z = -2 \end{pmatrix}$

45. $\begin{pmatrix} x + 2y + 3z = 2 \\ -3x - 4y + 3z = 0 \\ 2x + 4y - z = 4 \end{pmatrix}$

46. $\begin{pmatrix} x - 2y + 3z = -39 \\ -x + 3y - 2z = 40 \\ -2x + 6y + z = 45 \end{pmatrix}$

47. We can generate five systems of linear equations from the system

$$\begin{pmatrix} x + y + 2z = a \\ 2x + 3y - z = b \\ -3x + y - 2z = c \end{pmatrix}$$

by letting a, b, and c assume five different sets of values. Solve the system for each set of values. The inverse of the coefficient matrix of these systems is given in Example 2 of this section.

(a) $a = 7$, $b = 1$, and $c = -1$
(b) $a = -7$, $b = 5$, and $c = 1$
(c) $a = -9$, $b = -8$, and $c = 19$
(d) $a = -1$, $b = -13$, and $c = -17$
(e) $a = -2$, $b = 0$, and $c = -2$

■ ■ ■ **THOUGHTS INTO WORDS**

48. How would you describe row-by-column multiplication of matrices?

49. Give a step-by-step explanation of how to find the multiplicative inverse of the matrix $\begin{bmatrix} 1 & 3 \\ -2 & 4 \end{bmatrix}$ by using the technique of Section 12.3.

50. Explain how to find the multiplicative inverse of the matrix in Problem 49 by using the technique discussed in Section 12.2.

■ ■ ■ **FURTHER INVESTIGATIONS**

51. Matrices can be used to code and decode messages. For example, suppose that we set up a one-to-one correspondence between the letters of the alphabet and the first 26 counting numbers, as follows:

$$
\begin{array}{cccc}
A & B & C & Z \\
\updownarrow & \updownarrow & \updownarrow & \cdots \quad \updownarrow \\
1 & 2 & 3 & 26
\end{array}
$$

Now suppose that we want to code the message PLAY IT BY EAR. We can partition the letters of the message into groups of two. Because the last group will contain only one letter, let's arbitrarily stick in a Z to form a group of two. Let's also assign a number to each letter on the basis of the letter/number association we exhibited.

$$
\begin{array}{cccccccccccc}
P & L & A & Y & I & T & B & Y & E & A & R & Z \\
\updownarrow & \updownarrow & \updownarrow & \updownarrow & \updownarrow & \updownarrow & \updownarrow & \updownarrow & \updownarrow & \updownarrow & \updownarrow & \updownarrow \\
16 & 12 & 1 & 25 & 9 & 20 & 2 & 25 & 5 & 1 & 18 & 26
\end{array}
$$

Each pair of numbers can be recorded as columns in a 2×6 matrix B.

$$
B = \begin{bmatrix} 16 & 1 & 9 & 2 & 5 & 18 \\ 12 & 25 & 20 & 25 & 1 & 26 \end{bmatrix}
$$

Now let's choose a 2×2 matrix such that the matrix contains only integers and its inverse also contains only integers. For example, we can use $A = \begin{bmatrix} 3 & 1 \\ 5 & 2 \end{bmatrix}$; then $A^{-1} = \begin{bmatrix} 2 & -1 \\ -5 & 3 \end{bmatrix}$.

Next, let's find the product AB.

$$
AB = \begin{bmatrix} 3 & 1 \\ 5 & 2 \end{bmatrix}\begin{bmatrix} 16 & 1 & 9 & 2 & 5 & 18 \\ 12 & 25 & 20 & 25 & 1 & 26 \end{bmatrix}
$$

$$
= \begin{bmatrix} 60 & 28 & 47 & 31 & 16 & 80 \\ 104 & 55 & 85 & 60 & 27 & 142 \end{bmatrix}
$$

Now we have our coded message:

60 104 28 55 47 85 31 60 16 27 80 142

A person decoding the message would put the numbers back into a 2×6 matrix, multiply it on the left by A^{-1}, and convert the numbers back to letters.

Each of the following coded messages was formed by using the matrix $A = \begin{bmatrix} 2 & 3 \\ 1 & 2 \end{bmatrix}$. Decode each of the messages.

(a) 68 40 77 51 78 49 23 15 29 19 85 52 41 27

(b) 62 40 78 47 64 36 19 11 93 57 93 56 88 57

(c) 64 36 58 37 63 36 21 13 75 47 63 36 38 23 118 72

(d) 61 38 115 69 93 57 36 20 78 49 68 40 77 51 60 37 47 26 84 51 21 11

52. Suppose that the ordered pair (x, y) of a rectangular coordinate system is recorded as a 2×1 matrix and then multiplied on the left by the matrix $\begin{bmatrix} 1 & 0 \\ 0 & -1 \end{bmatrix}$. We would obtain

$$
\begin{bmatrix} 1 & 0 \\ 0 & -1 \end{bmatrix}\begin{bmatrix} x \\ y \end{bmatrix} = \begin{bmatrix} x \\ -y \end{bmatrix}
$$

The point $(x, -y)$ is an x axis reflection of the point (x, y). Therefore the matrix $\begin{bmatrix} 1 & 0 \\ 0 & -1 \end{bmatrix}$ performs an *x* axis reflection. What type of geometric transformation is performed by each of the following matrices?

(a) $\begin{bmatrix} -1 & 0 \\ 0 & 1 \end{bmatrix}$

(b) $\begin{bmatrix} -1 & 0 \\ 0 & -1 \end{bmatrix}$

(c) $\begin{bmatrix} 0 & -1 \\ 1 & 0 \end{bmatrix}$

(d) $\begin{bmatrix} 0 & 1 \\ -1 & 0 \end{bmatrix}$

[*Hint:* Check the slopes of lines through the origin.]

GRAPHING CALCULATOR ACTIVITIES

53. Use your calculator to check your answers for Problems 14, 18, 28, 30, 32, 34, 36, 42, 44, 46, and 47.

54. Use your calculator and the method of matrix inverses to solve each of the following systems. Be sure to check your solutions.

(a) $\begin{pmatrix} 2x - 3y + 4z = 54 \\ 3x + y - z = 32 \\ 5x - 4y + 3z = 58 \end{pmatrix}$

(b) $\begin{pmatrix} 17x + 15y - 19z = 10 \\ 18x - 14y + 16z = 94 \\ 13x + 19y - 14z = -23 \end{pmatrix}$

(c) $\begin{pmatrix} 1.98x + 2.49y + 3.15z = 45.72 \\ 2.29x + 1.95y + 2.75z = 42.05 \\ 3.15x + 3.20y + 1.85z = 42 \end{pmatrix}$

(d) $\begin{pmatrix} x_1 + 2x_2 - 4x_3 + 7x_4 = -23 \\ 2x_1 - 3x_2 + 5x_3 - x_4 = -22 \\ 5x_1 + 4x_2 - 2x_3 - 8x_4 = 59 \\ 3x_1 - 7x_2 + 8x_3 + 9x_4 = -103 \end{pmatrix}$

(e) $\begin{pmatrix} 2x_1 - x_2 + 3x_3 - 4x_4 + 12x_5 = 98 \\ x_1 + 2x_2 - x_3 - 7x_4 + 5x_5 = 41 \\ 3x_1 + 4x_2 - 7x_3 + 6x_4 - 9x_5 = -41 \\ 4x_1 - 3x_2 + x_3 - x_4 + x_5 = 4 \\ 7x_1 + 8x_2 - 4x_3 - 6x_4 - 6x_5 = 12 \end{pmatrix}$

12.4   Systems of Linear Inequalities: Linear Programming

Finding solution sets for **systems of linear inequalities** relies heavily on the graphing approach. (Recall that we discussed graphing of linear inequalities in Section 7.3.) The solution set of the system

$$\begin{pmatrix} x + y > 2 \\ x - y < 2 \end{pmatrix}$$

is the intersection of the solution sets of the individual inequalities. In Figure 12.1(a), we indicate the solution set for $x + y > 2$, and in Figure 12.1(b), we indicate the solution set for $x - y < 2$. The shaded region in Figure 12.1(c) represents the intersection of the two solution sets; therefore it is the graph of the system. Remember that dashed lines are used to indicate that the points on the lines are not included in the solution set. In the following examples, we indicate only the final solution set for the system.

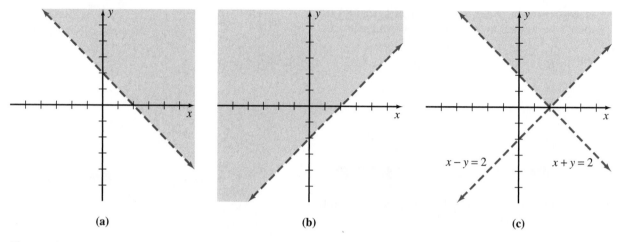

(a) (b) (c)

Figure 12.1

E X A M P L E   1

Solve the following system by graphing.

$$\begin{pmatrix} 2x - y \geq 4 \\ x + 2y < 2 \end{pmatrix}$$

Solution

The graph of $2x - y \geq 4$ consists of all points *on or below* the line $2x - y = 4$. The graph of $x + 2y < 2$ consists of all points *below* the line $x + 2y = 2$. The graph of the system is indicated by the shaded region in Figure 12.2. Note that all points in the shaded region are on or below the line $2x - y = 4$ and below the line $x + 2y = 2$.

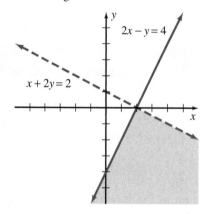

Figure 12.2 ■

E X A M P L E   2

Solve the following system by graphing:

$$\begin{pmatrix} x \leq 2 \\ y \geq -1 \end{pmatrix}$$

Solution

Remember that even though each inequality contains only one variable, we are working in a rectangular coordinate system involving ordered pairs. That is, the system could also be written

$$\begin{pmatrix} x + 0(y) \le 2 \\ 0(x) + y \ge -1 \end{pmatrix}$$

The graph of this system is the shaded region in Figure 12.3. Note that all points in the shaded region are *on or to the left* of the line $x = 2$ and *on or above* the line $y = -1$.

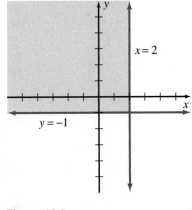

Figure 12.3 ■

A system may contain more than two inequalities, as the next example illustrates.

EXAMPLE 3

Solve the following system by graphing:

$$\begin{pmatrix} x \ge 0 \\ y \ge 0 \\ 2x + 3y \le 12 \\ 3x + y \le 6 \end{pmatrix}$$

Solution

The solution set for the system is the intersection of the solution sets of the four inequalities. The shaded region in Figure 12.4 indicates the solution set for the system. Note that all points in the shaded region are *on or to the right* of the y axis, *on or above* the x axis, *on or below* the line $2x + 3y = 12$, and *on or below* the line $3x + y = 6$.

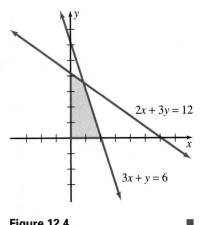

Figure 12.4 ■

■ Linear Programming: Another Look at Problem Solving

Throughout this text problem solving is a unifying theme. Therefore it seems appropriate at this time to give you a brief glimpse of an area of mathematics that was developed in the 1940s specifically as a problem-solving tool. Many applied problems involve the idea of *maximizing* or *minimizing* a certain function that is subject to various constraints; these can be expressed as linear inequalities. **Linear programming** was developed as one method for solving such problems.

Remark: The term *programming* refers to the distribution of limited resources in order to maximize or minimize a certain function, such as cost, profit, distance, and so on. Thus it does not mean the same thing that it means in computer programming. The constraints that govern the distribution of resources determine the linear inequalities and equations; thus the term *linear programming* is used.

Before we introduce a linear programming type of problem, we need to extend one mathematical concept a bit. A **linear function in two variables**, x and y, is a function of the form $f(x, y) = ax + by + c$, where a, b, and c are real numbers. In other words, with each ordered pair (x, y) we associate a third number by the rule $ax + by + c$. For example, suppose the function f is described by $f(x, y) = 4x + 3y + 5$. Then $f(2, 1) = 4(2) + 3(1) + 5 = 16$.

First, let's take a look at some mathematical ideas that form the basis for solving a linear programming problem. Consider the shaded region in Figure 12.5 and the following linear functions in two variables:

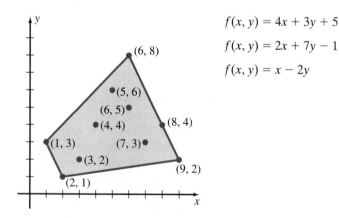

$$f(x, y) = 4x + 3y + 5$$

$$f(x, y) = 2x + 7y - 1$$

$$f(x, y) = x - 2y$$

Figure 12.5

Suppose that we need to find the maximum value and the minimum value achieved by each of the functions in the indicated region. The following chart summarizes the values for the ordered pairs indicated in Figure 12.5. Note that for each function, the maximum and minimum values are obtained at vertices of the region.

	Ordered pairs	Value of $f(x, y) = 4x + 3y = 5$	Value of $f(x, y) = 2x + 7y - 1$	Value of $f(x, y) = x - 2y$
Vertex	$(2, 1)$	16 (*minimum*)	10 (*minimum*)	0
	$(3, 2)$	23	19	-1
Vertex	$(9, 2)$	47	31	5 (*maximum*)
Vertex	$(1, 3)$	18	22	-5
	$(7, 3)$	42	34	1
	$(4, 4)$	33	35	-4
	$(8, 4)$	49	43	0
	$(6, 5)$	44	46	-4
	$(5, 6)$	43	51	-7
Vertex	$(6, 8)$	53 (*maximum*)	67 (*maximum*)	-10 (*minimum*)

We claim that for linear functions, maximum and minimum functional values are *always* obtained at vertices of the region. To substantiate this, let's consider the family of lines $x - 2y = k$, where k is an arbitrary constant. (We are now working only with the function $f(x, y) = x - 2y$.) In slope-intercept form, $x - 2y = k$ becomes $y = \frac{1}{2}x - \frac{1}{2}k$, so we have a family of parallel lines each having a slope of $\frac{1}{2}$. In Figure 12.6, we sketched some of these lines so that each line has at least one point in common with the given region. Note that $x - 2y$ reaches a minimum value of -10 at the vertex $(6, 8)$ and a maximum value of 5 at the vertex $(9, 2)$.

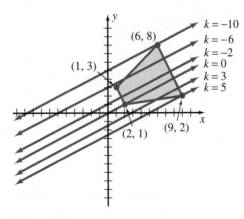

Figure 12.6

In general, suppose that f is a linear function in two variables x and y and that S is a region of the xy plane. If f attains a maximum (minimum) value in S, then that maximum (minimum) value is obtained at a vertex of S.

Remark: A subset of the xy plane is said to be **bounded** if there is a circle that contains all of its points; otherwise, the subset is said to be **unbounded**. A bounded set will contain maximum and minimum values for a function, but an unbounded set may not contain such values.

Now we will consider two examples that illustrate a general graphing approach to solving a linear programming problem in two variables. The first example gives us the general makeup of such a problem; the second example will illustrate the type of setting from which the function and inequalities evolve.

E X A M P L E   4

Find the maximum value and the minimum value of the function $f(x, y) = 9x + 13y$ in the region determined by the following system of inequalities:

$$\begin{pmatrix} x \geq 0 \\ y \geq 0 \\ 2x + 3y \leq 18 \\ 2x + y \leq 10 \end{pmatrix}$$

Solution

First, let's graph the inequalities to determine the region, as indicated in Figure 12.7. (Such a region is called the **set of feasible solutions**, and the inequalities are referred to as **constraints**.) The point $(3, 4)$ is determined by solving the system

$$\begin{pmatrix} 2x + 3y = 18 \\ 2x + y = 10 \end{pmatrix}$$

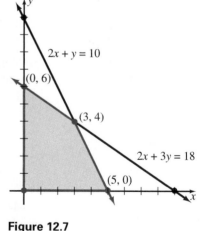

Figure 12.7

Next, we can determine the values of the given function at the vertices of the region. (Such a function to be maximized or minimized is called the **objective function**.)

Vertices	Value of $f(x, y) = 9x + 13y$
$(0, 0)$	0 (*minimum*)
$(5, 0)$	45
$(3, 4)$	79 (*maximum*)
$(0, 6)$	78

A minimum value of 0 is obtained at $(0, 0)$, and a maximum value of 79 is obtained at $(3, 4)$. ∎

PROBLEM 1

A company that manufactures gidgets and gadgets has the following production information available:

1. To produce a gidget requires 3 hours of working time on machine A and 1 hour on machine B.

2. To produce a gadget requires 2 hours on machine A and 1 hour on machine B.

3. Machine A is available for no more than 120 hours per week, and machine B is available for no more than 50 hours per week.

4. Gidgets can be sold at a profit of $3.75 each, and a profit of $3 can be realized on a gadget.

How many gidgets and how many gadgets should the company produce each week to maximize its profit? What would the maximum profit be?

Solution

Let x be the number of gidgets and y be the number of gadgets. Thus the profit function is $P(x, y) = 3.75x + 3y$. The constraints for the problem can be represented by the following inequalities:

$$3x + 2y \leq 120 \qquad \text{Machine A is available for no more than 120 hours.}$$

$$x + y \leq 50 \qquad \text{Machine B is available for no more than 50 hours.}$$

$$x \geq 0 \qquad \text{The number of gidgets and gadgets must be represented by a nonnegative number.}$$

$$y \geq 0$$

When we graph these inequalities, we obtain the set of feasible solutions indicated by the shaded region in Figure 12.8. Next, we find the value of the profit function at the vertices; this produces the chart that follows.

Vertices	Value of $P(x, y) = 3.75x + 3y$
$(0, 0)$	0
$(40, 0)$	150
$(20, 30)$	165 (*maximum*)
$(0, 50)$	150

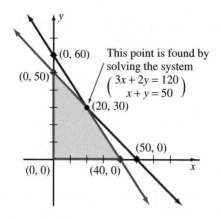

Figure 12.8

Thus a maximum profit of $165 is realized by producing 20 gidgets and 30 gadgets. ∎

Problem Set 12.4

For Problems 1–24, indicate the solution set for each system of inequalities by graphing the system and shading the appropriate region.

1. $\begin{pmatrix} x + y > 3 \\ x - y > 1 \end{pmatrix}$

2. $\begin{pmatrix} x - y < 2 \\ x + y < 1 \end{pmatrix}$

3. $\begin{pmatrix} x - 2y \le 4 \\ x + 2y > 4 \end{pmatrix}$

4. $\begin{pmatrix} 3x - y > 6 \\ 2x + y \le 4 \end{pmatrix}$

5. $\begin{pmatrix} 2x + 3y \le 6 \\ 3x - 2y \le 6 \end{pmatrix}$

6. $\begin{pmatrix} 4x + 3y \ge 12 \\ 3x - 4y \ge 12 \end{pmatrix}$

7. $\begin{pmatrix} 2x - y \ge 4 \\ x + 3y < 3 \end{pmatrix}$

8. $\begin{pmatrix} 3x - y < 3 \\ x + y \ge 1 \end{pmatrix}$

9. $\begin{pmatrix} x + 2y > -2 \\ x - y < -3 \end{pmatrix}$

10. $\begin{pmatrix} x - 3y < -3 \\ 2x - 3y > -6 \end{pmatrix}$

11. $\begin{pmatrix} y > x - 4 \\ y < x \end{pmatrix}$

12. $\begin{pmatrix} y \le x + 2 \\ y \ge x \end{pmatrix}$

13. $\begin{pmatrix} x - y > 2 \\ x - y > -1 \end{pmatrix}$

14. $\begin{pmatrix} x + y > 1 \\ x + y > 3 \end{pmatrix}$

15. $\begin{pmatrix} y \ge x \\ x > -1 \end{pmatrix}$

16. $\begin{pmatrix} y < x \\ y \le 2 \end{pmatrix}$

17. $\begin{pmatrix} y < x \\ y > x + 3 \end{pmatrix}$

18. $\begin{pmatrix} x \le 3 \\ y \le -1 \end{pmatrix}$

19. $\begin{pmatrix} y > -2 \\ x > 1 \end{pmatrix}$

20. $\begin{pmatrix} x + 2y > 4 \\ x + 2y < 2 \end{pmatrix}$

21. $\begin{pmatrix} x \ge 0 \\ y \ge 0 \\ x + y \le 4 \\ 2x + y \le 6 \end{pmatrix}$

22. $\begin{pmatrix} x \ge 0 \\ y \ge 0 \\ x - y \le 5 \\ 4x + 7y \le 28 \end{pmatrix}$

23. $\begin{pmatrix} x \ge 0 \\ y \ge 0 \\ 2x + y \le 4 \\ 2x - 3y \le 6 \end{pmatrix}$

24. $\begin{pmatrix} x \ge 0 \\ y \ge 0 \\ 3x + 5y \ge 15 \\ 5x + 3y \ge 15 \end{pmatrix}$

For Problems 25–28 (Figures 12.9 through 12.12), find the maximum value and the minimum value of the given function in the indicated region.

25. $f(x, y) = 3x + 5y$

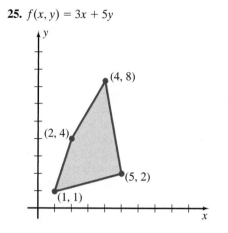

Figure 12.9

26. $f(x, y) = 8x + 3y$

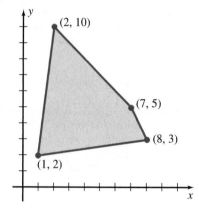

Figure 12.10

27. $f(x, y) = x + 4y$

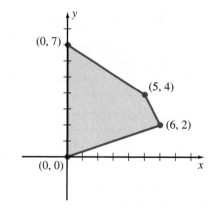

Figure 12.11

28. $f(x, y) = 2.5x + 3.5y$

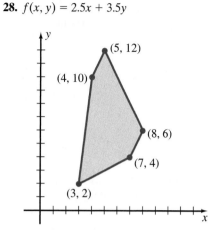

Figure 12.12

29. Maximize the function $f(x, y) = 3x + 7y$ in the region determined by the following constraints:

$$3x + 2y \leq 18$$
$$3x + 4y \geq 12$$
$$x \geq 0$$
$$y \geq 0$$

30. Maximize the function $f(x, y) = 1.5x + 2y$ in the region determined by the following constraints:

$$3x + 2y \leq 36$$
$$3x + 10y \leq 60$$
$$x \geq 0$$
$$y \geq 0$$

31. Maximize the function $f(x, y) = 40x + 55y$ in the region determined by the following constraints:

$$2x + y \leq 10$$
$$x + y \leq 7$$
$$2x + 3y \leq 18$$
$$x \geq 0$$
$$y \geq 0$$

32. Maximize the function $f(x, y) = 0.08x + 0.09y$ in the region determined by the following constraints:

$$x + y \leq 8000$$

$$y \leq \frac{1}{3}x$$

$$y \geq 500$$

$$x \leq 7000$$

$$x \geq 0$$

33. Minimize the function $f(x, y) = 0.2x + 0.5y$ in the region determined by the following constraints:

$$2x + y \geq 12$$

$$2x + 5y \geq 20$$

$$x \geq 0$$

$$y \geq 0$$

34. Minimize the function $f(x, y) = 3x + 7y$ in the region determined by the following constraints:

$$x + y \geq 9$$

$$6x + 11y \geq 84$$

$$x \geq 0$$

$$y \geq 0$$

35. Maximize the function $f(x, y) = 9x + 2y$ in the region determined by the following constraints:

$$5y - 4x \leq 20$$

$$4x + 5y \leq 60$$

$$x \geq 0$$

$$x \leq 10$$

$$y \geq 0$$

36. Maximize the function $f(x, y) = 3x + 4y$ in the region determined by the following constraints:

$$2y - x \leq 6$$

$$x + y \leq 12$$

$$x \geq 2$$

$$x \leq 8$$

$$y \geq 0$$

For Problems 37–42, solve each linear programming problem by using the graphing method illustrated in Problem 1 on page 677.

37. Suppose that an investor wants to invest up to $10,000. She plans to buy one speculative type of stock and one conservative type. The speculative stock is paying a 12% return, and the conservative stock is paying a 9% return. She has decided to invest at least $2000 in the conservative stock and no more than $6000 in the speculative stock. Furthermore, she does not want the speculative investment to exceed the conservative one. How much should she invest at each rate to maximize her return?

38. A manufacturer of golf clubs makes a profit of $50 per set on a model A set and $45 per set on a model B set. Daily production of the model A clubs is between 30 and 50 sets, inclusive, and that of the model B clubs is between 10 and 20 sets, inclusive. The total daily production is not to exceed 50 sets. How many sets of each model should be manufactured per day to maximize the profit?

39. A company makes two types of calculators. Type A sells for $12, and type B sells for $10. It costs the company $9 to produce one type A calculator and $8 to produce one type B calculator. In one month, the company is equipped to produce between 200 and 300, inclusive, of the type A calculator and between 100 and 250, inclusive, of the type B calculator, but not more than 300 altogether. How many calculators of each type should be produced per month to maximize the difference between the total selling price and the total cost of production?

40. A manufacturer of small copiers makes a profit of $200 on a deluxe model and $250 on a standard model. The company wants to produce at least 50 deluxe models per week and at least 75 standard models per week. However, the weekly production is not to exceed 150 copiers. How many copiers of each kind should be produced in order to maximize the profit?

41. Products A and B are produced by a company according to the following production information.
  (a) To produce one unit of product A requires 1 hour of working time on machine I, 2 hours on machine II, and 1 hour on machine III.
  (b) To produce one unit of product B requires 1 hour of working time on machine I, 1 hour on machine II, and 3 hours on machine III.

(c) Machine I is available for no more than 40 hours per week, machine II for no more than 40 hours per week, and machine III for no more than 60 hours per week.

(d) Product A can be sold at a profit of $2.75 per unit and product B at a profit of $3.50 per unit.

How many units each of product A and product B should be produced per week to maximize profit?

42. Suppose that the company we refer to in Problem 1 also manufactures widgets and wadgets and has the following production information available:

(a) To produce a widget requires 4 hours of working time on machine A and 2 hours on machine B.

(b) To produce a wadget requires 5 hours of working time on machine A and 5 hours on machine B.

(c) Machine A is available for no more than 200 hours per month, and machine B is available for no more than 150 hours per month.

(d) Widgets can be sold at a profit of $7 each and wadgets at a profit of $8 each.

How many widgets and how many wadgets should be produced per month in order to maximize profit?

■ ■ ■ THOUGHTS INTO WORDS

43. Describe in your own words the process of solving a system of inequalities.

44. What is linear programming? Write a paragraph or two answering this question in a way that elementary algebra students could understand.

Chapter 12   Summary

(12.1–12.3) Be sure that you understand the following ideas pertaining to the algebra of matrices.

1. Matrices of the same dimension are added by adding elements in corresponding positions.

2. Matrix addition is a commutative and an associative operation.

3. Matrices of any specific dimension have an additive identity element, which is the matrix of that same dimension containing all zeros.

4. Every matrix A has an additive inverse, $-A$, which can be found by changing the sign of each element of A.

5. Matrices of the same dimension can be subtracted by the definition $A - B = A + (-B)$.

6. The scalar product of a real number k and a matrix A can be found by multiplying each element of A by k.

7. The following properties hold for scalar multiplication and matrix addition.

$$k(A + B) = kA + kB$$

$$(k + l)A = kA + lA$$

$$(kl)A = k(lA)$$

8. If A is an $m \times n$ matrix and B is an $n \times p$ matrix, then the product AB is an $m \times p$ matrix. The general term, c_{ij}, of the product matrix $C = AB$ is determined by the equation

$$c_{ij} = a_{i1}b_{1j} + a_{i2}b_{2j} + \cdots + a_{in}b_{nj}$$

9. Matrix multiplication is not a commutative operation, but it is an associative operation.

10. Matrix multiplication has two distributive properties:

$$A(B+C) = AB + AC \quad \text{and} \quad (A+B)C = AC + BC$$

11. The general multiplicative identity element, I_n, for square $n \times n$ matrices contains only 1s in the main diagonal and 0s elsewhere. For example,

$$I_2 = \begin{bmatrix} 1 & 0 \\ 0 & 1 \end{bmatrix} \quad \text{and} \quad I_3 = \begin{bmatrix} 1 & 0 & 0 \\ 0 & 1 & 0 \\ 0 & 0 & 1 \end{bmatrix}$$

12. If a square matrix A has a multiplicative inverse A^{-1}, then $AA^{-1} = A^{-1}A = I_n$.

13. The multiplicative inverse of the 2×2 matrix

$$A = \begin{bmatrix} a_{11} & a_{12} \\ a_{21} & a_{22} \end{bmatrix}$$

is

$$A^{-1} = \frac{1}{|A|} \begin{bmatrix} a_{22} & -a_{12} \\ -a_{21} & a_{11} \end{bmatrix}$$

for $|A| \neq 0$. If $|A| = 0$, then the matrix A has no inverse.

14. A general technique for finding the inverse of a square matrix, when one exists, is described on page 665.

15. The solution set of a system of n linear equations in n variables can be found by multiplying the inverse of the coefficient matrix by the column matrix consisting of the constant terms. For example, the solution set of the system

$$\begin{pmatrix} 2x + 3y - z = 4 \\ 3x - y + 2z = 5 \\ 5x - 7y - 4z = -1 \end{pmatrix}$$

can be found by the product

$$\begin{bmatrix} 2 & 3 & -1 \\ 3 & -1 & 2 \\ 5 & -7 & -4 \end{bmatrix}^{-1} \begin{bmatrix} 4 \\ 5 \\ -1 \end{bmatrix}$$

(12.4) The solution set of a system of linear inequalities is the intersection of the solution sets of the individual inequalities. Such solution sets are easily determined by the graphing approach.

Linear programming problems deal with the idea of maximizing or minimizing a certain linear function that is subject to various constraints. The constraints are expressed as linear inequalities. Example 4 and Problem 1 are a good summary of the general approach to linear programming problems in this chapter.

Chapter 12   Review Problem Set

For Problems 1–10, compute the indicated matrix, if it exists, using the following matrices:

$$A = \begin{bmatrix} 2 & -4 \\ -3 & 8 \end{bmatrix} \qquad B = \begin{bmatrix} 5 & -1 \\ 0 & 2 \end{bmatrix}$$

$$C = \begin{bmatrix} 3 & -1 \\ -2 & 4 \\ 5 & -6 \end{bmatrix} \qquad D = \begin{bmatrix} -2 & -1 & 4 \\ 5 & 0 & -3 \end{bmatrix},$$

$$E = \begin{bmatrix} 1 \\ -3 \\ -7 \end{bmatrix} \qquad F = \begin{bmatrix} 1 & -2 \\ 4 & -4 \\ 7 & -8 \end{bmatrix}$$

1. $A + B$

2. $B - A$

3. $C - F$

4. $2A + 3B$

5. $3C - 2F$

6. CD

7. DC

8. $DC + AB$

9. DE

10. EF

11. Use A and B from the preceding problems and show that $AB \neq BA$.

12. Use C, D, and F from the preceding problems and show that $D(C + F) = DC + DF$.

13. Use C, D, and F from the preceding problems and show that $(C + F)D = CD + FD$.

For each matrix in Problems 14–23, find the multiplicative inverse, if it exists.

14. $\begin{bmatrix} 9 & 5 \\ 7 & 4 \end{bmatrix}$

15. $\begin{bmatrix} 9 & 4 \\ 7 & 3 \end{bmatrix}$

16. $\begin{bmatrix} -2 & 1 \\ 2 & 3 \end{bmatrix}$

17. $\begin{bmatrix} 4 & -6 \\ 2 & -3 \end{bmatrix}$

18. $\begin{bmatrix} -1 & -3 \\ -4 & -5 \end{bmatrix}$

19. $\begin{bmatrix} 0 & -3 \\ 7 & 6 \end{bmatrix}$

20. $\begin{bmatrix} 1 & -2 & 1 \\ 2 & -5 & 2 \\ -3 & 7 & 5 \end{bmatrix}$

21. $\begin{bmatrix} 1 & 3 & -2 \\ 4 & 13 & -7 \\ 5 & 16 & -8 \end{bmatrix}$

22. $\begin{bmatrix} -2 & 4 & 7 \\ 1 & -3 & 5 \\ 1 & -5 & 22 \end{bmatrix}$

23. $\begin{bmatrix} -1 & 2 & 3 \\ 2 & -5 & -7 \\ -3 & 5 & 11 \end{bmatrix}$

For Problems 24–28, use the multiplicative inverse matrix approach to solve each system. The required inverses were found in Problems 14–23.

24. $\begin{pmatrix} 9x + 5y = 12 \\ 7x + 4y = 10 \end{pmatrix}$

25. $\begin{pmatrix} -2x + y = -9 \\ 2x + 3y = 5 \end{pmatrix}$

26. $\begin{pmatrix} x - 2y + z = 7 \\ 2x - 5y + 2z = 17 \\ -3x + 7y + 5z = -32 \end{pmatrix}$

27. $\begin{pmatrix} x + 3y - 2z = -7 \\ 4x + 13y - 7z = -21 \\ 5x + 16y - 8z = -23 \end{pmatrix}$

28. $\begin{pmatrix} -x + 2y + 3z = 22 \\ 2x - 5y - 7z = -51 \\ -3x + 5y + 11z = 71 \end{pmatrix}$

For Problems 29–32, indicate the solution set for each system of linear inequalities by graphing the system and shading the appropriate region.

29. $\begin{pmatrix} 3x - 4y \geq 0 \\ 2x + 3y \leq 0 \end{pmatrix}$

30. $\begin{pmatrix} 3x - 2y < 6 \\ 2x - 3y < 6 \end{pmatrix}$

31. $\begin{pmatrix} x - 4y < 4 \\ 2x + y \geq 2 \end{pmatrix}$

32. $\begin{pmatrix} x \geq 0 \\ y \geq 0 \\ x + 2y \leq 4 \\ 2x - y \leq 4 \end{pmatrix}$

33. Maximize the function $f(x, y) = 8x + 5y$ in the region determined by the following constraints:

$$y \leq 4x$$
$$x + y \leq 5$$
$$x \geq 0$$
$$y \geq 0$$
$$x \leq 4$$

34. Maximize the function $f(x, y) = 2x + 7y$ in the region determined by the following constraints:

$$x \geq 0$$
$$y \geq 0$$
$$x + 2y \leq 16$$
$$x + y \leq 9$$
$$3x + 2y \leq 24$$

35. Maximize the function $f(x, y) = 7x + 5y$ in the region determined by the constraints of Problem 34.

36. Maximize the function $f(x, y) = 150x + 200y$ in the region determined by the constraints of Problem 34.

37. A manufacturer of electric ice cream freezers makes a profit of $4.50 on a one-gallon freezer and a profit of $5.25 on a two-gallon freezer. The company wants to produce at least 75 one-gallon and at least 100 two-gallon freezers per week. However, the weekly production is not to exceed a total of 250 freezers. How many freezers of each type should be produced per week in order to maximize the profit?

For Problems 1–10, compute the indicated matrix, if it exists, using the following matrices:

$$A = \begin{bmatrix} -1 & 3 \\ 4 & -2 \end{bmatrix} \quad B = \begin{bmatrix} 3 & -2 \\ 4 & -1 \end{bmatrix} \quad C = \begin{bmatrix} -3 \\ 5 \\ -6 \end{bmatrix}$$

$$D = \begin{bmatrix} 2 & -1 \\ 3 & -2 \\ 6 & 5 \end{bmatrix} \quad E = \begin{bmatrix} 2 & -1 & 4 \\ 5 & 1 & -3 \end{bmatrix}$$

$$F = \begin{bmatrix} -1 & 6 \\ 2 & -5 \\ 3 & 4 \end{bmatrix}$$

1. AB **2.** BA **3.** DE

4. BC **5.** EC **6.** $2A - B$

7. $3D + 2F$ **8.** $-3A - 2B$ **9.** EF

10. $AB - EF$

For Problems 11–16, find the multiplicative inverse, if it exists.

11. $\begin{bmatrix} 3 & -2 \\ 5 & -3 \end{bmatrix}$ **12.** $\begin{bmatrix} -2 & 5 \\ 3 & -7 \end{bmatrix}$ **13.** $\begin{bmatrix} 1 & -3 \\ -2 & 8 \end{bmatrix}$

14. $\begin{bmatrix} 3 & 5 \\ 1 & 4 \end{bmatrix}$ **15.** $\begin{bmatrix} -2 & 2 & 3 \\ 1 & -1 & 0 \\ 0 & 1 & 4 \end{bmatrix}$ **16.** $\begin{bmatrix} 1 & -2 & 4 \\ 0 & 1 & 3 \\ 0 & 0 & 1 \end{bmatrix}$

For Problems 17–19, use the multiplicative inverse matrix approach to solve each system.

17. $\begin{pmatrix} 3x - 2y = 48 \\ 5x - 3y = 76 \end{pmatrix}$ **18.** $\begin{pmatrix} x - 3y = 36 \\ -2x + 8y = -100 \end{pmatrix}$

19. $\begin{pmatrix} 3x + 5y = 92 \\ x + 4y = 61 \end{pmatrix}$

20. Solve the system

$$\begin{pmatrix} -x + 3y + z = 1 \\ 2x + 5y = 3 \\ 3x + y - 2z = -2 \end{pmatrix}$$

where the inverse of the coefficient matrix is

$$\begin{bmatrix} -\dfrac{10}{9} & \dfrac{7}{9} & -\dfrac{5}{9} \\[2mm] \dfrac{4}{9} & -\dfrac{1}{9} & \dfrac{2}{9} \\[2mm] -\dfrac{13}{9} & \dfrac{10}{9} & -\dfrac{11}{9} \end{bmatrix}$$

21. Solve the system

$$\begin{pmatrix} x + y + 2z = 3 \\ 2x + 3y - z = 3 \\ -3x + y - 2z = 3 \end{pmatrix}$$

where the inverse of the coefficient matrix is

$$\begin{bmatrix} -\dfrac{5}{24} & \dfrac{1}{6} & -\dfrac{7}{24} \\[2mm] \dfrac{7}{24} & \dfrac{1}{6} & \dfrac{5}{24} \\[2mm] \dfrac{11}{24} & -\dfrac{1}{6} & \dfrac{1}{24} \end{bmatrix}$$

For Problems 22–24, indicate the solution set for each system of inequalities by graphing the system and shading the appropriate region.

22. $\begin{pmatrix} 2x - y > 4 \\ x + 3y < 3 \end{pmatrix}$ **23.** $\begin{pmatrix} 2x - 3y \le 6 \\ x + 4y > 4 \end{pmatrix}$

24. $\begin{pmatrix} y \le 2x - 2 \\ y \ge x + 1 \end{pmatrix}$

25. Maximize the function $f(x, y) = 500x + 350y$ in the region determined by the following constraints:

$$3x + 2y \le 24$$

$$x + 2y \le 16$$

$$x + y \le 9$$

$$x \ge 0$$

$$y \ge 0$$

Conic Sections

13.1 Circles

13.2 Parabolas

13.3 Ellipses

13.4 Hyperbolas

13.5 Systems Involving Nonlinear Equations

Examples of conic sections, in particular, parabolas and ellipses, can be found in corporate logos throughout the world.

Circles, ellipses, parabolas, and hyperbolas can be formed by intersecting a plane and a right-circular conical surface as shown in Figure 13.1. These figures are often referred to as **conic sections**. In this chapter we will define each conic section as a set of points satisfying a set of conditions. Then we will use the definitions to develop standard forms for the equations of the conic sections. Next we will use the standard forms of the equations to (1) determine specific equations for specific conics, (2) determine graphs of specific equations, and (3) solve problems. Finally, we will consider some systems of equations involving the conic sections.

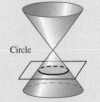
Circle

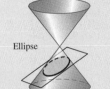
Ellipse

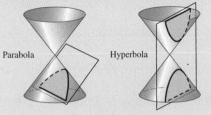
Parabola Hyperbola

Figure 13.1

13.1   Circles

The distance formula $d = \sqrt{(x_2 - x_1)^2 + (y_2 - y_1)^2}$, developed in Section 7.4 and applied to the definition of a circle, produces what is known as the **standard form of the equation of a circle**. We start with a precise definition of a circle.

Definition 13.1

A **circle** is the set of all points in a plane equidistant from a given fixed point called the **center**. A line segment determined by the center and any point on the circle is called a **radius**.

Now let's consider a circle having a radius of length r and a center at (h, k) on a coordinate system, as shown in Figure 13.2. For any point P on the circle with co-ordinates (x, y), the length of a radius, denoted by r, can be expressed as $r = \sqrt{(x - h)^2 + (y - k)^2}$. Thus, squaring both sides of the equation, we obtain the **standard form** of the equation of a circle:

$$(x - h)^2 + (y - k)^2 = r^2$$

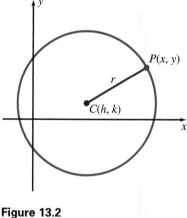

Figure 13.2

The standard form of the equation of a circle can be used to solve two basic kinds of problems: namely, (1) given the coordinates of the center and the length of a radius of a circle, find its equation; and (2) given the equation of a circle, determine its graph. Let's illustrate each of these types of problems.

E X A M P L E   1

Find the equation of a circle having its center at $(-3, 5)$ and a radius of length 4 units.

Solution

Substituting -3 for h, 5 for k, and 4 for r in the standard form and simplifying, we obtain

$$(x - h)^2 + (y - k)^2 = r^2$$
$$(x - (-3))^2 + (y - 5)^2 = 4^2$$
$$(x + 3)^2 + (y - 5)^2 = 4^2$$
$$x^2 + 6x + 9 + y^2 - 10y + 25 = 16$$
$$x^2 + y^2 + 6x - 10y + 18 = 0$$

■

Note in Example 1 that we simplified the equation to the form $x^2 + y^2 + Dx + Ey + F = 0$, where D, E, and F are constants. This is another form that we commonly use when working with circles.

E X A M P L E   2

Find the equation of a circle having its center at $(-5, -9)$ and a radius of length $2\sqrt{3}$ units. Express the final equation in the form $x^2 + y^2 + Dx + Ey + F = 0$.

Solution

In the standard form, substitute -5 for h, -9 for k, and $2\sqrt{3}$ for r.

$$(x - h)^2 + (y - k)^2 = r^2$$
$$(x - (-5))^2 + (y - (-9))^2 = (2\sqrt{3})^2$$
$$(x + 5)^2 + (y + 9)^2 = (2\sqrt{3})^2$$
$$x^2 + 10x + 25 + y^2 + 18y + 81 = 12$$
$$x^2 + y^2 + 10x + 18y + 94 = 0 \qquad \blacksquare$$

E X A M P L E   3

Find the equation of a circle having its center at the origin and a radius of length r units.

Solution

Substitute 0 for h, 0 for k, and r for r in the standard form of the equation of a circle.

$$(x - h)^2 + (y - k)^2 = r^2$$
$$(x - 0)^2 + (y - 0)^2 = r^2$$
$$x^2 + y^2 = r^2 \qquad \blacksquare$$

Note in Example 3 that

$$x^2 + y^2 = r^2$$

is the standard form of the equation of a circle that has its **center at the origin**. Therefore, by inspection, we can recognize that $x^2 + y^2 = 9$ is a circle with its center at the origin and radius of length 3 units. Likewise, the equation $5x^2 + 5y^2 = 10$ is equivalent to $x^2 + y^2 = 2$, and therefore its graph is a circle with its center at the origin and a radius of length $\sqrt{2}$ units. Furthermore, we can easily determine that the equation of the circle with its center at the origin and a radius of 8 units is $x^2 + y^2 = 64$.

EXAMPLE 4

Find the center and the length of a radius of the circle $x^2 + y^2 - 6x + 12y - 2 = 0$.

Solution

We can change the given equation into the standard form of the equation of a circle by completing the square on x and y as follows:

$$x^2 + y^2 - 6x + 12y - 2 = 0$$

$$(x^2 - 6x + \underline{\quad}) + (y^2 + 12y + \underline{\quad}) = 2$$

$$(x^2 - 6x + 9) + (y^2 + 12y + 36) = 2 + 9 + 36$$

| Add 9 to complete the square on x. | Add 36 to complete the square on y. | Add 9 and 36 to compensate for the 9 and 36 added on the left side. |

$$(x - 3)^2 + (y + 6)^2 = 47 \qquad \text{Factor.}$$

$$(x - 3)^2 + (y - (-6))^2 = (\sqrt{47})^2$$

$$\underset{h}{\uparrow} \qquad \underset{k}{\uparrow} \qquad \underset{r}{\uparrow}$$

The center is at $(3, -6)$, and the length of a radius is $\sqrt{47}$ units. ∎

EXAMPLE 5

Graph $x^2 + y^2 - 6x + 4y + 9 = 0$.

Solution

We can change the given equation into the standard form of the equation of a circle by completing the square on x and y as follows:

$$x^2 + y^2 - 6x + 4y + 9 = 0$$

$$(x^2 - 6x + \underline{\quad}) + (y^2 + 4y + \underline{\quad}) = -9$$

$$(x^2 - 6x + 9) + (y^2 + 4y + 4) = -9 + 9 + 4$$

| Add 9 to complete the square on x. | Add 4 to complete the square on y. | Add 9 and 4 to compensate for the 9 and 4 added on the left side. |

$$(x - 3)^2 + (y + 2)^2 = 2^2$$

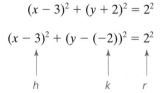

The center is at $(3, -2)$, and the length of a radius is 2 units. Thus the circle can be drawn as shown in Figure 13.3.

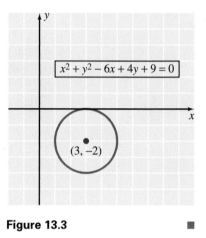

Figure 13.3 ∎

It should be evident that to determine the equation of a specific circle, we need the values of h, k, and r. To determine these values from a given set of conditions often requires the use of some of the following concepts from elementary geometry.

1. A tangent to a circle is a line that has one and only one point in common with the circle. This common point is called a *point of tangency*.

2. A radius drawn to the point of tangency is perpendicular to the tangent line.

3. Three noncollinear points in a plane determine a circle.

4. The perpendicular bisector of a chord contains the center of a circle.

Now let's consider two problems that use some of these concepts. We will offer an analysis of these problems but will leave the details for you to complete.

PROBLEM 1

Find the equation of the circle that has its center at $(2, 1)$ and is tangent to the line $x - 3y = 9$.

Analysis

Let's sketch a figure to help with the analysis of the problem (Figure 13.4). The point of tangency (a, b) is on the line $x - 3y = 9$, so we have $a - 3b = 9$. Also, the line determined by $(2, 1)$ and (a, b) is perpendicular to the line $x - 3y = 9$, so their slopes are negative reciprocals of each other. This relationship produces another

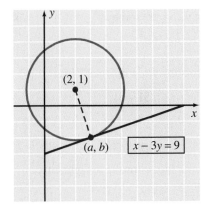

Figure 13.4

equation with the variables a and b. (This equation should be $3a + b = 7$.) Solving the system

$$\begin{pmatrix} a - 3b = 9 \\ 3a + b = 7 \end{pmatrix}$$

will produce the values for (a, b), and this point, along with the center of the circle, determines the length of a radius. Then the center along with the length of a radius determines the equation of the circle. (The equation is $x^2 + y^2 - 4x - 2y - 5 = 0$.) ∎

P R O B L E M   2

Find the equation of the circle that passes through the three points $(2, -4)$, $(-6, 4)$, and $(-2, -8)$.

Analysis

Three chords of the circle are determined by the three given points. (The points are noncollinear.) The center of the circle can be found at the intersection of the perpendicular bisectors of any two chords. Then the center and one of the given points can be used to find the length of a radius. From the center and the length of a radius, the equation of the circle can be determined. (The equation is $x^2 + y^2 + 8x + 4y - 20 = 0$.)

OR

Because three noncollinear points in a plane determine a circle, we could substitute the coordinates of the three given points into the general equation $x^2 + y^2 + Dx + Ey + F = 0$. This will produce a system of three linear equations in the three unknowns D, E, and F. (Perhaps you should do this and check your answer from the first method.) ∎

When using a graphing utility to graph circles, we need to solve the given equation for y in terms of x and then graph these two equations. Furthermore, it may be necessary to change the boundaries of the viewing rectangle so that a complete graph is shown. Let's consider an example.

EXAMPLE 6

Use a graphing utility to graph $x^2 - 40x + y^2 + 351 = 0$.

Solution

First we need to solve for y in terms of x.

$$x^2 - 40x + y^2 + 351 = 0$$

$$y^2 = -x^2 + 40x - 351$$

$$y = \pm\sqrt{-x^2 + 40x - 351}$$

Now we can make the following assignments:

$$Y_1 = \sqrt{-x^2 + 40x - 351}$$

$$Y_2 = -Y_1$$

(Note that we assigned Y_2 in terms of Y_1. By doing this, we avoid repetitive key strokes and thus reduce the chance for errors. You may need to consult your user's manual for instructions on how to keystroke $-Y_1$.) Figure 13.5 shows the graph.

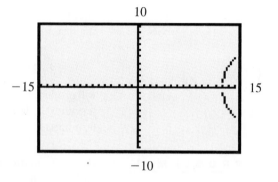

Figure 13.5

We know from the original equation that this graph is a circle, so we need to make some adjustments on the boundaries of the viewing rectangle in order to get a complete graph. This can be done by completing the square on the original equation to change its form to $(x - 20)^2 + y^2 = 49$, or simply by a trial-and-error process. By changing the boundaries on x so that $-15 \le x \le 30$, we obtain Figure 13.6.

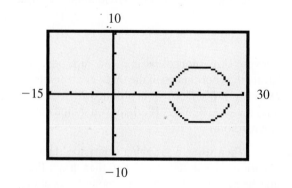

Figure 13.6

Problem Set 13.1

For Problems 1–14, write the equation of each of the circles that satisfies the stated conditions. In some cases there may be more than one circle that satisfies the conditions. Express the final equations in the form $x^2 + y^2 + Dx + Ey + F = 0$.

1. Center at $(2, 3)$ and $r = 5$

2. Center at $(-3, 4)$ and $r = 2$

3. Center at $(-1, -5)$ and $r = 3$

4. Center at $(4, -2)$ and $r = 1$

5. Center at $(3, 0)$ and $r = 3$

6. Center at $(0, -4)$ and $r = 6$

7. Center at the origin and $r = 7$

8. Center at the origin and $r = 1$

9. Tangent to the x axis, a radius of length 4, and abscissa of center is -3

10. Tangent to the y axis, a radius of length 5, and ordinate of center is 3

11. Tangent to both axes, a radius of 6, and the center in the third quadrant

12. x intercept of 6, y intercept of -4, and passes through the origin

13. Tangent to the y axis, x intercepts of 2 and 6

14. Tangent to the x axis, y intercepts of 1 and 5

For Problems 15–32, find the center and the length of a radius of each of the circles.

15. $(x - 5)^2 + (y - 7)^2 = 25$

16. $(x + 6)^2 + (y - 9)^2 = 49$

17. $(x + 1)^2 + (y + 8)^2 = 12$

18. $(x - 7)^2 + (y + 2)^2 = 24$

19. $3(x - 10)^2 + 3(y + 5)^2 = 9$

20. $5(x - 3)^2 + 5(y - 3)^2 = 30$

21. $x^2 + y^2 - 6x - 10y + 30 = 0$ $C(3,5)$

22. $x^2 + y^2 + 8x - 12y + 43 = 0$ $(-4 \quad -6)$

23. $x^2 + y^2 + 10x + 14y + 73 = 0$ $-5, -7$

24. $x^2 + y^2 + 6y - 7 = 0$ $0, -3$

25. $x^2 + y^2 - 10x = 0$ $5, 0$

26. $x^2 + y^2 + 7x - 2 = 0$

27. $x^2 + y^2 - 5y - 1 = 0$

28. $x^2 + y^2 - 4x + 2y = 0$ $2, -1$

29. $x^2 + y^2 = 8$

30. $4x^2 + 4y^2 = 1$

31. $4x^2 + 4y^2 - 4x - 8y - 11 = 0$

32. $36x^2 + 36y^2 + 48x - 36y - 11 = 0$

33. Find the equation of the line that is tangent to the circle $x^2 + y^2 - 2x + 3y - 12 = 0$ at the point $(4, 1)$.

34. Find the equation of the line that is tangent to the circle $x^2 + y^2 + 4x - 6y - 4 = 0$ at the point $(-1, -1)$.

35. Find the equation of the circle that passes through the origin and has its center at $(-3, -4)$.

36. Find the equation of the circle for which the line segment determined by $(-4, 9)$ and $(10, -3)$ is a diameter.

37. Find the equations of the circles that have their centers on the line $2x + 3y = 10$ and are tangent to both axes.

38. Find the equation of the circle that has its center at $(-2, -3)$ and is tangent to the line $x + y = -3$.

39. The point $(-1, 4)$ is the midpoint of a chord of a circle whose equation is $x^2 + y^2 + 8x + 4y - 30 = 0$. Find the equation of the chord.

40. Find the equation of the circle that is tangent to the line $3x - 4y = -26$ at the point $(-2, 5)$ and passes through the point $(5, -2)$.

41. Find the equation of the circle that passes through the three points $(1, 2)$, $(-3, -8)$, and $(-9, 6)$.

42. Find the equation of the circle that passes through the three points $(3, 0)$, $(6, -9)$ and $(10, -1)$.

■ ■ ■ **THOUGHTS INTO WORDS**

43. What is the graph of the equation $x^2 + y^2 = 0$? Explain your answer.

44. What is the graph of the equation $x^2 + y^2 = -4$? Explain your answer.

45. Your friend claims that the graph of an equation of the form $x^2 + y^2 + Dx + Ey + F = 0$, where $F = 0$, is a circle that passes through the origin. Is she correct? Explain why or why not.

■ ■ ■ **FURTHER INVESTIGATIONS**

46. Use a coordinate geometry approach to prove that an angle inscribed in a semicircle is a right angle. (See Figure 13.7.)

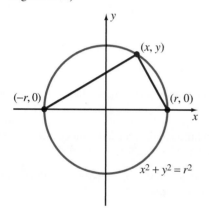

Figure 13.7

47. Use a coordinate geometry approach to prove that a line segment from the center of a circle bisecting a chord is perpendicular to the chord. [*Hint:* Let the ends of the chord be $(r, 0)$ and (a, b).]

48. By expanding $(x - h)^2 + (y - k)^2 = r^2$, we obtain $x^2 - 2hx + h^2 + y^2 - 2ky + k^2 - r^2 = 0$. When we compare this result to the form $x^2 + y^2 + Dx + Ey + F = 0$, we see that $D = -2h, E = -2k$, and $F = h^2 + k^2 - r^2$. Therefore, solving those equations respectively for h, k, and r, we can find the center and the length of a radius of a circle by using $h = \dfrac{D}{-2}, k = \dfrac{E}{-2}$, and $r = \sqrt{h^2 + k^2 - F}$. Use these relationships to find the center and the length of a radius of each of the following circles:

(a) $x^2 + y^2 - 2x - 8y + 8 = 0$
(b) $x^2 + y^2 + 4x - 14y + 49 = 0$
(c) $x^2 + y^2 + 12x + 8y - 12 = 0$
(d) $x^2 + y^2 - 16x + 20y + 115 = 0$
(e) $x^2 + y^2 - 12x - 45 = 0$
(f) $x^2 + y^2 + 14x = 0$

GRAPHING CALCULATOR ACTIVITIES

49. For each circle in Problems 15–32, you were asked to find the center and the length of a radius. Now use your graphing calculator and graph each of those circles. Be sure that your graph is consistent with the information you obtained earlier.

50. For each of the following, graph the two circles on the same set of axes and determine the coordinates of the points of intersection. Express the coordinates to the nearest tenth. If the circles do not intersect, so indicate.

(a) $x^2 + 4x + y^2 = 0$ and $x^2 - 2x + y^2 - 3 = 0$
(b) $x^2 + y^2 - 12y + 27 = 0$ and $x^2 + y^2 - 6y + 5 = 0$
(c) $x^2 - 4x + y^2 - 5 = 0$ and $x^2 - 14x + y^2 + 45.4 = 0$
(d) $x^2 - 6x + y^2 - 2y + 1 = 0$ and $x^2 - 6x + y^2 + 4y + 4 = 0$
(e) $x^2 - 4x + y^2 - 6y - 3 = 0$ and $x^2 - 8x + y^2 + 2y - 8 = 0$

13.2 Parabolas

We discussed parabolas as the graphs of quadratic functions in Sections 8.3 and 8.4. All parabolas in those sections had vertical lines as axes of symmetry. Furthermore, we did not state the definition for a parabola at that time. We shall now define a parabola and derive standard forms of equations for those that have either vertical or horizontal axes of symmetry.

Definition 13.2

A **parabola** is the set of all points in a plane such that the distance of each point from a fixed point F (the **focus**) is equal to its distance from a fixed line d (the **directrix**) in the plane.

Using Definition 13.2, we can sketch a parabola by starting with a fixed line d (directrix) and a fixed point F (focus) not on d. Then a point P is on the parabola if and only if $PF = PP'$, where $\overline{PP'}$ is perpendicular to the directrix d (Figure 13.8). The dashed curved line in Figure 13.8 indicates the possible positions of P; it is the parabola. The line l, through F and perpendicular to the directrix, is called the **axis of symmetry**. The point V, on the axis of symmetry halfway from F to the directrix d, is the **vertex** of the parabola.

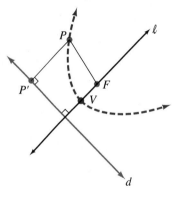

Figure 13.8

We can derive a standard form for the equation of a parabola by superimposing coordinates on the plane such that the origin is at the vertex of the parabola and the y axis is the axis of symmetry (Figure 13.9). If the focus is at $(0, p)$, where $p \neq 0$, then the equation of the directrix is $y = -p$. Therefore, for any point P on the parabola, $PF = PP'$,

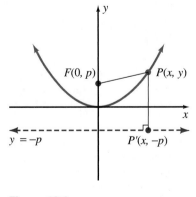

Figure 13.9

and using the distance formula yields

$$\sqrt{(x - 0)^2 + (y - p)^2} = \sqrt{(x - x)^2 + (y + p)^2}$$

Squaring both sides and simplifying, we obtain

$$(x - 0)^2 + (y - p)^2 = (x - x)^2 + (y + p)^2$$
$$x^2 + y^2 - 2py + p^2 = y^2 + 2py + p^2$$
$$x^2 = 4py$$

Thus the **standard form for the equation of a parabola** with its vertex at the origin and the y axis as its axis of symmetry is

$$x^2 = 4py$$

If $p > 0$, the parabola opens upward; if $p < 0$, the parabola opens downward.

A line segment that contains the focus and whose endpoints are on the parabola is called a **focal chord**. The specific focal chord that is parallel to the directrix we shall call the **primary focal chord**; this is line segment $\overline{QP}$ in Figure 13.10. Because $FP = PP' = |2p|$, the entire length of the primary focal chord is $|4p|$ units. You will see in a moment how we can use this fact when graphing parabolas.

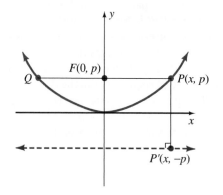

Figure 13.10

In a similar fashion, we can develop the standard form for the equation of a parabola with its vertex at the origin and the x axis as its axis of symmetry. By choosing a focus at $F(p, 0)$ and a directrix with an equation of $x = -p$ (see Figure 13.11), and by applying the definition of a parabola, we obtain the standard form for the equation:

$$y^2 = 4px$$

If $p > 0$, the parabola opens to the right, as in Figure 13.11; if $p < 0$, it opens to the left.

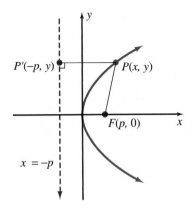

Figure 13.11

The concept of symmetry can be used to decide which of the two equations, $x^2 = 4py$ or $y^2 = 4px$, is to be used. The graph of $x^2 = 4py$ is symmetric with respect to the y axis because replacing x with $-x$ does not change the equation. Likewise, the graph of $y^2 = 4px$ is symmetric with respect to the x axis because replacing y with $-y$ leaves the equation unchanged. Let's summarize these ideas.

Standard Equations: Parabolas with Vertices at the Origin

> The graph of each of the following equations is a parabola that has its vertex at the origin and has the indicated focus, directrix, and symmetry.
>
> **1.** $x^2 = 4py$ focus $(0, p)$, directrix $y = -p$, y-axis symmetry
> **2.** $y^2 = 4px$ focus $(p, 0)$, directrix $x = -p$, x-axis symmetry

Now let's illustrate some uses of the equations $x^2 = 4py$ and $y^2 = 4px$.

EXAMPLE 1

Find the focus and directrix of the parabola $x^2 = -8y$ and sketch its graph.

Solution

Compare $x^2 = -8y$ to the standard form $x^2 = 4py$, and we have $4p = -8$. Therefore $p = -2$, and the parabola opens downward. The focus is at $(0, -2)$, and the equation of the directrix is $y = -(-2) = 2$. The primary focal chord is $|4p| = |-8| = 8$ units long. Therefore the endpoints of the primary focal chord are at $(4, -2)$ and $(-4, -2)$. The graph is sketched in Figure 13.12.

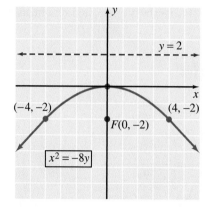

Figure 13.12

▪

EXAMPLE 2

Write the equation of the parabola that is symmetric with respect to the y axis, has its vertex at the origin, and contains the point $P(6, 3)$.

Solution

The standard form of the parabola is $x^2 = 4py$. Because P is on the parabola, the ordered pair $(6, 3)$ must satisfy the equation. Therefore

$$6^2 = 4p(3)$$
$$36 = 12p$$
$$3 = p$$

If $p = 3$, the equation becomes

$$x^2 = 4(3)y$$
$$x^2 = 12y$$

▪

EXAMPLE 3

Find the focus and directrix of the parabola $y^2 = 6x$ and sketch its graph.

Solution

Compare $y^2 = 6x$ to the standard form $y^2 = 4px$; we see that $4p = 6$ and therefore $p = \dfrac{3}{2}$. Thus the focus is at $\left(\dfrac{3}{2}, 0 \right)$, and the equation of the directrix is $x = -\dfrac{3}{2}$. Because $p > 0$, the parabola opens to the right. The primary focal chord is $|4p| = |6| = 6$ units long. Therefore the endpoints of the primary focal chord are at $\left(\dfrac{3}{2}, 3 \right)$ and $\left(\dfrac{3}{2}, -3 \right)$. The graph is sketched in Figure 13.13.

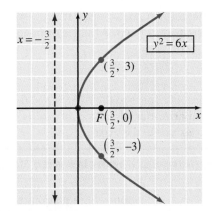

Figure 13.13

■ Other Parabolas

In much the same way, we can develop the standard form for the equation of a parabola that is symmetric with respect to a line parallel to a coordinate axis. In Figure 13.14 we have taken the vertex V at (h, k) and the focus F at $(h, k + p)$; the equation of the directrix is $y = k - p$. By the definition of a parabola, we know that $FP = PP'$. Therefore we can apply the distance formula as follows:

$$\sqrt{(x - h)^2 + (y - (k + p))^2} = \sqrt{(x - x)^2 + [y - (k - p)]^2}$$

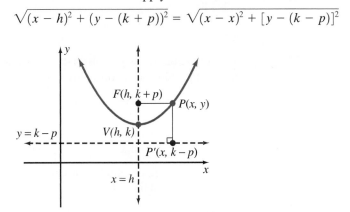

Figure 13.14

We leave it to the reader to show that this equation simplifies to

$$(x - h)^2 = 4p(y - k)$$

which is called the standard form of the equation of a parabola that has its vertex at (h, k) and is symmetric with respect to the line $x = h$. If $p > 0$, the parabola opens upward; if $p < 0$, the parabola opens downward.

In a similar fashion, we can show that the standard form of the equation of a parabola that has its vertex at (h, k) and is symmetric with respect to the line $y = k$ is

$$(y - k)^2 = 4p(x - h)$$

If $p > 0$, the parabola opens to the right; if $p < 0$, it opens to the left.

Let's summarize our discussion of parabolas that have lines of symmetry parallel to the x axis or to the y axis.

Standard Equations: Parabolas with Vertices Not at the Origin

The graph of each of the following equations is a parabola that has its vertex at (h, k) and has the indicated focus, directrix, and symmetry.

1. $(x - h)^2 = 4p(y - k)$ focus $(h, k + p)$, directrix $y = k - p$, line of symmetry $x = h$

2. $(y - k)^2 = 4p(x - h)$ focus $(h + p, k)$, directrix $x = h - p$, line of symmetry $y = k$

EXAMPLE 4

Find the vertex, focus, and directrix of the parabola $y^2 + 4y - 4x + 16 = 0$, and sketch its graph.

Solution

Write the equation as $y^2 + 4y = 4x - 16$, and we can complete the square on the left side by adding 4 to both sides.

$$y^2 + 4y + 4 = 4x - 16 + 4$$

$$(y + 2)^2 = 4x - 12$$

$$(y + 2)^2 = 4(x - 3)$$

Now let's compare this final equation to the form $(y - k)^2 = 4p(x - h)$:

$$[y - (-2)]^2 = 4(x - 3)$$

$$k = -2 \qquad 4p = 4 \qquad h = 3$$
$$p = 1$$

The vertex is at $(3, -2)$, and because $p > 0$, the parabola opens to the right and the focus is at $(4, -2)$. The equation of the directrix is $x = 2$. The primary focal chord is $|4p| = |4| = 4$ units long, and its endpoints are at $(4, 0)$ and $(4, -4)$. The graph is sketched in Figure 13.15.

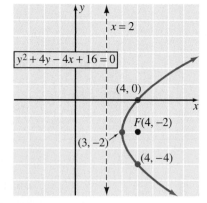

Figure 13.15

Remark: If we were using a graphing calculator to graph the parabola in Example 4, then after the step $(y + 2)^2 = 4x - 12$, we would solve for y to obtain $y = -2 \pm \sqrt{4x - 12}$. Then we could enter the two functions $Y_1 = -2 + \sqrt{4x - 12}$ and $Y_2 = -2 - \sqrt{4x - 12}$ and obtain a figure that closely resembles Figure 13.15. (You are asked to do this in the Graphing Calculator Activities.) Some graphing utilities can graph the equation in Example 4 without changing its form.

E X A M P L E   5

Write the equation of the parabola if its focus is at $(-4, 1)$ and the equation of its directrix is $y = 5$.

Solution

Because the directrix is a horizontal line, we know that the equation of the parabola is of the form $(x - h)^2 = 4p(y - k)$. The vertex is halfway between the focus and the directrix, so the vertex is at $(-4, 3)$. This means that $h = -4$ and $k = 3$. The parabola opens downward because the focus is below the directrix, and the distance between the focus and the vertex is 2 units; thus, $p = -2$. Substitute -4 for h, 3 for k, and -2 for p in the equation $(x - h)^2 = 4p(y - k)$ to obtain

$$(x - (-4))^2 = 4(-2)(y - 3)$$

which simplifies to

$$(x + 4)^2 = -8(y - 3)$$
$$x^2 + 8x + 16 = -8y + 24$$
$$x^2 + 8x + 8y - 8 = 0$$ ∎

Remark: For a problem such as Example 5, you may find it helpful to put the given information on a set of axes and draw a rough sketch of the parabola to assist in your analysis of the problem.

Parabolas possess various properties that make them very useful. For example, if a parabola is rotated about its axis, a parabolic surface is formed. The rays from a source of light placed at the focus of this surface reflect from the surface parallel to the axis. It is for this reason that parabolic reflectors are used on searchlights, as in Figure 13.16. Likewise, rays of light coming into a parabolic surface parallel to the axis are reflected through the focus. This property of parabolas is useful in the design of mirrors for telescopes (see Figure 13.17) and in the construction of radar antennas.

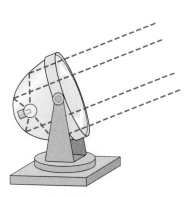

Figure 13.16

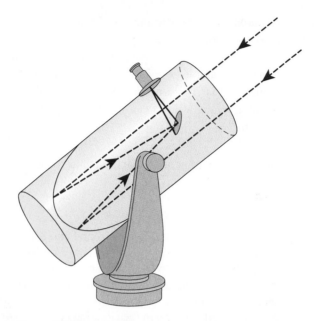

Figure 13.17

Problem Set 13.2

For Problems 1–30, find the vertex, focus, and directrix of the given parabola and sketch its graph.

1. $y^2 = 8x$

2. $y^2 = -4x$

3. $x^2 = -12y$

4. $x^2 = 8y$

5. $y^2 = -2x$

6. $y^2 = 6x$

7. $x^2 = 6y$

8. $x^2 = -7y$

9. $x^2 = 12(y + 1)$

10. $x^2 = -12(y - 2)$

11. $y^2 = -8(x - 3)$

12. $y^2 = 4(x + 1)$

13. $x^2 - 4y + 8 = 0$

14. $x^2 - 8y - 24 = 0$

15. $x^2 + 8y + 16 = 0$

16. $x^2 + 4y - 4 = 0$

17. $y^2 - 12x + 24 = 0$

18. $y^2 + 8x - 24 = 0$

19. $(x - 2)^2 = -4(y + 2)$

20. $(x + 3)^2 = 4(y - 4)$

21. $(y + 4)^2 = -8(x + 2)$

22. $(y - 3)^2 = 8(x - 1)$

23. $x^2 - 2x - 4y + 9 = 0$

24. $x^2 + 4x - 8y - 4 = 0$

25. $x^2 + 6x + 8y + 1 = 0$

26. $x^2 - 4x + 4y - 4 = 0$

27. $y^2 - 2y + 12x - 35 = 0$

28. $y^2 + 4y + 8x - 4 = 0$

29. $y^2 + 6y - 4x + 1 = 0$

30. $y^2 - 6y - 12x + 21 = 0$

For Problems 31–50, find an equation of the parabola that satisfies the given conditions.

31. Focus $(0, 3)$, directrix $y = -3$

32. Focus $\left(0, -\dfrac{1}{2}\right)$, directrix $y = \dfrac{1}{2}$

33. Focus $(-1, 0)$, directrix $x = 1$

34. Focus $(5, 0)$, directrix $x = 1$

35. Focus $(0, 1)$, directrix $y = 7$

36. Focus $(0, -2)$, directrix $y = -10$

37. Focus $(3, 4)$, directrix $y = -2$

38. Focus $(-3, -1)$, directrix $y = 7$

39. Focus $(-4, 5)$, directrix $x = 0$

40. Focus $(5, -2)$, directrix $x = -1$

41. Vertex $(0, 0)$, symmetric with respect to the x axis, and contains the point $(-3, 5)$

42. Vertex $(0, 0)$, symmetric with respect to the y axis, and contains the point $(-2, -4)$

43. Vertex $(0, 0)$, focus $\left(\dfrac{5}{2}, 0\right)$

44. Vertex $(0, 0)$, focus $\left(0, -\dfrac{7}{2}\right)$

45. Vertex $(7, 3)$, focus $(7, 5)$, and symmetric with respect to the line $x = 7$

46. Vertex $(-4, -6)$, focus $(-7, -6)$, and symmetric with respect to the line $y = -6$

47. Vertex $(8, -3)$, focus $(11, -3)$, and symmetric with respect to the line $y = -3$

48. Vertex $(-2, 9)$, focus $(-2, 5)$, and symmetric with respect to the line $x = -2$

49. Vertex $(-9, 1)$, symmetric with respect to the line $x = -9$, and contains the point $(-8, 0)$

50. Vertex $(6, -4)$, symmetric with respect to the line $y = -4$, and contains the point $(8, -3)$

For Problems 51–55, solve each problem.

51. One section of a suspension bridge hangs between two towers that are 40 feet above the surface and 300 feet apart, as shown in Figure 13.18. A cable strung between the tops of the two towers is in the shape of a parabola with its vertex 10 feet above the surface. With axes drawn as indicated in the figure, find the equation of the parabola.

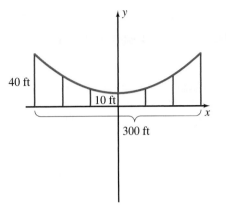

Figure 13.18

52. Suppose that five equally spaced vertical cables are used to support the bridge in Figure 13.18. Find the total length of these supports.

53. Suppose that an arch is shaped like a parabola. It is 20 feet wide at the base and 100 feet high. How wide is the arch 50 feet above the ground?

54. A parabolic arch 27 feet high spans a parkway. How wide is the arch if the center section of the parkway, a section that is 50 feet wide, has a minimum clearance of 15 feet?

55. A parabolic arch spans a stream 200 feet wide. How high above the stream must the arch be to give a minimum clearance of 40 feet over a channel in the center that is 120 feet wide?

■ ■ ■ THOUGHTS INTO WORDS

56. Give a step-by-step description of how you would go about graphing the parabola $x^2 - 2x - 4y - 7 = 0$.

57. Suppose that someone graphed the equation $y^2 - 6y - 2x + 11 = 0$ and obtained the graph in Figure 13.19. How do you know by looking at the equation that this graph is incorrect?

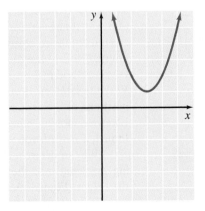

Figure 13.19

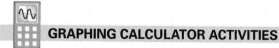

GRAPHING CALCULATOR ACTIVITIES

58. The parabola determined by the equation $x^2 + 4x - 8y - 4 = 0$ (Problem 24) is easy to graph using a graphing calculator because it can be expressed as a function of x without much computation. Let's solve the equation for y.

$$8y = x^2 + 4x - 4$$

$$y = \frac{x^2 + 4x - 4}{8}$$

Use your graphing calculator to graph this function.

As noted in the Remark that follows Example 4, solving the equation $y^2 + 4y - 4x + 16 = 0$ for y produces two functions: $Y_1 = -2 + \sqrt{4x - 12}$ and $Y_2 = -2 - \sqrt{4x - 12}$. Graph these two functions on the same set of axes. Your result should resemble Figure 13.15.

Use your graphing calculator to check your graphs for Problems 1–30.

13.3   Ellipses

Let's begin by defining an ellipse.

Definition 13.3

> An **ellipse** is the set of all points in a plane such that the sum of the distances of each point from two fixed points F and F' (the **foci**) in the plane is constant.

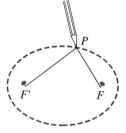

Figure 13.20

With two thumbtacks, a piece of string, and a pencil, it is easy to draw an ellipse by satisfying the conditions of Definition 13.3. First, insert two thumbtacks into a piece of cardboard at points F and F', and fasten the ends of the piece of string to the thumbtacks, as in Figure 13.20. Then loop the string around the point of a pencil and hold the pencil so that the string is taut. Finally, move the pencil around the tacks, always keeping the string taut. You will draw an ellipse. The two points F and F' are the foci referred to in Definition 13.3, and the sum of the distances FP and $F'P$ is constant because it represents the length of the piece of string. With the same piece of string, you can vary the shape of the ellipse by changing the positions of the foci. Moving F and F' farther apart will make the ellipse flatter. Likewise, moving F and F' closer together will cause the ellipse to resemble a circle. In fact, if $F = F'$, you will obtain a circle.

We can derive a standard form for the equation of an ellipse by superimposing coordinates on the plane such that the foci are on the x axis, equidistant from the origin (Figure 13.21). If F has coordinates $(c, 0)$, where $c > 0$, then F' has coordinates $(-c, 0)$, and the distance between F and F' is $2c$ units. We will let $2a$

represent the constant sum of $FP + F'P$. Note that $2a > 2c$ and therefore $a > c$. For any point P on the ellipse,

$$FP + F'P = 2a$$

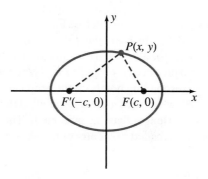

Figure 13.21

Use the distance formula to write this as

$$\sqrt{(x - c)^2 + (y - 0)^2} + \sqrt{(x + c)^2 + (y - 0)^2} = 2a$$

Let's change the form of this equation to

$$\sqrt{(x - c)^2 + y^2} = 2a - \sqrt{(x + c)^2 + y^2}$$

and square both sides:

$$(x - c)^2 + y^2 = 4a^2 - 4a\sqrt{(x + c)^2 + y^2} + (x + c)^2 + y^2$$

This can be simplified to

$$a^2 + cx = a\sqrt{(x + c)^2 + y^2}$$

Again, square both sides to produce

$$a^4 + 2a^2cx + c^2x^2 = a^2[(x + c)^2 + y^2]$$

which can be written in the form

$$x^2(a^2 - c^2) + a^2y^2 = a^2(a^2 - c^2)$$

Divide both sides by $a^2(a^2 - c^2)$, which yields the form

$$\frac{x^2}{a^2} + \frac{y^2}{a^2 - c^2} = 1$$

Letting $b^2 = a^2 - c^2$, where $b > 0$, produces the equation

$$\frac{x^2}{a^2} + \frac{y^2}{b^2} = 1$$

(1)

Because $c > 0$, $a > c$, and $b^2 = a^2 - c^2$, it follows that $a^2 > b^2$ and hence $a > b$. This equation that we have derived is called the **standard form of the equation of an ellipse** with its foci on the x axis and its center at the origin.

The x intercepts of equation (1) can be found by letting $y = 0$. Doing this produces $x^2/a^2 = 1$, or $x^2 = a^2$; consequently, the x intercepts are a and $-a$. The corresponding points on the graph (see Figure 13.22) are $A(a, 0)$ and $A'(-a, 0)$, and the line segment $\overline{A'A}$, which is of length $2a$, is called the **major axis** of the ellipse. The endpoints of the major axis are also referred to as the **vertices** of the ellipse. Similarly, letting $x = 0$ produces $y^2/b^2 = 1$ or $y^2 = b^2$; consequently the y intercepts are b and $-b$. The corresponding points on the graph are $B(0, b)$ and $B'(0, -b)$, and the line segment $\overline{BB'}$, which is of length $2b$, is called the **minor axis**. Because $a > b$, **the major axis is always longer than the minor axis**. The point of intersection of the major and minor axes is called the **center** of the ellipse.

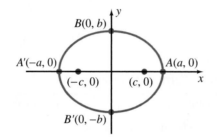

Figure 13.22

Standard Equation: Ellipse with Major Axis on the x Axis

The standard equation of an ellipse with its center at $(0, 0)$ and its major axis on the x axis is

$$\frac{x^2}{a^2} + \frac{y^2}{b^2} = 1$$

where $a > b$.

The vertices are $(-a, 0)$ and $(a, 0)$, and the length of the major axis is $2a$.

The endpoints of the minor axis are $(0, -b)$ and $(0, b)$, and the length of the minor axis is $2b$.

The foci are at $(-c, 0)$ and $(c, 0)$, where $c^2 = a^2 - b^2$.

Note that replacing y with $-y$, or x with $-x$, or both x and y with $-x$ and $-y$ leaves the equation unchanged. Thus the graph of

$$\frac{x^2}{a^2} + \frac{y^2}{b^2} = 1$$

is symmetric with respect to the x axis, the y axis, and the origin.

EXAMPLE 1

Find the vertices, the endpoints of the minor axis, and the foci of the ellipse $4x^2 + 9y^2 = 36$, and sketch the ellipse.

Solution

The given equation can be changed to standard form by dividing both sides by 36.

$$\frac{4x^2}{36} + \frac{9y^2}{36} = \frac{36}{36}$$

$$\frac{x^2}{9} + \frac{y^2}{4} = 1$$

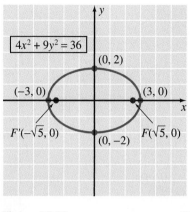

Therefore $a^2 = 9$ and $b^2 = 4$; hence the vertices are at $(3, 0)$ and $(-3, 0)$, and the endpoints of the minor axis are at $(0, 2)$ and $(0, -2)$. Because $c^2 = a^2 - b^2$, we have

$$c^2 = 9 - 4 = 5$$

Thus the foci are at $(\sqrt{5}, 0)$ and $(-\sqrt{5}, 0)$ The ellipse is sketched in Figure 13.23.

Figure 13.23 ■

EXAMPLE 2

Find the equation of the ellipse with vertices at $(\pm 6, 0)$ and foci at $(\pm 4, 0)$.

Solution

From the given information, we know that $a = 6$ and $c = 4$. Therefore

$$b^2 = a^2 - c^2 = 36 - 16 = 20$$

Substitute 36 for a^2 and 20 for b^2 in the standard form to produce

$$\frac{x^2}{36} + \frac{y^2}{20} = 1$$

Multiply both sides by 180 to get

$$5x^2 + 9y^2 = 180$$ ■

■ Ellipses with Foci on the *y* Axis

An ellipse with its center at the origin can also have its major axis on the *y* axis, as shown in Figure 13.24. In this case, the sum of the distances from any point *P* on the ellipse to the foci is set equal to the constant 2*b*.

$$\sqrt{(x - 0)^2 + (y - c)^2} + \sqrt{(x - 0)^2 + (y + c)^2} = 2b$$

With the conditions this time that $b > a$ and $c^2 = b^2 - a^2$, the equation simplifies to the same standard equation, $\dfrac{x^2}{a^2} + \dfrac{y^2}{b^2} = 1$. Let's summarize these ideas.

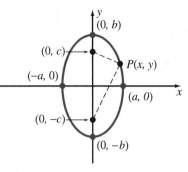

Figure 13.24

Standard Equation: Ellipse with Major Axis on the *y* Axis

The standard equation of an ellipse with its center at $(0, 0)$ and its major axis on the *y* axis is

$$\frac{x^2}{a^2} + \frac{y^2}{b^2} = 1$$

where $b > a$.

The vertices are $(0, -b)$ and $(0, b)$, and the length of the major axis is $2b$. The endpoints of the minor axis are $(-a, 0)$ and $(a, 0)$, and the length of the minor axis is $2a$.

The foci are at $(0, -c)$ and $(0, c)$, where $c^2 = b^2 - a^2$.

EXAMPLE 3

Find the vertices, the endpoints of the minor axis, and the foci of the ellipse $18x^2 + 4y^2 = 36$, and sketch the ellipse.

Solution

The given equation can be changed to standard form by dividing both sides by 36.

$$\frac{18x^2}{36} + \frac{4y^2}{36} = \frac{36}{36}$$

$$\frac{x^2}{2} + \frac{y^2}{9} = 1$$

Therefore $a^2 = 2$ and $b^2 = 9$; hence the vertices are at $(0, 3)$ and $(0, -3)$, and the endpoints of the minor axis are at $(\sqrt{2}, 0)$ and $(-\sqrt{2}, 0)$. From the relationship $c^2 = b^2 - a^2$, we obtain $c^2 = 9 - 2 = 7$; hence the foci are at $(0, \sqrt{7})$ and $(0, -\sqrt{7})$. The ellipse is sketched in Figure 13.25.

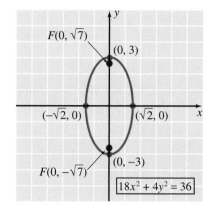

Figure 13.25

■ Other Ellipses

By applying the definition of an ellipse, we could also develop the standard equation of an ellipse whose center is not at the origin but whose major and minor axes are either on the coordinate axes or on lines parallel to the coordinate axes. In other words, we want to consider ellipses that are horizontal and vertical translations of the two basic ellipses. We will not show these developments in this text but will use Figures 13.26 (a) and (b) to indicate the basic facts needed to develop the standard equation. Note that in each figure, the center of the ellipse is at a point (h, k). Furthermore, the physical significance of a, b, and c is the same as before, but these values are used relative to the new center (h, k) to find the foci, vertices, and endpoints of the minor axis. Let's see how this works in a specific example.

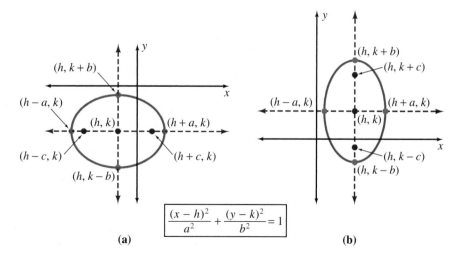

$$\frac{(x - h)^2}{a^2} + \frac{(y - k)^2}{b^2} = 1$$

(a)

(b)

Figure 13.26

EXAMPLE 4   Find the vertices, the endpoints of the minor axis, and the foci of the ellipse $9x^2 + 54x + 4y^2 - 8y + 49 = 0$, and sketch the ellipse.

Solution

First, we need to change to standard form by completing the square on both x and y.

$$9(x^2 + 6x + \underline{\ \ }) + 4(y^2 - 2y + \underline{\ \ }) = -49$$

$$9(x^2 + 6x + 9) + 4(y^2 - 2y + 1) = -49 + 9(9) + 4(1)$$

$$9(x + 3)^2 + 4(y - 1)^2 = 36$$

$$\frac{(x + 3)^2}{4} + \frac{(y - 1)^2}{9} = 1$$

From this equation, we can determine that $h = -3$, $k = 1$, $a = \sqrt{4} = 2$, and $b = \sqrt{9} = 3$. Because $b > a$, the foci and vertices are on the vertical line $x = -3$. The vertices are three units up and three units down from the center $(-3, 1)$, so they are at $(-3, 4)$ and $(-3, -2)$. The endpoints of the minor axis are two units to the right and two units to the left of the center, so they are at $(-1, 1)$ and $(-5, 1)$. From the relationship $c^2 = b^2 - a^2$, we obtain $c^2 = 9 - 4 = 5$. Thus the foci are at $(-3, 1 + \sqrt{5})$ and $(-3, 1 - \sqrt{5})$. The ellipse is sketched in Figure 13.27.

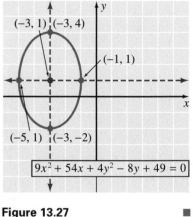

Figure 13.27   ∎

EXAMPLE 5   Write the equation of the ellipse that has vertices at $(-3, -5)$ and $(7, -5)$ and foci at $(-1, -5)$ and $(5, -5)$.

Solution

Because the vertices and foci are on the same horizontal line ($y = -5$), the equation of this ellipse is of the form

$$\frac{(x - h)^2}{a^2} + \frac{(y - k)^2}{b^2} = 1$$

where $a > b$. The center of the ellipse is at the midpoint of the major axis:

$$h = \frac{-3 + 7}{2} = 2 \quad \text{and} \quad k = \frac{-5 + (-5)}{2} = -5$$

The distance between the center $(2, -5)$ and a vertex $(7, -5)$ is 5 units; thus $a = 5$. The distance between the center $(2, -5)$ and a focus $(5, -5)$ is 3 units; thus $c = 3$. Using the relationship $c^2 = a^2 - b^2$, we obtain

$$b^2 = a^2 - c^2 = 25 - 9 = 16$$

Now let's substitute 2 for h, -5 for k, 25 for a^2, and 16 for b^2 in the standard form, and then we can simplify.

$$\frac{(x - 2)^2}{25} + \frac{(y + 5)^2}{16} = 1$$

$$16(x - 2)^2 + 25(y + 5)^2 = 400$$

$$16(x^2 - 4x + 4) + 25(y^2 + 10y + 25) = 400$$

$$16x^2 - 64x + 64 + 25y^2 + 250y + 625 = 400$$

$$16x^2 - 64x + 25y^2 + 250y + 289 = 0 \qquad \blacksquare$$

Remark: Again, for a problem such as Example 5, it might be helpful to start by recording the given information on a set of axes and drawing a rough sketch of the figure.

Like parabolas, ellipses possess properties that make them very useful. For example, the elliptical surface formed by rotating an ellipse about its major axis has the following property: Light or sound waves emitted at one focus reflect off the surface and converge at the other focus. This is the principle behind "whispering galleries," such as the Rotunda of the Capitol Building in Washington, D.C. In such buildings, two people standing at two specific spots that are the foci of the elliptical ceiling can whisper and yet hear each other clearly, even though they may be quite far apart.

One very important use of an elliptical surface is in the construction of a medical device called a lithotriptor. This device is used to break up kidney stones. A source that emits ultra-high-frequency shock waves is placed at one focus, and the kidney stone is placed at the other.

Ellipses also play an important role in astronomy. Johannes Kepler (1571–1630) showed that the orbit of a planet is an ellipse with the sun at one focus. For example, the orbit of earth is elliptical but nearly circular; at the same time, the moon moves about the earth in an elliptical path (see Figure 13.28).

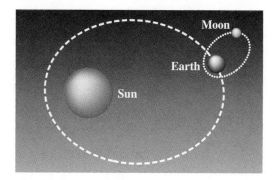

Figure 13.28

The arches for concrete bridges are sometimes elliptical. (One example is shown in Figure 13.30 in the next set of problems.) Also, elliptical gears are used in certain kinds of machinery that require a slow but powerful force at impact, such as a heavy-duty punch (see Figure 13.29).

Figure 13.29

Problem Set 13.3

For Problems 1–26, find the vertices, the endpoints of the minor axis, and the foci of the given ellipse, and sketch its graph.

1. $\dfrac{x^2}{4} + \dfrac{y^2}{1} = 1$

2. $\dfrac{x^2}{16} + \dfrac{y^2}{1} = 1$

3. $\dfrac{x^2}{4} + \dfrac{y^2}{9} = 1$

4. $\dfrac{x^2}{4} + \dfrac{y^2}{16} = 1$

5. $9x^2 + 3y^2 = 27$

6. $4x^2 + 3y^2 = 36$

7. $2x^2 + 5y^2 = 50$

8. $5x^2 + 36y^2 = 180$

9. $12x^2 + y^2 = 36$

10. $8x^2 + y^2 = 16$

11. $7x^2 + 11y^2 = 77$

12. $4x^2 + y^2 = 12$

13. $\dfrac{(x-2)^2}{9} + \dfrac{(y-1)^2}{4} = 1$

14. $\dfrac{(x+3)^2}{16} + \dfrac{(y-2)^2}{4} = 1$

15. $\dfrac{(x+1)^2}{9} + \dfrac{(y+2)^2}{16} = 1$

16. $\dfrac{(x-4)^2}{4} + \dfrac{(y+2)^2}{25} = 1$

17. $4x^2 - 8x + 9y^2 - 36y + 4 = 0$

18. $x^2 + 6x + 9y^2 - 36y + 36 = 0$

19. $4x^2 + 16x + y^2 + 2y + 1 = 0$

20. $9x^2 - 36x + 4y^2 + 16y + 16 = 0$

21. $x^2 - 6x + 4y^2 + 5 = 0$

22. $16x^2 + 9y^2 + 36y - 108 = 0$

23. $9x^2 - 72x + 2y^2 + 4y + 128 = 0$

24. $5x^2 + 10x + 16y^2 + 160y + 325 = 0$

25. $2x^2 + 12x + 11y^2 - 88y + 172 = 0$

26. $9x^2 + 72x + y^2 + 6y + 135 = 0$

For Problems 27–40, find an equation of the ellipse that satisfies the given conditions.

27. Vertices $(\pm 5, 0)$, foci $(\pm 3, 0)$

28. Vertices $(\pm 4, 0)$, foci $(\pm 2, 0)$

29. Vertices $(0, \pm 6)$, foci $(0, \pm 5)$

30. Vertices $(0, \pm 3)$, foci $(0, \pm 2)$

31. Vertices $(\pm 3, 0)$, length of minor axis is 2

32. Vertices $(0, \pm 5)$, length of minor axis is 4

33. Foci $(0, \pm 2)$, length of minor axis is 3

34. Foci $(\pm 1, 0)$, length of minor axis is 2

35. Vertices $(0, \pm 5)$, contains the point $(3, 2)$

36. Vertices $(\pm 6, 0)$, contains the point $(5, 1)$

37. Vertices $(5, 1)$ and $(-3, 1)$, foci $(3, 1)$ and $(-1, 1)$

38. Vertices $(2, 4)$ and $(2, -6)$, foci $(2, 3)$ and $(2, -5)$

39. Center $(0, 1)$, one focus at $(-4, 1)$, length of minor axis is 6

40. Center $(3, 0)$, one focus at $(3, 2)$, length of minor axis is 4

For Problems 41–44, solve each problem.

41. Find an equation of the set of points in a plane such that the sum of the distances between each point of the set and the points $(2, 0)$ and $(-2, 0)$ is 8 units.

42. Find an equation of the set of points in a plane such that the sum of the distances between each point of the set and the points $(0, 3)$ and $(0, -3)$ is 10 units.

43. An arch of the bridge shown in Figure 13.30 is semi-elliptical, and the major axis is horizontal. The arch is 30 feet wide and 10 feet high. Find the height of the arch 10 feet from the center of the base.

44. In Figure 13.30, how much clearance is there 10 feet from the bank?

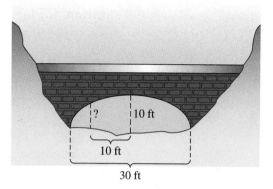

Figure 13.30

■ ■ ■ THOUGHTS INTO WORDS

45. What type of figure is the graph of the equation $x^2 + 6x + 2y^2 - 20y + 59 = 0$? Explain your answer.

46. Suppose that someone graphed the equation $4x^2 - 16x + 9y^2 + 18y - 11 = 0$ and obtained the graph shown in Figure 13.31. How do you know by looking at the equation that this is an incorrect graph?

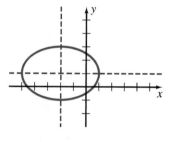

Figure 13.31

 GRAPHING CALCULATOR ACTIVITIES

47. Use your graphing calculator to check your graphs for Problems 17–26.

48. Use your graphing calculator to graph each of the following ellipses:

(a) $2x^2 - 40x + y^2 + 2y + 185 = 0$
(b) $x^2 - 4x + 2y^2 - 48y + 272 = 0$
(c) $4x^2 - 8x + y^2 - 4y - 136 = 0$
(d) $x^2 + 6x + 2y^2 + 56y + 301 = 0$

13.4 Hyperbolas

A hyperbola and an ellipse are similar by definition; however, an ellipse involves the *sum* of distances, and a hyperbola involves the *difference* of distances.

Definition 13.4

A **hyperbola** is the set of all points in a plane such that the difference of the distances of each point from two fixed points F and F' (the **foci**) in the plane is a positive constant.

Using Definition 13.4, we can sketch a hyperbola by starting with two fixed points F and F' as shown in Figure 13.32. Then we locate all points P such that $PF' - PF$ is a positive constant. Likewise, as shown in Figure 13.32, all points Q are located such that $QF - QF'$ is the same positive constant. The two dashed curved lines in Figure 13.32 make up the hyperbola. The two curves are sometimes referred to as the *branches* of the hyperbola.

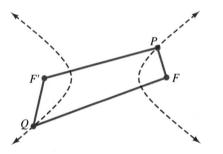

Figure 13.32

To develop a standard form for the equation of a hyperbola, let's superimpose coordinates on the plane such that the foci are located at $F(c, 0)$ and $F'(-c, 0)$, as indicated in Figure 13.33. Using the distance formula and setting $2a$ equal to the difference of the distances from any point P on the hyperbola to the foci, we have the following equation:

$$\left| \sqrt{(x - c)^2 + (y - 0)^2} - \sqrt{(x + c)^2 + (y - 0)^2} \right| = 2a$$

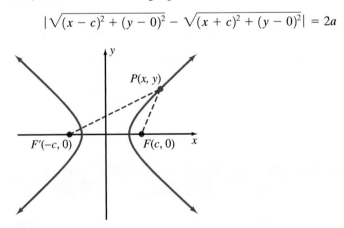

Figure 13.33

(The absolute-value sign is used to allow the point P to be on either branch of the hyperbola.) Using the same type of simplification procedure that we used for deriving the standard form for the equation of an ellipse, we find that this equation simplifies to

$$\frac{x^2}{a^2} - \frac{y^2}{c^2 - a^2} = 1$$

Letting $b^2 = c^2 - a^2$, where $b > 0$, we obtain the standard form

$$\frac{x^2}{a^2} - \frac{y^2}{b^2} = 1 \tag{1}$$

Equation (1) indicates that this hyperbola is symmetric with respect to both axes and the origin. Furthermore, by letting $y = 0$, we obtain $x^2/a^2 = 1$, or $x^2 = a^2$, so the x intercepts are a and $-a$. The corresponding points $A(a, 0)$ and $A'(-a, 0)$ are the **vertices** of the hyperbola, and the line segment $\overline{AA'}$ is called the **transverse axis**; it is of length $2a$ (see Figure 13.34). The midpoint of the transverse axis is called the **center** of the hyperbola; it is located at the origin. By letting $x = 0$ in equation (1), we obtain $-y^2/b^2 = 1$, or $y^2 = -b^2$. This implies that there are no y intercepts, as indicated in Figure 13.34.

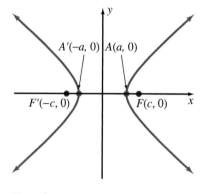

Figure 13.34

Standard Equation: Hyperbola with Transverse Axis on the *x* Axis

The standard equation of a hyperbola with its center at $(0,0)$ and its transverse axis on the x axis is

$$\frac{x^2}{a^2} - \frac{y^2}{b^2} = 1$$

where the foci are at $(-c, 0)$ and $(c, 0)$, the vertices are at $(-a, 0)$ and $(a, 0)$, and $c^2 = a^2 + b^2$.

In conjunction with every hyperbola, there are two intersecting lines that pass through the center of the hyperbola. These lines, referred to as *asymptotes*, are very helpful when we are sketching a hyperbola. Their equations are easily determined by using the following type of reasoning. Solving the equation

$$\frac{x^2}{a^2} - \frac{y^2}{b^2} = 1$$

for y produces $y = \pm\frac{b}{a}\sqrt{x^2 - a^2}$. From this form, it is evident that there are no points on the graph for $x^2 - a^2 < 0$ — that is, if $-a < x < a$. However, there are points on the graph if $x \geq a$ or $x \leq -a$. If $x \geq a$, then $y = \pm\frac{b}{a}\sqrt{x^2 - a^2}$ can be written

$$y = \pm\frac{b}{a}\sqrt{x^2\left(1 - \frac{a^2}{x^2}\right)}$$

$$= \pm\frac{b}{a}\sqrt{x^2}\sqrt{1 - \frac{a^2}{x^2}}$$

$$= \pm\frac{b}{a}x\sqrt{1 - \frac{a^2}{x^2}}$$

Now suppose that we are going to determine some y values for very large values of x. (Remember that a and b are arbitrary constants; they have specific values for a particular hyperbola.) When x is very large, a^2/x^2 will be close to zero, so the radicand will be close to 1. Therefore the y value will be close to either $(b/a)x$ or $-(b/a)x$. In other words, as x becomes larger and larger, the point $P(x, y)$ gets closer and closer to either the line $y = (b/a)x$ or the line $y = -(b/a)x$. A corresponding situation occurs when $x \leq a$. The lines with equations

$$y = \pm\frac{b}{a}x$$

are the **asymptotes** of the hyperbola.

As we mentioned earlier, the asymptotes are very helpful for sketching hyperbolas. An easy way to sketch the asymptotes is first to plot the vertices $A(a, 0)$ and $A'(-a, 0)$ and the points $B(0, b)$ and $B'(0, -b)$, as in Figure 13.35. The line

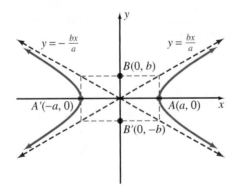

Figure 13.35

segment $\overline{BB'}$ is of length $2b$ and is called the **conjugate axis** of the hyperbola. The horizontal line segments drawn through B and B', together with the vertical line segments drawn through A and A', form a rectangle. The diagonals of this rectangle have slopes b/a and $-(b/a)$. Therefore, by extending the diagonals, we obtain the asymptotes $y = (b/a)x$ and $y = -(b/a)x$. The two branches of the hyperbola can be sketched by using the asymptotes as guidelines, as shown in Figure 13.35.

E X A M P L E   1

Find the vertices, the foci, and the equations of the asymptotes of the hyperbola $9x^2 - 4y^2 = 36$, and sketch the hyperbola.

Solution

Dividing both sides of the given equation by 36 and simplifying, we change the equation to the standard form

$$\frac{x^2}{4} - \frac{y^2}{9} = 1$$

where $a^2 = 4$ and $b^2 = 9$. Hence $a = 2$ and $b = 3$. The vertices are $(\pm 2, 0)$ and the endpoints of the conjugate axis are $(0, \pm 3)$; these points determine the rectangle whose diagonals extend to become the asymptotes. With $a = 2$ and $b = 3$, the equations of the asymptotes are $y = \frac{3}{2}x$ and $y = -\frac{3}{2}x$. Then, using the relationship $c^2 = a^2 + b^2$, we obtain $c^2 = 4 + 9 = 13$. Thus the foci are at $(\sqrt{13}, 0)$ and $(-\sqrt{13}, 0)$. (The foci are not shown in Figure 13.36.) Using the vertices and the asymptotes, we have sketched the hyperbola in Figure 13.36.

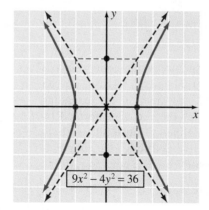

$9x^2 - 4y^2 = 36$

Figure 13.36 ■

E X A M P L E   2

Find the equation of the hyperbola with vertices at $(\pm 4, 0)$ and foci at $(\pm 2\sqrt{5}, 0)$.

Solution

From the given information, we know that $a = 4$ and $c = 2\sqrt{5}$. Then, using the relationship $b^2 = c^2 - a^2$, we obtain

$$b^2 = (2\sqrt{5})^2 - 4^2 = 20 - 16 = 4$$

Substituting 16 for a^2 and 4 for b^2 in the standard form produces

$$\frac{x^2}{16} - \frac{y^2}{4} = 1$$

Multiplying both sides of this equation by 16 yields

$$x^2 - 4y^2 = 16 \qquad \blacksquare$$

■ Hyperbolas with Foci on the *y* Axis

In a similar fashion, we could develop a standard form for the equation of a hyperbola whose foci are on the *y* axis. The following statement summarizes the results of such a development.

Standard Equation: Hyperbola with Transverse Axis on the *y* Axis

> The standard equation of a hyperbola with its center at $(0, 0)$ and its transverse axis on the *y* axis is
>
> $$\frac{y^2}{b^2} - \frac{x^2}{a^2} = 1$$
>
> where the foci are at $(0, -c)$ and $(0, c)$, the vertices are at $(0, -b)$ and $(0, b)$, and $c^2 = a^2 + b^2$.

The endpoints of the conjugate axis are at $(-a, 0)$ and $(a, 0)$. Again, we can determine the asymptotes by extending the diagonals of the rectangle formed by the horizontal lines through the vertices and the vertical lines through the endpoints of the conjugate axis. The equations of the asymptotes are again $y = \pm\dfrac{b}{a}x$. Let's summarize these ideas with Figure 13.37.

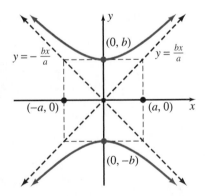

Figure 13.37

E X A M P L E   3

Find the vertices, the foci, and the equations of the asymptotes of the hyperbola $4y^2 - x^2 = 12$, and sketch the hyperbola.

Solution

Divide both sides of the given equation by 12 to change the equation to the standard form:

$$\frac{y^2}{3} - \frac{x^2}{12} = 1$$

where $b^2 = 3$ and $a^2 = 12$. Hence $b = \sqrt{3}$ and $a = 2\sqrt{3}$. The vertices, $(0, \pm\sqrt{3})$, and the endpoints of the conjugate axis, $(\pm 2\sqrt{3}, 0)$, determine the rectangle whose diagonals extend to become the asymptotes. With $b = \sqrt{3}$ and $a = 2\sqrt{3}$, the equations of the asymptotes are $y = \dfrac{\sqrt{3}}{2\sqrt{3}}x = \dfrac{1}{2}x$ and $y = -\dfrac{1}{2}x$. Then, using the relationship $c^2 = a^2 + b^2$, we obtain $c^2 = 12 + 3 = 15$. Thus the foci are at $(0, \sqrt{15})$ and $(0, -\sqrt{15})$. The hyperbola is sketched in Figure 13.38.

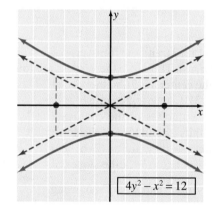

$$4y^2 - x^2 = 12$$

Figure 13.38   ■

■ Other Hyperbolas

In the same way, we can develop the standard form for the equation of a hyperbola that is symmetric with respect to a line parallel to a coordinate axis. We will not show such developments in this text but will simply state and use the results.

$\dfrac{(x - h)^2}{a^2} - \dfrac{(y - k)^2}{b^2} = 1$	A hyperbola with center at (h, k) and transverse axis on the horizontal line $y = k$
$\dfrac{(y - k)^2}{b^2} - \dfrac{(x - h)^2}{a^2} = 1$	A hyperbola with center at (h, k) and transverse axis on the vertical line $x = h$

The relationship $c^2 = a^2 + b^2$ still holds, and the physical significance of a, b, and c remains the same. However, these values are used relative to the center (h, k)

to find the endpoints of the transverse and conjugate axes and to find the foci. Furthermore, the slopes of the asymptotes are as before, but these lines now contain the new center, (h, k). Let's see how all of this works in a specific example.

E X A M P L E   4

Find the vertices, the foci, and the equations of the asymptotes of the hyperbola $9x^2 - 36x - 16y^2 + 96y - 252 = 0$, and sketch the hyperbola.

Solution

First, we need to change to the standard form by completing the square on both x and y.

$$9(x^2 - 4x + \underline{}) - 16(y^2 - 6y + \underline{}) = 252$$
$$9(x^2 - 4x + 4) - 16(y^2 - 6y + 9) = 252 + 9(4) - 16(9)$$
$$9(x - 2)^2 - 16(y - 3)^2 = 144$$
$$\frac{(x - 2)^2}{16} - \frac{(y - 3)^2}{9} = 1$$

The center is at $(2, 3)$, and the transverse axis is on the line $y = 3$. Because $a^2 = 16$, we know that $a = 4$. Therefore the vertices are four units to the right and four units to the left of the center, $(2, 3)$, so they are at $(6, 3)$ and $(-2, 3)$. Likewise, because $b^2 = 9$, or $b = 3$, the endpoints of the conjugate axis are three units up and three units down from the center, so they are at $(2, 6)$ and $(2, 0)$. With $a = 4$ and $b = 3$, the slopes of the asymptotes are $\frac{3}{4}$ and $-\frac{3}{4}$. Then, using the slopes, the center $(2, 3)$, and the point-slope form for writing the equation of a line, we can determine the equations of the asymptotes to be $3x - 4y = -6$ and $3x + 4y = 18$. From the relationship $c^2 = a^2 + b^2$, we obtain $c^2 = 16 + 9 = 25$. Thus the foci are at $(7, 3)$ and $(-3, 3)$. The hyperbola is sketched in Figure 13.39.

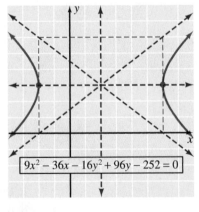

$9x^2 - 36x - 16y^2 + 96y - 252 = 0$

Figure 13.39 ■

E X A M P L E   5

Find the equation of the hyperbola with vertices at $(-4, 2)$ and $(-4, -4)$ and with foci at $(-4, 3)$ and $(-4, -5)$.

Solution

Because the vertices and foci are on the same vertical line ($x = -4$), this hyperbola has an equation of the form

$$\frac{(y - k)^2}{b^2} - \frac{(x - h)^2}{a^2} = 1$$

The center of the hyperbola is at the midpoint of the transverse axis. Therefore

$$h = \frac{-4 + (-4)}{2} = -4 \quad \text{and} \quad k = \frac{2 + (-4)}{2} = -1$$

The distance between the center, $(-4, -1)$, and a vertex, $(-4, 2)$, is three units, so $b = 3$. The distance between the center, $(-4, -1)$, and a focus, $(-4, 3)$, is four units, so $c = 4$. Then, using the relationship $c^2 = a^2 + b^2$, we obtain

$$a^2 = c^2 - b^2 = 16 - 9 = 7$$

Now we can substitute -4 for h, -1 for k, 9 for b^2, and 7 for a^2 in the general form and simplify.

$$\frac{(y + 1)^2}{9} - \frac{(x + 4)^2}{7} = 1$$

$$7(y + 1)^2 - 9(x + 4)^2 = 63$$

$$7(y^2 + 2y + 1) - 9(x^2 + 8x + 16) = 63$$

$$7y^2 + 14y + 7 - 9x^2 - 72x - 144 = 63$$

$$7y^2 + 14y - 9x^2 - 72x - 200 = 0 \qquad \blacksquare$$

The hyperbola also has numerous applications, including many you may not be aware of. For example, one method of artillery range-finding is based on the concept of a hyperbola. If each of two listening posts, P_1 and P_2 in Figure 13.40, records the time that an artillery blast is heard, then the difference between the times multiplied by the speed of sound gives the difference of the distances of the

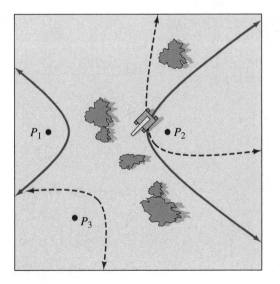

Figure 13.40

gun from the two fixed points. Thus the gun is located somewhere on the hyperbola whose foci are the two listening posts. By bringing in a third listening post, P_3, we can form another hyperbola with foci at P_2 and P_3. Then the location of the gun must be at one of the intersections of the two hyperbolas.

This same principle of intersecting hyperbolas is used in a long-range navigation system known as LORAN. Radar stations serve as the foci of the hyperbolas, and, of course, computers are used for the many calculations that are necessary to fix the location of a plane or ship. At the present time, LORAN is probably used mostly for coastal navigation in connection with small pleasure boats.

Some unique architectural creations have used the concept of a hyperbolic paraboloid, pictured in Figure 13.41. For example, the TWA building at Kennedy Airport is so designed. Some comets, upon entering the sun's gravitational field, follow a hyperbolic path, with the sun as one of the foci (see Figure 13.42).

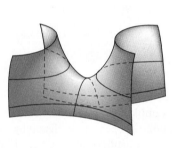

Figure 13.41

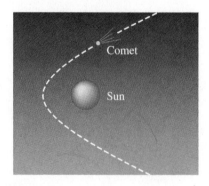

Figure 13.42

Problem Set 13.4

For Problems 1–26, find the vertices, the foci, and the equations of the asymptotes, and sketch each hyperbola.

1. $\dfrac{x^2}{9} - \dfrac{y^2}{4} = 1$

2. $\dfrac{x^2}{4} - \dfrac{y^2}{16} = 1$

3. $\dfrac{y^2}{4} - \dfrac{x^2}{9} = 1$

4. $\dfrac{y^2}{16} - \dfrac{x^2}{4} = 1$

5. $9y^2 - 16x^2 = 144$

6. $4y^2 - x^2 = 4$

7. $x^2 - y^2 = 9$

8. $x^2 - y^2 = 1$

9. $5y^2 - x^2 = 25$

10. $y^2 - 2x^2 = 8$

11. $y^2 - 9x^2 = -9$

12. $16y^2 - x^2 = -16$

13. $\dfrac{(x - 1)^2}{9} - \dfrac{(y + 1)^2}{4} = 1$

14. $\dfrac{(x + 2)^2}{9} - \dfrac{(y + 3)^2}{16} = 1$

15. $\dfrac{(y - 2)^2}{9} - \dfrac{(x - 1)^2}{16} = 1$

16. $\dfrac{(y + 1)^2}{1} - \dfrac{(x + 2)^2}{4} = 1$

17. $4x^2 - 24x - 9y^2 - 18y - 9 = 0$

18. $9x^2 + 72x - 4y^2 - 16y + 92 = 0$

19. $y^2 - 4y - 4x^2 - 24x - 36 = 0$

20. $9y^2 + 54y - x^2 + 6x + 63 = 0$

21. $2x^2 - 8x - y^2 + 4 = 0$

22. $x^2 + 6x - 3y^2 = 0$

23. $y^2 + 10y - 9x^2 + 16 = 0$

24. $4y^2 - 16y - x^2 + 12 = 0$

25. $x^2 + 4x - y^2 - 4y - 1 = 0$

26. $y^2 + 8y - x^2 + 2x + 14 = 0$

For Problems 27–42, find an equation of the hyperbola that satisfies the given conditions.

27. Vertices $(\pm 2, 0)$, foci $(\pm 3, 0)$

28. Vertices $(\pm 1, 0)$, foci $(\pm 4, 0)$

29. Vertices $(0, \pm 3)$, foci $(0, \pm 5)$

30. Vertices $(0, \pm 2)$, foci $(0, \pm 6)$

31. Vertices $(\pm 1, 0)$, contains the point $(2, 3)$

32. Vertices $(0, \pm 1)$, contains the point $(-3, 5)$

33. Vertices $(0, \pm \sqrt{3})$, length of conjugate axis is 4

34. Vertices $(\pm \sqrt{5}, 0)$, length of conjugate axis is 6

35. Foci $(\pm \sqrt{23}, 0)$, length of transverse axis is 8

36. Foci $(0, \pm 3\sqrt{2})$, length of conjugate axis is 4

37. Vertices $(6, -3)$ and $(2, -3)$, foci $(7, -3)$ and $(1, -3)$

38. Vertices $(-7, -4)$ and $(-5, -4)$, foci $(-8, -4)$ and $(-4, -4)$

39. Vertices $(-3, 7)$ and $(-3, 3)$, foci $(-3, 9)$ and $(-3, 1)$

40. Vertices $(7, 5)$ and $(7, -1)$, foci $(7, 7)$ and $(7, -3)$

41. Vertices $(0, 0)$ and $(4, 0)$, foci $(5, 0)$ and $(-1, 0)$

42. Vertices $(0, 0)$ and $(0, -6)$, foci $(0, 2)$ and $(0, -8)$

For Problems 43–52, identify the graph of each of the equations as a straight line, a circle, a parabola, an ellipse, or a hyperbola. Do not sketch the graphs.

43. $x^2 - 7x + y^2 + 8y - 2 = 0$ Circle

44. $x^2 - 7x - y^2 + 8y - 2 = 0$ Hyperbola

45. $5x - 7y = 9$ Linear

46. $4x^2 - x + y^2 + 2y - 3 = 0$ ellipse

47. $10x^2 + y^2 = 8$

48. $-3x - 2y = 9$

49. $5x^2 + 3x - 2y^2 - 3y - 1 = 0$

50. $x^2 + y^2 - 3y - 6 = 0$

51. $x^2 - 3x + y - 4 = 0$

52. $5x + y^2 - 2y - 1 = 0$

■ ■ ■ **THOUGHTS INTO WORDS**

53. What is the difference between the graphs of the equations $x^2 + y^2 = 0$ and $x^2 - y^2 = 0$?

54. What is the difference between the graphs of the equations $4x^2 + 9y^2 = 0$ and $9x^2 + 4y^2 = 0$?

55. A flashlight produces a "cone of light" that can be cut by the plane of a wall to illustrate the conic sections. Try

shining a flashlight against a wall (stand within a couple of feet of the wall) at different angles to produce a circle, an ellipse, a parabola, and one branch of a hyperbola. (You may find it difficult to distinguish between a parabola and a branch of a hyperbola.) Write a paragraph to someone else explaining this experiment.

GRAPHING CALCULATOR ACTIVITIES

56. Use a graphing calculator to check your graphs for Problems 17–26. Be sure to graph the asymptotes for each hyperbola.

57. Use a graphing calculator to check your answers for Problems 43–52.

13.5 Systems Involving Nonlinear Equations

In Chapters 11 and 12, we used several techniques to solve systems of linear equations. We will use two of those techniques in this section to solve some systems that contain at least one nonlinear equation. Furthermore, we will use our knowledge of graphing lines, circles, parabolas, ellipses, and hyperbolas to get a pictorial view of the systems. That will give us a basis for predicting approximate real number solutions if there are any. In other words, we have once again arrived at a topic that vividly illustrates the merging of mathematical ideas. Let's begin by considering a system that contains one linear and one nonlinear equation.

EXAMPLE 1 Solve the system $\left(\begin{array}{c} x^2 + y^2 = 13 \\ 3x + 2y = 0 \end{array} \right)$.

Solution

From our previous graphing experiences, we should recognize that $x^2 + y^2 = 13$ is a circle, and $3x + 2y = 0$ is a straight line. Thus the system can be pictured as in Figure 13.43. The graph indicates that the solution set of this system should consist of two ordered pairs of real numbers that represent the points of intersection in the second and fourth quadrants.

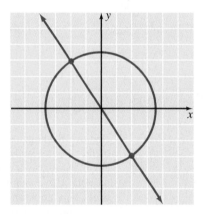

Figure 13.43

Now let's solve the system analytically by using the *substitution method*. Change the form of $3x + 2y = 0$ to $y = -3x/2$, and then substitute $-3x/2$ for y in the other equation to produce

$$x^2 + \left(-\frac{3x}{2} \right)^2 = 13$$

This equation can now be solved for x.

$$x^2 + \frac{9x^2}{4} = 13$$

$$4x^2 + 9x^2 = 52$$

$$13x^2 = 52$$

$$x^2 = 4$$

$$x = \pm 2$$

Substitute 2 for x and then -2 for x in the second equation of the system to produce two values for y.

$$3x + 2y = 0 \qquad\qquad 3x + 2y = 0$$

$$3(2) + 2y = 0 \qquad\qquad 3(-2) + 2y = 0$$

$$2y = -6 \qquad\qquad 2y = 6$$

$$y = -3 \qquad\qquad y = 3$$

Therefore the solution set of the system is $\{(2, -3), (-2, 3)\}$. ■

Remark: Don't forget that, as always, you can check the solutions by substituting them back into the original equations. Graphing the system permits you to approximate any possible real number solutions before solving the system. Then, after solving the system, you can use the graph again to check that the answers are reasonable.

E X A M P L E   2

Solve the system $\begin{pmatrix} x^2 + y^2 = 16 \\ y^2 - x^2 = 4 \end{pmatrix}$.

Solution

Graphing the system produces Figure 13.44. This figure indicates that there should be four ordered pairs of real numbers in the solution set of the system. Solving the system by using the *elimination method* works nicely. We can simply add the two equations, which eliminates the x's.

$$x^2 + y^2 = 16$$
$$\underline{-x^2 + y^2 = 4}$$
$$2y^2 = 20$$
$$y^2 = 10$$
$$y = \pm\sqrt{10}$$

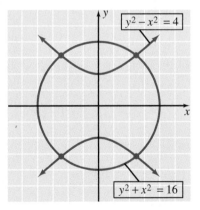

Figure 13.44

Substituting $\sqrt{10}$ for y in the first equation yields

$$x^2 + y^2 = 16$$

$$x^2 + (\sqrt{10})^2 = 16$$

$$x^2 + 10 = 16$$
$$x^2 = 6$$
$$x = \pm\sqrt{6}$$

Thus $(\sqrt{6},\sqrt{10})$ and $(-\sqrt{6},\sqrt{10})$ are solutions. Substituting $-\sqrt{10}$ for y in the first equation yields

$$x^2 + y^2 = 16$$
$$x^2 + (-\sqrt{10})^2 = 16$$
$$x^2 + 10 = 16$$
$$x^2 = 6$$
$$x = \pm\sqrt{6}$$

Thus $(\sqrt{6}, -\sqrt{10})$ and $(-\sqrt{6}, -\sqrt{10})$ are also solutions. The solution set is $\{(-\sqrt{6}, \sqrt{10}), (-\sqrt{6}, -\sqrt{10}), (\sqrt{6}, \sqrt{10}), (\sqrt{6}, -\sqrt{10})\}$. ∎

Sometimes a sketch of the graph of a system may not clearly indicate whether the system contains any real number solutions. The next example illustrates such a situation.

E X A M P L E   3   Solve the system $\begin{pmatrix} y = x^2 + 2 \\ 6x - 4y = -5 \end{pmatrix}$.

Solution

From our previous graphing experiences, we recognize that $y = x^2 + 2$ is the basic parabola shifted upward two units and that $6x - 4y = -5$ is a straight line (see Figure 13.45). Because of the close proximity of the curves, it is difficult to tell whether they intersect. In other words, the graph does not definitely indicate any real number solutions for the system.

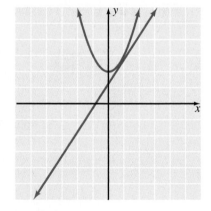

Figure 13.45

Let's solve the system by using the substitution method. We can substitute $x^2 + 2$ for y in the second equation, which produces two values for x.

$$6x - 4(x^2 + 2) = -5$$
$$6x - 4x^2 - 8 = -5$$
$$-4x^2 + 6x - 3 = 0$$
$$4x^2 - 6x + 3 = 0$$

$$x = \frac{6 \pm \sqrt{36 - 48}}{8}$$

$$= \frac{6 \pm \sqrt{-12}}{8}$$

$$= \frac{6 \pm 2i\sqrt{3}}{8}$$

$$= \frac{3 \pm i\sqrt{3}}{4}$$

It is now obvious that the system has no real number solutions. That is, the line and the parabola do not intersect in the real number plane. However, there will be two pairs of complex numbers in the solution set. We can substitute $(3 + i\sqrt{3})/4$ for x in the first equation.

$$y = \left(\frac{3 + i\sqrt{3}}{4}\right)^2 + 2$$

$$= \frac{6 + 6i\sqrt{3}}{16} + 2$$

$$= \frac{6 + 6i\sqrt{3} + 32}{16}$$

$$= \frac{38 + 6i\sqrt{3}}{16}$$

$$= \frac{19 + 3i\sqrt{3}}{8}$$

Likewise, we can substitute $(3 - i\sqrt{3})/4$ for x in the first equation.

$$y = \left(\frac{3 - i\sqrt{3}}{4}\right)^2 + 2$$

$$= \frac{6 - 6i\sqrt{3}}{16} + 2$$

$$= \frac{6 - 6i\sqrt{3} + 32}{16}$$

$$= \frac{38 - 6i\sqrt{3}}{16}$$

$$= \frac{19 - 3i\sqrt{3}}{8}$$

The solution set is $\left\{\left(\dfrac{3 + i\sqrt{3}}{4}, \dfrac{19 + 3i\sqrt{3}}{8}\right), \left(\dfrac{3 - i\sqrt{3}}{4}, \dfrac{19 - 3i\sqrt{3}}{8}\right)\right\}$ ∎

In Example 3 the use of a graphing utility may not, at first, indicate whether the system has any real number solutions. Suppose that we graph the system using a viewing rectangle such that $-15 \leq x \leq 15$ and $-10 \leq y \leq 10$. As shown in the display in Figure 13.46, we cannot tell whether the line and the parabola intersect. However, if we change the viewing rectangle so that $0 \leq x \leq 2$ and $0 \leq y \leq 4$, as shown in Figure 13.47, it becomes apparent that the two graphs do not intersect.

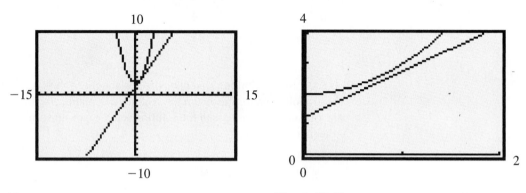

Figure 13.46   **Figure 13.47**

EXAMPLE 4   Find the real number solutions for the system $\begin{pmatrix} y = \log_2(x - 3) - 2 \\ y = -\log_2 x \end{pmatrix}$.

Solution

First, let's use a graphing calculator to obtain a graph of the system as shown in Figure 13.48. The two curves appear to intersect at approximately $x = 4$ and $y = -2$.

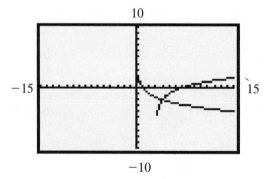

Figure 13.48

To solve the system algebraically, we can equate the two expressions for y and solve the resulting equation for x.

$$\log_2(x - 3) - 2 = -\log_2 x$$
$$\log_2 x + \log_2(x - 3) = 2$$
$$\log_2 x(x - 3) = 2$$

At this step, we can either change to exponential form or rewrite 2 as $\log_2 4$.

$$\log_2 x(x - 3) = \log_2 4$$

$$x(x - 3) = 4$$

$$x^2 - 3x - 4 = 0$$

$$(x - 4)(x + 1) = 0$$

$$x - 4 = 0 \quad \text{or} \quad x + 1 = 0$$

$$x = 4 \quad \text{or} \quad x = -1$$

Because logarithms are not defined for negative numbers, -1 is discarded. Therefore, if $x = 4$, then

$$y = -\log_2 x$$

becomes

$$y = -\log_2 4$$

$$= -2$$

Therefore the solution set is $\{(4, -2)\}$.   ∎

Problem Set 13.5

For Problems 1–30, (a) graph the system so that approximate real number solutions (if there are any) can be predicted, and (b) solve the system by the substitution or elimination method.

1. $\begin{pmatrix} x^2 + y^2 = 5 \\ x + 2y = 5 \end{pmatrix}$

2. $\begin{pmatrix} x^2 + y^2 = 13 \\ 2x + 3y = 13 \end{pmatrix}$

3. $\begin{pmatrix} x^2 + y^2 = 26 \\ x + y = -4 \end{pmatrix}$

4. $\begin{pmatrix} x^2 + y^2 = 10 \\ x + y = -2 \end{pmatrix}$

5. $\begin{pmatrix} x^2 + y^2 = 2 \\ x - y = 4 \end{pmatrix}$

6. $\begin{pmatrix} x^2 + y^2 = 3 \\ x - y = -5 \end{pmatrix}$

7. $\begin{pmatrix} y = x^2 + 6x + 7 \\ 2x + y = -5 \end{pmatrix}$

8. $\begin{pmatrix} y = x^2 - 4x + 5 \\ y - x = 1 \end{pmatrix}$

9. $\begin{pmatrix} 2x + y = -2 \\ y = x^2 + 4x + 7 \end{pmatrix}$

10. $\begin{pmatrix} 2x + y = 0 \\ y = -x^2 + 2x - 4 \end{pmatrix}$

11. $\begin{pmatrix} y = x^2 - 3 \\ x + y = -4 \end{pmatrix}$

12. $\begin{pmatrix} y = -x^2 + 1 \\ x + y = 2 \end{pmatrix}$

13. $\begin{pmatrix} x^2 + 2y^2 = 9 \\ x - 4y = -9 \end{pmatrix}$

14. $\begin{pmatrix} 2x - y = 7 \\ 3x^2 + y^2 = 21 \end{pmatrix}$

15. $\begin{pmatrix} x + y = -3 \\ x^2 + 2y^2 - 12y - 18 = 0 \end{pmatrix}$

16. $\begin{pmatrix} 4x^2 + 9y^2 = 25 \\ 2x + 3y = 7 \end{pmatrix}$

17. $\begin{pmatrix} x - y = 2 \\ x^2 - y^2 = 16 \end{pmatrix}$

18. $\begin{pmatrix} x^2 - 4y^2 = 16 \\ 2y - x = 2 \end{pmatrix}$

19. $\begin{pmatrix} y = -x^2 + 3 \\ y = x^2 + 1 \end{pmatrix}$

20. $\begin{pmatrix} y = x^2 \\ y = x^2 - 4x + 4 \end{pmatrix}$

21. $\begin{pmatrix} y = x^2 + 2x - 1 \\ y = x^2 + 4x + 5 \end{pmatrix}$

22. $\begin{pmatrix} y = -x^2 + 1 \\ y = x^2 - 2 \end{pmatrix}$

23. $\begin{pmatrix} x^2 - y^2 = 4 \\ x^2 + y^2 = 4 \end{pmatrix}$

24. $\begin{pmatrix} 2x^2 + y^2 = 8 \\ x^2 + y^2 = 4 \end{pmatrix}$

25. $\begin{pmatrix} 8y^2 - 9x^2 = 6 \\ 8x^2 - 3y^2 = 7 \end{pmatrix}$

26. $\begin{pmatrix} 2x^2 + y^2 = 11 \\ x^2 - y^2 = 4 \end{pmatrix}$

27. $\begin{pmatrix} 2x^2 - 3y^2 = -1 \\ 2x^2 + 3y^2 = 5 \end{pmatrix}$

28. $\begin{pmatrix} 4x^2 + 3y^2 = 9 \\ y^2 - 4x^2 = 7 \end{pmatrix}$

29. $\begin{pmatrix} xy = 3 \\ 2x + 2y = 7 \end{pmatrix}$

30. $\begin{pmatrix} x^2 + 4y^2 = 25 \\ xy = 6 \end{pmatrix}$

For Problems 31–36, solve each system for all real number solutions.

31. $\begin{pmatrix} y = \log_3(x - 6) - 3 \\ y = -\log_3 x \end{pmatrix}$ **32.** $\begin{pmatrix} y = \log_{10}(x - 9) - 1 \\ y = -\log_{10} x \end{pmatrix}$

33. $\begin{pmatrix} y = e^x - 1 \\ y = 2e^{-x} \end{pmatrix}$ **34.** $\begin{pmatrix} y = 28 - 11e^x \\ y = -e^{2x} \end{pmatrix}$

35. $\begin{pmatrix} y = x^3 \\ y = x^3 + 2x^2 + 5x - 3 \end{pmatrix}$

36. $\begin{pmatrix} y = 3(4^x) - 8 \\ y = 4^{2x} - 2(4^x) - 4 \end{pmatrix}$

■ ■ ■ **THOUGHTS INTO WORDS**

37. What happens if you try to graph the system

$$\begin{pmatrix} 7x^2 + 8y^2 = 36 \\ 11x^2 + 5y^2 = -4 \end{pmatrix}?$$

38. For what value(s) of k will the line $x + y = k$ touch the ellipse $x^2 + 2y^2 = 6$ in one and only one point? Defend your answer.

39. The system

$$\begin{pmatrix} x^2 - 6x + y^2 - 4y + 4 = 0 \\ x^2 - 4x + y^2 + 8y - 5 = 0 \end{pmatrix}$$

represents two circles that intersect in two points. An equivalent system can be formed by replacing the second equation with the result of adding -1 times the first equation to the second equation. Thus we obtain the system

$$\begin{pmatrix} x^2 - 6x + y^2 - 4y + 4 = 0 \\ 2x + 12y - 9 = 0 \end{pmatrix}$$

Explain why the linear equation in this system is the equation of the common chord of the original two intersecting circles.

▨ **GRAPHING CALCULATOR ACTIVITIES**

40. Graph the system of equations $\begin{pmatrix} y = x^2 + 2 \\ 6x - 4y = -5 \end{pmatrix}$, and use the TRACE and ZOOM features of your calculator to show that this system has no real number solutions.

41. Use a graphing calculator to graph the systems in Problems 31–36, and check the reasonableness of your answers to those problems.

For Problems 42–47, use a graphing calculator to approximate, to the nearest tenth, the real number solutions for each system of equations.

42. $\begin{pmatrix} y = e^x + 1 \\ y = x^3 + x^2 - 2x - 1 \end{pmatrix}$

43. $\begin{pmatrix} y = x^3 + 2x^2 - 3x + 2 \\ y = -x^3 - x^2 + 1 \end{pmatrix}$

44. $\begin{pmatrix} y = 2^x + 1 \\ y = 2^{-x} + 2 \end{pmatrix}$

45. $\begin{pmatrix} y = \ln(x - 1) \\ y = x^2 - 16x + 64 \end{pmatrix}$

46. $\begin{pmatrix} x = y^2 - 2y + 3 \\ x^2 + y^2 = 25 \end{pmatrix}$

47. $\begin{pmatrix} y^2 - x^2 = 16 \\ 2y^2 - x^2 = 8 \end{pmatrix}$

The following standard forms for the equations of conic sections were developed in this chapter.

(13.1) Circles

$x^2 + y^2 = r^2$

center at $(0, 0)$ and radius of length r

$(x - h)^2 + (y - k)^2 = r^2$

center at (h, k) and radius of length r

(13.2) Parabolas

$x^2 = 4py$

focus $(0, p)$, directrix $y = -p$, y axis symmetry

$(x - h)^2 = 4p(y - k)$

focus $(h, k + p)$, directrix $y = k - p$, symmetric with respect to the line $x = h$

$y^2 = 4px$

focus $(p, 0)$, directrix $x = -p$, x axis symmetry

$(y - k)^2 = 4p(x - h)$

focus $(h + p, k)$, directrix $x = h - p$, symmetric with respect to the line $y = k$

(13.3) Ellipses

$\dfrac{x^2}{a^2} + \dfrac{y^2}{b^2} = 1, \quad a^2 > b^2$

center $(0, 0)$, vertices $(\pm a, 0)$, endpoints of minor axis $(0, \pm b)$, foci $(\pm c, 0)$, $c^2 = a^2 - b^2$

$\dfrac{(x - h)^2}{a^2} + \dfrac{(y - k)^2}{b^2} = 1, \quad a^2 > b^2$

center (h, k), vertices $(h \pm a, k)$, endpoints of minor axis $(h, k \pm b)$, foci $(h \pm c, k)$, $c^2 = a^2 - b^2$

$\dfrac{x^2}{a^2} + \dfrac{y^2}{b^2} = 1, \quad b^2 > a^2$

center $(0, 0)$, vertices $(0, \pm b)$, endpoints of minor axis $(\pm a, 0)$, foci $(0, \pm c)$, $c^2 = b^2 - a^2$

$\dfrac{(x - h)^2}{a^2} + \dfrac{(y - k)^2}{b^2} = 1, \quad b^2 > a^2$

center (h, k), vertices $(h, k \pm b)$, endpoints of minor axis $(h \pm a, k)$, foci $(h, k \pm c)$, $c^2 = b^2 - a^2$

(13.4) Hyperbolas

$\dfrac{x^2}{a^2} - \dfrac{y^2}{b^2} = 1$

center $(0, 0)$, vertices $(\pm a, 0)$, endpoints of conjugate axis $(0, \pm b)$, foci $(\pm c, 0)$, $c^2 = a^2 + b^2$,

asymptotes $y = \pm\dfrac{b}{a}x$

$\dfrac{(x - h)^2}{a^2} - \dfrac{(y - k)^2}{b^2} = 1$

center (h, k), vertices $(h \pm a, k)$, endpoints of conjugate axis $(h, k \pm b)$, foci $(h \pm c, k)$, $c^2 = a^2 + b^2$,

asymptotes $y - k = \pm\dfrac{b}{a}(x - h)$

$\dfrac{y^2}{b^2} - \dfrac{x^2}{a^2} = 1$

center $(0, 0)$, vertices $(0, \pm b)$, endpoints of conjugate axis $(\pm a, 0)$, foci $(0, \pm c)$, $c^2 = a^2 + b^2$,

asymptotes $y = \pm\dfrac{b}{a}x$

$\dfrac{(y - k)^2}{b^2} - \dfrac{(x - h)^2}{a^2} = 1$

center (h, k), vertices $(h, k \pm b)$, endpoints of conjugate axis $(h \pm a, k)$, foci $(h, k \pm c)$, $c^2 = a^2 + b^2$,

asymptotes $y - k = \pm\dfrac{b}{a}(x - h)$

(13.5) Systems that contain at least one nonlinear equation can often be solved by substitution or by the elimination method. Graphing the system will often provide a basis for predicting approximate real number solutions if there are any.

Chapter 13   Review Problem Set

For Problems 1–14, (a) identify the conic section as a circle, a parabola, an ellipse, or a hyperbola. (b) If it is a circle, find its center and the length of a radius; if it is a parabola, find its vertex, focus, and directrix; if it is an ellipse, find its vertices, the endpoints of its minor axis, and its foci; if it is a hyperbola, find its vertices, the endpoints of its conjugate axis, its foci, and its asymptotes. (c) Sketch each of the curves.

1. $x^2 + 2y^2 = 32$ ℓ

2. $y^2 = -12x$?

3. $3y^2 - x^2 = 9$ P

4. $2x^2 - 3y^2 = 18$ ⌣

5. $5x^2 + 2y^2 = 20$ ℓ

6. $x^2 = 2y$ p

7. $x^2 + y^2 = 10$ c

8. $x^2 - 8x - 2y^2 + 4y + 10 = 0$ ⌣

9. $9x^2 - 54x + 2y^2 + 8y + 71 = 0$ ℓ

10. $y^2 - 2y + 4x + 9 = 0$ p

11. $x^2 + 2x + 8y + 25 = 0$ P

12. $x^2 + 10x + 4y^2 - 16y + 25 = 0$ ℓ

13. $3y^2 + 12y - 2x^2 - 8x - 8 = 0$ ⌣

14. $x^2 - 6x + y^2 + 4y - 3 = 0$ c

For Problems 15–28, find the equation of the indicated conic section that satisfies the given conditions.

15. Circle with center at $(-8, 3)$ and a radius of length $\sqrt{5}$ units

16. Parabola with vertex $(0, 0)$, focus $(-5, 0)$, directrix $x = 5$

17. Ellipse with vertices $(0, \pm 4)$, foci $(0, \pm \sqrt{15})$

18. Hyperbola with vertices $(\pm \sqrt{2}, 0)$, length of conjugate axis 10

19. Circle with center at $(5, -12)$, passes through the origin

20. Ellipse with vertices $(\pm 2, 0)$, contains the point $(1, -2)$

21. Parabola with vertex $(0, 0)$, symmetric with respect to the y axis, contains the point $(2, 6)$

22. Hyperbola with vertices $(0, \pm 1)$, foci $(0, \pm \sqrt{10})$

23. Ellipse with vertices $(6, 1)$ and $(6, 7)$, length of minor axis 2 units

24. Parabola with vertex $(4, -2)$, focus $(6, -2)$

25. Hyperbola with vertices $(-5, -3)$ and $(-5, -5)$, foci $(-5, -2)$ and $(-5, -6)$

26. Parabola with vertex $(-6, -3)$, symmetric with respect to the line $x = -6$, contains the point $(-5, -2)$

27. Ellipse with endpoints of minor axis $(-5, 2)$ and $(-5, -2)$, length of major axis 10 units

28. Hyperbola with vertices $(2, 0)$ and $(6, 0)$, length of conjugate axis 8 units

For Problems 29–34, (a) graph the system, and (b) solve the system by using the substitution or elimination method.

29. $\begin{pmatrix} x^2 + y^2 = 17 \\ x - 4y = -17 \end{pmatrix}$

30. $\begin{pmatrix} x^2 - y^2 = 8 \\ 3x - y = 8 \end{pmatrix}$

31. $\begin{pmatrix} x - y = 1 \\ y = x^2 + 4x + 1 \end{pmatrix}$

32. $\begin{pmatrix} 4x^2 - y^2 = 16 \\ 9x^2 + 9y^2 = 16 \end{pmatrix}$

33. $\begin{pmatrix} x^2 + 2y^2 = 8 \\ 2x^2 + 3y^2 = 12 \end{pmatrix}$

34. $\begin{pmatrix} y^2 - x^2 = 1 \\ 4x^2 + y^2 = 4 \end{pmatrix}$

1. Find the focus of the parabola $x^2 = -20y$.

2. Find the vertex of the parabola
$y^2 - 4y - 8x - 20 = 0$.

3. Find the equation of the directrix for the parabola
$2y^2 = 24x$.

4. Find the focus of the parabola $y^2 = 24x$.

5. Find the vertex of the parabola $x^2 + 4x - 12y - 8 = 0$.

6. Find the center of the circle $x^2 + 6x + y^2 + 18y + 87 = 0$.

7. Find the equation of the parabola that has its vertex at the origin, is symmetric with respect to the x axis, and contains the point $(-2, 4)$.

8. Find the equation of the parabola that has its vertex at $(3, 4)$ and its focus at $(3, 1)$.

9. Find the equation of the circle that has its center at $(-1, 6)$ and has a radius of length 5 units.

10. Find the length of the major axis of the ellipse $x^2 - 4x + 9y^2 - 18y + 4 = 0$.

11. Find the endpoints of the minor axis of the ellipse $9x^2 + 90x + 4y^2 - 8y + 193 = 0$.

12. Find the foci of the ellipse $x^2 + 4y^2 = 16$.

13. Find the center of the ellipse $3x^2 + 30x + y^2 - 16y + 79 = 0$.

14. Find the equation of the ellipse that has the endpoints of its major axis at $(0, \pm 10)$ and its foci at $(0, \pm 8)$.

15. Find the equation of the ellipse that has the endpoints of its major axis at $(2, -2)$ and $(10, -2)$ and the endpoints of its minor axis at $(6, 0)$ and $(6, -4)$.

16. Find the equations of the asymptotes of the hyperbola $4y^2 - 9x^2 = 32$.

17. Find the vertices of the hyperbola $y^2 - 6y - 3x^2 - 6x - 3 = 0$.

18. Find the foci of the hyperbola $5x^2 - 4y^2 = 20$.

19. Find the equation of the hyperbola that has its vertices at $(\pm 6, 0)$ and its foci at $(\pm 4\sqrt{3}, 0)$.

20. Find the equation of the hyperbola that has its vertices at $(0, 4)$ and $(-2, 4)$ and its foci at $(2, 4)$ and $(-4, 4)$.

21. How many real number solutions are there for the system $\begin{pmatrix} x^2 + y^2 = 16 \\ x^2 - 4y = 8 \end{pmatrix}$?

22. Solve the system $\begin{pmatrix} x^2 + 4y^2 = 25 \\ xy = 6 \end{pmatrix}$.

For Problems 23–25, graph each conic section.

23. $y^2 + 4y + 8x - 4 = 0$

24. $9x^2 - 36x + 4y^2 + 16y + 16 = 0$

25. $x^2 + 6x - 3y^2 = 0$

Sequences and Mathematical Induction

© Royalty-Free/CORBIS

When objects are arranged in a sequence, the total number of objects is the sum of the terms of the sequence.

Suppose that an auditorium has 35 seats in the first row, 40 seats in the second row, 45 seats in the third row, and so on, for ten rows. The numbers 35, 40, 45, 50, . . . , 80 represent the number of seats per row from row 1 through row 10. This list of numbers has a constant difference of 5 between any two successive numbers in the list; such a list is called an **arithmetic sequence**. (Used in this sense, the word arithmetic is pronounced with the accent on the syllable *met.*)

Suppose that a fungus culture growing under controlled conditions doubles in size each day. If today the size of the culture is 6 units, then the numbers 12, 24, 48, 96, 192 represent the size of the culture for the next 5 days. In this list of numbers, each number after the first is twice the previous number; such a list is called a **geometric sequence**. Arithmetic sequences and geometric sequences will be the center of our attention in this chapter.

14.1 Arithmetic Sequences

An **infinite sequence** is a function whose domain is the set of positive integers. For example, consider the function defined by the equation

$$f(n) = 5n + 1$$

where the domain is the set of positive integers. If we substitute the numbers of the domain in order, starting with 1, we can list the resulting ordered pairs:

$$(1, 6) \quad (2, 11) \quad (3, 16) \quad (4, 21) \quad (5, 26)$$

and so on. However, because we know we are using the domain of positive integers in order, starting with 1, there is no need to use ordered pairs. We can simply express the infinite sequence as

$$6, 11, 16, 21, 26, \ldots$$

Often the letter a is used to represent sequential functions, and the functional value of a at n is written a_n (this is read "a sub n") instead of $a(n)$. The sequence is then expressed as

$$a_1, a_2, a_3, a_4, \ldots$$

where a_1 is the **first term**, a_2 is the **second term**, a_3 is the **third term,** and so on. The expression a_n, which defines the sequence, is called the **general term** of the sequence. Knowing the general term of a sequence enables us to find as many terms of the sequence as needed and also to find any specific terms. Consider the following example.

EXAMPLE 1

Find the first five terms of the sequence where $a_n = 2n^2 - 3$; find the 20th term.

Solution

The first five terms are generated by replacing n with 1, 2, 3, 4, and 5.

$$a_1 = 2(1)^2 - 3 = -1 \qquad a_2 = 2(2)^2 - 3 = 5$$
$$a_3 = 2(3)^2 - 3 = 15 \qquad a_4 = 2(4)^2 - 3 = 29$$
$$a_5 = 2(5)^2 - 3 = 47$$

The first five terms are thus $-1, 5, 15, 29,$ and 47. The 20th term is

$$a_{20} = 2(20)^2 - 3 = 797 \qquad \blacksquare$$

■ Arithmetic Sequences

An **arithmetic sequence** (also called an **arithmetic progression**) is a sequence that has a common difference between successive terms. The following are examples of arithmetic sequences:

$$1, 8, 15, 22, 29, \ldots$$

$$4, 7, 10, 13, 16, \ldots$$

$$4, 1, -2, -5, -8, \ldots$$
$$-1, -6, -11, -16, -21, \ldots$$

The common difference in the first sequence is 7. That is, $8 - 1 = 7$, $15 - 8 = 7$, $22 - 15 = 7$, $29 - 22 = 7$, and so on. The common differences for the next three sequences are 3, -3, and -5, respectively.

In a more general setting, we say that the sequence

$$a_1, a_2, a_3, a_4, \ldots, a_n, \ldots$$

is an arithmetic sequence if and only if there is a real number d such that

$$a_{k+1} - a_k = d$$

for every positive integer k. The number d is called the **common difference**.

From the definition, we see that $a_{k+1} = a_k + d$. In other words, we can generate an arithmetic sequence that has a common difference of d by starting with a first term a_1 and then simply adding d to each successive term.

First term:	a_1	
Second term:	$a_1 + d$	
Third term:	$a_1 + 2d$	$(a_1 + d) + d = a_1 + 2d$
Fourth term:	$a_1 + 3d$	
.		
.		
.		
nth term:	$a_1 + (n - 1)d$	

Thus the **general term** of an arithmetic sequence is given by

$$a_n = a_1 + (n - 1)d$$

where a_1 is the first term, and d is the common difference. This formula for the general term can be used to solve a variety of problems involving arithmetic sequences.

E X A M P L E   2

Find the general term of the arithmetic sequence $6, 2, -2, -6, \ldots$.

Solution

The common difference, d, is $2 - 6 = -4$, and the first term, a_1, is 6. Substitute these values into $a_n = a_1 + (n - 1)d$ and simplify to obtain

$$a_n = a_1 + (n - 1)d$$
$$= 6 + (n - 1)(-4)$$
$$= 6 - 4n + 4$$
$$= -4n + 10$$

EXAMPLE 3

Find the 40th term of the arithmetic sequence $1, 5, 9, 13, \ldots$.

Solution

Using $a_n = a_1 + (n - 1)d$, we obtain

$$a_{40} = 1 + (40 - 1)4$$

$$= 1 + (39)(4)$$

$$= 157 \qquad\blacksquare$$

EXAMPLE 4

Find the first term of the arithmetic sequence where the fourth term is 26 and the ninth term is 61.

Solution

Using $a_n = a_1 + (n - 1)d$ with $a_4 = 26$ (the fourth term is 26) and $a_9 = 61$ (the ninth term is 61), we have

$$26 = a_1 + (4 - 1)d = a_1 + 3d$$

$$61 = a_1 + (9 - 1)d = a_1 + 8d$$

Solving the system of equations

$$\begin{pmatrix} a_1 + 3d = 26 \\ a_1 + 8d = 61 \end{pmatrix}$$

yields $a_1 = 5$ and $d = 7$. Thus the first term is 5. $\qquad\blacksquare$

■ Sums of Arithmetic Sequences

We often use sequences to solve problems, so we need to be able to find the sum of a certain number of terms of the sequence. Before we develop a general-sum formula for arithmetic sequences, let's consider an approach to a specific problem that we can then use in a general setting.

EXAMPLE 5

Find the sum of the first 100 positive integers.

Solution

We are being asked to find the sum of $1 + 2 + 3 + 4 + \cdots + 100$. Rather than adding in the usual way, we will find the sum in the following manner: Let's simply write the indicated sum forward and backward, and then add in a column fashion.

$$\begin{array}{ccccccccc} 1 & + & 2 & + & 3 & + & 4 & + \cdots + & 100 \\ 100 & + & 99 & + & 98 & + & 97 & + \cdots + & 1 \\ \hline 101 & + & 101 & + & 101 & + & 101 & + \cdots + & 101 \end{array}$$

We have produced 100 sums of 101. However, this result is double the amount we want because we wrote the sum twice. To find the sum of just the numbers 1 to 100, we need to multiply 100 by 101 and then divide by 2.

$$\frac{100(101)}{2} = \frac{\overset{50}{\cancel{100}}(101)}{\cancel{2}} = 5050$$

Thus the sum of the first 100 positive integers is 5050. ∎

The *forward–backward* approach we used in Example 5 can be used to develop a formula for finding the sum of the first n terms of any arithmetic sequence. Consider an arithmetic sequence $a_1, a_2, a_3, a_4, \ldots, a_n$ with a common difference of d. Use S_n to represent the sum of the first n terms, and proceed as follows:

$$S_n = a_1 + (a_1 + d) + (a_1 + 2d) + \cdots + (a_n - 2d) + (a_n - d) + a_n$$

Now write this sum in reverse.

$$S_n = a_n + (a_n - d) + (a_n - 2d) + \cdots + (a_1 + 2d) + (a_1 + d) + a_1$$

Add the two equations to produce

$$2S_n = (a_1 + a_n) + (a_1 + a_n) + (a_1 + a_n) + \cdots + (a_1 + a_n) + (a_1 + a_n) + (a_1 + a_n)$$

That is, we have n sums $a_1 + a_n$, so

$$2S_n = n(a_1 + a_n)$$

from which we obtain a **sum formula**:

$$S_n = \frac{n(a_1 + a_n)}{2}$$

Using the nth-term formula and/or the sum formula, we can solve a variety of problems involving arithmetic sequences.

E X A M P L E   6

Find the sum of the first 30 terms of the arithmetic sequence $3, 7, 11, 15, \ldots$.

Solution

To use the formula $S_n = \dfrac{n(a_1 + a_n)}{2}$, we need to know the number of terms (n), the first term (a_1), and the last term (a_n). We are given the number of terms and the first term, so we need to find the last term. Using $a_n = a_1 + (n - 1)d$, we can find the 30th term.

$$a_{30} = 3 + (30 - 1)4 = 3 + 29(4) = 119$$

Now we can use the sum formula.

$$S_{30} = \frac{30(3 + 119)}{2} = 1830$$

■

EXAMPLE 7

Find the sum $7 + 10 + 13 + \cdots + 157$.

Solution

To use the sum formula, we need to know the number of terms. Applying the nth-term formula will give us that information.

$$a_n = a_1 + (n - 1)d$$
$$157 = 7 + (n - 1)3$$
$$157 = 7 + 3n - 3$$
$$157 = 3n + 4$$
$$153 = 3n$$
$$51 = n$$

Now we can use the sum formula.

$$S_{51} = \frac{51(7 + 157)}{2} = 4182$$

■

Keep in mind that we developed the sum formula for an arithmetic sequence by using the forward–backward technique, which we had previously used on a specific problem. Now that we have the sum formula, we have two choices when solving problems. We can either memorize the formula and use it or simply use the forward–backward technique. If you choose to use the formula and some day you forget it, don't panic. Just use the forward–backward technique. In other words, understanding the development of a formula often enables you to do problems even when you forget the formula itself.

■ Summation Notation

Sometimes a special notation is used to indicate the sum of a certain number of terms of a sequence. The capital Greek letter *sigma*, Σ, is used as a **summation symbol**. For example,

$$\sum_{i=1}^{5} a_i$$

represents the sum $a_1 + a_2 + a_3 + a_4 + a_5$. The letter i is frequently used as the **index of summation**; the letter i takes on all integer values from the lower limit to the upper limit, inclusive. Thus

$$\sum_{i=1}^{4} b_i = b_1 + b_2 + b_3 + b_4$$

$$\sum_{i=3}^{7} a_i = a_3 + a_4 + a_5 + a_6 + a_7$$

$$\sum_{i=1}^{15} i^2 = 1^2 + 2^2 + 3^2 + \cdots + 15^2$$

$$\sum_{i=1}^{n} a_i = a_1 + a_2 + a_3 + \cdots + a_n$$

If $a_1, a_2, a_3, \ldots$ represents an arithmetic sequence, we can now write the sum formula

$$\sum_{i=1}^{n} a_i = \frac{n}{2}(a_1 + a_n)$$

EXAMPLE 8

Find the sum $\sum_{i=1}^{50} (3i + 4)$.

Solution

This indicated sum means

$$\sum_{i=1}^{50} (3i + 4) = [3(1) + 4] + [3(2) + 4] + [3(3) + 4] + \cdots + [3(50) + 4]$$
$$= 7 + 10 + 13 + \cdots + 154$$

Because this is an indicated sum of an arithmetic sequence, we can use our sum formula.

$$S_{50} = \frac{50}{2}(7 + 154) = 4025$$

∎

EXAMPLE 9

Find the sum $\sum_{i=2}^{7} 2i^2$

Solution

This indicated sum means

$$\sum_{i=2}^{7} 2i^2 = 2(2)^2 + 2(3)^2 + 2(4)^2 + 2(5)^2 + 2(6)^2 + 2(7)^2$$
$$= 8 + 18 + 32 + 50 + 72 + 98$$

This is not the indicated sum of an *arithmetic* sequence; therefore let's simply add the numbers in the usual way. The sum is 278.

∎

Example 9 suggests a word of caution. Be sure to analyze the sequence of numbers that is represented by the summation symbol. You may or may not be able to use a formula for adding the numbers.

Problem Set 14.1

For Problems 1–10, write the first five terms of the sequence that has the indicated general term.

1. $a_n = 3n - 7$ **2.** $a_n = 5n - 2$

3. $a_n = -2n + 4$ **4.** $a_n = -4n + 7$

5. $a_n = 3n^2 - 1$ **6.** $a_n = 2n^2 - 6$

7. $a_n = n(n - 1)$ **8.** $a_n = (n + 1)(n + 2)$

9. $a_n = 2^{n+1}$ **10.** $a_n = 3^{n-1}$

11. Find the 15th and 30th terms of the sequence where $a_n = -5n - 4$.

12. Find the 20th and 50th terms of the sequence where $a_n = -n - 3$.

13. Find the 25th and 50th terms of the sequence where $a_n = (-1)^{n+1}$.

14. Find the 10th and 15th terms of the sequence where $a_n = -n^2 - 10$.

For Problems 15–24, find the general term (the nth term) for each arithmetic sequence.

15. $11, 13, 15, 17, 19, \ldots$

16. $7, 10, 13, 16, 19, \ldots$

17. $2, -1, -4, -7, -10, \ldots$

18. $4, 2, 0, -2, -4, \ldots$

19. $\dfrac{3}{2}, 2, \dfrac{5}{2}, 3, \dfrac{7}{2}, \ldots$

20. $0, \dfrac{1}{2}, 1, \dfrac{3}{2}, 2, \ldots$

21. $2, 6, 10, 14, 18, \ldots$

22. $2, 7, 12, 17, 22, \ldots$

23. $-3, -6, -9, -12, -15, \ldots$

24. $-4, -8, -12, -16, -20, \ldots$

For Problems 25–30, find the required term for each arithmetic sequence.

25. The 15th term of $3, 8, 13, 18, \ldots$

26. The 20th term of $4, 11, 18, 25, \ldots$

27. The 30th term of $15, 26, 37, 48, \ldots$

28. The 35th term of $9, 17, 25, 33, \ldots$

29. The 52nd term of $1, \dfrac{5}{3}, \dfrac{7}{3}, 3, \ldots$

30. The 47th term of $\dfrac{1}{2}, \dfrac{5}{4}, 2, \dfrac{11}{4}, \ldots$

For Problems 31–42, solve each problem.

31. If the 6th term of an arithmetic sequence is 12 and the 10th term is 16, find the first term.

32. If the 5th term of an arithmetic sequence is 14 and the 12th term is 42, find the first term.

33. If the 3rd term of an arithmetic sequence is 20 and the 7th term is 32, find the 25th term.

34. If the 5th term of an arithmetic sequence is -5 and the 15th term is -25, find the 50th term.

35. Find the sum of the first 50 terms of the arithmetic sequence $5, 7, 9, 11, 13, \ldots$.

36. Find the sum of the first 30 terms of the arithmetic sequence $0, 2, 4, 6, 8, \ldots$.

37. Find the sum of the first 40 terms of the arithmetic sequence $2, 6, 10, 14, 18, \ldots$.

38. Find the sum of the first 60 terms of the arithmetic sequence $-2, 3, 8, 13, 18, \ldots$.

39. Find the sum of the first 75 terms of the arithmetic sequence $5, 2, -1, -4, -7, \ldots$.

40. Find the sum of the first 80 terms of the arithmetic sequence $7, 3, -1, -5, -9, \ldots$.

41. Find the sum of the first 50 terms of the arithmetic sequence $\dfrac{1}{2}, 1, \dfrac{3}{2}, 2, \dfrac{5}{2}, \ldots$.

42. Find the sum of the first 100 terms of the arithmetic sequence $-\dfrac{1}{3}, \dfrac{1}{3}, 1, \dfrac{5}{3}, \dfrac{7}{3}, \ldots$.

For Problems 43–50, find the indicated sum.

43. $1 + 5 + 9 + 13 + \cdots + 197$

44. $3 + 8 + 13 + 18 + \cdots + 398$

45. $2 + 8 + 14 + 20 + \cdots + 146$

46. $6 + 9 + 12 + 15 + \cdots + 93$

47. $(-7) + (-10) + (-13) + (-16) + \cdots + (-109)$

48. $(-5) + (-9) + (-13) + (-17) + \cdots + (-169)$

49. $(-5) + (-3) + (-1) + 1 + \cdots + 119$

50. $(-7) + (-4) + (-1) + 2 + \cdots + 131$

For Problems 51–58, solve each problem.

51. Find the sum of the first 200 odd whole numbers.

52. Find the sum of the first 175 positive even whole numbers.

53. Find the sum of all even numbers between 18 and 482, inclusive.

54. Find the sum of all odd numbers between 17 and 379, inclusive.

55. Find the sum of the first 30 terms of the arithmetic sequence with the general term $a_n = 5n - 4$.

56. Find the sum of the first 40 terms of the arithmetic sequence with the general term $a_n = 4n - 7$.

57. Find the sum of the first 25 terms of the arithmetic sequence with the general term $a_n = -4n - 1$.

58. Find the sum of the first 35 terms of the arithmetic sequence with the general term $a_n = -5n - 3$.

For Problems 59–70, find each sum.

59. $\sum_{i=1}^{45} (5i + 2)$

60. $\sum_{i=1}^{38} (3i + 6)$

61. $\sum_{i=1}^{30} (-2i + 4)$

62. $\sum_{i=1}^{40} (-3i + 3)$

63. $\sum_{i=4}^{32} (3i - 10)$

64. $\sum_{i=6}^{47} (4i - 9)$

65. $\sum_{i=10}^{20} 4i$

66. $\sum_{i=15}^{30} (-5i)$

67. $\sum_{i=1}^{5} i^2$

68. $\sum_{i=1}^{6} (i^2 + 1)$

69. $\sum_{i=3}^{8} (2i^2 + i)$

70. $\sum_{i=4}^{7} (3i^2 - 2)$

■■■ **THOUGHTS INTO WORDS**

71. Before developing the formula $a_n = a_1 + (n - 1)d$, we stated the equation $a_{k+1} - a_k = d$. In your own words, explain what this equation says.

72. Explain how to find the sum $1 + 2 + 3 + 4 + \cdots + 175$ without using the sum formula.

73. Explain in words how to find the sum of the first n terms of an arithmetic sequence.

74. Explain how one can tell that a particular sequence is an arithmetic sequence.

■■■ **FURTHER INVESTIGATIONS**

The general term of a sequence can consist of one expression for certain values of n and another expression (or expressions) for other values of n. That is, a **multiple description** of the sequence can be given. For example,

$$a_n = \begin{cases} 2n + 3 & \text{for } n \text{ odd} \\ 3n - 2 & \text{for } n \text{ even} \end{cases}$$

means that we use $a_n = 2n + 3$ for $n = 1, 3, 5, 7, \ldots$, and we use $a_n = 3n - 2$ for $n = 2, 4, 6, 8, \ldots$. The first six terms of this sequence are 5, 4, 9, 10, 13, and 16.

For Problems 75–78, write the first six terms of each sequence.

75. $a_n = \begin{cases} 2n + 1 & \text{for } n \text{ odd} \\ 2n - 1 & \text{for } n \text{ even} \end{cases}$

76. $a_n = \begin{cases} \dfrac{1}{n} & \text{for } n \text{ odd} \\ n^2 & \text{for } n \text{ even} \end{cases}$

77. $a_n = \begin{cases} 3n + 1 & \text{for } n \leq 3 \\ 4n - 3 & \text{for } n > 3 \end{cases}$

78. $a_n = \begin{cases} 5n - 1 & \text{for } n \text{ a multiple of 3} \\ 2n & \text{otherwise} \end{cases}$

The multiple-description approach can also be used to give a **recursive description** for a sequence. A sequence is said to be **described recursively** if the first n terms are stated, and then each succeeding term is defined as a function of one or more of the preceding terms. For example,

$$\begin{cases} a_1 = 2 \\ a_n = 2a_{n-1} & \text{for } n \geq 2 \end{cases}$$

means that the first term, a_1, is 2 and each succeeding term is 2 times the previous term. Thus the first six terms are 2, 4, 8, 16, 32, and 64.

For Problems 79–84, write the first six terms of each sequence.

79. $\begin{cases} a_1 = 4 \\ a_n = 3a_{n-1} & \text{for } n \geq 2 \end{cases}$

80. $\begin{cases} a_1 = 3 \\ a_n = a_{n-1} + 2 & \text{for } n \geq 2 \end{cases}$

81. $\begin{cases} a_1 = 1 \\ a_2 = 1 \\ a_n = a_{n-2} + a_{n-1} & \text{for } n \geq 3 \end{cases}$

82. $\begin{cases} a_1 = 2 \\ a_2 = 3 \\ a_n = 2a_{n-2} + 3a_{n-1} & \text{for } n \geq 3 \end{cases}$

83. $\begin{cases} a_1 = 3 \\ a_2 = 1 \\ a_n = (a_{n-1} - a_{n-2})^2 & \text{for } n \geq 3 \end{cases}$

84. $\begin{cases} a_1 = 1 \\ a_2 = 2 \\ a_3 = 3 \\ a_n = a_{n-1} + a_{n-2} + a_{n-3} & \text{for } n \geq 4 \end{cases}$

14.2   Geometric Sequences

A **geometric sequence** or **geometric progression** is a sequence in which we obtain each term after the first by multiplying the preceding term by a common multiplier called the **common ratio** of the sequence. We can find the common ratio of a geometric sequence by dividing any term (other than the first) by the preceding term. The following geometric sequences have common ratios of 3, 2, $\frac{1}{2}$, and -4, respectively:

$$1, 3, 9, 27, 81, \ldots$$

$$3, 6, 12, 24, 48, \ldots$$

$$16, 8, 4, 2, 1, \ldots$$

$$-1, 4, -16, 64, -256, \ldots$$

In a more general setting, we say that the sequence $a_1, a_2, a_3, \ldots, a_n, \ldots$ is a geometric sequence if and only if there is a nonzero real number r such that

$$a_{k+1} = ra_k$$

for every positive integer k. The nonzero real number r is called the common ratio of the sequence.

The previous equation can be used to generate a general geometric sequence that has a_1 as a first term and r as a common ratio. We can proceed as follows:

First term:	a_1	
Second term:	$a_1 r$	
Third term:	$a_1 r^2$	$(a_1 r)(r) = a_1 r^2$
Fourth term:	$a_1 r^3$	
$\cdot$		
$\cdot$		
$\cdot$		
nth term:	$a_1 r^{n-1}$	

Thus the **general term** of a geometric sequence is given by

$$a_n = a_1 r^{n-1}$$

where a_1 is the first term and r is the common ratio.

EXAMPLE 1   Find the general term for the geometric sequence $8, 16, 32, 64, \ldots$.

Solution

The common ratio (r) is $\dfrac{16}{8} = 2$, and the first term (a_1) is 8. Substitute these values into $a_n = a_1 r^{n-1}$ and simplify to obtain

$$a_n = 8(2)^{n-1} = (2^3)(2)^{n-1} = 2^{n+2}$$ ■

EXAMPLE 2   Find the ninth term of the geometric sequence $27, 9, 3, 1, \ldots$.

Solution

The common ratio (r) is $\dfrac{9}{27} = \dfrac{1}{3}$, and the first term (a_1) is 27. Using $a_n = a_1 r^{n-1}$, we obtain

$$a_9 = 27\left(\frac{1}{3}\right)^{9-1} = 27\left(\frac{1}{3}\right)^8$$

$$= \frac{3^3}{3^8}$$

$$= \frac{1}{3^5}$$

$$= \frac{1}{243}$$ ■

■ Sums of Geometric Sequences

As with arithmetic sequences, we often need to find the sum of a certain number of terms of a geometric sequence. Before we develop a general-sum formula for geometric sequences, let's consider an approach to a specific problem that we can then use in a general setting.

EXAMPLE 3

Find the sum of $1 + 3 + 9 + 27 + \cdots + 6561$.

Solution

Let S represent the sum and proceed as follows:

$$S = 1 + 3 + 9 + 27 + \cdots + 6561 \tag{1}$$

$$3S = \quad 3 + 9 + 27 + \cdots + 6561 + 19{,}683 \tag{2}$$

Equation (2) is the result of multiplying equation (1) by the common ratio 3. Subtracting equation (1) from equation (2) produces

$$2S = 19{,}683 - 1 = 19{,}682$$

$$S = 9841 \qquad\qquad ■$$

Now let's consider a general geometric sequence $a_1, a_1r, a_1r^2, \ldots, a_1r^{n-1}$. By applying a procedure similar to the one we used in Example 3, we can develop a formula for finding the sum of the first n terms of any geometric sequence. We let S_n represent the sum of the first n terms.

$$S_n = a_1 + a_1r + a_1r^2 + \cdots + a_1r^{n-1} \tag{3}$$

Next, we multiply both sides of equation (3) by the common ratio r.

$$rS_n = a_1r + a_1r^2 + a_1r^3 + \cdots + a_1r^n \tag{4}$$

We then subtract equation (3) from equation (4).

$$rS_n - S_n = a_1r^n - a_1$$

When we apply the distributive property to the left side and then solve for S_n, we obtain

$$S_n(r - 1) = a_1r^n - a_1$$

$$S_n = \frac{a_1r^n - a_1}{r - 1}, \qquad r \neq 1$$

Therefore the sum of the first n terms of a geometric sequence with a first term a_1 and a common ratio r is given by

$$S_n = \frac{a_1r^n - a_1}{r - 1}, \qquad r \neq 1$$

EXAMPLE 4

Find the sum of the first eight terms of the geometric sequence $1, 2, 4, 8, \ldots$.

Solution

To use the sum formula $S_n = \dfrac{a_1 r^n - a_1}{r - 1}$, we need to know the number of terms (n), the first term (a_1), and the common ratio (r). We are given the number of terms and the first term, and we can determine that $r = \dfrac{2}{1} = 2$. Using the sum formula, we obtain

$$S_8 = \frac{1(2)^8 - 1}{2 - 1} = \frac{2^8 - 1}{1} = 255$$ ∎

If the common ratio of a geometric sequence is less than 1, it may be more convenient to change the form of the sum formula. That is, the fraction

$$\frac{a_1 r^n - a_1}{r - 1}$$

can be changed to

$$\frac{a_1 - a_1 r^n}{1 - r}$$

by multiplying both the numerator and the denominator by -1. Thus by using

$$S_n = \frac{a_1 - a_1 r^n}{1 - r}$$

we can sometimes avoid unnecessary work with negative numbers when $r < 1$, as the next example illustrates.

EXAMPLE 5

Find the sum $1 + \dfrac{1}{2} + \dfrac{1}{4} + \cdots + \dfrac{1}{256}$.

Solution A

To use the sum formula, we need to know the number of terms, which can be found by counting them or by applying the nth-term formula, as follows:

$$a_n = a_1 r^{n-1}$$

$$\frac{1}{256} = 1\left(\frac{1}{2}\right)^{n-1}$$

$$\left(\frac{1}{2}\right)^8 = \left(\frac{1}{2}\right)^{n-1}$$

$$8 = n - 1 \qquad \text{If } b^n = b^m, \text{ then } n = m.$$

$$9 = n$$

Now we use $n = 9$, $a_1 = 1$, and $r = \dfrac{1}{2}$ in the sum formula of the form

$$S_n = \frac{a_1 - a_1 r^n}{1 - r}$$

$$S_9 = \frac{1 - 1\left(\frac{1}{2}\right)^9}{1 - \frac{1}{2}} = \frac{1 - \frac{1}{512}}{\frac{1}{2}} = \frac{\frac{511}{512}}{\frac{1}{2}} = 1\frac{255}{256}$$

We can also do a problem like Example 5 without finding the number of terms; we use the general approach illustrated in Example 3. Solution B demonstrates this idea.

Solution B

Let S represent the desired sum.

$$S = 1 + \frac{1}{2} + \frac{1}{4} + \cdots + \frac{1}{256}$$

Multiply both sides by the common ratio $\frac{1}{2}$.

$$\frac{1}{2}S = \frac{1}{2} + \frac{1}{4} + \frac{1}{8} + \cdots + \frac{1}{256} + \frac{1}{512}$$

Subtract the second equation from the first, and solve for S.

$$\frac{1}{2}S = 1 - \frac{1}{512} = \frac{511}{512}$$

$$S = \frac{511}{256} = 1\frac{255}{256}$$

Summation notation can also be used to indicate the sum of a certain number of terms of a geometric sequence.

EXAMPLE 6   Find the sum $\displaystyle\sum_{i=1}^{10} 2^i$.

Solution

This indicated sum means

$$\sum_{i=1}^{10} 2^i = 2^1 + 2^2 + 2^3 + \cdots + 2^{10}$$
$$= 2 + 4 + 8 + \cdots + 1024$$

This is the indicated sum of a geometric sequence, so we can use the sum formula with $a_1 = 2$, $r = 2$, and $n = 10$.

$$S_{10} = \frac{2(2)^{10} - 2}{2 - 1} = \frac{2(2^{10} - 1)}{1} = 2046$$

■ The Sum of an Infinite Geometric Sequence

Let's take the formula

$$S_n = \frac{a_1 - a_1 r^n}{1 - r}$$

and rewrite the right-hand side by applying the property

$$\frac{a - b}{c} = \frac{a}{c} - \frac{b}{c}$$

Thus we obtain

$$S_n = \frac{a_1}{1 - r} - \frac{a_1 r^n}{1 - r}$$

Now let's examine the behavior of r^n for $|r| < 1$, that is, for $-1 < r < 1$. For example, suppose that $r = \frac{1}{2}$. Then

$$r^2 = \left(\frac{1}{2}\right)^2 = \frac{1}{4} \qquad r^3 = \left(\frac{1}{2}\right)^3 = \frac{1}{8}$$

$$r^4 = \left(\frac{1}{2}\right)^4 = \frac{1}{16} \qquad r^5 = \left(\frac{1}{2}\right)^5 = \frac{1}{32}$$

and so on. We can make $\left(\frac{1}{2}\right)^n$ as close to zero as we please by choosing sufficiently large values for n. In general, for values of r such that $|r| < 1$, the expression r^n approaches zero as n gets larger and larger. Therefore the fraction $a_1 r^n / (1 - r)$ in equation (1) approaches zero as n increases. We say that the **sum of the infinite geometric sequence** is given by

$$S_\infty = \frac{a_1}{1 - r}, \qquad |r| < 1$$

E X A M P L E   7

Find the sum of the infinite geometric sequence

$$1, \frac{1}{2}, \frac{1}{4}, \frac{1}{8}, \dots$$

Solution

Because $a_1 = 1$ and $r = \frac{1}{2}$, we obtain

$$S_\infty = \frac{1}{1 - \dfrac{1}{2}} = \frac{1}{\dfrac{1}{2}} = 2$$

■

When we state that $S_\infty = 2$ in Example 7, we mean that as we add more and more terms, the sum approaches 2. Observe what happens when we calculate the sum up to five terms.

First term:	1
Sum of first two terms:	$1 + \dfrac{1}{2} = 1\dfrac{1}{2}$
Sum of first three terms:	$1 + \dfrac{1}{2} + \dfrac{1}{4} = 1\dfrac{3}{4}$
Sum of first four terms:	$1 + \dfrac{1}{2} + \dfrac{1}{4} + \dfrac{1}{8} = 1\dfrac{7}{8}$
Sum of first five terms:	$1 + \dfrac{1}{2} + \dfrac{1}{4} + \dfrac{1}{8} + \dfrac{1}{16} = 1\dfrac{15}{16}$

If $|r| > 1$, the absolute value of r^n increases without bound as n increases. In the next table, note the unbounded growth of the absolute value of r^n.

Let $r = 3$	Let $r = -2$	
$r^2 = 3^2 = 9$	$r^2 = (-2)^2 = 4$	
$r^3 = 3^3 = 27$	$r^3 = (-2)^3 = -8$	$|-8| = 8$
$r^4 = 3^4 = 81$	$r^4 = (-2)^4 = 16$	
$r^5 = 3^5 = 243$	$r^5 = (-2)^5 = -32$	$|-32| = 32$

If $r = 1$, then $S_n = na_1$, and as n increases without bound, $|S_n|$ also increases without bound. If $r = -1$, then S_n will be either a_1 or 0. Therefore we say that the sum of any infinite geometric sequence where $|r| \geq 1$ *does not exist*.

■ Repeating Decimals as Sums of Infinite Geometric Sequences

In Section 1.1, we defined rational numbers to be numbers that have either a terminating or a repeating decimal representation. For example,

$$2.23 \qquad 0.147 \qquad 0.\overline{3} \qquad 0.\overline{14} \qquad \text{and} \qquad 0.\overline{56}$$

are rational numbers. (Remember that $0.\overline{3}$ means $0.3333\ldots$.) Place value provides the basis for changing terminating decimals such as 2.23 and 0.147 to a/b form, where a and b are integers and $b \neq 0$.

$$2.23 = \frac{223}{100} \qquad \text{and} \qquad 0.147 = \frac{147}{1000}$$

However, changing repeating decimals to a/b form requires a different technique, and our work with sums of infinite geometric sequences provides the basis for one such approach. Consider the following examples.

EXAMPLE 8

Change $0.\overline{14}$ to a/b form, where a and b are integers and $b \neq 0$.

Solution

The repeating decimal $0.\overline{14}$ can be written as the indicated sum of an infinite geometric sequence with first term 0.14 and common ratio 0.01.

$$0.14 + 0.0014 + 0.000014 + \cdots$$

Using $S_\infty = a_1/(1 - r)$, we obtain

$$S_\infty = \frac{0.14}{1 - 0.01} = \frac{0.14}{0.99} = \frac{14}{99}$$

Thus $0.\overline{14} = \dfrac{14}{99}$. ∎

If the repeating block of digits does not begin immediately after the decimal point, as in $0.5\overline{6}$, we can make an adjustment in the technique we used in Example 8.

EXAMPLE 9

Change $0.5\overline{6}$ to a/b form, where a and b are integers and $b \neq 0$.

Solution

The repeating decimal $0.5\overline{6}$ can be written

$$(0.5) + (0.06 + 0.006 + 0.0006 + \cdots)$$

where

$$0.06 + 0.006 + 0.0006 + \cdots$$

is the indicated sum of the infinite geometric sequence with $a_1 = 0.06$ and $r = 0.1$. Therefore

$$S_\infty = \frac{0.06}{1 - 0.1} = \frac{0.06}{0.9} = \frac{6}{90} = \frac{1}{15}$$

Now we can add 0.5 and $\dfrac{1}{15}$.

$$0.5\overline{6} = 0.5 + \frac{1}{15} = \frac{1}{2} + \frac{1}{15} = \frac{15}{30} + \frac{2}{30} = \frac{17}{30}$$ ∎

Problem Set 14.2

For Problems 1–12, find the general term (the nth term) for each geometric sequence.

1. $3, 6, 12, 24, \ldots$

2. $2, 6, 18, 54, \ldots$

3. $3, 9, 27, 81, \ldots$

4. $2, 6, 18, 54, \ldots$

5. $\dfrac{1}{4}, \dfrac{1}{8}, \dfrac{1}{16}, \dfrac{1}{32}, \ldots$

6. $8, 4, 2, 1, \ldots$

7. $4, 16, 64, 256, \ldots$

8. $6, 2, \dfrac{2}{3}, \dfrac{2}{9}, \ldots$

9. $1, 0.3, 0.09, 0.027, \ldots$

10. $0.2, 0.04, 0.008, 0.0016, \ldots$

11. $1, -2, 4, -8, \ldots$

12. $-3, 9, -27, 81, \ldots$

For Problems 13–20, find the required term for each geometric sequence.

13. The 8th term of $\dfrac{1}{2}, 1, 2, 4, \ldots$

14. The 7th term of $2, 6, 18, 54, \ldots$

15. The 9th term of $729, 243, 81, 27, \ldots$

16. The 11th term of $768, 384, 192, 96, \ldots$

17. The 10th term of $1, -2, 4, -8, \ldots$

18. The 8th term of $-1, -\dfrac{3}{2}, -\dfrac{9}{4}, -\dfrac{27}{8}, \ldots$

19. The 8th term of $\dfrac{1}{2}, \dfrac{1}{6}, \dfrac{1}{18}, \dfrac{1}{54}, \ldots$

20. The 9th term of $\dfrac{16}{81}, \dfrac{8}{27}, \dfrac{4}{9}, \dfrac{2}{3}, \ldots$

For Problems 21–32, solve each problem.

21. Find the first term of the geometric sequence with 5th term $\dfrac{32}{3}$ and common ratio 2.

22. Find the first term of the geometric sequence with 4th term $\dfrac{27}{128}$ and common ratio $\dfrac{3}{4}$.

23. Find the common ratio of the geometric sequence with 3rd term 12 and 6th term 96.

24. Find the common ratio of the geometric sequence with 2nd term $\dfrac{8}{3}$ and 5th term $\dfrac{64}{81}$.

25. Find the sum of the first ten terms of the geometric sequence $1, 2, 4, 8, \ldots$.

26. Find the sum of the first seven terms of the geometric sequence $3, 9, 27, 81, \ldots$.

27. Find the sum of the first nine terms of the geometric sequence $2, 6, 18, 54, \ldots$.

28. Find the sum of the first ten terms of the geometric sequence $5, 10, 20, 40, \ldots$.

29. Find the sum of the first eight terms of the geometric sequence $8, 12, 18, 27, \ldots$.

30. Find the sum of the first eight terms of the geometric sequence $9, 12, 16, \dfrac{64}{3}, \ldots$.

31. Find the sum of the first ten terms of the geometric sequence $-4, 8, -16, 32, \ldots$.

32. Find the sum of the first nine terms of the geometric sequence $-2, 6, -18, 54, \ldots$.

For Problems 33–38, find each indicated sum.

33. $9 + 27 + 81 + \cdots + 729$

34. $2 + 8 + 32 + \cdots + 8192$

35. $4 + 2 + 1 + \cdots + \dfrac{1}{512}$

36. $1 + (-2) + 4 + \cdots + 256$

37. $(-1) + 3 + (-9) + \cdots + (-729)$

38. $16 + 8 + 4 + \cdots + \dfrac{1}{32}$

For Problems 39–44, find each indicated sum.

39. $\displaystyle\sum_{i=1}^{9} 2^{i-3}$

40. $\displaystyle\sum_{i=1}^{6} 3^{i}$

41. $\displaystyle\sum_{i=2}^{5} (-3)^{i+1}$

42. $\displaystyle\sum_{i=3}^{8} (-2)^{i-1}$

43. $\displaystyle\sum_{i=1}^{6} 3\left(\dfrac{1}{2}\right)^{i}$

44. $\displaystyle\sum_{i=1}^{5} 2\left(\dfrac{1}{3}\right)^{i}$

For Problems 45–56, find the sum of each infinite geometric sequence. If the sequence has no sum, so state.

45. $2, 1, \dfrac{1}{2}, \dfrac{1}{4}, \ldots$

46. $9, 3, 1, \dfrac{1}{3}, \ldots$

47. $1, \dfrac{2}{3}, \dfrac{4}{9}, \dfrac{8}{27}, \ldots$

48. $5, 3, \dfrac{9}{5}, \dfrac{27}{25}, \ldots$

49. $4, 8, 16, 32, \ldots$

50. $32, 16, 8, 4, \ldots$

51. $9, -3, 1, -\dfrac{1}{3}, \ldots$

52. $2, -6, 18, -54, \ldots$

53. $\dfrac{1}{2}, \dfrac{3}{8}, \dfrac{9}{32}, \dfrac{27}{128}, \ldots$

54. $4, -\dfrac{4}{3}, \dfrac{4}{9}, -\dfrac{4}{27}, \ldots$

55. $8, -4, 2, -1, \ldots$

56. $7, \dfrac{14}{5}, \dfrac{28}{25}, \dfrac{56}{125}, \ldots$

60. $0.\overline{18}$

61. $0.\overline{123}$

62. $0.\overline{273}$

63. $0.2\overline{6}$

64. $0.4\overline{3}$

65. $0.21\overline{4}$

For Problems 57–68, change each repeating decimal to a/b form, where a and b are integers and $b \neq 0$. Express a/b in reduced form.

66. $0.37\overline{1}$

67. $2.\overline{3}$

68. $3.\overline{7}$

57. $0.\overline{3}$

58. $0.\overline{4}$

59. $0.\overline{26}$

■ ■ ■ THOUGHTS INTO WORDS

69. Explain the difference between an arithmetic sequence and a geometric sequence.

70. What does it mean to say that the sum of the infinite geometric sequence $1, \dfrac{1}{2}, \dfrac{1}{4}, \dfrac{1}{8}, \ldots$ is 2?

71. What do we mean when we say that the infinite geometric sequence $1, 2, 4, 8, \ldots$ has no sum?

72. Why don't we discuss the sum of an infinite arithmetic sequence?

14.3 Another Look at Problem Solving

In the previous two sections, many of the exercises fell into one of the following four categories:

1. Find the nth term of an arithmetic sequence.

$$a_n = a_1 + (n-1)d$$

2. Find the sum of the first n terms of an arithmetic sequence.

$$S_n = \frac{n(a_1 + a_n)}{2}$$

3. Find the nth term of a geometric sequence.

$$a_n = a_1 r^{n-1}$$

4. Find the sum of the first n terms of a geometric sequence.

$$S_n = \frac{a_1 r^n - a_1}{r - 1}$$

In this section we want to use this knowledge of arithmetic sequences and geometric sequences to expand our problem-solving capabilities. Let's begin by restating some old problem-solving suggestions that continue to apply here; we will also consider some other suggestions that are directly related to problems that involve sequences of numbers. (We will indicate the new suggestions with an asterisk.)

Suggestions for Solving Word Problems

1. Read the problem carefully and make certain that you understand the meanings of all the words. Be especially alert for any technical terms used in the statement of the problem.

2. Read the problem a second time (perhaps even a third time) to get an overview of the situation being described and to determine the known facts, as well as what you are to find.

3. Sketch a figure, diagram, or chart that might be helpful in analyzing the problem.

*4. Write down the first few terms of the sequence to describe what is taking place in the problem. Be sure that you understand, term by term, what the sequence represents in the problem.

*5. Determine whether the sequence is arithmetic or geometric.

*6. Determine whether the problem is asking for a specific term of the sequence or for the sum of a certain number of terms.

7. Carry out the necessary calculations and check your answer for reasonableness.

As we solve some problems, these suggestions will become more meaningful.

PROBLEM 1

Domenica started to work in 1990 at an annual salary of $22,500. She received a $1200 raise each year. What was her annual salary in 1999?

Solution

The following sequence represents her annual salary beginning in 1990:

22,500,   23,700,   24,900,   26,100,   . . .

This is an arithmetic sequence, with $a_1 = 22,500$ and $d = 1200$. Her salary in 1990 is the first term of the sequence, and her salary in 1999 is the tenth term of the sequence. So, using $a_n = a_1 + (n - 1)d$, we obtain the tenth term of the arithmetic sequence.

$$a_{10} = 22,500 + (10 - 1)1200 = 22,500 + 9(1200) = 33,300$$

Her annual salary in 1999 was $33,300. ∎

PROBLEM 2

An auditorium has 20 seats in the front row, 24 seats in the second row, 28 seats in the third row, and so on, for 15 rows. How many seats are there in the auditorium?

Solution

The following sequence represents the number of seats per row, starting with the first row:

$$20, 24, 28, 32, \ldots$$

This is an arithmetic sequence, with $a_1 = 20$ and $d = 4$. Therefore the 15th term, which represents the number of seats in the 15th row, is given by

$$a_{15} = 20 + (15 - 1)4 = 20 + 14(4) = 76$$

The total number of seats in the auditorium is represented by

$$20 + 24 + 28 + \cdots + 76$$

Use the sum formula for an arithmetic sequence to obtain

$$S_{15} = \frac{15}{2}(20 + 76) = 720$$

There are 720 seats in the auditorium. ∎

PROBLEM 3

Suppose that you save 25 cents the first day of a week, 50 cents the second day, and one dollar the third day and that you continue to double your savings each day. How much will you save on the seventh day? What will be your total savings for the week?

Solution

The following sequence represents your savings per day, expressed in cents:

$$25, 50, 100, \ldots$$

This is a geometric sequence, with $a_1 = 25$ and $r = 2$. Your savings on the seventh day is the seventh term of this sequence. Therefore, using $a_n = a_1 r^{n-1}$, we obtain

$$a_7 = 25(2)^6 = 1600$$

You will save $16 on the seventh day. Your total savings for the seven days is given by

$$25 + 50 + 100 + \cdots + 1600$$

Use the sum formula for a geometric sequence to obtain

$$S_7 = \frac{25(2)^7 - 25}{2 - 1} = \frac{25(2^7 - 1)}{1} = 3175$$

Thus your savings for the entire week is $31.75. ∎

PROBLEM 4

A pump is attached to a container for the purpose of creating a vacuum. For each stroke of the pump, $\frac{1}{4}$ of the air that remains in the container is removed. To the nearest tenth of a percent, how much of the air remains in the container after six strokes?

Solution

Let's draw a chart to help with the analysis of this problem.

First stroke:	$\dfrac{1}{4}$ of the air is removed	$1 - \dfrac{1}{4} = \dfrac{3}{4}$ of the air remains
Second stroke:	$\dfrac{1}{4}\left(\dfrac{3}{4}\right) = \dfrac{3}{16}$ of the air is removed	$\dfrac{3}{4} - \dfrac{3}{16} = \dfrac{9}{16}$ of the air remains
Third stroke:	$\dfrac{1}{4}\left(\dfrac{9}{16}\right) = \dfrac{9}{64}$ of the air is removed	$\dfrac{9}{16} - \dfrac{9}{64} = \dfrac{27}{64}$ of the air remains

The diagram suggests two approaches to the problem.

Approach A   The sequence $\dfrac{1}{4}, \dfrac{3}{16}, \dfrac{9}{64}, \ldots$ represents, term by term, the fractional amount of air that is removed with each successive stroke. Therefore we can find the total amount removed and subtract it from 100%. The sequence is geometric with $a_1 = \dfrac{1}{4}$ and $r = \dfrac{\frac{3}{16}}{\frac{1}{4}} = \dfrac{3}{16} \cdot \dfrac{4}{1} = \dfrac{3}{4}$. Using the sum formula $S_n = \dfrac{a_1 - a_1 r^n}{1 - r}$, we obtain

$$S_6 = \frac{\dfrac{1}{4} - \dfrac{1}{4}\left(\dfrac{3}{4}\right)^6}{1 - \dfrac{3}{4}} = \frac{\dfrac{1}{4}\left[1 - \left(\dfrac{3}{4}\right)^6\right]}{\dfrac{1}{4}}$$

$$= 1 - \frac{729}{4096} = \frac{3367}{4096} = 82.2\%$$

Therefore $100\% - 82.2\% = 17.8\%$ of the air remains after six strokes.   ■

Approach B   The sequence

$$\frac{3}{4}, \frac{9}{16}, \frac{27}{64}, \ldots$$

represents, term by term, the amount of air that remains in the container after each stroke. Therefore when we find the sixth term of this geometric sequence, we will have the answer to the problem. Because $a_1 = \dfrac{3}{4}$ and $r = \dfrac{3}{4}$, we obtain

$$a_6 = \frac{3}{4}\left(\frac{3}{4}\right)^5 = \left(\frac{3}{4}\right)^6 = \frac{729}{4096} = 17.8\%$$

Therefore 17.8% of the air remains after six strokes. ∎

It will be helpful for you to take another look at the two approaches we used to solve Problem 4. Note that in Approach B, finding the sixth term of the sequence produced the answer to the problem without any further calculations. In Approach A, we had to find the sum of six terms of the sequence and then subtract that amount from 100%. As we solve problems that involve sequences, we must understand what each particular sequence represents on a term-by-term basis.

Problem Set 14.3

Use your knowledge of arithmetic sequences and geometric sequences to help solve Problems 1–28.

1. A man started to work in 1980 at an annual salary of $9500. He received a $700 raise each year. How much was his annual salary in 2001?

2. A woman started to work in 1985 at an annual salary of $13,400. She received a $900 raise each year. How much was her annual salary in 2000?

3. State University had an enrollment of 9600 students in 1992. Each year the enrollment increased by 150 students. What was the enrollment in 2005?

4. Math University had an enrollment of 12,800 students in 1998. Each year the enrollment decreased by 75 students. What was the enrollment in 2005?

5. The enrollment at University X is predicted to increase at the rate of 10% per year. If the enrollment for 2001 was 5000 students, find the predicted enrollment for 2005. Express your answer to the nearest whole number.

6. If you pay $12,000 for a car and it depreciates 20% per year, how much will it be worth in 5 years? Express your answer to the nearest dollar.

7. A tank contains 16,000 liters of water. Each day one-half of the water in the tank is removed and not replaced. How much water remains in the tank at the end of 7 days?

8. If the price of a pound of coffee is $3.20 and the projected rate of inflation is 5% per year, how much per pound should we expect coffee to cost in 5 years? Express your answer to the nearest cent.

9. A tank contains 5832 gallons of water. Each day one-third of the water in the tank is removed and not replaced. How much water remains in the tank at the end of 6 days?

10. A fungus culture growing under controlled conditions doubles in size each day. How many units will the culture contain after 7 days if it originally contains 4 units?

11. Sue is saving quarters. She saves 1 quarter the first day, 2 quarters the second day, 3 quarters the third day, and so on for 30 days. How much money will she have saved in 30 days?

12. Suppose you save a penny the first day of a month, 2 cents the second day, 3 cents the third day, and so on for 31 days. What will be your total savings for the 31 days?

13. Suppose you save a penny the first day of a month, 2 cents the second day, 4 cents the third day, and continue to double your savings each day. How much will you save on the 15th day of the month? How much will your total savings be for the 15 days?

14. Eric saved a nickel the first day of a month, a dime the second day, and 20 cents the third day and then continued to double his daily savings each day for 14 days. What were his daily savings on the 14th day? What were his total savings for the 14 days?

15. Ms. Bryan invested $1500 at 12% simple interest at the beginning of each year for a period of 10 years. Find the total accumulated value of all the investments at the end of the 10-year period.

16. Mr. Woodley invested $1200 at 11% simple interest at the beginning of each year for a period of 8 years. Find the total accumulated value of all the investments at the end of the 8-year period.

17. An object falling from rest in a vacuum falls approximately 16 feet the first second, 48 feet the second second, 80 feet the third second, 112 feet the fourth second, and so on. How far will it fall in 11 seconds?

18. A raffle is organized so that the amount paid for each ticket is determined by the number on the ticket. The tickets are numbered with the consecutive odd whole numbers 1, 3, 5, 7, ... Each contestant pays as many cents as the number on the ticket drawn. How much money will the raffle take in if 1000 tickets are sold?

19. Suppose an element has a half-life of 4 hours. This means that if n grams of it exist at a specific time, then only $\frac{1}{2}n$ grams remain 4 hours later. If at a particular moment we have 60 grams of the element, how many grams of it will remain 24 hours later?

20. Suppose an element has a half-life of 3 hours. (See Problem 19 for a definition of half-life.) If at a particular moment we have 768 grams of the element, how many grams of it will remain 24 hours later?

21. A rubber ball is dropped from a height of 1458 feet, and at each bounce it rebounds one-third of the height from which it last fell. How far has the ball traveled by the time it strikes the ground for the sixth time?

22. A rubber ball is dropped from a height of 100 feet, and at each bounce it rebounds one-half of the height from which it last fell. What distance has the ball traveled up to the instant it hits the ground for the eighth time?

23. A pile of logs has 25 logs in the bottom layer, 24 logs in the next layer, 23 logs in the next layer, and so on, until the top layer has 1 log. How many logs are in the pile?

24. A well driller charges $9.00 per foot for the first 10 feet, $9.10 per foot for the next 10 feet, $9.20 per foot for the next 10 feet, and so on, at a price increase of $0.10 per foot for succeeding intervals of 10 feet. How much does it cost to drill a well to a depth of 150 feet?

25. A pump is attached to a container for the purpose of creating a vacuum. For each stroke of the pump, one-third of the air remaining in the container is removed. To the nearest tenth of a percent, how much of the air remains in the container after seven strokes?

26. Suppose that in Problem 25, each stroke of the pump removes one-half of the air remaining in the container. What fractional part of the air has been removed after six strokes?

27. A tank contains 20 gallons of water. One-half of the water is removed and replaced with antifreeze. Then one-half of this mixture is removed and replaced with antifreeze. This process is continued eight times. How much water remains in the tank after the eighth replacement process?

28. The radiator of a truck contains 10 gallons of water. Suppose we remove 1 gallon of water and replace it with antifreeze. Then we remove 1 gallon of this mixture and replace it with antifreeze. This process is carried out seven times. To the nearest tenth of a gallon, how much antifreeze is in the final mixture?

■ ■ ■ **THOUGHTS INTO WORDS**

29. Your friend solves Problem 6 as follows: If the car depreciates 20% per year, then at the end of 5 years it will have depreciated 100% and be worth zero dollars. How would you convince him that his reasoning is incorrect?

30. A contractor wants you to clear some land for a housing project. He anticipates that it will take 20 working days to do the job. He offers to pay you one of two ways: (1) a fixed amount of $3000 or (2) a penny the first day, 2 cents the second day, 4 cents the third day, and so on, doubling your daily wages each day for the 20 days. Which offer should you take and why?

14.4   Mathematical Induction

Is $2^n > n$ for all positive integer values of n? In an attempt to answer this question, we might proceed as follows:

If $n = 1$, then $2^n > n$ becomes $2^1 > 1$, a true statement.

If $n = 2$, then $2^n > n$ becomes $2^2 > 2$, a true statement.

If $n = 3$, then $2^n > n$ becomes $2^3 > 3$, a true statement.

We can continue in this way as long as we want, but obviously we can never show in this manner that $2^n > n$ for *every* positive integer n. However, we do have a form of proof, called **proof by mathematical induction**, that can be used to verify the truth of many mathematical statements involving positive integers. This form of proof is based on the following principle.

Principle of Mathematical Induction

Let P_n be a statement in terms of n, where n is a positive integer. If

1. P_1 is true, and

2. the truth of P_k implies the truth of P_{k+1} for every positive integer k,

then P_n is true for every positive integer n.

The principle of mathematical induction, a proof that some statement is true for all positive integers, consists of two parts. First, we must show that the statement is true for the positive integer 1. Second, we must show that if the statement is true for some positive integer, then it follows that it is also true for the next positive integer. Let's illustrate what this means.

EXAMPLE 1

Prove that $2^n > n$ for all positive integer values of n.

Proof

Part 1   If $n = 1$, then $2^n > n$ becomes $2^1 > 1$, which is a true statement.

Part 2   We must prove that if $2^k > k$, then $2^{k+1} > k + 1$ for all positive integer values of k. In other words, we should be able to start with $2^k > k$ and from that deduce $2^{k+1} > k + 1$. This can be done as follows:

$$2^k > k$$

$$2(2^k) > 2(k) \qquad \text{Multiply both sides by 2.}$$

$$2^{k+1} > 2k$$

We know that $k \geq 1$ because we are working with positive integers. Therefore

$k + k \geq k + 1$ Add k to both sides.

$2k \geq k + 1$

Because $2^{k+1} > 2k$ and $2k \geq k + 1$, by the transitive property we conclude that

$2^{k+1} > k + 1$

Therefore, using parts 1 and 2, we proved that $2^n > n$ for *all* positive integers. ∎

It will be helpful for you to look back over the proof in Example 1. Note that in part 1, we established that $2^n > n$ is true for $n = 1$. Then, in part 2, we established that if $2^n > n$ is true for any positive integer, then it must be true for the next consecutive positive integer. Therefore, because $2^n > n$ is true for $n = 1$, it must be true for $n = 2$. Likewise, if $2^n > n$ is true for $n = 2$, then it must be true for $n = 3$, and so on, for *all* positive integers.

We can depict proof by mathematical induction with dominoes. Suppose that in Figure 14.1, we have infinitely many dominoes lined up. If we can push the first domino over (part 1 of a mathematical induction proof) and if the dominoes are spaced so that each time one falls over, it causes the next one to fall over (part 2 of a mathematical induction proof), then by pushing the first one over we will cause a chain reaction that will topple all of the dominoes (Figure 14.2).

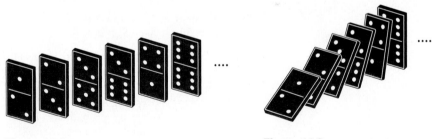

Figure 14.1   **Figure 14.2**

Recall that in the first three sections of this chapter, we used a_n to represent the nth term of a sequence and S_n to represent the sum of the first n terms of a sequence. For example, if $a_n = 2n$, then the first three terms of the sequence are $a_1 = 2(1) = 2$, $a_2 = 2(2) = 4$, and $a_3 = 2(3) = 6$. Furthermore, the kth term is $a_k = 2(k) = 2k$, and the $(k + 1)$ term is $a_{k+1} = 2(k + 1) = 2k + 2$. Relative to this same sequence, we can state that $S_1 = 2, S_2 = 2 + 4 = 6$, and $S_3 = 2 + 4 + 6 = 12$.

There are numerous sum formulas for sequences that can be verified by mathematical induction. For such proofs, the following property of sequences is used:

$S_{k+1} = S_k + a_{k+1}$

This property states that **the sum of the first $k + 1$ terms is equal to the sum of the first k terms plus the $(k + 1)$ term**. Let's see how this can be used in a specific example.

E X A M P L E   2

Prove that $S_n = n(n + 1)$ for the sequence $a_n = 2n$, where n is any positive integer.

Proof

Part 1   If $n = 1$, then $S_1 = 1(1 + 1) = 2$, and 2 is the first term of the sequence $a_n = 2n$, so $S_1 = a_1 = 2$.

Part 2   Now we need to prove that if $S_k = k(k + 1)$, then $S_{k+1} = (k + 1)(k + 2)$. Using the property $S_{k+1} = S_k + a_{k+1}$, we can proceed as follows:

$$S_{k+1} = S_k + a_{k+1}$$

$$= k(k + 1) + 2(k + 1)$$

$$= (k + 1)(k + 2)$$

Therefore, using parts 1 and 2, we proved that $S_n = n(n + 1)$ will yield the correct sum for any number of terms of the sequence $a_n = 2n$. ∎

E X A M P L E   3

Prove that $S_n = 5n(n + 1)/2$ for the sequence $a_n = 5n$, where n is any positive integer.

Proof

Part 1   Because $S_1 = 5(1)(1 + 1)/2 = 5$, and 5 is the first term of the sequence $a_n = 5n$, we have $S_1 = a_1 = 5$.

Part 2   We need to prove that if $S_k = 5k(k + 1)/2$, then $S_{k+1} = \dfrac{5(k + 1)(k + 2)}{2}$.

$$S_{k+1} = S_k + a_{k+1}$$

$$= \frac{5k(k + 1)}{2} + 5(k + 1)$$

$$= \frac{5k(k + 1)}{2} + 5k + 5$$

$$= \frac{5k(k + 1) + 2(5k + 5)}{2}$$

$$= \frac{5k^2 + 5k + 10k + 10}{2}$$

$$= \frac{5k^2 + 15k + 10}{2}$$

$$= \frac{5(k^2 + 3k + 2)}{2}$$

$$= \frac{5(k + 1)(k + 2)}{2}$$

Therefore, using parts 1 and 2, we proved that $S_n = 5n(n + 1)/2$ yields the correct sum for any number of terms of the sequence $a_n = 5n$. ∎

EXAMPLE 4   Prove that $S_n = (4^n - 1)/3$ for the sequence $a_n = 4^{n-1}$, where n is any positive integer.

Proof

Part 1   Because $S_1 = (4^1 - 1)/3 = 1$, and 1 is the first term of the sequence $a_n = 4^{n-1}$, we have $S_1 = a_1 = 1$.

Part 2   We need to prove that if $S_k = (4^k - 1)/3$, then $S_{k+1} = (4^{k+1} - 1)/3$.

$$S_{k+1} = S_k + a_{k+1}$$

$$= \frac{4^k - 1}{3} + 4^k$$

$$= \frac{4^k - 1 + 3(4^k)}{3}$$

$$= \frac{4^k + 3(4^k) - 1}{3}$$

$$= \frac{4^k(1 + 3) - 1}{3}$$

$$= \frac{4^k(4) - 1}{3}$$

$$= \frac{4^{k+1} - 1}{3}$$

Therefore, using parts 1 and 2, we proved that $S_n = (4^n - 1)/3$ yields the correct sum for any number of terms of the sequence $a_n = 4^{n-1}$. ∎

As our final example of this section, let's consider a proof by mathematical induction involving the concept of divisibility.

EXAMPLE 5   Prove that for all positive integers n, the number $3^{2n} - 1$ is divisible by 8.

Proof

Part 1   If $n = 1$, then $3^{2n} - 1$ becomes $3^{2(1)} - 1 = 3^2 - 1 = 8$, and of course 8 is divisible by 8.

Part 2 We need to prove that if $3^{2k} - 1$ is divisible by 8, then $3^{2k+2} - 1$ is divisible by 8 for all integer values of k. This can be verified as follows. If $3^{2k} - 1$ is divisible by 8, then for some integer x, we have $3^{2k} - 1 = 8x$. Therefore

$$3^{2k} - 1 = 8x$$

$$3^{2k} = 1 + 8x$$

$$3^2(3^{2k}) = 3^2(1 + 8x) \qquad \text{Multiply both sides by } 3^2.$$

$$3^{2k+2} = 9(1 + 8x)$$

$$3^{2k+2} = 9 + 9(8x)$$

$$3^{2k+2} = 1 + 8 + 9(8x) \qquad 9 = 1 + 8$$

$$3^{2k+2} = 1 + 8(1 + 9x)$$

$$3^{2k+2} - 1 = 8(1 + 9x) \qquad \begin{array}{l} \text{Apply distributive} \\ \text{property to } 8 + 9(8x). \end{array}$$

Therefore $3^{2k+2} - 1$ is divisible by 8.

Thus using parts 1 and 2, we proved that $3^{2n} - 1$ is divisible by 8 for all positive integers n. ∎

We conclude this section with a few final comments about proof by mathematical induction. Every mathematical induction proof is a two-part proof, and both parts are absolutely necessary. There can be mathematical statements that hold for one or the other of the two parts but not for both. For example, $(a + b)^n = a^n + b^n$ is true for $n = 1$, but it is false for every positive integer greater than 1. Therefore, if we were to attempt a mathematical induction proof for $(a + b)^n = a^n + b^n$, we could establish part 1 but not part 2. Another example of this type is the statement that $n^2 - n + 41$ produces a prime number for all positive integer values of n. This statement is true for $n = 1, 2, 3, 4, \ldots, 40$, but it is false when $n = 41$ (because $41^2 - 41 + 41 = 41^2$, which is not a prime number).

It is also possible that part 2 of a mathematical induction proof can be established but not part 1. For example, consider the sequence $a_n = n$ and the sum formula $S_n = (n + 3)(n - 2)/2$. If $n = 1$, then $a_1 = 1$ but $S_1 = (4)(-1)/2 = -2$, so part 1 does not hold. However, it is possible to show that $S_k = (k + 3)(k - 2)/2$ implies $S_{k+1} = (k + 4)(k - 1)/2$. We will leave the details of this for you to do.

Finally, it is important to realize that some mathematical statements are true for all positive integers greater than some fixed positive integer other than 1. (Back in Figure 14.1, perhaps we cannot knock down the first four dominoes, whereas we can knock down the fifth domino and every one thereafter.) For example, we can prove by mathematical induction that $2^n > n^2$ for all positive integers $n > 4$. It requires a slight variation in the statement of the principle of mathematical induction. We will not concern ourselves with such problems in this text, but we want you to be aware of their existence.

Problem Set 14.4

For Problems 1–10, use mathematical induction to prove each of the sum formulas for the indicated sequences. They are to hold for all positive integers n.

1. $S_n = \dfrac{n(n + 1)}{2}$ for $a_n = n$

2. $S_n = n^2$ for $a_n = 2n - 1$

3. $S_n = \dfrac{n(3n + 1)}{2}$ for $a_n = 3n - 1$

4. $S_n = \dfrac{n(5n + 9)}{2}$ for $a_n = 5n + 2$

5. $S_n = 2(2^n - 1)$ for $a_n = 2^n$

6. $S_n = \dfrac{3(3^n - 1)}{2}$ for $a_n = 3^n$

7. $S_n = \dfrac{n(n + 1)(2n + 1)}{6}$ for $a_n = n^2$

8. $S_n = \dfrac{n^2(n + 1)^2}{4}$ for $a_n = n^3$

9. $S_n = \dfrac{n}{n + 1}$ for $a_n = \dfrac{1}{n(n + 1)}$

10. $S_n = \dfrac{n(n + 1)(n + 2)}{3}$ for $a_n = n(n + 1)$

In Problems 11–20, use mathematical induction to prove that each statement is true for all positive integers n.

11. $3^n \geq 2n + 1$

12. $4^n \geq 4n$

13. $n^2 \geq n$

14. $2^n \geq n + 1$

15. $4^n - 1$ is divisible by 3

16. $5^n - 1$ is divisible by 4

17. $6^n - 1$ is divisible by 5

18. $9^n - 1$ is divisible by 4

19. $n^2 + n$ is divisible by 2

20. $n^2 - n$ is divisible by 2

■ ■ ■ THOUGHTS INTO WORDS

21. How would you describe proof by mathematical induction?

22. Compare inductive reasoning to prove by mathematical induction.

Chapter 14   Summary

There are four main topics in this chapter: arithmetic sequences, geometric sequences, problem solving, and mathematical induction.

(14.1) Arithmetic Sequences
The sequence $a_1, a_2, a_3, a_4, \ldots$ is called **arithmetic** if and only if

$$a_{k+1} - a_k = d$$

for every positive integer k. In other words, there is a **common difference**, d, between successive terms.

The **general term** of an arithmetic sequence is given by the formula

$$a_n = a_1 + (n-1)d$$

where a_1 is the first term, n is the number of terms, and d is the common difference.

The **sum** of the first n terms of an arithmetic sequence is given by the formula

$$S_n = \frac{n(a_1 + a_n)}{2}$$

Summation notation can be used to indicate the sum of a certain number of terms of a sequence. For example,

$$\sum_{i-1}^{5} 4^i = 4^1 + 4^2 + 4^3 + 4^4 + 4^5$$

(14.2) Geometric Sequences
The sequence $a_1, a_2, a_3, a_4, \ldots$ is called **geometric** if and only if

$$a_{k+1} = ra_k$$

for every positive integer k. There is a **common ratio**, r, between successive terms.

The **general term** of a geometric sequence is given by the formula

$$a_n = a_1 r^{n-1}$$

where a_1 is the first term, n is the number of terms, and r is the common ratio.

The **sum** of the first n terms of a geometric sequence is given by the formula

$$S_n = \frac{a_1 r^n - a_1}{r - 1} \qquad r \neq 1$$

The **sum of an infinite geometric sequence** is given by the formula

$$S_\infty = \frac{a_1}{1 - r} \qquad \text{for } |r| < 1$$

If $|r| \geq 1$, the sequence has no sum.

Repeating decimals (such as $0.\overline{4}$) can be changed to a/b form, where a and b are integers and $b \neq 0$, by treating them as the sum of an infinite geometric sequence. For example, the repeating decimal $0.\overline{4}$ can be written $0.4 + 0.04 + 0.004 + 0.0004 + \cdots$.

(14.3) Problem Solving
Many of the problem-solving suggestions offered earlier in this text are still appropriate when we are solving problems that deal with sequences. However, there are also some special suggestions pertaining to sequence problems.

1. Write down the first few terms of the sequence to describe what is taking place in the problem. Drawing a picture or diagram may help with this step.

2. Be sure that you understand, term by term, what the sequence represents in the problem.

3. Determine whether the sequence is arithmetic or geometric. (Those are the only kinds of sequences we are working with in this text.)

4. Determine whether the problem is asking for a specific term or for the sum of a certain number of terms.

(14.4) Mathematical Induction
Proof by mathematical induction relies on the following **principle of induction**: Let P_n be a statement in terms of n, where n is a positive integer. If

1. P_1 is true, and

2. the truth of P_k implies the truth of P_{k+1} for every positive integer k, then P_n is true for every positive integer n.

Chapter 14	**Review Problem Set**

For Problems 1–10, find the general term (the nth term) for each sequence. These problems include both arithmetic sequences and geometric sequences.

1. $3, 9, 15, 21, \ldots$

2. $\dfrac{1}{3}, 1, 3, 9, \ldots$

3. $10, 20, 40, 80, \ldots$

4. $5, 2, -1, -4, \ldots$

5. $-5, -3, -1, 1, \ldots$

6. $9, 3, 1, \dfrac{1}{3}, \ldots$

7. $-1, 2, -4, 8, \ldots$

8. $12, 15, 18, 21, \ldots$

9. $\dfrac{2}{3}, 1, \dfrac{4}{3}, \dfrac{5}{3}, \ldots$

10. $1, 4, 16, 64, \ldots$

For Problems 11–16, find the required term of each of the sequences.

11. The 19th term of $1, 5, 9, 13, \ldots$

12. The 28th term of $-2, 2, 6, 10, \ldots$

13. The 9th term of $8, 4, 2, 1, \ldots$

14. The 8th term of $\dfrac{243}{32}, \dfrac{81}{16}, \dfrac{27}{8}, \dfrac{9}{4}, \ldots$

15. The 34th term of $7, 4, 1, -2, \ldots$

16. The 10th term of $-32, 16, -8, 4, \ldots$

For Problems 17–29, solve each problem.

17. If the 5th term of an arithmetic sequence is -19 and the 8th term is -34, find the common difference of the sequence.

18. If the 8th term of an arithmetic sequence is 37 and the 13th term is 57, find the 20th term.

19. Find the first term of a geometric sequence if the third term is 5 and the sixth term is 135.

20. Find the common ratio of a geometric sequence if the second term is $\dfrac{1}{2}$ and the sixth term is 8.

21. Find the sum of the first nine terms of the sequence 81, 27, 9, 3, $\ldots$.

22. Find the sum of the first 70 terms of the sequence $-3, 0, 3, 6, \ldots$.

23. Find the sum of the first 75 terms of the sequence 5, 1, $-3, -7, \ldots$.

24. Find the sum of the first ten terms of the sequence where $a_n = 2^{5-n}$.

25. Find the sum of the first 95 terms of the sequence where $a_n = 7n + 1$.

26. Find the sum $5 + 7 + 9 + \cdots + 137$.

27. Find the sum $64 + 16 + 4 + \cdots + \dfrac{1}{64}$.

28. Find the sum of all even numbers between 8 and 384, inclusive.

29. Find the sum of all multiples of 3 between 27 and 276, inclusive.

For Problems 30–33, find each indicated sum.

30. $\displaystyle\sum_{i=1}^{45}(-2i + 5)$

31. $\displaystyle\sum_{i=1}^{5} i^3$

32. $\displaystyle\sum_{i=1}^{8} 2^{8-i}$

33. $\displaystyle\sum_{i=4}^{75}(3i - 4)$

For Problems 34–36, solve each problem.

34. Find the sum of the infinite geometric sequence 64, 16, 4, 1, $\ldots$.

35. Change $0.\overline{36}$ to reduced a/b form, where a and b are integers and $b \neq 0$.

36. Change $0.4\overline{5}$ to reduced a/b form, where a and b are integers and $b \neq 0$.

Solve each of Problems 37–40 by using your knowledge of arithmetic sequences and geometric sequences.

37. Suppose that your savings account contains $3750 at the beginning of a year. If you withdraw $250 per month from the account, how much will it contain at the end of the year?

38. Sonya decides to start saving dimes. She plans to save 1 dime the first day of April, 2 dimes the second day, 3 dimes the third day, 4 dimes the fourth day, and so on for the 30 days of April. How much money will she save in April?

39. Nancy decides to start saving dimes. She plans to save 1 dime the first day of April, 2 dimes the second day, 4 dimes the third day, 8 dimes the fourth day, and so on for the first 15 days of April. How much will she save in 15 days?

40. A tank contains 61,440 gallons of water. Each day one-fourth of the water is drained out. How much water remains in the tank at the end of 6 days?

For Problems 41–43, show a mathematical induction proof.

41. Prove that $5^n > 5n - 1$ for all positive integer values of n.

42. Prove that $n^3 - n + 3$ is divisible by 3 for all positive integer values of n.

43. Prove that

$$S_n = \frac{n(n + 3)}{4(n + 1)(n + 2)}$$

is the sum formula for the sequence

$$a_n = \frac{1}{n(n + 1)(n + 2)}$$

where n is any positive integer.

1. Find the 15th term of the sequence for which $a_n = -n^2 - 1$.

2. Find the fifth term of the sequence for which $a_n = 3(2)^{n-1}$.

3. Find the general term of the sequence $-3, -8, -13, -18, \ldots$.

4. Find the general term of the sequence $5, \dfrac{5}{2}, \dfrac{5}{4}, \dfrac{5}{8}, \ldots$.

5. Find the general term of the sequence $10, 16, 22, 28, \ldots$.

6. Find the seventh term of the sequence $8, 12, 18, 27, \ldots$.

7. Find the 75th term of the sequence $1, 4, 7, 10, \ldots$.

8. Find the number of terms in the sequence $7, 11, 15, \ldots, 243$.

9. Find the sum of the first 40 terms of the sequence $1, 4, 7, 10, \ldots$.

10. Find the sum of the first eight terms of the sequence $3, 6, 12, 24, \ldots$.

11. Find the sum of the first 45 terms of the sequence for which $a_n = 7n - 2$.

12. Find the sum of the first ten terms of the sequence for which $a_n = 3(2)^n$.

13. Find the sum of the first 150 positive even whole numbers.

14. Find the sum of the odd whole numbers between 11 and 193, inclusive.

15. Find the indicated sum $\displaystyle\sum_{i=1}^{50}(3i + 5)$.

16. Find the indicated sum $\displaystyle\sum_{i=1}^{10}(-2)^{i-1}$.

17. Find the sum of the infinite geometric sequence $3, \dfrac{3}{2}, \dfrac{3}{4}, \dfrac{3}{8}, \ldots$.

18. Find the sum of the infinite geometric sequence for which $a_n = 2\left(\dfrac{1}{3}\right)^{n+1}$.

19. Change $0.\overline{18}$ to reduced a/b form, where a and b are integers and $b \neq 0$.

20. Change $0.2\overline{6}$ to reduced a/b form, where a and b are integers and $b \neq 0$.

For Problems 21–23, solve each problem.

21. A tank contains 49,152 liters of gasoline. Each day, three-fourths of the gasoline remaining in the tank is pumped out and not replaced. How much gasoline remains in the tank at the end of 7 days?

22. Suppose that you save a dime the first day of a month, $0.20 the second day, and $0.40 the third day and that you continue to double your savings each day for 14 days. Find the total amount that you will save at the end of 14 days.

23. A woman invests $350 at 12% simple interest at the beginning of each year for a period of 10 years. Find the total accumulated value of all the investments at the end of the 10-year period.

For Problems 24 and 25, show a mathematical induction proof.

24. $S_n = \dfrac{n(3n - 1)}{2}$ for $a_n = 3n - 2$

25. $9^n - 1$ is divisible by 8 for all positive integer values for n.

Counting Techniques, Probability, and the Binomial Theorem

© AP/Wide World

Probability theory can determine the probability of winning a game of chance such as a lottery.

In a group of 30 people, there is approximately a 70% chance that at least 2 of them will have the same birthday (same month and same day of the month). In a group of 60 people, there is approximately a 99% chance that at least 2 of them will have the same birthday.

With an ordinary deck of 52 playing cards, there is 1 chance out of 54,145 that you will be dealt four aces in a five-card hand. The radio is predicting a 40% chance of locally severe thunderstorms by late afternoon. The odds in favor of the Cubs winning the pennant are 2 to 3. Suppose that in a box containing 50 light bulbs, 45 are good ones, and 5 are burned out. If 2 bulbs are chosen at random, the probability of getting at least 1 good bulb is $\frac{243}{245}$. Historically, many basic probability concepts have been developed as a result of studying various games of chance. However, in recent years, applications of probability have been surfacing at a phenomenal rate in a large variety of fields, such as physics, biology, psychology, economics, insurance, military science, manufacturing, and politics. It is our purpose in this chapter first to introduce some counting techniques and then to use those techniques to explore some basic concepts of probability. The last section of the chapter will be devoted to the binomial theorem.

15.1	**Fundamental Principle of Counting**

One very useful counting principle is referred to as the **fundamental principle of counting**. We will offer some examples, state the property, and then use it to solve a variety of counting problems. Let's consider two problems to lead up to the statement of the property.

PROBLEM 1

A woman has four skirts and five blouses. Assuming that each blouse can be worn with each skirt, how many different skirt–blouse outfits does she have?

Solution

For *each* of the four skirts, she has a choice of five blouses. Therefore she has $4(5) = 20$ different skirt–blouse outfits from which to choose. ∎

PROBLEM 2

Eric is shopping for a new bicycle and has two different models (5-speed or 10-speed) and four different colors (red, white, blue, or silver) from which to choose. How many different choices does he have?

Solution

His different choices can be counted with the help of a **tree diagram**.

Models	Colors	Choices
5-speed •	red	5-speed red
	white	5-speed white
	blue	5-speed blue
	silver	5-speed silver
10-speed •	red	10-speed red
	white	10-speed white
	blue	10-speed blue
	silver	10-speed silver

For each of the two model choices, there are four choices of color. Altogether, then, Eric has $2(4) = 8$ choices. ■

These two problems exemplify the following general principle:

Fundamental Principle of Counting

> If one task can be accomplished in x different ways and, following this task, a second task can be accomplished in y different ways, then the first task followed by the second task can be accomplished in $x \cdot y$ different ways. (This counting principle can be extended to any finite number of tasks.)

As you apply the fundamental principle of counting, it is often helpful to analyze a problem systematically in terms of the tasks to be accomplished. Let's consider some examples.

PROBLEM 3

How many numbers of three different digits each can be formed by choosing from the digits 1, 2, 3, 4, 5 and 6?

Solution

Let's analyze this problem in terms of three tasks.

Task 1 Choose the hundreds digit, for which there are six choices.

Task 2 Now choose the tens digit, for which there are only five choices because one digit was used in the hundreds place.

Task 3 Now choose the units digit, for which there are only four choices because two digits have been used for the other places.

Therefore task 1 followed by task 2 followed by task 3 can be accomplished in $(6)(5)(4) = 120$ ways. In other words, there are 120 numbers of three different digits that can be formed by choosing from the six given digits. ■

Now look back over the solution for Problem 3 and think about each of the following questions:

1. Can we solve the problem by choosing the units digit first, then the tens digit, and finally the hundreds digit?

2. How many three-digit numbers can be formed from 1, 2, 3, 4, 5, and 6 if we do not require each number to have three *different* digits? (Your answer should be 216.)

3. Suppose that the digits from which to choose are 0, 1, 2, 3, 4, and 5. Now how many numbers of three different digits each can be formed, assuming that we do not want zero in the hundreds place? (Your answer should be 100.)

4. Suppose that we want to know the number of *even* numbers with three different digits each that can be formed by choosing from 1, 2, 3, 4, 5, and 6. How many are there? (Your answer should be 60.)

PROBLEM   4

Employee ID numbers at a certain factory consist of one capital letter followed by a three-digit number that contains no repeat digits. For example, A-014 is an ID number. How many such ID numbers can be formed? How many can be formed if repeated digits *are* allowed?

Solution

Again, let's analyze the problem in terms of tasks to be completed.

Task 1   Choose the letter part of the ID number: there are 26 choices.

Task 2   Choose the first digit of the three-digit number: there are ten choices.

Task 3   Choose the second digit: there are nine choices.

Task 4   Choose the third digit: there are eight choices.

Therefore, applying the fundamental principle, we obtain $(26)(10)(9)(8) = 18{,}720$ possible ID numbers.

    If repeat digits were allowed, then there would be $(26)(10)(10)(10) = 26{,}000$ possible ID numbers.   ■

PROBLEM   5

In how many ways can Al, Barb, Chad, Dan, and Edna be seated in a row of five seats so that Al and Barb are seated side by side?

Solution

This problem can be analyzed in terms of three tasks.

Task 1   Choose the two adjacent seats to be occupied by Al and Barb. An illustration such as Figure 15.1 helps us to see that there are four choices for the two adjacent seats.

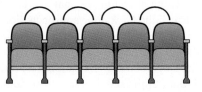

Figure 15.1

Task 2   Determine the number of ways in which Al and Barb can be seated. Because Al can be seated on the left and Barb on the right, or vice versa, there are two ways to seat Al and Barb for each pair of adjacent seats.

Task 3   The remaining three people must be seated in the remaining three seats. This can be done in $(3)(2)(1) = 6$ different ways.

Therefore, by the fundamental principle, task 1 followed by task 2 followed by task 3 can be done in $(4)(2)(6) = 48$ ways. ■

Suppose that in Problem 5, we wanted instead the number of ways in which the five people can sit so that Al and Barb are *not* side by side. We can determine this number by using either of two basically different techniques: (1) analyze and count the number of nonadjacent positions for Al and Barb, or (2) subtract the number of seating arrangements determined in Problem 5 from the total number of ways in which five people can be seated in five seats. Try doing this problem both ways, and see whether you agree with the answer of 72 ways.

As you apply the fundamental principle of counting, you may find that for certain problems, simply thinking about an appropriate tree diagram is helpful, even though the size of the problem may make it inappropriate to write out the diagram in detail. Consider the following problem.

PROBLEM 6

Suppose that the undergraduate students in three departments — geography, history, and psychology — are to be classified according to sex and year in school. How many categories are needed?

Solution

Let's represent the various classifications symbolically as follows:

M:	Male	1.	Freshman	G:	Geography
F:	Female	2.	Sophomore	H:	History
		3.	Junior	P:	Psychology
		4.	Senior		

We can mentally picture a tree diagram such that each of the two sex classifications branches into four school-year classifications, which in turn branch into three department classifications. Thus we have $(2)(4)(3) = 24$ different categories. ■

Another technique that works on certain problems involves what some people call the *back door* approach. For example, suppose we know that the classroom contains 50 seats. On some days, it may be easier to determine the number of students present by counting the number of empty seats and subtracting from 50 than by counting the number of students in attendance. (We suggested this back door approach as one way to count the nonadjacent seating arrangements in the discussion following Problem 5.) The next example further illustrates this approach.

PROBLEM 7

When rolling a pair of dice, in how many ways can we obtain a sum greater than 4?

Solution

For clarification purposes, let's use a red die and a white die. (It is not necessary to use different-colored dice, but it does help us analyze the different possible outcomes.) With a moment of thought, you will see that there are more ways to get a sum greater than 4 than there are ways to get a sum of 4 or less. Therefore let's determine the number of possibilities for getting a sum of 4 or less; then we'll subtract that number from the total number of possible outcomes when rolling a pair of dice. First, we can simply list and count the ways of getting a sum of 4 or less.

Red die	White die
1	1
1	2
1	3
2	1
2	2
3	1

There are six ways of getting a sum of 4 or less.

Second, because there are six possible outcomes on the red die and six possible outcomes on the white die, there is a total of $(6)(6) = 36$ possible outcomes when rolling a pair of dice.

Therefore, subtracting the number of ways of getting 4 or less from the total number of possible outcomes, we obtain $36 - 6 = 30$ ways of getting a sum greater than 4. ∎

Problem Set 15.1

Solve Problems 1–37.

1. If a woman has two skirts and ten blouses, how many different skirt–blouse combinations does she have?

2. If a man has eight shirts, five pairs of slacks, and three pairs of shoes, how many different shirt–slacks–shoe combinations does he have?

3. In how many ways can four people be seated in a row of four seats?

4. How many numbers of two different digits can be formed by choosing from the digits 1, 2, 3, 4, 5, 6, and 7?

5. How many *even* numbers of three different digits can be formed by choosing from the digits 2, 3, 4, 5, 6, 7, 8, and 9?

6. How many *odd* numbers of four different digits can be formed by choosing from the digits 1, 2, 3, 4, 5, 6, 7, and 8?

7. Suppose that the students at a certain university are to be classified according to their college (College of

Applied Science, College of Arts and Sciences, College of Business, College of Education, College of Fine Arts, College of Health and Physical Education), sex (female, male), and year in school (1, 2, 3, 4). How many categories are possible?

8. A medical researcher classifies subjects according to sex (female, male), smoking habits (smoker, non-smoker), and weight (below average, average, above average). How many different combined classifications are used?

9. A pollster classifies voters according to sex (female, male), party affiliation (Democrat, Republican, Independent), and family income (below $10,000, $10,000–$19,999, $20,000–$29,999, $30,000–$39,999, $40,000–$49,999, $50,000 and above). How many combined classifications does the pollster use?

10. A couple is planning to have four children. How many ways can this happen in terms of boy–girl classification? (For example, *BBBG* indicates that the first three children are boys and the last is a girl.)

11. In how many ways can three officers — president, secretary, and treasurer — be selected from a club that has 20 members?

12. In how many ways can three officers — president, secretary, and treasurer — be selected from a club with 15 female and 10 male members so that the president is female and the secretary and treasurer are male?

13. A disc jockey wants to play six songs once each in a half-hour program. How many different ways can he order these songs?

14. A state has agreed to have its automobile license plates consist of two letters followed by four digits. State officials do not want to repeat any letters or digits in any license numbers. How many different license plates will be available?

15. In how many ways can six people be seated in a row of six seats?

16. In how many ways can Al, Bob, Carlos, Don, Ed, and Fern be seated in a row of six seats if Al and Bob want to sit side by side?

17. In how many ways can Amy, Bob, Cindy, Dan, and Elmer be seated in a row of five seats so that neither Amy nor Bob occupies an end seat?

18. In how many ways can Al, Bob, Carlos, Don, Ed, and Fern be seated in a row of six seats if Al and Bob are not to be seated side by side? [*Hint*: Either Al and Bob will be seated side by side or they will not be seated side by side.]

19. In how many ways can Al, Bob, Carol, Dawn, and Ed be seated in a row of five chairs if Al is to be seated in the middle chair?

20. In how many ways can three letters be dropped in five mailboxes?

21. In how many ways can five letters be dropped in three mailboxes?

22. In how many ways can four letters be dropped in six mailboxes so that no two letters go in the same box?

23. In how many ways can six letters be dropped in four mailboxes so that no two letters go in the same box?

24. If five coins are tossed, in how many ways can they fall?

25. If three dice are tossed, in how many ways can they fall?

26. In how many ways can a sum less than ten be obtained when tossing a pair of dice?

27. In how many ways can a sum greater than five be obtained when tossing a pair of dice?

28. In how many ways can a sum greater than four be obtained when tossing three dice?

29. If no number contains repeated digits, how many numbers greater than 400 can be formed by choosing from the digits 2, 3, 4, and 5? [*Hint*: Consider both three-digit and four-digit numbers.]

30. If no number contains repeated digits, how many numbers greater than 5000 can be formed by choosing from the digits 1, 2, 3, 4, 5, and 6?

31. In how many ways can four boys and three girls be seated in a row of seven seats so that boys and girls occupy alternating seats?

32. In how many ways can three different mathematics books and four different history books be exhibited on a shelf so that all of the books in a subject area are side by side?

33. In how many ways can a true–false test of ten questions be answered?

34. If no number contains repeated digits, how many even numbers greater than 3000 can be formed by choosing from the digits 1, 2, 3, and 4?

35. If no number contains repeated digits, how many odd numbers greater than 40,000 can be formed by choosing from the digits 1, 2, 3, 4, and 5?

36. In how many ways can Al, Bob, Carol, Don, Ed, Faye, and George be seated in a row of seven seats so that Al, Bob, and Carol occupy consecutive seats in some order?

37. The license plates for a certain state consist of two letters followed by a four-digit number such that the first digit of the number is not zero. An example is PK-2446.
   (a) How many different license plates can be produced?
   (b) How many different plates do not have a repeated letter?
   (c) How many plates do not have any repeated digits in the number part of the plate?
   (d) How many plates do not have a repeated letter and also do not have any repeated digits?

38. How would you explain the fundamental principle of counting to a friend who missed class the day it was discussed?

39. Give two or three simple illustrations of the fundamental principle of counting.

40. Explain how you solved Problem 29.

15.2 Permutations and Combinations

As we develop the material in this section, **factorial notation** becomes very useful. The notation $n!$ (which is read "n factorial") is used with positive integers as follows:

$$1! = 1$$

$$2! = 2 \cdot 1 = 2$$

$$3! = 3 \cdot 2 \cdot 1 = 6$$

$$4! = 4 \cdot 3 \cdot 2 \cdot 1 = 24$$

Note that the factorial notation refers to an *indicated product*. In general, we write

$$n! = n(n - 1)(n - 2) \cdots 3 \cdot 2 \cdot 1$$

We also define $0! = 1$ so that certain formulas will be true for all nonnegative integers.

Now, as an introduction to the first concept of this section, let's consider a counting problem that closely resembles problems from the previous section.

PROBLEM 1

In how many ways can the three letters A, B, and C be arranged in a row?

Solution A

Certainly one approach to the problem is simply to list and count the arrangements.

ABC ACB BAC BCA CAB CBA

There are six arrangements of the three letters.

Solution B

Another approach, one that can be generalized for more difficult problems, uses the fundamental principle of counting. Because there are three choices for the first letter of an arrangement, two choices for the second letter, and one choice for the third letter, there are $(3)(2)(1) = 6$ arrangements. ■

■ Permutations

Ordered arrangements are called **permutations**. In general, a permutation of a set of n elements is an ordered arrangement of the n elements; we will use the symbol $P(n, n)$ to denote the number of such permutations. For example, from Problem 1, we know that $P(3, 3) = 6$. Furthermore, by using the same basic approach as in Solution B of Problem 1, we can obtain

$$P(1, 1) = 1 = 1!$$

$$P(2, 2) = 2 \cdot 1 = 2!$$

$$P(4, 4) = 4 \cdot 3 \cdot 2 \cdot 1 = 4!$$

$$P(5, 5) = 5 \cdot 4 \cdot 3 \cdot 2 \cdot 1 = 5!$$

In general, the following formula becomes evident:

$$P(n, n) = n!$$

Now suppose that we are interested in the number of two-letter permutations that can be formed by choosing from the four letters A, B, C, and D. (Some examples of such permutations are AB, BA, AC, BC, and CB.) In other words, we want to find the number of two-element permutations that can be formed from a set of four elements. We denote this number by $P(4, 2)$. To find $P(4, 2)$, we can reason as follows. First, we can choose any one of the four letters to occupy the first position in the permutation, and then we can choose any one of the remaining three letters for the second position. Therefore, by the fundamental principle of counting, we have $(4)(3) = 12$ different two-letter permutations; that is, $P(4, 2) = 12$. By using a similar line of reasoning, we can determine the following numbers. (Make sure that you agree with each of these.)

$$P(4, 3) = 4 \cdot 3 \cdot 2 = 24$$

$$P(5, 2) = 5 \cdot 4 = 20$$

$$P(6, 4) = 6 \cdot 5 \cdot 4 \cdot 3 = 360$$

$$P(7, 3) = 7 \cdot 6 \cdot 5 = 210$$

In general, we say that the number of r-element permutations that can be formed from a set of n elements is given by

$$P(n, r) = \underbrace{n(n - 1)(n - 2) \cdots}_{r \text{ factors}}$$

Note that the indicated product for $P(n, r)$ begins with n. Thereafter, each factor is 1 less than the previous one, and there is a total of r factors. For example,

$$P(6, 2) = 6 \cdot 5 = 30$$

$$P(8, 3) = 8 \cdot 7 \cdot 6 = 336$$

$$P(9, 4) = 9 \cdot 8 \cdot 7 \cdot 6 = 3024$$

Let's consider two problems that illustrate the use of $P(n, n)$ and $P(n, r)$.

PROBLEM 2   In how many ways can five students be seated in a row of five seats?

Solution

The problem is asking for the number of five-element permutations that can be formed from a set of five elements. Thus we can apply $P(n, n) = n!$.

$$P(5, 5) = 5! = 5 \cdot 4 \cdot 3 \cdot 2 \cdot 1 = 120$$ ■

PROBLEM 3   Suppose that seven people enter a swimming race. In how many ways can first, second, and third prizes be awarded?

Solution

This problem is asking for the number of three-element permutations that can be formed from a set of seven elements. Therefore, using the formula for $P(n, r)$, we obtain

$$P(7, 3) = 7 \cdot 6 \cdot 5 = 210$$ ■

It should be evident that both Problem 2 and Problem 3 could have been solved by applying the fundamental principle of counting. In fact, the formulas for $P(n, n)$ and $P(n, r)$ do not really give us much additional problem-solving power. However, as we will see in a moment, they do provide the basis for developing a formula that is very useful as a problem-solving tool.

■ Permutations Involving Nondistinguishable Objects

Suppose we have two identical H's and one T in an arrangement such as HTH. If we switch the two identical H's, the newly formed arrangement, HTH, will not be distinguishable from the original. In other words, there are fewer distinguishable permutations of n elements when some of those elements are identical than when the n elements are distinctly different.

To see the effect of identical elements on the number of distinguishable permutations, let's look at some specific examples:

2 identical H's	1 permutation (HH)
2 different letters	2! permutations (HT, TH)

Therefore, having two different letters affects the number of permutations by a *factor of* 2!.

$$\text{3 identical H's} \qquad \text{1 permutation (HHH)}$$

$$\text{3 different letters} \qquad \text{3! permutations}$$

Therefore, having three different letters affects the number of permutations by a *factor of* 3!.

$$\text{4 identical H's} \qquad \text{1 permutation (HHHH)}$$

$$\text{4 different letters} \qquad \text{4! permutations}$$

Therefore, having four different letters affects the number of permutations by a *factor of* 4!.

Now let's solve a specific problem.

PROBLEM 4

How many distinguishable permutations can be formed from three identical H's and two identical T's?

Solution

If we had five distinctly different letters, we could form 5! permutations. But the three identical H's affect the number of distinguishable permutations by a factor of 3!, and the two identical T's affect the number of permutations by a factor of 2!. Therefore we must divide 5! by 3! and 2!. We obtain

$$\frac{5!}{(3!)(2!)} = \frac{5 \cdot \overset{2}{4} \cdot 3 \cdot 2 \cdot 1}{3 \cdot 2 \cdot 1 \cdot 2 \cdot 1} = 10$$

distinguishable permutations of three H's and two T's. ∎

The type of reasoning used in Problem 4 leads us to the following general counting technique. If there are n elements to be arranged, where there are r_1 of one kind, r_2 of another kind, r_3 of another kind, . . . , r_k of a kth kind, then the total number of distinguishable permutations is given by the expression

$$\frac{n!}{(r_1!)(r_2!)(r_3!) \cdots (r_k!)}$$

PROBLEM 5

How many different 11-letter permutations can be formed from the 11 letters of the word MISSISSIPPI?

Solution

Because there are 4 I's, 4 S's, and 2 P's, we can form

$$\frac{11!}{(4!)(4!)(2!)} = \frac{11 \cdot 10 \cdot 9 \cdot 8 \cdot 7 \cdot 6 \cdot 5 \cdot 4 \cdot 3 \cdot 2 \cdot 1}{4 \cdot 3 \cdot 2 \cdot 1 \cdot 4 \cdot 3 \cdot 2 \cdot 1 \cdot 2 \cdot 1} = 34{,}650$$

distinguishable permutations. ∎

■ Combinations (Subsets)

Permutations are *ordered* arrangements; however, *order* is often not a consideration. For example, suppose that we want to determine the number of three-person committees that can be formed from the five people Al, Barb, Carol, Dawn, and Eric. Certainly the committee consisting of Al, Barb, and Eric is the same as the committee consisting of Barb, Eric, and Al. In other words, the order in which we choose or list the members is not important. Therefore we are really dealing with subsets; that is, we are looking for the number of three-element subsets that can be formed from a set of five elements. Traditionally in this context, subsets have been called **combinations**. Stated another way, then, we are looking for the number of combinations of five things taken three at a time. In general, *r*-element subsets taken from a set of *n* elements are called **combinations of *n* things taken *r* at a time**. The symbol $C(n, r)$ denotes the number of these combinations.

Now let's restate that committee problem and show a detailed solution that can be generalized to handle a variety of problems dealing with combinations.

PROBLEM 6

How many three-person committees can be formed from the five people Al, Barb, Carol, Dawn, and Eric?

Solution

Let's use the set {A, B, C, D, E} to represent the five people. Consider one possible three-person committee (subset), such as {A, B, C}; there are 3! permutations of these three letters. Now take another committee, such as {A, B, D}; there are also 3! permutations of these three letters. If we were to continue this process with all of the three-letter subsets that can be formed from the five letters, we would be counting all possible three-letter permutations of the five letters. That is, we would obtain $P(5, 3)$. Therefore, if we let $C(5, 3)$ represent the number of three-element subsets, then

$$(3!) \cdot C(5, 3) = P(5, 3)$$

Solving this equation for $C(5, 3)$ yields

$$C(5, 3) = \frac{P(5, 3)}{3!} = \frac{5 \cdot 4 \cdot 3}{3 \cdot 2 \cdot 1} = 10$$

Thus ten three-person committees can be formed from the five people. ■

In general, $C(n, r)$ times $r!$ yields $P(n, r)$. Thus

$$(r!) \cdot C(n, r) = P(n, r)$$

and solving this equation for $C(n, r)$ produces

$$C(n, r) = \frac{P(n, r)}{r!}$$

In other words, we can find the number of *combinations* of n things taken r at a time by dividing by $r!$, the number of permutations of n things taken r at a time. The following examples illustrate this idea:

$$C(7, 3) = \frac{P(7, 3)}{3!} = \frac{7 \cdot 6 \cdot 5}{3 \cdot 2 \cdot 1} = 35$$

$$C(9, 2) = \frac{P(9, 2)}{2!} = \frac{9 \cdot 8}{2 \cdot 1} = 36$$

$$C(10, 4) = \frac{P(10, 4)}{4!} = \frac{10 \cdot 9 \cdot 8 \cdot 7}{4 \cdot 3 \cdot 2 \cdot 1} = 210$$

P R O B L E M   7

How many different five-card hands can be dealt from a deck of 52 playing cards?

Solution

Because the order in which the cards are dealt is not an issue, we are working with a combination (subset) problem. Thus, using the formula for $C(n, r)$, we obtain

$$C(52, 5) = \frac{P(52, 5)}{5!} = \frac{52 \cdot 51 \cdot 50 \cdot 49 \cdot 48}{5 \cdot 4 \cdot 3 \cdot 2 \cdot 1} = 2{,}598{,}960$$

There are 2,598,960 different five-card hands that can be dealt from a deck of 52 playing cards.   ■

Some counting problems, such as Problem 8, can be solved by using the fundamental principle of counting along with the combination formula.

P R O B L E M   8

How many committees that consist of three women and two men can be formed from a group of five women and four men?

Solution

Let's think of this problem in terms of two tasks.

Task 1   Choose a subset of three women from the five women. This can be done in

$$C(5, 3) = \frac{P(5, 3)}{3!} = \frac{5 \cdot 4 \cdot 3}{3 \cdot 2 \cdot 1} = 10 \text{ ways}$$

Task 2   Choose a subset of two men from the four men. This can be done in

$$C(4, 2) = \frac{P(4, 2)}{2!} = \frac{4 \cdot 3}{2 \cdot 1} = 6 \text{ ways}$$

Task 1 followed by task 2 can be done in $(10)(6) = 60$ ways. Therefore there are 60 committees consisting of three women and two men that can be formed.   ■

Sometimes it takes a little thought to decide whether permutations or combinations should be used. Remember that **if order is to be considered, permutations should be used, but if order does not matter, then use combinations**. It is helpful to think of combinations as subsets.

| **PROBLEM 9** | A small accounting firm has 12 computer programmers. Three of these people are to be promoted to systems analysts. In how many ways can the firm select the three people to be promoted? |

Solution

Let's call the people A, B, C, D, E, F, G, H, I, J, K, and L. Suppose A, B, and C are chosen for promotion. Is this any different from choosing B, C, and A? Obviously not, so order does not matter, and we are being asked a question about combinations. More specifically, we need to find the number of combinations of 12 people taken three at a time. Thus there are

$$C(12, 3) = \frac{P(12, 3)}{3!} = \frac{12 \cdot 11 \cdot 10}{3 \cdot 2 \cdot 1} = 220$$

different ways to choose the three people to be promoted. ∎

| **PROBLEM 10** | A club is to elect three officers — president, secretary, and treasurer — from a group of six people, all of whom are willing to serve in any office. How many different ways can the officers be chosen? |

Solution

Let's call the candidates A, B, C, D, E, and F. Is electing A as president, B as secretary, and C as treasurer different from electing B as president, C as secretary, and A as treasurer? Obviously it is, so we are working with permutations. Thus there are

$$P(6, 3) = 6 \cdot 5 \cdot 4 = 120$$

different ways of filling the offices. ∎

Problem Set 15.2

In Problems 1–12, evaluate each.

1. $P(5, 3)$ **2.** $P(8, 2)$

3. $P(6, 4)$ **4.** $P(9, 3)$

5. $C(7, 2)$ **6.** $C(8, 5)$

7. $C(10, 5)$ **8.** $C(12, 4)$

9. $C(15, 2)$ **10.** $P(5, 5)$

11. $C(5, 5)$ **12.** $C(11, 1)$

For Problems 13–44, solve each problem.

13. How many permutations of the four letters A, B, C, and D can be formed by using all the letters in each permutation?

14. In how many ways can six students be seated in a row of six seats?

15. How many three-person committees can be formed from a group of nine people?

16. How many two-card hands can be dealt from a deck of 52 playing cards?

17. How many three-letter permutations can be formed from the first eight letters of the alphabet (a) if repetitions are not allowed? (b) if repetitions are allowed?

18. In a seven-team baseball league, in how many ways can the top three positions in the final standings be filled?

19. In how many ways can the manager of a baseball team arrange his batting order of nine starters if he wants his best hitters in the top four positions?

20. In a baseball league of nine teams, how many games are needed to complete the schedule if each team plays 12 games with each other team?

21. How many committees consisting of four women and four men can be chosen from a group of seven women and eight men?

22. How many three-element subsets containing one vowel and two consonants can be formed from the set {a, b, c, d, e, f, g, h, i}?

23. Five associate professors are being considered for promotion to the rank of full professor, but only three will be promoted. How many different combinations of three could be promoted?

24. How many numbers of four different digits can be formed from the digits 1, 2, 3, 4, 5, 6, 7, 8, and 9 if each number must consist of two odd and two even digits?

25. How many three-element subsets containing the letter A can be formed from the set {A, B, C, D, E, F}?

26. How many four-person committees can be chosen from five women and three men if each committee must contain at least one man?

27. How many different seven-letter permutations can be formed from four identical H's and three identical T's?

28. How many different eight-letter permutations can be formed from six identical H's and two identical T's?

29. How many different nine-letter permutations can be formed from three identical A's, four identical B's, and two identical C's?

30. How many different ten-letter permutations can be formed from five identical A's, four identical B's, and one C?

31. How many different seven-letter permutations can be formed from the seven letters of the word ALGEBRA?

32. How many different 11-letter permutations can be formed from the 11 letters of the word MATHEMATICS?

33. In how many ways can x^4y^2 be written without using exponents? [*Hint:* One way is *xxxxyy*.]

34. In how many ways can $x^3y^4z^3$ be written without using exponents?

35. Ten basketball players are going to be divided into two teams of five players each for a game. In how many ways can this be done?

36. Ten basketball players are going to be divided into two teams of five in such a way that the two best players are on opposite teams. In how many ways can this be done?

37. A box contains nine good light bulbs and four defective bulbs. How many samples of three bulbs contain one defective bulb? How many samples of three bulbs contain *at least* one defective bulb?

38. How many five-person committees consisting of two juniors and three seniors can be formed from a group of six juniors and eight seniors?

39. In how many ways can six people be divided into two groups so that there are four in one group and two in the other? In how many ways can six people be divided into two groups of three each?

40. How many five-element subsets containing A and B can be formed from the set {A, B, C, D, E, F, G, H}?

41. How many four-element subsets containing A or B but not both A and B can be formed from the set {A, B, C, D, E, F, G}?

42. How many different five-person committees can be selected from nine people if two of those people refuse to serve together on a committee?

43. How many different line segments are determined by five points? By six points? By seven points? By *n* points?

44. (a) How many five-card hands consisting of two kings and three aces can be dealt from a deck of 52 playing cards?

(b) How many five-card hands consisting of three kings and two aces can be dealt from a deck of 52 playing cards?

(c) How many five-card hands consisting of three cards of one face value and two cards of another face value can be dealt from a deck of 52 playing cards?

■ ■ ■ **THOUGHTS INTO WORDS**

45. Explain the difference between a permutation and a combination. Give an example of each one to illustrate your explanation.

46. Your friend is having difficulty distinguishing between permutations and combinations in problem-solving situations. What might you do to help her?

■ ■ ■ **FURTHER INVESTIGATIONS**

47. In how many ways can six people be seated at a circular table? [*Hint*: Moving each person one place to the right (or left) does not create a new seating.]

48. The quantity $P(8, 3)$ can be expressed completely in factorial notation as follows:

$$P(8, 3) = \frac{P(8, 3) \cdot 5!}{5!} = \frac{(8 \cdot 7 \cdot 6)(5 \cdot 4 \cdot 3 \cdot 2 \cdot 1)}{5!} = \frac{8!}{5!}$$

Express each of the following in terms of factorial notation.
(a) $P(7, 3)$
(b) $P(9, 2)$
(c) $P(10, 7)$
(d) $P(n, r)$, $r \leq n$ and $0!$ is defined to be 1

49. Sometimes the formula

$$C(n, r) = \frac{n!}{r!(n - r)!}$$

is used to find the number of combinations of n things taken r at a time. Use the result from part (d) of Problem 48 and develop this formula.

50. Compute $C(7, 3)$ and $C(7, 4)$. Compute $C(8, 2)$ and $C(8, 6)$. Compute $C(9, 8)$ and $C(9, 1)$. Now argue that $C(n, r) = C(n, n - r)$ for $r \leq n$.

▦ **GRAPHING CALCULATOR ACTIVITIES**

Before doing Problems 51–56, be sure that you can use your calculator to compute the number of permutations and combinations. Your calculator may possess a special sequence of keys for such computations. You may need to refer to your user's manual for this information.

51. Use your calculator to check your answers for Problems 1–12.

52. How many different five-card hands can be dealt from a deck of 52 playing cards?

53. How many different seven-card hands can be dealt from a deck of 52 playing cards?

54. How many different five-person committees can be formed from a group of 50 people?

55. How many different juries consisting of 11 people can be chosen from a group of 30 people?

56. How many seven-person committees consisting of three juniors and four seniors can be formed from 45 juniors and 53 seniors?

15.3   Probability

In order to introduce some terminology and notation, let's consider a simple experiment of tossing a regular six-sided die. There are six possible outcomes to this experiment: The 1, the 2, the 3, the 4, the 5, or the 6 will land up. This set of possible outcomes is called a "sample space," and the individual elements of the sample space are called "sample points." We will use S (sometimes with subscripts for identification purposes) to refer to a particular sample space of an experiment; then we will denote the number of sample points by $n(S)$. Thus for the experiment of tossing a die, $S = \{1, 2, 3, 4, 5, 6\}$ and $n(S) = 6$.

In general, the set of all possible outcomes of a given experiment is called the **sample space**, and the individual elements of the sample space are called **sample points**. (In this text, we will be working only with sample spaces that are finite.)

Now suppose we are interested in some of the various possible outcomes in the die-tossing experiment. For example, we might be interested in the event, *an even number comes up*. In this case we are satisfied if a 2, 4, or 6 appears on the top face of the die, and therefore the event, *an even number comes up*, is the subset $E = \{2, 4, 6\}$, where $n(E) = 3$. Perhaps, instead, we might be interested in the event, *a multiple of 3 comes up*. This event determines the subset $F = \{3, 6\}$, where $n(F) = 2$.

In general, any subset of a sample space is called an **event** or an **event space**. If the event consists of exactly one element of the sample space, then it is called a **simple event**. Any nonempty event that is not simple is called a **compound event**. A compound event can be represented as the union of simple events.

It is now possible to give a very simple definition for *probability* as we want to use the term in this text.

Definition 15.1

In an experiment where all possible outcomes in the sample space S are equally likely to occur, the **probability** of an event E is defined by

$$P(E) = \frac{n(E)}{n(S)}$$

where $n(E)$ denotes the number of elements in the event E, and $n(S)$ denotes the number of elements in the sample space S.

Many probability problems can be solved by applying Definition 15.1. Such an approach requires that we be able to determine the number of elements in the sample space and the number of elements in the event space. For example, in the die-tossing experiment, the probability of getting an even number with one toss of the die is given by

$$P(E) = \frac{n(E)}{n(S)} = \frac{3}{6} = \frac{1}{2}$$

Let's consider two examples where the number of elements in both the sample space and the event space are easy to determine.

P R O B L E M   1

A coin is tossed. Find the probability that a head turns up.

Solution

Let the sample space be $S = \{H, T\}$; then $n(S) = 2$. The event of a head turning up is the subset $E = \{H\}$, so $n(E) = 1$. Therefore the probability of getting a head with one flip of a coin is given by

$$P(E) = \frac{n(E)}{n(S)} = \frac{1}{2}$$

■

P R O B L E M   2

Two coins are tossed. What is the probability that *at least* one head will turn up?

Solution

For clarification purposes, let the coins be a penny and a nickel. The possible outcomes of this experiment are (1) a head on both coins, (2) a head on the penny and a tail on the nickel, (3) a tail on the penny and a head on the nickel, and (4) a tail on both coins. Using ordered-pair notation, where the first entry of a pair represents the penny and the second entry the nickel, we can write the sample space as

$$S = \{(H, H), (H, T), (T, H), (T, T)\}$$

and $n(S) = 4$.

Let E be the event of getting at least one head. Thus $E = \{(H, H), (H, T), (T, H)\}$ and $n(E) = 3$. Therefore the probability of getting at least one head with one toss of two coins is

$$P(E) = \frac{n(E)}{n(S)} = \frac{3}{4}$$

■

As you might expect, the counting techniques discussed in the first two sections of this chapter can frequently be used to solve probability problems.

P R O B L E M   3

Four coins are tossed. Find the probability of getting three heads and one tail.

Solution

The sample space consists of the possible outcomes for tossing four coins. Because there are two things that can happen on each coin, by the fundamental principle of counting there are $2 \cdot 2 \cdot 2 \cdot 2 = 16$ possible outcomes for tossing four coins. Thus we know that $n(S) = 16$ without taking the time to list all of the elements. The event of getting three heads and one tail is the subset $E = \{(H, H, H, T), (H, H, T, H), (H, T, H, H), (T, H, H, H)\}$, where $n(E) = 4$. Therefore the requested probability is

$$P(E) = \frac{n(E)}{n(S)} = \frac{4}{16} = \frac{1}{4}$$

■

P R O B L E M   4

Al, Bob, Chad, Dorcas, Eve, and Françoise are randomly seated in a row of six chairs. What is the probability that Al and Bob are seated in the end seats?

Solution

The sample space consists of all possible ways of seating six people in six chairs or, in other words, the permutations of six things taken six at a time. Thus $n(S) = P(6, 6) = 6! = 6 \cdot 5 \cdot 4 \cdot 3 \cdot 2 \cdot 1 = 720$.

The event space consists of all possible ways of seating the six people so that Al and Bob both occupy end seats. The number of these possibilities can be determined as follows:

Task 1   Put Al and Bob in the end seats. This can be done in two ways because Al can be on the left end and Bob on the right end, or vice versa.

Task 2   Put the other four people in the remaining four seats. This can be done in $4! = 4 \cdot 3 \cdot 2 \cdot 1 = 24$ different ways.

Therefore task 1 followed by task 2 can be done in $(2)(24) = 48$ different ways, so $n(E) = 48$. Thus the requested probability is

$$P(E) = \frac{n(E)}{n(S)} = \frac{48}{720} = \frac{1}{15}$$ ∎

Note that in Problem 3, by using the fundamental principle of counting to determine the number of elements in the sample space, we did not actually have to list all of the elements. For the event space, we listed the elements and counted them in the usual way. In Problem 4, we used the permutation formula $P(n, n) = n!$ to determine the number of elements in the sample space, and then we used the fundamental principle to determine the number of elements in the event space. There are no definite rules about when to list the elements and when to apply some sort of counting technique. In general, we suggest that if you do not immediately see a counting pattern for a particular problem, you should begin the listing process. If a counting pattern then emerges as you are listing the elements, use the pattern at that time.

The combination (subset) formula we developed in Section 15.2, $C(n, r) = P(n, r)/r!$, is also a very useful tool for solving certain kinds of probability problems. The next three examples illustrate some problems of this type.

P R O B L E M   5

A committee of three people is randomly selected from Alice, Bjorn, Chad, Dee, and Eric. What is the probability that Alice is on the committee?

Solution

The sample space, S, consists of all possible three-person committees that can be formed from the five people. Therefore

$$n(S) = C(5, 3) = \frac{P(5, 3)}{3!} = \frac{5 \cdot 4 \cdot 3}{3 \cdot 2 \cdot 1} = 10$$

The event space, E, consists of all the three-person committees that have Alice as a member. Each of those committees contains Alice and two other people chosen from the four remaining people. Thus the number of such committees is $C(4, 2)$, so we obtain

$$n(E) = C(4, 2) = \frac{P(4, 2)}{2!} = \frac{4 \cdot 3}{2 \cdot 1} = 6$$

The requested probability is

$$P(E) = \frac{n(E)}{n(S)} = \frac{6}{10} = \frac{3}{5}$$

∎

PROBLEM 6

A committee of four is chosen at random from a group of five seniors and four juniors. Find the probability that the committee will contain two seniors and two juniors.

Solution

The sample space, S, consists of all possible four-person committees that can be formed from the nine people. Thus

$$n(S) = C(9, 4) = \frac{P(9, 4)}{4!} = \frac{9 \cdot 8 \cdot 7 \cdot 6}{4 \cdot 3 \cdot 2 \cdot 1} = 126$$

The event space, E, consists of all four-person committees that contain two seniors and two juniors. They can be counted as follows.

Task 1 Choose two seniors from the five available seniors in $C(5, 2) = 10$ ways.

Task 2 Choose two juniors from the four available juniors in $C(4, 2) = 6$ ways.

Therefore there are $10 \cdot 6 = 60$ committees consisting of two seniors and two juniors. The requested probability is

$$P(E) = \frac{n(E)}{n(S)} = \frac{60}{126} = \frac{10}{21}$$

∎

PROBLEM 7

Eight coins are tossed. Find the probability of getting two heads and six tails.

Solution

Because either of two things can happen on each coin, the total number of possible outcomes, $n(S)$, is $2^8 = 256$.

We can select two coins, which are to fall heads, in $C(8, 2) = 28$ ways. For each of these ways, there is only one way to select the other six coins that are to fall tails. Therefore there are $28 \cdot 1 = 28$ ways of getting two heads and six tails, so $n(E) = 28$. The requested probability is

$$P(E) = \frac{n(E)}{n(S)} = \frac{28}{256} = \frac{7}{64}$$

∎

Problem Set 15.3

For Problems 1–4, *two* coins are tossed. Find the probability of tossing each of the following events:

1. One head and one tail

2. Two tails

3. At least one tail

4. No tails

For Problems 5–8, *three* coins are tossed. Find the probability of tossing each of the following events:

5. Three heads

6. Two heads and a tail

7. At least one head

8. Exactly one tail

For Problems 9–12, *four* coins are tossed. Find the probability of tossing each of the following events:

9. Four heads

10. Three heads and a tail

11. Two heads and two tails

12. At least one head

For Problems 13–16, *one* die is tossed. Find the probability of rolling each of the following events:

13. A multiple of 3

14. A prime number

15. An even number

16. A multiple of 7

For Problems 17–22, *two* dice are tossed. Find the probability of rolling each of the following events:

17. A sum of 6

18. A sum of 11

19. A sum less than 5

20. A 5 on exactly one die

21. A 4 on at least one die

22. A sum greater than 4

For Problems 23–26, *one* card is drawn from a standard deck of 52 playing cards. Find the probability of each of the following events:

23. A heart is drawn.

24. A king is drawn.

25. A spade or a diamond is drawn.

26. A red jack is drawn.

For Problems 27–30, suppose that 25 slips of paper numbered 1 to 25, inclusive, are put in a hat, and then one is drawn out at random. Find the probability of each of the following events:

27. The slip with the 5 on it is drawn.

28. A slip with an even number on it is drawn.

29. A slip with a prime number on it is drawn.

30. A slip with a multiple of 6 on it is drawn.

For Problems 31–34, suppose that a committee of two boys is to be chosen at random from the five boys Al, Bill, Carl, Dan, and Eli. Find the probability of each of the following events:

31. Dan is on the committee.

32. Dan and Eli are both on the committee.

33. Bill and Carl are not both on the committee.

34. Dan or Eli, but not both of them, is on the committee.

For Problems 35–38, suppose that a five-person committee is selected at random from the eight people Al, Barb, Chad, Dominique, Eric, Fern, George, and Harriet. Find the probability of each of the following events:

35. Al and Barb are both on the committee.

36. George is not on the committee.

37. Either Chad or Dominique, but not both, is on the committee.

38. Neither Al nor Barb is on the committee.

For Problems 39–41, suppose that a box of ten items from a manufacturing company is known to contain two defective and eight nondefective items. A sample of three items is selected at random. Find the probability of each of the following events:

39. The sample contains all nondefective items.

40. The sample contains one defective and two nondefective items.

41. The sample contains two defective and one nondefective item.

For Problems 42–60, solve each problem.

42. A building has five doors. Find the probability that two people, entering the building at random, will choose the same door.

43. Bill, Carol, and Alice are to be seated at random in a row of three seats. Find the probability that Bill and Carol will be seated side by side.

44. April, Bill, Carl, and Denise are to be seated at random in a row of four chairs. What is the probability that April and Bill will occupy the end seats?

45. A committee of four girls is to be chosen at random from the five girls Alice, Becky, Candy, Dee, and Elaine. Find the probability that Elaine is not on the committee.

46. Three boys and two girls are to be seated at random in a row of five seats. What is the probability that the boys and girls will be in alternating seats?

47. Four different mathematics books and five different history books are randomly placed on a shelf. What is the probability that all of the books on a subject are side by side?

48. Each of three letters is to be mailed in any one of five different mailboxes. What is the probability that all will be mailed in the same mailbox?

49. Randomly form a four-digit number by using the digits 2, 3, 4, and 6 once each. What is the probability that the number formed is greater than 4000?

50. Randomly select one of the 120 permutations of the letters a, b, c, d, and e. Find the probability that in the chosen permutation, the letter a precedes the b (the a is to the left of the b).

51. A committee of four is chosen at random from a group of six women and five men. Find the probability that the committee contains two women and two men.

52. A committee of three is chosen at random from a group of four women and five men. Find the probability that the committee contains at least one man.

53. Ahmed, Bob, Carl, Dan, Ed, Frank, Gino, Harry, Julio, and Mike are randomly divided into two five-man teams for a basketball game. What is the probability that Ahmed, Bob, and Carl are on the same team?

54. Seven coins are tossed. Find the probability of getting four heads and three tails.

55. Nine coins are tossed. Find the probability of getting three heads and six tails.

56. Six coins are tossed. Find the probability of getting at least four heads.

57. Five coins are tossed. Find the probability of getting no more than three heads.

58. Each arrangement of the 11 letters of the word MISSISSIPPI is put on a slip of paper and placed in a hat. One slip is drawn at random from the hat. Find the probability that the slip contains an arrangement of the letters with the four S's at the beginning.

59. Each arrangement of the seven letters of the word OSMOSIS is put on a slip of paper and placed in a hat. One slip is drawn at random from the hat. Find the probability that the slip contains an arrangement of the letters with an O at the beginning and an O at the end.

60. Consider all possible arrangements of three identical H's and three identical T's. Suppose that one of these arrangements is selected at random. What is the probability that the selected arrangement has the three H's in consecutive positions?

■ ■ ■ **THOUGHTS INTO WORDS**

61. Explain the concepts of sample space and event space.

62. Why must probability answers fall between 0 and 1, inclusive? Give an example of a situation for which the probability is 0. Also give an example for which the probability is 1.

■ ■ ■ **FURTHER INVESTIGATIONS**

In Problem 7 of Section 15.2, we found that there are 2,598,960 different five-card hands that can be dealt from a deck of 52 playing cards. Therefore, probabilities for certain kinds of five-card poker hands can be calculated by using 2,598,960 as the number of elements in the sample space. For Problems 63–71, determine the number of different five-card poker hands of the indicated type that can be obtained.

63. A straight flush (five cards in sequence and of the same suit; aces are both low and high, so A2345 and 10JQKA are both acceptable)

64. Four of a kind (four of the same face value, such as four kings)

65. A full house (three cards of one face value and two cards of another face value)

66. A flush (five cards of the same suit but not in sequence)

67. A straight (five cards in sequence but not all of the same suit)

68. Three of a kind (three cards of one face value and two cards of two different face values)

69. Two pairs

70. Exactly one pair

71. No pairs

15.4 Some Properties of Probability; Expected Values

There are several basic properties that are useful in the study of probability from both a theoretical and a computational viewpoint. We will discuss two of these properties at this time and some additional ones in the next section. The first property may seem to state the obvious, but it still needs to be mentioned.

Property 15.1

For all events E,

$$0 \le P(E) \le 1$$

Property 15.1 simply states that probabilities must fall in the range from 0 to 1, inclusive. This seems reasonable because $P(E) = n(E)/n(S)$, and E is a subset of S. The next two examples illustrate circumstances where $P(E) = 0$ and $P(E) = 1$.

PROBLEM 1   Toss a regular six-sided die. What is the probability of getting a 7?

Solution

The sample space is $S = \{1, 2, 3, 4, 5, 6\}$, thus $n(S) = 6$. The event space is $E = \varnothing$, so $n(E) = 0$. Therefore the probability of getting a 7 is

$$P(E) = \frac{n(E)}{n(S)} = \frac{0}{6} = 0$$ ∎

PROBLEM 2   What is the probability of getting a head or a tail with one flip of a coin?

Solution

The sample space is $S = \{H, T\}$, and the event space is $E = \{H, T\}$. Therefore $n(S) = n(E) = 2$, and

$$P(E) = \frac{n(E)}{n(S)} = \frac{2}{2} = 1$$ ∎

An event that has a probability of 1 is sometimes called **certain success**, and an event with a probability of 0 is called **certain failure**.

It should also be mentioned that Property 15.1 serves as a check for reasonableness of answers. In other words, when computing probabilities, we know that our answer must fall between 0 and 1, inclusive. Any other probability answer is simply not reasonable.

■ Complementary Events

Complementary events are complementary sets such that S, the sample space, serves as the universal set. The following examples illustrate this idea.

Sample space	Event space	Complement of event space
$S = \{1, 2, 3, 4, 5, 6\}$	$E = \{1, 2\}$	$E' = \{3, 4, 5, 6\}$
$S = \{H, T\}$	$E = \{T\}$	$E' = \{H\}$
$S = \{2, 3, 4, \ldots, 12\}$	$E = \{2, 3, 4\}$	$E' = \{5, 6, 7, \ldots, 12\}$
$S = \{1, 2, 3, \ldots, 25\}$	$E = \{3, 4, 5, \ldots, 25\}$	$E' = \{1, 2\}$

In each case, note that E' (the complement of E) consists of all elements of S that are *not* in E. Thus E and E' are called *complementary events*. Also note that for each example, $P(E) + P(E') = 1$. We can state the following general property:

Property 15.2

If E is any event of a sample space S, and E' is the complementary event, then

$$P(E) + P(E') = 1$$

From a computational viewpoint, Property 15.2 provides us with a double-barreled attack on some probability problems. That is, once we compute either $P(E)$ or $P(E')$, we can determine the other one simply by subtracting from 1. For example, suppose that for a particular problem we can determine that $P(E) = \dfrac{3}{13}$.

Then we immediately know that $P(E') = 1 - P(E) = 1 - \dfrac{3}{13} = \dfrac{10}{13}$. The following examples further illustrate the usefulness of Property 15.2.

PROBLEM 3

Two dice are tossed. Find the probability of getting a sum greater than 3.

Solution

Let S be the familiar sample space of ordered pairs for this problem, where $n(S) = 36$. Let E be the event of obtaining a sum greater than 3. Then E' is the event of obtaining a sum less than or equal to 3; that is, $E' = \{(1, 1), (1, 2), (2, 1)\}$. Thus

$$P(E') = \frac{n(E')}{n(S)} = \frac{3}{36} = \frac{1}{12}$$

From this, we conclude that

$$P(E) = 1 - P(E') = 1 - \frac{1}{12} = \frac{11}{12}$$

∎

P R O B L E M   4

Toss three coins and find the probability of getting at least one head.

Solution

The sample space, S, consists of all possible outcomes for tossing three coins. Using the fundamental principle of counting, we know that there are $(2)(2)(2) = 8$ outcomes, so $n(S) = 8$. Let E be the event of getting at least one head. Then E' is the complementary event of not getting any heads. The set E' is easy to list: $E' = \{(T, T, T)\}$. Thus $n(E') = 1$ and $P(E') = \frac{1}{8}$. From this, $P(E)$ can be determined to be

$$P(E) = 1 - P(E') = 1 - \frac{1}{8} = \frac{7}{8}$$

∎

P R O B L E M   5

A three-person committee is chosen at random from a group of five women and four men. Find the probability that the committee contains at least one woman.

Solution

Let the sample space, S, be the set of all possible three-person committees that can be formed from nine people. There are $C(9, 3) = 84$ such committees; therefore $n(S) = 84$.

Let E be the event, *the committee contains at least one woman*. Then E' is the complementary event, *the committee contains all men*. Thus E' consists of all three-man committees that can be formed from four men. There are $C(4, 3) = 4$ such committees; thus $n(E') = 4$. We have

$$P(E') = \frac{n(E')}{n(S)} = \frac{4}{84} = \frac{1}{21}$$

which determines $P(E)$ to be

$$P(E) = 1 - P(E') = 1 - \frac{1}{21} = \frac{20}{21}$$

∎

The concepts of **set intersection** and **set union** play an important role in the study of probability. If E and F are two events in a sample space S, then $E \cap F$ is the event consisting of all sample points of S that are in both E and F as indicated in Figure 15.2. Likewise, $E \cup F$ is the event consisting of all sample points of S that are in E or F, or both, as shown in Figure 15.3.

In Figure 15.4, there are 47 sample points in E, 38 sample points in F, and 15 sample points in $E \cap F$. How many sample points are there in $E \cup F$? Simply adding the number of points in E and F would result in counting the 15 points in

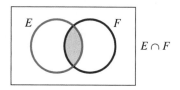

Figure 15.2

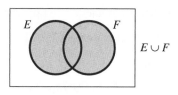

Figure 15.3

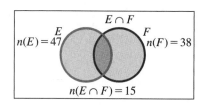

Figure 15.4

$E \cap F$ twice. Therefore, 15 must be subtracted from the total number of points in E and F, yielding $47 + 38 - 15 = 70$ points in $E \cup F$. We can state the following general counting property:

$$n(E \cup F) = n(E) + n(F) - n(E \cap F)$$

If we divide both sides of this equation by $n(S)$, we obtain the following probability property:

Property 15.3

For events E and F of a sample space S,

$$P(E \cup F) = P(E) + P(F) - P(E \cap F)$$

PROBLEM 6

What is the probability of getting an odd number or a prime number with one toss of a die?

Solution

Let $S = \{1, 2, 3, 4, 5, 6\}$ be the sample space, $E = \{1, 3, 5\}$ the event of getting an odd number, and $F = \{2, 3, 5\}$ the event of getting a prime number. Then $E \cap F = \{3, 5\}$, and using Property 15.3, we obtain

$$P(E \cup F) = \frac{3}{6} + \frac{3}{6} - \frac{2}{6} = \frac{4}{6} = \frac{2}{3}$$

■

PROBLEM 7

Toss three coins. What is the probability of getting at least two heads or exactly one tail?

Solution

Using the fundamental principle of counting, we know that there are $2 \cdot 2 \cdot 2 = 8$ possible outcomes of tossing three coins; thus $n(S) = 8$. Let

$$E = \{(H, H, H), (H, H, T), (H, T, H), (T, H, H)\}$$

be the event of getting at least two heads, and let

$$F = \{(H, H, T), (H, T, H), (T, H, H)\}$$

be the event of getting exactly one tail. Then

$$E \cap F = \{(H, H, T), (H, T, H), (T, H, H)\}$$

and we can compute $P(E \cup F)$ as follows.

$$P(E \cup F) = P(E) + P(F) - P(E \cap F)$$

$$= \frac{4}{8} + \frac{3}{8} - \frac{3}{8}$$

$$= \frac{4}{8} = \frac{1}{2}$$ ■

In Property 15.3, if $E \cap F = \varnothing$, then the events E and F are said to be **mutually exclusive**. In other words, mutually exclusive events are events that cannot occur at the same time. For example, when we roll a die, the event of getting a 4 and the event of getting a 5 are mutually exclusive; they cannot both happen on the same roll. If $E \cap F = \varnothing$, then $P(E \cap F) = 0$, and Property 15.3 becomes $P(E \cup F) = P(E) + P(F)$ **for mutually exclusive events**.

PROBLEM 8

Suppose we have a jar that contains five white, seven green, and nine red marbles. If one marble is drawn at random from the jar, find the probability that it is white or green.

Solution

The events of drawing a white marble and drawing a green marble are mutually exclusive. Therefore the probability of drawing a white or a green marble is

$$\frac{5}{21} + \frac{7}{21} = \frac{12}{21} = \frac{4}{7}$$ ■

Note that in the solution for Problem 8, we did not explicitly name and list the elements of the sample space or event spaces. It was obvious that the sample space contained 21 elements (21 marbles in the jar) and that the event spaces contained five elements (five white marbles) and seven elements (seven green marbles). Thus it was not necessary to name and list the sample space and event spaces.

PROBLEM 9

Suppose that the data in the following table represent the results of a survey of 1000 drivers after a holiday weekend.

	Rain (R)	No rain (R')	Total
Accident (A)	35	10	45
No accident (A')	450	505	955
Total	485	515	1000

If a person is selected at random, what is the probability that the person was in an accident or that it rained?

Solution

First, let's form a **probability table** by dividing each entry by 1000, the total number surveyed.

	Rain (R)	No rain (R')	Total
Accident (A)	0.035	0.010	0.045
No accident (A')	0.450	0.505	0.955
Total	0.485	0.515	1.000

Now we can use Property 15.3 and compute $P(A \cup R)$.

$$P(A \cup R) = P(A) + P(R) - P(A \cap R)$$
$$= 0.045 + 0.485 - 0.035$$
$$= 0.495 \quad \blacksquare$$

■ Expected Value

Suppose we toss a coin 500 times. We would expect to get approximately 250 heads. In other words, because the probability of getting a head with one toss of a coin is $\frac{1}{2}$, in 500 tosses we should get approximately $500\left(\frac{1}{2}\right) = 250$ heads. The word "approximately" conveys a key idea. As we know from experience, it is possible to toss a coin several times and get all heads. However, with a large number of tosses, things should average out so that we get about equal numbers of heads and tails.

As another example, consider the fact that the probability of getting a sum of 6 with one toss of a pair of dice is $\frac{5}{36}$. Therefore, if a pair of dice is tossed 360 times, we should expect to get a sum of 6 approximately $360\left(\frac{5}{36}\right) = 50$ times.

Let us now define the concept of *expected value*.

Definition 15.2

> If the k possible outcomes of an experiment are assigned the values x_1, x_2, $x_3, \ldots, x_k$, and if they occur with probabilities of $p_1, p_2, p_3, \ldots, p_k$, respectively, then the **expected value** of the experiment (E_v) is given by
>
> $$E_v = x_1 p_1 + x_2 p_2 + x_3 p_3 + \cdots + x_k p_k$$

The concept of expected value (also called **mathematical expectation**) is used in a variety of probability situations that deal with such things as fairness of games and decision making in business ventures. Let's consider some examples.

P R O B L E M   1 0

Suppose that you buy one ticket in a lottery where 1000 tickets are sold. Furthermore, suppose that three prizes are awarded: one of $500, one of $300, and one of $100. What is your mathematical expectation?

Solution

Because you bought one ticket, the probability of you winning $500 is $\dfrac{1}{1000}$; the probability of you winning $300 is $\dfrac{1}{1000}$; and the probability of you winning $100 is $\dfrac{1}{1000}$. Multiplying each of these probabilities by the corresponding prize money and then adding the results yields your mathematical expectation.

$$E_v = \$500\left(\frac{1}{100}\right) + \$300\left(\frac{1}{1000}\right) + \$100\left(\frac{1}{100}\right)$$

$$= \$0.50 + \$0.30 + \$0.10$$

$$= \$0.90 \qquad \blacksquare$$

In Problem 10, if you pay more than $0.90 for a ticket, then it is not a **fair game** from your standpoint. If the price of the game is included in the calculation of the expected value, then a fair game is defined to be one where the expected value is zero.

P R O B L E M   1 1

A player pays $5 to play a game where the probability of winning is $\dfrac{1}{5}$ and the probability of losing is $\dfrac{4}{5}$. If the player wins the game, he receives $25. Is this a fair game for the player?

Solution

From Definition 15.2, let $x_1 = \$20$, which represents the $25 won minus the $5 paid to play, and let $x_2 = -\$5$, the amount paid to play the game. We are also given that $p_1 = \dfrac{1}{5}$ and $p_2 = \dfrac{4}{5}$. Thus the expected value is

$$E_v = \$20\left(\frac{1}{5}\right) + (-\$5)\left(\frac{4}{5}\right)$$

$$= \$4 - \$4$$

$$= 0$$

Because the expected value is zero, it is a fair game. $\blacksquare$

P R O B L E M   1 2

Suppose you are interested in insuring a diamond ring for $2000 against theft. An insurance company charges a premium of $25 per year, claiming that there is a probability of 0.01 that the ring will be stolen during the year. What is your expected gain or loss if you take out the insurance?

Solution

From Definition 15.2, let $x_1 = \$1975$, which represents the $2000 minus the cost of the premium, $25, and let $x_2 = -\$25$. We also are given that $p_1 = 0.01$, so $p_2 = 1 - 0.01 = 0.99$. Thus the expected value is

$$E_v = \$1975(0.01) + (-\$25)(0.99)$$

$$= \$19.75 - \$24.75$$

$$= -\$5.00$$

This means that if you insure with this company over many years, and the circumstances remain the same, you will have an average net loss of $5 per year.   ■

Problem Set 15.4

For Problems 1–4, *two* dice are tossed. Find the probability of rolling each of the following events:

1. A sum of 6

2. A sum greater than 2

3. A sum less than 8

4. A sum greater than 1

For Problems 5–8, *three* dice are tossed. Find the probability of rolling each of the following events:

5. A sum of 3

6. A sum greater than 4

7. A sum less than 17

8. A sum greater than 18

For Problems 9–12, *four* coins are tossed. Find the probability of getting each of the following events:

9. Four heads

10. Three heads and a tail

11. At least one tail

12. At least one head

For Problems 13–16, *five* coins are tossed. Find the probability of getting each of the following events:

13. Five tails

14. Four heads and a tail

15. At least one tail

16. At least two heads

For Problems 17–23, solve each problem.

17. Toss a pair of dice. What is the probability of not getting a double?

18. The probability that a certain horse will win the Kentucky Derby is $\dfrac{1}{20}$. What is the probability that it will lose the race?

19. One card is randomly drawn from a deck of 52 playing cards. What is the probability that it is not an ace?

20. Six coins are tossed. Find the probability of getting at least two heads.

21. A subset of two letters is chosen at random from the set {a, b, c, d, e, f, g, h, i}. Find the probability that the subset contains at least one vowel.

22. A two-person committee is chosen at random from a group of four men and three women. Find the probability that the committee contains at least one man.

23. A three-person committee is chosen at random from a group of seven women and five men. Find the probability that the committee contains at least one man.

For Problems 24–27, one die is tossed. Find the probability of rolling each of the following events:

24. A 3 or an odd number

25. A 2 or an odd number

26. An even number or a prime number

27. An odd number or a multiple of 3

For Problems 28–31, two dice are tossed. Find the probability of rolling each of the following events:

28. A double or a sum of 6

29. A sum of 10 or a sum greater than 8

30. A sum of 5 or a sum greater than 10

31. A double or a sum of 7

For Problems 32–56, solve each problem.

32. Two coins are tossed. Find the probability of getting exactly one head or at least one tail.

33. Three coins are tossed. Find the probability of getting at least two heads or exactly one tail.

34. A jar contains seven white, six blue, and ten red marbles. If one marble is drawn at random from the jar, find the probability that (a) the marble is white or blue; (b) the marble is white or red; (c) the marble is blue or red.

35. A coin and a die are tossed. Find the probability of getting a head on the coin or a 2 on the die.

36. A card is randomly drawn from a deck of 52 playing cards. Find the probability that it is a red card or a face card. (Jacks, queens, and kings are the face cards.)

37. The data in the following table represent the results of a survey of 1000 drivers after a holiday weekend.

	Rain (R)	No rain (R')	Total
Accident (A)	45	15	60
No accident (A')	350	590	940
Total	395	605	1000

If a person is selected at random from those surveyed, find the probability of each of the following events. (Express the probabilities in decimal form.)
(a) The person was in an accident or it rained.
(b) The person was not in an accident or it rained.
(c) The person was not in an accident or it did not rain.

38. One hundred people were surveyed, and one question pertained to their educational background. The results of this question are given in the following table.

	Female (F)	Male (F')	Total
College degree (D)	30	20	50
No college degree (D')	15	35	50
Total	45	55	100

If a person is selected at random from those surveyed, find the probability of each of the following events. Express the probabilities in decimal form.
(a) The person is female or has a college degree.
(b) The person is male or does not have a college degree.
(c) The person is female or does not have a college degree.

39. In a recent election there were 1000 eligible voters. They were asked to vote on two issues, A and B. The results were as follows: 300 people voted for A, 400 people voted for B, and 175 voted for both A and B. If one person is chosen at random from the 1000 eligible voters, find the probability that the person voted for A or B.

40. A company has 500 employees among whom 200 are females, 15 are high-level executives, and 7 of the high-level executives are females. If one of the 500 employees is chosen at random, find the probability that the person chosen is female or is a high-level executive.

41. A die is tossed 360 times. How many times would you expect to get a 6?

42. Two dice are tossed 360 times. How many times would you expect to get a sum of 5?

43. Two dice are tossed 720 times. How many times would you expect to get a sum greater than 9?

44. Four coins are tossed 80 times. How many times would you expect to get one head and three tails?

45. Four coins are tossed 144 times. How many times would you expect to get four tails?

46. Two dice are tossed 300 times. How many times would you expect to get a double?

47. Three coins are tossed 448 times. How many times would you expect to get three heads?

48. Suppose 5000 tickets are sold in a lottery. There are three prizes: The first is $1000, the second is $500, and the third is $100. What is the mathematical expectation of winning?

49. Your friend challenges you with the following game: You are to roll a pair of dice, and he will give you $5 if you roll a sum of 2 or 12, $2 if you roll a sum of 3 or 11, $1 if you roll a sum of 4 or 10. Otherwise you are to pay him $1. Should you play the game?

50. A contractor bids on a building project. There is a probability of 0.8 that he can show a profit of $30,000 and a probability of 0.2 that he will have to absorb a loss of $10,000. What is his mathematical expectation?

51. Suppose a person tosses two coins and receives $5 if 2 heads come up, receives $2 if 1 head and 1 tail come up, and has to pay $2 if 2 tails come up. Is it a fair game for him?

52. A "wheel of fortune" is divided into four colors: red, white, blue, and yellow. The probability of the spinner landing on each of the colors and the money received is given by the following chart. The price to spin the wheel is $1.50. Is it a fair game?

Color	Probability of landing on the color	Money received for landing on the color
Red	$\frac{4}{10}$	$.50
White	$\frac{3}{10}$	1.00
Blue	$\frac{2}{10}$	2.00
Yellow	$\frac{1}{10}$	5.00

53. A contractor estimates a probability of 0.7 of making $20,000 on a building project and a probability of 0.3 of losing $10,000 on the project. What is his mathematical expectation?

54. A farmer estimates his corn crop at 30,000 bushels. On the basis of past experience, he also estimates a probability of $\frac{3}{5}$ that he will make a profit of $0.50 per bushel and a probability of $\frac{1}{5}$ of losing $0.30 per bushel. What is his expected income from the corn crop?

55. Bill finds that the annual premium for insuring a stereo system for $2500 against theft is $75. If the probability that the set will be stolen during the year is 0.02, what is Bill's expected gain or loss by taking out the insurance?

56. Sandra finds that the annual premium for a $2000 insurance policy against the theft of a painting is $100. If the probability that the painting will be stolen during the year is 0.01, what is Sandra's expected gain or loss in taking out the insurance?

■■■ THOUGHTS INTO WORDS

57. If the probability of some event happening is 0.4, what is the probability of the event not happening? Explain your answer.

58. Explain each of the following concepts to a friend who missed class the day this section was discussed: using

complementary events to determine probabilities, using union and intersection of sets to determine probabilities, and using expected value to determine the fairness of a game.

■■■ FURTHER INVESTIGATIONS

The term **odds** is sometimes used to express a probability statement. For example, we might say, "the odds in favor of

the Cubs winning the pennant are 5 to 1," or "the odds against the Mets winning the pennant are 50 to 1." *Odds in favor* and *odds against* for equally likely outcomes can be defined as follows:

$$\text{Odds in favor} = \frac{\text{Number of favorable outcomes}}{\text{Number of unfavorable outcomes}}$$

$$\text{Odds against} = \frac{\text{Number of unfavorable outcomes}}{\text{Number of favorable outcomes}}$$

We have used the fractional form to define odds; however, in practice, the *to* vocabulary is commonly used. Thus the odds in favor of rolling a 4 with one roll of a die are usually stated as *1 to 5* instead of $\frac{1}{5}$. The odds against rolling a 4 are stated as *5 to 1*.

The *odds in favor of* statement about the Cubs means that there are 5 favorable outcomes compared to 1 unfavorable, or a total of 6 possible outcomes, so the *5 to 1 in favor of* statement also means that the probability of the Cubs winning the pennant is $\frac{5}{6}$. Likewise, the *50 to 1 against* statement about the Mets means that the probability that the Mets will not win the pennant is $\frac{50}{51}$.

Odds are usually stated in reduced form. For example, odds of 6 to 4 are usually stated as 3 to 2. Likewise, a fraction representing probability is reduced before being changed to a statement about odds.

59. What are the odds in favor of getting three heads with a toss of three coins?

60. What are the odds against getting four tails with a toss of four coins?

61. What are the odds against getting three heads and two tails with a toss of five coins?

62. What are the odds in favor of getting four heads and two tails with a toss of six coins?

63. What are the odds in favor of getting a sum of 5 with one toss of a pair of dice?

64. What are the odds against getting a sum greater than 5 with one toss of a pair of dice?

65. Suppose that one card is drawn at random from a deck of 52 playing cards. Find the odds against drawing a red card.

66. Suppose that one card is drawn at random from a deck of 52 playing cards. Find the odds in favor of drawing an ace or a king.

67. If $P(E) = \frac{4}{7}$ for some event E, find the odds in favor of E happening.

68. If $P(E) = \frac{5}{9}$ for some event E, find the odds against E happening.

69. Suppose that there is a predicted 40% chance of freezing rain. State the prediction in terms of the odds against getting freezing rain.

70. Suppose that there is a predicted 20% chance of thunderstorms. State the prediction in terms of the odds in favor of getting thunderstorms.

71. If the odds against an event happening are 5 to 2, find the probability that the event will occur.

72. The odds against Belly Dancer winning the fifth race are 20 to 9. What is the probability of Belly Dancer winning the fifth race?

73. The odds in favor of the Mets winning the pennant are stated as 7 to 5. What is the probability of the Mets winning the pennant?

74. The following chart contains some poker-hand probabilities. Complete the last column, "Odds Against Being Dealt This Hand." Note that fractions are reduced before being changed to odds.

5-Card hand	Probability of being dealt this hand	Odds against being dealt this hand
Straight flush	$\frac{40}{2,598,960} = \frac{1}{64,974}$	64,973 to 1
Four of a kind	$\frac{624}{2,598,960} =$	
Full house	$\frac{3744}{2,598,960} =$	
Flush	$\frac{5108}{2,598,960} =$	
Straight	$\frac{10,200}{2,598,960} =$	
Three of a kind	$\frac{54,912}{2,598,960} =$	
Two pairs	$\frac{123,552}{2,598,960} =$	
One pair	$\frac{1,098,240}{2,598,960} =$	
No pairs	$\frac{1,302,540}{2,598,960} =$	

Conditional Probability: Dependent and Independent Events

Two events are often related in such a way that the probability of one of them may vary depending on whether the other event has occurred. For example, the probability of rain may change drastically if additional information is obtained indicating a front moving through the area. Mathematically, the additional information about the front changes the sample space for the probability of rain.

In general, the probability of the occurrence of an event E, given the occurrence of another event F, is called a **conditional probability** and is denoted $P(E|F)$. Let's look at a simple example and use it to motivate a definition for conditional probability.

What is the probability of rolling a prime number in one roll of a die? Let $S = \{1, 2, 3, 4, 5, 6\}$, so $n(S) = 6$; and let $E = \{2, 3, 5\}$, so $n(E) = 3$. Therefore

$$P(E) = \frac{n(E)}{n(S)} = \frac{3}{6} = \frac{1}{2}$$

Next, what is the probability of rolling a prime number in one roll of a die, *given that an odd number has turned up*? Let $F = \{1, 3, 5\}$ be the new sample space of odd numbers. Then $n(F) = 3$. We are now interested in only that part of E (rolling a prime number) that is also in F—in other words, $E \cap F$. Therefore, because $E \cap F = \{3, 5\}$, the probability of E given F is

$$P(E|F) = \frac{n(E \cap F)}{n(F)} = \frac{2}{3}$$

When we divide both the numerator and the denominator of $n(E \cap F)/n(F)$ by $n(S)$, we obtain

$$\frac{\dfrac{n(E \cap F)}{n(S)}}{\dfrac{n(F)}{n(S)}} = \frac{P(E \cap F)}{P(F)}$$

Therefore we can state the following general definition of the conditional probability of E given F for arbitrary events E and F:

Definition 15.3

$$P(E|F) = \frac{P(E \cap F)}{P(F)}, \qquad P(F) \neq 0$$

In a problem in the previous section, the following probability table was formed relative to car accidents and weather conditions on a holiday weekend.

	Rain (R)	No rain (R')	Total
Accident (A)	0.035	0.010	0.045
No accident (A')	0.450	0.505	0.955
Total	0.485	0.515	1.000

Some conditional probabilities that can be calculated from the table follow:

$$P(A|R) = \frac{P(A \cap R)}{P(R)} = \frac{0.035}{0.485} = \frac{35}{485} = \frac{7}{97}$$

$$P(A'|R) = \frac{P(A' \cap R)}{P(R)} = \frac{0.450}{0.485} = \frac{450}{485} = \frac{90}{97}$$

$$P(A|R') = \frac{P(A \cap R')}{P(R')} = \frac{0.010}{0.515} = \frac{10}{515} = \frac{2}{103}$$

Note that the probability of an accident given that it was raining, $P(A|R)$, is greater than the probability of an accident given that it was not raining, $P(A|R')$. This seems reasonable.

PROBLEM 1

A die is tossed. Find the probability that a 4 came up if it is known that an even number turned up.

Solution

Let E be the event of rolling a 4, and let F be the event of rolling an even number. Therefore $E = \{4\}$ and $F = \{2, 4, 6\}$, from which we obtain $E \cap F = \{4\}$. Using Definition 15.3, we obtain

$$P(E|F) = \frac{P(E \cap F)}{P(F)} = \frac{\frac{1}{6}}{\frac{3}{6}} = \frac{1}{3}$$

∎

PROBLEM 2

Suppose the probability that a student will enroll in a mathematics course is 0.45, the probability that he or she will enroll in a science course is 0.38, and the probability that he or she will enroll in both courses is 0.26. Find the probability that a student will enroll in a mathematics course, given that he or she is also enrolled in a science course. Also, find the probability that a student will enroll in a science course, given that he or she is enrolled in mathematics.

Solution

Let M be the event, *will enroll in mathematics*, and let S be the event, *will enroll in science*. Therefore, using Definition 10.3, we obtain

$$P(M|S) = \frac{P(M \cap S)}{P(S)} = \frac{0.26}{0.38} = \frac{26}{38} = \frac{13}{19}$$

and

$$P(S|M) = \frac{P(S \cap M)}{P(M)} = \frac{0.26}{0.45} = \frac{26}{45}$$ ∎

■ Independent and Dependent Events

Suppose that, when computing a conditional probability, we find that

$$P(E|F) = P(E)$$

This means that the probability of E is not affected by the occurrence or nonoccurrence of F. In such a situation, we say that event E is *independent* of event F. It can be shown that if event E is independent of event F, then F is also independent of E; thus E and F are referred to as **independent events**. Furthermore, from the equations

$$P(E|F) = \frac{P(E \cap F)}{P(F)} \qquad \text{and} \qquad P(E|F) = P(E)$$

we see that

$$\frac{P(E \cap F)}{P(F)} = P(E)$$

which can be written

$$P(E \cap F) = P(E)P(F)$$

Therefore we state the following general definition:

Definition 15.4

> Two events E and F are said to be **independent** if and only if
>
> $$P(E \cap F) = P(E)P(F)$$
>
> Two events that are not independent are called **dependent events**.

In the probability table preceding Problem 1, we see that $P(A) = 0.045$, $P(R) = 0.485$, and $P(A \cap R) = 0.035$. Because

$$P(A)P(R) = (0.045)(0.485) = 0.021825$$

and this does not equal $P(A \cap R)$, the events A (have a car accident) and R (rainy conditions) are not independent. This is not too surprising; we would certainly expect rainy conditions and automobile accidents to be related.

PROBLEM 3

Suppose we roll a white die and a red die. If we let E be the event, *we roll a 4 on the white die*, and if we let F be the event, *we roll a 6 on the red die*. Are E and F independent events?

Solution

The sample space for rolling a pair of dice has $(6)(6) = 36$ elements. Using ordered-pair notation, where the first entry represents the white die and the second entry the red die, we can list events E and F as follows:

$$E = \{(4, 1), (4, 2), (4, 3), (4, 4), (4, 5), (4, 6)\}$$

$$F = \{(1, 6), (2, 6), (3, 6), (4, 6), (5, 6), (6, 6)\}$$

Therefore $E \cap F = \{(4, 6)\}$. Because $P(F) = \dfrac{1}{6}$, $P(E) = \dfrac{1}{6}$, and $P(E \cap F) = \dfrac{1}{36}$, we see that $P(E \cap F) = P(E)P(F)$, and the events E and F are independent. ∎

PROBLEM 4

Two coins are tossed. Let E be the event, *toss not more than one head*, and let F be the event, *toss at least one of each face*. Are these events independent?

Solution

The sample space has $(2)(2) = 4$ elements. The events E and F can be listed as follows:

$$E = \{(H, T), (T, H), (T, T)\}$$

$$F = \{(H, T), (T, H)\}$$

Therefore $E \cap F = \{(H, T), (T, H)\}$. Because $P(E) = \dfrac{3}{4}$, $P(F) = \dfrac{1}{2}$, and $P(E \cap F) = \dfrac{1}{2}$, we see that $P(E \cap F) \neq P(E)P(F)$, so the events E and F are dependent. ∎

Sometimes the independence issue can be decided by the physical nature of the events in the problem. For instance, in Problem 3, it should seem evident that rolling a 4 on the white die is not affected by rolling a 6 on the red die. However, as in Problem 4, the description of the events may not clearly indicate whether the events are dependent.

From a problem-solving viewpoint, the following two statements are very helpful.

1. If E and F are independent events, then

$$P(E \cap F) = P(E)P(F)$$

(This property generalizes to any finite number of independent events.)

2. If E and F are dependent events, then

$$P(E \cap F) = P(E)P(F|E)$$

Let's analyze some problems using these ideas.

PROBLEM 5

A die is rolled three times. (This is equivalent to rolling three dice once each.) What is the probability of getting a 6 all three times?

Solution

The events of a 6 on the first roll, a 6 on the second roll, and a 6 on the third roll are independent events. Therefore the probability of getting three 6's is

$$\left(\frac{1}{6}\right)\left(\frac{1}{6}\right)\left(\frac{1}{6}\right) = \frac{1}{216}$$

∎

PROBLEM 6

A jar contains five white, seven green, and nine red marbles. If two marbles are drawn in succession *without replacement*, find the probability that both marbles are white.

Solution

Let E be the event of drawing a white marble on the first draw, and let F be the event of drawing a white marble on the second draw. Because the marble drawn first is not to be replaced before the second marble is drawn, we have dependent events. Therefore

$$P(E \cap F) = P(E)P(F|E)$$

$$= \left(\frac{5}{21}\right)\left(\frac{4}{20}\right) = \frac{20}{420} = \frac{1}{21}$$

$P(F|E)$ means the probability of drawing a white marble on the second draw, given that a white marble was obtained on the first draw.

∎

The concept of *mutually exclusive events* may also enter the picture when we are working with independent or dependent events. Our final problems of this section illustrate this idea.

PROBLEM 7

A coin is tossed three times. Find the probability of getting two heads and one tail.

Solution

Two heads and one tail can be obtained in three different ways: (1) HHT (head on first toss, head on second toss, and tail on third toss), (2) HTH, and (3) THH. Thus we have three *mutually exclusive* events, each of which can be broken into *independent* events: first toss, second toss, and third toss. Therefore the probability can be computed as follows:

$$\left(\frac{1}{2}\right)\left(\frac{1}{2}\right)\left(\frac{1}{2}\right) + \left(\frac{1}{2}\right)\left(\frac{1}{2}\right)\left(\frac{1}{2}\right) + \left(\frac{1}{2}\right)\left(\frac{1}{2}\right)\left(\frac{1}{2}\right) = \frac{3}{8}$$

∎

<table>
<tr><td>

</td><td>

A jar contains five white, seven green, and nine red marbles. If two marbles are drawn in succession *without replacement*, find the probability that one of them is white and the other is green.

</td></tr>
</table>

Solution

The drawing of a white marble and a green marble can occur in two different ways: (1) by drawing a white marble first and then a green, and (2) by drawing a green marble first and then a white. Thus we have two mutually exclusive events, each of which is broken into two *dependent* events: first draw and second draw. Therefore the probability can be computed as follows:

$$\left(\frac{5}{21}\right)\left(\frac{7}{20}\right) \quad + \quad \left(\frac{7}{21}\right)\left(\frac{5}{20}\right) = \frac{70}{420} = \frac{1}{6}$$

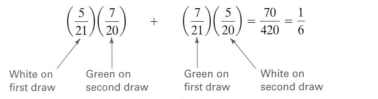

| White on first draw | Green on second draw | Green on first draw | White on second draw |

■

PROBLEM 9

Two cards are drawn in succession *with replacement* from a deck of 52 playing cards. Find the probability of drawing a jack and a queen.

Solution

Drawing a jack and a queen can occur in two different ways: (1) a jack on the first draw and a queen on the second and (2) a queen on the first draw and a jack on the second. Thus we have two mutually exclusive events, and each one is broken into the *independent* events of first draw and second draw with replacement. Therefore the probability can be computed as follows:

$$\left(\frac{4}{52}\right)\left(\frac{4}{52}\right) \quad + \quad \left(\frac{4}{52}\right)\left(\frac{4}{52}\right) = \frac{32}{2704} = \frac{2}{169}$$

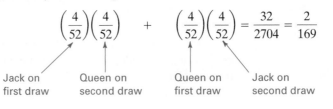

| Jack on first draw | Queen on second draw | Queen on first draw | Jack on second draw |

■

Problem Set 15.5

For Problems 1–22, solve each problem.

1. A die is tossed. Find the probability that a 5 came up if it is known that an odd number came up.

2. A die is tossed. Find the probability that a prime number was obtained, given that an even number came up.

Also find the probability that an even number came up, given that a prime number was obtained.

3. Two dice are rolled and someone indicates that the two numbers that come up are different. Find the probability that the sum of the two numbers is 6.

4. Two dice are rolled, and someone indicates that the two numbers that come up are identical. Find the probability that the sum of the two numbers is 8.

5. One card is randomly drawn from a deck of 52 playing cards. Find the probability that it is a jack, given that the card is a face card. (We are considering jacks, queens, and kings as face cards.)

6. One card is randomly drawn from a deck of 52 playing cards. Find the probability that it is a spade, given the fact that it is a black card.

7. A coin and a die are tossed. Find the probability of getting a 5 on the die, given that a head comes up on the coin.

8. A family has three children. Assume that each child is as likely to be a boy as it is to be a girl. Find the probability that the family has three girls if it is known that the family has at least one girl.

9. The probability that a student will enroll in a mathematics course is 0.7, the probability that he or she will enroll in a history course is 0.3, and the probability that he or she will enroll in both mathematics and history is 0.2. Find the probability that a student will enroll in mathematics, given that he or she is also enrolled in history. Also find the probability that a student will enroll in history, given that he or she is also enrolled in mathematics.

10. The following probability table contains data relative to car accidents and weather conditions on a holiday weekend.

	Rain (R)	No rain (R')	Total
Accident (A)	0.025	0.015	0.040
No accident (A')	0.400	0.560	0.960
Total	0.425	0.575	1.000

Find the probability that a person chosen at random from the survey was in an accident, given that it was raining. Also find the probability that a person was not in an accident, given that it was not raining.

11. One hundred people were surveyed, and one question pertained to their educational background. The responses to this question are given in the following table.

	Female (F)	Male (F')	Total
College degree (D)	30	20	50
No college degree (D')	15	35	50
Total	45	55	100

Find the probability that a person chosen at random from the survey has a college degree, given that the person is female. Also find the probability that a person chosen is male, given that the person has a college degree.

12. In a recent election there were 1000 eligible voters. They were asked to vote on two issues, A and B. The results were as follows: 200 people voted for A, 400 people voted for B, and 50 people voted for both A and B. If one person is chosen at random from the 100 eligible voters, find the probability that the person voted for A, given that he or she voted for B. Also find the probability that the person voted for B, given that he or she voted for A.

13. A small company has 100 employees; among them 75 are males, 7 are administrators, and 5 of the administrators are males. If a person is chosen at random from the employees, find the probability that the person is an administrator, given that he is male. Also find the probability that the person chosen is female, given that she is an administrator.

14. A survey claims that 80% of the households in a certain town have a high-definition TV, 10% have a microwave oven, and 2% have both a high-definition TV and a microwave oven. Find the probability that a randomly selected household will have a microwave oven, given that it has a high-definition TV.

15. Consider a family of three children. Let E be the event, *the first child is a boy*, and let F be the event, *the family has exactly one boy*. Are events E and F dependent or independent?

16. Roll a white die and a green die. Let E be the event, *roll a 2 on the white die*, and let F be the event, *roll a 4 on the green die*. Are E and F dependent or independent events?

17. Toss three coins. Let E be the event, *toss not more than one head*, and let F be the event, *toss at least one of each face*. Are E and F dependent or independent events?

18. A card is drawn at random from a standard deck of 52 playing cards. Let E be the event, *the card is a 2*, and let F be the event, *the card is a 2 or a 3*. Are the events E and F dependent or independent?

19. A coin is tossed four times. Find the probability of getting three heads and one tail.

20. A coin is tossed five times. Find the probability of getting four heads and one tail.

21. Toss a pair of dice three times. Find the probability that a double is obtained on all three tosses.

22. Toss a pair of dice three times. Find the probability that each toss will produce a sum of 4.

For Problems 23–26, suppose that two cards are drawn in succession *without replacement* from a deck of 52 playing cards. Find the probability of each of the following events:

23. Both cards are 4's.

24. One card is an ace and one card is a king.

25. One card is a spade and one card is a diamond.

26. Both cards are black.

For Problems 27–30, suppose that two cards are drawn in succession *with replacement* from a deck of 52 playing cards. Find the probability of each of the following events:

27. Both cards are spades.

28. One card is an ace and one card is a king.

29. One card is the ace of spades and one card is the king of spades.

30. Both cards are red.

For Problems 31 and 32, solve each problem.

31. A person holds three kings from a deck of 52 playing cards. If the person draws two cards without replacement from the 49 cards remaining in the deck, find the probability of drawing the fourth king.

32. A person removes two aces and a king from a deck of 52 playing cards and draws, without replacement, two more cards from the deck. Find the probability that the person will draw two aces, or two kings, or an ace and a king.

For Problems 33–36, a bag contains five red and four white marbles. Two marbles are drawn in succession *with*

replacement. Find the probability of each of the following events:

33. Both marbles drawn are red.

34. Both marbles drawn are white.

35. The first marble is red and the second marble is white.

36. At least one marble is red.

For Problems 37–40, a bag contains five white, four red, and four blue marbles. Two marbles are drawn in succession *with replacement*. Find the probability of each of the following events:

37. Both marbles drawn are white.

38. Both marbles drawn are red.

39. One red and one blue marble are drawn.

40. One white and one blue marble are drawn.

For Problems 41–44, a bag contains one red and two white marbles. Two marbles are drawn in succession *without replacement*. Find the probability of each of the following events:

41. One marble drawn is red, and one marble drawn is white.

42. The first marble drawn is red and the second is white.

43. Both marbles drawn are white.

44. Both marbles drawn are red.

For Problems 45–48, a bag contains five red and 12 white marbles. Two marbles are drawn in succession *without replacement*. Find the probability of each of the following events:

45. Both marbles drawn are red.

46. Both marbles drawn are white.

47. One red and one white marble are drawn.

48. At least one marble drawn is red.

For Problems 49–52, a bag contains two red, three white, and four blue marbles. Two marbles are drawn in succession *without replacement*. Find the probability of each of the following events:

49. Both marbles drawn are white.

50. One marble drawn is white, and one is blue.

51. Both marbles drawn are blue.

52. At least one red marble is drawn.

For Problems 53–56, a bag contains five white, one blue, and three red marbles. Three marbles are drawn in succession *with replacement*. Find the probability of each of the following events:

53. All three marbles drawn are blue.

54. One marble of each color is drawn.

55. One white and two red marbles are drawn.

56. One blue and two white marbles are drawn.

For Problems 57–60, a bag contains four white, one red, and two blue marbles. Three marbles are drawn in succession *without replacement*. Find the probability of each of the following events:

57. All three marbles drawn are white.

58. One red and two blue marbles are drawn.

59. One marble of each color is drawn.

60. One white and two red marbles are drawn.

For Problems 61 and 62, solve each problem.

61. Two boxes with red and white marbles are shown here. A marble is drawn at random from Box 1, and then a second marble is drawn from Box 2. Find the probability that both marbles drawn are white. Find the probability that both marbles drawn are red. Find the probability that one red and one white marble are drawn.

3 red 4 white	2 red 1 white
Box 1	Box 2

62. Three boxes containing red and white marbles are shown here. Randomly draw a marble from Box 1 and put it in Box 2. Then draw a marble from Box 2 and put it in Box 3. Then draw a marble from Box 3. What is the probability that the last marble drawn, from Box 3, is red? What is the probability that it is white?

2 red 2 white	3 red 1 white	3 white
Box 1	Box 2	Box 3

■ ■ ■ **THOUGHTS INTO WORDS**

63. How would you explain the concept of conditional probability to a classmate who missed the discussion of this section?

64. How would you give a nontechnical description of conditional probability to an elementary algebra student?

65. Explain in your own words the concept of independent events.

66. Suppose that a bag contains two red and three white marbles. Furthermore, suppose that two marbles are drawn from the bag in succession *with replacement*. Explain how the following tree diagram can be used to determine that the probability of drawing two white marbles is $\dfrac{9}{25}$.

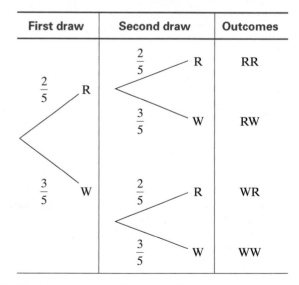

67. Explain how a tree diagram can be used to determine the probabilities for Problems 41–44.

15.6   Binomial Theorem

In Chapter 4, when multiplying polynomials, we developed patterns for squaring and cubing binomials. Now we want to develop a general pattern that can be used to raise a binomial to any positive integral power. Let's begin by looking at some specific expansions that can be verified by direct multiplication. (Note that the patterns for squaring and cubing a binomial are a part of this list.)

$$(x + y)^0 = 1$$
$$(x + y)^1 = x + y$$
$$(x + y)^2 = x^2 + 2xy + y^2$$
$$(x + y)^3 = x^3 + 3x^2y + 3xy^2 + y^3$$
$$(x + y)^4 = x^4 + 4x^3y + 6x^2y^2 + 4xy^3 + y^4$$
$$(x + y)^5 = x^5 + 5x^4y + 10x^3y^2 + 10x^2y^3 + 5xy^4 + y^5$$

First, note the pattern of the exponents for x and y on a term-by-term basis. The exponents of x begin with the exponent of the binomial and decrease by 1, term by term, until the last term has x^0, which is 1. The exponents of y begin with zero ($y^0 = 1$) and increase by 1, term by term, until the last term contains y to the power of the binomial. In other words, the variables in the expansion of $(x + y)^n$ have the following pattern.

$$x^n, \quad x^{n-1}y, \quad x^{n-2}y^2, \quad x^{n-3}y^3, \quad \ldots, \quad xy^{n-1}, \quad y^n$$

Note that for each term, the sum of the exponents of x and y is n.

Now let's look for a pattern for the coefficients by examining specifically the expansion of $(x + y)^5$.

$$(x + y)^5 = x^5 + 5x^4y^1 + 10x^3y^2 + 10x^2y^3 + 5x^1y^4 + 1y^5$$

$$C(5, 1) \quad C(5, 2) \quad C(5, 3) \quad C(5, 4) \quad C(5, 5)$$

As indicated by the arrows, the coefficients are numbers that arise as different-sized combinations of five things. To see why this happens, consider the coefficient for the term containing x^3y^2. The two y's (for y^2) come from two of the factors of $(x + y)$, and therefore the three x's (for x^3) must come from the other three factors of $(x + y)$. In other words, the coefficient is $C(5, 2)$.

We can now state a general expansion formula for $(x + y)^n$; this formula is often called the **binomial theorem**. But before stating it, let's make a small switch in notation. Instead of $C(n, r)$, we shall write $\binom{n}{r}$, which will prove to be a little more convenient at this time. The symbol $\binom{n}{r}$, still refers to the number of combinations of n things taken r at a time, but in this context it is often called a **binomial coefficient**.

Binomial Theorem

For any binomial $(x + y)$ and any natural number n,

$$(x + y)^n = x^n + \binom{n}{1}x^{n-1}y + \binom{n}{2}x^{n-2}y^2 + \cdots + \binom{n}{n}y^n$$

The binomial theorem can be proved by mathematical induction, but we will not do that in this text. Instead, we'll consider a few examples that put the binomial theorem to work.

E X A M P L E   1   Expand $(x + y)^7$.

Solution

$$(x + y)^7 = x^7 + \binom{7}{1}x^6y + \binom{7}{2}x^5y^2 + \binom{7}{3}x^4y^3 + \binom{7}{4}x^3y^4$$

$$+ \binom{7}{5}x^2y^5 + \binom{7}{6}xy^6 + \binom{7}{7}y^7$$

$$= x^7 + 7x^6y + 21x^5y^2 + 35x^4y^3 + 35x^3y^4 + 21x^2y^5 + 7xy^6 + y^7 \quad \blacksquare$$

E X A M P L E   2   Expand $(x - y)^5$.

Solution

We shall treat $(x - y)^5$ as $[x + (-y)]^5$.

$$[x + (-y)]^5 = x^5 + \binom{5}{1}x^4(-y) + \binom{5}{2}x^3(-y)^2 + \binom{5}{3}x^2(-y)^3$$

$$+ \binom{5}{4}x(-y)^4 + \binom{5}{5}(-y)^5$$

$$= x^5 - 5x^4y + 10x^3y^2 - 10x^2y^3 + 5xy^4 - y^5 \quad \blacksquare$$

E X A M P L E   3   Expand $(2a + 3b)^4$.

Solution

Let $x = 2a$ and $y = 3b$ in the binomial theorem.

$$(2a + 3b)^4 = (2a)^4 + \binom{4}{1}(2a)^3(3b) + \binom{4}{2}(2a)^2(3b)^2$$

$$+ \binom{4}{3}(2a)(3b)^3 + \binom{4}{4}(3b)^4$$

$$= 16a^4 + 96a^3b + 216a^2b^2 + 216ab^3 + 81b^4 \quad \blacksquare$$

E X A M P L E   4

Expand $\left(a + \dfrac{1}{n}\right)^5$.

Solution

$$\left(a + \frac{1}{n}\right)^5 = a^5 + \binom{5}{1}a^4\left(\frac{1}{n}\right) + \binom{5}{2}a^3\left(\frac{1}{n}\right)^2 + \binom{5}{3}a^2\left(\frac{1}{n}\right)^3 + \binom{5}{4}a\left(\frac{1}{n}\right)^4 + \binom{5}{5}\left(\frac{1}{n}\right)^5$$

$$= a^5 + \frac{5a^4}{n} + \frac{10a^3}{n^2} + \frac{10a^2}{n^3} + \frac{5a}{n^4} + \frac{1}{n^5}$$
■

E X A M P L E   5

Expand $(x^2 - 2y^3)^6$.

Solution

$$[x^2 + (-2y^3)]^6 = (x^2)^6 + \binom{6}{1}(x^2)^5(-2y^3) + \binom{6}{2}(x^2)^4(-2y^3)^2$$

$$+ \binom{6}{3}(x^2)^3(-2y^3)^3 + \binom{6}{4}(x^2)^2(-2y^3)^4$$

$$+ \binom{6}{5}(x^2)(-2y^3)^5 + \binom{6}{6}(-2y^3)^6$$

$$= x^{12} - 12x^{10}y^3 + 60x^8y^6 - 160x^6y^9 + 240x^4y^{12} - 192x^2y^{15}$$
$$+ 64y^{18}$$
■

■ Finding Specific Terms

Sometimes it is convenient to be able to write down the specific term of a binomial expansion without writing out the entire expansion. For example, suppose that we want the sixth term of the expansion $(x + y)^{12}$. We can proceed as follows: The sixth term will contain y^5. (Note in the binomial theorem that the **exponent of y is always one less than the number of the term**.) Because the sum of the exponents for x and y must be 12 (the exponent of the binomial), the sixth term will also contain x^7. The coefficient is $\binom{12}{5}$, where the 5 agrees with the exponent of y^5. Therefore the sixth term of $(x + y)^{12}$ is

$$\binom{12}{5}x^7y^5 = 792x^7y^5$$

E X A M P L E   6

Find the fourth term of $(3a + 2b)^7$.

Solution

The fourth term will contain $(2b)^3$, and therefore it will also contain $(3a)^4$. The coefficient is $\binom{7}{3}$. Thus the fourth term is

$$\binom{7}{3}(3a)^4(2b)^3 = (35)(81a^4)(8b^3) = 22{,}680a^4b^3$$
■

E X A M P L E 7   Find the sixth term of $(4x - y)^9$.

Solution

The sixth term will contain $(-y)^5$, and therefore it will also contain $(4x)^4$. The coefficient is $\binom{9}{5}$. Thus the sixth term is

$$\binom{9}{5}(4x)^4(-y)^5 = (126)(256x^4)(-y^5) = -32{,}256x^4y^5$$

∎

Problem Set 15.6

For Problems 1–26, expand and simplify each binomial.

1. $(x + y)^8$

2. $(x + y)^9$

3. $(x - y)^6$

4. $(x - y)^4$

5. $(a + 2b)^4$

6. $(3a + b)^4$

7. $(x - 3y)^5$

8. $(2x - y)^6$

9. $(2a - 3b)^4$

10. $(3a - 2b)^5$

11. $(x^2 + y)^5$

12. $(x + y^3)^6$

13. $(2x^2 - y^2)^4$

14. $(3x^2 - 2y^2)^5$

15. $(x + 3)^6$

16. $(x + 2)^7$

17. $(x - 1)^9$

18. $(x - 3)^4$

19. $\left(1 + \dfrac{1}{n}\right)^4$

20. $\left(2 + \dfrac{1}{n}\right)^5$

21. $\left(a - \dfrac{1}{n}\right)^6$

22. $\left(2a - \dfrac{1}{n}\right)^5$

23. $(1 + \sqrt{2})^4$

24. $(2 + \sqrt{3})^3$

25. $(3 - \sqrt{2})^5$

26. $(1 - \sqrt{3})^4$

For Problems 27–36, write the first four terms of each expansion.

27. $(x + y)^{12}$

28. $(x + y)^{15}$

29. $(x - y)^{20}$

30. $(a - 2b)^{13}$

31. $(x^2 - 2y^3)^{14}$

32. $(x^3 - 3y^2)^{11}$

33. $\left(a + \dfrac{1}{n}\right)^9$

34. $\left(2 - \dfrac{1}{n}\right)^6$

35. $(-x + 2y)^{10}$

36. $(-a - b)^{14}$

For Problems 37–46, find the specified term for each binomial expansion.

37. The fourth term of $(x + y)^8$

38. The seventh term of $(x + y)^{11}$

39. The fifth term of $(x - y)^9$

40. The fourth term of $(x - 2y)^6$

41. The sixth term of $(3a + b)^7$

42. The third term of $(2x - 5y)^5$

43. The eighth term of $(x^2 + y^3)^{10}$

44. The ninth term of $(a + b^3)^{12}$

45. The seventh term of $\left(1 - \dfrac{1}{n}\right)^{15}$

46. The eighth term of $\left(1 - \dfrac{1}{n}\right)^{13}$

■ ■ ■ **THOUGHTS INTO WORDS**

47. How would you explain binomial expansions to an elementary algebra student?

48. Explain how to find the fifth term of the expansion of $(2x + 3y)^9$ without writing out the entire expansion.

49. Is the tenth term of the expansion $(1 - 2)^{15}$ positive or negative? Explain how you determined the answer to this question.

■ ■ ■ **FURTHER INVESTIGATIONS**

For Problems 50–53, expand and simplify each complex number.

50. $(1 + 2i)^5$

51. $(2 + i)^6$

52. $(2 - i)^6$

53. $(3 - 2i)^5$

We can summarize this chapter with three main topics: counting techniques, probability, and the binomial theorem.

(15.1) Counting Techniques

The **fundamental principle of counting** states that if a first task can be accomplished in x ways and, following this task, a second task can be accomplished in y ways, then task 1 followed by task 2 can be accomplished in $x \cdot y$ ways. The principle extends to any finite number of tasks. As you solve problems involving the fundamental principle of counting, it is often helpful to analyze the problem in terms of the tasks to be completed.

(15.2) Ordered arrangements are called **permutations**. The number of permutations of n things taken n at a time is given by

$$P(n, n) = n!$$

The number of r-element permutations that can be formed from a set of n elements is given by

$$P(n, r) = \underbrace{n(n - 1)(n - 2) \cdots}_{r \text{ factors}}$$

If there are n elements to be arranged, where there are r_1 of one kind, r_2 of another kind, r_3 of another kind, ..., r_k of a kth kind, then the number of distinguishable permutations is given by

$$\frac{n!}{(r_1!)(r_2!)(r_3!) \ldots (r_k!)}$$

Combinations are subsets; the order in which the elements appear does not make a difference. The number of r-element combinations (subsets) that can be formed from a set of n elements is given by

$$C(n, r) = \frac{P(n, r)}{r!}$$

Does the order in which the elements appear make any difference? This is a key question to consider when trying to decide whether a particular problem involves permutations or combinations. If the answer to the question is yes, then it is a permutation problem; if the answer is no, then it is a combination problem. Don't forget that combinations are subsets.

(15.3–15.5) Probability

In an experiment where all possible outcomes in the sample space S are equally likely to occur, the **probability** of an event E is defined by

$$P(E) = \frac{n(E)}{n(S)}$$

where $n(E)$ denotes the number of elements in the event E, and $n(S)$ denotes the number of elements in the sample space S. The numbers $n(E)$ and $n(S)$ can often be determined by using one or more of the previously listed counting techniques. For all events E, it is always true that $0 \leq P(E) \leq 1$. That is, all probabilities fall in the range from 0 to 1, inclusive.

If E and E' are **complementary events**, then $P(E) + P(E') = 1$. Therefore, if we can calculate either $P(E)$ or $P(E')$, then we can find the other one by subtracting from 1.

For two events E and F, the probability of E or F is given by

$$P(E \cup F) = P(E) + P(F) - P(E \cap F)$$

If $E \cap F = \varnothing$, then E and F are **mutually exclusive events**.

The probability that an event E occurs, given that another event F has already occurred, is called **conditional probability**. It is given by the equation

$$P(E|F) = \frac{P(E \cap F)}{P(F)}$$

Two events E and F are said to be **independent** if and only if

$$P(E \cap F) = P(E)P(F)$$

Two events that are not independent are called **dependent events**, and the probability of two dependent events is given by

$$P(E \cap F) = P(E)P(F|E)$$

(15.6) The Binomial Theorem

For any binomial $(x + y)$ and any natural number n,

$$(x + y)^n = x^n + \binom{n}{1}x^{n-1}y + \binom{n}{2}x^{n-2}y^2$$

$$+ \cdots + \binom{n}{n}y^n$$

Note the following patterns in a binomial expansion:

1. In each term, the sum of the exponents of x and y is n.
2. The exponents of x begin with the exponent of the binomial and decrease by 1, term by term, until the last term has x^0, which is 1. The exponents of y begin with zero ($y^0 = 1$) and increase by 1, term by term, until the last term contains y to the power of the binomial.
3. The coefficient of any term is given by $\binom{n}{r}$, where the value of r agrees with the exponent of y for that term. For example, if the term contains y^3, then the coefficient of that term is $\binom{n}{3}$.
4. The expansion of $(x + y)^n$ contains $n + 1$ terms.

Chapter 15   Review Problem Set

Problems 1–14 are counting type problems.

1. How many different arrangements of the letters A, B, C, D, E, and F can be made?

2. How many different nine-letter arrangements can be formed from the nine letters of the word APPARATUS?

3. How many odd numbers of three different digits each can be formed by choosing from the digits 1, 2, 3, 5, 7, 8, and 9?

4. In how many ways can Arlene, Brent, Carlos, Dave, Ernie, Frank, and Gladys be seated in a row of seven seats so that Arlene and Carlos are side by side?

5. In how many ways can a committee of three people be chosen from six people?

6. How many committees consisting of three men and two women can be formed from seven men and six women?

7. How many different five-card hands consisting of all hearts can be formed from a deck of 52 playing cards?

8. If no number contains repeated digits, how many numbers greater than 500 can be formed by choosing from the digits 2, 3, 4, 5, and 6?

9. How many three-person committees can be formed from four men and five women so that each committee contains at least one man?

10. How many different four-person committees can be formed from eight people if two particular people refuse to serve together on a committee?

11. How many four-element subsets containing A or B but not both A and B can be formed from the set {A, B, C, D, E, F, G, H}?

12. How many different six-letter permutations can be formed from four identical H's and two identical T's?

13. How many four-person committees consisting of two seniors, one sophomore, and one junior can be formed from three seniors, four juniors, and five sophomores?

14. In a baseball league of six teams, how many games are needed to complete a schedule if each team plays eight games with each other team?

Problems 15–35 pose some probability questions.

15. If three coins are tossed, find the probability of getting two heads and one tail.

16. If five coins are tossed, find the probability of getting three heads and two tails.

17. What is the probability of getting a sum of 8 with one roll of a pair of dice?

18. What is the probability of getting a sum greater than 5 with one roll of a pair of dice?

19. Aimée, Brenda, Chuck, Dave, and Eli are randomly seated in a row of five seats. Find the probability that Aimée and Chuck are not seated side by side.

20. Four girls and three boys are to be randomly seated in a row of seven seats. Find the probability that the girls and boys will be seated in alternating seats.

21. Six coins are tossed. Find the probability of getting at least two heads.

22. Two cards are randomly chosen from a deck of 52 playing cards. What is the probability that two jacks are drawn?

23. Each arrangement of the six letters of the word CYCLIC is put on a slip of paper and placed in a hat. One slip is drawn at random. Find the probability that the slip contains an arrangement with the Y at the beginning.

24. A committee of three is randomly chosen from one man and six women. What is the probability that the man is not on the committee?

25. A four-person committee is selected at random from the eight people Alice, Bob, Carl, Dee, Enrique, Fred, Gina, and Hilda. Find the probability that Alice or Bob, but not both, is on the committee.

26. A committee of three is chosen at random from a group of five men and four women. Find the probability that the committee contains two men and one woman.

27. A committee of four is chosen at random from a group of six men and seven women. Find the probability that the committee contains at least one woman.

28. A bag contains five red and eight white marbles. Two marbles are drawn in succession *with replacement.* What is the probability that at least one red marble is drawn?

29. A bag contains four red, five white, and three blue marbles. Two marbles are drawn in succession *with replacement.* Find the probability that one red and one blue marble are drawn.

30. A bag contains four red and seven blue marbles. Two marbles are drawn in succession *without replacement.* Find the probability of drawing one red and one blue marble.

31. A bag contains three red, two white, and two blue marbles. Two marbles are drawn in succession *without*

replacement. Find the probability of drawing at least one red marble.

32. Each of three letters is to be mailed in any one of four different mailboxes. What is the probability that all three letters will be mailed in the same mailbox?

33. The probability that a customer in a department store will buy a blouse is 0.15, the probability that she will buy a pair of shoes is 0.10, and the probability that she will buy both a blouse and a pair of shoes is 0.05. Find the probability that the customer will buy a blouse, given that she has already purchased a pair of shoes. Also find the probability that she will buy a pair of shoes, given that she has already purchased a blouse.

34. A survey of 500 employees of a company produced the following information.

Employment level	College degree	No college degree
Managerial	45	5
Nonmanagerial	50	400

Find the probability that an employee chosen at random (a) is working in a managerial position, given that he or she has a college degree; and (b) has a college degree, given that he or she is working in a managerial position.

35. From a survey of 1000 college students, it was found that 450 of them owned cars, 700 of them owned sound systems, and 200 of them owned both a car and a sound system. If a student is chosen at random from the 1000 students, find the probability that the student (a) owns a car, given the fact that he or she owns a sound system, and (b) owns a sound system, given the fact that he or she owns a car.

For Problems 36–41, expand each binomial and simplify.

36. $(x + 2y)^5$ **37.** $(x - y)^8$ **38.** $(a^2 - 3b^3)^4$

39. $\left(x + \dfrac{1}{n}\right)^6$ **40.** $(1 - \sqrt{2})^5$ **41.** $(-a + b)^3$

42. Find the fourth term of the expansion of $(x - 2y)^{12}$.

43. Find the tenth term of the expansion of $(3a + b^2)^{13}$.

For Problems 1–21, solve each problem.

1. In how many ways can Abdul, Barb, Corazon, and Doug be seated in a row of four seats so that Abdul occupies an end seat?

2. How many even numbers of four different digits each can be formed by choosing from the digits 1, 2, 3, 5, 7, 8, and 9?

3. In how many ways can three letters be mailed in six mailboxes?

4. In a baseball league of ten teams, how many games are needed to complete the schedule if each team plays six games against each other team?

5. In how many ways can a sum greater than 5 be obtained when tossing a pair of dice?

6. In how many ways can six different mathematics books and three different biology books be placed on a shelf so that all of the books in a subject area are side by side?

7. How many four-element subsets containing A or B, but not both A and B, can be formed from the set {A, B, C, D, E, F, G}?

8. How many five-card hands consisting of two aces, two kings, and one queen can be dealt from a deck of 52 playing cards?

9. How many different nine-letter arrangements can be formed from the nine letters of the word SASSAFRAS?

10. How many committees consisting of four men and three women can be formed from a group of seven men and five women?

11. What is the probability of rolling a sum less than 9 with a pair of dice?

12. Six coins are tossed. Find the probability of getting three heads and three tails.

13. All possible numbers of three different digits each are formed from the digits 1, 2, 3, 4, 5, and 6. If one number is then chosen at random, find the probability that it is greater than 200.

14. A four-person committee is selected at random from Anwar, Barb, Chad, Dick, Edna, Fern, and Giraldo. What is the probability that neither Anwar nor Barb is on the committee?

15. From a group of three men and five women, a three-person committee is selected at random. Find the probability that the committee contains at least one man.

16. A box of 12 items is known to contain one defective and 11 nondefective items. If a sample of three items is selected at random, what is the probability that all three items are nondefective?

17. Five coins are tossed 80 times. How many times should you expect to get three heads and two tails?

18. Suppose 3000 tickets are sold in a lottery. There are three prizes: The first prize is $500, the second is $300, and the third is $100. What is the mathematical expectation of winning?

19. A bag contains seven white and 12 green marbles. Two marbles are drawn in succession, *with replacement*. Find the probability that one marble of each color is drawn.

20. A bag contains three white, five green, and seven blue marbles. Two marbles are drawn *without replacement*. Find the probability that two green marbles are drawn.

21. In an election there were 2000 eligible voters. They were asked to vote on two issues, A and B. The results were as follows: 500 people voted for A, 800 people voted for B, and 250 people voted for both A and B. If one person is chosen at random from the 2000 eligible voters, find the probability that this person voted for A, given that he or she voted for B.

22. Expand and simplify $\left(2 - \dfrac{1}{n}\right)^6$.

23. Expand and simplify $(3x + 2y)^5$.

24. Find the ninth term of the expansion of $\left(x - \dfrac{1}{2}\right)^{12}$.

25. Find the fifth term of the expansion of $(x + 3y)^7$.

Appendix

A   Prime Numbers and Operations with Fractions

This appendix reviews the operations with rational numbers in common fraction form. Throughout this section, we will speak of "multiplying fractions." Be aware that this phrase means multiplying rational numbers in common fraction form. A strong foundation here will simplify your later work in rational expressions. Because prime numbers and prime factorization play an important role in the operations with fractions, let's begin by considering two special kinds of whole numbers, prime numbers and composite numbers.

Definition 4.1

A **prime number** is a whole number greater than 1 that has no factors (divisors) other than itself and 1. Whole numbers greater than 1 that are not prime numbers are called **composite numbers**.

The prime numbers less than 50 are 2, 3, 5, 7, 11, 13, 17, 19, 23, 29, 31, 37, 41, 43, and 47. Note that each of these has no factors other than itself and 1. We can express every composite number as the indicated product of prime numbers. Consider the following examples:

$$4 = 2 \cdot 2 \qquad 6 = 2 \cdot 3 \qquad 8 = 2 \cdot 2 \cdot 2 \qquad 10 = 2 \cdot 5 \qquad 12 = 2 \cdot 2 \cdot 3$$

In each case we express a composite number as the indicated product of prime numbers. The indicated-product form is called the prime-factored form of the number. There are various procedures to find the prime factors of a given composite number. For our purposes, the simplest technique is to factor the given composite number into any two easily recognized factors and then continue to factor each of these until we obtain only prime factors. Consider these examples:

$$18 = 2 \cdot 9 = 2 \cdot 3 \cdot 3 \qquad\qquad\qquad 27 = 3 \cdot 9 = 3 \cdot 3 \cdot 3$$

$$24 = 4 \cdot 6 = 2 \cdot 2 \cdot 2 \cdot 3 \qquad\qquad\qquad 150 = 10 \cdot 15 = 2 \cdot 5 \cdot 3 \cdot 5$$

It does not matter which two factors we choose first. For example, we might start by expressing 18 as $3 \cdot 6$ and then factor 6 into $2 \cdot 3$, which produces a final result of $18 = 3 \cdot 2 \cdot 3$. Either way, 18 contains two prime factors of 3 and one prime factor of 2. The order in which we write the prime factors is not important.

■ Least Common Multiple

It is sometimes necessary to determine the smallest common nonzero multiple of two or more whole numbers. We call this nonzero number the **least common multiple**. In our work with fractions, there will be problems where it will be necessary to find the least common multiple of some numbers, usually the denominators of fractions. So let's review the concepts of multiples. We know that 35 is a multiple of 5 because $5 \cdot 7 = 35$. The set of all whole numbers that are multiples of 5 consists of 0, 5, 10, 15, 20, 25, and so on. In other words, 5 times each successive whole number ($5 \cdot 0 = 0, 5 \cdot 1 = 5, 5 \cdot 2 = 10, 5 \cdot 3 = 15$, etc.) produces the multiples of 5. In a like manner, the set of multiples of 4 consists of 0, 4, 8, 12, 16, and so on. We can illustrate the concept of least common multiple and find the least common multiple of 5 and 4 by using a simple listing of the multiples of 5 and the multiples of 4.

Multiples of 5 are 0, 5, 10, 15, 20, 25, 30, 35, 40, 45, . . .

Multiples of 4 are 0, 4, 8, 12, 16, 20, 24, 28, 32, 36, 40, 44, 48, . . .

The nonzero numbers in common on the lists are 20 and 40. The least of these, 20, is the least common multiple. Stated another way, 20 is the smallest nonzero whole number that is divisible by both 4 and 5.

Often, from your knowledge of arithmetic, you will be able to determine the least common multiple by inspection. For instance, the least common multiple of 6 and 8 is 24. Therefore, 24 is the smallest nonzero whole number that is divisible by both 6 and 8. If we cannot determine the least common multiple by inspection, then using the prime-factorized form of composite numbers is helpful. The procedure is as follows.

Step 1 Express each number as a product of prime factors.

Step 2 The least common multiple contains each different prime factor as many times as the most times it appears in any one of the factorizations from step 1.

The following examples illustrate this technique for finding the least common multiple of two or more numbers.

E X A M P L E   1   Find the least common multiple of 24 and 36.

Solution

Let's first express each number as a product of prime factors.

$24 = 2 \cdot 2 \cdot 2 \cdot 3$

$36 = 2 \cdot 2 \cdot 3 \cdot 3$

The prime factor 2 occurs the most times (three times) in the factorization of 24. Because the factorization of 24 contains three 2s, the least common multiple must have three 2s. The prime factor 3 occurs the most times (two times) in the factorization of 36. Because the factorization of 36 contains two 3s, the least common multiple must have two 3s. The least common multiple of 24 and 36 is therefore $2 \cdot 2 \cdot 2 \cdot 3 \cdot 3 = 72$. ■

EXAMPLE 2

Find the least common multiple of 48 and 84.

Solution

$$48 = 2 \cdot 2 \cdot 2 \cdot 2 \cdot 3$$

$$84 = 2 \cdot 2 \cdot 3 \cdot 7$$

We need four 2s in the least common multiple because of the four 2s in 48. We need one 3 because of the 3 in each of the numbers, and we need one 7 because of the 7 in 84. The least common multiple of 48 and 84 is $2 \cdot 2 \cdot 2 \cdot 2 \cdot 3 \cdot 7 = 336$. ■

EXAMPLE 3

Find the least common multiple of 12, 18, and 28.

Solution

$$28 = 2 \cdot 2 \cdot 7$$

$$18 = 2 \cdot 3 \cdot 3$$

$$12 = 2 \cdot 2 \cdot 3$$

The least common multiple is $2 \cdot 2 \cdot 3 \cdot 3 \cdot 7 = 252$. ■

EXAMPLE 4

Find the least common multiple of 8 and 9.

Solution

$$9 = 3 \cdot 3$$

$$8 = 2 \cdot 2 \cdot 2$$

The least common multiple is $2 \cdot 2 \cdot 2 \cdot 3 \cdot 3 = 72$. ■

■ Multiplying Fractions

We can define the multiplication of fractions in common fractional form as follows:

Multiplying Fractions

If a, b, c, and d are integers, with b and d not equal to zero, then $\dfrac{a}{b} \cdot \dfrac{c}{d} = \dfrac{a \cdot c}{b \cdot d}$.

To multiply fractions in common fractional form, we simply multiply numerators and multiply denominators. The following examples illustrate the multiplying of fractions.

$$\frac{1}{3} \cdot \frac{2}{5} = \frac{1 \cdot 2}{3 \cdot 5} = \frac{2}{15}$$

$$\frac{3}{4} \cdot \frac{5}{7} = \frac{3 \cdot 5}{4 \cdot 7} = \frac{15}{28}$$

$$\frac{3}{5} \cdot \frac{5}{3} = \frac{15}{15} = 1$$

The last of these examples is a very special case. If the product of two numbers is 1, then the numbers are said to be reciprocals of each other.

Before we proceed too far with multiplying fractions, we need to learn about reducing fractions. The following property is applied throughout our work with fractions. We call this property the fundamental property of fractions.

Fundamental Property of Fractions

If b and k are nonzero integers, and a is any integer, then $\dfrac{a \cdot k}{b \cdot k} = \dfrac{a}{b}$.

The fundamental property of fractions provides the basis for what is often called reducing fractions to lowest terms, or expressing fractions in simplest or reduced form. Let's apply the property to a few examples.

EXAMPLE 5

Reduce $\dfrac{12}{18}$ to lowest terms.

Solution

$$\frac{12}{18} = \frac{2 \cdot 6}{3 \cdot 6} = \frac{2}{3}$$

A common factor of 6 has been divided out of both numerator and denominator. ∎

EXAMPLE 6

Change $\dfrac{14}{35}$ to simplest form.

Solution

$$\frac{14}{35} = \frac{2 \cdot 7}{5 \cdot 7} = \frac{2}{5}$$

A common factor of 7 has been divided out of both numerator and denominator. ∎

EXAMPLE 7   Reduce $\dfrac{72}{90}$.

Solution

$$\frac{72}{90} = \frac{2 \cdot 2 \cdot 2 \cdot 3 \cdot 3}{2 \cdot 3 \cdot 3 \cdot 5} = \frac{4}{5}$$

The prime-factored forms of the numerator and denominator may be used to find common factors.

We are now ready to consider multiplication problems with the understanding that the final answer should be expressed in reduced form. Study the following examples carefully; we use different methods to simplify the problems.

EXAMPLE 8   Multiply $\left(\dfrac{9}{4}\right)\left(\dfrac{14}{15}\right)$.

Solution

$$\left(\frac{9}{4}\right)\left(\frac{14}{15}\right) = \frac{3 \cdot 3 \cdot 2 \cdot 7}{2 \cdot 2 \cdot 3 \cdot 5} = \frac{21}{10}$$

EXAMPLE 9   Find the product of $\dfrac{8}{9}$ and $\dfrac{18}{24}$.

Solution

$$\frac{\overset{1}{\cancel{8}}}{\underset{1}{\cancel{9}}} \cdot \frac{\overset{2}{\cancel{18}}}{\underset{3}{\cancel{24}}} = \frac{2}{3}$$

A common factor of 8 has been divided out of 8 and 24, and a common factor of 9 has been divided out of 9 and 18.

■ Dividing Fractions

The next example motivates a definition for division of rational numbers in fractional form:

$$\frac{\frac{3}{4}}{\frac{2}{3}} = \left(\frac{\frac{3}{4}}{\frac{2}{3}}\right)\left(\frac{\frac{3}{2}}{\frac{3}{2}}\right) = \frac{\left(\frac{3}{4}\right)\left(\frac{3}{2}\right)}{1} = \left(\frac{3}{4}\right)\left(\frac{3}{2}\right) = \frac{9}{8}$$

Note that $\left(\dfrac{\frac{3}{2}}{\frac{3}{2}}\right)$ is a form of 1, and $\dfrac{3}{2}$ is the reciprocal of $\dfrac{2}{3}$. In other words, $\dfrac{3}{4}$ divided by $\dfrac{2}{3}$ is equivalent to $\dfrac{3}{4}$ times $\dfrac{3}{2}$. The following definition for division now should seem reasonable.

Division of Fractions

If b, c, and d are nonzero integers, and a is any integer, then $\dfrac{a}{b} \div \dfrac{c}{d} = \dfrac{a}{b} \cdot \dfrac{d}{c}$.

Note that to divide $\dfrac{a}{b}$ by $\dfrac{c}{d}$, we multiply $\dfrac{a}{b}$ times the reciprocal of $\dfrac{c}{d}$, which is $\dfrac{d}{c}$. The next examples demonstrate the important steps of a division problem.

$$\frac{2}{3} \div \frac{1}{2} = \frac{2}{3} \cdot \frac{2}{1} = \frac{4}{3}$$

$$\frac{5}{6} \div \frac{3}{4} = \frac{5}{6} \cdot \frac{4}{3} = \frac{5 \cdot 4}{6 \cdot 3} = \frac{5 \cdot \cancel{2} \cdot 2}{\cancel{2} \cdot 3 \cdot 3} = \frac{10}{9}$$

$$\frac{\frac{6}{7}}{\frac{3}{2}} = \frac{\cancel{6}^{3}}{7} \cdot \frac{1}{\cancel{2}_{1}} = \frac{3}{7}$$

■ Adding and Subtracting Fractions

Suppose that it is one-fifth of a mile between your dorm and the union and two-fifths of a mile between the union and the library along a straight line as indicated in Figure A.1. The total distance between your dorm and the library is three-fifths of a mile, and we write $\dfrac{1}{5} + \dfrac{2}{5} = \dfrac{3}{5}$.

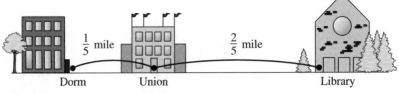

$\frac{1}{5}$ mile $\frac{2}{5}$ mile

Dorm Union Library

Figure A.1

Figure A.2

A pizza is cut into seven equal pieces and you eat two of the pieces (see Figure A.2). How much of the pizza remains? We represent the whole pizza by $\frac{7}{7}$ and conclude that $\frac{7}{7} - \frac{2}{7} = \frac{5}{7}$ of the pizza remains.

These examples motivate the following definition for addition and subtraction of rational numbers in $\frac{a}{b}$ form.

Addition and Subtraction of Fractions

If a, b, and c are integers, and b is not zero, then

$$\frac{a}{b} + \frac{c}{b} = \frac{a + c}{b}$$ Addition

$$\frac{a}{b} - \frac{c}{b} = \frac{a - c}{b}$$ Subtraction

We say that fractions with common denominators can be added or subtracted by adding or subtracting the numerators and placing the results over the common denominator. Consider the following examples:

$$\frac{3}{7} + \frac{2}{7} = \frac{3 + 2}{7} = \frac{5}{7}$$

$$\frac{7}{8} - \frac{2}{8} = \frac{7 - 2}{8} = \frac{5}{8}$$

$$\frac{5}{6} - \frac{1}{6} = \frac{5 - 1}{6} = \frac{4}{6} = \frac{2}{3}$$ We agree to reduce the final answer.

How do we add or subtract if the fractions do not have a common denominator? We use the fundamental principle of fractions, $\dfrac{a \cdot k}{b \cdot k} = \dfrac{a}{b}$, to get equivalent fractions that have a common denominator. **Equivalent fractions** are fractions that name the same number. Consider the next example, which shows the details.

E X A M P L E 1 0	Add $\dfrac{1}{4} + \dfrac{2}{5}$.

Solution

$$\frac{1}{4} = \frac{1 \cdot 5}{4 \cdot 5} = \frac{5}{20} \qquad \frac{1}{4} \text{ and } \frac{5}{10} \text{ are equivalent fractions.}$$

$$\frac{2}{5} = \frac{2 \cdot 4}{5 \cdot 4} = \frac{8}{20} \qquad \frac{2}{5} \text{ and } \frac{8}{20} \text{ are equivalent fractions.}$$

$$\frac{5}{20} + \frac{8}{20} = \frac{13}{20} \qquad \qquad \blacksquare$$

Note that in Example 10 we chose 20 as the common denominator, and 20 is the least common multiple of the original denominators 4 and 5. (Recall that the least common multiple is the smallest nonzero whole number divisible by the given numbers.) In general, we use the least common multiple of the denominators of the fractions to be added or subtracted as a **least common denominator** (LCD).

Recall that the least common multiple may be found either by inspection or by using prime factorization forms of the numbers. Consider some examples involving these procedures.

E X A M P L E 1 1	Subtract $\dfrac{5}{8} - \dfrac{7}{12}$.

Solution

By inspection the LCD is 24.

$$\frac{5}{8} - \frac{7}{12} = \frac{5 \cdot 3}{8 \cdot 3} - \frac{7 \cdot 2}{12 \cdot 2} = \frac{15}{24} - \frac{14}{24} = \frac{1}{24} \qquad \qquad \blacksquare$$

If the LCD is not obvious by inspection, then we can use the technique of prime factorization to find the least common multiple.

E X A M P L E 1 2	Add $\dfrac{5}{18} + \dfrac{7}{24}$.

Solution

If we cannot find the LCD by inspection, then we can use the prime-factorized forms.

$$\left. \begin{aligned} 18 &= 2 \cdot 3 \cdot 3 \\ 24 &= 2 \cdot 2 \cdot 2 \cdot 3 \end{aligned} \right\} \longrightarrow \text{LCD} = 2 \cdot 2 \cdot 2 \cdot 3 \cdot 3 = 72$$

$$\frac{5}{18} + \frac{7}{24} = \frac{5 \cdot 4}{18 \cdot 4} + \frac{7 \cdot 3}{24 \cdot 3} = \frac{20}{72} + \frac{21}{72} = \frac{41}{72} \qquad \qquad \blacksquare$$

EXAMPLE   13

Marcey put $\dfrac{5}{8}$ pound of chemicals in the spa to adjust the water quality. Michael, not realizing Marcey had already put in chemicals, put $\dfrac{3}{14}$ pound of chemicals in the spa. The chemical manufacturer states that you should never add more than 1 pound of chemicals. Have Marcey and Michael together put in more than 1 pound of chemicals?

Solution

Add $\dfrac{5}{8} + \dfrac{3}{14}$.

$$\left.\begin{array}{l} 8 = 2 \cdot 2 \cdot 2 \\ 14 = 2 \cdot 7 \end{array}\right\} \longrightarrow \text{LCD} = 2 \cdot 2 \cdot 2 \cdot 7 = 56$$

$$\frac{5}{8} + \frac{3}{14} = \frac{5 \cdot 7}{8 \cdot 7} + \frac{3 \cdot 4}{14 \cdot 4} = \frac{35}{56} + \frac{12}{56} = \frac{47}{56}$$

No, Marcey and Michael have not added more than 1 pound of chemicals. ■

■ Simplifying Numerical Expressions

We now consider simplifying numerical expressions that contain fractions. In agreement with the order of operations, first multiplications and divisions are done as they appear from left to right, and then additions and subtractions are performed as they appear from left to right. In these next examples, we show only the major steps. Be sure you can fill in all the details.

EXAMPLE   14

Simplify $\dfrac{3}{4} + \dfrac{2}{3} \cdot \dfrac{3}{5} - \dfrac{1}{2} \cdot \dfrac{1}{5}$.

Solution

$$\frac{3}{4} + \frac{2}{3} \cdot \frac{3}{5} - \frac{1}{2} \cdot \frac{1}{5} = \frac{3}{4} + \frac{2}{5} - \frac{1}{10}$$

$$= \frac{15}{20} + \frac{8}{20} - \frac{2}{20} = \frac{15 + 8 - 2}{20} = \frac{21}{20} \qquad ■$$

EXAMPLE   15

Simplify $\dfrac{5}{8}\left(\dfrac{1}{2} + \dfrac{1}{3}\right)$.

Solution

$$\frac{5}{8}\left(\frac{1}{2} + \frac{1}{3}\right) = \frac{5}{8}\left(\frac{3}{6} + \frac{2}{6}\right) = \frac{5}{8}\left(\frac{5}{6}\right) = \frac{25}{48} \qquad ■$$

Practice Exercises

For Problems 1–12, factor each composite number into a product of prime numbers; for example, $18 = 2 \cdot 3 \cdot 3$.

1. 26   **2.** 16

3. 36   **4.** 80

5. 49   **6.** 92

7. 56   **8.** 144

9. 120   **10.** 84

11. 135   **12.** 98

For Problems 13–24, find the least common multiple of the given numbers.

13. 6 and 8   **14.** 8 and 12

15. 12 and 16   **16.** 9 and 12

17. 28 and 35   **18.** 42 and 66

19. 49 and 56   **20.** 18 and 24

21. 8, 12, and 28   **22.** 6, 10, and 12

23. 9, 15, and 18   **24.** 8, 14, and 24

For Problems 25–30, reduce each fraction to lowest terms.

25. $\dfrac{8}{12}$ **26.** $\dfrac{12}{16}$

27. $\dfrac{16}{24}$ **28.** $\dfrac{18}{32}$

29. $\dfrac{15}{9}$ **30.** $\dfrac{48}{36}$

For Problems 31–36, multiply or divide as indicated, and express answers in reduced form.

31. $\dfrac{3}{4} \cdot \dfrac{5}{7}$ **32.** $\dfrac{4}{5} \cdot \dfrac{3}{11}$

33. $\dfrac{2}{7} \div \dfrac{3}{5}$ **34.** $\dfrac{5}{6} \div \dfrac{11}{13}$

35. $\dfrac{3}{8} \cdot \dfrac{12}{15}$ **36.** $\dfrac{4}{9} \cdot \dfrac{3}{2}$

37. A certain recipe calls for $\dfrac{3}{4}$ cup of milk. To make half of the recipe, how much milk is needed?

38. John is adding a diesel fuel additive to his fuel tank, which is half full. The directions say to add $\dfrac{1}{3}$ of the bottle to a full fuel tank. What portion of the bottle should he add to the fuel tank?

39. Mark shares a computer with his roommates. He has partitioned the hard drive in such a way that he gets $\dfrac{1}{3}$ of the disk space. His part of the hard drive is currently $\dfrac{2}{3}$ full. What portion of the computer's hard drive space is he currently taking up?

40. Angelina teaches $\dfrac{2}{3}$ of the deaf children in her local school. Her local school educates $\dfrac{1}{2}$ of the deaf children in the school district. What portion of the school district's deaf children is Angelina teaching?

For Problems 41–57, add or subtract as indicated and express answers in lowest terms.

41. $\dfrac{2}{7} + \dfrac{3}{7}$ **42.** $\dfrac{3}{11} + \dfrac{5}{11}$

43. $\dfrac{7}{9} - \dfrac{2}{9}$ **44.** $\dfrac{11}{13} - \dfrac{6}{13}$

45. $\dfrac{3}{4} + \dfrac{9}{4}$ **46.** $\dfrac{5}{6} + \dfrac{7}{6}$

47. $\dfrac{11}{12} - \dfrac{3}{12}$ **48.** $\dfrac{13}{16} - \dfrac{7}{16}$

49. $\dfrac{5}{24} + \dfrac{11}{24}$ **50.** $\dfrac{7}{36} + \dfrac{13}{36}$

51. $\dfrac{1}{3} + \dfrac{1}{5}$ **52.** $\dfrac{1}{6} + \dfrac{1}{8}$

53. $\dfrac{15}{16} - \dfrac{3}{8}$ **54.** $\dfrac{13}{12} - \dfrac{1}{6}$

55. $\dfrac{7}{10} + \dfrac{8}{15}$ **56.** $\dfrac{7}{12} + \dfrac{5}{8}$

57. $\dfrac{11}{24} + \dfrac{5}{32}$

58. Alicia and her brother Jeff shared a pizza. Alicia ate $\frac{1}{8}$ of the pizza, while Jeff ate $\frac{2}{3}$ of the pizza. How much of the pizza has been eaten?

59. Rosa has $\frac{1}{3}$ pound of blueberries, $\frac{1}{4}$ pound of strawberries, and $\frac{1}{2}$ pound of raspberries. If she combines these for a fruit salad, how many pounds of these berries will be in the salad?

60. A chemist has $\frac{11}{16}$ of an ounce of dirt residue to perform crime lab tests. He needs $\frac{3}{8}$ of an ounce to perform a test for iron content. How much of the dirt residue will be left for the chemist to use in other testing?

For Problems 61–68, simplify each numerical expression, expressing answers in reduced form.

61. $\frac{1}{4} - \frac{3}{8} + \frac{5}{12} - \frac{1}{24}$

62. $\frac{3}{4} + \frac{2}{3} - \frac{1}{6} + \frac{5}{12}$

63. $\frac{5}{6} + \frac{2}{3} \cdot \frac{3}{4} - \frac{1}{4} \cdot \frac{2}{5}$

64. $\frac{2}{3} + \frac{1}{2} \cdot \frac{2}{5} - \frac{1}{3} \cdot \frac{1}{5}$

65. $\frac{3}{4} \cdot \frac{6}{9} - \frac{5}{6} \cdot \frac{8}{10} + \frac{2}{3} \cdot \frac{6}{8}$

66. $\frac{3}{5} \cdot \frac{5}{7} + \frac{2}{3} \cdot \frac{3}{5} - \frac{1}{7} \cdot \frac{2}{5}$

67. $\frac{7}{13}\left(\frac{2}{3} - \frac{1}{6}\right)$

68. $48\left(\frac{5}{12} - \frac{1}{6} + \frac{3}{8}\right)$

69. Blake Scott leaves $\frac{1}{4}$ of his estate to the Boy Scouts, $\frac{2}{5}$ to the local cancer fund, and the rest to his church. What fractional part of the estate does the church receive?

70. Franco has $\frac{7}{8}$ of an ounce of gold. He wants to give $\frac{3}{16}$ of an ounce to his friend Julie. He plans to divide the remaining amount of his gold in half to make two rings. How much gold will he have for each ring?

Answers to Odd-Numbered Problems and All Chapter Review, Chapter Test, and Cumulative Review Problems

Problem Set 1.1 (page 10)

1. True **3.** False **5.** True **7.** False **9.** True

11. 0 and 14 **13.** $0, 14, \frac{2}{3}, -\frac{11}{14}, 2.34, 3.2\bar{1}, \frac{55}{8}, -19$, and

-2.6 **15.** 0 and 14 **17.** All of them **19.** $\not\subseteq$ **21.** $\subseteq$

23. $\not\subseteq$ **25.** $\subseteq$ **27.** $\not\subseteq$ **29.** Real, rational, an integer,

and negative **31.** Real, irrational, and negative

33. $\{1, 2\}$ **35.** $\{0, 1, 2, 3, 4, 5\}$ **37.** $\{\ldots, -1, 0, 1, 2\}$

39. $\varnothing$ **41.** $\{0, 1, 2, 3, 4\}$ **43.** -6 **45.** 2 **47.** $3x + 1$

49. $5x$ **51.** 26 **53.** 84 **55.** 23 **57.** 65 **59.** 60 **61.** 33

63. 1320 **65.** 20 **67.** 119 **69.** 18 **71.** 4 **73.** 31

Problem Set 1.2 (page 20)

1. -7 **3.** -19 **5.** -22 **7.** -7 **9.** 108 **11.** -70

13. 14 **15.** -7 **17.** $3\frac{1}{2}$ **19.** $5\frac{1}{2}$ **21.** $-\frac{2}{15}$ **23.** -4

25. 0 **27.** Undefined **29.** -60 **31.** -4.8 **33.** 14.13

35. -6.5 **37.** -38.88 **39.** 0.2 **41.** $-\frac{13}{12}$ **43.** $-\frac{3}{4}$

45. $-\frac{13}{9}$ **47.** $-\frac{3}{5}$ **49.** $-\frac{3}{2}$ **51.** -12 **53.** -24 **55.** $\frac{35}{4}$

57. 15 **59.** -17 **61.** $\frac{47}{12}$ **63.** 5 **65.** 0 **67.** 26 **69.** 6

71. 25 **73.** 78 **75.** -10 **77.** 5 **79.** -5 **81.** 10.5

83. -3.3 **85.** 19.5 **87.** $\frac{3}{4}$ **89.** $\frac{5}{2}$ **93.** 10 over par

95. Lost $16.50 **97.** A gain of 0.88 dollar

99. No; they made it 49.1 pounds lighter

Problem Set 1.3 (page 28)

1. Associative property of addition

3. Commutative property of addition

5. Additive inverse property

7. Multiplication property of negative one

9. Commutative property of multiplication

11. Distributive property

13. Associative property of multiplication

15. 18 **17.** 2 **19.** -1300 **21.** 1700 **23.** -47

25. 3200 **27.** -19 **29.** -41 **31.** -17 **33.** -39

35. 24 **37.** 20 **39.** 55 **41.** 16 **43.** 49 **45.** -216

47. -14 **49.** -8 **51.** $\frac{3}{16}$ **53.** $-\frac{10}{9}$ **57.** 2187

59. -2048 **61.** $-15,625$ **63.** 3.9525416

Problem Set 1.4 (page 37)

1. $4x$ **3.** $-a^2$ **5.** $-6n$ **7.** $-5x + 2y$ **9.** $6a^2 + 5b^2$

11. $21x - 13$ **13.** $-2a^2b - ab^2$ **15.** $8x + 21$

17. $-5a + 2$ **19.** $-5n^2 + 11$ **21.** $-7x^2 + 32$

23. $22x - 3$ **25.** $-14x - 7$ **27.** $-10n^2 + 4$

29. $4x - 30y$ **31.** $-13x - 31$ **33.** $-21x - 9$ **35.** -17

37. 12 **39.** 4 **41.** 3 **43.** -38 **45.** -14 **47.** 64

49. 104 **51.** 5 **53.** 4 **55.** $-\frac{22}{3}$ **57.** $\frac{29}{4}$

59. 221.6 **61.** 1092.4 **63.** 1420.5 **65.** $n + 12$

67. $n - 5$ **69.** $50n$ **71.** $\frac{1}{2}n - 4$ **73.** $\frac{n}{8}$ **75.** $2n - 9$

77. $10(n - 6)$ **79.** $n + 20$ **81.** $2t - 3$ **83.** $n + 47$

85. $8y$ **87.** 25 cm **89.** $\frac{c}{25}$ **91.** $n + 2$ **93.** $\frac{c}{5}$

95. $12d$ **97.** $3y + f$ **99.** $5280m$

Chapter 1 Review Problem Set (page 41)

1. (a) 67 **(b)** $0, -8$, and 67 **(c)** 0 and 67

(d) $0, \frac{3}{4}, -\frac{5}{6}, \frac{25}{3}, -8, 0.34, 0.2\bar{3}, 67$, and $\frac{9}{7}$

(e) $\sqrt{2}$ and $-\sqrt{3}$

2. Associative property for addition

3. Substitution property of equality

4. Multiplication property of negative one

5. Distributive property
6. Associative property for multiplication
7. Commutative property for addition
8. Distributive property
9. Multiplicative inverse property
10. Symmetric property of equality
11. $-6\frac{1}{2}$ **12.** $-6\frac{1}{6}$ **13.** -8 **14.** -15 **15.** 20 **16.** 49
17. -56 **18.** -24 **19.** 6 **20.** 4 **21.** 100 **22.** 8
23. $-4a^2 - 5b^2$ **24.** $3x - 2$ **25.** ab^2 **26.** $-\frac{7}{3}x^2y$
27. $10n^2 - 17$ **28.** $-13a + 4$ **29.** $-2n + 2$
30. $-7x - 29y$ **31.** $-7a - 9$ **32.** $-9x^2 + 7$ **33.** $-6\frac{1}{2}$
34. $-\frac{5}{16}$ **35.** -55 **36.** 144 **37.** -16 **38.** -44
39. 19.4 **40.** 59.6 **41.** $-\frac{59}{3}$ **42.** $\frac{9}{2}$ **43.** $4 + 2n$
44. $3n - 50$ **45.** $\frac{2}{3}n - 6$ **46.** $10(n - 14)$ **47.** $5n - 8$
48. $\frac{n}{n - 3}$ **49.** $5(n + 2) - 3$ **50.** $\frac{3}{4}(n + 12)$
51. $37 - n$ **52.** $\frac{w}{60}$ **53.** $2y - 7$ **54.** $n + 3$
55. $p + 5n + 25q$ **56.** $\frac{i}{48}$ **57.** $24f + 72y$ **58.** $10d$
59. $12f + i$ **60.** $25 - c$

Chapter 1 Test (page 43)

1. Symmetric property **2.** Distributive property **3.** -3
4. -23 **5.** $-\frac{23}{6}$ **6.** 11 **7.** 8 **8.** -94 **9.** -4 **10.** 960
11. -32 **12.** $-x^2 - 8x - 2$ **13.** $-19n - 20$ **14.** 27
15. $\frac{11}{16}$ **16.** $\frac{2}{3}$ **17.** 77 **18.** -22.5 **19.** 93 **20.** -5
21. $6n - 30$ **22.** $3n + 28$ or $3(n + 8) + 4$ **23.** $\frac{72}{n}$
24. $5n + 10d + 25q$ **25.** $6x + 2y$

CHAPTER 2

Problem Set 2.1 (page 51)

1. $\{4\}$ **3.** $\{-3\}$ **5.** $\{-14\}$ **7.** $\{6\}$ **9.** $\left\{\frac{19}{3}\right\}$ **11.** $\{1\}$
13. $\left\{-\frac{10}{3}\right\}$ **15.** $\{4\}$ **17.** $\left\{-\frac{13}{3}\right\}$ **19.** $\{3\}$ **21.** $\{8\}$
23. $\{-9\}$ **25.** $\{-3\}$ **27.** $\{0\}$ **29.** $\left\{-\frac{7}{2}\right\}$ **31.** $\{-2\}$

33. $\left\{-\frac{5}{3}\right\}$ **35.** $\left\{\frac{33}{2}\right\}$ **37.** $\{-35\}$ **39.** $\left\{\frac{1}{2}\right\}$ **41.** $\left\{\frac{1}{6}\right\}$
43. $\{5\}$ **45.** $\{-1\}$ **47.** $\left\{-\frac{21}{16}\right\}$ **49.** $\left\{\frac{12}{7}\right\}$ **51.** 14
53. $13, 14,$ and 15 **55.** $9, 11,$ and 13 **57.** 14 and 81
59. $\$11$ per hour **61.** 30 pennies, 50 nickels, and 70 dimes
63. $\$300$ **65.** 20 three-bedroom, 70 two-bedroom, and
140 one-bedroom **73.** **(a)** $\varnothing$ **(c)** $\{0\}$ **(e)** $\varnothing$

Problem Set 2.2 (page 59)

1. $\{12\}$ **3.** $\left\{-\frac{3}{5}\right\}$ **5.** $\{3\}$ **7.** $\{-2\}$ **9.** $\{-36\}$ **11.** $\left\{\frac{20}{9}\right\}$
13. $\{3\}$ **15.** $\{3\}$ **17.** $\{-2\}$ **19.** $\left\{\frac{8}{5}\right\}$ **21.** $\{-3\}$
23. $\left\{\frac{48}{17}\right\}$ **25.** $\left\{\frac{103}{6}\right\}$ **27.** $\{3\}$ **29.** $\left\{\frac{40}{3}\right\}$ **31.** $\left\{-\frac{20}{7}\right\}$
33. $\left\{\frac{24}{5}\right\}$ **35.** $\{-10\}$ **37.** $\left\{-\frac{25}{4}\right\}$ **39.** $\{0\}$ **41.** 18
43. 16 inches long and 5 inches wide **45.** $14, 15,$ and 16
47. 8 feet **49.** Angie is 22 and her mother is 42.
51. Sydney is 18 and Marcus is 36. **53.** $80, 90,$ and 94
55. $48°$ and $132°$ **57.** $78°$

Problem Set 2.3 (page 67)

1. $\{20\}$ **3.** $\{50\}$ **5.** $\{40\}$ **7.** $\{12\}$ **9.** $\{6\}$ **11.** $\{400\}$
13. $\{400\}$ **15.** $\{38\}$ **17.** $\{6\}$ **19.** $\{3000\}$ **21.** $\{3000\}$
23. $\{400\}$ **25.** $\{14\}$ **27.** $\{15\}$ **29.** $\$90$ **31.** $\$54.40$
33. $\$48$ **35.** $\$400$ **37.** 65% **39.** 62.5% **41.** $\$32,500$
43. $\$3000$ at 10% and $\$4500$ at 11% **45.** $\$53,000$
47. 8 pennies, 15 nickels, and 18 dimes
49. 15 dimes, 45 quarters, and 10 half-dollars **55.** $\{7.5\}$
57. $\{-4775\}$ **59.** $\{8.7\}$ **61.** $\{17.1\}$ **63.** $\{13.5\}$

Problem Set 2.4 (page 77)

1. $\$120$ **3.** 3 years **5.** 6% **7.** $\$800$ **9.** $\$1600$
11. 8% **13.** $\$200$ **15.** 6 feet; 14 feet; 10 feet; 20 feet;
7 feet; 2 feet **17.** $h = \dfrac{V}{B}$ **19.** $h = \dfrac{V}{\pi r^2}$ **21.** $r = \dfrac{C}{2\pi}$
23. $C = \dfrac{100M}{I}$ **25.** $C = \dfrac{5}{9}(F - 32)$ or $C = \dfrac{5F - 160}{9}$
27. $x = \dfrac{y - b}{m}$ **29.** $x = \dfrac{y - y_1 + mx_1}{m}$
31. $x = \dfrac{ab + bc}{b - a}$ **33.** $x = a + bc$ **35.** $x = \dfrac{3b - 6a}{2}$
37. $x = \dfrac{5y + 7}{2}$ **39.** $y = -7x - 4$ **41.** $x = \dfrac{6y + 4}{3}$

43. $x = \dfrac{cy - ac - b^2}{b}$ **45.** $y = \dfrac{x - a + 1}{a - 3}$

47. 22 meters long and 6 meters wide **49.** $11\frac{1}{9}$ years

51. $11\frac{1}{9}$ years **53.** 4 hours **55.** 3 hours **57.** 40 miles

59. 15 quarts of 30% solution and 5 quarts of 70% solution

61. 25 milliliters **67.** \$596.25 **69.** 1.5 years **71.** 14.5%

73. \$1850

Problem Set 2.5 (page 86)

1. $(1, \infty)$

3. $[-1, \infty)$

5. $(-\infty, -2)$

7. $(-\infty, 2]$

9. $x < 4$ **11.** $x \le -7$ **13.** $x > 8$ **15.** $x \ge -7$

17. $(1, \infty)$

19. $(-\infty, -4]$

21. $(-\infty, -2]$

23. $(-\infty, 2)$

25. $(-1, \infty)$

27. $[-1, \infty)$

29. $(-2, \infty)$

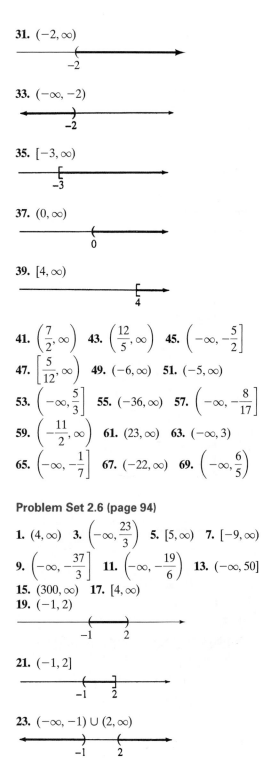

31. $(-2, \infty)$

33. $(-\infty, -2)$

35. $[-3, \infty)$

37. $(0, \infty)$

39. $[4, \infty)$

41. $\left(\dfrac{7}{2}, \infty\right)$ **43.** $\left(\dfrac{12}{5}, \infty\right)$ **45.** $\left(-\infty, -\dfrac{5}{2}\right]$

47. $\left[\dfrac{5}{12}, \infty\right)$ **49.** $(-6, \infty)$ **51.** $(-5, \infty)$

53. $\left(-\infty, \dfrac{5}{3}\right]$ **55.** $(-36, \infty)$ **57.** $\left(-\infty, -\dfrac{8}{17}\right]$

59. $\left(-\dfrac{11}{2}, \infty\right)$ **61.** $(23, \infty)$ **63.** $(-\infty, 3)$

65. $\left(-\infty, -\dfrac{1}{7}\right]$ **67.** $(-22, \infty)$ **69.** $\left(-\infty, \dfrac{6}{5}\right)$

Problem Set 2.6 (page 94)

1. $(4, \infty)$ **3.** $\left(-\infty, \dfrac{23}{3}\right)$ **5.** $[5, \infty)$ **7.** $[-9, \infty)$

9. $\left(-\infty, -\dfrac{37}{3}\right]$ **11.** $\left(-\infty, -\dfrac{19}{6}\right)$ **13.** $(-\infty, 50]$

15. $(300, \infty)$ **17.** $[4, \infty)$

19. $(-1, 2)$

21. $(-1, 2]$

23. $(-\infty, -1) \cup (2, \infty)$

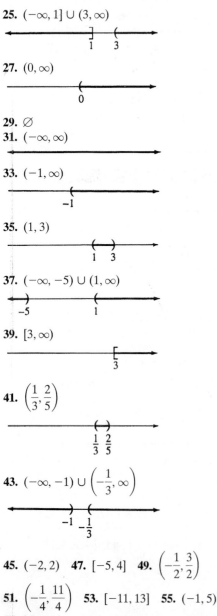

25. $(-\infty, 1] \cup (3, \infty)$

27. $(0, \infty)$

29. $\varnothing$
31. $(-\infty, \infty)$

33. $(-1, \infty)$

35. $(1, 3)$

37. $(-\infty, -5) \cup (1, \infty)$

39. $[3, \infty)$

41. $\left(\dfrac{1}{3}, \dfrac{2}{5}\right)$

43. $(-\infty, -1) \cup \left(-\dfrac{1}{3}, \infty\right)$

45. $(-2, 2)$ **47.** $[-5, 4]$ **49.** $\left(-\dfrac{1}{2}, \dfrac{3}{2}\right)$

51. $\left(-\dfrac{1}{4}, \dfrac{11}{4}\right)$ **53.** $[-11, 13]$ **55.** $(-1, 5)$

57. More than 10% **59.** 5 feet and 10 inches or better
61. 168 or better **63.** 77 or less **65.** $163°F \le C \le 218°F$
67. $6.3 \le M \le 11.25$

Problem Set 2.7 (page 101)
1. $(-5, 5)$

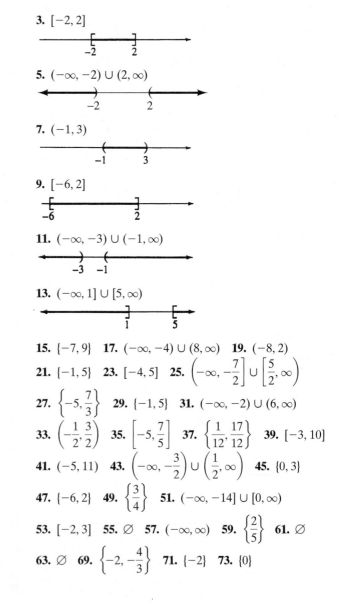

3. $[-2, 2]$

5. $(-\infty, -2) \cup (2, \infty)$

7. $(-1, 3)$

9. $[-6, 2]$

11. $(-\infty, -3) \cup (-1, \infty)$

13. $(-\infty, 1] \cup [5, \infty)$

15. $\{-7, 9\}$ **17.** $(-\infty, -4) \cup (8, \infty)$ **19.** $(-8, 2)$
21. $\{-1, 5\}$ **23.** $[-4, 5]$ **25.** $\left(-\infty, -\dfrac{7}{2}\right] \cup \left[\dfrac{5}{2}, \infty\right)$
27. $\left\{-5, \dfrac{7}{3}\right\}$ **29.** $\{-1, 5\}$ **31.** $(-\infty, -2) \cup (6, \infty)$
33. $\left(-\dfrac{1}{2}, \dfrac{3}{2}\right)$ **35.** $\left[-5, \dfrac{7}{5}\right]$ **37.** $\left\{\dfrac{1}{12}, \dfrac{17}{12}\right\}$ **39.** $[-3, 10]$
41. $(-5, 11)$ **43.** $\left(-\infty, -\dfrac{3}{2}\right) \cup \left(\dfrac{1}{2}, \infty\right)$ **45.** $\{0, 3\}$
47. $\{-6, 2\}$ **49.** $\left\{\dfrac{3}{4}\right\}$ **51.** $(-\infty, -14] \cup [0, \infty)$
53. $[-2, 3]$ **55.** $\varnothing$ **57.** $(-\infty, \infty)$ **59.** $\left\{\dfrac{2}{5}\right\}$ **61.** $\varnothing$
63. $\varnothing$ **69.** $\left\{-2, -\dfrac{4}{3}\right\}$ **71.** $\{-2\}$ **73.** $\{0\}$

Chapter 2 Review Problem Set (page 104)

1. $\{18\}$ **2.** $\{-14\}$ **3.** $\{0\}$ **4.** $\left\{\dfrac{1}{2}\right\}$ **5.** $\{10\}$ **6.** $\left\{\dfrac{7}{3}\right\}$
7. $\left\{\dfrac{28}{17}\right\}$ **8.** $\left\{-\dfrac{1}{38}\right\}$ **9.** $\left\{\dfrac{27}{17}\right\}$ **10.** $\left\{-\dfrac{10}{3}, 4\right\}$
11. $\{50\}$ **12.** $\left\{-\dfrac{39}{2}\right\}$ **13.** $\{200\}$ **14.** $\{-8\}$
15. $\left\{-\dfrac{7}{2}, \dfrac{1}{2}\right\}$ **16.** $x = \dfrac{2b + 2}{a}$ **17.** $x = \dfrac{c}{a - b}$

18. $x = \dfrac{pb - ma}{m - p}$ 19. $x = \dfrac{11 + 7y}{5}$

20. $x = \dfrac{by + b + ac}{c}$ 21. $s = \dfrac{A - \pi r^2}{\pi r}$

22. $b_2 = \dfrac{2A - hb_1}{h}$ 23. $n = \dfrac{2S_n}{a_1 + a_2}$ 24. $R = \dfrac{R_1 R_2}{R_1 + R_2}$

25. $[-5, \infty)$ 26. $(4, \infty)$ 27. $\left(-\dfrac{7}{3}, \infty\right)$ 28. $\left[\dfrac{17}{2}, \infty\right)$

29. $\left(-\infty, \dfrac{1}{3}\right)$ 30. $\left(\dfrac{53}{11}, \infty\right)$ 31. $[6, \infty)$ 32. $(-\infty, 100]$

33. $(-5, 6)$ 34. $\left(-\infty, -\dfrac{11}{3}\right) \cup (3, \infty)$ 35. $(-\infty, -17)$

36. $\left(-\infty, -\dfrac{15}{4}\right)$

37. (number line: -1, 1)

38. (number line: -3, 2)

39. (number line: 3)

40. (number line)

41. (number line: -2, 1)

42. (number line: -2, 1)

43. (number line: $\dfrac{1}{2}$, 3)

44. $\varnothing$ 45. The length is 15 meters and the width is 7 meters.   46. $200 at 7% and $300 at 8%   47. 88 or better   48. 4, 5, and 6   49. $10.50 per hour   50. 20 nickels, 50 dimes, and 75 quarters   51. 80°   52. $45.60

53. $6\dfrac{2}{3}$ pints   54. 55 miles per hour   55. Sonya for $3\dfrac{1}{4}$ hours and Rita for $4\dfrac{1}{2}$ hours   56. $6\dfrac{1}{4}$ cups

Chapter 2 Test (page 107)

1. $\{-3\}$ 2. $\{5\}$ 3. $\left\{\dfrac{1}{2}\right\}$ 4. $\left\{\dfrac{16}{5}\right\}$ 5. $\left\{-\dfrac{14}{5}\right\}$ 6. $\{-1\}$

7. $\left\{-\dfrac{3}{2}, 3\right\}$ 8. $\{3\}$ 9. $\left\{\dfrac{31}{3}\right\}$ 10. $\{650\}$

11. $y = \dfrac{8x - 24}{9}$ 12. $h = \dfrac{S - 2\pi r^2}{2\pi r}$ 13. $(-2, \infty)$

14. $[-4, \infty)$ 15. $(-\infty, -35]$ 16. $(-\infty, 10)$ 17. $(3, \infty)$

18. $(-\infty, 200]$ 19. $\left(-1, \dfrac{7}{3}\right)$ 20. $\left(-\infty, -\dfrac{11}{4}\right] \cup \left[\dfrac{1}{4}, \infty\right)$

21. $72   22. 19 centimeters   23. $\dfrac{2}{3}$ of a cup

24. 97 or better   25. 70°

CHAPTER 3

Problem Set 3.1 (page 113)

1. 2 3. 3 5. 2 7. 6 9. 0 11. $10x - 3$
13. $-11t + 5$ 15. $-x^2 + 2x - 2$ 17. $17a^2b^2 - 5ab$
19. $-9x + 7$ 21. $-2x + 6$ 23. $10a + 7$
25. $4x^2 + 10x + 6$ 27. $-6a^2 + 12a + 14$
29. $3x^3 + x^2 + 13x - 11$ 31. $7x + 8$ 33. $-3x - 16$
35. $2x^2 - 2x - 8$ 37. $-3x^3 + 5x^2 - 2x + 9$
39. $5x^2 - 4x + 11$ 41. $-6x^2 + 9x + 7$
43. $-2x^2 + 9x + 4$ 45. $-10n^2 + n + 9$ 47. $8x - 2$
49. $8x - 14$ 51. $-9x^2 - 12x + 4$ 53. $10x^2 + 13x - 18$
55. $-n^2 - 4n - 4$ 57. $-x + 6$ 59. $6x^2 - 4$
61. $-7n^2 + n + 6$ 63. $t^2 - 4t + 8$ 65. $4n^2 - n - 12$
67. $-4x - 2y$ 69. $-x^3 - x^2 + 3x$ 71. (a) $8x + 4$
(c) $12x + 6$ 73. $8\pi h + 32\pi$ (a) 226.1 (c) 452.2

Problem Set 3.2 (page 120)

1. $36x^4$ 3. $-12x^5$ 5. $4a^3b^4$ 7. $-3x^3y^2z^6$ 9. $-30xy^4$

11. $27a^4b^5$ 13. $-m^3n^3$ 15. $\dfrac{3}{10}x^3y^6$ 17. $-\dfrac{3}{20}a^3b^4$

19. $-\dfrac{1}{6}x^3y^4$ 21. $30x^6$ 23. $-18x^9$ 25. $-3x^6y^6$

27. $-24y^9$ 29. $-56a^4b^2$ 31. $-18a^3b^3$ 33. $-10x^7y^7$
35. $50x^5y^2$ 37. $27x^3y^6$ 39. $-32x^{10}y^5$ 41. $x^{16}y^{20}$
43. $a^6b^{12}c^{18}$ 45. $64a^{12}b^{18}$ 47. $81x^2y^8$ 49. $81a^4b^{12}$
51. $-16a^4b^4$ 53. $-x^6y^{12}z^{18}$ 55. $-125a^6b^6c^3$
57. $-x^7y^{28}z^{14}$ 59. $3x^3y^3$ 61. $-5x^3y^2$ 63. $9bc^2$
65. $-18xyz^4$ 67. $-a^2b^3c^2$ 69. 9 71. $-b^2$ 73. $-18x^3$
75. $6x^{3n}$ 77. a^{5n+3} 79. x^{4n} 81. a^{5n+1} 83. $-10x^{2n}$
85. $12a^{n+4}$ 87. $6x^{3n+2}$ 89. $12x^{n+2}$ 91. $22x^2; 6x^3$
93. $\pi r^2 - 36\pi$

Problem Set 3.3 (page 127)

1. $10x^2y^3 + 6x^3y^4$ 3. $-12a^3b^3 + 15a^5b$
5. $24a^4b^5 - 16a^4b^6 + 32a^5b^6$ 7. $-6x^3y^3 - 3x^4y^4 + x^5y^2$
9. $ax + ay + 2bx + 2by$ 11. $ac + 4ad - 3bc - 12bd$
13. $x^2 + 16x + 60$ 15. $y^2 + 6y - 55$ 17. $n^2 - 5n - 14$

19. $x^2 - 36$ **21.** $x^2 - 12x + 36$ **23.** $x^2 - 14x + 48$
25. $x^3 - 4x^2 + x + 6$ **27.** $x^3 - x^2 - 9x + 9$
29. $t^2 + 18t + 81$ **31.** $y^2 - 14y + 49$
33. $4x^2 + 33x + 35$ **35.** $9y^2 - 1$ **37.** $14x^2 + 3x - 2$
39. $5 + 3t - 2t^2$ **41.** $9t^2 + 42t + 49$ **43.** $4 - 25x^2$
45. $49x^2 - 56x + 16$ **47.** $18x^2 - 39x - 70$
49. $2x^2 + xy - 15y^2$ **51.** $25x^2 - 4a^2$ **53.** $t^3 - 14t - 15$
55. $x^3 + x^2 - 24x + 16$ **57.** $2x^3 + 9x^2 + 2x - 30$
59. $12x^3 - 7x^2 + 25x - 6$
61. $x^4 + 5x^3 + 11x^2 + 11x + 4$
63. $2x^4 - x^3 - 12x^2 + 5x + 4$
65. $x^3 + 6x^2 + 12x + 8$ **67.** $x^3 - 12x^2 + 48x - 64$
69. $8x^3 + 36x^2 + 54x + 27$ **71.** $64x^3 - 48x^2 + 12x - 1$
73. $125x^3 + 150x^2 + 60x + 8$ **75.** $x^{2n} - 16$
77. $x^{2a} + 4x^a - 12$ **79.** $6x^{2n} + x^n - 35$
81. $x^{4a} - 10x^{2a} + 21$ **83.** $4x^{2n} + 20x^n + 25$
87. $2x^2 + 6$ **89.** $4x^3 - 64x^2 + 256x; 256 - 4x^2$
93. (a) $a^6 + 6a^5b + 15a^4b^2 + 20a^3b^3 + 15a^2b^4 + 6ab^5 + b^6$
(c) $a^8 + 8a^7b + 28a^6b^2 + 56a^5b^3 + 70a^4b^4 + 56a^3b^5 + 28a^2b^6 + 8ab^7 + b^8$

Problem Set 3.4 (page 135)

1. Composite **3.** Prime **5.** Composite **7.** Composite
9. Prime **11.** $2 \cdot 2 \cdot 7$ **13.** $2 \cdot 2 \cdot 11$ **15.** $2 \cdot 2 \cdot 2 \cdot 7$
17. $2 \cdot 2 \cdot 2 \cdot 3 \cdot 3$ **19.** $3 \cdot 29$ **21.** $3(2x + y)$
23. $2x(3x + 7)$ **25.** $4y(7y - 1)$ **27.** $5x(4y - 3)$
29. $x^2(7x + 10)$ **31.** $9ab(2a + 3b)$ **33.** $3x^3y^3(4y - 13x)$
35. $4x^2(2x^2 + 3x - 6)$ **37.** $x(5 + 7x + 9x^3)$
39. $5xy^2(3xy + 4 + 7x^2y^2)$ **41.** $(y + 2)(x + 3)$
43. $(2a + b)(3x - 2y)$ **45.** $(x + 2)(x + 5)$
47. $(a + 4)(x + y)$ **49.** $(a - 2b)(x + y)$
51. $(a - b)(3x - y)$ **53.** $(a + 1)(2x + y)$
55. $(a - 1)(x^2 + 2)$ **57.** $(a + b)(2c + 3d)$
59. $(a + b)(x - y)$ **61.** $(x + 9)(x + 6)$
63. $(x + 4)(2x + 1)$ **65.** $\{-7, 0\}$ **67.** $\{0, 1\}$
69. $\{0, 5\}$ **71.** $\left\{-\dfrac{1}{2}, 0\right\}$ **73.** $\left\{-\dfrac{7}{3}, 0\right\}$ **75.** $\left\{0, \dfrac{5}{4}\right\}$
77. $\left\{0, \dfrac{1}{4}\right\}$ **79.** $\{-12, 0\}$ **81.** $\left\{0, \dfrac{3a}{5b}\right\}$ **83.** $\left\{-\dfrac{3a}{2b}, 0\right\}$
85. $\{a, -2b\}$ **87.** 0 or 7 **89.** 6 units **91.** $\dfrac{4}{\pi}$ units
93. The square is 100 feet by 100 feet, and the rectangle is 50 feet by 100 feet.
95. 6 units **101.** $x^a(2x^a - 3)$ **103.** $y^{2m}(y^m + 5)$
105. $x^{4a}(2x^{2a} - 3x^a + 7)$

Problem Set 3.5 (page 142)

1. $(x + 1)(x - 1)$ **3.** $(4x + 5)(4x - 5)$
5. $(3x + 5y)(3x - 5y)$ **7.** $(5xy + 6)(5xy - 6)$
9. $(2x + y^2)(2x - y^2)$ **11.** $(1 + 12n)(1 - 12n)$

13. $(x + 2 + y)(x + 2 - y)$ **15.** $(2x + y + 1)(2x - y - 1)$
17. $(3a + 2b + 3)(3a - 2b - 3)$ **19.** $-5(2x + 9)$
21. $9(x + 2)(x - 2)$ **23.** $5(x^2 + 1)$ **25.** $8(y + 2)(y - 2)$
27. $ab(a + 3)(a - 3)$ **29.** Not factorable
31. $(n + 3)(n - 3)(n^2 + 9)$ **33.** $3x(x^2 + 9)$
35. $4xy(x + 4y)(x - 4y)$ **37.** $6x(1 + x)(1 - x)$
39. $(1 + xy)(1 - xy)(1 + x^2y^2)$ **41.** $4(x + 4y)(x - 4y)$
43. $3(x + 2)(x - 2)(x^2 + 4)$ **45.** $(a - 4)(a^2 + 4a + 16)$
47. $(x + 1)(x^2 - x + 1)$
49. $(3x + 4y)(9x^2 - 12xy + 16y^2)$
51. $(1 - 3a)(1 + 3a + 9a^2)$ **53.** $(xy - 1)(x^2y^2 + xy + 1)$
55. $(x + y)(x - y)(x^2 - xy + y^2)(x^2 + xy + y^2)$
57. $\{-5, 5\}$ **59.** $\left\{-\dfrac{7}{3}, \dfrac{7}{3}\right\}$ **61.** $\{-2, 2\}$ **63.** $\{-1, 0, 1\}$
65. $\{-2, 2\}$ **67.** $\{-3, 3\}$ **69.** $\{0\}$ **71.** $-3, 0$, or 3
73. 4 centimeters and 8 centimeters **75.** 10 meters long and 5 meters wide **77.** 6 inches **79.** 8 yards

Problem Set 3.6 (page 150)

1. $(x + 5)(x + 4)$ **3.** $(x - 4)(x - 7)$ **5.** $(a + 9)(a - 4)$
7. $(y + 6)(y + 14)$ **9.** $(x - 7)(x + 2)$ **11.** Not factorable **13.** $(6 - x)(1 + x)$ **15.** $(x + 3y)(x + 12y)$
17. $(a - 8b)(a + 7b)$ **19.** $(3x + 1)(5x + 6)$
21. $(4x - 3)(3x + 2)$ **23.** $(a + 3)(4a - 9)$
25. $(n - 4)(3n + 5)$ **27.** Not factorable
29. $(2n - 7)(5n + 3)$ **31.** $(4x - 5)(2x + 9)$
33. $(1 - 6x)(6 + x)$ **35.** $(5y + 9)(4y - 1)$
37. $(12n + 5)(2n - 1)$ **39.** $(5n + 3)(n + 6)$
41. $(x + 10)(x + 15)$ **43.** $(n - 16)(n - 20)$
45. $(t + 15)(t - 12)$ **47.** $(t^2 - 3)(t^2 - 2)$
49. $(2x^2 - 1)(5x^2 + 4)$ **51.** $(x + 1)(x - 1)(x^2 - 8)$
53. $(3n + 1)(3n - 1)(2n^2 + 3)$
55. $(x + 1)(x - 1)(x + 4)(x - 4)$ **57.** $2(t + 2)(t - 2)$
59. $(4x + 5y)(3x - 2y)$ **61.** $3n(2n + 5)(3n - 1)$
63. $(n - 12)(n - 5)$ **65.** $(6a - 1)^2$ **67.** $6(x^2 + 9)$
69. Not factorable **71.** $(x + y - 7)(x - y + 7)$
73. $(1 + 4x^2)(1 + 2x)(1 - 2x)$ **75.** $(4n + 9)(n + 4)$
77. $n(n + 7)(n - 7)$ **79.** $(x - 8)(x + 1)$
81. $3x(x - 3)(x^2 + 3x + 9)$ **83.** $(x^2 + 3)^2$
85. $(x + 3)(x - 3)(x^2 + 4)$ **87.** $(2w - 7)(3w + 5)$
89. Not factorable **91.** $2n(n^2 + 7n - 10)$
93. $(2x + 1)(y + 3)$ **99.** $(x^a + 3)(x^a + 7)$
101. $(2x^a + 5)^2$ **103.** $(5x^n - 1)(4x^n + 5)$
105. $(x - 4)(x - 2)$ **107.** $(3x - 11)(3x + 2)$
109. $(3x + 4)(5x + 9)$

Problem Set 3.7 (page 156)

1. $\{-3, -1\}$ **3.** $\{-12, -6\}$ **5.** $\{4, 9\}$ **7.** $\{-6, 2\}$
9. $\{-1, 5\}$ **11.** $\{-13, -12\}$ **13.** $\left\{-5, \dfrac{1}{3}\right\}$

15. $\left\{-\dfrac{7}{2}, -\dfrac{2}{3}\right\}$ **17.** $\{0, 4\}$ **19.** $\left\{\dfrac{1}{6}, 2\right\}$ **21.** $\{-6, 0, 6\}$

23. $\{-4, 6\}$ **25.** $\{-4, 4\}$ **27.** $\{-11, 4\}$ **29.** $\{-5, 5\}$

31. $\left\{-\dfrac{5}{3}, -\dfrac{3}{5}\right\}$ **33.** $\left\{-\dfrac{1}{8}, 6\right\}$ **35.** $\left\{\dfrac{3}{7}, \dfrac{5}{4}\right\}$ **37.** $\left\{-\dfrac{2}{7}, \dfrac{4}{5}\right\}$

39. $\left\{-7, \dfrac{2}{3}\right\}$ **41.** $\{-20, 18\}$ **43.** $\left\{-2, -\dfrac{1}{3}, \dfrac{1}{3}, 2\right\}$

45. $\left\{-\dfrac{2}{3}, 16\right\}$ **47.** $\left\{-\dfrac{3}{2}, 1\right\}$ **49.** $\left\{-\dfrac{5}{2}, -\dfrac{4}{3}, 0\right\}$

51. $\left\{-1, \dfrac{5}{3}\right\}$ **53.** $\left\{-\dfrac{3}{2}, \dfrac{1}{2}\right\}$ **55.** 8 and 9 or -9 and -8

57. 7 and 15   **59.** 10 inches by 6 inches
61. -7 and -6 or 6 and 7   **63.** 4 centimeters by
4 centimeters and 6 centimeters by 8 centimeters
65. 3, 4, and 5 units   **67.** 9 inches and 12 inches
69. An altitude of 4 inches and a side 14 inches long
77. (a) 0.28 and 3.73   **(c)** 2.27 and 5.76   **(e)** 0.71

Chapter 3 Review Problem Set (page 160)

1. $5x - 3$ **2.** $3x^2 + 12x - 2$ **3.** $12x^2 - x + 5$
4. $-20x^5y^7$ **5.** $-6a^5b^5$ **6.** $15a^4 - 10a^3 - 5a^2$
7. $24x^2 + 2xy - 15y^2$ **8.** $3x^3 + 7x^2 - 21x - 4$
9. $256x^8y^{12}$ **10.** $9x^2 - 12xy + 4y^2$ **11.** $-8x^6y^9z^3$
12. $-13x^2y$ **13.** $2x + y - 2$
14. $x^4 + x^3 - 18x^2 - x + 35$ **15.** $21 + 26x - 15x^2$
16. $-12a^5b^7$ **17.** $-8a^7b^3$ **18.** $7x^2 + 19x - 36$
19. $6x^3 - 11x^2 - 7x + 2$ **20.** $6x^{4n}$
21. $4x^2 + 20xy + 25y^2$ **22.** $x^3 - 6x^2 + 12x - 8$
23. $8x^3 + 60x^2 + 150x + 125$ **24.** $(x + 7)(x - 4)$
25. $2(t + 3)(t - 3)$ **26.** Not factorable
27. $(4n - 1)(3n - 1)$ **28.** $x^2(x^2 + 1)(x + 1)(x - 1)$
29. $x(x - 12)(x + 6)$ **30.** $2a^2b(3a + 2b - c)$
31. $(x - y + 1)(x + y - 1)$ **32.** $4(2x^2 + 3)$
33. $(4x + 7)(3x - 5)$ **34.** $(4n - 5)^2$ **35.** $4n(n - 2)$
36. $3w(w^2 + 6w - 8)$ **37.** $(5x + 2y)(4x - y)$
38. $16a(a - 4)$ **39.** $3x(x + 1)(x - 6)$
40. $(n + 8)(n - 16)$ **41.** $(t + 5)(t - 5)(t^2 + 3)$
42. $(5x - 3)(7x + 2)$ **43.** $(3 - x)(5 - 3x)$
44. $(4n - 3)(16n^2 + 12n + 9)$
45. $2(2x + 5)(4x^2 - 10x + 25)$ **46.** $\{-3, 3\}$
47. $\{-6, 1\}$ **48.** $\left\{\dfrac{2}{7}\right\}$ **49.** $\left\{-\dfrac{2}{5}, \dfrac{1}{3}\right\}$

50. $\left\{-\dfrac{1}{3}, 3\right\}$ **51.** $\{-3, 0, 3\}$ **52.** $\{-1, 0, 1\}$

53. $\{-7, 9\}$ **54.** $\left\{-\dfrac{4}{7}, \dfrac{2}{7}\right\}$ **55.** $\left\{-\dfrac{4}{5}, \dfrac{5}{6}\right\}$

56. $\{-2, 2\}$ **57.** $\left\{\dfrac{5}{3}\right\}$ **58.** $\{-8, 6\}$

59. $\left\{-5, \dfrac{2}{7}\right\}$ **60.** $\{-8, 5\}$ **61.** $\{-12, 1\}$ **62.** $\varnothing$

63. $\left\{-5, \dfrac{6}{5}\right\}$ **64.** $\{0, 1, 8\}$ **65.** $\left\{-10, \dfrac{1}{4}\right\}$

66. 8, 9, and 10 or -1, 0, and 1   **67.** -6 and 8
68. 13 and 15   **69.** 12 miles and 16 miles   **70.** 4 meters by
12 meters   **71.** 9 rows and 16 chairs per row
72. The side is 13 feet long and the altitude is 6 feet.
73. 3 feet   **74.** 5 centimeters by 5 centimeters and
8 centimeters by 8 centimeters   **75.** 6 inches

Chapter 3 Test (page 162)

1. $2x - 11$ **2.** $-48x^4y^4$ **3.** $-27x^6y^{12}$
4. $20x^2 + 17x - 63$ **5.** $6n^2 - 13n + 6$
6. $x^3 - 12x^2y + 48xy^2 - 64y^3$ **7.** $2x^3 + 11x^2 - 11x - 30$
8. $-14x^3y$ **9.** $(6x - 5)(x + 4)$ **10.** $3(2x + 1)(2x - 1)$
11. $(4 + t)(16 - 4t + t^2)$ **12.** $2x(3 - 2x)(5 + 4x)$
13. $(x - y)(x + 4)$ **14.** $(3n + 8)(8n - 3)$ **15.** $\{-12, 4\}$

16. $\left\{0, \dfrac{1}{4}\right\}$ **17.** $\left\{\dfrac{3}{2}\right\}$ **18.** $\{-4, -1\}$ **19.** $\{-9, 0, 2\}$

20. $\left\{-\dfrac{3}{7}, \dfrac{4}{5}\right\}$ **21.** $\left\{-\dfrac{1}{3}, 2\right\}$ **22.** $\{-2, 2\}$ **23.** 9 inches

24. 15 rows   **25.** 8 feet

Cumulative Review Problem Set (page 163)

1. 4   **2.** -19 **3.** 9   **4.** 21   **5.** -78 **6.** -33 **7.** -43
8. -11 **9.** -39 **10.** 57   **11.** $2x - 11$ **12.** $36a^2b^6$
13. $30x^2 - 37x - 7$ **14.** $-2x^2 - 11x - 12$ **15.** $-64a^6b^9$
16. $5x^3 - 6x^2 - 20x + 24$ **17.** $x^3 - 4x^2 - x + 12$
18. $2x^4 - x^3 - 2x^2 - 19x - 28$ **19.** $7(x + 1)(x - 1)$
20. $(2a - b)^2$ **21.** $(3x + 7)(x - 8)$
22. $(1 - x)(1 + x + x^2)$ **23.** $(y - 5)(x + 2)$
24. $3(x - 4)^2$ **25.** $(4n^2 + 3)(n + 1)(n - 1)$
26. $4x(2x + 3)(4x^2 - 6x + 9)$ **27.** $4(x^2 + 9)$
28. $(3x + 4)(2x - 1)$ **29.** $(3x - 5)^2$
30. $(x + 3y)(2x + 1)$ **31.** $(2a + 3b)(4a^2 - 6ab + 9b^2)$
32. $(x^2 + 4)(x + 2)(x - 2)$ **33.** $2m^2n^2(5m^2 - mn - 2n^2)$
34. $(2y + 7z)(5x - 12)$ **35.** $(3x + 5)(x - 2)$
36. $(5 - 2a)(5 + 2a)$ **37.** $(6x + 5)^2$

38. $(4y + 1)(16y^2 - 4y + 1)$ **39.** $x = \dfrac{2y + 6}{5}$

40. $y = \dfrac{12 - 3x}{4}$ **41.** $h = \dfrac{V - 2\pi r^2}{2\pi r}$ **42.** $R_1 = \dfrac{RR_2}{R_2 - R}$

43. 10.5%   **44.** $-15°$ **45.** $\{-6, 3\}$ **46.** $\left\{-\dfrac{7}{3}, \dfrac{2}{5}\right\}$

47. $\{-4\}$ **48.** $\{-9, 2\}$ **49.** $\{-1, 1\}$ **50.** $\{-10\}$

51. $\{15\}$ **52.** $\left\{-\dfrac{25}{4}\right\}$ **53.** $\left\{-\dfrac{5}{3}, 3\right\}$ **54.** $\{-1, 5\}$

55. $\{400\}$ **56.** $\{-4, 10\}$ **57.** $\{-4, 0, 4\}$

58. $\{-2, 3\}$ **59.** $\left\{-\dfrac{1}{4}, \dfrac{2}{3}\right\}$ **60.** $\{-6, 5\}$ **61.** $\{-5, 0, 2\}$

62. $\left\{\dfrac{17}{12}\right\}$ **63.** $(-22, \infty)$ **64.** $(23, \infty)$

65. $(-\infty, -3) \cup (4, \infty)$ **66.** $\left(-7, \dfrac{7}{3}\right)$ **67.** $(300, \infty)$

68. $\left(-\infty, \dfrac{7}{8}\right]$ **69.** $\left[\dfrac{5}{2}, \infty\right)$ **70.** $\left(-\infty, \dfrac{32}{31}\right)$

71. 7, 9, and 11 **72.** 8 nickels, 15 dimes, 25 quarters
73. 12 and 34 **74.** $62°$ and $118°$ **75.** \$400 at 8% and \$600
at 9% **76.** 35 pennies, 40 nickels, 70 dimes
77. 1 hour and 40 minutes **78.** 25 milliliters
79. 40% **80.** Better than 88 **81.** 4 inches
82. 7 meters by 14 meters **83.** 8 rows and 12 chairs
per row **84.** 9 feet, 12 feet, and 15 feet

CHAPTER 4

Problem Set 4.1 (page 170)

1. $\dfrac{3}{4}$ **3.** $\dfrac{5}{6}$ **5.** $-\dfrac{2}{5}$ **7.** $\dfrac{2}{7}$ **9.** $\dfrac{2x}{7}$ **11.** $\dfrac{2a}{5b}$ **13.** $-\dfrac{y}{4x}$

15. $-\dfrac{9c}{13d}$ **17.** $\dfrac{5x^2}{3y^3}$ **19.** $\dfrac{x-2}{x}$ **21.** $\dfrac{3x+2}{2x-1}$ **23.** $\dfrac{a+5}{a-9}$

25. $\dfrac{n-3}{5n-1}$ **27.** $\dfrac{5x^2+7}{10x}$ **29.** $\dfrac{3x+5}{4x+1}$

31. $\dfrac{3x}{x^2+4x+16}$ **33.** $\dfrac{x+6}{3x-1}$ **35.** $\dfrac{x(2x+7)}{y(x+9)}$

37. $\dfrac{y+4}{5y-2}$ **39.** $\dfrac{3x(x-1)}{x^2+1}$ **41.** $\dfrac{2(x+3y)}{3x(3x+y)}$

43. $\dfrac{3n-2}{7n+2}$ **45.** $\dfrac{4-x}{5+3x}$ **47.** $\dfrac{9x^2+3x+1}{2(x+2)}$

49. $\dfrac{-2(x-1)}{x+1}$ **51.** $\dfrac{y+b}{y+c}$ **53.** $\dfrac{x+2y}{2x+y}$ **55.** $\dfrac{x+1}{x-6}$

57. $\dfrac{2s+5}{3s+1}$ **59.** -1 **61.** $-n-7$ **63.** $-\dfrac{2}{x+1}$

65. -2 **67.** $-\dfrac{n+3}{n+5}$

Problem Set 4.2 (page 176)

1. $\dfrac{1}{10}$ **3.** $-\dfrac{4}{15}$ **5.** $\dfrac{3}{16}$ **7.** $-\dfrac{5}{6}$ **9.** $-\dfrac{2}{3}$

11. $\dfrac{10}{11}$ **13.** $-\dfrac{5x^3}{12y^2}$ **15.** $\dfrac{2a^3}{3b}$ **17.** $\dfrac{3x^3}{4}$

19. $\dfrac{25x^3}{108y^2}$ **21.** $\dfrac{ac^2}{2b^2}$ **23.** $\dfrac{3x}{4y}$ **25.** $\dfrac{3(x^2+4)}{5y(x+8)}$

27. $\dfrac{5(a+3)}{a(a-2)}$ **29.** $\dfrac{3}{2}$ **31.** $\dfrac{3xy}{4(x+6)}$

33. $\dfrac{5(x-2y)}{7y}$ **35.** $\dfrac{5+n}{3-n}$ **37.** $\dfrac{x^2+1}{x^2-10}$

39. $\dfrac{6x+5}{3x+4}$ **41.** $\dfrac{2t^2+5}{2(t^2+1)(t+1)}$ **43.** $\dfrac{t(t+6)}{4t+5}$

45. $\dfrac{n+3}{n(n-2)}$ **47.** $\dfrac{25x^3y^3}{4(x+1)}$ **49.** $\dfrac{2(a-2b)}{a(3a-2b)}$

Problem Set 4.3 (page 184)

1. $\dfrac{13}{12}$ **3.** $\dfrac{11}{40}$ **5.** $\dfrac{19}{20}$ **7.** $\dfrac{49}{75}$ **9.** $\dfrac{17}{30}$ **11.** $-\dfrac{11}{84}$

13. $\dfrac{2x+4}{x-1}$ **15.** 4 **17.** $\dfrac{7y-10}{7y}$ **19.** $\dfrac{5x+3}{6}$

21. $\dfrac{12a+1}{12}$ **23.** $\dfrac{n+14}{18}$ **25.** $-\dfrac{11}{15}$ **27.** $\dfrac{3x-25}{30}$

29. $\dfrac{43}{40x}$ **31.** $\dfrac{20y-77x}{28xy}$ **33.** $\dfrac{16y+15x-12xy}{12xy}$

35. $\dfrac{21+22x}{30x^2}$ **37.** $\dfrac{10n-21}{7n^2}$ **39.** $\dfrac{45-6n+20n^2}{15n^2}$

41. $\dfrac{11x-10}{6x^2}$ **43.** $\dfrac{42t+43}{35t^3}$ **45.** $\dfrac{20b^2-33a^3}{96a^2b}$

47. $\dfrac{14-24y^3+45xy}{18xy^3}$ **49.** $\dfrac{2x^2+3x-3}{x(x-1)}$

51. $\dfrac{a^2-a-8}{a(a+4)}$ **53.** $\dfrac{-41n-55}{(4n+5)(3n+5)}$

55. $\dfrac{-3x+17}{(x+4)(7x-1)}$ **57.** $\dfrac{-x+74}{(3x-5)(2x+7)}$

59. $\dfrac{38x+13}{(3x-2)(4x+5)}$ **61.** $\dfrac{5x+5}{2x+5}$ **63.** $\dfrac{x+15}{x-5}$

65. $\dfrac{-2x-4}{2x+1}$ **67. (a)** -1 **(c)** 0

Problem Set 4.4 (page 193)

1. $\dfrac{7x+20}{x(x+4)}$ **3.** $\dfrac{-x-3}{x(x+7)}$ **5.** $\dfrac{6x-5}{(x+1)(x-1)}$

7. $\dfrac{1}{a+1}$ **9.** $\dfrac{5n+15}{4(n+5)(n-5)}$ **11.** $\dfrac{x^2+60}{x(x+6)}$

13. $\dfrac{11x+13}{(x+2)(x+7)(2x+1)}$ **15.** $\dfrac{-3a+1}{(a-5)(a+2)(a+9)}$

17. $\dfrac{3a^2+14a+1}{(4a-3)(2a+1)(a+4)}$ **19.** $\dfrac{3x^2+20x-111}{(x^2+3)(x+7)(x-3)}$

21. $\dfrac{x+6}{(x-3)^2}$ **23.** $\dfrac{14x-4}{(x-1)(x+1)^2}$

25. $\dfrac{-7y-14}{(y+8)(y-2)}$ **27.** $\dfrac{-2x^2-4x+3}{(x+2)(x-2)}$

29. $\dfrac{2x^2+14x-19}{(x+10)(x-2)}$ **31.** $\dfrac{2n+1}{n-6}$

33. $\dfrac{2x^2 - 32x + 16}{(x + 1)(2x - 1)(3x - 2)}$ **35.** $\dfrac{1}{(n^2 + 1)(n + 1)}$

37. $\dfrac{-16x}{(5x - 2)(x - 1)}$ **39.** $\dfrac{t + 1}{t - 2}$ **41.** $\dfrac{2}{11}$

43. $-\dfrac{7}{27}$ **45.** $\dfrac{x}{4}$ **47.** $\dfrac{3y - 2x}{4x - 7}$ **49.** $\dfrac{6ab^2 - 5a^2}{12b^2 + 2a^2b}$

51. $\dfrac{2y - 3xy}{3x + 4xy}$ **53.** $\dfrac{3n + 14}{5n + 19}$ **55.** $\dfrac{5n - 17}{4n - 13}$

57. $\dfrac{-x + 5y - 10}{3y - 10}$ **59.** $\dfrac{-x + 15}{-2x - 1}$ **61.** $\dfrac{3a^2 - 2a + 1}{2a - 1}$

63. $\dfrac{-x^2 + 6x - 4}{3x - 2}$

Problem Set 4.5 (page 200)

1. $3x^3 + 6x^2$ **3.** $-6x^4 + 9x^6$ **5.** $3a^2 - 5a - 8$
7. $-13x^2 + 17x - 28$ **9.** $-3xy + 4x^2y - 8xy^2$

11. $x - 13$ **13.** $x + 20$ **15.** $2x + 1 - \dfrac{3}{x - 1}$

17. $5x - 1$ **19.** $3x^2 - 2x - 7$ **21.** $x^2 + 5x - 6$

23. $4x^2 + 7x + 12 + \dfrac{30}{x - 2}$ **25.** $x^3 - 4x^2 - 5x + 3$

27. $x^2 + 5x + 25$ **29.** $x^2 - x + 1 + \dfrac{63}{x + 1}$

31. $2x^2 - 4x + 7 - \dfrac{20}{x + 2}$ **33.** $4a - 4b$

35. $4x + 7 + \dfrac{23x - 6}{x^2 - 3x}$ **37.** $8y - 9 + \dfrac{8y + 5}{y^2 + y}$

39. $2x - 1$ **41.** $x - 3$ **43.** $5a - 8 + \dfrac{42a - 41}{a^2 + 3a - 4}$

45. $2n^2 + 3n - 4$ **47.** $x^4 + x^3 + x^2 + x + 1$

49. $x^3 - x^2 + x - 1$ **51.** $3x^2 + x + 1 + \dfrac{7}{x^2 - 1}$

53. $x - 6$ **55.** $x + 6, R = 14$ **57.** $x^2 - 1$
59. $x^2 - 2x - 3$ **61.** $2x^2 - x - 6, R = -6$
63. $x^3 + 7x^2 + 21x + 56, R = 167$

Problem Set 4.6 (page 208)

1. $\{2\}$ **3.** $\{-3\}$ **5.** $\{6\}$ **7.** $\left\{-\dfrac{85}{18}\right\}$ **9.** $\left\{\dfrac{7}{10}\right\}$

11. $\{5\}$ **13.** $\{58\}$ **15.** $\left\{\dfrac{1}{4}, 4\right\}$ **17.** $\left\{-\dfrac{2}{5}, 5\right\}$ **19.** $\{-16\}$

21. $\left\{-\dfrac{13}{3}\right\}$ **23.** $\{-3, 1\}$ **25.** $\left\{-\dfrac{5}{2}\right\}$ **27.** $\{-51\}$

29. $\left\{-\dfrac{5}{3}, 4\right\}$ **31.** $\varnothing$ **33.** $\left\{-\dfrac{11}{8}, 2\right\}$ **35.** $\{-29, 0\}$

37. $\{-9, 3\}$ **39.** $\left\{-2, \dfrac{23}{8}\right\}$ **41.** $\left\{\dfrac{11}{23}\right\}$ **43.** $\left\{3, \dfrac{7}{2}\right\}$

45. \$750 and \$1000   **47.** $48°$ and $72°$ **49.** $\dfrac{2}{7}$ or $\dfrac{7}{2}$

51. \$3500   **53.** \$69 for Tammy and \$51.75 for Laura
55. 8 and 82   **57.** 14 feet and 6 feet
59. 690 females and 460 males

Problem Set 4.7 (page 217)

1. $\{-21\}$ **3.** $\{-1, 2\}$ **5.** $\{2\}$ **7.** $\left\{\dfrac{37}{15}\right\}$ **9.** $\{-1\}$

11. $\{-1\}$ **13.** $\left\{0, \dfrac{13}{2}\right\}$ **15.** $\left\{-2, \dfrac{19}{2}\right\}$ **17.** $\{-2\}$

19. $\left\{-\dfrac{1}{5}\right\}$ **21.** $\varnothing$ **23.** $\left\{\dfrac{7}{2}\right\}$ **25.** $\{-3\}$ **27.** $\left\{-\dfrac{7}{9}\right\}$

29. $\left\{-\dfrac{7}{6}\right\}$ **31.** $x = \dfrac{18y - 4}{15}$ **33.** $y = \dfrac{-5x + 22}{2}$

35. $M = \dfrac{IC}{100}$ **37.** $R = \dfrac{ST}{S + T}$

39. $y = \dfrac{bx - x - 3b + a}{a - 3}$ **41.** $y = \dfrac{ab - bx}{a}$

43. $y = \dfrac{-2x - 9}{3}$

45. 50 miles per hour for Dave and 54 miles per hour for Kent
47. 60 minutes
49. 60 words per minute for Connie and 40 words per minute for Katie
51. Plane B could travel at 400 miles per hour for 5 hours and plane A at 350 miles per hour for 4 hours, or plane B could travel at 250 miles per hour for 8 hours and plane A at 200 miles per hour for 7 hours.
53. 60 minutes for Nancy and 120 minutes for Amy
55. 3 hours
57. 16 miles per hour on the way out and 12 miles per hour on the way back, or 12 miles per hour out and 8 miles per hour back

Chapter 4 Review Problem Set (page 221)

1. $\dfrac{2y}{3x^2}$ **2.** $\dfrac{a - 3}{a}$ **3.** $\dfrac{n - 5}{n - 1}$ **4.** $\dfrac{x^2 + 1}{x}$ **5.** $\dfrac{2x + 1}{3}$

6. $\dfrac{x^2 - 10}{2x^2 + 1}$ **7.** $\dfrac{3}{22}$ **8.** $\dfrac{18y + 20x}{48y - 9x}$ **9.** $\dfrac{3x + 2}{3x - 2}$

10. $\dfrac{x - 1}{2x - 1}$ **11.** $\dfrac{2x}{7y^2}$ **12.** $3b$ **13.** $\dfrac{n(n + 5)}{n - 1}$

14. $\dfrac{x(x - 3y)}{x^2 + 9y^2}$ **15.** $\dfrac{23x - 6}{20}$ **16.** $\dfrac{57 - 2n}{18n}$

840 Answers to Odd-Numbered Problems

17. $\dfrac{3x^2 - 2x - 14}{x(x + 7)}$ 18. $\dfrac{2}{x - 5}$

19. $\dfrac{5n - 21}{(n - 9)(n + 4)(n - 1)}$ 20. $\dfrac{6y - 23}{(2y + 3)(y - 6)}$

21. $6x - 1$ 22. $3x^2 - 7x + 22 - \dfrac{90}{x + 4}$ 23. $\left\{\dfrac{4}{13}\right\}$

24. $\left\{\dfrac{3}{16}\right\}$ 25. $\varnothing$ 26. $\{-17\}$ 27. $\left\{\dfrac{2}{7}, \dfrac{7}{2}\right\}$ 28. $\{22\}$

29. $\left\{-\dfrac{6}{7}, 3\right\}$ 30. $\left\{\dfrac{3}{4}, \dfrac{5}{2}\right\}$ 31. $\left\{\dfrac{9}{7}\right\}$ 32. $\left\{-\dfrac{5}{4}\right\}$

33. $y = \dfrac{3x + 27}{4}$ 34. $y = \dfrac{bx - ab}{a}$ 35. \$525 and \$875

36. 20 minutes for Julio and 30 minutes for Dan
37. 50 miles per hour and 55 miles per hour or 8⅓ miles per hour and 13⅓ miles per hour
38. 9 hours 39. 80 hours 40. 13 miles per hour

Chapter 4 Test (page 223)

1. $\dfrac{13y^2}{24x}$ 2. $\dfrac{3x - 1}{x(x - 6)}$ 3. $\dfrac{2n - 3}{n + 4}$ 4. $-\dfrac{2x}{x + 1}$ 5. $\dfrac{3y^2}{8}$

6. $\dfrac{a - b}{4(2a + b)}$ 7. $\dfrac{x + 4}{5x - 1}$ 8. $\dfrac{13x + 7}{12}$ 9. $\dfrac{3x}{2}$

10. $\dfrac{10n - 26}{15n}$ 11. $\dfrac{3x^2 + 2x - 12}{x(x - 6)}$ 12. $\dfrac{11 - 2x}{x(x - 1)}$

13. $\dfrac{13n + 46}{(2n + 5)(n - 2)(n + 7)}$ 14. $3x^2 - 2x - 1$

15. $\dfrac{18 - 2x}{8 + 9x}$ 16. $y = \dfrac{4x + 20}{3}$ 17. $\{1\}$ 18. $\left\{\dfrac{1}{10}\right\}$

19. $\{-35\}$ 20. $\{-1, 5\}$ 21. $\left\{\dfrac{5}{3}\right\}$ 22. $\left\{-\dfrac{9}{13}\right\}$ 23. $\dfrac{27}{72}$

24. 1 hour 25. 15 miles per hour

CHAPTER 5

Problem Set 5.1 (page 231)

1. $\dfrac{1}{27}$ 3. $-\dfrac{1}{100}$ 5. 81 7. -27 9. -8 11. 1 13. $\dfrac{9}{49}$

15. 16 17. $\dfrac{1}{1000}$ 19. $\dfrac{1}{1000}$ 21. 27 23. $\dfrac{1}{125}$ 25. $\dfrac{9}{8}$

27. $\dfrac{256}{25}$ 29. $\dfrac{2}{25}$ 31. $\dfrac{81}{4}$ 33. 81 35. $\dfrac{1}{10,000}$ 37. $\dfrac{13}{36}$

39. $\dfrac{1}{2}$ 41. $\dfrac{72}{17}$ 43. $\dfrac{1}{x^6}$ 45. $\dfrac{1}{a^3}$ 47. $\dfrac{1}{a^8}$ 49. $\dfrac{y^6}{x^2}$

51. $\dfrac{c^8}{a^4 b^{12}}$ 53. $\dfrac{y^{12}}{8x^9}$ 55. $\dfrac{x^3}{y^{12}}$ 57. $\dfrac{4a^4}{9b^2}$ 59. $\dfrac{1}{x^2}$ 61. $a^5 b^2$

63. $\dfrac{6y^3}{x}$ 65. $7b^2$ 67. $\dfrac{7x}{y^2}$ 69. $-\dfrac{12b^3}{a}$ 71. $\dfrac{x^5 y^5}{5}$

73. $\dfrac{b^{20}}{81}$ 75. $\dfrac{x + 1}{x^3}$ 77. $\dfrac{y - x^3}{x^3 y}$ 79. $\dfrac{3b + 4a^2}{a^2 b}$

81. $\dfrac{1 - x^2 y}{xy^2.}$ 83. $\dfrac{2x - 3}{x^2}$

Problem Set 5.2 (page 242)

1. 8 3. -10 5. 3 7. -4 9. 3 11. $\dfrac{4}{5}$ 13. $-\dfrac{6}{7}$

15. $\dfrac{1}{2}$ 17. $\dfrac{3}{4}$ 19. 8 21. $3\sqrt{3}$ 23. $4\sqrt{2}$ 25. $4\sqrt{5}$

27. $4\sqrt{10}$ 29. $12\sqrt{2}$ 31. $-12\sqrt{5}$ 33. $2\sqrt{3}$

35. $3\sqrt{6}$ 37. $-\dfrac{5}{3}\sqrt{7}$ 39. $\dfrac{\sqrt{19}}{2}$ 41. $\dfrac{3\sqrt{3}}{4}$ 43. $\dfrac{5\sqrt{3}}{9}$

45. $\dfrac{\sqrt{14}}{7}$ 47. $\dfrac{\sqrt{6}}{3}$ 49. $\dfrac{\sqrt{15}}{6}$ 51. $\dfrac{\sqrt{66}}{12}$ 53. $\dfrac{\sqrt{6}}{3}$

55. $\sqrt{5}$ 57. $\dfrac{2\sqrt{21}}{7}$ 59. $-\dfrac{8\sqrt{15}}{5}$ 61. $\dfrac{\sqrt{6}}{4}$ 63. $-\dfrac{12}{25}$

65. $2\sqrt[3]{2}$ 67. $6\sqrt[3]{3}$ 69. $\dfrac{2\sqrt[3]{3}}{3}$ 71. $\dfrac{3\sqrt[3]{2}}{2}$ 73. $\dfrac{\sqrt[3]{12}}{2}$

75. 42 miles per hour; 49 miles per hour; 65 miles per hour
77. 107 square centimeters 79. 140 square inches
85. (a) 1.414 (c) 12.490 (e) 57.000 (g) 0.374
(i) 0.930

Problem Set 5.3 (page 248)

1. $13\sqrt{2}$ 3. $54\sqrt{3}$ 5. $-30\sqrt{2}$ 7. $-\sqrt{5}$ 9. $-21\sqrt{6}$

11. $-\dfrac{7\sqrt{7}}{12}$ 13. $\dfrac{37\sqrt{10}}{10}$ 15. $\dfrac{41\sqrt{2}}{20}$ 17. $-9\sqrt[3]{3}$

19. $10\sqrt[3]{2}$ 21. $4\sqrt{2x}$ 23. $5x\sqrt{3}$ 25. $2x\sqrt{5y}$

27. $8xy^3\sqrt{xy}$ 29. $3a^2 b\sqrt{6b}$ 31. $3x^3 y^4 \sqrt{7}$

33. $4a\sqrt{10a}$ 35. $\dfrac{8y}{3}\sqrt{6xy}$ 37. $\dfrac{\sqrt{10xy}}{5y}$ 39. $\dfrac{\sqrt{15}}{6x^2}$

41. $\dfrac{5\sqrt{2y}}{6y}$ 43. $\dfrac{\sqrt{14xy}}{4y^3}$ 45. $\dfrac{3y\sqrt{2xy}}{4x}$ 47. $\dfrac{2\sqrt{42ab}}{7b^2}$

49. $2\sqrt[3]{3y}$ 51. $2x\sqrt[3]{2x}$ 53. $2x^2 y^2\sqrt[3]{7y^2}$ 55. $\dfrac{\sqrt[3]{21x}}{3x}$

57. $\dfrac{\sqrt[3]{12x^2 y}}{4x^2}$ 59. $\dfrac{\sqrt[3]{4x^2 y^2}}{xy^2}$ 61. $2\sqrt{2x + 3y}$

63. $4\sqrt{x + 3y}$ 65. $33\sqrt{x}$ 67. $-30\sqrt{2x}$ 69. $7\sqrt{3n}$

71. $-40\sqrt{ab}$ 73. $-7x\sqrt{2x}$ 79. (a) $5|x|\sqrt{5}$ (b) $4x^2$
(c) $2b\sqrt{2b}$ (d) $y^2\sqrt{3y}$ (e) $12|x^3|\sqrt{2}$ (f) $2m^4\sqrt{7}$

(g) $8|c^5|\sqrt{2}$ **(h)** $3d^3\sqrt{2d}$ **(i)** $7|x|$ **(j)** $4n^{10}\sqrt{5}$
(k) $9h\sqrt{h}$

Problem Set 5.4 (page 254)

1. $6\sqrt{2}$ **3.** $18\sqrt{2}$ **5.** $-24\sqrt{10}$ **7.** $24\sqrt{6}$ **9.** 120
11. 24 **13.** $56\sqrt[3]{3}$ **15.** $\sqrt{6}+\sqrt{10}$
17. $6\sqrt{10}-3\sqrt{35}$ **19.** $24\sqrt{3}-60\sqrt{2}$
21. $-40-32\sqrt{15}$ **23.** $15\sqrt{2x}+3\sqrt{xy}$
25. $5xy-6x\sqrt{y}$ **27.** $2\sqrt{10xy}+2y\sqrt{15y}$
29. $-25\sqrt{6}$ **31.** $-25-3\sqrt{3}$ **33.** $23-9\sqrt{5}$
35. $6\sqrt{35}+3\sqrt{10}-4\sqrt{21}-2\sqrt{6}$
37. $8\sqrt{3}-36\sqrt{2}+6\sqrt{10}-18\sqrt{15}$
39. $11+13\sqrt{30}$ **41.** $141-51\sqrt{6}$ **43.** -10 **45.** -8
47. $2x-3y$ **49.** $10\sqrt[3]{12}+2\sqrt[3]{18}$ **51.** $12-36\sqrt[3]{2}$
53. $\dfrac{\sqrt{7}-1}{3}$ **55.** $\dfrac{-3\sqrt{2}-15}{23}$ **57.** $\dfrac{\sqrt{7}-\sqrt{2}}{5}$
59. $\dfrac{2\sqrt{5}+\sqrt{6}}{7}$ **61.** $\dfrac{\sqrt{15}-2\sqrt{3}}{2}$ **63.** $\dfrac{6\sqrt{7}+4\sqrt{6}}{13}$
65. $\sqrt{3}-\sqrt{2}$ **67.** $\dfrac{2\sqrt{x}-8}{x-16}$ **69.** $\dfrac{x+5\sqrt{x}}{x-25}$
71. $\dfrac{x-8\sqrt{x}+12}{x-36}$ **73.** $\dfrac{x-2\sqrt{xy}}{x-4y}$ **75.** $\dfrac{6\sqrt{xy}+9y}{4x-9y}$

Problem Set 5.5 (page 260)

1. $\{20\}$ **3.** $\varnothing$ **5.** $\left\{\dfrac{25}{4}\right\}$ **7.** $\left\{\dfrac{4}{9}\right\}$ **9.** $\{5\}$ **11.** $\left\{\dfrac{39}{4}\right\}$
13. $\left\{\dfrac{10}{3}\right\}$ **15.** $\{-1\}$ **17.** $\varnothing$ **19.** $\{1\}$ **21.** $\left\{\dfrac{3}{2}\right\}$ **23.** $\{3\}$
25. $\left\{\dfrac{61}{25}\right\}$ **27.** $\{-3,3\}$ **29.** $\{-9,-4\}$ **31.** $\{0\}$ **33.** $\{3\}$
35. $\{4\}$ **37.** $\{-4,-3\}$ **39.** $\{12\}$ **41.** $\{25\}$ **43.** $\{29\}$
45. $\{-15\}$ **47.** $\left\{-\dfrac{1}{3}\right\}$ **49.** $\{-3\}$ **51.** $\{0\}$ **53.** $\{5\}$
55. $\{2,6\}$ **57.** 56 feet; 106 feet; 148 feet
59. 3.2 feet; 5.1 feet; 7.3 feet

Problem Set 5.6 (page 266)

1. 9 **3.** 3 **5.** -2 **7.** -5 **9.** $\dfrac{1}{6}$ **11.** 3 **13.** 8 **15.** 81

17. -1 **19.** -32 **21.** $\dfrac{81}{16}$ **23.** 4 **25.** $\dfrac{1}{128}$ **27.** -125

29. 625 **31.** $\sqrt[3]{x^4}$ **33.** $3\sqrt{x}$ **35.** $\sqrt[3]{2y}$ **37.** $\sqrt{2x-3y}$
39. $\sqrt[3]{(2a-3b)^2}$ **41.** $\sqrt[3]{x^2 y}$ **43.** $-3\sqrt[5]{xy^2}$ **45.** $5^{\frac{1}{2}}y^{\frac{1}{2}}$
47. $3y^{\frac{1}{2}}$ **49.** $x^{\frac{1}{3}}y^{\frac{2}{3}}$ **51.** $a^{\frac{1}{2}}b^{\frac{3}{4}}$ **53.** $(2x-y)^{\frac{3}{5}}$ **55.** $5xy^{\frac{1}{2}}$

57. $-(x+y)^{\frac{1}{3}}$ **59.** $12x^{\frac{13}{20}}$ **61.** $y^{\frac{5}{12}}$ **63.** $\dfrac{4}{x^{\frac{1}{10}}}$ **65.** $16xy^2$

67. $2x^2 y$ **69.** $4x^{\frac{4}{15}}$ **71.** $\dfrac{4}{b^{\frac{5}{12}}}$ **73.** $\dfrac{36x^{\frac{4}{5}}}{49y^{\frac{4}{3}}}$ **75.** $\dfrac{y^{\frac{3}{2}}}{x}$ **77.** $4x^{\frac{1}{6}}$

79. $\dfrac{16}{a^{\frac{11}{10}}}$ **81.** $\sqrt[6]{243}$ **83.** $\sqrt[4]{216}$ **85.** $\sqrt[12]{3}$ **87.** $\sqrt{2}$

89. $\sqrt[4]{3}$ **93. (a)** 12 **(c)** 7 **(e)** 11 **95. (a)** 1024
(c) 512 **(e)** 49

Problem Set 5.7 (page 271)

1. $(8.9)(10)^1$ **3.** $(4.29)(10)^3$ **5.** $(6.12)(10)^6$ **7.** $(4)(10)^7$
9. $(3.764)(10)^2$ **11.** $(3.47)(10)^{-1}$ **13.** $(2.14)(10)^{-2}$
15. $(5)(10)^{-5}$ **17.** $(1.94)(10)^{-9}$ **19.** 23 **21.** 4190
23. $500,000,000$ **25.** $31,400,000,000$ **27.** 0.43
29. 0.000914 **31.** 0.00000005123 **33.** 0.000000074
35. 0.77 **37.** $300,000,000,000$ **39.** 0.000000004
41. 1000 **43.** 1000 **45.** 3000 **47.** 20 **49.** $27,000,000$
51. $(6.02)(10^{23})$ **53.** 831 **55.** $(2.07)(10^4)$ dollars
57. $(1.99)(10^{-26})$ kg **59.** 1833 **63. (a)** 7000 **(c)** 120
(e) 30 **65. (a)** $(4.385)(10)^{14}$ **(c)** $(2.322)(10)^{17}$
(e) $(3.052)(10)^{12}$

Chapter 5 Review Problem Set (page 275)

1. $\dfrac{1}{64}$ **2.** $\dfrac{9}{4}$ **3.** 3 **4.** -2 **5.** $\dfrac{2}{3}$ **6.** 32 **7.** 1 **8.** $\dfrac{4}{9}$
9. -64 **10.** 32 **11.** 1 **12.** 27 **13.** $3\sqrt{6}$
14. $4x\sqrt{3xy}$ **15.** $2\sqrt{2}$ **16.** $\dfrac{\sqrt{15x}}{6x^2}$ **17.** $2\sqrt[3]{7}$
18. $\dfrac{\sqrt[3]{6}}{3}$ **19.** $\dfrac{3\sqrt{5}}{5}$ **20.** $\dfrac{x\sqrt{21x}}{7}$ **21.** $3xy^2\sqrt[3]{4xy^2}$
22. $\dfrac{15\sqrt{6}}{4}$ **23.** $2y\sqrt{5xy}$ **24.** $2\sqrt{x}$ **25.** $24\sqrt{10}$
26. 60 **27.** $24\sqrt{3}-6\sqrt{14}$ **28.** $x-2\sqrt{x}-15$
29. 17 **30.** $12-8\sqrt{3}$ **31.** $6a-5\sqrt{ab}-4b$ **32.** 70
33. $\dfrac{2(\sqrt{7}+1)}{3}$ **34.** $\dfrac{2\sqrt{6}-\sqrt{15}}{3}$ **35.** $\dfrac{3\sqrt{5}-2\sqrt{3}}{11}$
36. $\dfrac{6\sqrt{3}+3\sqrt{5}}{7}$ **37.** $\dfrac{x^6}{y^8}$ **38.** $\dfrac{27a^3b^{12}}{8}$ **39.** $20x^{\frac{7}{10}}$
40. $7a^{\frac{5}{12}}$ **41.** $\dfrac{y^{\frac{4}{3}}}{x}$ **42.** $\dfrac{x^{12}}{9}$ **43.** $\sqrt{5}$ **44.** $5\sqrt[3]{3}$
45. $\dfrac{29\sqrt{6}}{5}$ **46.** $-15\sqrt{3x}$ **47.** $\dfrac{y+x^2}{x^2 y}$ **48.** $\dfrac{b-2a}{a^2 b}$
49. $\left\{\dfrac{19}{7}\right\}$ **50.** $\{4\}$ **51.** $\{8\}$ **52.** $\varnothing$ **53.** $\{14\}$
54. $\{-10,1\}$ **55.** $\{2\}$ **56.** $\{8\}$ **57.** 0.000000006

58. 36,000,000,000 **59.** 6 **60.** 0.15 **61.** 0.000028
62. 0.002 **63.** 0.002 **64.** 8,000,000,000

Chapter 5 Test (page 277)

1. $\dfrac{1}{32}$ **2.** -32 **3.** $\dfrac{81}{16}$ **4.** $\dfrac{1}{4}$ **5.** $3\sqrt{7}$ **6.** $3\sqrt[3]{4}$

7. $2x^2y\sqrt{13y}$ **8.** $\dfrac{5\sqrt{6}}{6}$ **9.** $\dfrac{\sqrt{42x}}{12x^2}$ **10.** $72\sqrt{2}$

11. $-5\sqrt{6}$ **12.** $-38\sqrt{2}$ **13.** $\dfrac{3\sqrt{6}+3}{10}$ **14.** $\dfrac{9x^2y^2}{4}$

15. $-\dfrac{12}{a^{\frac{3}{10}}}$ **16.** $\dfrac{y^3+x}{xy^3}$ **17.** $-12x^{\frac{1}{4}}$ **18.** 33 **19.** 600

20. 0.003 **21.** $\left\{\dfrac{8}{3}\right\}$ **22.** $\{2\}$ **23.** $\{4\}$ **24.** $\{5\}$

25. $\{4, 6\}$

CHAPTER 6

Problem Set 6.1 (page 285)

1. False **3.** True **5.** True **7.** True **9.** $10 + 8i$
11. $-6 + 10i$ **13.** $-2 - 5i$ **15.** $-12 + 5i$ **17.** $-1 - 23i$

19. $-4 - 5i$ **21.** $1 + 3i$ **23.** $\dfrac{5}{3} - \dfrac{5}{12}i$ **25.** $-\dfrac{17}{9} + \dfrac{23}{30}i$

27. $9i$ **29.** $i\sqrt{14}$ **31.** $\dfrac{4}{5}i$ **33.** $3i\sqrt{2}$ **35.** $5i\sqrt{3}$

37. $6i\sqrt{7}$ **39.** $-8i\sqrt{5}$ **41.** $36i\sqrt{10}$ **43.** -8
45. $-\sqrt{15}$ **47.** $-3\sqrt{6}$ **49.** $-5\sqrt{3}$ **51.** $-3\sqrt{6}$

53. $4i\sqrt{3}$ **55.** $\dfrac{5}{2}$ **57.** $2\sqrt{2}$ **59.** $2i$ **61.** $-20 + 0i$

63. $42 + 0i$ **65.** $15 + 6i$ **67.** $-42 + 12i$ **69.** $7 + 22i$
71. $40 - 20i$ **73.** $-3 - 28i$ **75.** $-3 - 15i$
77. $-9 + 40i$ **79.** $-12 + 16i$ **81.** $85 + 0i$

83. $5 + 0i$ **85.** $\dfrac{3}{5} + \dfrac{3}{10}i$ **87.** $\dfrac{5}{17} - \dfrac{3}{17}i$ **89.** $2 + \dfrac{2}{3}i$

91. $0 - \dfrac{2}{7}i$ **93.** $\dfrac{22}{25} - \dfrac{4}{25}i$ **95.** $-\dfrac{18}{41} + \dfrac{39}{41}i$ **97.** $\dfrac{9}{2} - \dfrac{5}{2}i$

99. $\dfrac{4}{13} - \dfrac{1}{26}i$ **101.** (a) $-2 - i\sqrt{3}$ (c) $\dfrac{-1 - 3i\sqrt{2}}{2}$

(e) $\dfrac{10 + 3i\sqrt{5}}{4}$

Problem Set 6.2 (page 293)

1. $\{0, 9\}$ **3.** $\{-3, 0\}$ **5.** $\{-4, 0\}$ **7.** $\left\{0, \dfrac{9}{5}\right\}$ **9.** $\{-6, 5\}$

11. $\{7, 12\}$ **13.** $\left\{-8, -\dfrac{3}{2}\right\}$ **15.** $\left\{-\dfrac{7}{3}, \dfrac{2}{5}\right\}$ **17.** $\left\{\dfrac{3}{5}\right\}$

19. $\left\{-\dfrac{3}{2}, \dfrac{7}{3}\right\}$ **21.** $\{1, 4\}$ **23.** $\{8\}$ **25.** $\{12\}$ **27.** $\{0, 5k\}$

29. $\{0, 16k^2\}$ **31.** $\{5k, 7k\}$ **33.** $\left\{\dfrac{k}{2}, -3k\right\}$ **35.** $\{\pm 1\}$

37. $\{\pm 6i\}$ **39.** $\{\pm\sqrt{14}\}$ **41.** $\{\pm 2\sqrt{7}\}$ **43.** $\{\pm 3\sqrt{2}\}$

45. $\left\{\pm\dfrac{\sqrt{14}}{2}\right\}$ **47.** $\left\{\pm\dfrac{2\sqrt{3}}{3}\right\}$ **49.** $\left\{\pm\dfrac{2i\sqrt{30}}{5}\right\}$

51. $\left\{\pm\dfrac{\sqrt{6}}{2}\right\}$ **53.** $\{-1, 5\}$ **55.** $\{-8, 2\}$ **57.** $\{-6 \pm 2i\}$

59. $\{1, 2\}$ **61.** $\{4 \pm \sqrt{5}\}$ **63.** $\{-5 \pm 2\sqrt{3}\}$

65. $\left\{\dfrac{2 \pm 3i\sqrt{3}}{3}\right\}$ **67.** $\{-12, -2\}$ **69.** $\left\{\dfrac{2 \pm \sqrt{10}}{5}\right\}$

71. $2\sqrt{13}$ centimeters **73.** $4\sqrt{5}$ inches **75.** 8 yards
77. $6\sqrt{2}$ inches **79.** $a = b = 4\sqrt{2}$ meters
81. $b = 3\sqrt{3}$ inches and $c = 6$ inches
83. $a = 7$ centimeters and $b = 7\sqrt{3}$ centimeters

85. $a = \dfrac{10\sqrt{3}}{3}$ feet and $c = \dfrac{20\sqrt{3}}{3}$ feet **87.** 17.9 feet

89. 38 meters **91.** 53 meters **95.** 10.8 centimeters
97. $h = s\sqrt{2}$

Problem Set 6.3 (page 299)

1. $\{-6, 10\}$ **3.** $\{4, 10\}$ **5.** $\{-5, 10\}$ **7.** $\{-8, 1\}$

9. $\left\{-\dfrac{5}{2}, 3\right\}$ **11.** $\left\{-3, \dfrac{2}{3}\right\}$ **13.** $\{-16, 10\}$

15. $\{-2 \pm \sqrt{6}\}$ **17.** $\{-3 \pm 2\sqrt{3}\}$ **19.** $\{5 \pm \sqrt{26}\}$
21. $\{4 \pm i\}$ **23.** $\{-6 \pm 3\sqrt{3}\}$ **25.** $\{-1 \pm i\sqrt{5}\}$

27. $\left\{\dfrac{-3 \pm \sqrt{17}}{2}\right\}$ **29.** $\left\{\dfrac{-5 \pm \sqrt{21}}{2}\right\}$ **31.** $\left\{\dfrac{7 \pm \sqrt{37}}{2}\right\}$

33. $\left\{\dfrac{-2 \pm \sqrt{10}}{2}\right\}$ **35.** $\left\{\dfrac{3 \pm i\sqrt{6}}{3}\right\}$ **37.** $\left\{\dfrac{-5 \pm \sqrt{37}}{6}\right\}$

39. $\{-12, 4\}$ **41.** $\left\{\dfrac{4 \pm \sqrt{10}}{2}\right\}$ **43.** $\left\{-\dfrac{9}{2}, \dfrac{1}{3}\right\}$

45. $\{-3, 8\}$ **47.** $\{3 \pm 2\sqrt{3}\}$ **49.** $\left\{\dfrac{3 \pm i\sqrt{3}}{3}\right\}$

51. $\{-20, 12\}$ **53.** $\left\{-1, -\dfrac{2}{3}\right\}$ **55.** $\left\{\dfrac{1}{2}, \dfrac{3}{2}\right\}$

57. $\{-6 \pm 2\sqrt{10}\}$ **59.** $\left\{\dfrac{-1 \pm \sqrt{3}}{2}\right\}$

61. $\left\{\dfrac{-b \pm \sqrt{b^2 - 4ac}}{2a}\right\}$ **65.** $x = \dfrac{a\sqrt{b^2 - y^2}}{b}$

67. $r = \dfrac{\sqrt{A\pi}}{\pi}$ **69.** $\{2a, 3a\}$ **71.** $\left\{\dfrac{a}{2}, -\dfrac{2a}{3}\right\}$

73. $\left\{\dfrac{2b}{3}\right\}$

Problem Set 6.4 (page 307)

1. Two real solutions; $\{-7, 3\}$ **3.** One real solution; $\left\{\dfrac{1}{3}\right\}$

5. Two complex solutions; $\left\{\dfrac{7 \pm i\sqrt{3}}{2}\right\}$

7. Two real solutions; $\left\{-\dfrac{4}{3}, \dfrac{1}{5}\right\}$

9. Two real solutions; $\left\{\dfrac{-2 \pm \sqrt{10}}{3}\right\}$ **11.** $\{-1 \pm \sqrt{2}\}$

13. $\left\{\dfrac{-5 \pm \sqrt{37}}{2}\right\}$ **15.** $\{4 \pm 2\sqrt{5}\}$ **17.** $\left\{\dfrac{-5 \pm i\sqrt{7}}{2}\right\}$

19. $\{8, 10\}$ **21.** $\left\{\dfrac{9 \pm \sqrt{61}}{2}\right\}$ **23.** $\left\{\dfrac{-1 \pm \sqrt{33}}{4}\right\}$

25. $\left\{\dfrac{-1 \pm i\sqrt{3}}{4}\right\}$ **27.** $\left\{\dfrac{4 \pm \sqrt{10}}{3}\right\}$ **29.** $\left\{-1, \dfrac{5}{2}\right\}$

31. $\left\{-5, -\dfrac{4}{3}\right\}$ **33.** $\left\{\dfrac{5}{6}\right\}$ **35.** $\left\{\dfrac{1 \pm \sqrt{13}}{4}\right\}$

37. $\left\{0, \dfrac{13}{5}\right\}$ **39.** $\left\{\pm\dfrac{\sqrt{15}}{3}\right\}$ **41.** $\left\{\dfrac{-1 \pm \sqrt{73}}{12}\right\}$

43. $\{-18, -14\}$ **45.** $\left\{\dfrac{11}{4}, \dfrac{10}{3}\right\}$ **47.** $\left\{\dfrac{2 \pm i\sqrt{2}}{2}\right\}$

49. $\left\{\dfrac{1 \pm \sqrt{7}}{6}\right\}$ **55.** $\{-1.381, 17.381\}$

57. $\{-13.426, 3.426\}$ **59.** $\{-0.347, -8.653\}$
61. $\{0.119, 1.681\}$ **63.** $\{-0.708, 4.708\}$
65. $k = 4$ or $k = -4$

Problem Set 6.5 (page 317)

1. $\{2 \pm \sqrt{10}\}$ **3.** $\left\{-9, \dfrac{4}{3}\right\}$ **5.** $\{9 \pm 3\sqrt{10}\}$

7. $\left\{\dfrac{3 \pm i\sqrt{23}}{4}\right\}$ **9.** $\{-15, -9\}$ **11.** $\{-8, 1\}$

13. $\left\{\dfrac{2 \pm i\sqrt{10}}{2}\right\}$ **15.** $\{9 \pm \sqrt{66}\}$ **17.** $\left\{-\dfrac{5}{4}, \dfrac{2}{5}\right\}$

19. $\left\{\dfrac{-1 \pm \sqrt{2}}{2}\right\}$ **21.** $\left\{\dfrac{3}{4}, 4\right\}$ **23.** $\left\{\dfrac{11 \pm \sqrt{109}}{2}\right\}$

25. $\left\{\dfrac{3}{7}, 4\right\}$ **27.** $\left\{\dfrac{7 \pm \sqrt{129}}{10}\right\}$ **29.** $\left\{-\dfrac{10}{7}, 3\right\}$

31. $\{1 \pm \sqrt{34}\}$ **33.** $\{\pm\sqrt{6}, \pm 2\sqrt{3}\}$

35. $\left\{\pm 3, \pm\dfrac{2\sqrt{6}}{3}\right\}$ **37.** $\left\{\pm\dfrac{i\sqrt{15}}{3}, \pm 2i\right\}$

39. $\left\{\pm\dfrac{\sqrt{14}}{2}, \pm\dfrac{2\sqrt{3}}{3}\right\}$ **41.** 8 and 9   **43.** 9 and 12

45. $5 + \sqrt{3}$ and $5 - \sqrt{3}$ **47.** 3 and 6
49. 9 inches and 12 inches   **51.** 1 meter
53. 8 inches by 14 inches   **55.** 20 miles per hour for
Lorraine and 25 miles per hour for Charlotte, or 45 miles
per hour for Lorraine and 50 miles per hour for Charlotte

57. 55 miles per hour   **59.** 6 hours for Tom and 8 hours
for Terry   **61.** 2 hours   **63.** 8 students   **65.** 40 shares at
$20 per share   **67.** 50 numbers   **69.** 9%   **75.** $\{9, 36\}$

77. $\{1\}$ **79.** $\left\{-\dfrac{8}{27}, \dfrac{27}{8}\right\}$ **81.** $\left\{-4, \dfrac{3}{5}\right\}$ **83.** $\{4\}$

85. $\{\pm 4\sqrt{2}\}$ **87.** $\left\{\dfrac{3}{2}, \dfrac{5}{2}\right\}$ **89.** $\{-1, 3\}$

Problem Set 6.6 (page 325)

1. $(-\infty, -2) \cup (1, \infty)$

3. $(-4, -1)$

5. $\left(-\infty, -\dfrac{7}{3}\right] \cup \left[\dfrac{1}{2}, \infty\right)$

7. $\left[-2, \dfrac{3}{4}\right]$

9. $(-1, 1) \cup (3, \infty)$

11. $(-\infty, -2] \cup [0, 4]$

13. $(-\infty, -1) \cup (2, \infty)$

15. $(-2, 3)$

17. $(-\infty, 0) \cup \left[\dfrac{1}{2}, \infty\right)$

19. $(-\infty, 1) \cup [2, \infty)$

21. $(-7, 5)$ **23.** $(-\infty, 4) \cup (7, \infty)$ **25.** $\left[-5, \frac{2}{3}\right]$

27. $\left(-\infty, -\frac{5}{2}\right] \cup \left[-\frac{1}{4}, \infty\right)$ **29.** $\left(-\infty, -\frac{4}{5}\right) \cup (8, \infty)$

31. $(-\infty, \infty)$ **33.** $\left\{-\frac{5}{2}\right\}$ **35.** $(-1, 3) \cup (3, \infty)$

37. $(-\infty, -2) \cup (2, \infty)$ **39.** $(-6, 6)$ **41.** $(-\infty, \infty)$
43. $(-\infty, 0] \cup [2, \infty)$ **45.** $(-4, 0) \cup (0, \infty)$ **47.** $(-6, -3)$

49. $(-\infty, 5) \cup [9, \infty)$ **51.** $\left(-\infty, \frac{4}{3}\right) \cup (3, \infty)$

53. $(-4, 6]$ **55.** $(-\infty, 2)$

Chapter 6 Review Problem Set (page 328)
1. $2 - 2i$ **2.** $-3 - i$ **3.** $30 + 15i$ **4.** $86 - 2i$

5. $-32 + 4i$ **6.** $25 + 0i$ **7.** $\frac{9}{20} + \frac{13}{20}i$ **8.** $-\frac{3}{29} + \frac{7}{29}i$

9. One real solution with a multiplicity of 2.
10. Two nonreal complex solutions
11. Two unequal real solutions
12. Two unequal real solutions   **13.** $\{0, 17\}$ **14.** $\{-4, 8\}$
15. $\left\{\frac{1 \pm 8i}{2}\right\}$ **16.** $\{-3, 7\}$ **17.** $\{-1 \pm \sqrt{10}\}$

18. $\{3 \pm 5i\}$ **19.** $\{25\}$ **20.** $\left\{-4, \frac{2}{3}\right\}$ **21.** $\{-10, 20\}$

22. $\left\{\frac{-1 \pm \sqrt{61}}{6}\right\}$ **23.** $\left\{\frac{1 \pm i\sqrt{11}}{2}\right\}$

24. $\left\{\frac{5 \pm i\sqrt{23}}{4}\right\}$ **25.** $\left\{\frac{-2 \pm \sqrt{14}}{2}\right\}$ **26.** $\{-9, 4\}$

27. $\{-2 \pm i\sqrt{5}\}$ **28.** $\{-6, 12\}$ **29.** $\{1 \pm \sqrt{10}\}$

30. $\left\{\pm\frac{\sqrt{14}}{2}, \pm 2\sqrt{2}\right\}$ **31.** $\left\{\frac{-3 \pm \sqrt{97}}{2}\right\}$

32. $(-\infty, -5) \cup (2, \infty)$ **33.** $\left[-\frac{7}{2}, 3\right]$

34. $(-\infty, -6) \cup [4, \infty)$ **35.** $\left(-\frac{5}{2}, -1\right)$

36. $3 + \sqrt{7}$ and $3 - \sqrt{7}$ **37.** 20 shares at \$15 per share
38. 45 miles per hour and 52 miles per hour   **39.** 8 units
40. 8 and 10   **41.** 7 inches by 12 inches   **42.** 4 hours for
Reena and 6 hours for Billy   **43.** 10 meters

Chapter 6 Test (page 330)
1. $39 - 2i$ **2.** $-\frac{6}{25} - \frac{17}{25}i$ **3.** $\{0, 7\}$ **4.** $\{-1, 7\}$

5. $\{-6, 3\}$ **6.** $\{1 - \sqrt{2}, 1 + \sqrt{2}\}$ **7.** $\left\{\frac{1 - 2i}{5}, \frac{1 + 2i}{5}\right\}$

8. $\{-16, -14\}$ **9.** $\left\{\frac{1 - 6i}{3}, \frac{1 + 6i}{3}\right\}$ **10.** $\left\{-\frac{7}{4}, \frac{6}{5}\right\}$

11. $\left\{-3, \frac{19}{6}\right\}$ **12.** $\left\{-\frac{10}{3}, 4\right\}$ **13.** $\{-2, 2, -4i, 4i\}$

14. $\left\{-\frac{3}{4}, 1\right\}$ **15.** $\left\{\frac{1 - \sqrt{10}}{3}, \frac{1 + \sqrt{10}}{3}\right\}$

16. Two equal real solutions   **17.** Two nonreal complex

solutions   **18.** $[-6, 9]$ **19.** $(-\infty, -2) \cup \left(\frac{1}{3}, \infty\right)$

20. $[-10, -6)$ **21.** 20.8 feet   **22.** 29 meters

23. 150 shares   **24.** $6\frac{1}{2}$ inches   **25.** $3 + \sqrt{5}$

Cumulative Review Problem Set (page 331)
1. $\frac{64}{15}$ **2.** $\frac{11}{3}$ **3.** $\frac{1}{6}$ **4.** $-\frac{44}{5}$ **5.** -7 **6.** $-24a^4b^5$

7. $2x^3 + 5x^2 - 7x - 12$ **8.** $\frac{3x^2y^2}{8}$ **9.** $\frac{a(a + 1)}{2a - 1}$

10. $\frac{-x + 14}{18}$ **11.** $\frac{5x + 19}{x(x + 3)}$ **12.** $\frac{2}{n + 8}$

13. $\frac{x - 14}{(5x - 2)(x + 1)(x - 4)}$ **14.** $y^2 - 5y + 6$

15. $x^2 - 3x - 2$ **16.** $20 + 7\sqrt{10}$

17. $2x - 2\sqrt{xy} - 12y$ **18.** $-\frac{3}{8}$ **19.** $-\frac{2}{3}$ **20.** 0.2

21. $\frac{1}{2}$ **22.** $\frac{13}{9}$ **23.** -27 **24.** $\frac{16}{9}$ **25.** $\frac{8}{27}$

26. $3x(x + 3)(x^2 - 3x + 9)$ **27.** $(6x - 5)(x + 4)$
28. $(4 + 7x)(3 - 2x)$ **29.** $(3x + 2)(3x - 2)(x^2 + 8)$
30. $(2x - y)(a - b)$ **31.** $(3x - 2y)(9x^2 + 6xy + 4y^2)$

32. $\left\{-\frac{12}{7}\right\}$ **33.** $\{150\}$ **34.** $\{25\}$ **35.** $\{0\}$ **36.** $\{-2, 2\}$

37. $\{-7\}$ **38.** $\left\{-6, \frac{4}{3}\right\}$ **39.** $\left\{\frac{5}{4}\right\}$ **40.** $\{3\}$ **41.** $\left\{\frac{4}{5}, 1\right\}$

42. $\left\{-\frac{10}{3}, 4\right\}$ **43.** $\left\{\frac{1}{4}, \frac{2}{3}\right\}$ **44.** $\left\{-\frac{3}{2}, 3\right\}$ **45.** $\left\{\frac{1}{5}\right\}$

46. $\left\{\frac{5}{7}\right\}$ **47.** $\{-2, 2\}$ **48.** $\{0\}$ **49.** $\{-6, 19\}$

50. $\left\{-\frac{3}{4}, \frac{2}{3}\right\}$ **51.** $\{1 \pm 5i\}$ **52.** $\{1, 3\}$ **53.** $\left\{-4, \frac{1}{3}\right\}$

54. $\{-2 \pm 4i\}$ **55.** $\left\{\frac{1 \pm \sqrt{33}}{4}\right\}$ **56.** $(-\infty, -2]$

57. $\left(-\infty, \frac{19}{5}\right)$ **58.** $\left(\frac{1}{4}, \infty\right)$ **59.** $(-2, 3)$

60. $\left(-\infty, -\frac{13}{3}\right) \cup (3, \infty)$ **61.** $(-\infty, 29]$ **62.** $[-2, 4]$

63. $(-\infty, -5) \cup \left(\dfrac{1}{3}, \infty\right)$ **64.** $(-\infty, -2] \cup (7, \infty)$

65. $(-3, 4)$ **66.** 6 liters   **67.** \$900 and \$1350
68. 12 inches by 17 inches   **69.** 5 hours   **70.** 7 golf balls
71. 12 minutes   **72.** 7%   **73.** 15 chairs per row
74. 140 shares

CHAPTER 7

Problem Set 7.1 (page 346)

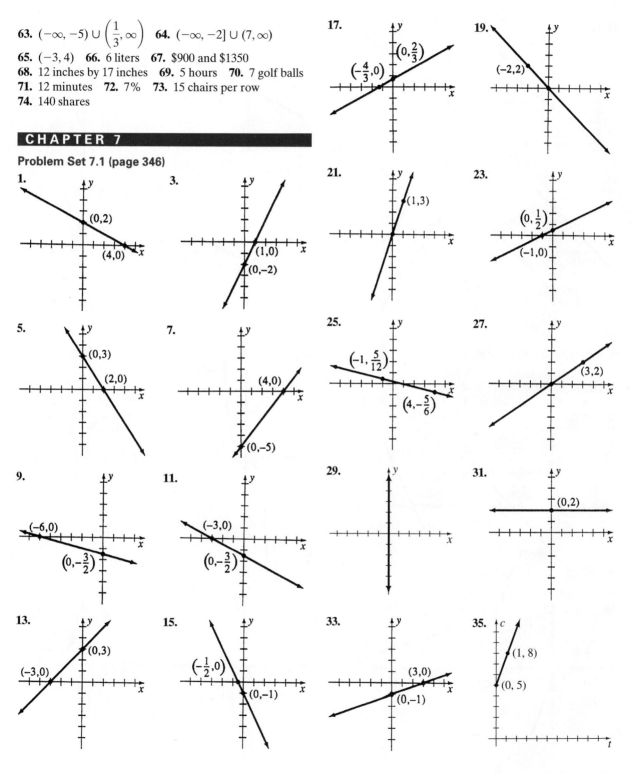

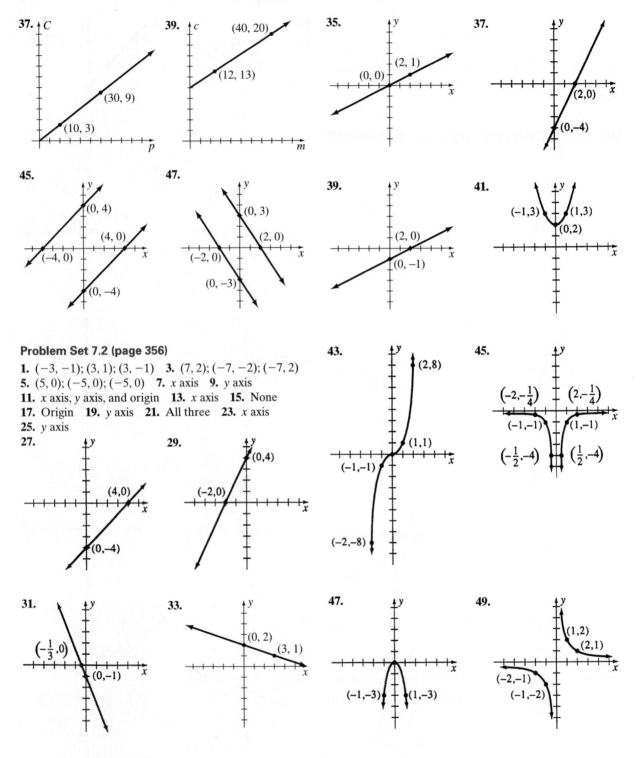

37.

(30, 9)

(10, 3)

p

39. (40, 20)

(12, 13)

m

35. (2, 1)

(0, 0)

x

37. (2,0)

(0,−4)

x

45. (0, 4)

(4, 0)

(−4, 0)

(0, −4)

x

47. (0, 3)

(2, 0)

(−2, 0)

(0, −3)

x

39. (2, 0)

(0, −1)

x

41. (−1,3)   (1,3)

(0,2)

x

Problem Set 7.2 (page 356)

1. $(-3, -1); (3, 1); (3, -1)$ **3.** $(7, 2); (-7, -2); (-7, 2)$
5. $(5, 0); (-5, 0); (-5, 0)$ **7.** x axis   **9.** y axis
11. x axis, y axis, and origin   **13.** x axis   **15.** None
17. Origin   **19.** y axis   **21.** All three   **23.** x axis
25. y axis

27. (4,0)

(0,−4)

x

29. (0,4)

(−2,0)

x

43. (2,8)

(1,1)

(−1,−1)

(−2,−8)

x

45. $\left(-2,-\frac{1}{4}\right)$ $\left(2,-\frac{1}{4}\right)$

(−1,−1)   (1,−1)

$\left(-\frac{1}{2},-4\right)$ $\left(\frac{1}{2},-4\right)$

x

31. $\left(-\frac{1}{3},0\right)$

(0,−1)

x

33. (0, 2)

(3, 1)

x

47. (−1,−3)   (1,−3)

x

49. (1,2)

(2,1)

(−2,−1)

(−1,−2)

x

51.

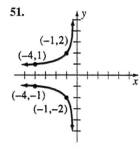

53.

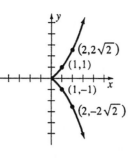

5.

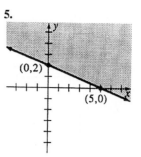

7.

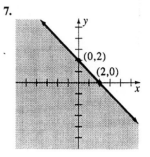

55.

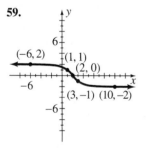

57.

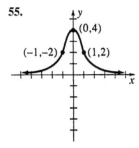

9.

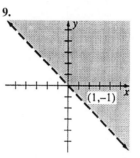

11.

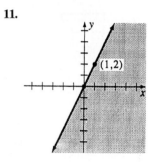

59.

13.

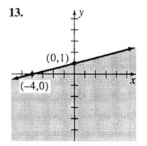

15.

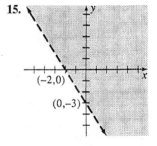

Problem Set 7.3 (page 361)

1.   **3.**   **17.**   **19.**

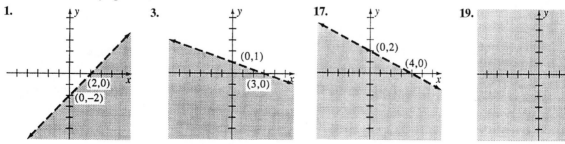

21.

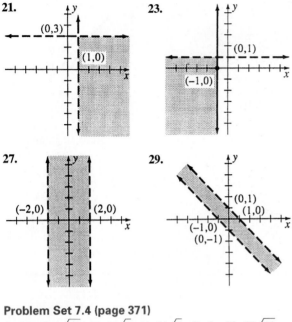

23.

27.

29.

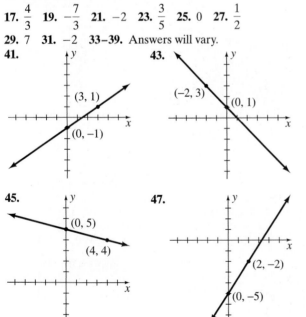

49. $-\dfrac{2}{3}$ **51.** $\dfrac{1}{2}$ **53.** $\dfrac{4}{7}$ **55.** 0 **57.** -5 **59.** 105.6 feet
61. 8.1% **63.** 19 centimeters **69. (a)** $(3, 5)$ **(c)** $(2, 5)$
(e) $\left(\dfrac{17}{8}, -7\right)$

Problem Set 7.5 (page 383)
1. $x - 2y = -7$ **3.** $3x - y = -10$ **5.** $3x + 4y = -15$
7. $5x - 4y = 28$ **9.** $x - y = 1$ **11.** $5x - 2y = -4$
13. $x + 7y = 11$ **15.** $x + 2y = -9$ **17.** $7x - 5y = 0$
19. $y = \dfrac{3}{7}x + 4$ **21.** $y = 2x - 3$ **23.** $y = -\dfrac{2}{5}x + 1$
25. $y = 0(x) - 4$ **27.** $2x - y = 4$ **29.** $5x + 8y = -15$
31. $x + 0(y) = 2$ **33.** $0(x) + y = 6$ **35.** $x + 5y = 16$
37. $4x - 7y = 0$ **39.** $x + 2y = 5$ **41.** $3x + 2y = 0$
43. $m = -3$ and $b = 7$ **45.** $m = -\dfrac{3}{2}$ and $b = \dfrac{9}{2}$
47. $m = \dfrac{1}{5}$ and $b = -\dfrac{12}{5}$

Problem Set 7.4 (page 371)
1. 15 **3.** $\sqrt{13}$ **5.** $3\sqrt{2}$ **7.** $3\sqrt{5}$ **9.** 6 **11.** $3\sqrt{10}$
13. The lengths of the sides are $10, 5\sqrt{5}$, and 5. Because $10^2 + 5^2 = (5\sqrt{5})^2$, it is a right triangle.
15. The distances between $(3, 6)$, and $(7, 12)$, between $(7, 12)$ and $(11, 18)$, and between $(11, 18)$ and $(15, 24)$ are all $2\sqrt{13}$ units.
17. $\dfrac{4}{3}$ **19.** $-\dfrac{7}{3}$ **21.** -2 **23.** $\dfrac{3}{5}$ **25.** 0 **27.** $\dfrac{1}{2}$
29. 7 **31.** -2 **33–39.** Answers will vary.

41. **43.** **45.** **47.**

49. **51.** **53.** **55.** **57.** **59.**

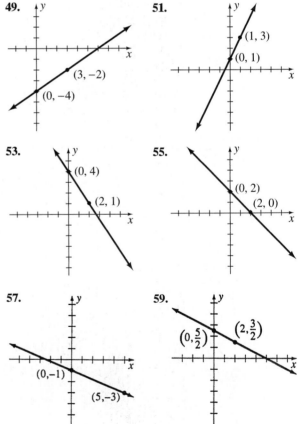

61.

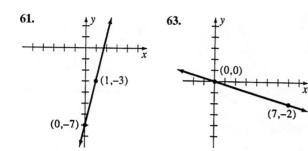

63.

18.

19.

65.

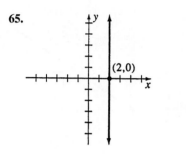

20.

21.

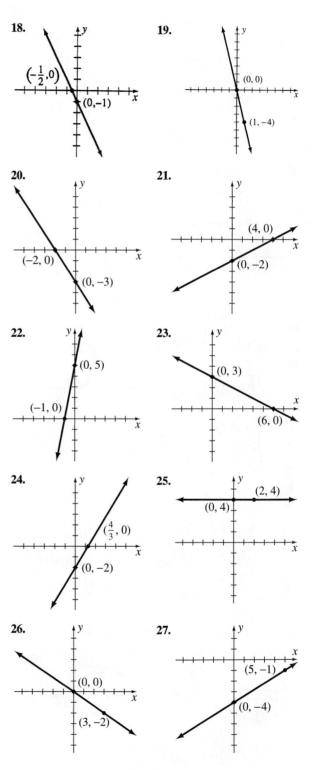

67. $y = \dfrac{1}{1000}x + 2$ **69.** $y = \dfrac{9}{5}x + 32$

77. (a) $2x - y = 1$ **(b)** $5x - 6y = 29$ **(c)** $x + y = 2$
(d) $3x - 2y = 18$

Chapter 7 Review Problem Set (page 388)

1. (a) $\dfrac{6}{5}$ **(b)** $-\dfrac{2}{3}$ **2.** 5   **3.** -1

4. (a) $m = -4$ **(b)** $m = \dfrac{2}{7}$ **5.** 5, 10, and $\sqrt{97}$

6. (a) $2\sqrt{10}$ **(b)** $\sqrt{58}$ **8.** $7x + 4y = 1$
9. $3x + 7y = 28$ **10.** $2x - 3y = 16$
11. $x - 2y = -8$ **12.** $2x - 3y = 14$
13. $x - y = -4$ **14.** $x + y = -2$
15. $4x + y = -29$

16.

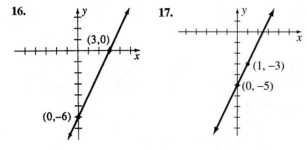

17.

26.

27.

22.

23.

24.

25.

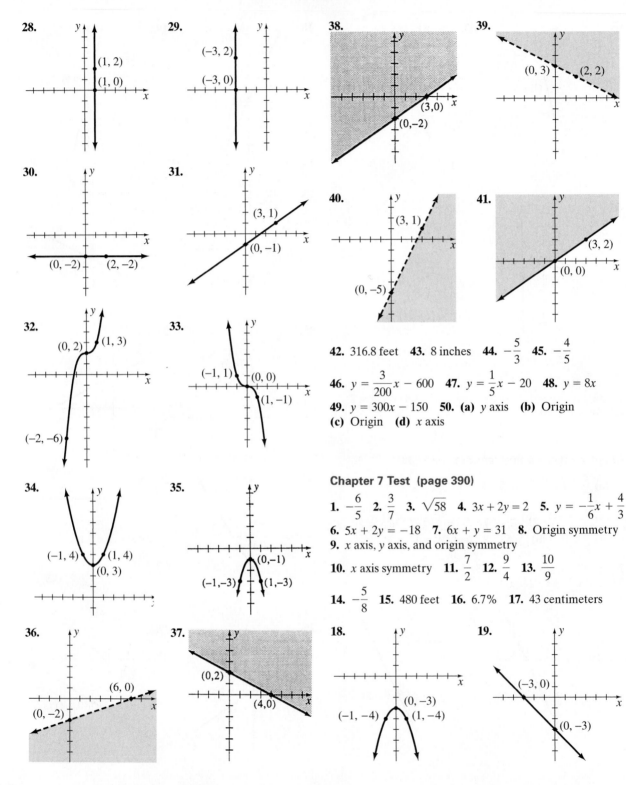

28. (1, 2) (1, 0)

29. (−3, 2) (−3, 0)

30. (0, −2) (2, −2)

31. (3, 1) (0, −1)

32. (0, 2) (1, 3) (−2, −6)

33. (−1, 1) (0, 0) (1, −1)

34. (−1, 4) (1, 4) (0, 3)

35. (0, −1) (−1, −3) (1, −3)

36. (6, 0) (0, −2)

37. (0, 2) (4, 0)

38. (3, 0) (0, −2)

39. (0, 3) (2, 2)

40. (3, 1) (0, −5)

41. (3, 2) (0, 0)

42. 316.8 feet   **43.** 8 inches   **44.** $-\dfrac{5}{3}$ **45.** $-\dfrac{4}{5}$

46. $y = \dfrac{3}{200}x - 600$ **47.** $y = \dfrac{1}{5}x - 20$ **48.** $y = 8x$

49. $y = 300x - 150$ **50. (a)** y axis   **(b)** Origin
(c) Origin   **(d)** x axis

Chapter 7 Test (page 390)

1. $-\dfrac{6}{5}$ **2.** $\dfrac{3}{7}$ **3.** $\sqrt{58}$ **4.** $3x + 2y = 2$ **5.** $y = -\dfrac{1}{6}x + \dfrac{4}{3}$

6. $5x + 2y = -18$ **7.** $6x + y = 31$ **8.** Origin symmetry

9. x axis, y axis, and origin symmetry

10. x axis symmetry   **11.** $\dfrac{7}{2}$ **12.** $\dfrac{9}{4}$ **13.** $\dfrac{10}{9}$

14. $-\dfrac{5}{8}$ **15.** 480 feet   **16.** 6.7%   **17.** 43 centimeters

18. (0, −3) (−1, −4) (1, −4)

19. (−3, 0) (0, −3)

20.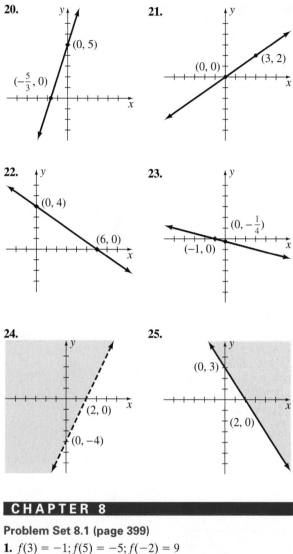

21.

22.

23.

24.

25.

21. -7 **23.** $-2a - h + 4$ **25.** $6a + 3h - 1$
27. $3a^2 + 3ah + h^2 - 2a - h + 2$
29. $-\dfrac{2}{(a-1)(a+h-1)}$ **31.** $-\dfrac{2a+h}{a^2(a+h)^2}$ **33.** Yes
35. No   **37.** Yes   **39.** Yes   **41.** $D = \left\{x \mid x \geq \dfrac{4}{3}\right\}$;
$\qquad R = \{f(x) \mid f(x) \geq 0\}$
43. $D = \{x \mid x \text{ is any real number}\}$;
$\qquad R = \{f(x) \mid f(x) \geq -2\}$
45. $D = \{x \mid x \text{ is any real number}\}$;
$\qquad R = \{f(x) \mid f(x) \text{ is any nonnegative real number}\}$
47. $D = \{x \mid x \text{ is any nonnegative real number}\}$;
$\qquad R = \{f(x) \mid f(x) \text{ is any nonpositive real number}\}$
49. $D = \{x \mid x \neq -2\}$
51. $D = \left\{x \mid x \neq \dfrac{1}{2} \text{ and } x \neq -4\right\}$
53. $D = \{x \mid x \neq 2 \text{ and } x \neq -2\}$
55. $D = \{x \mid x \neq -3 \text{ and } x \neq 4\}$
57. $D = \left\{x \mid x \neq -\dfrac{5}{2} \text{ and } x \neq \dfrac{1}{3}\right\}$
59. $(-\infty, -4] \cup [4, \infty)$ **61.** $(-\infty, \infty)$
63. $(-\infty, -5] \cup [8, \infty)$ **65.** $\left(-\infty, -\dfrac{5}{2}\right] \cup \left[\dfrac{7}{4}, \infty\right)$
67. $[-1, 1]$ **69.** 12.57; 28.27; 452.39; 907.92
71. 48; 64; 48; 0
73. \$55; \$60; \$67.50; \$75   **75.** 125.66; 301.59; 804.25

CHAPTER 8

Problem Set 8.1 (page 399)
1. $f(3) = -1; f(5) = -5; f(-2) = 9$
3. $g(3) = -20; g(-1) = -8; g(2a) = -8a^2 + 2a - 5$
5. $h(3) = \dfrac{5}{4}; h(4) = \dfrac{23}{12}; h\left(-\dfrac{1}{2}\right) = -\dfrac{13}{12}$
7. $f(5) = 3; f\left(\dfrac{1}{2}\right) = 0; f(23) = 3\sqrt{5}$
9. $-2a + 7, -2a + 3, -2a - 2h + 7$
11. $a^2 + 4a + 10, a^2 - 12a + 42,$
$\qquad a^2 + 2ah + h^2 - 4a - 4h + 10$
13. $-a^2 - 3a + 5, -a^2 - 9a - 13, -a^2 - a + 7$
15. $f(4) = 4; f(10) = 10; f(-3) = 9; f(-5) = 25$
17. $f(3) = 6; f(5) = 10; f(-3) = 6; f(-5) = 10$
19. $f(2) = 1; f(0) = 0; f\left(-\dfrac{1}{2}\right) = 0; f(-4) = -1$

Problem Set 8.2 (page 408)
1.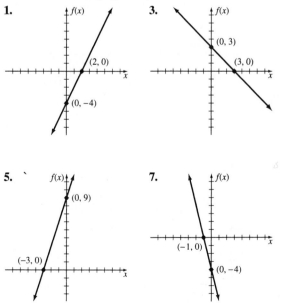

3.

5.

7.

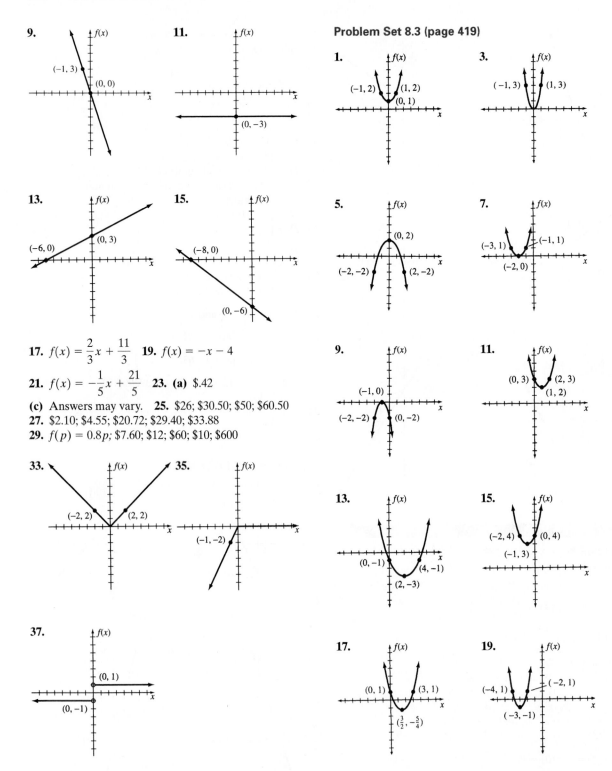

9. (−1, 3) (0, 0)

11. (0, −3)

13. (−6, 0) (0, 3)

15. (−8, 0) (0, −6)

17. $f(x) = \dfrac{2}{3}x + \dfrac{11}{3}$ **19.** $f(x) = -x - 4$

21. $f(x) = -\dfrac{1}{5}x + \dfrac{21}{5}$ **23. (a)** \$.42

(c) Answers may vary.   **25.** \$26; \$30.50; \$50; \$60.50
27. \$2.10; \$4.55; \$20.72; \$29.40; \$33.88
29. $f(p) = 0.8p$; \$7.60; \$12; \$60; \$10; \$600

33. (−2, 2) (2, 2)

35. (−1, −2)

37. (0, 1) (0, −1)

Problem Set 8.3 (page 419)

1. (−1, 2) (1, 2) (0, 1)

3. (−1, 3) (1, 3)

5. (0, 2) (−2, −2) (2, −2)

7. (−3, 1) (−1, 1) (−2, 0)

9. (−1, 0) (−2, −2) (0, −2)

11. (0, 3) (2, 3) (1, 2)

13. (0, −1) (4, −1) (2, −3)

15. (−2, 4) (0, 4) (−1, 3)

17. (0, 1) (3, 1) $\left(\frac{3}{2}, -\frac{5}{4}\right)$

19. (−4, 1) (−2, 1) (−3, −1)

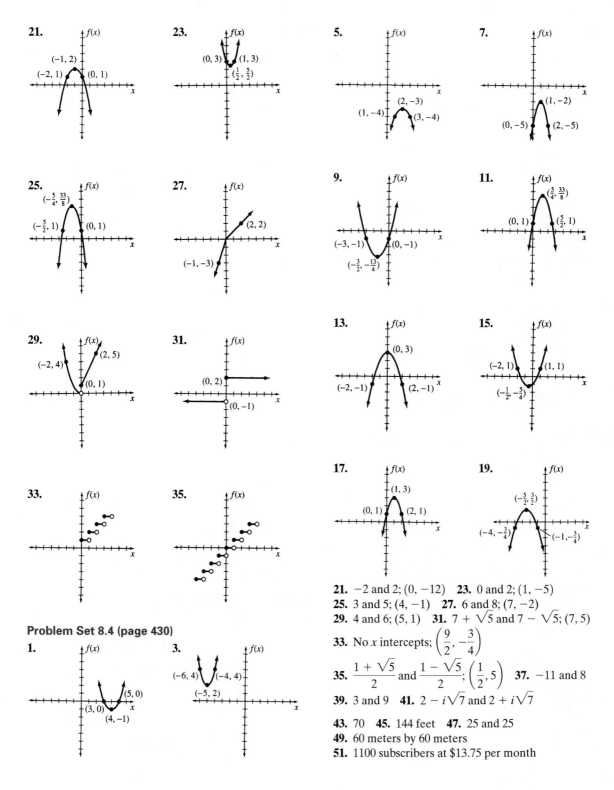

21. -2 and 2; $(0, -12)$ **23.** 0 and 2; $(1, -5)$

25. 3 and 5; $(4, -1)$ **27.** 6 and 8; $(7, -2)$

29. 4 and 6; $(5, 1)$ **31.** $7 + \sqrt{5}$ and $7 - \sqrt{5}$; $(7, 5)$

33. No x intercepts; $\left(\dfrac{9}{2}, -\dfrac{3}{4} \right)$

35. $\dfrac{1 + \sqrt{5}}{2}$ and $\dfrac{1 - \sqrt{5}}{2}$; $\left(\dfrac{1}{2}, 5 \right)$ **37.** -11 and 8

39. 3 and 9 **41.** $2 - i\sqrt{7}$ and $2 + i\sqrt{7}$

43. 70 **45.** 144 feet   **47.** 25 and 25

49. 60 meters by 60 meters

51. 1100 subscribers at $\$13.75$ per month

Problem Set 8.5 (page 441)

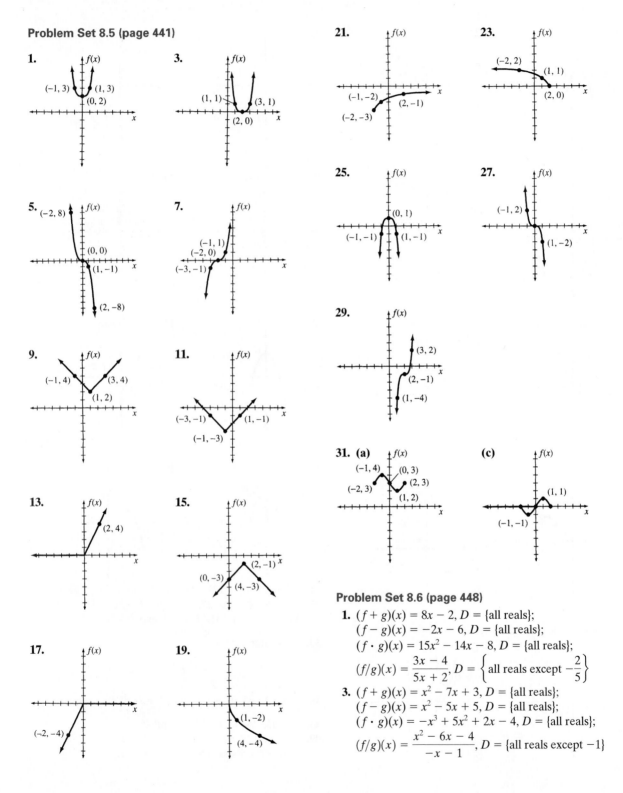

Problem Set 8.6 (page 448)

1. $(f + g)(x) = 8x - 2, D = \{\text{all reals}\};$
   $(f - g)(x) = -2x - 6, D = \{\text{all reals}\};$
   $(f \cdot g)(x) = 15x^2 - 14x - 8, D = \{\text{all reals}\};$
   $(f/g)(x) = \dfrac{3x - 4}{5x + 2}, D = \left\{\text{all reals except } -\dfrac{2}{5}\right\}$

3. $(f + g)(x) = x^2 - 7x + 3, D = \{\text{all reals}\};$
   $(f - g)(x) = x^2 - 5x + 5, D = \{\text{all reals}\};$
   $(f \cdot g)(x) = -x^3 + 5x^2 + 2x - 4, D = \{\text{all reals}\};$
   $(f/g)(x) = \dfrac{x^2 - 6x - 4}{-x - 1}, D = \{\text{all reals except } -1\}$

5. $(f + g)(x) = 2x^2 + 3x - 6, D = \{$all reals$\}$;
$(f - g)(x) = -5x + 4, D = \{$all reals$\}$;
$(f \cdot g)(x) = x^4 + 3x^3 - 10x^2 + x + 5, D = \{$all reals$\}$;
$(f/g)(x) = \dfrac{x^2 - x - 1}{x^2 + 4x - 5}, D = \{$all reals except -5 and $1\}$

7. $(f + g)(x) = \sqrt{x - 1} + \sqrt{x}, D = \{x | x \geq 1\}$;
$(f - g)(x) = \sqrt{x - 1} - \sqrt{x}, D = \{x | x \geq 1\}$;
$(f \cdot g)(x) = \sqrt{x^2 - x}, D = \{x | x \geq 1\}$;
$(f/g)(x) = \dfrac{\sqrt{x - 1}}{\sqrt{x}}, D = \{x | x \geq 1\}$

9. $(f \circ g)(x) = 6x - 2, D = \{$all reals$\}$;
$(g \circ f)(x) = 6x - 1, D = \{$all reals$\}$

11. $(f \circ g)(x) = 10x + 2, D = \{$all reals$\}$;
$(g \circ f)(x) = 10x - 5, D = \{$all reals$\}$

13. $(f \circ g)(x) = 3x^2 + 7, D = \{$all reals$\}$;
$(g \circ f)(x) = 9x^2 + 24x + 17, D = \{$all reals$\}$

15. $(f \circ g)(x) = 3x^2 + 9x - 16, D = \{$all reals$\}$;
$(g \circ f)(x) = 9x^2 - 15x, D = \{$all reals$\}$

17. $(f \circ g)(x) = \dfrac{1}{2x + 7}, D = \left\{ x \Big| x \neq -\dfrac{7}{2} \right\}$;
$(g \circ f)(x) = \dfrac{7x + 2}{x}, D = \{x | x \neq 0\}$

19. $(f \circ g)(x) = \sqrt{3x - 3}, D = \{x | x \geq 1\}$;
$(g \circ f)(x) = 3\sqrt{x - 2} - 1, D = \{x | x \geq 2\}$

21. $(f \circ g)(x) = \dfrac{x}{2 - x}, D = \{x | x \neq 0 \text{ and } x \neq 2\}$;
$(g \circ f)(x) = 2x - 2, D = \{x | x \neq 1\}$

23. $(f \circ g)(x) = 2\sqrt{x - 1} + 1, D = \{x | x \geq 1\}$;
$(g \circ f)(x) = \sqrt{2x}, D = \{x | x \geq 0\}$

25. $(f \circ g)(x) = x, D = \{x | x \neq 0\}$;
$(g \circ f)(x) = x, D = \{x | x \neq 1\}$

27. $4; 50$ **29.** $9; 0$ **31.** $\sqrt{11}; 5$

Problem Set 8.7 (page 456)

1. $y = kx^3$ **3.** $A = klw$ **5.** $V = \dfrac{k}{P}$ **7.** $V = khr^2$

9. 24 **11.** $\dfrac{22}{7}$ **13.** $\dfrac{1}{2}$ **15.** 7 **17.** 6 **19.** 8 **21.** 96

23. 5 hours   **25.** 2 seconds   **27.** 24 days   **29.** 28
31. $\$2400$ **37.** 2.8 seconds   **39.** 1.4

Chapter 8 Review Problem Set (page 460)

1. $7; 4; 32$ **2. (a)** -5 **(b)** $4a + 2h - 1$
(c) $-6a - 3h + 2$

3. $D = \{x | x \text{ is any real number}\}; R = \{f(x) | f(x) \geq 5\}$

4. $D = \left\{ x \Big| x \neq \dfrac{1}{2}, x \neq -4 \right\}$

5. $(-\infty, 2] \cup [5, \infty)$

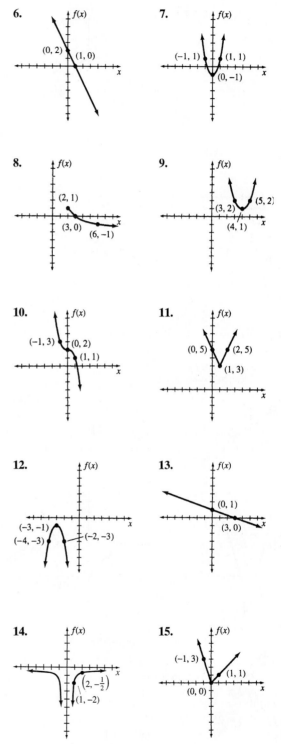

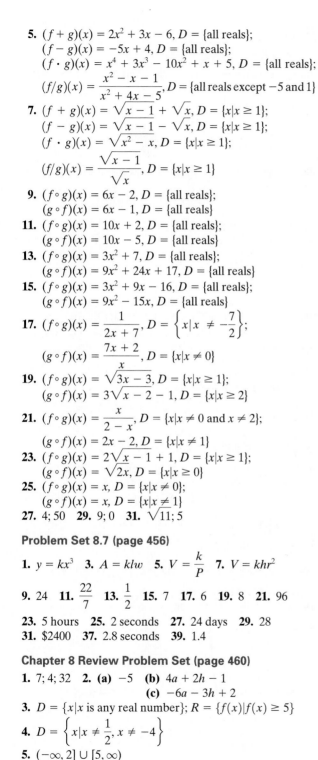

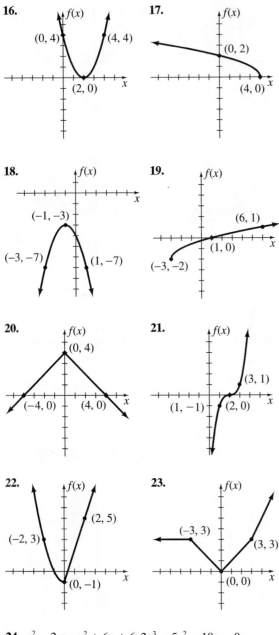

16.

(0, 4)   (4, 4)

(2, 0)

17.

(0, 2)

(4, 0)

18.

(−1, −3)

(−3, −7)   (1, −7)

19.

(6, 1)

(1, 0)

(−3, −2)

20.

(0, 4)

(−4, 0)   (4, 0)

21.

(3, 1)

(1, −1)   (2, 0)

22.

(2, 5)

(−2, 3)

(0, −1)

23.

(−3, 3)

(3, 3)

(0, 0)

24. $x^2 - 2x; -x^2 + 6x + 6; 2x^3 - 5x^2 - 18x - 9;$

$$\frac{2x + 3}{x^2 - 4x - 3}$$

25. $(f \circ g)(x) = -6x + 12, D = \{\text{all reals}\};$
$(g \circ f)(x) = -6x + 25, D = \{\text{all reals}\}$

26. $(f \circ g)(x) = 25x^2 - 40x + 11, D = \{\text{all reals}\};$
$(g \circ f)(x) = 5x^2 - 29, D = \{\text{all reals}\}$

27. $(f \circ g)(x) = \sqrt{x - 3}, D = \{x | x \geq 3\};$
$(g \circ f)(x) = \sqrt{x - 5} + 2, D = \{x | x \geq 5\}$

28. $(f \circ g)(x) = \dfrac{1}{x^2 - x - 6}, D = \{x | x \neq 3 \text{ and } x \neq -2\};$

$(g \circ f)(x) = \dfrac{1 - x - 6x^2}{x^2}, D = \{x | x \neq 0\}$

29. $(f \circ g)(x) = x - 1, D = \{x | x \geq 1\};$
$(g \circ f)(x) = \sqrt{x^2 - 1}, D = \{x | x \leq -1 \text{ or } x \geq 1\}$

30. $(f \circ g)(x) = \dfrac{x + 2}{-3x - 5}, D = \left\{x | x \neq -2 \text{ and } x \neq -\dfrac{5}{3}\right\};$

$(g \circ f)(x) = \dfrac{x - 3}{2x - 5}, D = \left\{x | x \neq 3 \text{ and } x \neq \dfrac{5}{2}\right\}$

31. $f(5) = 23; f(0) = -2; f(-3) = 13$
32. $f(g(6)) = -2; g(f(-2)) = 0$

33. $f(g(1)) = 1; g(f(-3)) = 5$ **34.** $f(x) = \dfrac{2}{3}x - \dfrac{16}{3}$

35. $f(x) = 2x + 15$ **36.** \$.72
37. $f(x) = 0.7x$; \$45.50; \$33.60; \$10.85
38. −4 and 2; (−1, −27)   **39.** $3 \pm \sqrt{14}$; (3, −14)
40. No x intercepts; (7, 3)   **41.** 2 and 8   **42.** 112 students
43. 9   **44.** 441   **45.** 128 pounds   **46.** 15 hours

Chapter 8 Test (page 462)

1. $\dfrac{11}{6}$ **2.** 11   **3.** $6a + 3h + 2$

4. $\left\{x | x \neq -4 \text{ and } x \neq \dfrac{1}{2}\right\}$ **5.** $\left\{x | x \leq \dfrac{5}{3}\right\}$

6. $(f + g)(x) = 2x^2 + 2x - 6; (f - g)(x) = -2x^2 + 4x + 4; (f \cdot g)(x) = 6x^3 - 5x^2 - 14x + 5$
7. $(f \circ g)(x) = -21x - 2$ **8.** $(g \circ f)(x) = 8x^2 + 38x + 48$

9. $(f \circ g)(x) = \dfrac{3x}{2 - 2x}$ **10.** 12; 7   **11.** $f(x) = -\dfrac{5}{6}x - \dfrac{14}{3}$

12. $\{x | x \neq 0 \text{ and } x \neq 1\}$ **13.** 18; 10; 0

14. $(f \cdot g)(x) = x^3 + 4x^2 - 11x + 6; \left(\dfrac{f}{g}\right)(x) = x + 6$

15. 6 and 54   **16.** 15   **17.** −4   **18.** \$96
19. $s(c) = 1.35c$; \$17.55   **20.** −2 and 6; (2, −64)

21.

(3, −2)
(2, −3)
(1, −4)

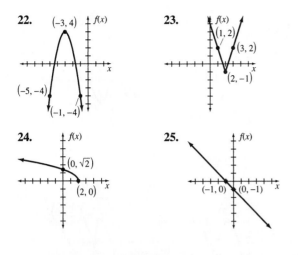

22. $(-3, 4)$, $f(x)$, $(-5, -4)$, $(-1, -4)$

23. $f(x)$, $(1, 2)$, $(3, 2)$, $(2, -1)$

24. $f(x)$, $(0, \sqrt{2})$, $(2, 0)$

25. $f(x)$, $(-1, 0)$, $(0, -1)$

CHAPTER 9

Problem Set 9.1 (page 468)

1. $Q: 4x + 3; R: 0$ **3.** $Q: 2x - 7; R: 0$ **5.** $Q: 3x - 4; R: 1$
7. $Q: 4x - 5; R: -2$ **9.** $Q: x^2 + 3x - 4; R: 0$
11. $Q: 3x^2 + 2x - 4; R: 0$ **13.** $Q: 5x^2 + x - 1; R: -4$
15. $Q: x^2 - x - 1; R: 8$ **17.** $Q: -x^2 + 4x - 2; R: 0$
19. $Q: -3x^2 + 4x - 2; R: 4$ **21.** $Q: 3x^2 + 6x + 10; R: 15$
23. $Q: 2x^3 - x^2 + 4x - 2; R: 0$
25. $Q: x^3 + 7x^2 + 21x + 56; R: 167$
27. $Q: x^3 - x + 5; R: 0$ **29.** $Q: x^3 + 2x^2 + 4x + 8; R: 0$
31. $Q: x^4 - x^3 + x^2 - x + 1; R: -2$
33. $Q: x^4 - x^3 + x^2 - x + 1; R: 0$
35. $Q: x^4 - x^3 - x^2 + x - 1; R: 0$
37. $Q: 4x^4 - 2x^3 + 2x - 3; R: -1$

39. $Q: 9x^2 - 3x + 2; R: -\dfrac{10}{3}$

41. $Q: 3x^3 - 3x^2 + 6x - 3; R: 0$

Problem Set 9.2 (page 472)

1. $f(3) = 9$ **3.** $f(-1) = -7$ **5.** $f(2) = 19$ **7.** $f(6) = 74$
9. $f(-2) = -65$ **11.** $f(-1) = -1$ **13.** $f(8) = -83$
15. $f(3) = 8751$ **17.** $f(-6) = 31$ **19.** $f(4) = -1113$
21. Yes **23.** No **25.** Yes **27.** Yes **29.** No **31.** Yes
33. Yes **35.** $f(x) = (x - 2)(x + 3)(x - 7)$
37. $f(x) = (x + 2)(4x - 1)(3x + 2)$
39. $f(x) = (x + 1)^2(x - 4)$
41. $f(x) = (x - 6)(x + 2)(x - 2)(x^2 + 4)$
43. $f(x) = (x + 5)(3x - 4)^2$ **45.** $k = 1$ or $k = -4$
47. $k = 6$ **49.** $f(c) > 0$ for all values of c
51. Let $f(x) = x^n - 1$. Because $(-1)^n = 1$ for all even positive integral values of n, $f(-1) = 0$ and $x - (-1) = x + 1$ is a factor.

53. **(a)** Let $f(x) = x^n - y^n$. Therefore $f(y) = y^n - y^n = 0$ and $x - y$ is a factor of $f(x)$.   **(c)** Let $f(x) = x^n + y^n$. Therefore $f(-y) = (-y)^n + y^n = -y^n + y^n = 0$ when n is odd, and $x - (-y) = x + y$ is a factor of $f(x)$.
57. $f(1 + i) = 2 + 6i$
61. **(a)** $f(4) = 137; f(-5) = 11; f(7) = 575$
   (c) $f(4) = -79; f(5) = -162; f(-3) = 110$

Problem Set 9.3 (page 483)

1. $\{-3, 1, 4\}$ **3.** $\left\{-1, -\dfrac{1}{3}, \dfrac{2}{5}\right\}$ **5.** $\left\{-2, -\dfrac{1}{4}, \dfrac{5}{2}\right\}$

7. $\{-3, 2\}$ **9.** $\{2, 1 \pm \sqrt{5}\}$ **11.** $\{-3, -2, -1, 2\}$
13. $\{-2, 3, -1 \pm 2i\}$ **15.** $\{1, \pm i\}$ **17.** $\left\{-\dfrac{5}{2}, 1, \pm\sqrt{3}\right\}$

19. $\left\{-2, \dfrac{1}{2}\right\}$ **27.** $\{-3, -1, 2\}$ **29.** $\left\{-\dfrac{5}{2}, \dfrac{1}{3}, 3\right\}$

31. 1 positive and 1 negative solution
33. 1 positive and 2 nonreal complex solutions
35. 1 negative and 2 positive solutions *or* 1 negative and 2 nonreal complex solutions
37. 5 positive solutions *or* 3 positive and 2 nonreal complex solutions *or* 1 positive solutions and 4 nonreal complex solutions
39. 1 negative and 4 nonreal complex solutions
47. **(a)** $\{4, -3 \pm i\}$ **(c)** $\{-2, 6, 1 \pm \sqrt{3}\}$
   (e) $\{12, 1 \pm i\}$

Problem Set 9.4 (page 494)

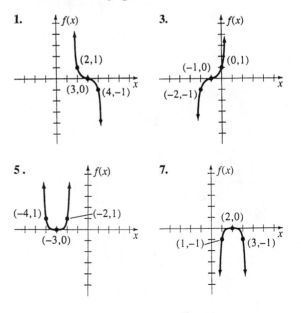

1. $f(x)$, $(2, 1)$, $(3, 0)$, $(4, -1)$

3. $f(x)$, $(-1, 0)$, $(0, 1)$, $(-2, -1)$

5. $f(x)$, $(-4, 1)$, $(-2, 1)$, $(-3, 0)$

7. $f(x)$, $(2, 0)$, $(1, -1)$, $(3, -1)$

9.

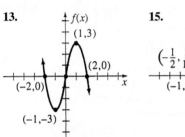

11.

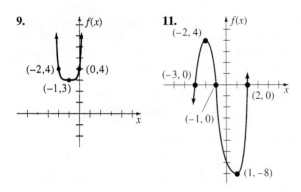

13.

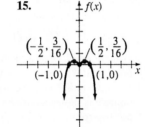

15.

17.

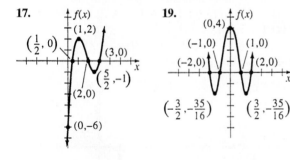

19.

21.

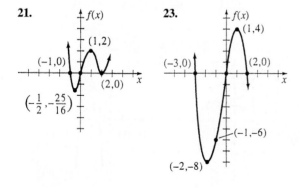

23.

25.

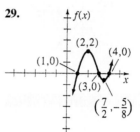

27.

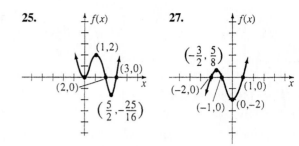

29.

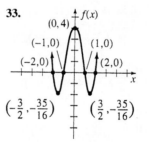

31.

33.

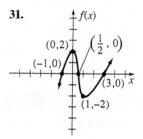

35. **(a)** -144 **(b)** $-3, 6$, and 8
(c) $f(x) > 0$ for $\{x|x < -3$ or $6 < x < 8\}$; $f(x) < 0$ for
$\{x|-3 < x < 6$ or $x > 8\}$

37. **(a)** -81 **(b)** -3 and 1 **(c)** $f(x) > 0$ for
$\{x|x > 1\}$; $f(x) < 0$ for $\{x|x < -3$ or $-3 < x < 1\}$

39. **(a)** 0 **(b)** $-4, 0$, and 6
(c) $f(x) > 0$ for $\{x|x < -4$ or $0 < x < 6$ or $x > 6\}$;
$f(x) < 0$ for $\{x|-4 < x < 0\}$

41. **(a)** 0 **(b)** $-3, 0$, and 2
(c) $f(x) > 0$ for $\{x|-3 < x < 0$ or $0 < x < 2\}$;
$f(x) < 0$ for $\{x|x < -3$ or $x > 2\}$

45. **(a)** 1.6 **(c)** 6.1 **(e)** 2.5

51. **(a)** $-2, 1$, and 4; $f(x) > 0$ for $(-2, 1) \cup (4, \infty)$; $f(x) < 0$
for $(-\infty, -2) \cup (1, 4)$
(c) 2 and 3; $f(x) > 0$ for $(3, \infty)$; $f(x) < 0$ for $(2, 3) \cup$
$(-\infty, -2)$ **(e)** $-3, -1$, and 2; $f(x) > 0$ for
$(-\infty, -3) \cup (2, \infty)$; $f(x) < 0$ for $(-3, -1) \cup (-1, 2)$

53. (a) -3.3; $(0.5, 3.1)$, $(-1.9, 10.1)$
  (c) $-2.2, 2.2$; $(-1.4, -8.0)$, $(0, -4.0)$, $(1.4, -8.0)$
55. 32 units

Problem Set 9.5 (page 506)

1. **3.**

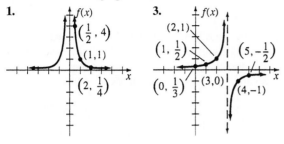

13.

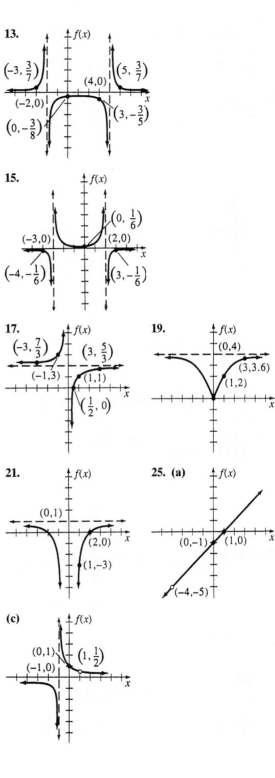

5. **7.**

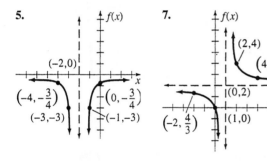

15.

17. **19.**

9.

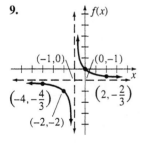

21. **25. (a)**

11. **(c)**

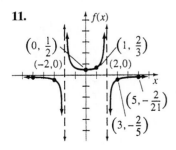

Problem Set 9.6 (page 515)

1.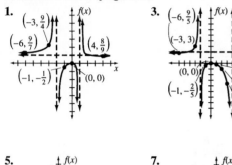

3.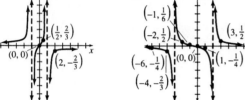

5.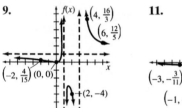

7.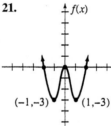

9.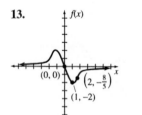

11.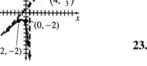

13.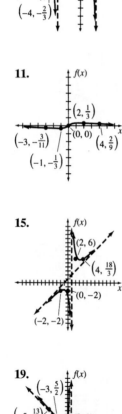

15.

17.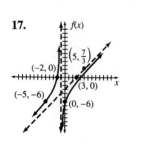

19.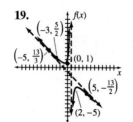

Chapter 9 Review Problem Set (page 518)

1. $Q: 3x^2 - x + 5; R: 3$ **2.** $Q: 5x^2 - 3x - 3; R: 16$
3. $Q: -2x^3 + 9x^2 - 38x + 151; R: -605$
4. $Q: -3x^3 - 9x^2 - 32x - 96; R: -279$ **5.** $f(1) = 1$
6. $f(-3) = -197$ **7.** $f(-2) = 20$ **8.** $f(8) = 0$
9. Yes **10.** No **11.** Yes **12.** Yes **13.** $\{-3, 1, 5\}$
14. $\left\{-\dfrac{7}{2}, -1, \dfrac{5}{4}\right\}$ **15.** $\{1, 2, 1 \pm 5i\}$ **16.** $\{-2, 3 \pm \sqrt{7}\}$
17. 2 positive and 2 negative solutions *or* 2 positive and 2 nonreal complex solutions *or* 2 negative and 2 nonreal complex solutions *or* 4 nonreal complex solutions
18. 1 negative and 4 nonreal complex solutions

19.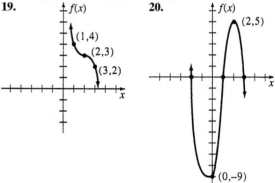

20.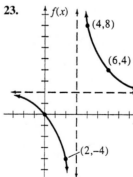

21.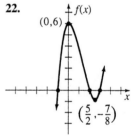

22.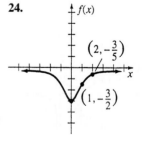

23.

24.

25.
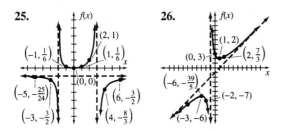

Chapter 9 Test (page 519)

1. $Q: 3x^2 - 4x - 2; R: 0$
2. $Q: 4x^3 + 8x^2 + 9x + 17; R: 38$ **3.** -24 **4.** 5
5. 39 **6.** No **7.** No **8.** Yes **9.** No **10.** $\{-4, 1, 3\}$
11. $\left\{-4, \dfrac{3 \pm \sqrt{17}}{4}\right\}$ **12.** $\{-3, 1, 3 \pm i\}$
13. $\left\{-4, 1, \dfrac{3}{2}\right\}$ **14.** $\left\{-\dfrac{5}{3}, 2\right\}$
15. 1 positive, 1 negative, and 2 nonreal complex solutions
16. $-7, 0,$ and $\dfrac{2}{3}$ **17.** $x = -3$ **18.** $f(x) = 5$ or $y = 5$
19. y axis symmetry **20.** Origin symmetry

21. 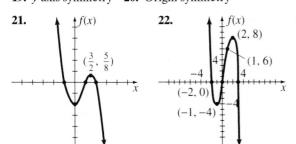 **22.**

23. 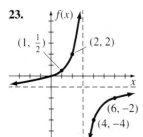 **24.**

25.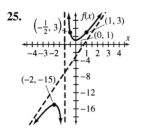

CHAPTER 10

Problem Set 10.1 (page 528)

1. $\{6\}$ **3.** $\left\{\dfrac{3}{2}\right\}$ **5.** $\{7\}$ **7.** $\{5\}$ **9.** $\{1\}$ **11.** $\{-3\}$
13. $\left\{\dfrac{3}{2}\right\}$ **15.** $\left\{\dfrac{1}{5}\right\}$ **17.** $\{0\}$ **19.** $\{-1\}$ **21.** $\left\{\dfrac{5}{2}\right\}$
23. $\{3\}$ **25.** $\left\{\dfrac{1}{2}\right\}$

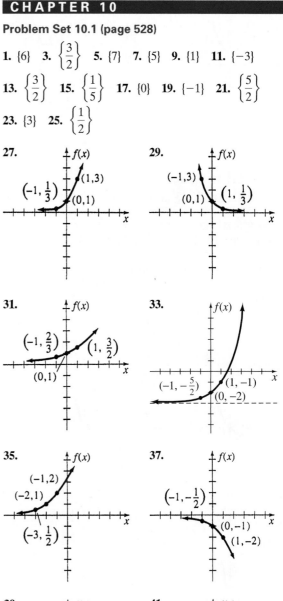
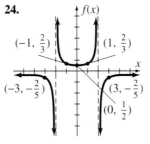
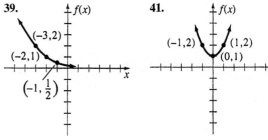

43.

45.

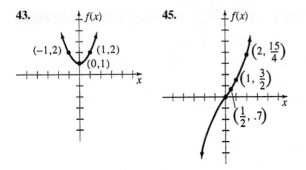

51.

	8%	10%	12%	14%
Compounded annually	$2159	2594	3106	3707
Compounded semiannually	2191	2653	3207	3870
Compounded quarterly	2208	2685	3262	3959
Compounded monthly	2220	2707	3300	4022
Compounded continuously	2226	2718	3320	4055

53.

55.

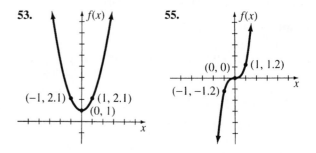

Problem Set 10.2 (page 538)

1. (a) $0.87 **(c)** $2.33 **(e)** $21,900 **(g)** $658
3. $283.70 **5.** $865.84 **7.** $1782.25 **9.** $2725.05
11. $16,998.71 **13.** $22,553.65 **15.** $567.63
17. $1422.36 **19.** $8963.38 **21.** $17,547.35
23. $32,558.88 **25.** 5.9% **27.** 8.06%
29. 8.25% compounded quarterly
31. 50 grams; 37 grams **33.** 2226; 3320; 7389
35. 2000 **37. (a)** 6.5 pounds per square inch
(c) 13.6 pounds per square inch

39.

41.

43.

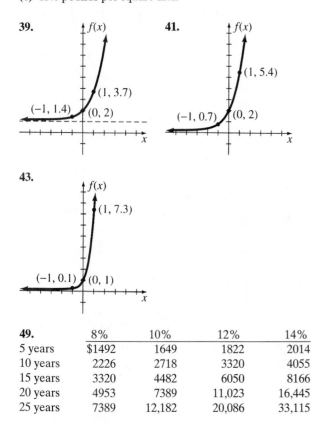

Problem Set 10.3 (page 549)

1. Yes **3.** No **5.** Yes **7.** Yes **9.** Yes
11. No **13.** No
15. (a) Domain of f: $\{1, 2, 5\}$; Range of f: $\{5, 9, 21\}$
(b) $f^{-1} = \{(5, 1), (9, 2), (21, 5)\}$
(c) Domain of f^{-1}: $\{5, 9, 21\}$; Range of f^{-1}: $\{1, 2, 5\}$
17. (a) Domain of f: $\{0, 2, -1, -2\}$;
Range of f: $\{0, 8, -1, -8\}$
(b) f^{-1}: $\{(0, 0), (8, 2), (-1, -1), (-8, -2)\}$
(c) Domain of f^{-1}: $\{0, 8, -1, -8\}$;
Range of f^{-1}: $\{0, 2, -1, -2\}$
27. No **29.** Yes **31.** No **33.** Yes **35.** Yes
37. $f^{-1}(x) = x + 4$ **39.** $f^{-1}(x) = \dfrac{-x - 4}{3}$
41. $f^{-1}(x) = \dfrac{12x + 10}{9}$ **43.** $f^{-1}(x) = -\dfrac{3}{2}x$
45. $f^{-1}(x) = x^2$ for $x \geq 0$ **47.** $f^{-1}(x) = \sqrt{x - 4}$ for $x \geq 4$
49. $f^{-1}(x) = \dfrac{1}{x - 1}$ for $x > 1$
51. $f^{-1}(x) = \dfrac{1}{3}x$ **53.** $f^{-1}(x) = \dfrac{x - 1}{2}$

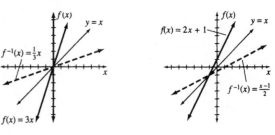

49.

	8%	10%	12%	14%
5 years	$1492	1649	1822	2014
10 years	2226	2718	3320	4055
15 years	3320	4482	6050	8166
20 years	4953	7389	11,023	16,445
25 years	7389	12,182	20,086	33,115

55. $f^{-1}(x) = \dfrac{x + 2}{x}$ for $x > 0$

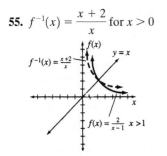

57. $f^{-1}(x) = \sqrt{x + 4}$ for $x \geq -4$

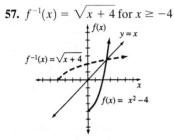

59. Increasing on $[0, \infty)$ and decreasing on $(-\infty, 0]$
61. Decreasing on $(-\infty, \infty)$
63. Increasing on $(-\infty, -2]$ and decreasing on $[-2, \infty)$
65. Increasing on $(-\infty, -4]$ and increasing on $[-4, \infty)$
71. (a) $f^{-1}(x) = \dfrac{x + 9}{3}$ **(c)** $f^{-1}(x) = -x + 1$
(e) $f^{-1}(x) = -\dfrac{1}{5}x$

Problem Set 10.4 (page 560)
1. $\log_2 128 = 7$ **3.** $\log_5 125 = 3$ **5.** $\log_{10} 1000 = 3$
7. $\log_2 \left(\dfrac{1}{4}\right) = -2$ **9.** $\log_{10} 0.1 = -1$ **11.** $3^4 = 81$

13. $4^3 = 64$ **15.** $10^4 = 10,000$ **17.** $2^{-4} = \dfrac{1}{16}$

19. $10^{-3} = 0.001$ **21.** 4   **23.** 4   **25.** 3   **27.** $\dfrac{1}{2}$ **29.** 0
31. -1 **33.** 5   **35.** -5 **37.** 1   **39.** 0   **41.** $\{49\}$
43. $\{16\}$ **45.** $\{27\}$ **47.** $\left\{\dfrac{1}{8}\right\}$ **49.** $\{4\}$ **51.** 5.1293
53. 6.9657   **55.** 1.4037   **57.** 7.4512   **59.** 6.3219
61. -0.3791 **63.** 0.5766   **65.** 2.1531   **67.** 0.3949
69. $\log_b x + \log_b y + \log_b z$ **71.** $\log_b y - \log_b z$
73. $3 \log_b y + 4 \log_b z$ **75.** $\dfrac{1}{2} \log_b x + \dfrac{1}{3} \log_b y - 4 \log_b z$
77. $\dfrac{2}{3} \log_b x + \dfrac{1}{3} \log_b z$ **79.** $\dfrac{3}{2} \log_b x - \dfrac{1}{2} \log_b y$

81. $\log_b \left(\dfrac{x^2}{y^4}\right)$ **83.** $\log_b \left(\dfrac{xz}{y}\right)$ **85.** $\log_b \left(\dfrac{x^2 z^4}{z^3}\right)$
87. $\log_b \left(\dfrac{y^4 \sqrt{x}}{x}\right)$ **89.** $\left\{\dfrac{9}{4}\right\}$ **91.** $\{25\}$ **93.** $\{4\}$ **95.** $\{-2\}$
97. $\left\{-\dfrac{4}{3}\right\}$ **99.** $\left\{\dfrac{19}{8}\right\}$ **101.** $\{9\}$ **103.** $\varnothing$ **105.** $\{1\}$

Problem Set 10.5 (page 568)
1. 0.8597   **3.** 1.7179   **5.** 3.5071   **7.** -0.1373
9. -3.4685 **11.** 411.43   **13.** 90,095   **15.** 79.543
17. 0.048440   **19.** 0.0064150   **21.** 1.6094
23. 3.4843   **25.** 6.0638   **27.** -0.7765 **29.** -3.4609
31. 1.6034   **33.** 3.1346   **35.** 108.56   **37.** 0.48268
39. 0.035994

41.   **43.**

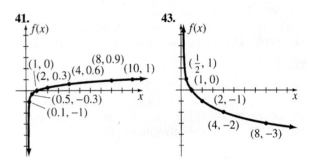

45.   **47.**

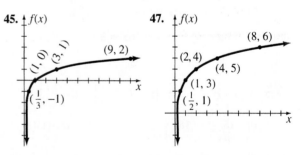

49.   **51.**

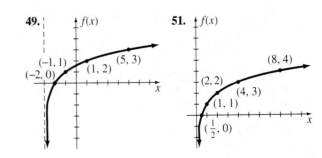

53.

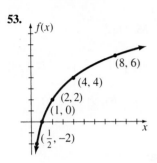

55. 0.36   **57.** 0.73   **59.** 23.10   **61.** 7.93

Problem Set 10.6 (page 578)

1. {2.33}   **3.** {2.56}   **5.** {5.43}   **7.** {4.18}   **9.** {0.12}
11. {3.30}   **13.** {4.57}   **15.** {1.79}   **17.** {3.32}   **19.** {2.44}
21. {4}   **23.** $\left\{\dfrac{19}{47}\right\}$ **25.** $\left\{\dfrac{-1 + \sqrt{33}}{4}\right\}$ **27.** {1}
29. {8}   **31.** {1, 10,000}   **33.** 5.322   **35.** 2.524   **37.** 0.339
39. −0.837   **41.** 3.194   **43.** 2.4 years   **45.** 5.3 years
47. 5.9%   **49.** 6.8 hours   **51.** 6100 feet   **53.** 3.5 hours
55. 6.7   **57.** Approximately 8 times   **65.** {1.13}
67. $x = \ln(y + \sqrt{y^2 + 1})$

Chapter 10 Review Problem (page 581)

1. 32   **2.** −125   **3.** 81   **4.** 3   **5.** −2   **6.** $\dfrac{1}{3}$ **7.** $\dfrac{1}{4}$
8. −5   **9.** 1   **10.** 12   **11.** {5}   **12.** $\left\{\dfrac{1}{9}\right\}$ **13.** $\left\{\dfrac{7}{2}\right\}$
14. {3.40}   **15.** {8}   **16.** $\left\{\dfrac{1}{11}\right\}$ **17.** {1.95}   **18.** {1.41}
19. {1.56}   **20.** {20}   **21.** {10^{100}}   **22.** {2}   **23.** $\left\{\dfrac{11}{2}\right\}$
24. {0}   **25.** 0.3680   **26.** 1.3222   **27.** 1.4313   **28.** 0.5634
29. (a) $\log_b x - 2\log_b y$ (b) $\dfrac{1}{4}\log_b x + \dfrac{1}{2}\log_b y$
(c) $\dfrac{1}{2}\log_b x - 3\log_b y$ **30.** (a) $\log_b x^3 y^2$ (b) $\log_b\left(\dfrac{\sqrt{y}}{x^4}\right)$
(c) $\log_b\left(\dfrac{\sqrt{xy}}{z^2}\right)$ **31.** 1.585   **32.** 0.631
33. 3.789   **34.** −2.120

35. (a)

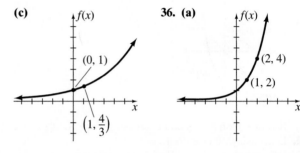

(b)

(c)

36. (a)

(b)

(c)

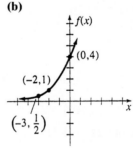

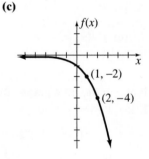

37. (a)

(b)

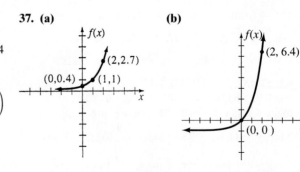

(c)

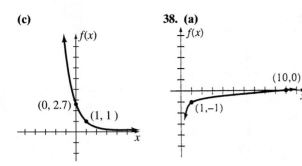

38. (a)

42.

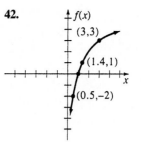

43. $2219.91   **44.** $4797.55   **45.** $15,999.31   **46.** Yes

47. No   **48.** Yes   **49.** Yes   **50.** $f^{-1}(x) = \dfrac{x - 5}{4}$

(b)

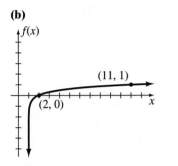

51. $f^{-1}(x) = \dfrac{-x - 7}{3}$ **52.** $f^{-1}(x) = \dfrac{6x + 2}{5}$

53. $f^{-1}(x) = \sqrt{-2 - x}$

54. Increasing on $(-\infty, 4]$ and decreasing on $[4, \infty)$

55. Increasing on $[3, \infty)$ **56.** Approximately 5.3 years

57. Approximately 12.1 years   **58.** Approximately 8.7%

59. 61,070; 67,493; 74,591   **60.** Approximately 4.8 hours

61. 133 grams   **62.** 8.1

(c)

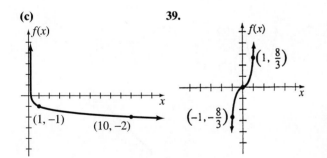

39.

Chapter 10 Test (page 584)

1. $\dfrac{1}{2}$ **2.** 1   **3.** 1   **4.** -1 **5.** $\{-3\}$ **6.** $\left\{-\dfrac{3}{2}\right\}$ **7.** $\left\{\dfrac{8}{3}\right\}$

8. $\{243\}$ **9.** $\{2\}$ **10.** $\left\{\dfrac{2}{5}\right\}$ **11.** 4.1919   **12.** 0.2031

13. 0.7325   **14.** $f^{-1}(x) = \dfrac{-6 - x}{3}$ **15.** $\{5.17\}$

16. $\{10.29\}$ **17.** 4.0069   **18.** $f^{-1}(x) = \dfrac{3}{2}x + \dfrac{9}{10}$

19. $6342.08   **20.** 13.5 years   **21.** 7.8 hours

22. 4813 grams

40.   **41.**   **23.**   **24.**

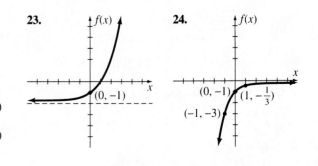

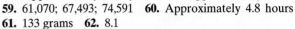

25.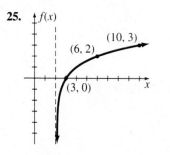

Cumulative Review Problem Set (page 585)

1. -6 **2.** -8 **3.** $\dfrac{13}{24}$ **4.** 56 **5.** $\dfrac{13}{6}$ **6.** $-90\sqrt{2}$

7. $2x + 5\sqrt{x} - 12$ **8.** $-18 + 22\sqrt{3}$

9. $2x^3 + 11x^2 - 14x + 4$ **10.** $\dfrac{x + 4}{x(x + 5)}$ **11.** $\dfrac{16x^2}{27y}$

12. $\dfrac{16x + 43}{90}$ **13.** $\dfrac{35a - 44b}{60a^2b}$ **14.** $\dfrac{2}{x - 4}$

15. $2x^2 - x - 4$ **16.** $\dfrac{5y^2 - 3xy^2}{x^2y + 2x^2}$ **17.** $\dfrac{2y - 3xy}{3x + 4xy}$

18. $\dfrac{(2n - 5)(n + 3)}{(n - 2)(3n + 13)}$ **19.** $\dfrac{3a^2 - 2a + 1}{2a - 1}$

20. $(5x - 2)(4x + 3)$ **21.** $2(2x + 3)(4x^2 - 6x + 9)$
22. $(2x + 3)(2x - 3)(x + 2)(x - 2)$
23. $4x(3x + 2)(x - 5)$ **24.** $(y - 6)(x + 3)$

25. $(5 - 3x)(2 + 3x)$ **26.** $\dfrac{81}{16}$ **27.** 4 **28.** $-\dfrac{3}{4}$

29. -0.3 **30.** $\dfrac{1}{81}$ **31.** $\dfrac{21}{16}$ **32.** $\dfrac{9}{64}$ **33.** 72 **34.** 6

35. -2 **36.** $\dfrac{-12}{x^3y}$ **37.** $\dfrac{8y}{x^5}$ **38.** $-\dfrac{a^3}{9b}$ **39.** $4\sqrt{5}$

40. $-6\sqrt{6}$ **41.** $\dfrac{5\sqrt{3}}{9}$ **42.** $\dfrac{2\sqrt{3}}{3}$ **43.** $2\sqrt[3]{7}$ **44.** $\dfrac{\sqrt[3]{6}}{2}$

45. $8xy\sqrt{13x}$ **46.** $\dfrac{\sqrt{6xy}}{3y}$ **47.** $11\sqrt{6}$ **48.** $-\dfrac{169\sqrt{2}}{12}$

49. $-16\sqrt[3]{3}$ **50.** $\dfrac{-3\sqrt{2} - 2\sqrt{6}}{2}$

51. $\dfrac{6\sqrt{15} - 3\sqrt{35} - 6 + \sqrt{21}}{5}$ **52.** 0.021 **53.** 300

54. 0.0003 **55.** $32 + 22i$ **56.** $-17 + i$ **57.** $0 - \dfrac{5}{4}i$

58. $-\dfrac{19}{53} + \dfrac{40}{53}i$ **59.** $-\dfrac{10}{3}$ **60.** $\dfrac{4}{7}$ **61.** $2\sqrt{13}$

62. $5x - 4y = 19$ **63.** $4x + 3y = -18$
64. $(-2, 6)$ and $r = 3$ **65.** $(-5, -4)$ **66.** 8 units

67.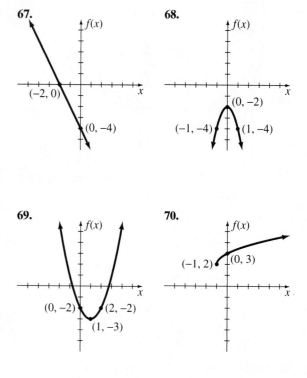

68.

69.

70.

71. 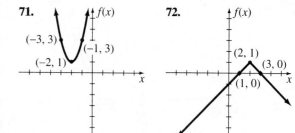 **72.**

73. 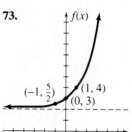 **74.**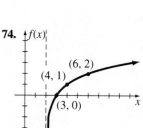

75.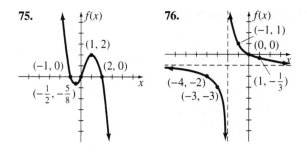

76.

77. $(g \circ f)(x) = 2x^2 - 13x + 20; (f \circ g)(x) = 2x^2 - x - 4$

78. $f^{-1}(x) = \dfrac{x + 7}{3}$ **79.** $f^{-1}(x) = -2x + \dfrac{4}{3}$

80. $k = -3$ **81.** $y = 1$ **82.** 12 cubic centimeters

83. $\left\{-\dfrac{21}{16}\right\}$ **84.** $\left\{\dfrac{40}{3}\right\}$ **85.** $\{6\}$ **86.** $\left\{-\dfrac{5}{2}, 3\right\}$

87. $\left\{0, \dfrac{7}{3}\right\}$ **88.** $\{-6, 0, 6\}$ **89.** $\left\{-\dfrac{5}{6}, \dfrac{2}{5}\right\}$

90. $\left\{-3, 0, \dfrac{3}{2}\right\}$ **91.** $\{\pm 1, \pm 3i\}$ **92.** $\{-5, 7\}$ **93.** $\{-29, 0\}$

94. $\left\{\dfrac{7}{2}\right\}$ **95.** $\{12\}$ **96.** $\{-3\}$ **97.** $\left\{\dfrac{1 \pm 3\sqrt{5}}{3}\right\}$

98. $\left\{\dfrac{-5 \pm 4i\sqrt{2}}{2}\right\}$ **99.** $\left\{\dfrac{3 \pm i\sqrt{23}}{4}\right\}$

100. $\left\{\dfrac{3 \pm \sqrt{3}}{3}\right\}$ **101.** $\{1 \pm \sqrt{34}\}$

102. $\left\{\pm\dfrac{\sqrt{5}}{2}, \pm\dfrac{\sqrt{3}}{3}\right\}$ **103.** $\left\{\dfrac{-5 \pm i\sqrt{15}}{4}\right\}$

104. $\{-4, 1, 7\}$ **105.** $\left\{-\dfrac{1}{2}, \dfrac{5}{3}, 2\right\}$ **106.** $\left\{\dfrac{3}{2}\right\}$

107. $\{81\}$ **108.** $\{4\}$ **109.** $\{6\}$ **110.** $\left\{\dfrac{1}{5}\right\}$

111. $(-\infty, 3)$ **112.** $(-\infty, 50]$

113. $\left(-\infty, -\dfrac{11}{5}\right) \cup (3, \infty)$ **114.** $\left(-\dfrac{5}{3}, 1\right)$

115. $\left[-\dfrac{9}{11}, \infty\right)$ **116.** $[-4, 2]$

117. $\left(-\infty, \dfrac{1}{3}\right) \cup (4, \infty)$ **118.** $(-8, 3)$

119. $(-\infty, 3] \cup (7, \infty)$ **120.** $(-6, -3)$
121. 17, 19, and 21
122. 14 nickels, 20 dimes, and 29 quarters
123. $48°$ and $132°$ **124.** $600
125. $1700 at 8% and $2000 at 9%
126. 66 miles per hour and 76 miles per hour
127. 4 quarts **128.** 69 or less **129.** $-3, 0,$ or 3

130. 1-inch strip **131.** $1050 and $1400 **132.** 3 hours
133. 30 shares at $10 per share **134.** 37
135. $10°, 60°,$ and $110°$

CHAPTER 11

Problem Set 11.1 (page 598)

1. $\{(3, 2)\}$ **3.** $\{(2, 1)\}$ **5.** Dependent **7.** $\{(4, -3)\}$
9. Inconsistent **11.** $\{(7, 9)\}$ **13.** $\{(-4, 7)\}$ **15.** $\{(6, 3)\}$
17. $a = -3$ and $b = -4$

19. $\left\{\left(k, \dfrac{2}{3}k - \dfrac{4}{3}\right)\right\}$, a dependent system

21. $u = 5$ and $t = 7$ **23.** $\{(2, -5)\}$

25. $\varnothing$, an inconsistent system **27.** $\left\{\left(-\dfrac{3}{4}, -\dfrac{6}{5}\right)\right\}$

29. $\{(3, -4)\}$ **31.** $\{(2, 8)\}$ **33.** $\{(-1, -5)\}$

35. $\varnothing$, an inconsistent system **37.** $a = 2$ and $b = -\dfrac{1}{3}$

39. $s = -6$ and $t = 12$ **41.** $\left\{\left(-\dfrac{1}{2}, \dfrac{1}{3}\right)\right\}$

43. $\left\{\left(\dfrac{3}{4}, -\dfrac{2}{3}\right)\right\}$ **45.** $\{(-4, 2)\}$ **47.** $\{(5, 5)\}$

49. $\varnothing$, an inconsistent system **51.** $\{(12, -24)\}$
53. $t = 8$ and $u = 3$ **55.** $\{(200, 800)\}$ **57.** $\{(400, 800)\}$
59. $\{(3.5, 7)\}$ **61.** 17 and 36 **63.** $15°, 75°$ **65.** 72
67. 34 **69.** 8 single rooms and 15 double rooms
71. 2500 student tickets and 500 nonstudent tickets
73. $500 at 9% and $1500 at 11%
75. 3 miles per hour
77. $1.25 per tennis ball and $1.75 per golf ball
79. 30 five-dollar bills and 18 ten-dollar bills

85. $\{(4, 6)\}$ **87.** $\{(2, -3)\}$ **89.** $\left\{\left(\dfrac{1}{4}, -\dfrac{2}{3}\right)\right\}$

Problem Set 11.2 (page 608)

1. $\{(-4, -2, 3)\}$ **3.** $\{(-2, 5, 2)\}$ **5.** $\{(4, -1, -2)\}$
7. $\{(3, 1, 2)\}$ **9.** $\{(-1, 3, 5)\}$ **11.** $\{(-2, -1, 3)\}$
13. $\{(0, 2, 4)\}$ **15.** $\{(4, -1, -2)\}$ **17.** $\{(-4, 0, -1)\}$
19. $\{(2, 2, -3)\}$
21. 4 pounds of pecans, 4 pounds of almonds, and
    12 pounds of peanuts
23. 7 nickels, 13 dimes, and 22 quarters
25. $40°, 60°,$ and $80°$
27. $500 at 12%, $1000 at 13%, and $1500 at 14%
29. 50 of type A, 75 of type B, and 150 of type C

Problem Set 11.3 (page 618)

1. Yes **3.** Yes **5.** No **7.** No **9.** Yes **11.** $\{(-1, -5)\}$
13. $\{(3, -6)\}$ **15.** $\varnothing$ **17.** $\{(-2, -9)\}$ **19.** $\{(-1, -2, 3)\}$

21. $\{(3, -1, 4)\}$ **23.** $\{(0, -2, 4)\}$
25. $\{(-7k + 8, -5k + 7, k)\}$ **27.** $\{(-4, -3, -2)\}$
29. $\{(4, -1, -2)\}$ **31.** $\{(1, -1, 2, -3)\}$
33. $\{(2, 1, 3, -2)\}$ **35.** $\{(-2, 4, -3, 0)\}$
37. $\varnothing$ **39.** $\{(-3k + 5, -1, -4k + 2, k)\}$
41. $\{(-3k + 9, k, 2, -3)\}$
45. $\{(17k - 6, 10k - 5, k)\}$
47. $\left\{\left(-\dfrac{1}{2}k + \dfrac{34}{11}, \dfrac{1}{2}k - \dfrac{5}{11}, k\right)\right\}$ **49.** $\varnothing$

Problem Set 11.4 (page 627)

1. 22 **3.** -29 **5.** 20 **7.** 5 **9.** -2 **11.** $-\dfrac{2}{3}$ **13.** -25
15. 58 **17.** 39 **19.** -12 **21.** -41 **23.** -8 **25.** 1088
27. -140 **29.** 81 **31.** 146 **33.** Property 11.3
35. Property 11.2 **37.** Property 11.4 **39.** Property 11.3
41. Property 11.5

Problem Set 11.5 (page 635)

1. $\{(1, 4)\}$ **3.** $\{(3, -5)\}$ **5.** $\{(2, -1)\}$ **7.** $\varnothing$
9. $\left\{\left(-\dfrac{1}{4}, \dfrac{2}{3}\right)\right\}$ **11.** $\left\{\left(\dfrac{2}{17}, \dfrac{52}{17}\right)\right\}$ **13.** $\{(9, -2)\}$
15. $\left\{\left(2, -\dfrac{5}{7}\right)\right\}$ **17.** $\{(0, 2, -3)\}$ **19.** $\{(2, 6, 7)\}$
21. $\{(4, -4, 5)\}$ **23.** $\{(-1, 3, -4)\}$
25. Infinitely many solutions **27.** $\left\{\left(-2, \dfrac{1}{2}, -\dfrac{2}{3}\right)\right\}$
29. $\left\{\left(3, \dfrac{1}{2}, -\dfrac{1}{3}\right)\right\}$ **31.** $(-4, 6, 0)$ **37.** $(0, 0, 0)$
39. Infinitely many solutions

Problem Set 11.6 (page 642)

1. $\dfrac{4}{x - 2} + \dfrac{7}{x + 1}$ **3.** $\dfrac{3}{x + 1} - \dfrac{5}{x - 1}$
5. $\dfrac{1}{3x - 1} + \dfrac{6}{2x + 3}$ **7.** $\dfrac{2}{x - 1} + \dfrac{3}{x + 2} - \dfrac{4}{x - 3}$
9. $\dfrac{-1}{x} + \dfrac{2}{2x - 1} - \dfrac{3}{4x + 1}$ **11.** $\dfrac{2}{x - 2} + \dfrac{5}{(x - 2)^2}$
13. $\dfrac{4}{x} + \dfrac{7}{x^2} - \dfrac{10}{x + 3}$ **15.** $\dfrac{-3}{x^2 + 1} - \dfrac{2}{x - 4}$
17. $\dfrac{3}{x + 2} - \dfrac{2}{(x + 2)^2} + \dfrac{1}{(x + 2)^3}$
19. $\dfrac{2}{x} + \dfrac{3x + 5}{x^2 - x + 3}$ **21.** $\dfrac{2x}{x^2 + 1} + \dfrac{3 - x}{(x^2 + 1)^2}$

Chapter 11 Review Problem Set (page 644)

1. $\{(3, -7)\}$ **2.** $\{(-1, -3)\}$ **3.** $\{(0, -4)\}$
4. $\left\{\left(\dfrac{23}{3}, -\dfrac{14}{3}\right)\right\}$ **5.** $\{(4, -6)\}$ **6.** $\left\{\left(-\dfrac{6}{7}, -\dfrac{15}{7}\right)\right\}$
7. $\{(-1, 2, -5)\}$ **8.** $\{(2, -3, -1)\}$ **9.** $\{(5, -4)\}$
10. $\{(2, 7)\}$ **11.** $\{(-2, 2, -1)\}$ **12.** $\{(0, -1, 2)\}$
13. $\{(-3, -1)\}$ **14.** $\{(4, 6)\}$ **15.** $\{(2, -3, -4)\}$
16. $\{(-1, 2, -5)\}$ **17.** $\{(5, -5)\}$ **18.** $\{(-12, 12)\}$
19. $\left\{\left(\dfrac{5}{7}, \dfrac{4}{7}\right)\right\}$ **20.** $\{(-10, -7)\}$ **21.** $\{(1, 1, -4)\}$
22. $\{(-4, 0, 1)\}$ **23.** $\varnothing$ **24.** $\{(-2, -4, 6)\}$ **25.** -34
26. 13 **27.** -40 **28.** 16 **29.** 51 **30.** 125 **31.** 72
32. $900 at 10% and $1600 at 12%
33. 20 nickels, 32 dimes, and 54 quarters
34. 25°, 45°, and 110°

Chapter 11 Test (page 646)

1. III **2.** I **3.** III **4.** II **5.** 8 **6.** $-\dfrac{7}{12}$ **7.** -18
8. 112 **9.** Infinitely many **10.** $\{(-2, 4)\}$ **11.** $\{(3, -1)\}$
12. $x = -12$ **13.** $y = -\dfrac{13}{11}$ **14.** $x = 14$ **15.** $y = 13$
16. Infinitely many **17.** None **18.** $\left\{\left(\dfrac{11}{5}, 6, -3\right)\right\}$
19. $\{(-2, -1, 0)\}$ **20.** $x = 1$ **21.** $y = 4$ **22.** 2 liters
23. 22 quarters **24.** 5 batches of cream puffs, 4 batches of eclairs, and 10 batches of Danish rolls
25. 100°, 45°, and 35°

CHAPTER 12

Problem Set 12.1 (page 653)

1. $\begin{bmatrix} 3 & -5 \\ 8 & 3 \end{bmatrix}$ **3.** $\begin{bmatrix} -2 & 21 \\ -7 & 2 \end{bmatrix}$ **5.** $\begin{bmatrix} -2 & 1 \\ -3 & 19 \end{bmatrix}$
7. $\begin{bmatrix} -1 & -5 \\ 2 & 3 \end{bmatrix}$ **9.** $\begin{bmatrix} -12 & -14 \\ -18 & -20 \end{bmatrix}$ **11.** $\begin{bmatrix} 2 & -11 \\ -7 & 0 \end{bmatrix}$
13. $AB = \begin{bmatrix} 4 & -6 \\ 8 & -12 \end{bmatrix}, BA = \begin{bmatrix} -5 & 5 \\ 3 & -3 \end{bmatrix}$
15. $AB = \begin{bmatrix} -5 & -18 \\ -4 & 42 \end{bmatrix}, BA = \begin{bmatrix} 19 & -39 \\ -16 & 18 \end{bmatrix}$
17. $AB = \begin{bmatrix} 14 & -28 \\ 7 & -14 \end{bmatrix}, BA = \begin{bmatrix} 0 & 0 \\ 0 & 0 \end{bmatrix}$
19. $AB = \begin{bmatrix} -14 & -7 \\ -12 & -1 \end{bmatrix}, BA = \begin{bmatrix} -2 & -3 \\ -32 & -13 \end{bmatrix}$
21. $AB = \begin{bmatrix} 1 & 0 \\ 0 & 1 \end{bmatrix}, BA = \begin{bmatrix} 1 & 0 \\ 0 & 1 \end{bmatrix}$

23. $AB = \begin{bmatrix} 0 & -\dfrac{5}{3} \\ \dfrac{17}{6} & -3 \end{bmatrix}$, $BA = \begin{bmatrix} 0 & -\dfrac{17}{6} \\ \dfrac{5}{3} & -3 \end{bmatrix}$

25. $AB = \begin{bmatrix} 1 & 0 \\ 0 & 1 \end{bmatrix}$, $BA = \begin{bmatrix} 1 & 0 \\ 0 & 1 \end{bmatrix}$

27. $AB = \begin{bmatrix} 3 & -2 \\ 4 & 5 \end{bmatrix}$, $BA = \begin{bmatrix} 5 & 4 \\ -2 & 3 \end{bmatrix}$

29. $AD = \begin{bmatrix} 1 & 1 \\ 9 & 9 \end{bmatrix}$, $DA = \begin{bmatrix} 3 & 7 \\ 3 & 7 \end{bmatrix}$

49. $A^2 = \begin{bmatrix} -1 & -4 \\ 8 & 7 \end{bmatrix}$, $A^3 = \begin{bmatrix} -9 & -11 \\ 22 & 13 \end{bmatrix}$

Problem Set 12.2 (page 660)

1. $\begin{bmatrix} 3 & -7 \\ -2 & 5 \end{bmatrix}$ **3.** $\begin{bmatrix} -5 & 8 \\ 2 & -3 \end{bmatrix}$ **5.** $\begin{bmatrix} -\dfrac{2}{5} & \dfrac{1}{5} \\ \dfrac{3}{10} & \dfrac{1}{10} \end{bmatrix}$

7. Does not exist   **9.** $\begin{bmatrix} -\dfrac{5}{7} & \dfrac{2}{7} \\ -\dfrac{4}{7} & \dfrac{3}{7} \end{bmatrix}$ **11.** $\begin{bmatrix} -\dfrac{3}{5} & \dfrac{1}{5} \\ 1 & 0 \end{bmatrix}$

13. $\begin{bmatrix} -\dfrac{4}{5} & \dfrac{3}{5} \\ \dfrac{1}{5} & -\dfrac{2}{5} \end{bmatrix}$ **15.** $\begin{bmatrix} 2 & -\dfrac{5}{3} \\ 1 & -\dfrac{2}{3} \end{bmatrix}$ **17.** $\begin{bmatrix} \dfrac{1}{2} & \dfrac{1}{2} \\ \dfrac{1}{2} & -\dfrac{1}{2} \end{bmatrix}$

19. $\begin{bmatrix} 30 \\ 36 \end{bmatrix}$ **21.** $\begin{bmatrix} 0 \\ 5 \end{bmatrix}$ **23.** $\begin{bmatrix} -4 \\ 13 \end{bmatrix}$ **25.** $\begin{bmatrix} -4 \\ -13 \end{bmatrix}$ **27.** $\{(2, 3)\}$

29. $\{(-2, 5)\}$ **31.** $\{(0, -1)\}$ **33.** $\{(-1, -1)\}$ **35.** $\{(4, 7)\}$

37. $\left\{\left(-\dfrac{1}{3}, \dfrac{1}{2}\right)\right\}$ **39.** $\{(-9, 20)\}$

Problem Set 12.3 (page 668)

1. $\begin{bmatrix} 1 & 3 & -3 \\ 3 & -6 & 7 \end{bmatrix}$; $\begin{bmatrix} 3 & -5 & 11 \\ -7 & 6 & 3 \end{bmatrix}$; $\begin{bmatrix} 1 & 10 & -13 \\ 11 & -18 & 16 \end{bmatrix}$; $\begin{bmatrix} 10 & -12 & 30 \\ -18 & 12 & 16 \end{bmatrix}$

3. $[-1 \ -7 \ 13 \ 7]$; $[5 \ 5 \ -5 \ 17]$; $[-5 \ -20 \ 35 \ 9]$; $[14 \ 8 \ -2 \ 58]$

5. $\begin{bmatrix} 8 & -3 & -2 \\ 9 & 2 & -3 \\ 7 & 5 & 21 \end{bmatrix}$; $\begin{bmatrix} -2 & -1 & 4 \\ -11 & 6 & -11 \\ -7 & 5 & -3 \end{bmatrix}$; $\begin{bmatrix} 21 & -7 & -7 \\ 28 & 2 & 2 \\ 21 & 10 & 54 \end{bmatrix}$; $\begin{bmatrix} 2 & -6 & 10 \\ -24 & 20 & -36 \\ -14 & 20 & 12 \end{bmatrix}$

7. $\begin{bmatrix} 0 & 2 \\ -1 & 10 \\ 1 & -9 \\ 2 & 9 \end{bmatrix}$; $\begin{bmatrix} -2 & -2 \\ 5 & -4 \\ -11 & 1 \\ -16 & 13 \end{bmatrix}$; $\begin{bmatrix} 1 & 6 \\ -5 & 27 \\ 8 & -23 \\ 13 & 16 \end{bmatrix}$; $\begin{bmatrix} -6 & -4 \\ 14 & -2 \\ -32 & -6 \\ -46 & 48 \end{bmatrix}$

9. $AB = \begin{bmatrix} 11 & -8 & 14 \\ 4 & -16 & 8 \\ -28 & 22 & -36 \end{bmatrix}$; $BA = \begin{bmatrix} -20 & 21 \\ 8 & -21 \end{bmatrix}$

11. $AB = \begin{bmatrix} 22 & -8 & 1 & 3 \\ -42 & 36 & -26 & -20 \end{bmatrix}$; BA does not exist.

13. $AB = \begin{bmatrix} -12 & 5 & -5 \\ 14 & -2 & 4 \\ -10 & 13 & -5 \end{bmatrix}$; $BA = \begin{bmatrix} -1 & 0 & -6 \\ 10 & -2 & 16 \\ -8 & 5 & -16 \end{bmatrix}$

15. $AB = [-9]$; $BA = \begin{bmatrix} -2 & 1 & -3 & -4 \\ -6 & 3 & -9 & -12 \\ 4 & -2 & 6 & 8 \\ -8 & 4 & -12 & -16 \end{bmatrix}$

17. AB does not exist; $BA = \begin{bmatrix} 20 \\ 2 \\ -30 \end{bmatrix}$

19. $AB = \begin{bmatrix} 9 & -12 \\ -12 & 16 \\ 6 & -8 \end{bmatrix}$; BA does not exist.

21. $\begin{bmatrix} -\dfrac{1}{5} & \dfrac{3}{10} \\ \dfrac{2}{5} & -\dfrac{1}{10} \end{bmatrix}$ **23.** $\begin{bmatrix} 4 & -1 \\ -7 & 2 \end{bmatrix}$ **25.** $\begin{bmatrix} \dfrac{4}{5} & -\dfrac{1}{5} \\ -\dfrac{3}{5} & \dfrac{2}{5} \end{bmatrix}$

27. $\begin{bmatrix} \dfrac{7}{2} & -3 & \dfrac{1}{2} \\ -\dfrac{1}{2} & 0 & \dfrac{1}{2} \\ -\dfrac{1}{2} & 1 & -\dfrac{1}{2} \end{bmatrix}$ **29.** $\begin{bmatrix} -50 & -9 & 11 \\ -23 & -4 & 5 \\ 5 & 1 & -1 \end{bmatrix}$

31. Does not exist   **33.** $\begin{bmatrix} \dfrac{4}{7} & -1 & -\dfrac{9}{7} \\ -\dfrac{3}{14} & \dfrac{1}{2} & \dfrac{6}{7} \\ \dfrac{2}{7} & 0 & -\dfrac{1}{7} \end{bmatrix}$

35. $\begin{bmatrix} \dfrac{1}{2} & 0 & 0 \\ 0 & \dfrac{1}{4} & 0 \\ 0 & 0 & \dfrac{1}{10} \end{bmatrix}$ **37.** $\{(-3, 2)\}$ **39.** $\{(2, 5)\}$

41. $\{(-1, -2, 1)\}$ **43.** $\{(-2, 3, 5)\}$ **45.** $\{(-4, 3, 0)\}$

47. (a) $\{(-1, 2, 3)\}$ **(c)** $\{(-5, 0, -2)\}$ **(e)** $\{(1, -1, -1)\}$

Problem Set 12.4 (page 678)

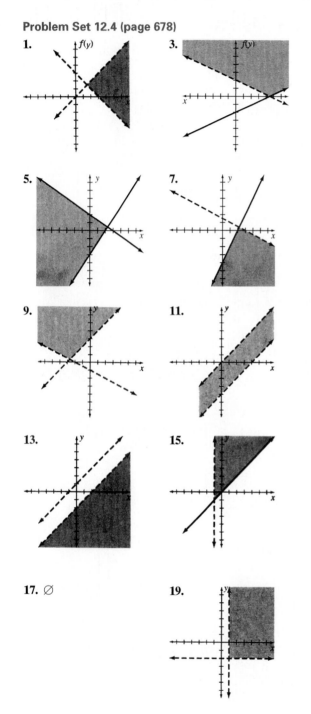

1.

3.

5.

7.

9.

11.

13.

15.

17. ∅

19.

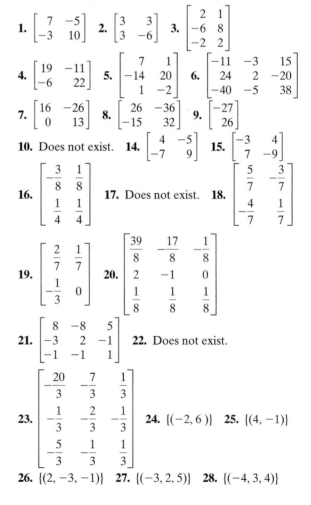

21.

23.

25. Minimum of 8 and maximum of 52
27. Minimum of 0 and maximum of 28
29. 63 **31.** 340 **33.** 2 **35.** 98
37. $5000 at 9% and $5000 at 12%
39. 300 of type A and 200 of type B
41. 12 units of A and 16 units of B

Chapter 12 Review Problem Set (page 683)

1. $\begin{bmatrix} 7 & -5 \\ -3 & 10 \end{bmatrix}$ **2.** $\begin{bmatrix} 3 & 3 \\ 3 & -6 \end{bmatrix}$ **3.** $\begin{bmatrix} 2 & 1 \\ -6 & 8 \\ -2 & 2 \end{bmatrix}$

4. $\begin{bmatrix} 19 & -11 \\ -6 & 22 \end{bmatrix}$ **5.** $\begin{bmatrix} 7 & 1 \\ -14 & 20 \\ 1 & -2 \end{bmatrix}$ **6.** $\begin{bmatrix} -11 & -3 & 15 \\ 24 & 2 & -20 \\ -40 & -5 & 38 \end{bmatrix}$

7. $\begin{bmatrix} 16 & -26 \\ 0 & 13 \end{bmatrix}$ **8.** $\begin{bmatrix} 26 & -36 \\ -15 & 32 \end{bmatrix}$ **9.** $\begin{bmatrix} -27 \\ 26 \end{bmatrix}$

10. Does not exist. **14.** $\begin{bmatrix} 4 & -5 \\ -7 & 9 \end{bmatrix}$ **15.** $\begin{bmatrix} -3 & 4 \\ 7 & -9 \end{bmatrix}$

16. $\begin{bmatrix} -\dfrac{3}{8} & \dfrac{1}{8} \\ \dfrac{1}{4} & \dfrac{1}{4} \end{bmatrix}$ **17.** Does not exist. **18.** $\begin{bmatrix} \dfrac{5}{7} & -\dfrac{3}{7} \\ \dfrac{4}{7} & \dfrac{1}{7} \end{bmatrix}$

19. $\begin{bmatrix} \dfrac{2}{7} & \dfrac{1}{7} \\ -\dfrac{1}{3} & 0 \end{bmatrix}$ **20.** $\begin{bmatrix} \dfrac{39}{8} & -\dfrac{17}{8} & -\dfrac{1}{8} \\ 2 & -1 & 0 \\ \dfrac{1}{8} & \dfrac{1}{8} & \dfrac{1}{8} \end{bmatrix}$

21. $\begin{bmatrix} 8 & -8 & 5 \\ -3 & 2 & -1 \\ -1 & -1 & 1 \end{bmatrix}$ **22.** Does not exist.

23. $\begin{bmatrix} -\dfrac{20}{3} & -\dfrac{7}{3} & \dfrac{1}{3} \\ -\dfrac{1}{3} & -\dfrac{2}{3} & -\dfrac{1}{3} \\ -\dfrac{5}{3} & \dfrac{1}{3} & \dfrac{1}{3} \end{bmatrix}$ **24.** $\{(-2, 6)\}$ **25.** $\{(4, -1)\}$

26. $\{(2, -3, -1)\}$ **27.** $\{(-3, 2, 5)\}$ **28.** $\{(-4, 3, 4)\}$

29.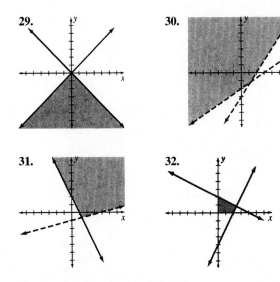

30.

31.

32.

22.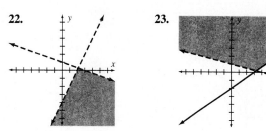

23.

24.

33. 37 **34.** 56 **35.** 57 **36.** 1700
37. 75 one-gallon and 175 two-gallon freezers

25. 4050

Chapter 12 Test (page 685)

1. $\begin{bmatrix} 9 & -1 \\ 4 & -6 \end{bmatrix}$ **2.** $\begin{bmatrix} -11 & 13 \\ -8 & 14 \end{bmatrix}$ **3.** $\begin{bmatrix} -1 & -3 & 11 \\ -4 & -5 & 18 \\ 37 & -1 & 9 \end{bmatrix}$

4. Does not exist **5.** $\begin{bmatrix} -35 \\ 8 \end{bmatrix}$ **6.** $\begin{bmatrix} -5 & 8 \\ 4 & -3 \end{bmatrix}$

7. $\begin{bmatrix} 4 & 9 \\ 13 & -16 \\ 24 & 23 \end{bmatrix}$ **8.** $\begin{bmatrix} -3 & -5 \\ -20 & 8 \end{bmatrix}$ **9.** $\begin{bmatrix} 8 & 33 \\ -12 & 13 \end{bmatrix}$

10. $\begin{bmatrix} 1 & -34 \\ 16 & -19 \end{bmatrix}$ **11.** $\begin{bmatrix} -3 & 2 \\ -5 & 3 \end{bmatrix}$ **12.** $\begin{bmatrix} 7 & 5 \\ 3 & 2 \end{bmatrix}$

13. $\begin{bmatrix} 4 & \dfrac{3}{2} \\ 1 & \dfrac{1}{2} \end{bmatrix}$ **14.** $\begin{bmatrix} \dfrac{4}{7} & -\dfrac{5}{7} \\ -\dfrac{1}{7} & \dfrac{3}{7} \end{bmatrix}$

15. $\begin{bmatrix} -\dfrac{4}{3} & -\dfrac{5}{3} & 1 \\ -\dfrac{4}{3} & -\dfrac{8}{3} & 1 \\ \dfrac{1}{3} & \dfrac{2}{3} & 0 \end{bmatrix}$ **16.** $\begin{bmatrix} 1 & 2 & -10 \\ 0 & 1 & -3 \\ 0 & 0 & 1 \end{bmatrix}$

17. $\{(8, -12)\}$ **18.** $\{(-6, -14)\}$ **19.** $\{(9, 13)\}$
20. $\left\{\left(\dfrac{7}{3}, -\dfrac{1}{3}, \dfrac{13}{3}\right)\right\}$ **21.** $\{(-1, 2, 1)\}$

CHAPTER 13

Problem Set 13.1 (page 693)

1. $x^2 + y^2 - 4x - 6y - 12 = 0$
3. $x^2 + y^2 + 2x + 10y + 17 = 0$
5. $x^2 + y^2 - 6x = 0$ **7.** $x^2 + y^2 = 49$
9. $x^2 + y^2 + 6x - 8y + 9 = 0$ and
$x^2 + y^2 + 6x + 8y + 9 = 0$
11. $x^2 + y^2 + 12x + 12y + 36 = 0$
13. $x^2 + y^2 - 8x + 4\sqrt{3}y + 12 = 0$ and
$x^2 + y^2 - 8x - 4\sqrt{3}y + 12 = 0$
15. $(5, 7); r = 5$ **17.** $(-1, -8); r = 2\sqrt{3}$
19. $(10, -5); r = \sqrt{3}$
21. $(3, 5), r = 2$ **23.** $(-5, -7), r = 1$ **25.** $(5, 0), r = 5$
27. $\left(0, \dfrac{5}{2}\right); r = \dfrac{\sqrt{29}}{2}$ **29.** $(0, 0), r = 2\sqrt{2}$
31. $\left(\dfrac{1}{2}, 1\right), r = 2$ **33.** $6x + 5y = 29$
35. $x^2 + y^2 + 6x + 8y = 0$
37. $x^2 + y^2 - 4x - 4y + 4 = 0$ and
$x^2 + y^2 + 20x - 20y + 100 = 0$
39. $x + 2y = 7$ **41.** $x^2 + y^2 + 12x + 2y - 21 = 0$

Problem Set 13.2 (page 702)

1. $V(0, 0)$, $F(2, 0)$, $x = -2$

3. $V(0, 0)$, $F(0, -3)$, $y = 3$

5. $V(0, 0)$, $F\left(-\dfrac{1}{2}, 0\right)$, $x = \dfrac{1}{2}$

7. $V(0, 0)$, $F\left(0, \dfrac{3}{2}\right)$, $y = -\dfrac{3}{2}$

9. $V(0, -1)$, $F(0, 2)$, $y = -4$

11. $V(3, 0)$, $F(1, 0)$, $x = 5$

13. $V(0, 2)$, $F(0, 3)$, $y = 1$

15. $V(0, -2)$, $F(0, -4)$, $y = 0$

17. $V(2, 0)$, $F(5, 0)$, $x = -1$

19. $V(2, -2)$, $F(2, -3)$, $y = -1$

21. $V(-2, -4)$, $F(-4, -4)$, $x = 0$

23. $V(1, 2)$, $F(1, 3)$, $y = 1$

25. $V(-3, 1)$, $F(-3, -1)$, $y = 3$

27. $V(3, 1)$, $F(0, 1)$, $x = 6$

29. $V(-2, -3)$, $F(-1, -3)$, $x = -3$

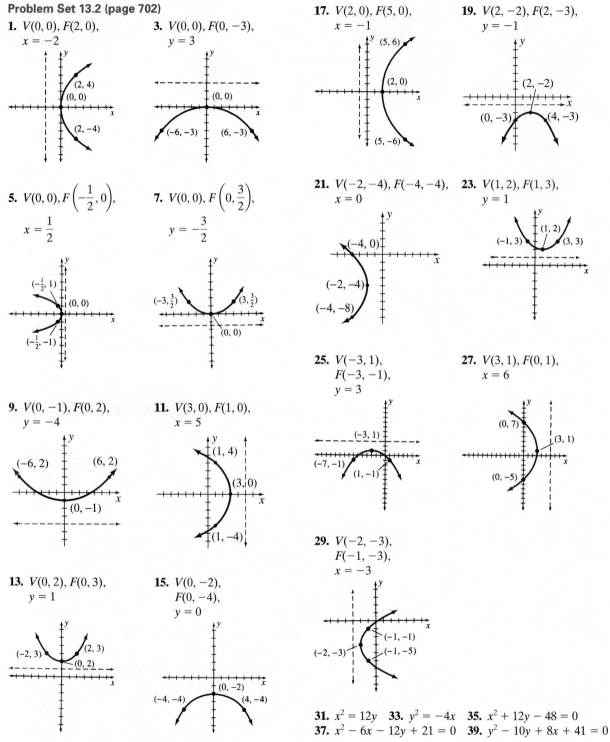

31. $x^2 = 12y$ **33.** $y^2 = -4x$ **35.** $x^2 + 12y - 48 = 0$
37. $x^2 - 6x - 12y + 21 = 0$ **39.** $y^2 - 10y + 8x + 41 = 0$

41. $y^2 = \dfrac{-25}{3}x$ **43.** $y^2 = 10x$

45. $x^2 - 14x - 8y + 73 = 0$ **47.** $y^2 + 6y - 12x + 105 = 0$

49. $x^2 + 18x + y + 80 = 0$ **51.** $x^2 = 750(y - 10)$

53. $10\sqrt{2}$ feet   **55.** 62.5 feet

Problem Set 13.3 (page 712)

For Problems 1–21, the foci are indicated above the graph, and the vertices and endpoints of the minor axes are indicated on the graph.

1. $F(\sqrt{3}, 0)$,
   $F'(-\sqrt{3}, 0)$

3. $F(0, \sqrt{5})$,
   $F'(0, -\sqrt{5})$

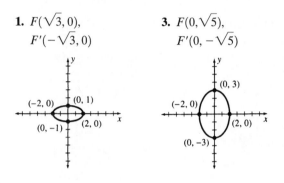

5. $F(0, \sqrt{6})$
   $F'(0, -\sqrt{6})$

7. $F(\sqrt{15}, 0)$
   $F'(-\sqrt{15}, 0)$

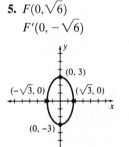
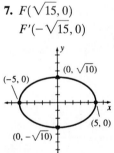

9. $F(0, \sqrt{33})$
   $F'(0, -\sqrt{33})$

11. $F(2, 0)$
   $F'(-2, 0)$

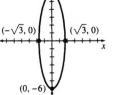

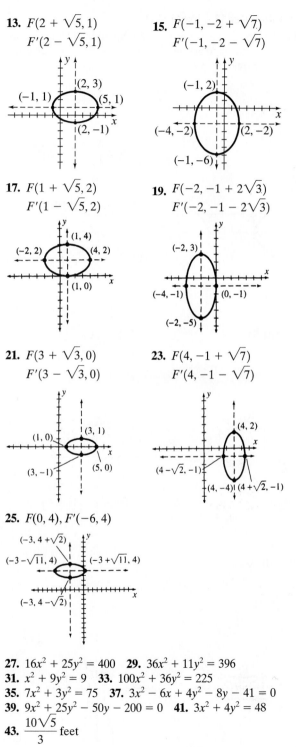

13. $F(2 + \sqrt{5}, 1)$
   $F'(2 - \sqrt{5}, 1)$

15. $F(-1, -2 + \sqrt{7})$
   $F'(-1, -2 - \sqrt{7})$

17. $F(1 + \sqrt{5}, 2)$
   $F'(1 - \sqrt{5}, 2)$

19. $F(-2, -1 + 2\sqrt{3})$
   $F'(-2, -1 - 2\sqrt{3})$

21. $F(3 + \sqrt{3}, 0)$
   $F'(3 - \sqrt{3}, 0)$

23. $F(4, -1 + \sqrt{7})$
   $F'(4, -1 - \sqrt{7})$

25. $F(0, 4)$, $F'(-6, 4)$

27. $16x^2 + 25y^2 = 400$ **29.** $36x^2 + 11y^2 = 396$

31. $x^2 + 9y^2 = 9$ **33.** $100x^2 + 36y^2 = 225$

35. $7x^2 + 3y^2 = 75$ **37.** $3x^2 - 6x + 4y^2 - 8y - 41 = 0$

39. $9x^2 + 25y^2 - 50y - 200 = 0$ **41.** $3x^2 + 4y^2 = 48$

43. $\dfrac{10\sqrt{5}}{3}$ feet

Problem Set 13.4 (page 722)

For Problems 1–22, the foci and equations of the asymptotes are indicated above the graphs. The vertices are given on the graphs.

1. $F(\sqrt{13}, 0)$,
  $F'(-\sqrt{13}, 0)$
  $y = \pm\dfrac{2}{3}x$

3. $F(0, \sqrt{13})$,
  $F'(0, -\sqrt{13})$
  $y = \pm\dfrac{2}{3}x$

5. $F(0, 5)$,
  $F(0, -5)$
  $y = \pm\dfrac{4}{3}x$

7. $F(3\sqrt{2}, 0)$
  $F(-3\sqrt{2}, 0)$
  $y = \pm x$

9. $F(0, \sqrt{30})$,
  $F(0, -\sqrt{30})$
  $y = \pm\dfrac{\sqrt{5}}{5}x$

11. $F(\sqrt{10}, 0)$,
  $F'(-\sqrt{10}, 0)$
  $y = \pm 3x$

13. $F(1 - \sqrt{13}, -1)$,
  $F'(1 - \sqrt{13}, -1)$
  $2x - 3y = 5$ and
  $2x + 3y = -1$

15. $F(1, 7)$,
  $F'(1, -3)$
  $3x - 4y = -5$ and
  $3x + 4y = 11$

17. $F(13 + \sqrt{13}, -1)$,
  $F'(3 - \sqrt{13}, -1)$
  $2x - 3y = 9$ and
  $2x + 3y = 3$

19. $F(-3, 2 + \sqrt{5})$,
  $F'(-3, 2 - \sqrt{5})$
  $2x - y = -8$ and
  $2x + y = -4$

21. $F(2 + \sqrt{6}, 0)$,
  $F'(2 - \sqrt{6}, 0)$
  $\sqrt{2}x - y = 2\sqrt{2}$ and
  $\sqrt{2}x + y = 2\sqrt{2}$

23. $F(0, -5 + \sqrt{10})$,
  $F'(0, -5 - \sqrt{10})$
  $3x - y = 5$ and
  $3x + y = -5$

25. $F(-2 + \sqrt{2}, -2)$, $F'(-2 - \sqrt{2}, -2)$
  $x - y = 0$ and $x + y = -4$

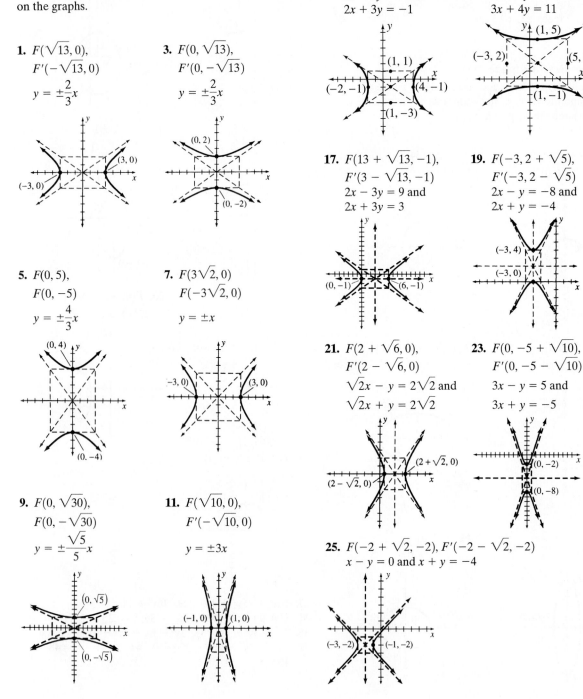

27. $5x^2 - 4y^2 = 20$ **29.** $16y^2 - 9x^2 = 144$
31. $3x^2 - y^2 = 3$ **33.** $4y^2 - 3x^2 = 12$ **35.** $7x^2 - 16y^2 = 112$
37. $5x^2 - 40x - 4y^2 - 24y + 24 = 0$
39. $3y^2 - 30y - x^2 - 6x + 54 = 0$
41. $5x^2 - 20x - 4y^2 = 0$ **43.** Circle   **45.** Straight line
47. Ellipse   **49.** Hyperbola   **51.** Parabola

Problem Set 13.5 (page 729)
1. $\{(1, 2)\}$ **3.** $\{(1, -5), (-5, 1)\}$
5. $\{(2 + i\sqrt{3}, -2 + i\sqrt{3}), (2 - i\sqrt{3}, -2 -i\sqrt{3})\}$
7. $\{(-6, 7), (-2, -1)\}$ **9.** $\{(-3, 4)\}$
11. $\left\{\left(\dfrac{-1 + i\sqrt{3}}{2}, \dfrac{-7 - i\sqrt{3}}{2}\right),\right.$
$\left.\left(\dfrac{-1 - i\sqrt{3}}{2}, \dfrac{-7 + i\sqrt{3}}{2}\right)\right\}$
13. $\{(-1, 2)\}$ **15.** $\{(-6, 3), (-2, -1)\}$ **17.** $\{(5, 3)\}$
19. $\{(1, 2), (-1, 2)\}$ **21.** $\{(-3, 2)\}$ **23.** $\{(2, 0), (-2, 0)\}$
25. $\{(\sqrt{2}, \sqrt{3}), (\sqrt{2}, -\sqrt{3}), (-\sqrt{2}, \sqrt{3}), (-\sqrt{2}, -\sqrt{3})\}$
27. $\{(1, 1), (1, -1), (-1, 1), (-1, -1)\}$
29. $\left\{\left(2, \dfrac{3}{2}\right), \left(\dfrac{3}{2}, 2\right)\right\}$ **31.** $\{(9, -2)\}$ **33.** $\{(\ln 2, 1)\}$
35. $\left\{\left(\dfrac{1}{2}, \dfrac{1}{8}\right), (-3, -27)\right\}$ **43.** $\{(-2.3, 7.4)\}$
45. $\{(6.7, 1.7), (9.5, 2.1)\}$ **47.** None

Chapter 13 Review Problem Set (page 732)
1. $F(4, 0), F'(-4, 0)$ **2.** $F(-3, 0)$

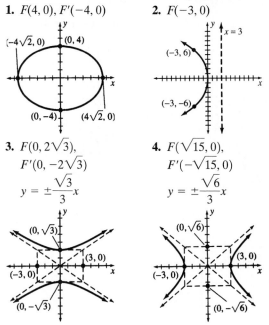

3. $F(0, 2\sqrt{3}),$
$F'(0, -2\sqrt{3})$
$y = \pm\dfrac{\sqrt{3}}{3}x$

4. $F(\sqrt{15}, 0),$
$F'(-\sqrt{15}, 0)$
$y = \pm\dfrac{\sqrt{6}}{3}x$

5. $F(0, \sqrt{6}),$
$F'(0, -\sqrt{6})$

6. $F\left(0, \dfrac{1}{2}\right)$

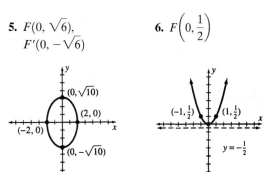

7.

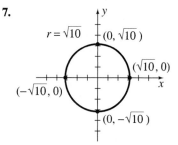

8. $F(4 + \sqrt{6}, 1), F'(4 - \sqrt{6}, 1)$
$\sqrt{2}x - 2y = 4\sqrt{2} - 2$ and $\sqrt{2}x + 2y = 4\sqrt{2} + 2$

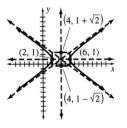

9. $F(3, -2 + \sqrt{7}), F'(3, -2 -\sqrt{7})$

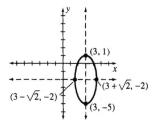

10. $F(-3, 1), x = -1$

11. $F(-1, -5), y = -1$

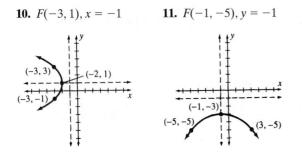

12. $F(-5 + 2\sqrt{3}, 2), F'(-5 - 2\sqrt{3}, 2)$

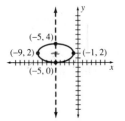

13. $F(-2, -2 + \sqrt{10}), F'(-2, -2 - \sqrt{10})$

$$\sqrt{6}x - 3y = 6 - 2\sqrt{6} \text{ and } \sqrt{6}x + 3y = -6 - 2\sqrt{6}$$

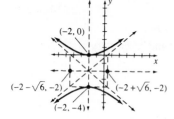

14. Center at $(3, -2)$ and $r = 4$

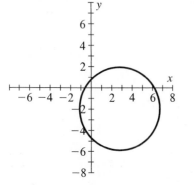

15. $x^2 + 16x + y^2 - 6y + 68 = 0$ **16.** $y^2 = -20x$
17. $16x^2 + y^2 = 16$ **18.** $25x^2 - 2y^2 = 50$

19. $x^2 - 10x + y^2 + 24y = 0$ **20.** $4x^2 + 3y^2 = 16$
21. $x^2 = \dfrac{2}{3}y$ **22.** $9y^2 - x^2 = 9$
23. $9x^2 - 108x + y^2 - 8y + 331 = 0$
24. $y^2 + 4y - 8x + 36 = 0$
25. $3y^2 + 24y - x^2 - 10x + 20 = 0$
26. $x^2 + 12x - y + 33 = 0$ **27.** $4x^2 + 40x + 25y^2 = 0$
28. $4x^2 - 32x - y^2 + 48 = 0$ **29.** $\{(-1, 4)\}$ **30.** $\{(3, 1)\}$
31. $\{(-1, -2), (-2, -3)\}$
32. $\left\{\left(\dfrac{4\sqrt{2}}{3}, \dfrac{4}{3}i\right), \left(\dfrac{4\sqrt{2}}{3}, -\dfrac{4}{3}i\right), \left(-\dfrac{4\sqrt{2}}{3}, \dfrac{4}{3}i\right),\right.$
$\left.\left(-\dfrac{4\sqrt{2}}{3}, -\dfrac{4}{3}i\right)\right\}$ **33.** $\{(0, 2), (0, -2)\}$

34. $\left\{\left(\dfrac{\sqrt{15}}{5}, \dfrac{2\sqrt{10}}{5}\right), \left(\dfrac{\sqrt{15}}{5}, -\dfrac{2\sqrt{10}}{5}\right),\right.$
$\left.\left(-\dfrac{\sqrt{15}}{5}, \dfrac{2\sqrt{10}}{5}\right), \left(-\dfrac{\sqrt{15}}{5}, -\dfrac{2\sqrt{10}}{5}\right)\right\}$

Chapter 13 Test (page 733)

1. $(0, -5)$ **2.** $(-3, 2)$ **3.** $x = -3$
4. $(6, 0)$ **5.** $(-2, -1)$ **6.** $(-3, -9)$
7. $y^2 + 8x = 0$ **8.** $x^2 - 6x + 12y - 39 = 0$
9. $x^2 + 2x + y^2 - 12y + 12 = 0$ **10.** 6 units
11. $(-7, 1)$ and $(-3, 1)$ **12.** $(-2\sqrt{3}, 0)$ and $(2\sqrt{3}, 0)$
13. $(-5, 8)$ **14.** $25x^2 + 9y^2 = 900$

15. $x^2 - 12x + 4y^2 + 16y + 36 = 0$ **16.** $y = \pm\dfrac{3}{2}x$

17. $(-1, 6)$ and $(-1, 0)$ **18.** $(\pm 3, 0)$ **19.** $x^2 - 3y^2 = 36$
20. $8x^2 + 16x - y^2 + 8y - 16 = 0$ **21.** 2

22. $\left\{(3, 2), (-3, -2), \left(4, \dfrac{3}{2}\right), \left(-4, -\dfrac{3}{2}\right)\right\}$

23.

24.

25.

CHAPTER 14

Problem Set 14.1 (page 741)

1. $-4, -1, 2, 5, 8$ **3.** $2, 0, -2, -4, -6$ **5.** $2, 11, 26, 47, 74$
7. $0, 2, 6, 12, 20$ **9.** $4, 8, 16, 32, 64$
11. $a_{15} = -79; a_{30} = -154$ **13.** $a_{25} = 1; a_{50} = -1$
15. $2n + 9$ **17.** $-3n + 5$ **19.** $\dfrac{n+2}{2}$ **21.** $4n - 2$
23. $-3n$ **25.** 73 **27.** 334 **29.** 35 **31.** 7 **33.** 86
35. 2700 **37.** 3200 **39.** -7950 **41.** 637.5 **43.** 4950
45. 1850 **47.** -2030 **49.** 3591 **51.** $40,000$ **53.** $58,250$
55. 2205 **57.** -1325 **59.** 5265 **61.** -810 **63.** 1276
65. 660 **67.** 55 **69.** 431 **75.** $3, 3, 7, 7, 11, 11$
77. $4, 7, 10, 13, 17, 21$ **79.** $4, 12, 36, 108, 324, 972$
81. $1, 1, 2, 3, 5, 8$ **83.** $3, 1, 4, 9, 25, 256$

Problem Set 14.2 (page 750)

1. $3(2)^{n-1}$ **3.** 3^n **5.** $\left(\dfrac{1}{2}\right)^{n+1}$ **7.** 4^n **9.** $(0.3)^{n-1}$
11. $(-2)^{n-1}$ **13.** 64 **15.** $\dfrac{1}{9}$ **17.** -512 **19.** $\dfrac{1}{4374}$
21. $\dfrac{2}{3}$ **23.** 2 **25.** 1023 **27.** $19,682$ **29.** $394\dfrac{1}{16}$
31. 1364 **33.** 1089 **35.** $7\dfrac{511}{512}$ **37.** -547 **39.** $127\dfrac{3}{4}$
41. 540 **43.** $2\dfrac{61}{64}$ **45.** 4 **47.** 3 **49.** No sum   **51.** $\dfrac{27}{4}$
53. 2 **55.** $\dfrac{16}{3}$ **57.** $\dfrac{1}{3}$ **59.** $\dfrac{26}{99}$ **61.** $\dfrac{41}{333}$ **63.** $\dfrac{4}{15}$
65. $\dfrac{106}{495}$ **67.** $\dfrac{7}{3}$

Problem Set 14.3 (page 756)

1. $\$24,200$ **3.** $11,550$ **5.** 7320 **7.** 125 liters
9. 512 gallons   **11.** $\$116.25$ **13.** $\$163.84; \$327.67$
15. $\$24,900$ **17.** 1936 feet   **19.** $\dfrac{15}{16}$ of a gram
21. 2910 feet   **23.** 325 logs   **25.** 5.9%
27. $\dfrac{5}{64}$ of a gallon

Problem Set 14.4 (page 763)

These problems call for proof by mathematical induction
and require class discussion.

Chapter 14 Review Problem Set (page 765)

1. $6n - 3$ **2.** 3^{n-2} **3.** $5(2^n)$ **4.** $-3n + 8$ **5.** $2n - 7$
6. 3^{3-n} **7.** $-(-2)^{n-1}$ **8.** $3n + 9$ **9.** $\dfrac{n+1}{3}$ **10.** 4^{n-1}

11. 73 **12.** 106 **13.** $\dfrac{1}{32}$ **14.** $\dfrac{4}{9}$ **15.** -92 **16.** $\dfrac{1}{16}$
17. -5 **18.** 85 **19.** $\dfrac{5}{9}$ **20.** 2 or -2 **21.** $121\dfrac{40}{81}$
22. 7035 **23.** $-10,725$ **24.** $31\dfrac{31}{32}$ **25.** $32,015$ **26.** 4757
27. $85\dfrac{21}{64}$ **28.** $37,044$ **29.** $12,726$ **30.** -1845
31. 225 **32.** 255 **33.** 8244 **34.** $85\dfrac{1}{3}$ **35.** $\dfrac{4}{11}$ **36.** $\dfrac{41}{90}$
37. $\$750$ **38.** $\$46.50$ **39.** $\$3276.70$ **40.** $10,935$ gallons

Chapter 14 Test (page 767)

1. -226 **2.** 48 **3.** $-5n + 2$ **4.** $5(2)^{1-n}$ **5.** $6n + 4$
6. $\dfrac{729}{8}$ or $91\dfrac{1}{8}$ **7.** 223 **8.** 60 terms   **9.** 2380 **10.** 765
11. 7155 **12.** 6138 **13.** $22,650$ **14.** 9384 **15.** 4075
16. -341 **17.** 6 **18.** $\dfrac{1}{3}$ **19.** $\dfrac{2}{11}$ **20.** $\dfrac{4}{15}$ **21.** 3 liters
22. $\$1638.30$ **23.** $\$5810$
24. and **25.** Instructor supplies proof.

CHAPTER 15

Problem Set 15.1 (page 773)

1. 20 **3.** 24 **5.** 168 **7.** 48 **9.** 36 **11.** 6840 **13.** 720
15. 720 **17.** 36 **19.** 24 **21.** 243 **23.** Impossible
25. 216 **27.** 26 **29.** 36 **31.** 144 **33.** 1024 **35.** 30
37. (a) $6,084,000$ **(c)** $3,066,336$

Problem Set 15.2 (page 781)

1. 60 **3.** 360 **5.** 21 **7.** 252 **9.** 105 **11.** 1 **13.** 24
15. 84 **17. (a)** 336 **19.** 2880 **21.** 2450 **23.** 10
25. 10 **27.** 35 **29.** 1260 **31.** 2520 **33.** 15 **35.** 126
37. $144; 202$ **39.** $15; 20$ **41.** 20
43. $10; 15; 21; \dfrac{n(n-1)}{2}$ **47.** 120 **53.** $133,784,560$
55. $54,627,300$

Problem Set 15.3 (page 788)

1. $\dfrac{1}{2}$ **3.** $\dfrac{3}{4}$ **5.** $\dfrac{1}{8}$ **7.** $\dfrac{7}{8}$ **9.** $\dfrac{1}{16}$ **11.** $\dfrac{3}{8}$ **13.** $\dfrac{1}{3}$ **15.** $\dfrac{1}{2}$
17. $\dfrac{5}{36}$ **19.** $\dfrac{1}{6}$ **21.** $\dfrac{11}{36}$ **23.** $\dfrac{1}{4}$ **25.** $\dfrac{1}{2}$ **27.** $\dfrac{1}{25}$ **29.** $\dfrac{9}{25}$
31. $\dfrac{2}{5}$ **33.** $\dfrac{9}{10}$ **35.** $\dfrac{5}{14}$ **37.** $\dfrac{15}{28}$ **39.** $\dfrac{7}{15}$ **41.** $\dfrac{1}{15}$ **43.** $\dfrac{2}{3}$

45. $\dfrac{1}{5}$ **47.** $\dfrac{1}{63}$ **49.** $\dfrac{1}{2}$ **51.** $\dfrac{5}{11}$ **53.** $\dfrac{1}{6}$ **55.** $\dfrac{21}{128}$ **57.** $\dfrac{13}{16}$

59. $\dfrac{1}{21}$ **63.** 40   **65.** 3744   **67.** 10,200   **69.** 123,552

71. 1,302,540

Problem Set 15.4 (page 797)

1. $\dfrac{5}{36}$ **3.** $\dfrac{7}{12}$ **5.** $\dfrac{1}{216}$ **7.** $\dfrac{53}{54}$ **9.** $\dfrac{1}{16}$ **11.** $\dfrac{15}{16}$ **13.** $\dfrac{1}{32}$

15. $\dfrac{31}{32}$ **17.** $\dfrac{5}{6}$ **19.** $\dfrac{12}{13}$ **21.** $\dfrac{7}{12}$ **23.** $\dfrac{37}{44}$ **25.** $\dfrac{2}{3}$ **27.** $\dfrac{2}{3}$

29. $\dfrac{5}{18}$ **31.** $\dfrac{1}{3}$ **33.** $\dfrac{1}{2}$ **35.** $\dfrac{7}{12}$ **37. (a)** 0.410   **(c)** 0.955

39. 0.525   **41.** 60   **43.** 120   **45.** 9   **47.** 56
49. It is a fair game.   **51.** Yes   **53.** $11,000   **55.** −$25
59. 1 to 7   **61.** 11 to 5   **63.** 1 to 8   **65.** 1 to 1   **67.** 4 to 3

69. 3 to 2   **71.** $\dfrac{2}{7}$ **73.** $\dfrac{7}{12}$

Problem Set 15.5 (page 806)

1. $\dfrac{1}{3}$ **3.** $\dfrac{2}{15}$ **5.** $\dfrac{1}{3}$ **7.** $\dfrac{1}{6}$ **9.** $\dfrac{2}{3};\dfrac{2}{7}$ **11.** $\dfrac{2}{3};\dfrac{2}{5}$ **13.** $\dfrac{1}{5};\dfrac{2}{7}$

15. Dependent   **17.** Independent   **19.** $\dfrac{1}{4}$ **21.** $\dfrac{1}{216}$

23. $\dfrac{1}{221}$ **25.** $\dfrac{13}{102}$ **27.** $\dfrac{1}{16}$ **29.** $\dfrac{1}{1352}$ **31.** $\dfrac{2}{49}$ **33.** $\dfrac{25}{81}$

35. $\dfrac{20}{81}$ **37.** $\dfrac{25}{169}$ **39.** $\dfrac{32}{169}$ **41.** $\dfrac{2}{3}$ **43.** $\dfrac{1}{3}$ **45.** $\dfrac{5}{68}$

47. $\dfrac{15}{34}$ **49.** $\dfrac{1}{12}$ **51.** $\dfrac{1}{6}$ **53.** $\dfrac{1}{729}$ **55.** $\dfrac{5}{27}$ **57.** $\dfrac{4}{35}$

59. $\dfrac{8}{35}$ **61.** $\dfrac{4}{21};\dfrac{2}{7};\dfrac{11}{21}$

Problem Set 15.6 (page 813)

1. $x^8 + 8x^7y + 28x^6y^2 + 56x^5y^3 + 70x^4y^4 + 56x^3y^5 + 28x^2y^6 + 8xy^7 + y^8$
3. $x^6 - 6x^5y + 15x^4y^2 - 20x^3y^3 + 15x^2y^4 - 6xy^5 + y^6$
5. $a^4 + 8a^3b + 24a^2b^2 + 32ab^3 + 16b^4$
7. $x^5 - 15x^4y + 90x^3y^2 - 270x^2y^3 + 405xy^4 - 243y^5$
9. $16a^4 - 96a^3b + 216a^2b^2 - 216ab^3 + 81b^4$
11. $x^{10} + 5x^8y + 10x^6y^2 + 10x^4y^3 + 5x^2y^4 + y^5$
13. $16x^8 - 32x^6y^2 + 24x^4y^4 - 8x^2y^6 + y^8$
15. $x^6 + 18x^5 + 135x^4 + 540x^3 + 1215x^2 + 1458x + 729$
17. $x^9 - 9x^8 + 36x^7 - 84x^6 + 126x^5 - 126x^4 + 84x^3 - 36x^2 + 9x - 1$

19. $1 + \dfrac{4}{n} + \dfrac{6}{n^2} + \dfrac{4}{n^3} + \dfrac{1}{n^4}$

21. $a^6 - \dfrac{6a^5}{n} + \dfrac{15a^4}{n^2} - \dfrac{20a^3}{n^3} + \dfrac{15a^2}{n^4} - \dfrac{6a}{n^5} + \dfrac{1}{n^6}$
23. $17 + 12\sqrt{2}$ **25.** $843 - 589\sqrt{2}$
27. $x^{12} + 12x^{11}y + 66x^{10}y^2 + 220x^9y^3$
29. $x^{20} - 20x^{19}y + 190x^{18}y^2 - 1140x^{17}y^3$
31. $x^{28} - 28x^{26}y^3 + 364x^{24}y^6 - 2912x^{22}y^9$

33. $a^9 + \dfrac{9a^8}{n} + \dfrac{36a^7}{n^2} + \dfrac{84a^6}{n^3}$

35. $x^{10} - 20x^9y + 180x^8y^2 - 960x^7y^3$ **37.** $56x^5y^3$

39. $126x^5y^4$ **41.** $189a^2b^5$ **43.** $120x^6y^{21}$ **45.** $\dfrac{5005}{n^6}$

51. $-117 + 44i$ **53.** $-597 - 122i$

Chapter 15 Review Problem Set (page 816)

1. 720   **2.** 30,240   **3.** 150   **4.** 1440   **5.** 20   **6.** 525
7. 1287   **8.** 264   **9.** 74   **10.** 55   **11.** 40   **12.** 15

13. 60   **14.** 120   **15.** $\dfrac{3}{8}$ **16.** $\dfrac{5}{16}$ **17.** $\dfrac{5}{36}$ **18.** $\dfrac{13}{18}$ **19.** $\dfrac{3}{5}$

20. $\dfrac{1}{35}$ **21.** $\dfrac{57}{64}$ **22.** $\dfrac{1}{221}$ **23.** $\dfrac{1}{6}$ **24.** $\dfrac{4}{7}$ **25.** $\dfrac{4}{7}$

26. $\dfrac{10}{21}$ **27.** $\dfrac{140}{143}$ **28.** $\dfrac{105}{169}$ **29.** $\dfrac{1}{6}$ **30.** $\dfrac{28}{55}$ **31.** $\dfrac{5}{7}$

32. $\dfrac{1}{16}$ **33.** $\dfrac{1}{2};\dfrac{1}{3}$ **34. (a)** $\dfrac{9}{19}$ **(b)** $\dfrac{9}{10}$ **35. (a)** $\dfrac{2}{7}$

(b) $\dfrac{4}{9}$ **36.** $x^5 + 10x^4y + 40x^3y^2 + 80x^2y^3 + 80xy^4 + 32y^5$
37. $x^8 - 8x^7y + 28x^6y^2 - 56x^5y^3 + 70x^4y^4 - 56x^3y^5 + 28x^2y^6 - 8xy^7 + y^8$
38. $a^8 - 12a^6b^3 + 54a^4b^6 - 108a^2b^9 + 81b^{12}$

39. $x^6 + \dfrac{6x^5}{n} + \dfrac{15x^4}{n^2} + \dfrac{20x^3}{n^3} + \dfrac{15x^2}{n^4} + \dfrac{6x}{n^5} + \dfrac{1}{n^6}$
40. $41 - 29\sqrt{2}$ **41.** $-a^3 + 3a^2b - 3ab^2 + b^3$
42. $-1760x^9y^3$ **43.** $57,915a^4b^{18}$

Chapter 15 Test (page 818)

1. 12   **2.** 240   **3.** 216   **4.** 270   **5.** 26   **6.** 8640   **7.** 20

8. 144   **9.** 2520   **10.** 350   **11.** $\dfrac{13}{18}$ **12.** $\dfrac{5}{16}$ **13.** $\dfrac{5}{6}$

14. $\dfrac{1}{7}$ **15.** $\dfrac{23}{28}$ **16.** $\dfrac{3}{4}$ **17.** 25   **18.** $0.30   **19.** $\dfrac{168}{361}$

20. $\dfrac{2}{21}$ **21.** $\dfrac{5}{16}$

22. $64 - \dfrac{192}{n} + \dfrac{240}{n^2} - \dfrac{160}{n^3} + \dfrac{60}{n^4} - \dfrac{12}{n^5} + \dfrac{1}{n^6}$
23. $243x^5 + 810x^4y + 1080x^3y^2 + 720x^2y^3 + 240xy^4 + 32y^5$

24. $\dfrac{495}{256}x^4$ **25.** $2835x^3y^4$

APPENDIX A

Practice Exercises (page 828)

1. $2 \cdot 13$ **2.** $2 \cdot 2 \cdot 2 \cdot 2$ **3.** $2 \cdot 2 \cdot 3 \cdot 3$
4. $2 \cdot 2 \cdot 2 \cdot 2 \cdot 5$ **5.** $7 \cdot 7$ **6.** $2 \cdot 2 \cdot 23$
7. $2 \cdot 2 \cdot 2 \cdot 7$ **8.** $2 \cdot 2 \cdot 2 \cdot 2 \cdot 3 \cdot 3$
9. $2 \cdot 2 \cdot 2 \cdot 3 \cdot 5$ **10.** $2 \cdot 2 \cdot 3 \cdot 7$ **11.** $3 \cdot 3 \cdot 3 \cdot 5$
12. $2 \cdot 7 \cdot 7$ **13.** 24 **14.** 24 **15.** 48 **16.** 36 **17.** 140
18. 462 **19.** 392 **20.** 72 **21.** 168 **22.** 60 **23.** 90
24. 168 **25.** $\dfrac{2}{3}$ **26.** $\dfrac{3}{4}$ **27.** $\dfrac{2}{3}$ **28.** $\dfrac{9}{16}$ **29.** $\dfrac{5}{3}$ **30.** $\dfrac{4}{3}$
31. $\dfrac{15}{28}$ **32.** $\dfrac{12}{55}$ **33.** $\dfrac{10}{21}$ **34.** $\dfrac{65}{66}$ **35.** $\dfrac{3}{10}$ **36.** $\dfrac{2}{3}$

37. $\dfrac{3}{8}$ cup **38.** $\dfrac{1}{6}$ of the bottle **39.** $\dfrac{2}{9}$ of the disk space
40. $\dfrac{1}{3}$ **41.** $\dfrac{5}{7}$ **42.** $\dfrac{8}{11}$ **43.** $\dfrac{5}{9}$ **44.** $\dfrac{5}{13}$ **45.** 3 **46.** 2
47. $\dfrac{2}{3}$ **48.** $\dfrac{3}{8}$ **49.** $\dfrac{2}{3}$ **50.** $\dfrac{5}{9}$ **51.** $\dfrac{8}{15}$ **52.** $\dfrac{7}{24}$ **53.** $\dfrac{9}{16}$
54. $\dfrac{11}{12}$ **55.** $\dfrac{37}{30}$ **56.** $\dfrac{29}{24}$ **57.** $\dfrac{59}{96}$ **58.** $\dfrac{19}{24}$ **59.** $\dfrac{13}{12}$
60. $\dfrac{5}{16}$ **61.** $\dfrac{1}{4}$ **62.** $\dfrac{5}{3}$ **63.** $\dfrac{37}{30}$ **64.** $\dfrac{4}{5}$ **65.** $\dfrac{1}{3}$ **66.** $\dfrac{27}{35}$
67. $\dfrac{7}{26}$ **68.** 30 **69.** $\dfrac{7}{20}$ **70.** $\dfrac{11}{32}$

Index

Properties of Absolute Value

$|a| \geq 0$

$|a| = |-a|$

$|a - b| = |b - a|$

$|a^2| = |a|^2 = a^2$

Multiplication Patterns

$(a + b)^2 = a^2 + 2ab + b^2$

$(a - b)^2 = a^2 - 2ab + b^2$

$(a + b)(a - b) = a^2 - b^2$

$(a + b)^3 = a^3 + 3a^2b + 3ab^2 + b^3$

$(a - b)^3 = a^3 - 3a^2b + 3ab^2 - b^3$

Properties of Exponents

$b^n \cdot b^m = b^{n+m}$ $\dfrac{b^n}{b^m} = b^{n-m}$

$(b^n)^m = b^{mn}$

$(ab)^n = a^n b^n$

$\left(\dfrac{a}{b}\right)^n = \dfrac{a^n}{b^n}$

Equations Determining Functions

Linear function: $f(x) = ax + b$

Quadratic function: $f(x) = ax^2 + bx + c$

Polynomial function: $f(x) = a_n x^n + a_{n-1} x^{n-1} + \ldots + a_1 x + a_0,$ where n is a nonnegative integer

Rational function: $f(x) = \dfrac{g(x)}{h(x)},$ where g and h are polynomials; $h(x) \neq 0$

Exponential function: $f(x) = b^x,$ where $b > 0$ and $b \neq 1$

Logarithmic function: $f(x) = \log_b x,$ where $b > 0$ and $b \neq 1$

Interval Notation   **Set Notation**

(a, ∞) $\{x \mid x > a\}$

$(-\infty, b)$ $\{x \mid x < b\}$

(a, b) $\{x \mid a < x < b\}$

$[a, \infty)$ $\{x \mid x \geq a\}$

$(-\infty, b]$ $\{x \mid x \leq b\}$

$(a, b]$ $\{x \mid a < x \leq b\}$

$[a, b)$ $\cdot$ $\{x \mid a \leq x < b\}$

$[a, b]$ $\{x \mid a \leq x \leq b\}$

Properties of Logarithms

$\log_b b = 1$

$\log_b 1 = 0$

$\log_b rs = \log_b r + \log_b s$

$\log_b\left(\dfrac{r}{s}\right) = \log_b r - \log_b s$

$\log_b r^p = p(\log_b r)$

Factoring Patterns

$a^2 - b^2 = (a + b)(a - b)$

$a^3 - b^3 = (a - b)(a^2 + ab + b^2)$

$a^3 + b^3 = (a + b)(a^2 - ab + b^2)$